PREALGEBRA

TOM CARSON

Midlands Technical College

Addison Wesley

Boston   San Francisco   New York
London   Toronto   Sydney   Tokyo   Singapore   Madrid
Mexico City   Munich   Paris   Cape Town   Hong Kong   Montreal

Publisher: Jason A. Jordan
Acquisitions Editor: Jennifer P. Crum
Editorial Project Manager: Ruth Berry
Managing Editor: Ron Hampton
Text and Cover Design Supervision: Susan C. Raymond
Production Coordinator: Sheila Spinney
Production Services: Pre-Press Company
Editorial Assistant: Gregory Erb
Media Producer: Lorie Reilly
Marketing Manager: Dona Kenly
Marketing Coordinator: Jennifer Berman
Prepress Services Buyer: Caroline Fell
First Print Buyer: Evelyn Beaton
Cover Image: © Index Stock Imagery/Charles Schoffner, photographer

Library of Congress Cataloging-in-Publication Data
Carson, Tom, 1967–
    Prealgebra/Tom Carson.
      p. cm.
    Includes index.
    ISBN 0-201-65553-5
    1. Mathematics. I. Title.
  QA39.2.C353 2000
510—dc21   00-023163

ISBN: 0-201-65553-5

1 2 3 4 5 6 7 8 9 10   DOW   03 02 01 00

Contents

Preface

*P*realgebra is designed for the student who needs a brush up in arithmetic and basic algebra concepts, or for people who are encountering algebra for the first time. Written in a relaxed, nonthreatening style, this text takes great care to ensure that students who have struggled with math in the past will be comfortable with the subject matter. Explanations are carefully developed to provide a sense of why a mathematical process works the way it does, instead of just an explanation of how to follow the process. Problems from science, engineering, accounting, health fields, the arts, and every-day life link mathematics to the real world. The link to real-world problem solving is further developed through two project portfolios available in a separate workbook (see pages ix and xiii for a complete description of the Project Portfolios Workbook). In addition, a complete study system with a learning-styles inventory provides further guidance for students. (See pages xiv–xxiv.)

Upon completing the material in this text, a student should be able to proceed successfully in an introductory algebra course or a survey math course. This text is designed to be versatile enough for use in a standard lecture format, a self-paced lab, or even in an independent study format. A strong ancillary package provides a wealth of supplemental resources for both instructors and students.

Key Features

Real, relevant, and interesting applications. Nearly every application problem is a real situation taken from science, engineering, health, finance, the arts, or just every-day life. The real-world applications not only illustrate the uses of basic arithmetic and algebra concepts, but they also expose students to the wonders of the world in which we live. Often the problems follow up with open-ended discussion questions where there is no "correct" answer. These questions help students to think beyond just getting a numeric answer by encouraging them to apply mathematical results to solve problems (see pages 32, 325, and 573).

Thorough explanations. This text explains not only how to do the math, but also why the math works the way it does, where it comes from, and how it is relevant to students' every-day lives. Knowing all of this helps the students remember the concept.

Study system. A study system is presented in the *TO THE STUDENT* section on pages xv–xxii, and this system is reinforced throughout the text. The system recommends using color codes for taking notes. The color codes are consistent in the text itself: red for definitions, blue for outlines, rules, and things to remember, and black for notes and examples. In addition, the study system presents strategies for succeeding in the course, and these strategies are revisited in the chapter openers throughout the text (see pages 1, 149, and 237).

Learning Styles Inventory. A learning styles inventory is presented on page xxii to help students assess their particular style of learning. Throughout the text, explanations are presented with different learning styles in mind (see pages 87, 95, and 318).

Problem-solving process. On page 65 of Section 1.7, a problem-solving outline is developed with the following headings:

1. Understand
2. Plan
3. Execute
4. Answer
5. Check

Every application example throughout the rest of the text follows the steps given in this outline, presenting the headings to show the thinking process clearly (see pages 69, 219, and 570).

Historical notes and trivia bites. The historical notes and trivia bites are designed to enhance the learning process by making concepts fun and interesting (see pages 248, 342, and 573).

Connection boxes. Connection boxes bridge concepts so that students see how the concepts build on each other and are interrelated (see pages 180, 325, and 410).

Calculator explanations and exercises. The relevant functions on a scientific calculator are explained and illustrated throughout the text. An occasional calculator icon in the exercise sets indicates problems that are appropriate to be solved using a scientific calculator (see pages 98, 101, and 506). Note that the icon does not mean that a calculator is required for those problems.

Work text format. The worktext format leaves space for working problems in the margins and exercise sets, encouraging students to become actively involved in their learning.

Examples and margin problems that correlate. Each example has at least one margin problem for students to complete as they read through the text so that they can practice concepts continually. Answers to all margin problems are provided at the bottom of the page, giving students immediate feedback (see pages 5, 44, and 245).

Puzzle Problems. These mathematical brain-teasers appear at the end of selected exercise sets to encourage creative and critical thinking (see pages 20, 178, and 652).

Continuous review problems. Problems that review previous concepts appear at the end of each exercise set (see pages 168, 242, and 420). These problems also review concepts that are foundational to the discussion in the next section. Answers to these review exercises appear in the back of the text.

Chapter Summaries and Review Exercises. A summary at the end of each chapter provides a list of key terms referenced by section, a two-column summary of key concepts (procedures and rules on the left and examples on the right), and a list of key formulas. A set of review exercises is also provided (see pages 75, 76, and 81). Answers to all chapter review exercises appear in the back of the text.

Practice tests. A practice test follows each set of chapter review exercises (see pages 85, 145, and 541). The problem types in the practice tests correlate to the short-answer tests in the Printed Test Bank. This is especially comforting for people who are math-anxious or experience test anxiety. Answers to all practice test questions are provided in the back of the text.

Cumulative reviews. Beginning with Chapter 2, cumulative review exercises appear after each practice test (see pages 147, 287, and 673). These help students stay current with *all* the material they have learned. Answers to all cumulative review exercises appear in the back of the text.

Supplements

FOR THE INSTRUCTOR:

Instructor's Solutions Manual Contains complete solutions to all of the even-numbered section exercises and solutions to all puzzle problems. ISBN 0-201-66049-0

Printed Test Bank/Instructor's Resource Guide The Printed Test Bank portion of this manual contains a diagnostic pretest for each chapter; four free-response test forms for each chapter, one of which contains higher-level questions; one multiple-choice test per chapter; one free-response midterm exam; two free-response final exams; and two multiple-choice final exams. The *Instructor's Resource Guide* portion of the manual contains sample syllabi, chapter-by-chapter teaching strategies, extra practice worksheets for the algebra topics in the text and answers and teaching suggestions for the project portfolios (see *For the Student*). ISBN 0-201-66050-4

Answer Book This supplement contains answers to all exercises in the text. ISBN 0-201-66051-2

TestGen-EQ with QuizMaster-EQ Available on a dual-platform Windows/Macintosh CD-ROM, this fully networkable software enables professors to create, edit, and administer tests using a computerized test bank of questions organized according to the chapter content of the text. Six question formats are available, and a built-in question editor allows the user to create graphs, import graphics, and insert mathematical symbols and templates, variable numbers, or text. An "Export to HTML" feature allows practice tests to be posted to the Internet, and instructors can use QuizMaster-EQ to post quizzes to a local computer network so that students can take them online. QuizMaster-EQ automatically grades the quizzes, stores results, and lets the instructor view or print a variety of reports for individual students or for an entire class or section. ISBN 0-201-66053-9

InterAct Math® Plus This networkable software provides course-management capabilities and on-line test administration for Addison Wesley Longman's InterAct Math Tutorial Software (see *For the Student*). InterAct Math Plus enables instructors to create and administer on-line tests, summarize students' results, and monitor students' progress in the tutorial software, providing an invaluable teaching and tracking resource.
InterAct Math Instructor Package for Windows: ISBN 0-201-63555-0
InterAct Math Instructor Package for Macintosh: ISBN 0-201-64805-9

FOR THE STUDENT:

Student's Solutions Manual This manual contains complete solutions to the odd-numbered exercises in each section and to all the section-level review exercises, chapter review exercises, practice tests, and cumulative review exercises. ISBN 0-201-66048-2

Project Portfolios Workbook Two unique project portfolios—one on building a house and one on starting and operating a business—are available in a student workbook that can be bundled with the text. Each portfolio contains problems keyed to the relevant sections of the text, so that as they work through each project over the course of the semester, students apply the skills and concepts they've learned in a realistic and integrated context. The project problems foster skill-building in a creative setting while also leading students to think critically, synthesize concepts, and apply mathematics to real-life situations. The projects also provide a wonderful resource for collaborative work, writing across the curriculum, and Internet research. ISBN 0-201-70404-8

InterAct Math® Tutorial Software Available on a dual-platform Windows/Macintosh CD-ROM, this interactive tutorial software provides algorithmically generated practice exercises that are correlated at the objective level to the content of the text. Every exercise in the program is accompanied by an example and a guided solution designed to involve students in the solution process. The software recognizes common student errors and provides appropriate feedback. It also tracks student activity and scores and can generate printed summaries of students' progress. Instructors can use the InterAct Math Plus course management software (see *For the Instructor*) to create, administer, and track on-line tests and monitor student performance during their practice sessions in InterAct Math. ISBN 0-201-66052-0

InterAct MathXL InterAct MathXL is a Web-based tutorial system that helps students prepare for tests by allowing them to take practice tests and receive a personalized study plan based on their results. Practice tests are correlated directly to the section objectives in the text, and once a student has taken an on-line practice test, the software scores the test and generates a study plan that identifies strengths, pinpoints topics where more review is needed, and links directly to the appropriate section(s) of the InterAct Math® tutorial software for additional practice and review. Students gain access to the InterAct MathXL Web site through a password-protected subscription; subscriptions can either be bundled with new copies of *Prealgebra* or purchased separately.
InterAct MathXL subscription bundled with the text: ISBN 0-201-71430-2
InterAct MathXL subscription purchased separately: ISBN 0-201-71630-5

Videotapes Designed specifically for *Prealgebra*, this set of videotapes presents interactive lectures that cover every section of the text. Featuring engaging mathematics instructors who use examples and problems from the text itself, these videos provide an ideal form of supplemental instruction that supports visualization and problem-solving techniques. ISBN 0-201-66054-7

Digital Video Tutor This innovative supplement provides the entire set of *Prealgebra* video tapes in digital format on CD-ROMs, making it easy and convenient for students to view the videos from a computer, either at home or on campus. Available for purchase with the text at a minimal cost, the Digital Video Tutor is ideal for distance learning and supplemental instruction. ISBN 0-201-70407-2

AWL Math Tutor Center The AWL Math Tutor Center is staffed by qualified mathematics instructors who tutor students via toll-free telephone, fax, or e-mail on examples and exercises from their text. The Tutor Center is open five days a week, seven hours a day, and is accessed through a registration number that may be bundled with a new textbook or purchased separately with a used book.
Tutor Center registration bundled with the text: ISBN 0-201-71429-9
Tutor Center registration purchased separately: ISBN 0-201-44461-5

Web Site: http://www.carsonmath.com Designed specifically for *Prealgebra*, this Web site features InterAct Math® tutorial exercises as well as chapter quizzes. A Learning Styles Inventory section helps students determine their own personal learning style. Other features include study skills, links to additional resources, and a way for students to contact the author with feedback or questions about the text.

MathPass, Version 2.2 for Windows MathPass helps students succeed in their developmental mathematics courses by creating customized study plans based on diagnostic test results from ACT, Inc.'s Computer-Adaptive Placement Assessment and Support System (COMPASS®). MathPass pinpoints topics where the student needs in-depth study or targeted review and correlates these topics with the student's textbook and related supplements (such as videos, student's solutions manuals, Web sites, and tutorial software). The MathPass Learning System provides diagnostic assessment, focused instruction, and exit placement all in one package. Contact your local Addison Wesley Longman sales consultant for more information about MathPass.

MathPass bundled with the text: ISBN 0-201-72384-0

MathPass purchased separately: ISBN 0-201-71728-X

Project Portfolios Workbook

In order to bridge mathematical concepts and real-world problem solving, two projects have been developed to accompany the text. The Project Portfolios Workbook can be bundled with student copies of the text. The two projects included are:
1. Building a House
2. Starting and Operating a Business

Both projects are correlated to the chapter content of the text and build and develop as students progress throughout the semester. Students are guided through a series of questions that reference relevant sections in the text, and in working through the problems, students are able to apply the skills and concepts they have learned in a realistic and integrated context. Following is a little more about each project.

Building a House

Students are given the following resources to get them started:
- A financial profile with relevant income, credit, and debt information
- A floor plan
- A pricing option sheet
- Closing costs

This project takes students through the process of building a house. As the project progresses, students will have to think through budget constraints, financing, and construction. The mathematical topics developed in this project include computations, geometry (area, perimeter, surface area, and the Pythagorean theorem), algebraic expressions, solving equations, ratios, percents, and graphs.

Starting and Operating a Business

Students are given the following resources to get them started:
- Profile with initial capital and start-up costs
- Monthly production information, including material costs, labor costs, and plant lease

This project takes students through the process of starting and operating a business. As the project progresses, students will be asked to work through budgeting, accounting, and employee benefits and to follow the stock market. The mathematical topics developed in this project include computations, algebraic expressions, solving equations, ratios, percents, and graphs.

Note to instructors: The Printed Test Bank/Instructors Resource Guide includes tips and strategies for incorporating the project portfolios into your course.

Acknowledgments

So many people have helped me in so many ways that I could write another book in saying thank you. Though the words of thanks to follow may be few, no amount of space can contain the genuine gratitude that I feel toward each and every person that gave of themselves to make this work the best that it can be.

I would like to thank the following people who gave of their time in reviewing the text. Their thoughtful input was vital to the development of the text.

Carla K. Ainsworth, *Salt Lake Community College*

James R. Barr, *Laramie County Community College*

Kenneth Benson, *University of Illinois at Urbana-Champaign*

Rebecca Benson-Beaver, *Valencia Community College*

Alberto Beron, *Moorpark College*

Abraham K. Biggs, *Broward Community College*

Vernon L. Bridges, *Durham Technical Community College*

John F. Close, *Salt Lake Community College*

Cheryl B. Davids, *Central Carolina Technical College*

Donna Foster, *Piedmont Technical College*

Miguel San Miguel Gonzalez, *Texas A&M International University*

Lisa Grenier, *Pima Community College, Downtown Campus*

Loretta Griffy, *Austin Peay State University*

Paige Hamersma, *Manatee Community College, Bradenton Campus*

Beth E. Hempleman, *Miracosta College*

Karen Jensen, *Southeastern Community College*

Nancy Johnson, *Broward Community College, North Campus*

Steve Kahn, *Anne Arundel Community College*

Robert Kaiden, *Lorain County Community College*

Helen C. Kirk, *Palo Alto College*

Kristi Laird, *Jackson Community College*

Robert Maynard, *Tidewater Community College*

Paul Wayne Lee, *St. Philip's College*

Nenette Loftsgaarden, *University of Montana*

Cherry Mauk, *Kutztown University*

Frank L. Miller, *Orange Coast College*

Linda J. Murphy, *Northern Essex Community College*

Donna E. Nordstrom, *Pasadena City College*

Jody Rooney, *Jackson Community College*

Hassan Saffari, *Prestonburg Community College*

Ellen Sawyer, *College of DuPage*

Jolene Schoenherr, *Concord High School*

Sounny Slitine, *Palo Alto College*

Marcia Swope, *Broward Community College*

Dr. Sharon Testone, *Onondaga Community College*

Lucio Della Vecchia, *Daytona Beach Community College*

Kathy Wagner, *Prestonburg Community College*

Johanna Wirbel, *Washtenaw Community College*

Kevin Yokoyama, *College of the Redwoods*

My deepest thanks to the following teachers and their students for their willingness to class test the text while in manuscript form. Kenneth Benson, University of Illinois at Urbana-Champaign; Vera Brennan, Ulster County Community College; Karen Jensen, Southeastern Community College; Dave Pritchard, Midlands Technical College; Sounny Slitine, Palo Alto College; and Ellyn Webb, Midlands Technical College.

I would like to extend special thanks to Aimee Tait, who encouraged me to pursue this endeavor and put me in touch with Jennifer Crum and Jason Jordan, who gave me the opportunity. I am forever indebted to Jennifer Crum for believing in me and encouraging me throughout the entire process. My heartfelt thanks to Ruth Berry, who kept me on track, and to Bobbie Lewis, whose keen eyes and superb advice were crucial during the development of the manuscript. I would be remiss without also saying thank you to John Hornsby, who answered all my questions about the business.

A very special thank you to Pamela Watkins and Carol Nessmith for their inspiration and contribution to the development of the study system. Special thanks to Cheryl Davids, Donna Foster, Paul Lorczak, Susanne Lynch, Linda Murphy, Nancy Nickerson, and Steve Ouellette for accuracy checking the manuscript.

Thank you to Susan Raymond, who created the beautiful design, and to Ron Hampton, who supervised the production process. Thanks to Jennifer Carley and all the folks with Pre-Press Company who put together the finished product. A big thank you to Michelle Lulos, whose copyedits were fabulous.

To Ruth Berry, Sheila Spinney, Lorie Reilly, Kimberly Bock-Schutz, Sharon Smith, Greg Erb, and all the people involved in the development of the supplements package, my deepest thanks. To my colleague, Ellyn Webb, thank you for the exceptional quality of the solutions manuals, and thank you to Abby Tanenbaum and Sheri Minkner for the superb Printed Test Bank.

Finally, I'd like to dedicate this work in thanks to my parents, Tom and Janice, who gave me all that I am, and to my wife Laura for her unwavering support and love.

Tom Carson

To the Student

This is one of the first questions people often ask when they find out they must take an algebra course, especially when they believe that they will never use the math again. For the most part, it is true that you probably will not use algebra directly in daily life. However, if you think about it, we do not use very much of any general education subject in our daily lives. To be functional, we just need to be able to read a little, write a little, and know enough arithmetic to balance a check book. Some might argue that even these skills are not really necessary! So why do we study anything at all? What is the real point of education? Why don't colleges just train us for the job we want? The purpose of education is not training but exercise—mental exercise. An analogy that illustrates this quite well is the physical training of athletes.

During the off-season, athletes will usually develop an exercise routine that may involve weight lifting, running, swimming, aerobics, or maybe even dance lessons. Athletes will often seek out a professional trainer to push them further than they might push themselves. The trainer's job is not to teach an athlete better technique in his or her sport, but to work on the athlete's raw material—to work the body for more strength, stamina, balance, etc. Educators are like physical trainers, and going to college is like going to the gym. An educator's job is to push people mentally and work the "muscle" of the mind. A physical trainer will design an exercise program to develop the raw material the athlete will need to be competitive in his or her sport. Likewise, a college program is designed to develop the "raw material" of intellect to be competitive in the job market. After the athlete completes the off-season exercise program, he or she will return to the coach and receive specific technique training. Similarly, when students complete their college education and begin a job, they receive specific training to do that job. If we trainers, physical and mental, have done our jobs well with hard-working clients, the coaching or job training should be absorbed easily.

Take this analogy a step further. I think it is safe to say that all athletes need strength, stamina, coordination, and balance. The best athletes have a good dose of all of these things. Employers want employees that have good communication skills, people skills, and problem-solving abilities. The courses in a college curriculum are designed to develop these areas, and the assignments in the courses are merely exercising the parts of the mind that are responsible for these skills. Writing and speaking activities develop the communication skills; social and behavioral science courses develop relational and people skills; math, logic, and computer programming courses develop problem-solving skills.

Now consider the exercises themselves. Why do difficult assignments? Why does your algebra teacher want you to do things the "long way?" Why does your English teacher want you to write a huge research paper with a bibliography? These are valid questions. Consider the athlete again. If the physical trainer gave him or her a one-pound hand weight and said to do only ten curls with it, the athlete wouldn't get much out of the exercise. A good physical trainer will find the athlete's weaknesses and design exercises that the athlete has never performed before, then push him or her accordingly. Teachers do the same thing. The assignments are difficult so that they work the mind effectively. When I hear students groan about doing things the "long way," it lets me know I am working something, just like when the athlete groans a bit from the strain on his or her muscles. So keep in mind as you go through your courses that if you feel "brain-strained," that's a good sign that you are working something, and you should keep it up.

Following is a study system that is designed to help you in your academic workouts. I find that most students that struggle with mathematics have never really studied math. The thinking is, "If I pay attention in class then that should be all I need to do." However, when you watch a teacher do math, keep in mind you are watching a pro. Going back to the athlete analogy, you can't expect to be able to shoot perfect 3-pointers by watching Michael Jordan play basketball. Nor could you expect to shoot a round of 68 in golf by watching Tiger Woods. You have to work at things yourself in order to learn and improve. The study system outlined in the following pages is designed to get you organized and make efficient use of your time so that you can maximize the benefits of your course work.

What do I need to do to succeed?

I think there are four prerequisites that one must have or acquire in order to succeed in college:

1. Positive Attitude
2. Commitment
3. Discipline
4. Time

A **Positive Attitude** is most important because commitment and discipline flow naturally from it. Consider Thomas Edison, inventor of the lightbulb. He tried over 2000 different combinations of materials for the filament before he found the right combination that worked the light bulb. When asked by a reporter about all his failed attempts Edison replied, "I didn't fail once, I invented the lightbulb. It was just a 2000-step process." Now that's how to put a positive spin on struggling towards success! Recognize that learning can be uncomfortable and difficult, and mistakes are part of the process. So erase the "*t*" from "*can't*" and embrace the learning process with all its discomforts and difficulties and you'll see how easy it is to be committed and disciplined.

Commitment means giving everything you've got with no turning back. Consider Edison again. Imagine the doubts and frustrations he must have felt trying material after material for the filament of his lightbulb without success. Yet he forged ahead. Quitting was simply not an option. In Edison's own words, "Our greatest weakness lies in giving up. The most certain way to succeed is always to try just one more time."

Discipline means doing the things you should be doing even when you don't want to. According to W. K. Hope, "Self discipline is when your conscience tells you to do something and you don't talk back." Staying disciplined can be difficult given all the distractions in our society. The best way to develop discipline is to create a schedule and stick to it.

With respect to **time**, make sure you have enough time to study properly, and also make sure that you manage that time wisely. Too often, students try to fit school into an already full schedule. Take a moment to complete the exercise that follows to make sure you haven't committed yourself to too much. Once you have a sense of how much time school requires, read on about the study system that will help you maximize the benefits of your study time.

To put it all together, the elements work something like this:
A positive attitude will lead you to commit yourself whole-heartedly and discipline yourself so that you manage your time wisely.

How do I do it all?

Now that we know a little about what it takes to be successful, let's make sure that you have enough time for school. In general humans have a maximum of about 60 hours of productivity per week. A 60-hour work week (ten hours a day, six days a week) is about the most a person can handle. Therefore, as a guide, let's set the maximum number of "work" hours, which means job(s) and school combined, at 60 hours per week. Use the following exercise to figure your combined job/school commitment.

Exercise: Figure your job–school commitment.

1. Calculate the total hours you work in one week.
2. Calculate the number of hours you are in class each week.

   NOTE: The number of credits that a course is worth is usually the number of hours that the class meets each week.

3. Estimate the number of hours you should expect to spend outside of class studying. A general rule is to double the number of hours spent in class. For example, if a class meets three hours per week, expect to spend six hours outside of class each week.

4. Add your work hours, in-class hours, and estimated outside-of-class hours to get your total time commitment.

5. Evaluate the results.

Evaluating the Results:

a. If you are over 60 hours total, you will probably find yourself overwhelmed. It may not occur at first, but doing that much for extended periods of time will eventually catch up with you. Eventually, something may suffer, whether it is school, job, or family. If you are over 60 hours, it is in your best interest to cut back on work or school until you get your time commitment under 60 hours.

b. If you are under the 60 hours, good. However, no matter how many hours of commitment you have, consider other circumstances in your life, such as your children, spouse, health problems, and anything that could make demands on your time. Make sure that you have enough time for everything you put in your life. Above all, never let the desire to carry through your program by taking full loads every semester overrule your sensibility. Many people find themselves in a bad position because they feel they have to be a full-time student and end up failing courses because they have a heavier load and do poorly. This type of situation not only hurts their GPA but it also slows their academic progress because they have to repeat courses.

How do I make the best of my time? How should I study?

I've seen many students who had been making Ds and Fs in mathematics transform their grades to As and Bs by using the study system that follows. Here are a few testimonials from former students:

"I had never done well in math courses in high school. Using this study system, I made a 100 on a math test for the first time in my life."

—Allen Creel

"I never knew how to study mathematics. Using this study system, I've made As and Bs in each math course since taking the prealgebra-level course, and I'm about to go into trigonometry and calculus."

—Lonnie Villarino

"After being out of school for 26 years, I returned to college with a fear of mathematics. I had never done well in math in high school and continued to struggle in the college Prealgebra course, but as I began to understand how to use the study system, my grades improved. On the last test in the course, I made the highest grade in the class. Though I didn't pass the course because of my low test scores from earlier in the semester, I knew I had overcome my math struggles. When I repeated the course and used the study system, I not only made an A, I learned enough to exempt the next course—Beginning Algebra. Using these study techniques, I've made As in every course since."

—Ruth Gilbert

YOUR NOTEBOOK:

1. Get a loose-leaf binder so that you can put papers in and take them out without ripping.

2. Organize the notebook into four parts:
   a. Class notes
   b. Homework
   c. Study sheets (a single piece of paper for each chapter onto which you will transfer procedures from your notes)
   d. Practice test

IN CLASS: Involve your mind completely.

1. **Take good notes.** Use three different colors. Most students like using red, blue, and black (pencil).
   - Use the red pen to write definitions. Also, use this color to mark problems or items that the instructor indicates will be covered on a test.
   - Use the blue pen to write procedures and rules.
   - Use the pencil to write problems and explanations.

   When taking notes, don't just write the solutions to the problems that the instructor works out, but write the explanations as well. Out to the side of the math, make notes on the steps so that you understand the significance of each step. Also, pay attention to examples or issues the instructor seems to be stressing. Mark these because they will usually appear on a test. Also, include common errors that the instructor points out or any words of caution. If you have difficulty writing everything and paying attention at the same time, ask your instructor whether you can record the lectures with a small tape recorder. Also, if your instructor follows the text closely, when he or she points out definitions or procedures in the text, highlight them or write a page reference in your notes. You can then write these referenced items in their proper place in your notes after class.

2. **Answer the instructor's questions.** This does not mean you have to answer every question verbally, but you should think through every question and answer in your mind, on your paper, or out loud if you are comfortable.

3. **Ask questions.** You may find it uncomfortable to ask questions in front of other people, but keep in mind that if you have a question, then it is very likely that someone else has the same question. If you still don't feel like asking in class, then be sure to ask as soon as class is over. The main thing is to get that question answered as soon as possible because in mathematics, one misconception can fester and become a major problem in the future.

AFTER CLASS: Prepare for the next class meeting as if you are going to have a test on everything covered so far.

To make the most of your time, set aside a specific time that is reserved for math. Since there are often too many distractions at home, do your math study while on campus in a quiet place such as the library or tutorial lab. Also, staying on campus allows you to visit your instructor or tutorial services if you have a question that you cannot resolve. Here is a systematic approach to use in organizing your math study time outside of class:

1. As soon as possible, go over your notes. Clarify any sentences that weren't quite complete. Fill in any page-referenced material.

2. Read through the relevant section(s) in the text again, and make sure you understand all the examples.

3. Transfer each new procedure or rule to your study sheet for that chapter. You might also write down important terms and their definitions. A good way to approach your study sheets is to make headings for each objective in the section(s) you covered that day. Write the procedures and definitions in your own words. Don't feel like you have to duplicate the wording in the text or your instructor's words. The important thing is to write in language that makes sense to you.

4. Study the examples worked in class. Transfer each example to the practice test section of your notebook. Do not write the solution. Just write the problem and its instructions and leave room to work it out later.

5. Use your outline to do the assigned practice problems. As soon as you finish each problem, check your answer in the back of the book or in the solutions manual. If you did not get it correct, then troubleshoot it immediately. If you are asked to do even-numbered problems, then work odd-numbered problems that are just like the even problems. That way you can check your answers for the odd-numbered problems and then work the even-numbered problems with confidence.

Troubleshooting: For the problems you do not get correct, first look for simple arithmetic errors. If you find no arithmetic errors, then make sure you followed the outline and rules correctly. If you followed the rules correctly and are still getting it wrong, this usually means you have interpreted something incorrectly either with the problem or with the rules. Go back and read the instructions again carefully and try to find similar examples in your notes or the book. If after this you still can't find the mistake, then it is best to walk away or go on to something else for a while. Many times we just need a break, and upon taking a fresh look we see the mistake right away. If all these tips fail to resolve the problem, then mark it as a question for the next class meeting.

---

6. After completing the homework, prepare a quiz for yourself. Select from the homework one of each type of problem. Don't just pick the easy ones! Set the quiz aside for later.

7. After making the quiz, study your study sheet. To test your understanding, write the objective headings on a blank piece of paper and see whether you can write an explanation of each objective. Again, do not focus on memorizing wording. The important thing is for you to evaluate how well you understand what you are learning. If you can explain each objective in writing then you have a good understanding of the material.

8. Now it is time to begin preparing for the next class meeting. Read the next section(s) to be covered. Don't worry if you do not understand everything. The idea is to get some feeling for the topics to be discussed so that the class discussion will actually be the second time you encounter with the material, not the first. While reading, you might mark points that you find difficult so that if the instructor does not clear them up, you can ask about them. Also, attempt to work through the examples. Again, the

idea is for you to try to do as much as possible on your own before class so that the in-class discussion merely ties loose ends together and solidifies the material.

9. After you have finished preparing for the next day, go back and do the quiz that you made. If you can do the quiz without having to ponder each problem and if you get all the answers correct, then you have the material down. If you have to stop and think about how to do any of the problems, then you don't have it solid. If it isn't solid, return to your study sheet and repeat the exercise of writing explanations for each objective.

How do I ace the test?

PREPARING FOR A TEST: If you have followed all of the suggestions above, then preparing for a test should actually be quite easy.

1. **Read.** In one sitting (no breaks), read through all of your notes on the material to be tested. In the same sitting, read through the book, observing what the instructor has highlighted in class. The example problems the instructor does will usually reflect what is to be on the test.

2. **Study.** Compare your study sheet to the summary in the book at the end of the chapter. Use both to guide you in your preparation, but keep in mind that the one you've made came from your notes and reflects what the instructor has emphasized. Make sure you understand everything on your study sheet. Write explanations of the objectives until you eliminate all hesitation about how to approach an objective. The rules and procedures should become second-nature.

3. **Practice.** When you have the rules and procedures down, do the practice test at the end of the chapter, as well as the practice test you've made. As you complete the tests, take note of how quickly you recall how to do each problem. Did you launch right into the procedure or did you pause a moment to remember what to do? If you paused, then you need to work on the rules and procedures a little more.

4. **Evaluate.** Once you have completed the practice tests, check them. The answers to the practice tests in the book are in the Appendix. Check the practice test that you made using your notes. (Remember, you made the practice test from the examples done in class.)

5. **Repeat.** Keep repeating steps 2–4 until you get every problem right on the practice tests.

TAKING A TEST:

1. When the test hits your desk, don't look at it. Instead, do a memory dump. A memory dump is just what it sounds like: dump everything you think you might forget on paper. Write out rules, procedures, notes to yourself, things to watch out for, special instructions from the instructor, etc. This will allow you to relax and cruise through the test stress-free.

2. If you get to a problem that you cannot figure out, skip it and come back to it after finishing the rest. Do all the problems you are certain of first and come back to any you were not sure about afterwards.

3. Use all the time given. If you finish early, check to make sure you have answered every problem. Even if you cannot figure out a problem, at least guess. Any remaining time should be used to check as many problems as possible by doing them over on separate paper.

IF YOU ARE NOT GETTING GOOD RESULTS

Evaluate the situation. What are you doing or not doing in the course? Are you doing all the homework and taking the time to prepare as suggested? Sometimes people misjudge how well they have prepared. If you are doing all the homework regularly and preparing for the tests yet are still not getting good results, then evaluate how you are preparing. Are you really preparing as thoroughly as suggested? Just like an athlete, to excel, you will need to prepare beyond the minimum requirements.

Here are some suggestions:

1. **Go to your instructor.** Ask your instructor for help in evaluating what is wrong. Remember that the instructor's office hours are for helping students. Don't feel like you are imposing on the instructor's time, because it is actually your time with him or her.

2. **Get a tutor.** Most schools have tutorial services. If your school does have a tutorial service, then go there to do your homework so that if you have a question, tutors will be right there to help.

3. **Use Addison Wesley Longman's Support materials.** Use the support materials that are available with your text, which include a *Student's Solutions Manual,* the AWL Tutor Center, videotapes (available on CD-ROM as well), tutorial software, and a Web site. Full descriptions of these supplements are provided on pages ix and x of this book.

4. **Join a study group.** Get together regularly with a few people from class and go over material together. Quiz each other and answer questions. Also, the study group can be of help if you miss a class. You can just contact someone in your group for the assignments and notes. Be careful how you use the study group, though. Don't let it be a session where one person does the homework and the others just follow along. You should meet with the group only after you have done your own preparation. Then you can compare notes or discuss problems with which you had difficulty.

Learning Styles Inventory

What is your personal learning style?

A learning style is the way in which a person processes new information. Knowing your learning style can help you make choices in the way you focus on and study new material. Below are fifteen statements that will help you assess your learning style. After reading each statement, rate your response to the statement using the scale below. There are no right or wrong answers.

3 = Often applies   **2** = Sometimes applies   **1** = Never or almost never applies

_____ **1.** I remember information better if I write it down or draw a picture of it.

_____ **2.** I remember things better when I hear them instead of just reading or seeing them.

_____ **3.** When I receive something that has to be assembled, I just start doing it. I don't read the directions.

_____ **4.** If I am taking a test, I can "see" the page of text or lecture notes where the answer is located.

_____ **5.** I would rather the professor explain a graph, chart, or diagram to me instead of just showing it to me.

_____ **6.** When learning new things, I want to "do it" rather than hear about it.

_____ **7.** I would rather the instructor write the information on the board or overhead instead of just lecturing.

_____ **8.** I would rather listen to a book on tape than read it.

_____ **9.** I enjoy making things, putting things together, and working with my hands.

_____ **10.** I am able to conceptualize quickly and visualize information.

_____ **11.** I learn best by hearing words.

_____ **12.** I have been called hyperactive by my parents, spouse, partner, or professor.

_____ **13.** I have no trouble reading maps, charts, or diagrams.

_____ **14.** I can usually pick up on small sounds like bells, crickets, frogs, or distant sounds like train whistles.

_____ **15.** I use my hands and gesture a lot when I speak to others.

Write your score for each statement beside the appropriate statement number below. Then add the scores in each column to get a total score for that column.

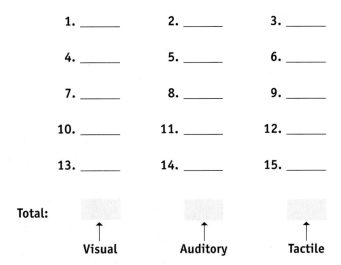

1. _____   2. _____   3. _____

4. _____   5. _____   6. _____

7. _____   8. _____   9. _____

10. _____   11. _____   12. _____

13. _____   14. _____   15. _____

Total:

↑   ↑   ↑

Visual   **Auditory**   **Tactile**

The largest total of the three columns indicates your dominant learning style.

Visual learners learn best by seeing. If this is your dominant learning style then you should focus on study strategies that involve seeing. The color coding in the study system will be especially important. The same color coding is used in the text. Draw lots of diagrams, arrows, and pictures in your notes to help you see what is happening. Reading your notes, study sheets, and text repeatedly will be an important strategy.

Auditory learners learn best by hearing. If this is your dominant learning style then you should use study strategies that involve hearing. Bring a tape recorder to class to record the discussion. When you study your notes, play back the tape. Also, when you learn rules, say the rule over and over. As you work problems, say the rule before you do the problem. You may also find the video tapes beneficial in that you can hear explanations of problems taken from the text.

Tactile (also known as Kinesthetic) learners learn best by touching or doing. If this is your dominant learning style, you should use study strategies that involve doing. Doing lots of practice problems will be important. Make use of the margin exercises. Watch for the "Do Margin" prompts in the text. These are designed so that as soon as a topic is developed, you are given an opportunity to do problems like the examples. Writing out your study sheets and doing your practice tests repeatedly will be important strategies for you.

Note that the study system developed in this text is for all learners. Your learning style will help you decide what aspects and strategies in the study system to focus on, but being predominantly an auditory learner does not mean that you shouldn't read the textbook, do lots of practice problems, or use the color-coding system in your notes. Auditory learners can benefit from seeing and doing, and tactile learners can benefit from seeing and hearing. In other words, do not use your dominant learning style as a reason for not doing things that are beneficial to the learning process.

This learning styles inventory is adapted from *Cornerstone: Building on Your Best* by Montgomery/Moody/Sherfield, © 2000. Reprinted by permission of Prentice-Hall, Inc., Upper Saddle River, NJ.

Whole Numbers

Taking Notes

Good note taking is essential to success in college. If you have not read through the study system in the *To the Student* section at the beginning of this text, take a moment to do so now. You will find that the text follows the color-coding suggested in the note-taking discussion. Definitions will appear in red. Procedures and rules will appear in blue.

When taking notes, do your best to record not only what your instructor writes but also what your instructor says about example problems. Outline the thought process involved in solving the problems. Note the instructor's cautions and warnings about common errors. If your instructor follows the text, then be sure to include page references in your notes and highlight important points in the text.

Of course, it is difficult to try to understand everything as it's happening and to write it all down at the same time. Try to write down the essence of the sentence and leave out nonessential words like "the." Also develop codes for words that are used frequently. For example, $\therefore$ could mean "therefore" and $+$ could mean "and." It is also a good idea to tape lectures.

As soon as possible after class, tidy up your notes or even rewrite them. If you tape the lecture, you can listen to the tape as you rewrite the notes. This will allow you to fill in those nonessential words and replace any symbols with actual words, if needed. Taping will also allow you to write more slowly so that the notes will be more legible.

"The strokes of the pen need deliberation as much as the sword needs swiftness."
—Julia Ward Howe

"As to the adjective, when in doubt, strike it out."
—Mark Twain

"Good writing is clear thinking made visible."
—Bill Wheeler

1.1 Introduction to Numbers, Notation, and Rounding

OBJECTIVES

1 Name the digit in a specified place.

2 Write numbers in standard and expanded form.

3 Write the word name for a given number.

4 Use <, >, or = to make a true statement.

5 Round numbers to a specified place.

What are numbers? They're not things we can see or touch. So what are they?

DEFINITION   **Numbers:** Amounts or quantities.

What do we do with numbers, especially if we cannot hold them or see them? We can use numbers to describe measurements, such as how much liquid is in a container or how fast an object is traveling or how much distance is between objects. If a number is used to describe a measurement, it will have a unit attached, such as feet (ft.) or meters (m) or seconds (sec).

For some numbers and ideas in mathematics, no practical use has yet been found. Many people merely play with mathematics for the sheer beauty and enjoyment of it. In this way, mathematics is like art.

Let's be a little more specific about some numbers. Numbers are classified into groups called **sets.** In this first chapter, we'll focus on the set of **whole numbers.** Within the set of whole numbers is a **subset** of numbers called **natural numbers.**

DEFINITIONS   **Set:** A group of elements.

Subset: A set within a set.

Natural numbers: The natural numbers are 1, 2, 3, . . .

Whole numbers: The whole numbers are 0, 1, 2, 3, . . .

> Three periods mean that the numbers continue forever.

All natural numbers are also whole numbers. This is why the set of natural numbers is a subset of the set of whole numbers. But not all whole numbers are natural. The number 0 is a whole number but not a natural number. In a box diagram it looks something like this:

We use symbols to represent the numbers. We call these symbols *numerals* or *digits*. There are 10 numerals in our numeral system:

$$0, 1, 2, 3, 4, 5, 6, 7, 8, 9$$

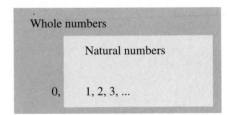

Whole numbers

Natural numbers

0,   1, 2, 3, ...

Our system is a base-10 system, which means we have only 10 symbols. To write numbers beyond 9, we use combinations of numerals to represent the larger numbers, called a *place value system*.

Historical Note

Numerals have been written many different ways throughout history. It is generally accepted that our modern numerals are derived from Hindu-Arabic forms. The Babylonians used wedge-shaped marks in clay tablets to represent numbers. The ancient Greeks used the letters of their alphabet to represent numbers. We still find Roman numerals in use in clocks, in architecture, in books, and even in numbering the Super Bowls.

Roman

Greek

Egyptian

Chinese

OBJECTIVE 1 *Name the digit in a specified place.*

Our Place Value System

Figure 1.1 Place Values

Place Values			
Billions period	Millions period	Thousands period	Ones period
Hundred billions — Ten billions — Billions	Hundred millions — Ten millions — Millions	Hundred thousands — Ten thousands — Thousands	Hundreds — Tens — Ones

Notice how the places are grouped in threes. These groups are called *periods*. When we write numbers using the place value system, we say we are writing in *standard form*. When writing numbers in standard form, we separate the periods with commas. The chart extends to the left indefinitely.

> **Tip**
> Think of the place value chart as the money tray in a cash register. A digit in a particular place is like having that many bills in that tray of the register. For example, if we place a numeral like 7 in the thousands place, it's the same as having 7 one-thousand-dollar bills in the register, which is worth a total of $7000.

EXAMPLE 1 What digit is in the thousands place in 209,812?

Answer: 9

Explanation: The digit in the thousands place is the fourth digit from the right.

Do Margin 1. ▶

OBJECTIVE 2 *Write numbers in standard and expanded form.*

When we write a number in standard form such as 430, what are we saying?

Think of the money tray analogy.

The 4 in the hundreds place means 4 hundreds.

The 3 in the tens place means 3 tens.

The 0 in the ones place means 0 ones.

If we add those amounts together we have 430.

4 hundreds + 3 tens + 0 ones = 430

or

4 hundreds + 3 tens = 430

We do not need to represent the 0 ones because it means there is a lack of bills in that slot of the money tray and mathematically the 0 will not affect the total.

◆ **Margin 1**

a. What digit is in the hundred thousands place in 62,407,981?

b. What digit is in the ten millions place in 417,290,006?

Answers: **a.** 4 **b.** 1

Write in expanded form.

a. 82,469

b. 7,082,049

c. 410,159,200

Mathematically 430 is the result of multiplying the 4 times 100, the 3 times 10, and the 0 times 1 and then adding it all up.

$$4 \times 100 + 3 \times 10 + 0 \times 1 \qquad \text{or just} \qquad 4 \times 100 + 3 \times 10$$

A number written this way is in *expanded notation* or *expanded form*.

Procedure *To write a number in expanded form:*
1. Write each digit multiplied by its place value.
2. Express it all as a sum.

EXAMPLE 2 Write 57,483 in expanded form.

Answer: $5 \times 10,000 + 7 \times 1000 + 4 \times 100 + 8 \times 10 + 3 \times 1$

Explanation: We simply wrote each digit multiplied by its respective place value and expressed it all as a sum. (Notice if we were to calculate, we'd get 57,483.)

EXAMPLE 3 Write 4,705,208 in expanded form.

Answer: $4 \times 1,000,000 + 7 \times 100,000 + 5 \times 1000 + 2 \times 100 + 8 \times 1$

Explanation: We simply wrote each digit (except 0's) multiplied by its respective place value and expressed it all as a sum.

◀ **Do Margin 2.**

Let's write expanded form in standard form.

◆ **Margin 3**

Write in standard form.

a. $2 \times 100,000 + 3 \times 10,000 + 1 \times 1000 + 5 \times 100 + 9 \times 10 + 8 \times 1$

b. 3 hundred thousands + 5 thousands + 9 tens + 7 ones

c. $9 \times 100,000,000 + 1 \times 10,000 + 4 \times 1000 + 7 \times 100$

Procedure *To change a number from expanded form to standard form, write each digit in the place indicated by the corresponding place value.*

EXAMPLE 4 Write $6 \times 10,000 + 9 \times 1000 + 2 \times 100 + 5 \times 10 + 3 \times 1$ in standard form.

Answer: 69,253

Explanation: We wrote each digit in the place indicated by the corresponding place value.

EXAMPLE 5 Write $9 \times 1,000,000 + 2 \times 10,000 + 7 \times 100 + 9 \times 10 + 3 \times 1$ in standard form.

Answer: 9,020,793

Explanation: Notice that some places were skipped. There were no hundred thousands and no thousands in the expanded form so we placed 0's in the hundred thousands and thousands places in the standard form.

◀ **Do Margin 3.**

Answers to Margin 2: **a.** $8 \times 10,000 + 2 \times 1000 + 4 \times 100 + 6 \times 10 + 9 \times 1$ **b.** $7 \times 1,000,000 + 8 \times 10,000 + 2 \times 1000 + 4 \times 10 + 9 \times 1$ **c.** $4 \times 100,000,000 + 1 \times 10,000,000 + 1 \times 100,000 + 5 \times 10,000 + 9 \times 1000 + 2 \times 100$

Answers to Margin 3: **a.** 231,598 **b.** 305,097 **c.** 900,014,700

OBJECTIVE 3 *Write the word name for a given number.*

The word name for a number is the way we speak the number.

> **Procedure** *To write a word name, work from left to right through the periods.*
> **1.** Write the name of the digits in the left-most period.
> **2.** Write the period name followed by a comma.
> **3.** Repeat steps 1 and 2 until you get to the ones period. We do not follow the ones period with its name.

WARNING Do not write the word *and* anywhere in a word name for a whole number. The word *and* takes the place of a decimal point. We'll see this in Chapter 6.

EXAMPLE 6 The Earth is on average 92,958,349 miles from the sun. Write the word name for 92,958,349.

Answer: Ninety-two million, nine hundred fifty-eight thousand, three hundred forty-nine.

Explanation: Starting at the left, we wrote 92 as "ninety-two" and then stated the period name, "million." Next we wrote 958 as "nine hundred fifty-eight" with its period name, "thousand." Finally, we wrote 349 as "three hundred forty-nine." Since this was the ones period we did not say the period name.

Do Margin 4.

OBJECTIVE 4 *Use <, >, or = to make a true statement.*

When we use an equal sign to show that two amounts are equal, we call the statement an **equation.**

> **DEFINITION** **Equation:** A mathematical statement that contains an equal sign.

$12 = 12$ is an equation. It is read as *twelve is equal to twelve.* Because 12 is equal to 12, we say the equation is true. An equation can be false as well. $12 = 9$ is an equation that is false because 12 and 9 are not equal.

When two amounts are not equal, we can use inequality symbols to indicate which amount is larger. A statement that contains an inequality symbol is an **inequality.**

> **DEFINITION** **Inequality:** A mathematical statement that contains an inequality symbol.

The standard inequality symbols are:

Greater than symbol: $>$

Less than symbol: $<$

In comparing 12 and 9 we can write a true statement by using an appropriate symbol.

We can say: 12 <u>is greater than</u> 9   9 <u>is less than</u> 12

Translation: $12 > 9$ $9 < 12$

> **Tip**
> Notice the inequality symbol always opens to the larger value and points at the smaller value.

♦ **Margin 4**

Write the word name.

a. 44,568 (1997 median family income; U.S. Census Bureau)

b. 8,509,000,000 (1998 Gross Domestic Product; U.S. Census Bureau)

c. 847,716 (Diameter of the sun in miles)

Answers: **a.** Forty-four thousand, five hundred sixty-eight. **b.** Eight billion, five hundred nine million **c.** Eight hundred forty-seven thousand, seven hundred sixteen.

EXAMPLE 7 Use $<$, $>$, or $=$ to make a true statement.

a. 159,208 ? 161,000

Answer: 159,208 $<$ 161,000

Explanation: Because 159,208 is smaller than 161,000 we need to use the less than symbol. The inequality is read as "159,208 is less than 161,000."

b. 48,090 ? 12,489

Answer: 48,090 $>$ 12,489

Explanation: Because 48,090 is larger than 12,489 we need to use a greater than symbol.

◀ **Do Margin 5.**

OBJECTIVE 5 *Round numbers to a specified place.*

The speed of light in a vacuum is 299,792,458 meters per second (abbreviated m/s). However, because this value is rather tedious to say and work with, it is usually rounded to 300,000,000 m/s. We round numbers to make them easier to communicate. We will see later that we can round numbers to work out an estimate of a calculation.

When we round, we must decide on a place value or have a place value specified. We then have to determine whether the given number is closer to the nearest whole value above or below it in the specified place.

EXAMPLE 8 Round 46,357,204 to the nearest million.

Answer: 46,000,000

> **WARNING** Be sure to include all 0's after the rounded place. Instead of 46, be sure to write 46,000,000. Think about money. $46 is very different from $46,000,000.

Explanation: The nearest whole millions to 46,357,204 are 46,000,000 and 47,000,000. We had to decide which whole million it was closer to. Because 46,357,204 is below the halfway point, the number 46,500,000, we rounded down to 46,000,000.

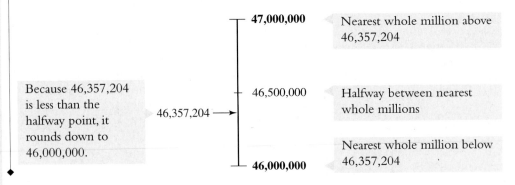

Because 46,357,204 is less than the halfway point, it rounds down to 46,000,000.

46,357,204 →

47,000,000 — Nearest whole million above 46,357,204

46,500,000 — Halfway between nearest whole millions

46,000,000 — Nearest whole million below 46,357,204

Notice that the digit to the right of the place to be rounded determines whether the number is above or below the halfway point. What if this digit were a 5, that is, at the halfway point? Since the digit is no closer to the number above than the number below, we could round either way. Mathematicians agree to always round up in these cases so that everyone rounds in a consistent way.

◆ Margin 5

Use $<$, $>$, or $=$ to make a true statement.

a. 15,907 ? 15,906

b. 1,291,304 ? 1,291,309

c. 64,108 ? 64,108

d. 24,300 ? 25,300

Answers: **a.** 15,907 $>$ 15,906
b. 1,291,304 $<$ 1,291,309
c. 64,108 $=$ 64,108
d. 24,300 $<$ 25,300

Conclusion: Because we can look at the digits to round, we don't have to draw a scale anymore. We can simply use the following procedure:

Procedure *To round a number to a given place value, consider the digit to the right of the desired place value.*

If this digit is 5 or greater, round up.

If this digit is 4 or less, round down.

EXAMPLE 9   Round 43,572,991 to the specified place:

a. hundred thousands

Solution: 43,**5**72,991

hundred thousands ↗ Because this digit is greater than 5 we round up.

Answer: 43,600,000

b. ten thousands

Solution: 43,5**7**2,991

ten thousands ↗ Because this digit is less than 4 we round down.

Answer: 43,570,000

c. millions

Solution: 43,**5**72,991

millions ↗ Because this digit is 5 we agree to round up.

Answer: 44,000,000

d. hundreds

Solution: 43,572,**9**91

hundreds ↗ Because this digit is greater than 5 we round up. Note that when we round the 9 in the hundreds place up, it becomes 10. Since 10 hundreds is 1000, we must increase the thousands place by 1 so that the 2 becomes a 3. This process of adding to the next place to the left is often called *carrying*.

◆ **Answer:** 43,573,000

Do Margin 6. ▶

Margin 6

Round 602,549,961 to the specified place.

a. Ten thousands

b. Tens

c. Millions

d. Thousands

e. Hundreds

Answers: a. 602,550,000 **b.** 602,549,960 **c.** 603,000,000 **d.** 602,550,000 **e.** 602,550,000

In the real world, we are not told what place to round to. The idea is to round to the place that makes sense for the situation. Some questions to consider are:

How precise must the numbers be in this situation?

How accurately can the amounts involved be measured?

Are others depending on what we do?

Are lives at stake?

For estimation purposes, a guideline is to round so that there is only one digit that is not 0. This means to round to the place farthest to the left.

EXAMPLE 10   Round each so that there is only one nonzero digit.

a. 36,568

Solution: 36,568

Farthest ↗
left

Because this digit is greater than 5 we round up.

Answer: 40,000

b. 621,905

Solution: 621,905

Farthest ↗
left

Because this digit is less than 4 we round down.

Answer: 600,000

Reminder
When rounding down, the digit in the rounded place remains the same.

◀ **Do Margin 7.**

◆ Margin 7

Round each number so that there is only one nonzero digit.

a. 27,502,341

b. 6,128,200

c. 453,219

Answers: **a.** 30,000,000
b. 6,000,000 **c.** 500,000

1.1 Exercises

FOR EXTRA HELP

 Videotape 1

 InterAct Math Tutorial Software

 www.carsonmath.com

 AWL Math Tutor Center

 InterAct MathXL www.mathxl.com

 Student's Solutions Manual

For Exercises 1–4, name the digit in the requested place in the number 56,324,092.

1. Hundreds

2. Thousands

3. Ten millions

4. Hundred thousands

For Exercises 5–8, name the place held by the 7.

5. 2,457,502

6. 57,414

7. 9,706,541

8. 70,412,581

For Exercises 9–14, write the number in expanded form.

9. 24,319

10. 78,625

11. 5,213,304

12. 2,410,512

13. 93,014,008

14. 86,300,905

For Exercises 15–20, write the number in standard form.

15. 8 thousands + 7 hundreds + 9 tens + 2 ones

16. $9 \times 10{,}000 + 2 \times 1000 + 5 \times 100 + 8 \times 10 + 1 \times 1$

17. $6 \times 1{,}000{,}000 + 3 \times 10{,}000 + 9 \times 1000 + 2 \times 10$

18. 8 hundred millions + 7 thousands + 2 hundreds + 3 ones

19. $4 \times 10{,}000{,}000 + 9 \times 100{,}000 + 8 \times 10{,}000 + 1 \times 100 + 9 \times 1$

20. $7 \times 100{,}000{,}000 + 2 \times 1{,}000{,}000 + 5 \times 10{,}000 + 8 \times 10$

For Exercises 21–26, write the word name.

21. 7768 (diameter of Earth in miles)

22. 29,028 (height of the peak of Mt. Everest in feet)

23. 272,190,000 (estimated resident population of the United States as of April 1, 1999, according to the U.S. Census Bureau)

24. 5,479,000,000,000 (U.S. gross outstanding national debt in 1998)

25. 186,171 (speed of light in mi./s)

26. 299,792,458 (speed of light in m/s)

Trivia Bite

The Sun is 92,958,349 miles from Earth. The light leaving the Sun travels at a speed of 186,171 miles per second. Even at this speed, it takes the light from the Sun a little over 8 minutes to reach Earth.

For Exercises 27–32, use $<$, $>$, or $=$ to make a true statement.

27. 599 ? 899

28. 88,332 ? 88,332

29. 4,299,308 ? 4,298,308

30. 89,900 ? 89,902

31. 609,001 ? 609,001

32. 9911 ? 9199

For Exercises 33–40, round 5,652,992,481 to the specified place.

33. Thousands

34. Hundred millions

35. Millions

36. Ten millions

37. Billions

38. Hundred thousands

39. Hundreds

40. Ten thousands

For Exercises 41–46, round each number so that there is only one nonzero digit.

41. 32,607

42. 281,506

43. 851,220

44. 4,513,541

45. 8723

46. 54,298

47. The distance from Earth to the Sun is 92,958,349 miles. Round the distance to a reasonable place. Why did you round to the place you chose?

48. The U.S. gross national debt was $5,479,000,000,000 in 1998. Round the debt to a reasonable place. Why did you round to the place you chose?

1.2 Adding, Subtracting, and Solving Equations with Whole Numbers

OBJECTIVES

1 Add whole numbers.

2 Estimate sums.

3 Solve applications involving addition.

4 Subtract whole numbers.

5 Solve equations containing a missing addend.

6 Solve applications involving subtraction.

7 Solve applications involving addition and subtraction.

OBJECTIVE 1 *Add whole numbers.*

What are we really doing when we add numbers?

DEFINITION	**Addition:** The arithmetic operation that combines amounts.

When we write an addition sentence, the *addends* are the parts that are added and the *sum* is the answer.

Notation: $3 + 5 = 8$

$\uparrow \quad \uparrow \quad \uparrow$

Addends Sum

Notice how the order of addends can be changed without affecting the outcome.

$5 + 3$ and $3 + 5$ both make a sum of 8.

This fact about addition is called the *commutative property*. The word commutative comes from the root word commute, which means to move from one place to another. That's exactly what happened when we moved the addends—they commuted.

> **Commutative Property of Addition** Changing the order of addends does not affect the sum.
>
> ***In math language*** $a + b = b + a$, where a and b are any numbers.

How do we add? We must add according to place value.

> **Procedure** *To add whole numbers:*
> **1.** Stack with corresponding place values aligned.
> **2.** Add the digits.

EXAMPLE 1 $423 + 64 =$

Solution:
$$
\begin{array}{r} 423 \\ +64 \\ \hline 487 \end{array}
\quad \text{or} \quad
\begin{array}{r} 64 \\ +423 \\ \hline 487 \end{array}
$$

Note: The commutative property allows us to stack either way.

Connection
When we added the 2 and 6 digits we were adding 2 tens and 6 tens, which makes 8 tens.
In expanded form:

$$
\begin{array}{r}
4 \text{ hundreds} + 2 \text{ tens} + 3 \text{ ones} \\
+ 6 \text{ tens} + 4 \text{ ones} \\
\hline
4 \text{ hundreds} + 8 \text{ tens} + 7 \text{ ones} = 487
\end{array}
$$

Explanation: We stacked the numbers with the ones and tens places aligned, then added the corresponding digits. Notice the answer was the same whether we stacked 423 over 64 or 64 over 423.

Sometimes when we add digits, the sum is more than 9, so we have to carry the extra over to the next column.

◆ **Margin 1**

Add.

a. $527 + 41 =$

b. $5802 + 549 =$

c. $59,481 + 8574 =$

EXAMPLE 2 $5408 + 916 =$

Solution:
$$\begin{array}{r} {\scriptstyle 1\ 1} \\ 5408 \\ +916 \\ \hline 6324 \end{array}$$

Explanation: Notice when we add the 8 and 6, we get 14. The 4 is placed in the ones place. The extra 10 in 14 is expressed as a 1 over the tens column of digits and is added with the other tens digits. The same thing happens with the 9 and 4 in the hundreds column. Finally we added the 1 and 5 in the thousands column.

◀ **Do Margin 1.**

When we have three or more addends, we can group them any way we wish and get the same answer each time. Consider $2 + 3 + 4$.

We can add the 2 and 3 first, which is 5, and then add the 4 for a total of 9. Or, we can add the 3 and 4 first, which is 7, then add the 2 to again total 9.

This fact about addition is called the *associative property* because to associate is to group. To write the associative property symbolically, we use parentheses to indicate the different ways we can group the addends.

$(2 + 3) + 4$ indicates add the $2 + 3$ first, then add 4 to the result.
$2 + (3 + 4)$ indicates add the $3 + 4$ first, then add that result to 2.

Either way we get 9.

Associative Property of Addition Grouping three or more addends differently does not affect the sum.

In math language $(a + b) + c = a + (b + c)$, where a, b, and c are any numbers.

The commutative and associative properties allow flexibility in the way we add. We can change the order of the addends or group the addends any way we wish without affecting the sum.

◆ **Margin 2**

Add

a. $48 + 70 + 62 =$

b. $114 + 85 + 30 + 79 =$

c. $35,604 + 907 + 3215 + 42,008 =$

EXAMPLE 3 $35 + 60 + 18 =$

Solution:
$$\begin{array}{r} {\scriptstyle 1} \\ 35 \\ 60 \\ +18 \\ \hline 113 \end{array}$$

Connection
The addition properties tell us we could have added $35 + 60 + 18$ differently and still had 113 as the sum. We could have added 35 and 60 then added the resulting sum of 95 to 18 to get 113. Or we could have added 60 with 18 and then added the resulting sum of 78 to 35 to get 113.

Explanation: We stacked the numbers by place value and added the digits.

◀ **Do Margin 2.**

Answers to Margin 1: **a.** 568
b. 6351 **c.** 68,055

Answers to Margin 2: **a.** 180
b. 308 **c.** 81,734

OBJECTIVE 2 *Estimate sums.*

Often we aren't really concerned about the actual answer to a calculation; we just need a quick approximation. This is called an *estimate*.

To Estimate a Calculation
1. Round the numbers
2. Perform the calculation with the rounded numbers.

An estimate can also serve as a check. The idea is to do the estimate first, then do the actual calculation and make sure the actual amount is reasonable based on the estimate.

EXAMPLE 4 Estimate 68,214 + 4,318 by rounding so that there is only one nonzero digit, then calculate the actual sum.

Estimate: 70,000   **Actual:** 68,214
          +4,000   +4,318
           74,000   72,532

Explanation: To do the estimate, we rounded 68,214 to 70,000 and 4,318 to 4,000. We then added the rounded numbers. To get the actual answer, we added the numbers as they were given. Notice how the estimate is quite close to the actual answer, close enough to know the actual answer is reasonable.

Do Margin 3. ▶

OBJECTIVE 3 *Solve applications involving addition.*

Often we can focus on words in a situation that help us recognize that the situation is addition. We refer to those words as *key words*. Here are some key words and key questions that help us recognize addition:

Key Words for *Addition* Add, plus, sum, total, increased by, in all, altogether, perimeter.
Key Questions   "How much in all?"
                     "How much altogether?"
                     "What is the total?"

EXAMPLE 5 Bob and Kim find out that the flat cost of building the home they have chosen is $112,500. They decide they want to add some extra features. A wood-burning fireplace costs an additional $980. They also want to upgrade the fixtures and appliances at a cost of $2158. What will be the total price of the house?

Solution: 112,500   **Check: We can estimate:** 113,000
                  2,158   2,000
                 +980   +1,000
                $115,638   116,000   This is reasonably close
                                                                    to the actual sum.

Explanation: Because the upgrades are costs in *addition* to the base price, we must add. Also notice the word *total* in the question.

Do Margin 4. ▶

◆ **Margin 3**

Estimate each sum by rounding so that there is only one nonzero digit, then find the actual sum.

a. 67,482 + 8,190 =
  Estimate: Actual:

b. 4,586 + 62 + 871 =
  Estimate: Actual:

◆ **Margin 4**

Below are the base price and prices for upgrades. Estimate the total cost of the house with the upgrades by rounding so that the numbers are easy to work with but are still fairly accurate. (Answers may vary.)

House base price = $125,480

Deck = $1280

Fireplace = $420

Kitchen upgrade = $675

Answers to Margin 3: **a.** Estimate: 78,000, Actual: 75,672
b. Estimate: 5960, Actual: 5519

Answer to Margin 4: 127,000

◆ **Margin 5**

Sarah is planning her budget for the week. She has $642 in her checking account and $283 in her savings account. She owes rent of $450. She knows groceries will be around $100, and she also needs to get her car serviced, which should be $125 according to the advertisement. Will she have enough money to cover these expenses?

Suppose we want to construct a fence around some property. To get a sense of how much fencing material we need, we would measure or calculate the total distance around the space to be fenced. The total distance around a shape is called **perimeter.**

◀ **Do Margin 5.**

DEFINITION | **Perimeter:** The total distance around a shape.

Procedure *To find the perimeter of a shape, add the lengths of all the sides of the shape.*

EXAMPLE 6 The Jensons want to put a border on the walls of their son's room at the ceiling. The room is a 10 ft. by 12 ft. rectangle. How much border material must they buy?

Solution: To find the total amount of material, we need the total distance around the room. It is helpful to draw the picture.

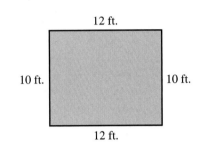

Perimeter = total distance around

$P = 12 \text{ ft.} + 10 \text{ ft.} + 12 \text{ ft.} + 10 \text{ ft.}$

$P = 44 \text{ ft.}$

The Jensons will need at least 44 ft. of border material. Why *at least* 44 ft? Because buying a little more gives them room for error.

Check: Estimate: $10 + 10 + 10 + 10 = 40$ ft.

You probably noticed that we used the letter P in our example. Writing the word *perimeter* each time becomes tedious so we abbreviate to P. A letter or symbol used to represent an

DEFINITION | **Variable:** A symbol that can vary in value.

unknown value is called a **variable.**

The word *variable* comes from the root word *vary*. The word was chosen because the same symbol may have different values in different problems or may even mean something totally different in a different context. For example, the letter P in a geometry context may mean perimeter while in a financial context it might mean principle.

Some symbols do not vary in value. We call these symbols **constants.**

DEFINITION | **Constant:** Any symbol that does not vary in value.

All numerals are constants. Remember 3 is merely a symbol that means *three*. The symbol 3 always means three, so it is a constant.

◀ **Do Margin 6.**

OBJECTIVE **4** *Subtract whole numbers.*

Subtraction is the *inverse* of addition. Addition and subtraction are inverse operations because they *undo* each other. If $5 + 4 = 9$, then we can undo the addition with subtraction like this: $9 - 4 = 5$. There are actually three ways of interpreting subtraction.

DEFINITION | **Subtraction:** 1. Take away
2. Difference
3. Missing addend

◆ **Margin 6**

Mr. Marcus needs to fence a space in his yard for a garden. It is a rectangular space 45 ft. by 60 ft. How much fencing material will he need?

Answer to Margin 4: 210 ft.

Answer to Margin 5: Yes

We interpret subtraction to mean *take away* when we remove an amount from another amount.

We interpret subtraction as *difference* when we must find the distance between two positions. For example, on a highway if you are at mile marker 42 and later you see you are at mile marker 60, you can use subtraction to find the difference between the two positions and thus determine how far you have traveled.

We interpret subtraction as a *missing addend* when an addend, which is part of an addition statement, is missing. For example, in $?$ + 4 = 9, the first addend is missing. This corresponds to having $4, needing $9, and wishing to know how much more is needed to bring us up to that total of $9. We use subtraction to find that missing addend.

Historical Note

Fifteenth-century writers generally used a lowercase m for minus. In the same way that *et* became a + sign, the m was simplified to a − sign.

Source: D. E. Smith, *History of Mathematics*.

Notation: $9 - 4 = 5$

minuend subtrahend difference

Connection
Since addition and subtraction are inverse operations we can check the accuracy of $9 - 4 = 5$ by adding $5 + 4 = 9$.

Let's look at how we subtract. We stack according to place value, just like in addition. However, subtraction is *not* commutative, which means we cannot change the order of the numbers in the minuend and subtrahend, positions. In subtraction if we change the order of the minuend and subtrahend, we get different results.

$7 - 4$ is 3, but what does $4 - 7$ produce? Certainly not 3. When we discuss integers in Chapter 2, we'll be able to answer this question.

Conclusion: To subtract whole numbers, we stack the larger number on top, align the place values, then subtract the corresponding digits.

EXAMPLE 7 $795 - 32 =$

Solution:
$$\begin{array}{r} 795 \\ -32 \\ \hline 763 \end{array}$$

Check:
$$\begin{array}{r} 763 \\ +32 \\ \hline 795 \end{array}$$

Note: We can use the inverse operation to check. Adding the difference to the subtrahend should equal the minuend.

◆ **Explanation:** Because 795 was the larger number, it went on top, and we subtracted the digits.

When one of the digits above is smaller than the digit directly below, we cannot subtract the digit below from the digit above. In these cases, we must rename the number above.

EXAMPLE 8 $8982 - 765 =$

Solution:
$$\begin{array}{r} {}^{7\,1} \\ 89\cancel{8}2 \\ -765 \\ \hline 8217 \end{array}$$

Connection
Think about the cash register analogy. Suppose there are two $1 bills and eight $10 bills in the register, which amount to $82. To give more than two $1 bills out in change, the cashier must break one of the $10 bills. The manager gives the cashier ten $1 bills in exchange for one $10 bill. There are now seven $10 and twelve $1 in the register, which still amount to $82.

Explanation: Because the digit 5 cannot be subtracted from the 2, we had to rename 2 by taking 10 from the 8 tens and adding the 10 to 2 to make 12. This process is often called *borrowing*. When we take 1 ten from the 8 tens it becomes 7 tens. Notice this does not change the value of 8982. We merely moved 1 ten from the tens place over to the ones place so that we
◆ could carry out the subtraction.

Procedure *To subtract whole numbers:*
1. Stack the larger number on top of the smaller number aligning the place values.
2. Subtract the digits in the bottom number from the digits directly above.
3. If the digit above is smaller than the digit below, then rename the digit above.

EXAMPLE 9 $45,002 - 8,473 =$

Solution:

$$
\begin{array}{r}
{\scriptstyle 3}{\overset{1}{\scriptstyle 4}}\,{\overset{9}{\scriptstyle \cancel{0}}}\,{\overset{9}{\scriptstyle \cancel{0}}}\,{\overset{1\,1}{\scriptstyle 2}} \\
\cancel{45,002} \\
-\,8,473 \\
\hline
36,529
\end{array}
$$

Check:

$$
\begin{array}{r}
{\scriptstyle 1}\ {\scriptstyle 11} \\
36,529 \\
+\,8,473 \\
\hline
45,002
\end{array}
$$

Explanation: Like Example 8 we had to rename. The challenge is the fact that the tens and hundreds places contain 0's. When we renamed the 2 we had to take 1000 from the 5 thousands to make 10 hundreds, then take 100 from the 10 hundreds to make 10 tens. This, in turn, allowed us to take 10 from the 10 tens to make 12 in the ones place.

◀ **Do Margin 7.**

OBJECTIVE 5 *Solve equations containing a missing addend.*

Suppose you have $200 saved and you wish to buy an appliance that costs $350. The question becomes how much more you need in order to have enough to make the purchase. In mathematical form it would look like this:

$$\$200 + (?) = \$350$$

This is a missing addend equation. Because the amount you would need is the difference between the $350 price and the $200 that you have, you can write a **related subtraction sentence** to calculate the needed amount.

DEFINITION   **Related Sentence:** A mathematical equation that relates the same pieces of a given equation using the inverse operation.

We can say: $(?) = \text{sum} - \text{known addend}$
$(?) = \$350 - \$200$
$(?) = \$150$

Answer: You will need $150.

Discussion For an addition problem we write a related subtraction sentence. What type of related sentence would we write given a multiplication problem?

Procedure *To find a missing addend, write a related subtraction sentence. Subtract the known addend from the sum.*

Earlier we defined a variable as a symbol or letter that can vary in value, and we said that we use these symbols to represent unknown amounts. Because (?) indicates an unknown amount, it is a variable. However, it is more common to use a letter such as x, y, or t for a variable. From here on we will use letter variables.

Instead of: $200 + (?) = 350$
$(?) = 350 - 200$
$(?) = 150$

we write: $200 + x = 350$
$x = 350 - 200$
$x = 150$

The number 150 is the **solution** to the equation because it can replace the variable and make the equation true. The act of finding this solution is called *solving* the equation.

DEFINITION   **Solution:** A number that can replace the variable(s) in an equation and make the equation true.

The definition for *solution* suggests a method for checking. If 150 is correct, it should replace the x in $200 + x = 350$ and make the equation true.

◆ **Margin 7**

Estimate the difference by rounding so that there is only one nonzero digit, then calculate the actual difference.

a. $48,975 - 6241 =$
Estimate: Actual:

b. $5941 - 218 =$
Estimate: Actual:

c. $520,048 - 63,793 =$
Estimate: Actual:

Answers: **a.** Estimate: 44,000, Actual: 42,734 **b.** Estimate: 5800, Actual: 5723 **c.** Estimate: 440,000, Actual: 456,255

Check: $200 + 150 \overset{?}{=} 350$

$350 = 350$

The symbol ? is used to indicate that we are asking whether $200 + 150$ is equal to 350.

Because $200 + 150 = 350$, 150 is in fact the solution to $200 + x = 350$.

EXAMPLE 10 Solve and check. $45 + x = 73$

Solution: Recall $45 + x = 73$ is the same as $45 + (?) = 73$. To solve for the missing addend, we write a related subtraction sentence.

$x = 73 - 45$ **Check:** $45 + 28 \overset{?}{=} 73$

$x = 28$ $73 = 73$ It checks.

Explanation: Think about the missing addend statement in money terms. We have $45 and we want to end up with $73. We must subtract to find out how much more is needed.

Do Margin 8. ▷

OBJECTIVE 6 *Solve applications involving subtraction.*

Like addition, there are key words and key questions that help us recognize subtraction situations.

> **Key Words** Subtract, minus, remove, decreased by, difference.
> **Key Questions** How much is left?   (take away)
>   How much more/higher/warmer?   (difference)
>   How much more is needed?   (missing addend)

EXAMPLE 11 A small computer business has $5678 in the bank. During the month, $2985 is spent on parts for production. How much is left?

Solution: Notice from the clues *spent,* indicating *to remove,* and the key question, *How much is left?* we can conclude that we must subtract.

$$\begin{array}{r} 4\overset{1}{5}1 \\ \cancel{5}678 \\ -2985 \\ \hline 2693 \end{array}$$ **Check:** $$\begin{array}{r} 11 \\ 2693 \\ +2985 \\ \hline 5678 \end{array}$$

◆ **Answer:** $2693 will be left.

EXAMPLE 12 A room temperature (78°F) mixture of water and sugar is heated to boiling, which is 212°F, so that more sugar may be added to supersaturate the mixture. How much did the temperature change?

Solution: Because we want to know the amount of change, we are looking for the difference. This means we must subtract.

$$\begin{array}{r} 1\overset{10}{}1 \\ 2\cancel{1}2 \\ -78 \\ \hline 134 \end{array}$$ **Check:** $$\begin{array}{r} 11 \\ 134 \\ +78 \\ \hline 212 \end{array}$$

◆ **Answer:** 134°F

EXAMPLE 13 A charity organization has collected $43,587 through a fundraiser. The goal for the year is to raise $125,000. How much more is needed?

Solution: Because we must increase the original amount of $43,587 by an unknown amount to end up with $125,000, this is a missing addend situation.

We could set it up like this: $43,587 + x = $125,000

In order to find that missing addend, we can subtract the addend we know, which is the 43,587, from the total we want to end up with, 125,000.

♦ **Margin 8**

Solve and check.

a. $14 + x = 20$

b. $y + 32 = 60$

c. $140 + t = 216$

d. $m + 89 = 191$

♦ **Margin 9**

A food bank has 3452 boxes of food. They distribute 928 boxes in one neighborhood. How many boxes do they have left?

♦ **Margin 10**

A plane is flying at 24,500 ft. The plane experiences a downdraft, which causes it to abruptly drop to an altitude of 22,750 ft. How much altitude did the plane lose?

♦ **Margin 11**

Debbie sells cars at a very competitive dealership. As an incentive, the first salesperson to sell $150,000 worth of cars during the year will win a cruise to the Caribbean. Debbie has sold $78,520 so far. How much more does she need to sell to win? Can you see a flaw in the incentive idea?

Answers to Margin 8: **a.** 6 **b.** 28 **c.** 76 **d.** 102

Answer to Margin 9: 2524

Answer to Margin 10: 1750 ft.

Answer to Margin 11: $71,480

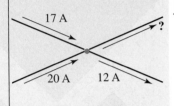

$$\begin{array}{r} 0\ \ 4\ 9\ 9 \\ {}_{1}\ {}_{1}{}_{1}{}_{1} \\ \cancel{1\,2\,5,\!0\,0\,0} \\ -\ 4\,3,\!5\,8\,7 \\ \hline 8\,1,\!4\,1\,3 \end{array}$$

Check:
$$\begin{array}{r} 1\ \ 11 \\ 8\,1,\!4\,1\,3 \\ +\,4\,3,\!5\,8\,7 \\ \hline 1\,2\,5,\!0\,0\,0 \end{array}$$

Connection

We could have written our solution to the problem in an algebraic form:

$$43,587 + x = 125,000$$
$$x = 125,000 - 43,587$$
$$x = 81,413$$

Check: $43,587 + 81,413 \overset{?}{=} 125,000$
$$125,000 = 125,000$$

♦ **Answer:** $81,413 is needed.

◀ **Do Margin 9–11.**

OBJECTIVE **7** *Solve applications involving addition and subtraction.*

Much of the time the problems we encounter in life involve more than one step to get to a solution. Let's consider some problems that involve addition and subtraction in the same problem.

EXAMPLE 14 In circuits, a wire connection is often referred to as a *node.* Current is a measure of electricity moving through a wire and is measured in amperes (or amps). One property of circuits is that all the current entering a node must equal all the current exiting the node. In the circuit diagram below, how much must the missing current be?

Solution: We can calculate the total current entering the node and then subtract the known current that is leaving the node.

Entering current: $9\,A + 15\,A = 24\,A$

If 24 amps are entering and we know 13 amps are leaving, then we can subtract 13 from 24 to find the unknown current.

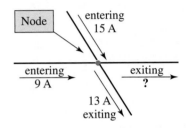

♦ **Unknown current:** $24\,A - 13\,A = 11\,A$

◀ **Do Margin 12.**

EXAMPLE 15 An accountant is given the following spreadsheet of expenses and income for a company. What is the final balance?

Description of expenses	Amount	Description of assets/income	Amount
Payroll	$10,548	Checking account	$ 5,849
Utilities	$ 329	Income	$19,538
Water	$ 87	Donations	$ 1,200
Waste disposal	$ 285		
New inventory	$12,243		

Solution: We must calculate the total expenses and total income/assets, then find the difference of the amounts.

Expenses	Income/Assets
$10,548	$ 5,849
$ 329	$19,538
$ 87	+$ 1,200
$ 285	Total $26,587
+ $12,243	
Total $23,492	

Now that we have the total expenses and income/assets figured, we need to find the difference to determine the final balance.

Balance
$26,587
−$23,492
$ 3,095

◀ **Do Margin 13.**

♦ **Margin 13**

Lynda is a checkout clerk. At the end of her shift, the management makes sure the amount of money in the register equals her sales. The printout breaks the sales down into three categories: food, clothing, and nonperishable items. Lynda must count the money in the register and her credit card sales. Finally $200 always stays in the register. The following chart shows the amounts. Based on the information in the chart, how did Lynda do? Is her register in balance? Does she have too little or too much and if so, how much?

Category	Amount
Food	$1587
Clothing	$2768
Nonperishables	$ 988

Totals	Amount
Register count	$4192
Credit cards	$1339

Answer to Margin 12: 25 A

Answer to Margin 13: There should be $12 more in the register.

1.2 Exercises

For Exercises 1–6, estimate each sum by rounding to the nearest thousand, then find the actual sum.

1. 6,051
      +2,798

Estimate: Actual:

2. 52,407
      +31,596

Estimate: Actual:

3. 91,512
      +8,756

Estimate: Actual:

4. 82,098
      +7971

Estimate: Actual:

5. 10,516
         982
      +4,516

Estimate: Actual:

6. 150,412
         258
      +6,239

Estimate: Actual:

For Exercises 7–10, estimate each sum by rounding so that there is only one nonzero digit, then find the actual sum.

7. $9319 + 519 + 5408 =$

Estimate: Actual:

8. $6809 + 398 + 2087 =$

Estimate: Actual:

9. $43,210 + 135,569 + 2088 + 516 =$

Estimate: Actual:

10. $128,402 + 4480 + 93,095 + 98 =$

Estimate: Actual:

For Exercises 11–14, estimate by rounding to the nearest hundred, then find the actual difference.

11. 5873
       −521

Estimate: Actual:

12. 9478
       −253

Estimate: Actual:

13. 40,302
       −6,141

Estimate: Actual:

14. 510,304
       −42,183

Estimate: Actual:

For Exercises 15–20, estimate by rounding so that there is only one nonzero digit, then find the actual difference.

15. 50,016
       −4,682

Estimate: Actual:

16.   42,003
       −23,567

Estimate: Actual:

17. $51,980 − 25,461 =$

Estimate: Actual:

18. $413,609 − 20,724 =$

Estimate: Actual:

19. $210,007 − 43,519 =$

Estimate: Actual:

20. $6,005,002 − 258,496 =$

Estimate: Actual:

For Exercises 21–30, solve and check.

21. $8 + x = 12$

22. $12 + y = 19$

23. $t + 9 = 30$

24. $m + 16 = 28$

25. $17 + n = 35$

26. $r + 27 = 70$

27. $125 + u = 280$

28. $88 + a = 115$

29. $b + 76 = 301$

30. $470 + c = 611$

For Exercises 31–39, solve.

31. Tamika has to keep track of total ticket sales at a box office. There are four attendants selling tickets. The first attendant sold 548 tickets, the second sold 354, the third 481, and the fourth 427. How many tickets were sold in all?

32. Margo Incorporated produces a motor that turns the fan in air conditioners. The annual cost of materials last year was $75,348. The labor costs were $284,568. Utilities and operations costs for the plant were $54,214. The company received $548,780 in revenue from sales of the motors. What was the profit?

33. Mr. Grayson has just acquired some new land that he wants to fence in for a cow herd. To the right is a copy of the plot. How many feet of fencing will he need in all?

200 ft.

50 ft.

70 ft.

45 ft.

225 ft.

34. Jose is raising funds to run for a seat in the United States Senate. Currently he has raised $280,540 and he intends to add $28,580 of his own money. His competitor has raised a total of $485,300. How much more does Jose need to raise to match his competitor's funds?

35. The photosphere, which is the outer layer of the Sun, has an average temperature of about 5500 K. A typical atomic blast at ground zero has a temperature of about 3400 K. How much warmer is the Sun's temperature?

36. The Sun is 92,958,349 miles from Earth. The Moon is 615,042 miles from Earth. During a solar eclipse, the Moon is directly between the Sun and Earth. How far is the Moon from the Sun during a solar eclipse?

37. An accountant is given the following spreadsheet of expenses and income for a company. What is the final balance?

Description of expenses	Amount	Description of assets/income	Amount
Payroll	$18,708	Checking account	$10,489
Utilities	$ 678	Income	$32,500
Water	$ 126		
Waste disposal	$ 1,245		
New inventory	$15,621		

38. An accountant is given the following spreadsheet of expenses and income for a company. What is the final balance?

Description of expenses	Amount	Description of assets/income	Amount
Payroll	$22,512	Checking account	$15,489
Utilities	$ 1,265	Income	$82,569
Water	$ 240		
Waste disposal	$ 1,350		
New inventory	$24,882		

39. Find the missing current in the circuit shown.

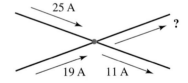

40. A family is purchasing a home in Florida that is said to be 2400 sq. ft. They later find out that this figure includes an unfinished bonus room over the garage and the garage itself. The garage is 440 sq. ft. and the bonus room is 240 sq. ft. What is the area of the actual living space?

Trivia Bite

The symbol K represents Kelvin temperature units. The scale was proposed by William Thomson Kelvin in 1848 and is used extensively in sciences. Increments on the Kelvin scale are the same as the Celsius scale. However, 0 K is equivalent to −273°C.

Puzzle Problem

Eight people are at a group-therapy session. Everyone hugs everyone once. How many hugs take place?

REVIEW EXERCISES

1. What digit is in the millions place in 456,028,549?

2. Write expanded form for 307,491,024.

3. Write the word name for 1,472,359.

4. Use $<$, $>$, or $=$ to make a true statement.

12,305 ? 12,350

5. Round 23,405,172 to the nearest ten thousand.

1.3 Multiplying Whole Numbers and Exponents

OBJECTIVES

1 Multiply whole numbers.

2 Solve applications involving multiplication.

3 Evaluate numbers in exponential form.

4 Write in exponential form.

5 Solve applications.

OBJECTIVE 1 *Multiply whole numbers.*

Imagine a merchant in ancient times selling baskets of fruit. He has four baskets each containing five apples. If he wanted to know how many apples he had in all, he could add four sets of five, but after a while he'd simply memorize that when he has four sets of five, the total is always 20. This is probably how multiplication was born.

| **DEFINITION** | **Multiplication:** Repeated addition of the same number. |

Notation: $4 \times 5 = 20$

Factors   Product

We can also write: $4 \cdot 5 = 20$

Discussion When would we avoid using $\cdot$ or $\times$ to indicate multiplication?

Or we can use parentheses with no $\cdot$ or $\times$ in between: $(4)(5) = 20$
or $4(5) = 20$
or $(4)5 = 20$

If we write a multiplication statement with the product first, such as $20 = 4 \cdot 5$, we say it is in *factored form*.

There are many real situations that involve multiplication: area, volume, voltage, and force, just to name a few. We'll consider these later.

To make multiplication useful, you must memorize a certain number of multiplication facts. The more facts you memorize, the faster you'll be at multiplying. Here is a table of multiplication facts through 12.

Figure 1.2 Multiplication Table

×	0	1	2	3	4	5	6	7	8	9	10	11	12
0	0	0	0	0	0	0	0	0	0	0	0	0	0
1	0	1	2	3	4	5	6	7	8	9	10	11	12
2	0	2	4	6	8	10	12	14	16	18	20	22	24
3	0	3	6	9	12	15	18	21	24	27	30	33	36
4	0	4	8	12	16	20	24	28	32	36	40	44	48
5	0	5	10	15	20	25	30	35	40	45	50	55	60
6	0	6	12	18	24	30	36	42	48	54	60	66	72
7	0	7	14	21	28	35	42	49	56	63	70	77	84
8	0	8	16	24	32	40	48	56	64	72	80	88	96
9	0	9	18	27	36	45	54	63	72	81	90	99	108
10	0	10	20	30	40	50	60	70	80	90	100	110	120
11	0	11	22	33	44	55	66	77	88	99	110	121	132
12	0	12	24	36	48	60	72	84	96	108	120	132	144

Like addition, multiplication is commutative. The *commutative property of multiplication* says that changing the order of factors will not affect the product.

$$(2)(5) = 10 \text{ and } (5)(2) = 10$$

Commutative Property of Multiplication Changing factor order will not affect the product.

In math language $a \cdot b = b \cdot a$, where a and b are any numbers.

Also like addition, multiplication is associative. The *associative property of multiplication* says if three or more factors occur, grouping them differently will not affect the product.

$$(2 \cdot 3) \cdot 5 = 30 \text{ or } 2 \cdot (3 \cdot 5) = 30$$

Associative Property of Multiplication Grouping three or more factors differently will not affect the product.

In math language $(a \cdot b) \cdot c = a \cdot (b \cdot c)$ where a, b, and c are any numbers.

Note that the product of 0 and a number is always 0 and the product of 1 and a number is always the number.

Multiplicative Property of 0 The product of 0 and a number is always 0.

In math language $0 \cdot n = 0$ and $n \cdot 0 = 0$, where n is any number.

Multiplicative Property of 1 The product of 1 and a number is always the number.

In math language $1 \cdot n = n$ and $n \cdot 1 = n$, where n is any number.

Another property exists called the *distributive property*. The distributive property can be used when a number multiplies a sum or difference, such as 2(3 + 5). Think about the word distribute. When we distribute weight evenly in the seats of a car, we place the same weight in each seat. Similarly, when we apply the distributive property we multiply each number inside parentheses by the number outside the parentheses.

$$2(3 + 5) = 2 \cdot 3 + 2 \cdot 5$$

Both expressions have the same value.

$$
\begin{array}{ll}
2(3 + 5) & \qquad 2 \cdot 3 + 2 \cdot 5 \\
= 2 \cdot 8 & \qquad = 6 + 10 \\
= 16 & \qquad = 16
\end{array}
$$

Distributive Property If a sum or difference is multiplied by a number, then each number inside parentheses may be multiplied by the number outside the parentheses.

In math language $a(b + c) = ab + ac$ and $a(b - c) = ab - ac$, where a, b, and c are any numbers.

We apply the distributive property in our method for multiplying numbers having several digits. Let's see how this works.

EXAMPLE 1 Multiply 2 × 34.

Solution:
$$
\begin{array}{r}
34 \\
\times 2 \\
\hline
68
\end{array}
$$
Multiply 2 times 4, then 2 times 3.

Explanation: We are applying the distributive property here. Since 34 is the same as 30 + 4 in expanded form, we could set the problem up in the distributive property format this way:

$$
\begin{aligned}
2(30 + 4) &= 2 \cdot 30 + 2 \cdot 4 \\
&= 60 + 8 \\
&= 68
\end{aligned}
$$

We are simply multiplying the 2 times the 4 in the ones place and since the 3 in the tens place means 30, we're really multiplying 2 times 30.

Do Margin 1. ▷

EXAMPLE 2 Multiply 503 × 62

Solution:
$$
\begin{array}{r}
\overset{1}{5}03 \\
\times 62 \\
\hline
1\,006 \\
+30\,18 \\
\hline
31{,}186
\end{array}
$$
Multiply 2 times 503.
Multiply 6 (tens) times 503.
Add.

Explanation: Notice what happened when we multiplied 6 times 503. Because it was actually 6 tens times 3 ones we got 180. We wrote 80 underneath. Note that we just wrote the 8 in the tens place and left out the 0. The 100 part of 180 was expressed as a small digit 1 over the 0 in 503. When we multiplied 6 tens times 0 tens, we got 0 hundreds but had to add the extra 1 hundred. We brought down this 1 hundred to the left of the 8.

Do Margin 2. ▷

◆ **Margin 1**
Multiply.

a. 5 × 31

b. 3 × 402

c. 18 × 0

d. 248 × 1

◆ **Margin 2**
Multiply.

a. 32 × 41

b. (51)(23)

c. 802 × 74

d. 512 × 61

Answers to Margin 1: **a.** 155
b. 1206 **c.** 0 **d.** 248

Answers to Margin 2: **a.** 1312
b. 1173 **c.** 59,348 **d.** 31,232

EXAMPLE 3 Estimate 42,109 × 7104 by rounding so that there is only one nonzero digit, then calculate the actual product.

Estimate: To estimate, we round 42,109 to 40,000 and 7104 becomes 7000. Because the numbers now contain lots of 0's, we can simply multiply the 7 times 4 mentally and write the total number of 0's from both numbers after the product.

```
                              Four 0's
    42,109  →   40,000                  total of seven 0's.
   ×7,104   →  ×7,000  ← Three 0's
              Think 7 × 4 = 28 and add seven 0's.
```

Answer: 280,000,000

Actual:
```
              1  6
               3
          42,109
         ×7,104
         168 436      Multiply 4 times 42,109.
         000 00       Multiply 0 times 42,109.
        4 210 9       Multiply 1 times 42,109.
       +294 763       Multiply 7 times 42,109.
       299,142,336
```

◆ **Answer:** 299,142,336. Notice how the estimate and actual answer are in reasonable agreement.

◀ **Do Margin 3.**

OBJECTIVE 2 *Solve applications involving multiplication.*

There are certain key words you can look for that indicate multiplication. Since multiplication is repeated addition, you can expect to see words that indicate addition as well.

> **Key Words** Multiply, times, product, each, of, by

EXAMPLE 4 The human heart averages about 70 beats each minute. How many times would the heart beat in an hour? In a week?

Solution: a. For the number of heartbeats in an hour, since there are 60 minutes in an hour and each minute is 70 beats, we multiply 60 × 70.

```
        60
       ×70
        00
       420
      4200
```

Answer: The heart beats about 4200 times per hour.

Solution: b. For the number of heartbeats in a week, we need to first figure out how many hours there are in a week. Since there are 24 hours in a day, and 7 days in a week, we multiply 24 × 7 to get the number of hours in a week.

```
        2
       24
       ×7
      168
```

◆ Margin 3

Estimate each product by rounding to only one nonzero digit, then calculate the actual product.

a. 56,045 × 6,714
Estimate: Actual:

b. 24 × 365 × 8
Estimate: Actual:

Answers: **a.** Estimate: 420,000,000, Actual: 376,286,130 **b.** Estimate: 80,000, Actual: 70,080

There are 168 hours in a week. From part (a) we found there were 4200 beats in an hour. To calculate the number of beats in a week, we multiply 168 × 4200.

$$\begin{array}{r} 4{,}200 \\ \times 168 \\ \hline 33\ 600 \\ 252\ 00 \\ \underline{420\ 0} \\ 705{,}600 \end{array}$$

Discussion What things affect the heart rate? Measure your own heart rate and do the same calculation to estimate the number of times your own heart beats in a week.

◆ **Answer:** The heart beats about 705,600 times in a week (truly an amazing muscle).

Do Margin 4. ▶

We can use multiplication to count objects arranged in a **rectangular array.**

DEFINITION **Rectangular array:** A rectangle formed by a pattern of neatly arranged rows and columns.

The buttons on a telephone form a rectangular array.

Notice we could count the buttons individually to get 12 total buttons.

A faster way would be to multiply. Four rows with three buttons in each row is a rectangular array.

$$4 \cdot 3 = 12$$

Conclusion: To calculate the total number of items in a rectangular array, multiply the number of rows by the number of columns.

EXAMPLE 5 A section in the upper deck of a football stadium has 42 rows, each containing 40 seats. There are 8 sections in the upper deck. How many people can be seated in the upper deck altogether?

Solution: The 8 sections of 42 rows with 40 seats in each row form a large rectangular array so we multiply.

◆ Margin 4

a. If the heart beats about 70 beats per minute, then how many times does the heart beat in a 30-day month?

b. A manager decides to monitor the copy machine use in her office. She finds that the people in her area are using two boxes of paper each week. Each box contains 10 reams of paper and each ream contains 500 pieces of paper. How many pages are being copied each week? Each month?

Discussion What are some ways the manager could cut back on copying?

Answers: **a.** 3,024,000 **b.** 10,000 each week; 40,000 each month

Multiply 42 × 40 × 8

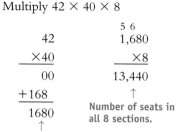

Answer: The upper deck can seat 13,440 people in all.

Explanation: Notice that when the example says, "42 rows *each* containing 40 seats," the key word *each* tells us we have a repeated addition situation, so we multiply. Since there are 8 identical sections, we are once again repeatedly adding, so once again we multiply.

EXAMPLE 6 In order to get into a certain restricted building, a code must be entered. The code box has two digital windows (see illustration). The first window can contain any number from 1 to 5. The second window can contain any letter from A to Z (only uppercase letters appear). How many possible codes are there in all?

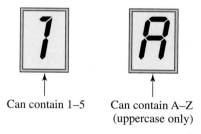

Can contain 1–5   Can contain A–Z
(uppercase only)

Solution: Each number in the first window could be paired with all 26 letters of the alphabet. In other words, the 1 could be paired with all 26 letters, then 2 with all 26 letters, and so on (see below). That means there could be a total of 5 sets of 26. Notice how we can make a rectangular array out of the combinations. Therefore we multiply 5 times 26.

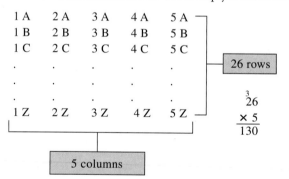

Answer: There are 130 possible codes.

Another way to analyze this problem is to think of each window as a place value or a *slot*, like on a slot machine. Figure the number of possible values each slot could contain and multiply.

Can contain 1–5 so there   Can contain A–Z (uppercase
are 5 possibilities   only) so there are 26
in this slot.   possibilities in this slot.

Solution: Multiply $5 \times 26 = 130$

The branch of mathematics that deals with counting total combinations and arrangements of items is called *combinatorics*.

◆

Procedure *To get the total number of possible combinations in a problem with multiple slots to fill, multiply the number of items that can fill the first slot by the number of items that can fill the next, and so on.*

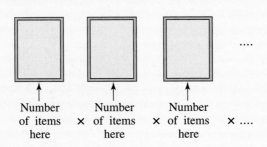

Number of items here × Number of items here × Number of items here × ....

Do Margin 6. ▶

OBJECTIVE 3 *Evaluate numbers in exponential form.*

A computer translates everything into *binary* code. This means that numbers are written so that each place value either contains a 1 or a 0 digit. If a certain computer chip has 7 place values, then how many different numbers can be coded?

Suppose each box below represents a bit or place value on the chip.

0 or 1   0 or 1   0 or 1   0 or 1   0 or 1   0 or 1   0 or 1

Since each box could have 2 different values, we must multiply seven 2's together.

$$2 \times 2 \times 2 \times 2 \times 2 \times 2 \times 2 = 128$$

Notice that we had to multiply the same number repeatedly. This suggests the need for a notation to indicate repeated multiplication. We use an **exponent** to indicate repeated multiplication of a **base** number.

DEFINITIONS **Exponent:** A symbol written to the upper right of a base number that indicates how many times to use the base as a factor.
Base: The number that is repeatedly multiplied.

Notation: When we write a number with an exponent, we say it is in *exponential form*.

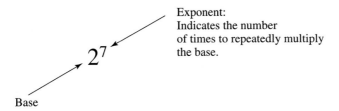

Exponent:
Indicates the number
of times to repeatedly multiply
the base.

2^7

Base

2^7 is read "two to the seventh power," or, simply, "two to the seventh."

We could answer the computer-coding question by saying the computer can code a total of 2^7 different numbers.

◆ **Margin 6**

A certain combination lock has three dials. The first dial has the letters A–F inscribed. The second dial has the numbers 0–9. The third dial has the names of the planets. How many combinations are possible?

Trivia Bite

In computer lingo, the place values are called *bits*. Bit is short for binary digit.

Answer: 540

To *evaluate* 2^7, we must multiply seven 2's together. It is helpful to first write the multiplication in factored form.

$$2^7 = \underbrace{2 \cdot 2 \cdot 2 \cdot 2 \cdot 2 \cdot 2 \cdot 2}_{} = 128$$

exponential form   factored form   standard form

 WARNING 2^7 does not mean $2 \cdot 7$

EXAMPLE 7 Write 3^4 in factored form then evaluate.

Solution:
$$\begin{aligned} 3^4 &= 3 \cdot 3 \cdot 3 \cdot 3 \\ &= 9 \cdot 3 \cdot 3 \\ &= 27 \cdot 3 \\ &= 81 \end{aligned}$$

Explanation: The exponent 4 means to multiply four 3's together. The first two 3's multiply to make 9. The 9 times the next 3 is 27, then 27 times the last 3 makes 81.

◀ **Do Margin 7 and 8.**

OBJECTIVE 4 *Write in exponential form.*

EXAMPLE 8 Write $5 \cdot 5 \cdot 5 \cdot 5 \cdot 5 \cdot 5$ in exponential form.

Answer: 5^6

Explanation: Since there were six 5's multiplied, the exponent is 6.

◀ **Do Margin 9 and 10.**

When an exponential form has a base of 10 we say it is a *power* of 10. Consider the pattern:

$10^2 = 10 \cdot 10$	$= 100$
$10^3 = 10 \cdot 10 \cdot 10$	$= 1000$
$10^4 = 10 \cdot 10 \cdot 10 \cdot 10$	$= 10,000$
$10^5 = 10 \cdot 10 \cdot 10 \cdot 10 \cdot 10$	$= 100,000$
$10^6 = 10 \cdot 10 \cdot 10 \cdot 10 \cdot 10 \cdot 10$	$= 1,000,000$

Conclusion: Because the exponent indicates the number of 10's to be multiplied, and each 10 contributes an additional 0 after the 1, the number of 0's in the product matches the exponent.

Notice we can write expanded notation more simply with the powers of 10. Instead of writing all the zeros in the place values, we can simply express the place values with powers of 10.

EXAMPLE 9 Write 1,239,405 in expanded notation using powers of 10.

Solution: $1,239,405 =$
$$1 \times 1,000,000 + 2 \times 100,000 + 3 \times 10,000 + 9 \times 1,000 + 4 \times 100 + 0 \times 10 + 5 \times 1$$
$$= 1 \times 10^6 + 2 \times 10^5 + 3 \times 10^4 + 9 \times 10^3 + 4 \times 10^2 + 0 \times 10^1 + 5 \times 10^0$$
$$= 1 \times 10^6 + 2 \times 10^5 + 3 \times 10^4 + 9 \times 10^3 + 4 \times 10^2 + 5 \times 1$$

Reminder
We did not write the tens place in the expanded notations because it contained a 0 digit.

What does 10^0 mean?

Notice the pattern of exponents. Start with 10^3. If $10^3 = 1000$, $10^2 = 100$, and $10^1 = 10$, then 10^0 has to be 1 for the pattern to hold true.

$$10^0 = 1$$

Strange but true.

◆ **Margin 7**

Write each in factored form, then evaluate.

a. 7^2

b. 10^4

c. 1^9

d. 0^4

◆ **Margin 8**

Use a calculator to evaluate each of the following.

a. 2^{12}

b. 5^7

Answers to Margin 7: **a.** 49 **b.** 10,000 **c.** 1 **d.** 0

Answers to Margin 8: **a.** 4096 **b.** 78,125

Each power of 10 represents a different place value. Remember that the period names change every three places. So every third power of 10 will have a new name. The chart below illustrates the connection between powers of 10 and the period names.

Figure 1.3 Powers of 10 and Period Names

Period Names

$10^3 = 1000$ = one thousand
$10^6 = 1,000,000$ = one million
$10^9 = 1,000,000,000$ = one billion
$10^{12} = 1,000,000,000,000$ = one quadrillion
$10^{15} = $ (1 with 15 zeros) = one quintillion

The names continue using the above pattern. Some names are rather colorful.

$10^{100} = $ (1 with 100 zeros) = googol
$10^{googol} = $ (1 with a googol of zeros) = googolplex

Do Margin 11. ▶

OBJECTIVE 5 *Solve applications.*

Suppose we wanted to know how much surface space is within the borders or edges of a shape. We measure the space in **square units** and call the measurement **area**.

DEFINITION   **Square unit:** A 1 × 1 square.

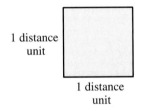

1 distance unit

1 distance unit

Examples of some square units:

Square foot:

1 ft.

1 ft.

A square foot is a 1 ft. by 1 ft. square and is represented mathematically by ft².

Square meter:

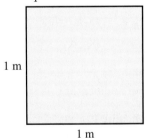

1 m

1 m

A square meter is a 1 m by 1 m square and is represented mathematically by m²

Margin 9

Write each in exponential form.

a. $14 \cdot 14 \cdot 14 \cdot 14$

b. $2 \cdot 2 \cdot 2 \cdot 2 \cdot 2 \cdot 2 \cdot 2 \cdot 2 \cdot 2 \cdot 2$

c. $9 \cdot 9 \cdot 9 \cdot 9 \cdot 9 \cdot 9 \cdot 9$

Margin 10

🖩 How many different values can be represented by a 16–bit computer chip? Write your answer in exponential form, then evaluate.

Margin 11

Write each number in expanded form using powers of 10.

a. 9,210,984

b. 46,009

c. 1,204,918,205

Answers to Margin 9: **a.** 14^4
b. 2^{10} **c.** 9^7

Answer to Margin 10: $2^{16} = 65336$

Answers to Margin 11: **a.** $9 \times 10^6 + 2 \times 10^5 + 1 \times 10^4 + 9 \times 10^2 + 8 \times 10 + 4 \times 1$
b. $4 \times 10^4 + 6 \times 10^3 + 9 \times 1$
c. $1 \times 10^9 + 2 \times 10^8 + 4 \times 10^6 + 9 \times 10^5 + 1 \times 10^4 + 8 \times 10^3 + 2 \times 10^2 + 5 \times 1$

DEFINITION | **Area:** The total number of square units that completely fill a shape.

If we were to measure the area of a rectangular room, we could place square tiles down until we completely covered the floor with tiles and then count the total number of tiles. Suppose we wanted to measure the floor area of a 4 ft. × 8 ft. bathroom.

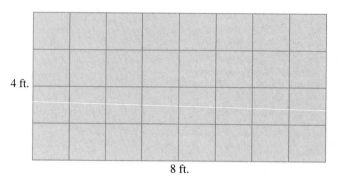

4 ft.

8 ft.

If we tiled the floor with 1 ft. × 1 ft. square tiles, it would take 32 of those square feet to cover the floor.

Of course, it's much easier to calculate the area simply by multiplying the length of the rectangle by the width, rather than actually measuring the area with square tiles.

Conclusion: Area of a rectangle = length × width

Or we can abbreviate like so:

$$A = l \cdot w$$

This abbreviation is called a **formula.**

DEFINITION | **Formula:** An equation that describes a procedure.

EXAMPLE 10   A standard football field measures 100 yd. by 50 yd., not counting the end zones. Find the area of a standard football field.

Solution: Since the field is a rectangle, to calculate the area, we multiply length times width.

$A = l \cdot w$

$A = 100 \text{ yd} \cdot 50 \text{ yd}$

$A = 5000 \text{ yd}^2$

> This unit of area is read "yards squared" or "square yards." All area calculations will involve multiplying two distance measurements. Area units can always be expressed as the distance unit with an exponent of 2.

Explanation: Notice how the square units are written with an exponent of 2. It's like multiplication of numbers. It is as if yards times yards are multiplied to equal yd^2.

◀ **Do Margin 12.**

◆ **Margin 12**

A room measures 15 ft. by 24 ft. How many square feet of carpet will be needed to cover the floor of the room?

Answer: 360 ft.^{2}

1.3 Exercises

For Exercises 1–4, find the product.

1. $352 \times 1 =$

2. $0 \times 495 =$

3. $8 \times 9 \times 0 \times 7 =$

4. $12 \times 3 \times 1 =$

For Exercises 5–10, estimate each product by rounding to a single nonzero digit, then find the actual product.

5. $24 \times 18 =$
Estimate: Actual:

6. $63 \times 42 =$
Estimate: Actual:

7. $54 \times 91 =$
Estimate: Actual:

8. $47 \times 86 =$
Estimate: Actual:

9. $246 \times 381 =$
Estimate: Actual:

10. $384 \times 25 =$
Estimate: Actual:

For Exercises 11–20, multiply.

11. $642 \times 70 =$

12. $408 \times 92 =$

13. $2{,}065 \times 482 =$

14. $9{,}402 \times 608 =$

15. $207 \times 41{,}308 =$

16. $361 \times 80{,}290 =$

17. $60{,}309 \times 4002 =$

18. $314 \times 12 \times 47 =$

19. $205 \times 23 \times 70 =$

20. $14 \times 1 \times 289 \times 307 =$

For Exercises 21–28, write each number in factored form and evaluate.

21. 2^4

22. 5^3

23. 1^6

24. 13^2

25. 3^5

26. 10^5

27. 10^7

28. 0^7

📓 *For Exercises 29–32, use a scientific calculator to evaluate.*

29. 6536^2 　　　　　**30.** 279^3 　　　　　**31.** 42^5 　　　　　**32.** 34^6

For Exercises 33–38, write in exponential form.

33. $9 \cdot 9 \cdot 9 \cdot 9$ 　　　　　**34.** $12 \cdot 12 \cdot 12$ 　　　　　**35.** $7 \cdot 7 \cdot 7 \cdot 7 \cdot 7$

36. $10 \cdot 10 \cdot 10 \cdot 10 \cdot 10$ 　　　　　**37.** $14 \cdot 14 \cdot 14 \cdot 14 \cdot 14$ 　　　　　**38.** $2 \cdot 2 \cdot 2 \cdot 2 \cdot 2 \cdot 2 \cdot 2 \cdot 2$

For Exercises 39–44, write in expanded form using powers of 10.

39. 24,902 　　　　　**40.** 604,057 　　　　　**41.** 9,128,020

42. 10,945 　　　　　**43.** 407,210,925 　　　　　**44.** 3,029,408

For Exercises 45–69, solve.

45. In a certain city, avenues run north/south, while streets run east/west. If there are 97 streets and 82 avenues, how many intersections are there? If the city must put up 8 traffic lights at each intersection, then how many traffic lights are required in all?

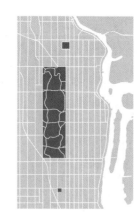

46. A parking lot for a department store has 23 rows of spaces. Each row can hold 36 cars. What is the maximum number of cars that can be parked in the lot? Now, suppose the lot is full. Estimate the number of people in the store if we assume that there are two people in the store for each car in the lot.

📓 **47.** On average, the human heart beats 70 times in one minute. How many beats is this in a year? If the average male in the United States lives to be 76, estimate the number of heartbeats in an average male's lifetime.

📓 **48.** The average female life span in the United States is about 82 years. Using the information in Exercise 47, estimate the number of heartbeats in the average female's lifetime.

49. A prescription indicates that a patient is to receive 10 mg of aminophylline per kg that the person weighs. If the person weighs 74 kg, then how many mg should the patient receive?

50. A nurse sets the drip rate for an IV at 26 drops each minute. How many drops does the patient receive in one hour?

51. Michael works 5 days every week and buys lunch each workday. On average, he spends $7 each day. If he has 10 vacation days each year, then how much does he spend on lunch in a year?

52. A flyer advertising telemarketing job openings promises that a person can make up to $800 per week. If a person made $800 each week, how much would they make in a year?

> **Discussion** If you got the job and did not make $800 per week, can you argue that the flyer was false advertising? Why or why not?

53. It is recommended that a person drink eight 8-oz glasses of water every day. If every one of the 270 million Americans drank eight 8-oz glasses of water in the same day, how many ounces of water would be consumed in that day?

54. On the evening news you hear a conversation between the anchor and a commentator. The commentator says that a computer company sold nearly eight hundred thousand copies of a new software package at $40 each. The anchor then says, "So that's about $320 billion. Wow." Is the anchor's response accurate? Explain.

55. A combination lock is opened by a combination of three things. You must turn a key to one of two positions, left or right; set a dial to a day of the week; and set another dial to a letter from A to H. How many possible combinations are there?

56. If a license plate consists of three capital letters and a three-digit number, how many license plates are possible? How many are possible in the entire United States if each state uses the same system of three letters and a three-digit number?

57. A certain computer chip processes 5-digit binary numbers, meaning each of the 5 places could contain a 1 or 0. How many different binary numbers can it process? If each letter, A–Z, and each numeral, 0–9, corresponds to a different binary number, would the 5-digit chip be able to recognize them all? Explain.

58. How many codes are possible using an 12-bit binary coder?

59. MIDI stands for musical instrument data interface and is the common coding for digital instruments, such as synthesizers. Most MIDI devices have a 7-bit binary memory chip. How many memory locations are possible?

60. A psychology quiz has 10 true-false questions. How many different answer keys are possible? Express the answer in exponential form, then calculate the actual number.

61. A man walks into a carpet store and says he has a 24 ft. by 26 ft. room and therefore needs about 4000 square feet of carpet. Is his estimate reasonable? Exactly how many square feet does he need?

62. You wish to tile the kitchen floor in your home. The kitchen is a 16 ft. by 12 ft. rectangle. If you used 1 ft. by 1 ft. square tiles, how many would be needed?

63. Delia plans to stain a new wood deck. The deck is 15 ft. by 20 ft. What is the area of the deck?

64. Taylor plans to paint a wall in his living room. The wall is 18 ft. by 9 ft. What is the area of the wall?

Puzzle Problem

A certain bacterium reproduces by dividing itself every minute. This causes the total population to double every minute. If we begin with a single specimen on a petri dish, how many will exist after 5 minutes? 10 Minutes? Write a formula that describes the pattern.

REVIEW EXERCISES

1. Write the word name for 16,507,309.

2. Write expanded notation for 23,506.

3. Calculate: $54,391 + 2079 + 518$.

4. Calculate: $901,042 - 69,318$.

5. Solve and check. $n + 19 = 32$

1.4 Dividing, Square Roots, and Solving Equations with Whole Numbers

OBJECTIVES

1 Divide whole numbers.

2 Solve equations containing a missing factor.

3 Solve applications involving division.

4 Find the square root of a perfect square.

5 Solve applications involving square roots.

OBJECTIVE **1** *Divide whole numbers.*

Like addition and subtraction, multiplication and division are inverse operations. This means that division *undoes* multiplication. Because multiplication means to repeatedly add, it follows that division means to repeatedly subtract.

DEFINITION | **Division:** Repeated subtraction of the same number.

Notation: $20 \div 5 = 4$

$ \uparrow \qquad \uparrow \qquad \uparrow$

 Dividend Divisor Quotient

> **Connection**
> Because multiplication and division are inverse operations, we can check $20 \div 5 = 4$ by multiplying:
> $$4 \cdot 5 = 20$$

Like multiplication, there are several ways to indicate division. We can write $20 \div 5 = 4$ using a slash, fraction bar, or long division.

Slash: $20/5 = 4$

Fraction bar: $\dfrac{20}{5} = 4$

Long division: $5 \overline{)20}^{\,4}$

At one time people divided by repeatedly subtracting the same number until they couldn't subtract anymore.

To calculate $20 \div 5$ by repeated subtraction, we must repeatedly subtract 5 from 20 until we can no longer subtract 5. The answer is the number of times we have to subtract 5.

$$
\begin{array}{r}
20 \\
-5 \\
\hline
15 \\
-5 \\
\hline
10 \\
-5 \\
\hline
5 \\
-5 \\
\hline
0
\end{array}
$$

1st subtraction

2nd subtraction

3rd subtraction

4th subtraction

Because we subtracted 5 from 20 four times, the answer is 4.

The amount left after dividing is called the **remainder.**

DEFINITION | **Remainder:** The amount left over after dividing two whole numbers.

In the case of $20 \div 5$ there are exactly four 5's in 20, so the remainder is 0. Sometimes the division does not end in 0. For example, if we had to split $21 into $5 bills we could only get four $5 bills with $1 left over. The $1 left over is the remainder and we would write $21 \div 5 = 4$ r1, using the letter r to indicate the remainder.

Let's explore a few properties of division.

Like subtraction, division is neither commutative nor associative.

Division is not commutative: $20 \div 5 = 4$ while $5 \div 20 \neq 4$ Note: $\neq$ means "is *not* equal to"

Division is not associative: $(32 \div 8) \div 2$ while $32 \div (8 \div 2)$
$$= 4 \div 2 \qquad = 32 \div 4$$
$$= 2 \qquad = 8$$

Different results

What if the divisor is 1?

$24 \div 1 = ?$

To check the quotient, we must be able to multiply the quotient by the divisor to get the dividend.

$? \cdot 1 = 24$

In Section 1.3 we learned that any number times 1 is the number. Therefore $24 \cdot 1 = 24$, which means $24 \div 1 = 24$.

Conclusion: Any number divided by 1 is that number.

> **Division Property** When 1 is the divisor, the quotient is equal to the dividend.
>
> **In math language** $n \div 1 = n$, where n is any number

What if the divisor is 0?

$14 \div 0 = ?$

To check the quotient, we must be able to multiply the quotient by the divisor to get the dividend.

$? \cdot 0 = 14$

This is impossible. Any number times 0 should make 0.

Conclusion: It is impossible to divide by 0.

When 0 is a divisor, we say the quotient is *undefined*, so $14 \div 0$ is undefined.

> **Division Property** When 0 is the divisor with any dividend other than 0, the quotient is undefined.
>
> **In math language** $n \div 0$ is undefined, where $n \neq 0$

What if the dividend is 0?

$$0 \div 12 = \text{?}$$

Again, we must be able to multiply the quotient by the divisor to get the dividend.

$$\text{?} \cdot 12 = 0$$

Notice the missing factor must be 0, therefore $0 \div 12 = 0$. This is true as long as the divisor is not also 0. 0 divided by 0 deserves some special attention.

$$0 \div 0 = \text{?}$$

To check we must be able to multiply the quotient by the divisor to get the dividend.

$$\text{?} \cdot 0 = 0$$

It is tempting to say the missing factor is still 0, but actually, it could be any number because all numbers multiply by 0 to make 0. Because we cannot determine a solution for $0 \div 0$, we say the solution is *indeterminate*.

Conclusion: When 0 is divided by any number other than 0, the quotient is 0. When 0 is divided by itself, the quotient is indeterminate.

> **Division Properties** When 0 is the dividend, the quotient is 0 as long as the divisor is not also 0.
>
> ***In math language*** $0 \div n = 0$, as long as $n \neq 0$
>
> If both dividend and divisor are 0, the quotient is indeterminate.
>
> ***In math language*** $0 \div 0$ is indeterminate

What if we divide a number by itself?

$$65 \div 65 = \text{?}$$

Again, we must be able to check by multiplying the quotient times the divisor to get the dividend.

$$\text{?} \cdot 65 = 65$$

The only number that works here is 1. $1 \cdot 65 = 65$, which means $65 \div 65 = 1$.

Conclusion: When a number other than 0 is divided by itself, the quotient is 1.

> **Division Property** When a number (other than 0) is divided by itself, the quotient is 1.
>
> ***In math language*** $n \div n = 1$, as long as $n \neq 0$

In division when the remainder is 0, we say the divisor is an *exact divisor*. For example, because $20 \div 5 = 4$, we say that 5 is an exact divisor of 20. There are ways to tell whether certain numbers are exact divisors without actually dividing. We call these *divisibility rules*. There are many divisibility rules, but, for our purposes, the following rules are the most useful.

Divisibility Rules

1. 2 is an exact divisor for all even numbers. Even numbers have 0, 2, 4, 6, or 8 in the ones place.

Example: 2 goes evenly into 3596 because it is an even number. 3596 has the digit 6 in the ones place.

2. To determine whether 3 is an exact divisor for a given number:
a. Add the digits in the dividend.
b. If the resulting sum is a number that is evenly divisible by 3, then so is the dividend.

Example: 3 goes evenly into 58,014 because if we add the digits, $5 + 8 + 0 + 1 + 4$, we get 18. Because 18 is divisible by 3, so is 58,014.

3. 5 is an exact divisor for numbers that have 0 or 5 in the ones place.

Example: 5 goes evenly into 91,285 because it has a 5 in the ones place.

EXAMPLE 1   Use the divisibility rules to determine whether the given number is divisible by 2.

a. 45,091

Answer: 45,091 is not divisible by 2 because it is not even. In other words it does not end in 0, 2, 4, 6, or 8.

b. 691,134

◆ **Answer:** 691,134 is divisible by 2 because it is an even number.

◀ **Do Margin 1.**

EXAMPLE 2   Use divisibility rules to determine whether the given number is divisible by 3.

a. 76,413

Solution: Add the digits. $7 + 6 + 4 + 1 + 3 = 21$

Because the sum, 21, is evenly divisible by 3, so is the number.

Answer: 76,413 is divisible by 3.

b. 4256

Solution: Add the digits. $4 + 2 + 5 + 6 = 17$

Because the sum, 17, is not evenly divisible by 3, neither is the number.

◆ **Answer:** 4256 is not divisible by 3.

◀ **Do Margin 2.**

EXAMPLE 3   Use divisibility rules to determine whether the given number is divisible by 5.

a. 2,380,956

Answer: 2,380,956 is not divisible by 5 because it does not end in 0 or 5.

b. 49,360

◆ **Answer:** 49,360 is divisible by 5 because it ends in 0.

◀ **Do Margin 3.**

Now that we have explored some properties and rules of division, we can turn our attention to more complex division. Because repeated subtraction can be quite tedious, the common method for dividing is *long division*.

In a completed long division it can be difficult to comprehend everything that happens, so in our first example of long division, we will break the process into steps. Thereafter, we will only show the completed long division.

EXAMPLE 4   Divide $6408 \div 7$

Solution: $7\overline{)6408}$

Because 6 is not larger than 7, we include the 4 with the 6. Because 64 is larger than 7, we are ready to divide.

1. Write the problem in long division form.
2. Compare the left-most digit in the dividend with the divisor. Is it larger? If not, then include the next digit. Continue in this way until the digits name a number that is larger than the divisor.

$$\begin{array}{r} 9 \\ 7\overline{)6408} \\ -63 \\ \hline 1 \end{array}$$

3. Divide 7 into 64 and place the quotient, which is 9, over the 4.
4. Multiply 9 times 7 to get 63, then subtract the 63 from 64 to get 1.

$$\begin{array}{r} 9 \\ 7\overline{)6408} \\ -63\downarrow \\ \hline 10 \end{array}$$

5. Write the next digit in the dividend, 0, beside the 1 remainder. It is common to say that we are *bringing down* the 0.

$$\begin{array}{r} 91 \\ 7\overline{)6408} \\ -63 \\ \hline 10 \\ -7 \\ \hline 3 \end{array}$$

6. Divide 7 into 10 and place the quotient, 1, over the 0 digit in the dividend.
7. Multiply 1 times 7 to get 7. Subtract 7 from 10 to get 3.

$$\begin{array}{r} 915 \\ 7\overline{)6408} \\ -63 \\ \hline 10 \\ -7\downarrow \\ \hline 38 \\ -35 \\ \hline 3 \end{array}$$

8. Write the last digit, 8, beside the 3 to make the number 38.
9. Divide 7 into 38 to get 5 and write the 5 over the 8.
10. Multiply 5 times 7 to get 35, then subtract the 35 from 38 to get a remainder of 3.

Answer: 915 r3

Check: When checking results that have a remainder, we multiply the quotient by the divisor, then add the remainder to get the dividend.

$$\begin{array}{r} {\scriptstyle 1\ 3} \\ 915 \quad \text{Quotient} \\ \times 7 \quad \text{Divisor} \\ \hline 6405 \\ +3 \quad \text{Remainder} \\ \hline 6408 \quad \text{Dividend} \end{array}$$

Do Margin 4. ▷

◆ **Margin 4**

Divide.

a. $5028 \div 3$

b. $6409 \div 5$

c. $12{,}059 \div 23$

Answers: **a.** 1676 **b.** 1281 r4 **c.** 524 r7

Divide.

a. $10{,}091 \div 97$

b. $2617 \div 65$

c. $493{,}231 \div 205$

EXAMPLE 5 Divide. $42{,}017 \div 41$

$$
\begin{array}{r}
1{,}024 \\
41\overline{)42{,}017} \\
-41 \\
\hline
10 \\
-0 \\
\hline
101 \\
-82 \\
\hline
197 \\
-164 \\
\hline
33
\end{array}
$$

Because 41 does not divide 10, we place a 0 in the quotient.

Answer: 1024 r33

We can check by reversing the process. We multiply the quotient by the divisor then add the remainder to the resulting product. We should get the dividend.

Check:
$$
\begin{array}{rl}
\overset{1}{1024} & \text{Quotient} \\
\times 41 & \text{Divisor} \\
\hline
1\,024 & \\
40\,96 & \\
\hline
41{,}984 & \\
+33 & \text{Remainder} \\
\hline
42{,}017 & \text{Dividend}
\end{array}
$$

◀ **Do Margin 5.**

OBJECTIVE 2 *Solve equations containing a missing factor.*

Suppose we have to design a room in a house, and we know we want the area to be 150 ft.^{2} and the length to be 15 ft. What must the width be?

We know the area of a rectangle is found by multiplying the length by the width, so we can write an equation with a missing factor:

$$(15 \text{ ft.}) \cdot w = 150 \text{ ft.}^2$$

We could have said $(15 \text{ ft.}) \cdot (?) = 150 \text{ ft.}^2$ but using the variable w for the unknown amount makes more sense because we are looking for width.

We've seen in the past that multiplying by 10 places a zero after the digits in the other factor, so this missing factor must be 10 ft. If we replace the missing factor with 10 ft. it works:

$$15 \text{ ft.} \cdot 10 \text{ ft.} = 150 \text{ ft.}^2$$

Notice we could divide 150 by 15 to get this value. In Section 1.2 we used a related subtraction sentence to solve for a missing addend.

Conclusion: To solve for a missing factor, we write a related division sentence and divide the product by the known factor. We can say:

$$w = 150 \text{ ft.}^2 \div 10 \text{ ft.}$$
$$w = 15 \text{ ft.}$$

Procedure *To solve for a missing factor, write a related division sentence, dividing the product by the known factor.*

Answers: **a.** 104 r3 **b.** 40 r17
c. 2406 r1

EXAMPLE 6 Solve and check.

a. $x \cdot 16 = 208$

Solution: $x = 208 \div 16$

$\qquad\qquad x = 13$

$$\begin{array}{r} 13 \\ 16\overline{)208} \\ -16 \\ \hline 48 \\ -48 \\ \hline 0 \end{array}$$

Check: We can check the result by replacing the missing factor with what we found. We want to make sure that $13 \cdot 16$ does produce 208.

$$13 \cdot 16 \text{ ? } 208$$
$$208 = 208$$

$$\begin{array}{r} 16 \\ \times 13 \\ \hline 48 \\ +16 \\ \hline 208 \end{array}$$

b. $14 \cdot n = 0$

Solution: $n = 0 \div 14$

$\qquad\qquad n = 0$

Explanation: The related division statement has 0 as a dividend. When 0 is a dividend with a divisor other than 0, the quotient is 0.

Check: Replace the missing factor with the number we found and make sure it works. Verify that $14 \cdot 0$ is equal to 0.

$$14 \cdot 0 \stackrel{?}{=} 0$$
$$0 = 0$$

c. $y \cdot 0 = 5$

Solution: $y = 5 \div 0$

$\qquad\qquad y$ is undefined

Explanation: The related division statement has 0 as a divisor. We learned earlier that when 0 is a divisor with a dividend other than 0, the quotient is undefined.

Check: We cannot check this problem because the answer is undefined. No number can replace the missing factor to make the statement true.

Do Margin 6. ▶

OBJECTIVE 3 *Solve applications involving division.*

Like the other operations, division problems have key words or phrases that we can look for to help us identify them in the context of a real problem.

Key Words Divide, distribute, each, split

◆ Margin 6

Find the missing factor.

a. $6 \cdot x = 54$

b. $n \cdot 18 = 378$

c. $28 \cdot a = 0$

d. $r \cdot 0 = 37$

Answers: **a.** 9 **b.** 21 **c.** 0
d. undefined

Margin 7

a. Alicia wants to place a wooden fence along the back of her property. She measures the distance to be 180 ft. She decides to buy prefabricated 8-ft. sections of fencing. How many sections must she purchase?

b. Juan is to set an IV drip so that a patient receives 300 units of heparin, an anticoagulant medication, in D/W solution during one hour. How many units should the patient receive each minute?

Trivia Bite

D/W is an abbreviation for Dextrose in Water. Dextrose is a type of sugar in animal and plant tissues.

EXAMPLE 7 An egg farmer has 4394 eggs to distribute into packages of a dozen each. How many packages can be made? How many eggs will be left?

Solution: We must split up the total number of eggs into groups of a dozen.

$$
\begin{array}{r}
366 \\
12\overline{)4394} \\
-36 \\
\hline
79 \\
-72 \\
\hline
74 \\
-72 \\
\hline
2
\end{array}
$$

Answer: 366 packages, each holding a dozen eggs, can be made with 2 eggs left over.

Check:
$$
\begin{array}{r}
366 \\
\times 12 \\
\hline
732 \\
+366 \\
\hline
4392 \\
+2 \\
\hline
4394
\end{array}
$$

◀ **Do Margin 7.**

OBJECTIVE 4 *Find the square root of a perfect square.*

Suppose we want to design a fenced area for our pet dog. The veterinarian tells us that the dog needs 400 square feet of area to stay healthy. If we make the fenced area a square, how long must each side be?

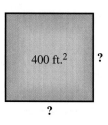

400 ft.² ?

?

A **square** is a special rectangle in which the length and width are the same. Recall that to get the area of a rectangle we must multiply the length by the width. In the case of a square, since the length and width are the same, we would have to multiply a number by itself, or *square* the number to get the area. Note the two meanings of the word *square*.

DEFINITION **Square:** a. Geometrical: A four-sided figure with all sides equal in length.
b. Algebraic: To multiply a number by itself.

Answers: **a.** 23 whole sections
b. 5 units

42 ■ Chapter 1 Whole Numbers

So in designing the fence, to find the length of each side, we must find a number that can be squared to make 400.

$$(\boxed{?})^2 = 400$$

This is exactly what it means to find a **square root.** In fact, roots are the inverse of exponents just as division is the inverse of multiplication.

DEFINITION **Square root:** The base number that can be squared to make a given number.

Because exponents mean repeated multiplication and because roots and exponents are inverses, then *roots* mean repeated division.

To answer our question, since 20 can be squared to make 400, the fenced area should be a 20 ft. by 20 ft. square. In other words, the square root of 400 is 20.

The symbol for square root is the radical sign. The number we wish to find the square root of is called the *radicand.*

$$\text{Radical sign} \rightarrow \sqrt{radicand}$$

In this section, we will only consider radicands that are **perfect squares.** Perfect squares are numbers that have whole number square roots. In other words, perfect squares come from squaring whole numbers.

DEFINITION **Perfect Square:** A number that has a whole number square root.

It is helpful to memorize a certain number of perfect squares. Table 1.1 lists the first 20 perfect squares and their roots.

Table 1.1 Roots and Their Squares

Root	Perfect square	Root	Perfect square
0	$0^2 = 0$	10	$10^2 = 100$
1	$1^2 = 1$	11	$11^2 = 121$
2	$2^2 = 4$	12	$12^2 = 144$
3	$3^2 = 9$	13	$13^2 = 169$
4	$4^2 = 16$	14	$14^2 = 196$
5	$5^2 = 25$	15	$15^2 = 225$
6	$6^2 = 36$	16	$16^2 = 256$
7	$7^2 = 49$	17	$17^2 = 289$
8	$8^2 = 64$	18	$18^2 = 324$
9	$9^2 = 81$	19	$19^2 = 361$

Discussion There is a pattern to the ones places in the perfect squares. Do you see it?

Hint:
$0^2 = \mathbf{0}$ and $10^2 = 100$
$1^2 = \mathbf{1}$ and $11^2 = 121$
$2^2 = \mathbf{4}$ and $12^2 = 144$

What is the pattern and why is it like that?

Procedure *To find a square root of a given number, find a number that can be squared to equal the given number.*

EXAMPLE 8 $\sqrt{169} =$

Answer: 13

Explanation: The square root of 169 is 13 because $13^2 = 169$.

Note: We can check by multiplying 13×13 to be sure we get 169.

Check:
$$
\begin{array}{r}
13 \\
\times 13 \\
\hline
39 \\
+13 \\
\hline
169
\end{array}
$$

◀ **Do Margin 8.**

OBJECTIVE 5 *Solve applications involving square roots.*

EXAMPLE 9   In building your new home, you decide to include an office. The plans can be developed with an office that has an area between 180 ft.² and 200 ft.². What size perfect-square office would fit within the specified area?

Solution: The perfect square that is between 180 and 200 is 196. To get the dimensions of this square, we must get the square root of 196.

$$\sqrt{196} = 14$$

Answer: To make the room a perfect square with an area in the specified range, the dimensions should be 14 ft. $\times$ 14 ft.

◆ Check: $14^2 = 196$

◀ **Do Margin 9.**

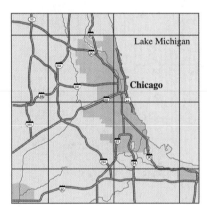

◆ **Margin 8**

Find each square root.

a. $\sqrt{196} =$

b. $\sqrt{324} =$

◆ **Margin 9**

a. A package of grass seed indicates that the whole package will cover 2500 ft.². What would be the dimensions of a square area that could be covered by the whole package?

b. A grid is drawn on a map with a scale where 1 in.² = 196 mi.². What are the dimensions of each square in the grid?

Smith's Premium Grass Seed

Covers 2500 ft²

Answers to Margin 8: **a.** 14 **b.** 18

Answers to Margin 9: **a.** 50 ft. $\times$ 50 ft. **b.** 14 mi. $\times$ 14 mi.

1.4 Exercises

FOR EXTRA HELP

 Videotape 1

 InterAct Math Tutorial Software

 www.carsonmath.com

 AWL Math Tutor Center

InterAct MathXL www.mathxl.com

 Student's Solutions Manual

For Exercises 1–8, determine the quotient and explain your answer.

1. $26 \div 26$ **2.** $381 \div 1$ **3.** $0 \div 49$ **4.** $29 \div 0$

5. $0 \div 0$ **6.** $462 \div 462$ **7.** $22 \div 0$ **8.** $0 \div 95$

For Exercises 9–14, use divisibility rules to determine whether the given number is divisible by 2.

9. 19,761   **10.** 24,978   **11.** 143,706

12. 801,907   **13.** 431,970   **14.** 8445

For Exercises 15–20, use divisibility rules to determine whether the given number is divisible by 3.

15. 19,704   **16.** 1101   **17.** 26,093

18. 450,917   **19.** 98,757   **20.** 241,080

For Exercises 21–26, use divisibility rules to determine whether the given number is divisible by 5.

21. 27,005   **22.** 148,070   **23.** 64,320

24. 704,995   **25.** 4,195,786   **26.** 319,424

For Exercises 27–44, divide.

27. $3834 \div 9 =$ **28.** $2166 \div 6 =$ **29.** $1038 \div 5 =$

30. $24,083 \div 4 =$ **31.** $3472 \div 16 =$ **32.** $10,836 \div 28 =$

33. $6399 \div 26 =$ **34.** $9770 \div 19 =$ **35.** $12,592 \div 41 =$

36. $19,076 \div 38 =$ **37.** $27,600 \div 120 =$ **38.** $21,600 \div 270 =$

39. $235,600 \div 124 =$ **40.** $3,912,517 \div 93 =$ **41.** $14,780 \div 0 =$

42. $174,699 \div 87 =$ **43.** $331,419 \div 207 =$ **44.** $1,031,096 \div 514 =$

For Exercises 45–56, solve and check.

45. $9 \cdot x = 54$

46. $y \cdot 4 = 40$

47. $m \cdot 11 = 88$

48. $15 \cdot a = 45$

49. $24 \cdot t = 0$

50. $b \cdot 21 = 105$

51. $17 \cdot n = 119$

52. $u \cdot 0 = 409$

53. $v \cdot 2 \cdot 13 = 1092$

54. $16 \cdot 5 \cdot k = 560$

55. $29 \cdot 6 \cdot h = 3480$

56. $18 \cdot c \cdot 41 = 6642$

For Exercises 57–70, solve.

57. Dedra's annual salary is $34,248. She gets paid once per month. What is her gross monthly salary?

58. A financial planner is asked to split $16,800 evenly among 7 investments. How much does she put into each investment?

59. Carl is to set an IV drip so that a patient receives 840 ml of 5% D/W solution in an hour. How many ml should the patient receive each minute?

60. The federal government grants a state $5,473,000 to be distributed equally among the 13 technical colleges in the state. How much does each college receive?

61. An employee of a copying company needs to make small fliers for a client. He can make 4 fliers out of each piece of paper. The client needs 500 fliers. How many pieces of paper must be used?

62. A printing company is asked to make 800 business cards for a client. They can print 12 cards on each sheet of card stock. How many sheets will be used?

63. How many 33¢ stamps can be bought with $15? Explain.

64. A long-distance company charges 19¢ per minute. How long can you talk for $5? Explain.

65. A cereal factory produces 153,600 oz. of cereal each day. Each box is to contain 16 oz. The boxes are packaged in bundles of 8 boxes per bundle, then stacked on pallets, 24 bundles to a pallet. Finally, the pallets are loaded onto trucks, 28 pallets to each truck.
  a. How many boxes of cereal are produced each day?
  b. How many bundles are produced?
  c. How many pallets are needed?
  d. How many trucks are needed? Explain.

66. A bottling company produces 48,000 bottles each day. The bottles are packaged first in six-packs, then bundled into cases of 4 six-packs per case. The cases are then loaded onto pallets, 36 cases to a pallet.
  a. How many six-packs are produced?
  b. How many cases?
  c. How many pallets are needed? Explain.

67. In designing a room for a house, the desired length is 24 ft. The desired area is 432 ft.². What must the width be? (Tip: Draw a picture.)

68. A landscaper needs to cover a 4000-ft.² yard with sod. Sod is delivered in pallets. Each pallet holds enough sod to cover 504 ft.². How many pallets must be purchased if the company that sells the pallets will only sell whole pallets?

69. Barry wants to fence in his backyard. He decides to use prefabricated sections of wooden fencing. Each section is 8 ft. in length. How many sections must be purchased to complete 170 ft?

70. Building code says that electrical outlets must be spaced no more than 8 ft. apart along the wall perimeter in each room. If a room has a perimeter of 60 ft., then how many electrical outlets will be needed to meet the code requirements?

For Exercises 71–82, find the square root. You should have these perfect squares and their roots memorized.

71. $\sqrt{100}$ **72.** $\sqrt{81}$ **73.** $\sqrt{169}$ **74.** $\sqrt{121}$

75. $\sqrt{0}$ **76.** $\sqrt{289}$ **77.** $\sqrt{1}$ **78.** $\sqrt{25}$

79. $\sqrt{49}$ **80.** $\sqrt{144}$ **81.** $\sqrt{196}$ **82.** $\sqrt{225}$

📱 *For Exercises 83–86, use a calculator to find the square root.*

83. $\sqrt{1369}$ **84.** $\sqrt{7056}$ **85.** $\sqrt{45{,}796}$ **86.** $\sqrt{21{,}316}$

For Exercises 87–90, solve.

87. A machinist is hired to design a square plate cover for an engine. The hole to be covered is 36 square inches. What must be the dimensions of the plate?

88. A search-and-rescue team is flying in a helicopter over the ocean. They must search a square area of 784 square miles. What are the dimensions of this square?

89. The Jacksons wish to build a deck onto the back of their house. The builder charges $13 per ft.². They have $1872 in savings to spend.
a. How many square feet can they afford?
b. If they build a square deck, what would be the dimensions?

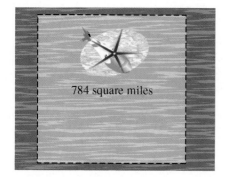

784 square miles

90. Michael is building a concrete patio. He has purchased enough concrete to cover an area of 324 ft.². If he built a square patio, what would be its dimensions?

REVIEW EXERCISES ───────────────────────────────────

1. Solve and check. $184 + t = 361$

2. Connie is placing a decorative border strip near the ceiling in her bedroom. The room is 14 ft. by 16 ft. How much will she need?

3. Estimate the product by rounding to the nearest ten, then find the actual product.

$42 \times 367 =$

4. Evaluate 5^4.

5. Write 49,602 in expanded notation with powers of 10.

1.5 Order of Operations

OBJECTIVES

1 Simplify numerical expressions by following the order of operations agreement.

OBJECTIVE **1** *Simplify numerical expressions by following the order of operations agreement.*

How do we handle problems like $16 - 3 \cdot 2$ that involve different operations? Notice we get different answers depending on the order in which we perform the operations.

If we subtract first, we get:

$$16 - 3 \cdot 2$$
$$= 13 \cdot 2$$
$$= 26$$

If we multiply first, we get:

$$16 - 3 \cdot 2$$
$$= 16 - 6$$
$$= 10$$

Which is correct? It turns out that 10 is the correct answer because we must multiply before subtracting.

The correct order is determined by an agreement. Mathematicians simply agreed upon how to do problems involving more than one operation. Because the specific order of operations that we follow is an agreement, it cannot be derived or proven.

> **Order of Operations Agreement** *Perform operations in the following order:*
> **1.** Parentheses ( ), brackets [ ], and braces { }
> **2.** Exponents/Roots
> **3.** Multiplication/Division from left to right
> **4.** Addition/Subtraction from left to right

Many people use a sentence to help remember the order of operations. Here's one that is popular:

Please	**E**xcuse	**M**y	**D**ear	**A**unt	**S**ally.
a	x	u	i	d	u
r	p	l	v	d	b
e	o	t	i	i	t
n	n	i	s	t	r
t	e	p	i	i	a
h	n	l	o	o	c
e	t	i	n	n	t
s	s	c			i
e		a			o
s		t			n
		i			
		o			
		n			

Another one:

Please **E**ducate **M**y **D**aughter **A**nd **S**on.

◆ Margin 1

Simplify. Show your steps.

a. $15 + 4 \cdot 6$

b. $20 - 18 \div 3$

c. $12 + 14 \div 2 \cdot 5$

◆ Margin 2

Simplify. Show your steps.

a. $12 + 2^4 - 9 \cdot 2$

b. $2 \cdot 6 \div 4 + \sqrt{49}$

c. $3^3 - 5 \cdot 3 + \sqrt{64} - 24 \div 3$

◆ Margin 3

Simplify. Show your steps.

a. $7(9 - 5) + 16 \div (3 + 5)$

b. $3^4 - 5(8 - 2) \div 6 + (7 - 3)^2$

Answers to Margin 1: **a.** 39 **b.** 14 **c.** 47

Answers to Margin 2: **a.** 10 **b.** 10 **c.** 12

Answers to Margin 3: **a.** 30 **b.** 92

Other people form a word from the initial letters:

PEMDAS

WARNING Some people make the mistake of thinking that multiplication always comes before division and addition always comes before subtraction. However, proper order is to multiply or divide from *left to right* in the order they occur. Then add or subtract from *left to right* in the order they occur.

EXAMPLE 1 Simplify. $8 + 6 \cdot 2$

Solution: $8 + 6 \cdot 2$ Multiply $6 \cdot 2$ first to get 12.

$\qquad\quad = 8 + 12$ Add $8 + 12$ to get 20.

$\qquad\quad = 20$

Explanation: The agreement is to multiply before adding, so we multiplied $6 \cdot 2$ first to get 12. Then we added 8 and 12 to get 20.

Note: Notice how we are writing steps underneath instead of outwards to the right. This is an algebraic form of writing mathematics. The purpose for writing steps underneath is that it makes the problem easier to follow. You can see very clearly where the results come from.

Discussion Explain the mistake: $8 + 6 \cdot 2$

$\qquad\qquad\qquad\qquad\qquad = 14 \cdot 2$

$\qquad\qquad\qquad\qquad\qquad = 28$

◀ **Do Margin 1.**

EXAMPLE 2 Simplify. $16 - 3^2 + 15 \div \sqrt{25}$

Solution: $16 - 3^2 + 15 \div \sqrt{25}$ Evaluate: $3^2 = 9$ and $\sqrt{25} = 5$

$\qquad\quad = 16 - 9 + 15 \div 5$ Divide $15 \div 5$ to get 3.

$\qquad\quad = 16 - 9 + 3$ Subtract $16 - 9$ to get 7.

$\qquad\quad = 7 + 3$ Add $7 + 3$ to get 10.

$\qquad\quad = 10$

In $16 - 9 + 3$ our choices are subtraction or addition. We are to add/subtract from left to right, so we subtract first.

◀ **Do Margin 2.**

EXAMPLE 3 Simplify $3(12 - 9) + 5^2 - (18 + 2) \div 5$

Solution:

$\qquad 3(12 - 9) + 5^2 - (18 + 2) \div 5$ Perform operations within parentheses.

A number next to parentheses means multiply. ↗

$\qquad = 3 \cdot 3 + 5^2 - 20 \div 5$ Evaluate: $5^2 = 25$.

$\qquad = 3 \cdot 3 + 25 - 20 \div 5$ Multiply and divide from left to right. $3 \cdot 3 = 9$ and $20 \div 5 = 4$.

$\qquad = 9 + 25 - 4$ Add and subtract from left to right. $9 + 25 = 34$ and $34 - 4 = 30$

$\qquad = 34 - 4$

$\qquad = 30$

Note: Once the calculations within parentheses, brackets, or braces have been completed we can drop the parentheses, brackets, or braces.

Note: Once you get used to the order, you'll find that you can do some of the operations in the same step. For example, we could have performed the operations within the parentheses and evaluated 5^2 in one step. That would have saved writing an extra line.

◀ **Do Margin 3.**

What if we have parentheses, brackets, or braces inside other parentheses, brackets, or braces? When these grouping symbols appear within other grouping symbols, we say they are *embedded*. When we have embedded parentheses, brackets, or braces, we must work from the innermost symbols outward. We perform the inner computations first.

EXAMPLE 4 Simplify $[26 - 3(2 + 5)] + 30 \div \sqrt{9 + 16}$

Solution: $[26 - 3(2 + 5)] + 30 \div \sqrt{9 + 16}$ Start with the operation in the innermost parentheses. $2 + 5 = 7$

$= [26 - 3 \cdot 7] + 30 \div \sqrt{9 + 16}$ Multiply within the brackets. $3 \cdot 7 = 21$

$= [26 - 21] + 30 \div \sqrt{9 + 16}$ Finish within the brackets. $26 - 21 = 5$

$= 5 + 30 \div \sqrt{9 + 16}$ Add within the radical.

$= 5 + 30 \div \sqrt{25}$ Find the square root. $\sqrt{25} = 5$

$= 5 + 30 \div 5$ Divide. $30 \div 5 = 6$

$= 5 + 6$ Add. $5 + 6 = 11$

$= 11$

Notice that $\sqrt{9 + 16}$ is not the same as $\sqrt{9} + \sqrt{16}$.

$\sqrt{9 + 16}$ means to find the square root of the sum of 9 and 16 while $\sqrt{9} + \sqrt{16}$ means to find the sum of the square root of 9 and the square root of 16.

Look at the difference in answers.

$$\sqrt{9} + \sqrt{16} \qquad\qquad \sqrt{9 + 16}$$
$$= 3 + 4 \qquad\qquad\quad = \sqrt{25}$$
$$= 7 \qquad\qquad\qquad\;\; = 5$$

Conclusion: When getting the square root of a sum or difference, we must add or subtract first, then get the root of the sum or difference.

EXAMPLE 5 Simplify $\{3^2 + 5[(12 - 8) \div 2]\} + 2\sqrt{16 \cdot 9}$

Solution: $\{3^2 + 5[(12 - 8) \div 2]\} + 2\sqrt{16 \cdot 9}$ Innermost parentheses first. $12 - 8 = 4$

$= \{3^2 + 5[4 \div 2]\} + 2\sqrt{16 \cdot 9}$ Divide within the brackets. $4 \div 2 = 2$

$= \{3^2 + 5 \cdot 2\} + 2\sqrt{16 \cdot 9}$ Evaluate within the braces. $3^2 = 9$

$= \{9 + 5 \cdot 2\} + 2\sqrt{16 \cdot 9}$ Multiply inside the braces. $5 \cdot 2 = 10$

$= \{9 + 10\} + 2\sqrt{16 \cdot 9}$ Finish within the braces. $9 + 10 = 19$

$= 19 + 2\sqrt{16 \cdot 9}$ Multiply inside the radical. $16 \cdot 9 = 144$

$= 19 + 2\sqrt{144}$ Find the square root. $\sqrt{144} = 12$

$= 19 + 2 \cdot 12$ Multiply. $2 \cdot 12 = 24$

$= 19 + 24$ Add. $19 + 24 = 43$

$= 43$

A number next to a radical sign means multiply.

Working with Scientific Calculators

Scientific calculators are programmed to follow the proper order of operations. This means you merely type in the problem as it is written.

Example: $(12 + 5)^3 + 360 \div [3^2 + (40 - 31)] + \sqrt{841}$

Enter:

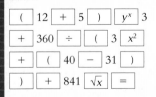

Answer: 186

Margin 4

Simplify.

a. $\{5 + 2[18 \div (2 + 7)]\}$
   $+ 5\sqrt{25 \cdot 4}$

b. $5^3 - 3[(6 + 14) \div$
   $(3 + 2)] + \sqrt{36 \div 9}$

c. $4\sqrt{169 - 25} +$
   $\{[30 - 3(5 + 2)] +$
   $[70 - (3 + 5)^2]\}$

We saw in Example 5 that $\sqrt{9 + 16}$ is not the same as $\sqrt{9} + \sqrt{16}$. What about $\sqrt{16 \cdot 9}$ compared to $\sqrt{16} \cdot \sqrt{9}$? Let's have a look:

Multiplying first: $\sqrt{16 \cdot 9}$
$$= \sqrt{144}$$
$$= 12$$

Roots first: $\sqrt{16} \cdot \sqrt{9}$
$$= 4 \cdot 3$$
$$= 12$$

Notice we get the same answer either way. This is always true with square roots of products or quotients.

Conclusion: If multiplication or division occurs under a radical, we can multiply or divide first, then find the square root of the product or quotient. Or we can find the square roots first, then multiply or divide the roots.

◀ **Do Margin 4.**

What if a fraction bar is used like $\dfrac{7^2 - 4}{29 - 3(8 - 6)^3}$? The fraction line means to divide. However, before we can divide, we must work out the top to have a dividend and the bottom to have a divisor. In fraction terminology the top is called the *numerator* and the bottom is called the *denominator*. Think of $\dfrac{7^2 - 4}{29 - 3(8 - 6)^3}$ as having implied parentheses in the numerator and denominator. In other words, the order of operations is the same as $(7^2 - 4) \div [29 - 3(8 - 6)^3]$.

EXAMPLE 6 Simplify.

$$\frac{7^2 - 4}{29 - 3(8 - 6)^3}$$

Solution: Work the numerator and denominator separately.

$$\frac{7^2 - 4}{29 - 3(8 - 6)^3}$$ In the top: 7^2 5 49. At the bottom: 8 2 6 5 2.

$$= \frac{49 - 4}{29 - 3(2)^3}$$ In the top: 49 2 4 5 45. At the bottom: 2^3 5 8.

$$= \frac{45}{29 - 3 \cdot 8}$$ Multiply in the bottom $3 \cdot 8 = 24$.

$$= \frac{45}{29 - 24}$$ Finish the bottom: $29 - 24 = 5$.

$$= \frac{45}{5}$$ Divide 45/5 = 9.

$$= 9$$

◀ **Do Margin 5.**

1.5 Exercises

For Exercises 1–36, simplify using the order of operations. Show your steps.

1. $7 + 5 \cdot 3$

2. $35 - 3 \cdot 7$

3. $18 + 36 \div 4 \cdot 3$

4. $20 \div 2 \cdot 5 + 6$

5. $12^2 - 6 \cdot 4 \div 8$

6. $36 - 3^2 \cdot 4 + 14$

7. $39 \div 13 + \sqrt{36} - 2^3$

8. $29 - 4^2 + 21 \div \sqrt{49}$

9. $12 + 8\sqrt{25} \div 4$

10. $7\sqrt{64} - 40 \div 5 \cdot 2 + 9$

11. $4^2 - 6 \cdot 2 + \sqrt{81} - 42 \div 7$

12. $2^6 - 18 \div 3 \cdot 5 - \sqrt{100}$

13. $24 \div 8 \cdot 6 + \sqrt{121} - 5^2$

14. $3^2 \cdot 4 \div 6 - 5 + \sqrt{16}$

15. $2(14 + 3) - 8 \cdot 2 + (9 - 2)$

16. $5(13 - 7) + 12 \div (4 + 2)$

17. $2^5 \div (14 - 6) + 7(23 - 12)$

18. $2(7 - 4) + 8^2 - (16 + 5) \div 7$

19. $(13 - 8)3^2 - 48 \div (15 - 9)$

20. $(19 - 16)\sqrt{169} + 54 \div 9(7 - 5)$

21. $[14 - (3 + 2)] \div 3 + 4(17 - 6)$

22. $31 - 3[(20 - 6) - 3 \cdot 2] + 2^4$

23. $\{18 - 4[21 \div (3 + 4)]\} + 3\sqrt{16 \cdot 4}$

24. $\sqrt{25 \cdot 9} + 2\{[15 - (3 + 9)] \cdot 6\}$

25. $4^3 - 5[(28 - 4) \div 2^3] + \sqrt{100 \div 25}$

26. $2[(2 + 9)\sqrt{36 \div 9}] + 28 \div 7$

27. $4\{[25 - (19 + 2)] \cdot (3 + 1)\} - (3 + 5)^2$

28. $3\sqrt{25 - 16} + \{[25 + 2(14 - 9)] \div [43 - (2 + 4)^2]\}$

29. $4\{[49 - 5(2 + 6)] + [21 - (3 + 1)^2]\} - \sqrt{169 - 144}$ **30.** $(8 - 3)\sqrt{36 + 64} - 3\{[24 + 4 \cdot 8] - (11 - 9)[19 - 16]^3\}$

31. $\dfrac{9^2 - 21}{56 - 2(4 + 1)^2}$

32. $\dfrac{22 + 3^3}{3(14 - 6) - (8 + 9)}$

33. $\dfrac{38 - 4(15 - 12)}{(3 + 5)^2 - 2(39 - 8)}$

34. $\dfrac{(12 - 5)^2 + 2^3}{10 \div 2 - (11 - 9)}$

35. $[485 - (68 + 39)] + 4^5 - 24 \cdot 16 \div 8$

36. $(25 - 9)^3 + 420 \div (28 - 7) + \sqrt{2209}$

For Exercises 37–40, explain the mistake, then work the problem correctly.

37. $48 - 6(9 - 4)$
$= 48 - 6(5)$
$= 42(5)$
$= 210$

38. $23 + \sqrt{100 - 64}$
$= 23 + 10 - 8$
$= 33 - 8$
$= 25$

39. $(3 + 5)^2 - 2\sqrt{49}$
$= 9 + 25 - 2\sqrt{49}$
$= 9 + 25 - 2(7)$
$= 9 + 25 - 14$
$= 34 - 14$
$= 20$

40. $[12 - 2 \cdot 3] + 4(3)^2$
$= [12 - 6] + 4(3)^2$
$= 6 + 4(3)^2$
$= 6 + 12^2$
$= 6 + 144$
$= 150$

REVIEW EXERCISES

1. Estimate $42,320 + 25,015$ by rounding to the nearest ten thousand.

2. $498,503 \times 209 =$

3. $21,253 \div 17 =$

4. A 12 ft. by 16 ft. room is to have ceiling molding installed. How much molding is needed? If the carpenter charges \$2 per foot to install the molding, how much will it cost?

5. The floor of the same 12 ft. by 16 ft. room is to be covered with square 1 ft. by 1 ft. tiles. How many tiles must be purchased?

1.6 Variables, Formulas, and Solving Equations

OBJECTIVES

1 Use the formula $P = 2l + 2w$ to find the perimeter of a rectangle.

2 Use the formula $A = bh$ to find the area of a parallelogram.

3 Use the formula $V = lwh$ to find the volume of a box.

4 Solve for a missing number in a formula.

OBJECTIVE 1 *Use the formula $P = 2l + 2w$ to find the perimeter of a rectangle.*

We have solved for missing pieces of information in the preceding problems. Rather than use words or (?) to describe the missing value, we began using variables. Let's recall the definition of a variable.

DEFINITION   **Variable:** A symbol that can vary in value.

Most of the time, variables are letters of our alphabet such as x, y, or z. However, a variable does not have to be a letter of the English alphabet. Often you will encounter Greek letters for variables, such as λ (lambda), θ (theta), or τ (tau).

We use variables when we don't know or weren't given actual numbers. For example, in Section 1.3 when we discovered that to get the area of a rectangle we multiply length times width, we condensed that relationship to a formula: $A = l \cdot w$. Let's recall the definition of a formula.

DEFINITION   **Formula:** An equation that describes a procedure.

Some symbols do not vary in value. We call these symbols **constants.**

DEFINITION   **Constant:** Any symbol that does not vary in value.

All of our numerals are constants. Remember, 3 is merely a symbol that means three. The symbol 3 always means three, so it is a constant. Sometimes letters are used as constants. This usually happens when a value is very large or unruly because of having lots of decimal places to it. An example is the Greek letter π (*pi*: pronounced pī). The numerical value of π happens to be a decimal number that we cannot express exactly because the decimal digits continue on forever without repetition. Since we cannot write all those decimal digits, we chose a symbol to represent the exact value.

When scientists or mathematicians discover some new phenomenon, they try to write a formula that describes what is happening. The formula then allows them to predict what might happen if the conditions of the situation are altered.

Let's see how we work with formulas.

> **Procedure:** *To use a formula:*
>
> **1.** Replace the variables with the corresponding given values.
> **2.** Solve for the missing variable.

In Section 1.2 we defined perimeter to be the total distance along the edge of a shape. To get perimeter we need to add the lengths of all sides. Let's look at a rectangle. We could label the sides as *length* for the longer sides and *width* for the shorter sides. However, using these full words is tedious, so let's use letters. Use *l* for length and *w* for width.

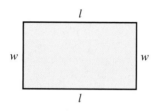

Perimeter of a rectangle = length + width + length + width
$$P = l + w + l + w$$

Notice we can simplify the relationship even further. Since we are repeatedly adding two lengths and two widths, we can multiply the length by 2 and multiply the width by 2 then add. We now have a formula for the perimeter of a rectangle.

$$P = 2l + 2w$$

Note: When we write a number next to a variable as in $2l$ or $2w$, it means to multiply 2 times the numbers that l and w represent. It is the same when variables are written next to each other. For example, we've been writing the formula for the area of a rectangle as $A = l \cdot w$. We can simply write $A = lw$.

◆ **Margin 1**

a. A rectangle has a length of 19 m and width of 42 m. Find the perimeter.

b. A rectangle has a length of 7 mi. and a width of 6 mi. Find the perimeter.

EXAMPLE 1 A school practice field is to be enclosed with a chain-link fence. The field is 400 ft. by 280 ft. How much fencing is needed?

Solution: Because we must find the total distance around a rectangular field, we use the formula for the perimeter of a rectangle.

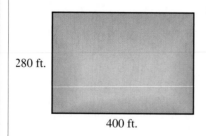

280 ft.

400 ft.

$P = 2l + 2w$
$P = 2(400 \text{ ft.}) + 2(280 \text{ ft.})$
$P = 800 \text{ ft.} + 560 \text{ ft.}$
$P = 1360 \text{ ft.}$

Replace *l* with 400 and *w* with 280 then calculate.

Discussion What would happen if we replaced *l* with 280 and *w* with 400? Would the answer be different?

◀ **Do Margin 1.**

Answers: **a.** 122 m **b.** 26 mi.

OBJECTIVE 2 *Use the formula A = bh to find the area of a parallelogram.*

In Section 1.3 we discussed that area is surface space measured in square units. We also developed the formula for calculating the area of a rectangle, $A = lw$. Squares and rectangles are special forms of a more general class of figures called **parallelograms.** Parallelogram comes from the root word **parallel.**

> **DEFINITION** **Parallelogram:** A four-sided figure with two pair of parallel sides.
>
> **Parallel lines:** Straight lines that never intersect.

Examples of parallelograms:

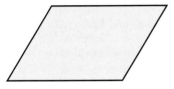 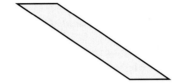

To calculate the area of a parallelogram, we need the length of the base and the height. The height is measured along a line that makes a 90° angle with the base. An angle that measures 90° is called a **right angle.**

> **DEFINITION** **Right angle:** An angle that measures 90°.

We will use the letter b for the length of the base and h for the height.

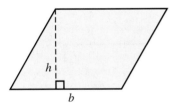

Note The right angle is indicated by the small square where the height line meets the base.

Notice if we cut along the height line we would cut off a triangle. If we move the triangle around to the right side and place the parallel sides together, the resulting figure is a rectangle.

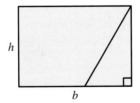

The rectangle we have made has the same area as the original parallelogram. The length of this rectangle is b and the width is h. So the formula to calculate the area of a parallelogram is:

$$A = bh$$

Also, because squares and rectangles are special parallelograms, we can use $A = bh$ for them as well.

Margin 2

Use the formula A = bh to find the area.

a.

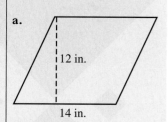

12 in.

14 in.

b.

8 ft.

4 ft.

EXAMPLE 2 Find the area.

Solution: We have a formula for the area of a parallelogram: $A = bh$.

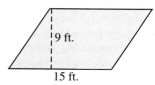

9 ft.

15 ft.

$$A = bh$$
$$A = (15)(9)$$
$$A = 135 \text{ ft.}^2$$

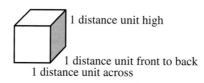 Replace b with 15 and h with 9 then multiply.

Discussion What would happen if we replaced b with 9 and h with 15? Would the answer be different?

◀ **Do Margin 2.**

OBJECTIVE 3 *Use the formula V = lwh to find the volume of a box.*

Volume is a measure of the amount of space inside a three-dimensional object. We measure volume using **cubic units.**

DEFINITION **Cubic unit:** A 1 × 1 × 1 cube.

1 distance unit high

1 distance unit front to back

1 distance unit across

Examples of cubic units:

Cubic Inch:

1 in.
1 in.
1 in.

A cubic inch is a 1 in. by 1 in. by 1 in. cube and is represented mathematically by $in.^3$.

Cubic meter:

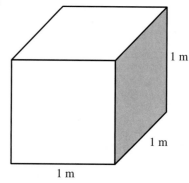

1 m

1 m

1 m

A cubic meter is a 1 m by 1 m by 1 m cube and is represented mathematically by m^3.

DEFINITION **Volume:** The total number of cubic units that completely fill an object.

Answers: **a.** 168 in.^{2} **b.** 32 ft.^{2}

Let's develop a formula for calculating the volume of a box. Consider a box that is 3 m by 2 m by 4 m.

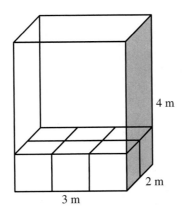

If we begin to fill the box with cubic meters, we can fit 6 cubic meters in a layer at the bottom. Because each layer is 1 m high, we can stack a total of four layers inside 4 m the box. This makes a total of 24 cubic meters inside the box.

$$\text{Volume} = 24 \text{ cubic meters}$$

Notice we can calculate the volume by multiplying.

$$\text{Volume} = 3 \text{ m} \cdot 2 \text{ m} \cdot 4 \text{ m}$$
$$V = 24 \text{ m}^3$$

This is read *cubic meters* or *meters cubed*. All volume calculations will involve multiplying three distance measurements. Therefore, volume units will always be expressed as the distance unit with an exponent of 3.

Conclusion: Volume of a box = length · width · height

$$V = lwh$$

EXAMPLE 3   An engine is a block that is 25 cm by 20 cm by 10 cm. What is the volume of the engine block?

Solution: Because the engine block is a box, we can use the formula $V = lwh$.

$$V = lwh$$
$$V = (25 \text{ cm})(20 \text{ cm})(10 \text{ cm})$$
$$V = 5000 \text{ cm}^3$$

◆ **Explanation:** We replaced the letters with the corresponding values and did the calculation.

▸ **Discussion** What would happen if we interchanged the length value with the width value (or height)? Would the answer be different?

Do Margin 3. ▷

◆ **Margin 3**

Find the volume of each box.

a.

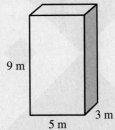

9 m

5 m

3 m

b.

3 ft.

5 ft.

2 ft.

Answers: **a.** 135 m³ **b.** 30 ft.³

OBJECTIVE 4 *Solve for a missing number in a formula.*

What if we were given the area and length of a rectangle and asked to find the width?

EXAMPLE 4 The area of a parallelogram is 40 ft.² and the base is 8 ft. Find the height.

Solution: Use the formula $A = bh$. Replace A with 40 and b with 8.

$$A = bh$$
$$40 = 8h$$
$$40 \div 8 = h$$
$$5 = h$$

$40 = 8h$ is a missing factor equation. We learned to solve missing factor equations in Section 1.4 by dividing the product by the known factor.

Answer: The height is 5 ft.

Check: Verify that a parallelogram with a base of 8 ft. and height of 5 ft. has an area of 40 ft.².

$$A = (8 \text{ ft.})(5 \text{ ft.})$$
$$A = 40 \text{ ft.}^2$$

◀ **Do Margin 4.**

We can use the volume formula in a similar way. If we are given the volume of a box and two of the three measurements (length, width, and height), we can solve for the missing measurement.

EXAMPLE 5 The volume of a box is 480 cm³. If the length is 10 cm and the width is 8 cm, then find the height.

Solution: Use the formula $V = lwh$. Replace V with 480, l with 10, and w with 8, then solve for h.

$$V = lwh$$
$$480 = (10)(8)h$$
$$480 = 80h$$
$$480 \div 80 = h$$
$$6 = h$$

Because we know two out of the three factors, to simplify the problem we multiply the two known factors.

Answer: The width is 6 cm.

Check: Verify that a box with a length of 10 cm, width of 8 cm, and height of 6 cm has a volume of 480 cm³.

$$V = (10 \text{ cm})(8 \text{ cm})(6 \text{ cm})$$
$$V = 480 \text{ cm}^3$$

◀ **Do Margin 5.**

◆ Margin 4

Find the missing number.

a. The area of a parallelogram is 72 in.². If the height is 9 in., find the base.

b. The area of a rectangle is 60 m². If the width is 5 m, find the length.

◆ Margin 5

Solve for the missing number.

a. The volume of a box is 336 cm³. If the length is 12 cm and the width is 7 cm, find the height.

b. The volume of a box is 810 in.³. If the width is 9 in. and the height is 5 in., find the length.

Answers to Margin 4: **a.** 8 in. **b.** 12 m

Answers to Margin 5: **a.** 4 cm **b.** 18 in.

1.6 Exercises

For Exercises 1–6, find the perimeter.

1.

29 cm

18 cm

2.
10 ft.

15 ft.

3.
32 in.

32 in.

4.
40 m

7 m

5.
24 km

8 km

6.
60 mi.

60 mi.

For Exercises 7–12, find the area.

7.
19 m

12 m

8.
9 ft.

13 ft.
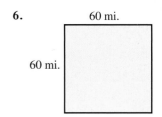

9.
12 in.

12 in.

10.
26 m

5 m
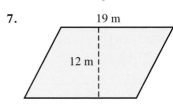

11.

16 km
7 km

12.

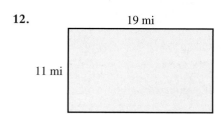
19 mi
11 mi

For Exercises 13–18, find the volume.

13.

10 ft.
4 ft.
7 ft.

14.

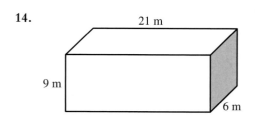
21 m
9 m
6 m

15.

6 in.
6 in.
6 in.

16.

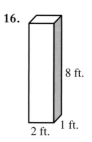
8 ft.
1 ft.
2 ft.

17.

14 km
5 km
3 km

18.

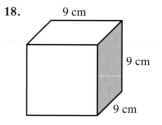
9 cm
9 cm
9 cm

For Exercises 19–36, find the missing number.

19. A parallelogram has an area of 144 m². If the base is 16 m, find the height.

20. A parallelogram has an area of 399 ft.². If the height is 19 ft., find the base.

21. A parallelogram has an area of 1922 in.². If the height is 31 in., find the base.

22. A parallelogram has an area of 2408 cm². If the base is 56 cm, find the height.

23. A canning company feeds sheets of metal into a machine that molds them into cans. The area of a rectangular piece of sheet metal must be 180 in.². To fit into the machine the width of a sheet of metal must be 9 in. What must the length be?

24. A search-and-rescue helicopter is assigned to search an 3584 mi.² area of the Atlantic Ocean. They have flown north in a straight path of 64 mi. and will now turn east (90° turn). How far must they travel along this new course to make a rectangle that will have the desired area?

25. A box has a volume of 12,320 ft.³. If the length is 22 ft. and the width is 35 ft., find the height.

26. A box has a volume of 1368 m³. If the width is 19 m and the height is 3 m, find the length.

27. A box has a volume of 22,320 in.³. If the length is 31 in. and the height is 40 in., find the width.

28. A box has a volume of 30,600 cm³. If the height is 60 cm and the length is 15 cm, find the width.

29. An engineer is designing a refrigerator. The desired volume is to be 24 ft.³. If the length must be 3 ft. and width must be 2 ft., what must the height be?

30. The research and development department for an electronics company has calculated that 18 ft.^{3} are necessary to house a new high-resolution T.V. screen with all its components. Most entertainment centers allow for a maximum space of 2 ft. by 3 ft. by 3 ft. Will the new design fit into this space?

REVIEW EXERCISES

1. An accountant is given the following spreadsheet of expenses and income for a company. What is the final balance?

Description of expense	Amount	Description of assets/income	Amount
Payroll	$16,980	Checking account	$ 2,359
Utilities	$ 1,250	Income	$41,300
Water	$ 158		
Waste disposal	$ 97		
New inventory	$12,341		

2. Find the area of a parallelogram with a base of 14 ft. and a height of 16 ft.

3. A contractor charges $16 per ft.^{2} to put in hardwood floors. How much would it cost to have the contractor cover a 32 ft.^{2}-entry space with hardwood flooring?

4. To open a lock, you must use a combination of three things. You must turn a key to one of two positions, left or right; set a dial to a digit 0–9; and set another dial to a letter from A to F. How many possible combinations are there?

5. An employee of a copying company needs to make leaflets for a client. She can make 6 leaflets out of each piece of paper. The client needs 1800 leaflets. How many pieces of paper must be used?

1.7 Problem Solving and Applications

OBJECTIVES

1 Use a problem-solving process to solve problems involving several operations.

OBJECTIVE 1 *Use a problem-solving process to solve problems involving several operations.*

As the problems we solve get more complex, it is helpful to follow a problem-solving outline. The purpose of the outline is to suggest strategies for solving problems and to offer insight into the process of problem solving. Throughout the rest of the text, all application problems will be solved using the outline.

Problem-Solving Outline

1. Develop an *understanding* of the problem.
   a. Read the question(s) (not the whole problem, just the question at the end) and write a note to yourself about what it is you are to find.
   b. Now read the whole problem.
   c. Underline key words.
   d. Draw a picture (if possible).
   e. Make a chart or list.
   f. Simulate the situation.
   g. Search for a related example problem.
2. Develop a *plan*.
   a. Translate the key words to an equation.
   b. Search for a formula.
3. *Execute* your plan by solving the equation or formula.
4. *Answer* the question. Look back at the note you made to yourself about what it was you were to find, and make sure you answer that question. Be sure to include appropriate units.
5. *Check* your results. Make sure your answer is reasonable.
   a. Repeat the solution in a different way.
   b. Reverse the process.
   c. Estimate the answer and make sure the estimate and actual answer are reasonably close.

Discussion Why are units important in your answer? What are the pros and cons of each different method of checking?

Let's use the problem-solving outline to solve problems involving composite shapes. Composite shapes are shapes formed by putting together two or more fundamental shapes, like rectangles or parallelograms. Or, a composite shape can be a shape within a shape.

EXAMPLE 1 Below is a base plan for a new building. What is the total area occupied by the building?

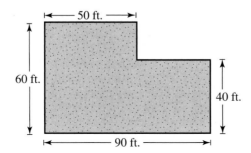

Solution: Follow the problem-solving outline.

Develop an understanding: First, read the question and note what it is we are to find.

What is the total area occupied by the building?

Note: We must find the total area of the building.

The key words that we should underline are *total* and *area*. From Section 1.2 we recall that *total* indicates we must add. From Section 1.3 we recall that *area* indicates that we must multiply.

When we calculated area earlier, we always had a rectangle. But the shape of this building is a composite shape.

There are several ways we can look at this composite shape. We could look at it as a rectangle in the upper right-hand corner that was removed from a larger rectangle. However, most people find it easier to look at it as two rectangles put together.

We could picture a horizontal line separating the shape into two rectangles:

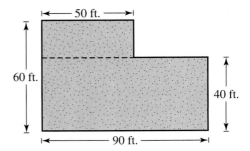

Or a vertical line separating the shape:

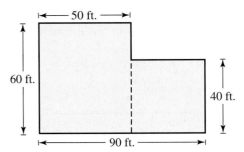

Visualize the rectangles in the way that is easiest for you. You will get the same result as everyone else. Let's work with drawing a vertical line to divide the shape into two rectangles as shown previously.

Develop a plan: To get the total area of the shape, we calculate the area of each rectangle, then add.

Execute the plan: To calculate the area of a rectangle, we multiply length by width.

The larger rectangle is 50 ft. by 60 ft.

$$\text{Area of the larger rectangle} = 50 \text{ ft.} \times 60 \text{ ft.} = 3000 \text{ ft.}^2$$

Notice we were not given the length of the smaller rectangle. However, we can figure it out from the other dimensions. If it is 90 ft. across the bottom, then the total distance across the top must also be 90 ft. We know 50 ft. and the missing amount together must be 90 ft. total. This means that the missing amount must be 40 ft.

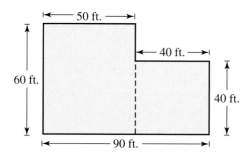

So the smaller rectangle is actually a 40 ft. by 40 ft. square. Now we can get its area:

$$\text{Area of the square} = 40 \text{ ft.} \times 40 \text{ ft.} = 1600 \text{ ft.}^2$$

The total area of the building is the sum of the two areas.

Total area = 3000 ft.^{2} + 1600 ft.^{2} = 4600 ft.^{2}

Answer the question: The total area is 4600 ft.^{2}. ◄——

Check: If we calculate the total area using a horizontal line to separate the two rectangles, we should get the same answer.

> **Reminder**
> Be sure to include the units.

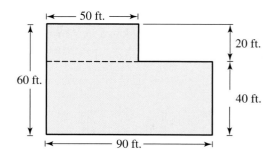

$$\text{Area of top rectangle} = 50 \text{ ft.} \cdot 20 \text{ ft.}$$
$$A = 1000 \text{ ft.}^2$$

$$\text{Area of bottom rectangle} = 90 \text{ ft.} \cdot 40 \text{ ft.}$$
$$A = 3600 \text{ ft.}^2$$

$$\text{Total area} = 1000 \text{ ft.}^2 + 3600 \text{ ft.}^2 = 4600 \text{ ft.}^2$$

Because we get the same answer we can feel secure that our calculation for the total area of
◆ this shape is correct.

Do Margin 1. ▶

EXAMPLE 2 After building a new home, the builder is required to sod the yard. The lot is a rectangle 75 ft. by 125 ft. The house occupies 2500 ft.^{2} of the lot. Each pallet of sod costs $85 and covers 500 ft.^{2}. Partial pallets may not be purchased. How much will it cost to sod the yard?

Solution: Follow the problem-solving outline.

Understand: First, read the question and note what it is we are to find.

How much will it cost to sod the yard?

Note: We must calculate the cost.

◆ **Margin 1**

a. The picture below shows two bedrooms and a connecting hallway that are to be carpeted. How many square feet of carpet will be needed?

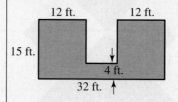

b. The picture below shows the driveway and walk of a new house. The builder needs to know how much concrete to plan for. What is the total area of the driveway and walk?

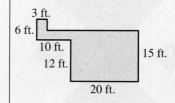

Answers: **a.** 392 ft.^{2} **b.** 339 ft.^{2}

Now let's underline key words and important information.

After building a new home, the builder is required to sod the yard. The lot is a <u>rectangle 75 ft. by 125 ft.</u> The house <u>occupies 2500 ft.² of the lot.</u> <u>Each</u> pallet of sod <u>costs $85</u> and <u>covers 500 ft.²</u> Partial pallets may not be purchased. How much will it cost to sod the yard?

Because the problem involves rectangles, let's draw a picture.

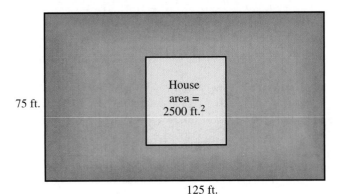

Develop a *Plan:* We ask ourselves questions starting with the question in the problem.

What do we need to know in order to get the cost of the sod?

Answer: If each pallet costs $85, then we need to know how many pallets will be used.

How can we figure the number of pallets required?

Answer: If each pallet covers 500 ft.², then we need to know how many square feet of the lot will be covered with sod.

How can we figure the number of square feet to be covered?

Answer: Because the whole lot, except for the house, is to be covered, we need to subtract the area occupied by the house from the total area of the lot.

We were given the area occupied by the house, but not the lot. How can we get the area of the lot?

Answer: Because the lot is a rectangle and the area of a rectangle is found by the formula $A = lw$, we need to multiply the length and width of the lot.

Plan: Now we have our starting point. We'll get the area of the lot, then subtract the area occupied by the house, and this will give us the area to be covered with sod. From there we can calculate how many pallets will be needed. Once we get the number of pallets, we can figure the cost.

Execute:

$$\text{Calculate the area of the lot: } A = lw$$
$$A = (125 \text{ ft.})(75 \text{ ft.})$$
$$A = 9375 \text{ ft.}^2$$

Now we can calculate how much area will be covered with sod by subtracting the area occupied by the house.

$$\text{Area to sod} = 9375 \text{ ft.}^2 - 2500 \text{ ft.}^2$$
$$\text{Area to sod} = 6875 \text{ ft.}^2$$

Next we need to figure out how many pallets will be needed to cover 6875 ft.². Because each pallet covers 500 ft.², we need to divide 500 into 6875 to see how many sets of 500 are in 6875.

$$\text{Number of pallets} = 6875 \div 500$$
$$\text{Number of pallets} = 13 \text{ r } 375$$

This means 13 whole pallets are needed with 375 ft.² uncovered by sod. Because partial pallets cannot be purchased, the builder must buy one more pallet to cover that 375 ft.², even though not all the sod will be used from that extra pallet. So, a total of 14 pallets must be purchased.

Because each of the 14 pallets costs $85, we must multiply to get the total cost.

$$\text{Total cost} = 14 \times \$85 = \$1190$$

Answer: The total cost to sod the yard is $1190

Check: Let's do a quick estimate of the area. We'll round the length and width of the lot so that we have only one digit other than zero and calculate the area.

$$\text{Actual area} = 125 \text{ ft.} \cdot 75 \text{ ft.}$$
$$\text{Estimate} = 100 \text{ ft.} \cdot 80 \text{ ft.} = 8000 \text{ ft.}^2$$

The house occupies 2500 ft.², so let's round that to 3000 ft.². We can now estimate the amount of land to be covered in sod by subtracting the area of the house from our estimate for the whole lot.

$$\text{Approximate area of coverage: } 8000 - 3000 = 5000 \text{ ft.}^2$$

If each pallet covers 500 ft.², then about 10 pallets would be needed to cover 5000 ft.².

If we round the cost of each pallet to about $90, then the total cost is $90 · 10 pallets = $900. So we should expect the cost to be somewhere in the neighborhood of $900. Our actual cal-
◆ culation of $1190 is reasonable given the amount of rounding we did in our estimates.

Do Margin 2. ▶

EXAMPLE 3 Approximate the volume of the automobile shown.

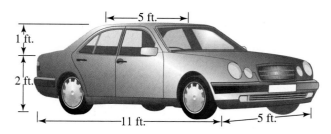

Solution: Follow the problem-solving outline.

Understand: The key words here are *volume* and *approximate*. We know that *volume* means we want to calculate the total number of cubic units that completely fill the object. The only object we've seen so far is a box and the formula is $V = lwh$. But the car isn't exactly a perfect box.

However, since we only need an approximation, we can consider the automobile as two boxes stacked this way:

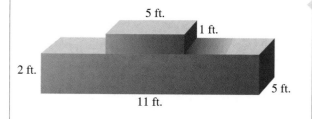

Connection

Two boxes combined is similar to two rectangles combined. To find the area made by the two combined rectangles we add the areas of the rectangles. To find the volume of two boxes combined we add the volumes of the two boxes.

Plan: To get the total volume, we need to get the volume of both boxes and add them together.

$$\text{Total Volume} = \text{Volume of bottom box} + \text{Volume of top box}$$

Execute:

$$V = l \cdot w \cdot h \qquad\qquad + l \cdot w \cdot h$$
$$V = (11 \text{ ft.})(5 \text{ ft.})(2 \text{ ft.}) \qquad + (4 \text{ ft.})(5 \text{ ft.})(1 \text{ ft.})$$
$$V = 110 \text{ ft.}^3 \qquad\qquad + 20 \text{ ft.}^3$$

Answer: $V = 130 \text{ ft.}^3$

Check: Let's reverse the process. We start with 130 ft.³ and see if we can work our way back to the original dimensions. The last step in the preceding process was $110 + 20 = 130$, so we can check by subtracting.

$$130 - 20 = 110 \qquad \text{This is true, so our addition checks.}$$

Now let's check the 110 and 20. The 110 ft.³ came from multiplying (11 ft.)(5 ft.)(2 ft.), so we can reverse this step and divide $110 \div 2 \div 5$ and see if we get 11.

$$110 \div 2 \div 5$$
$$= 55 \div 5$$
$$= 11 \qquad \text{It checks.}$$

Now check the 20. Since we multiplied (4 ft.)(5 ft.)(1 ft.) to get 20, we can reverse this step and divide $20 \div 1 \div 5$ and see if we get 4.

$$20 \div 1 \div 5$$
$$= 20 \div 5$$
$$= 4 \qquad \text{It checks.}$$

> **Tip**
> Whenever dealing with a composite form, ask yourself if it is two or more shapes that have been put together or a shape within another?
>
> When two or more shapes or objects are put together, we calculate the individual areas or volumes and add them.
>
> When we have a shape or object within a shape and must calculate the frame, we calculate the individual areas or volumes and subtract.

◀ **Do Margin 3.**

♦ Margin 3

Engineers are designing a clean room for a computer chip manufacturer. The room is to be L-shaped (see diagram). In designing the air recycling and filter system, the engineers need to know the volume of the room. Find the volume of the room if the whole room has a 10 ft. ceiling.

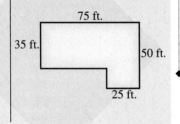

Answer: 30,000 ft.³

1.7 Exercises

For Exercises 1–20, solve.

1. Crown molding is to be installed in a 16 ft. by 20 ft. living room. The subcontractor charges $4 per foot to install the molding. What will be the total cost of molding for the room?

2. Angela plans to put a wallpaper border just below the ceiling in her bedroom. The room is 12 ft. by 14 ft. If the border costs $2 per foot, what will be the total cost?

3. Baseboard molding is to be placed where the walls connect to the floor in the room below. An 8-ft. section of baseboard molding costs $4. If no partial sections are sold, how much will the total baseboard cost?

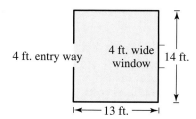

4. Chair railing is to be placed on the walls around the room mentioned in Exercise 3. The railing will be at a height of 3 ft. from the floor. If the railing costs $5 for an 8-ft. section and no partial sections are sold, how much will the chair railing cost for the room?

5. Mr. Williams wishes to fence in a large pasture. The plot plan for the land is shown below. The fencing company charges $5 per foot for fencing and $75 for the 10 ft. gate. What will be the total cost?

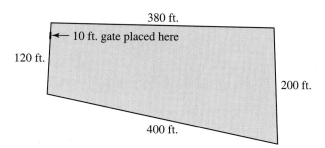

6. Mr. Williams decides to investigate the cost of building the fence in Exercise 3 himself.
   a. If he places wood posts every 8 ft. (except where the gate goes), then how many posts must he purchase? The posts cost $2 each. What will be the cost for all the posts?
   b. He will nail four rows of barbed wire along the posts (except where the gate goes). How many feet of barb wire must he have in all? A roll of barbed wire contains 1320 ft. and costs $45.
   c. Because he'll be nailing four rows of wire to each post, he'll need 4 U-nails for each post. How many nails will be needed? If the nails come in boxes of 100, how many boxes will he need? If the boxes cost $3 each, how much will the nails cost?

d. The gate costs $75. What will be the total cost of materials?

e. Compare the total cost of materials for Mr. Williams to build the fence himself with the total cost of having the fencing company build it. What should Mr. Williams do? What factors should he consider?

7. The Henry family wishes to put a fence around their backyard. The plan is shown below. The fencing company charges $8 per foot for the wood picket fence and $50 for each 4-ft.–wide gate. What will be the total cost?

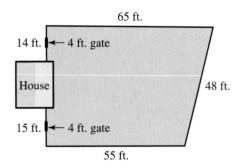

8. Mr. Henry from Exercise 7 decides to investigate the cost of building the fence himself. He decides that 8-ft. prefabricated panels are the least expensive option. Each 8-ft. panel costs $25. Posts are placed at each end of each section and at each end of each gate. The posts cost $3 each. Rental of an auger to dig the post holes costs $50. The gates cost $45 each. What will be the total cost? (Hint: Draw the plan for post placement.)

9. Find the area of the metal (shaded) in the plate shown. The hole in the center is an 8 cm by 8 cm square.

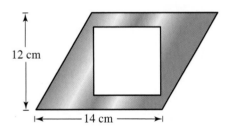

10. A 84 ft. by 72 ft. building is on a lot that is 245 ft. by 170 ft. The entire area surrounding the building is to be paved for parking. Calculate the area that will be paved.

11. The floor of a house is to be made from 4 ft. by 8 ft. sheets of plywood. The floor plan is shown below. How many sheets of plywood will be needed? If each sheet of wood costs $15, what will be the total cost?

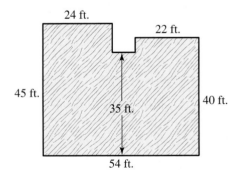

12. The area shown below is to be carpeted. How many square feet will be needed in all? If the chosen carpet costs $3 for each square foot, then how much will it cost for carpet for the area?

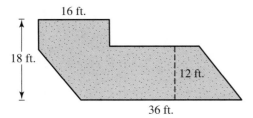

16 ft.

18 ft.

12 ft.

36 ft.

13. A 24 ft. by 15 ft. room with a 9 ft. ceiling is to be painted. The room has a 6 ft. by 7 ft. entryway and two windows that each measure 3 ft. by 2 ft. How many square feet are to be painted? If a single gallon of paint covers 400 square feet, how many gallons will be needed?

14. Nine offices in an office building are to be painted. Each office is 10 ft. by 9 ft. and has 8 ft. ceilings. Each office has a 3 ft. by 7 ft. doorway and a 3 ft. by 2 ft. window.
   a. How many square feet will be painted?
   b. If a gallon of paint will cover 400 square feet, then how many cans will be needed?
   c. If each gallon costs $14, then what will be the total cost of the paint?

15. Three floors in an office building are to have new carpet installed. To estimate the amount of carpet needed the manager decides to use the dimensions of the building, which is 60 ft. by 80 ft. How many square feet of carpet will be needed? If the carpet company charges $2 per square foot, then what will be the cost of installing the carpet? What are the problems with his estimate?

16. The outside of a large building is to be covered with 4 ft. by 8 ft. glass panels. The building is a box that is 75 ft. by 60 ft. by 150 ft. How many glass panels will be needed? (Note: The roof will not be covered with the panels.)

17. The NASA assembling building where NASA constructs its rockets and shuttles is 120 ft. by 140 ft. by 500 ft. Find the volume of the building.

Trivia Bite

The NASA assembling building is actually a large shell. Because the open space inside is so large, clouds could sometimes form in the upper part of the building. NASA had to design a humidity/temperature-control system to keep storms from developing inside the building.

18. A bus is a box that is 35 ft. by 9 ft. by 8 ft. What is the volume of a bus?

19. The World Trade Center in New York City comprises twin skyscrapers. Tower One is 208 ft. by 208 ft. by 1368 ft. and Tower Two is 208 ft. by 208 ft. by 1362 ft. What is the combined volume of the towers?

20. Ziggurats are terraced pyramids. Find the volume of the ziggurat shown if the bottom level is 297 ft. by 297 ft. by 12 ft., the middle level is 260 ft. by 260 ft. by 10 ft., and the top level is 145 ft. by 145 ft. by 9 ft.

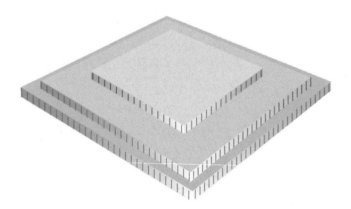

Puzzle Problem

An *equilateral triangle* is a triangle with all three sides equal in length. Suppose we construct an equilateral triangle out of three equal length pencils like so:

We can construct larger equilateral triangles out of the smaller version like so:

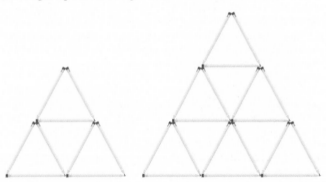

How many pencils would be needed to construct an equilateral triangle like those above that has fifteen pencils along each side?

REVIEW EXERCISES

1. Write 2,408,073 in expanded form using powers of 10.

2. Solve and check. $x + 175 = 2104$

3. Estimate by rounding so that there is only one nonzero digit, and then calculate the actual product.

$$452 \cdot 71{,}203 =$$

4. Solve and check. $17y = 3451$

5. Simplify using the order of operations. $5^3 - [(9 + 3)^2 - 2^6] + \sqrt{100 - 36}$

Summary

Defined Terms

Review the following terms, and for those you do not know, study its definition on the page number next to it.

Section 1.1
Numbers *p. 2*
Set *p. 2*
Subset *p. 2*
Natural numbers *p. 2*
Whole numbers *p. 2*
Equation *p. 5*
Inequality *p. 5*

Section 1.2
Perimeter *p. 14*
Variable *p. 14, 55*

Constants *p. 14*
Related sentence *p. 16*
Solution *p. 16*

Section 1.3
Rectangular array *p. 25*
Exponent *p. 27*
Base *p. 27*
Square unit *p. 29*
Area *p. 30*
Formula *p. 30, 55*

Section 1.4
Remainder *p. 35*
Square *p. 42*
Square root *p. 43*
Perfect square *p. 43*

Section 1.6
Constant *p. 55*
Parallelogram *p. 57*
Parallel *p. 57*
Right angle *p. 57*
Cubic unit *p. 58*
Volume *p. 58*

Our Place Value System

Place Values			
Billions period	**Millions period**	**Thousands period**	**Ones period**
Hundred billions / Ten billions / Billions	Hundred millions / Ten millions / Millions	Hundred thousands / Ten thousands / Thousands	Hundreds / Tens / Ones

Arithmetic Summary Diagram

Each operation has an inverse operation. In the following diagram, the operations build from the top down. Addition leads to multiplication, which leads to exponents. Subtraction leads to division, which leads to roots.

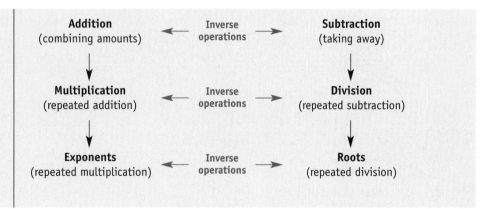

Properties of arithmetic:

Commutative of addition
$a + b = b + a$

Associative of addition
$(a + b) + c = a + (b + c)$

Commutative of multiplication
$a \cdot b = b \cdot a$

Associative of multiplication
$(a \cdot b) \cdot c = a \cdot (b \cdot c)$

Distributive law
$a(b + c) = ab + ac$

Addition (combining amounts) ← Inverse operations → **Subtraction** (taking away)

Multiplication (repeated addition) ← Inverse operations → **Division** (repeated subtraction)

Exponents (repeated multiplication) ← Inverse operations → **Roots** (repeated division)

Procedures, Rules, and Key Examples

Procedures/Rules	Key Example(s)
Section 1.1	
To write a number in expanded form: 1. Write each digit multiplied by its place value. 2. Express it all as a sum.	$70,916 = 7 \times 10,000 + 9 \times 100 + 1 \times 10 + 6 \times 1$
To change a number from expanded form to standard form: Write each digit in the place indicated by the corresponding place value.	$2 \times 1,000,000 + 8 \times 10,000 + 5 \times 1000 + 4 \times 10 + 7 = 2,085,047$
To write a word name, work from left to right through the periods. 1. Write the name of the digits in the left-most period. 2. Write the period name followed by a comma. 3. Repeat steps 1 and 2 until you get to the ones period. We do not follow the ones period with its name. Warning: Do not write the word *and* anywhere in a word name for a whole number. The word *and* takes the place of a decimal point.	$14,907,156 =$ fourteen million, nine hundred seven thousand, one hundred fifty-six.
To round a number to a given place value, consider the digit to the right of the desired place value. If this digit is 5 or greater, round up. If this digit is 4 or less, round down.	Round to the nearest thousand. **a.** 54,259 Answer: 54,000 **b.** 127,502 Answer: 128,000 **c.** 4,801 Answer: 5,000
Section 1.2 **To add whole numbers:** 1. Stack by place value. 2. Add the digits.	$5408 + 913 =$ $\begin{array}{r} \overset{1\ \ 1}{5408} \\ +913 \\ \hline 6321 \end{array}$
To estimate a calculation: 1. Round the numbers. 2. Perform the calculation with the rounded numbers.	Estimate by rounding to the nearest thousand. $\begin{array}{ll} 12,591 & \text{Estimate:}\ \ 13,000 \\ +4,298 & \text{Estimate:}\ +4,000 \\ \hline & \hphantom{\text{Estimate:}\ \ }17,000 \end{array}$
To find the perimeter of a shape: Add the lengths of all the sides of the shape.	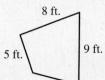 8 ft. 5 ft. 9 ft. 6 ft. $P = 5\ \text{ft.} + 8\ \text{ft.} + 9\ \text{ft.} + 6\ \text{ft.}$ $P = 28\ \text{ft.}$

Procedures/Rules (cont.)	Key Example(s)
To subtract whole numbers: 1. Stack the larger number on top of the smaller number aligning the place values. 2. Subtract the digits in the bottom number from the digits directly above. 3. If the digit above is smaller than the digit below then rename.	$8982 - 765 =$ $89\overset{7}{\cancel{8}}\overset{1}{2}$ $\underline{-765}$ 8217
To find a missing addend, write a related subtraction sentence, subtracting the known addend from the sum.	Solve. $23 + x = 57$ $x = 57 - 23$ $x = 34$

Section 1.3

To multiply: Stack and apply the distributive law.	$503 \times 62 =$ 1 503 $\underline{\times 62}$ $1\,006 \qquad$ Multiply 2 times 503. $\underline{+30\,18} \qquad$ Multiply 6 times 503. $31{,}186 \qquad$ Add.
Multiplicative Property of 0: The product of 0 and a number is always 0. In math language: $0 \cdot n = 0$ and $n \cdot 0 = 0$, where n is any number.	$0 \cdot 15 = 0$ $271 \cdot 0 = 0$
Multiplicative Property of 1: The product of 1 and a number is always the number. In math language: $1 \cdot n = n$ and $n \cdot 1 = n$, where n is any number.	$1 \cdot 94 = 94$ $68 \cdot 1 = 68$
Exponential form when the exponent is a natural number: Repeated multiplication of the base.	Write in factored form, then calculate. $2^5 = 2 \cdot 2 \cdot 2 \cdot 2 \cdot 2 = 32$ Write in exponential form. $7 \cdot 7 \cdot 7 \cdot 7 \cdot 7 \cdot 7 = 7^6$
In general, to get the total number of possible combinations in a situation where one is filling slots, count the number of ways to fill each slot and multiply; that is, multiply the number of items that can fill the first slot by the number of items that can fill the next, and so on.	In order to get into a certain restricted building, a code must be entered. The code box has two digital windows. The first window can contain any number from 1 to 6. The second window can contain any letter from A to J (only capitals appear). How many possible codes are there in all?

Number of items here $\times$ Number of items here $\times$ Number of items here $\times$ ....

Number of items in first window	$\times$	Number of items in second window	
6	$\times$	10	$= 60$

Procedures/Rules (cont.)	Key Example(s)

Section 1.4

Division property: When 1 is the divisor, the quotient is equal to the dividend.

In math language: $n \div 1 = n$, where n is any number.

$78 \div 1 = 78$

Division properties:

When 0 is the divisor with any dividend other than 0, the quotient is undefined.

In math language: $n \div 0$ is undefined, where $n \neq 0$

When 0 is the dividend, the quotient is 0 as long as the divisor is not also 0.

In math language: $0 \div n = 0$, as long as $n \neq 0$.

If both dividend and divisor are 0, the quotient is indeterminate.

In math language: $0 \div 0$ is indeterminate.

$25 \div 0$ is undefined
(because no number checks)

$0 \div 19 = 0$

$0 \div 0$ is indeterminate
(because every number checks)

Division property: When a number (other than 0) is divided by itself, the quotient is 1.

In math language: $n \div n = 1$, as long as $n \neq 0$.

$245 \div 245 = 1$

Divisibility Rules:

1. 2 is an exact divisor for all even numbers. Even numbers have 0, 2, 4, 6, or 8 in the ones place.

24,806 is divisible by 2 because it is an even number. It has a 6 in the ones place.

617 is not divisible by 2 because it is not an even number. It has a digit other than 0, 2, 4, 6, or 8 in the ones place.

2. To determine whether 3 is an exact divisor for a given number:
   a. Add the digits in the dividend.
   b. If the resulting sum is a number that is divisible by 3, then so is the dividend.

5142 is divisible by 3 because the sum of its digits is divisible by 3.

$$5 + 1 + 4 + 2 = 12$$

215 is not divisible by 3 because the sum of its digits is not divisible by 3.

$$2 + 1 + 5 = 8$$

3. 5 is an exact divisor for numbers that have 0 or 5 in the ones place.

945 and 260 are both divisible by 5 because they both have 0 or 5 in the ones place.

612 is not divisible by 5 because it has a digit other than 0 or 5 in the ones place.

Procedures/Rules (cont.)	Key Example(s)

To do long division:

1. Write the problem in long division form.
2. Compare the left-most digit in the dividend with the divisor. Is it larger? If not, then include the next digit. Continue in this way until the digits name a number that is larger than the divisor.
3. Ask yourself how many times the divisor goes into those digits and place that result over the last digit you included from step 2.
4. Multiply the result by the divisor and align the product under the digits from step 3.
5. Subtract. The remainder you get should be less than the divisor. If not, then add 1 to your quotient digit and repeat steps 4 and 5.
6. Write the next digit in the quotient directly beside the last digit in the remainder from step 5.
7. Divide the divisor into the number made by the remainder and the digit you brought down.
8. Repeat steps 3–7 until you have brought down and divided the digit in the ones place of the dividend. When you subtract this last time, if you get zero, then we say the number has no remainder or that it goes evenly. If you get a remainder other than zero, then list it as a remainder in your answer.

$350 \div 2 =$

$$
\begin{array}{r}
175 \\
2\overline{)350} \\
-2 \\
\hline
15 \\
-14 \\
\hline
10 \\
-10 \\
\hline
0
\end{array}
$$

Answer: 175 remainder 0

$6409 \div 7 =$

$$
\begin{array}{r}
915 \\
7\overline{)6409} \\
-63 \\
\hline
10 \\
-7 \\
\hline
39 \\
-35 \\
\hline
4
\end{array}
$$

Answer: 915 r 4

To solve for a missing factor, write a related division sentence, dividing the product by the known factor.

Solve: $x \cdot 12 = 180$

$x = 180 \div 12$

$x = 15$

Square Roots:
To find the square root of a given number, find a number that can be squared to equal the given number.

$\sqrt{81} = 9$ because $(9)^2 = 81$

Order of operations agreement:
Perform operations in the following order:

1. Parentheses, brackets, and braces
2. Exponents
3. Multiplication/Division from left to right
4. Addition/Subtraction from left to right

Memory tips:
Please Excuse My Dear Aunt Sally
or
Please Educate My Daughter And Son.
or
PEMDAS

Simplify.

$26 - [19 - 2(15 - 9)] + 4^2$

$= 26 - [19 - 2(6)] + 4^2$

$= 26 - [19 - 12] + 4^2$

$= 26 - 7 + 4^2$

$= 26 - 7 + 16$

$= 19 + 16$

$= 35$

Section 1.6	Find the perimeter of a rectangle with a length of 24 cm and a width of 17 cm.
To use a formula:	
1. Replace the variables with the corresponding given values.	$P = 2l + 2w$
2. Solve for the missing variable.	$P = 2(24 \text{ cm}) + 2(17 \text{ cm})$
	$P = 48 \text{ cm} + 34 \text{ cm}$
	$P = 82 \text{ cm}$

Section 1.7

Problem-Solving Outline:
1. Develop an **understanding** of the problem.
   a. Read the question(s) (not the whole problem, just the question at the end), and write a note to yourself about what it is you are to find.
   b. Now read the whole problem.
   c. Underline key words.
   d. Draw a picture (if possible).
   e. Make a chart or list.
   f. Simulate the situation.
   g. Search for a related example problem.
2. Develop a **plan.**
   a. Translate the key words to an equation.
   b. Search for a formula.
3. **Execute** your plan by solving the equation or formula.
4. **Answer** the question. Look back at the note you made to yourself about what it was you were to find, and make sure you answer that question. Be sure to include appropriate units.
5. **Check** your results. Make sure your answer is reasonable.
   a. Repeat the solution in a different way.
   b. Reverse the process.
   c. Estimate the answer and make sure the estimate and actual answer are reasonably close.

Formulas

Perimeter of a rectangle:	$P = 2l + 2w$
Area of a parallelogram:	$A = bh$
Volume of a box:	$V = lwh$

CHAPTER 1

Review Exercises

For Exercises 1–6, answer true or false.

1. 0 is a natural number.

2. All natural numbers are also whole numbers.

3. $16 - 2 = 2 - 16$.

4. $2(3 + 5) = 2 \cdot 3 + 2 \cdot 5$

5. 3 is an exact divisor for 6201.

6. $0 \div 0 = 0$

For Exercises 7–10, solve as indicated.

7. Explain in your own words how to check that 8 is the solution for $x + 9 = 17$.

8. Explain or show why $14 \div 0$ is undefined.

9. Explain the mistake in the problem.
$$17 + 3(9 - 2)$$
$$= 17 + 3(7)$$
$$= 20(7)$$
$$= 140$$

10. Explain in your own words the difference between the square of a number and the square root of a number.

For Exercises 11 and 12, write in expanded form.

11. 5,680,901

12. 42,519

For Exercises 13–15, write in standard form.

13. $9 \times 10,000 + 8 \times 1000 + 2 \times 100 + 7 \times 10 + 4 \times 1$

14. $8 \times 1,000,000 + 2 \times 10,000 + 9 \times 10 + 6 \times 1$

15. $7 \times 10^8 + 9 \times 10^5 + 2 \times 10^4 + 8 \times 10^3 + 6 \times 1$

For Exercises 16–18, write the word name.

16. 47,609,204

17. 9421

18. 123,405,600

For Exercises 19 and 20, use < or > to make a true statement.

19. 14 ? 19

20. 2930 ? 2899

For Exercises 21 and 22, round to the specified place.

21. 5,689,412 to the nearest ten thousand

22. 2,512,309 to the nearest million

For Exercises 23 and 24, estimate by rounding so that there is only one nonzero digit, then calculate the actual sum.

23. 45,902 + 6819

24. 545 + 9091 + 28 + 30,009

For Exercises 25 and 26, estimate by rounding so that there is only one nonzero digit, then calculate the actual difference.

25. 541,908 − 56,192

26. 8002 − 295

For Exercises 27 and 28, solve and check.

27. 29 + x = 54

28. y + 14 = 203

For Exercises 29 and 30, solve.

29. An accountant is given the following spreadsheet of expenses and income for a company. What is the final balance?

Description of expense	Amount	Description of assets/income	Amount
Payroll	$19,445	Checking account	$ 9,876
Utilities	$ 704	Income	$48,984
Water	$ 155		
Waste disposal	$ 1,282		
New inventory	$21,931		

30. Find the missing current in the circuit shown.

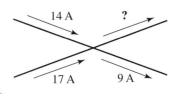

For Exercises 31 and 32, estimate by rounding so that there is only one nonzero digit, then calculate the actual product.

31. 4591×307

32. $16,140 \times 25$

For Exercises 33 and 34, evaluate.

33. 2^7

34. 5^3

For Exercises 35 and 36, write in exponential form.

35. $10 \cdot 10 \cdot 10 \cdot 10 \cdot 10$

36. $7 \cdot 7 \cdot 7 \cdot 7 \cdot 7 \cdot 7 \cdot 7 \cdot 7$

For Exercises 37 and 38, solve.

37. A computer translates everything into binary code. This means that numbers are written so that each place value either contains a 1 or a 0 digit. If a certain computer chip has 6 place values (bits), then how many different numbers can be coded?

38. A room measures 16 ft. by 18 ft. How many square feet of carpet will be needed to cover the floor of the room?

For Exercises 39 and 40, estimate by rounding so that there is only one nonzero digit. Then calculate the actual quotient.

39. $78,413 \div 19 =$

40. $83,451 \div 26 =$

For Exercises 41 and 42, solve and check.

41. $19b = 456$

42. $8k = 2448$

For Exercises 43 and 44, solve.

43. Jana is to administer 120 ml of Pitocin in an IV drip over the course of an hour. How many ml of Pitocin should the patient receive each minute?

44. Andre has an annual salary of $29,256. What is his gross monthly salary?

For Exercises 45 and 46, find the square root.

45. $\sqrt{196} =$

46. $\sqrt{225} =$

For Exercises 47–50, simplify.

47. $24 \div 8 \cdot 6 + \sqrt{121} - 5^2$

48. $4^3 - 5[(28 - 4) \div 2^3] + \sqrt{100 \div 25}$

49. $3\sqrt{25 - 16} + \{[25 + 2(14 - 9)] \div [43 - (2 + 4)^2]\}$

50. $\dfrac{9^2 - 21}{56 - 2(4 + 1)^2}$

51. Find the area of the figure.

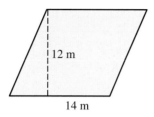

12 m

14 m

52. Find the volume of a 7 ft. by 6 ft. by 9 ft. box.

For Exercises 53–58, solve.

53. In designing a room for a house, the desired length is 18 ft. and the desired area is 288 ft.² What must the width be?

54. The Dobsons are told that a 225-ft.² square patio is included in the price of the house they are having built. What are the dimensions of the patio?

55. A desk manufacturer has an order to make 500 desks (shown below). Each desk is to have a plastic binder placed along the edge. How many feet of binder must be purchased?

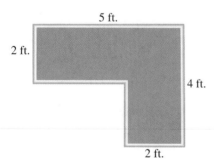

5 ft.

2 ft.

4 ft.

2 ft.

56. The picture below shows the plans for a department store and parking lot. What is the area of the parking lot?

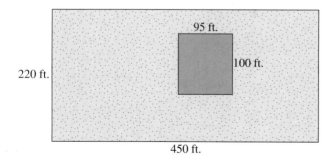

95 ft.

100 ft.

220 ft.

450 ft.

Practice Test

1. What digit is in the ten thousands place in 1,258,987?

2. Write 48,210,907 in expanded form using powers of 10.

3. Use $<$ or $>$ to make a true statement.
$$19,304 \quad ? \quad 19,204$$

4. Write the word name for 67,194,210.

5. Estimate the sum by rounding to the nearest hundred thousand.
$$2,346,502 + 481,901 =$$

6. Add. $63,209 + 4981 =$

7. Subtract. $480,091 - 54,382 =$

8. Solve and check. $\quad 27 + x = 35$

9. Multiply. $357 \times 691 =$

10. $4^3 =$

11. Divide. $45,019 \div 28 =$

12. Solve and check. $\quad 7y = 126$

13. $\sqrt{196} =$

14. $24 + 40 \div 4$

15. $14 + 2^5 - 8 \cdot 3 =$

1. _____

2. _____

3. _____

4. _____

5. _____

6. _____

7. _____

8. _____

9. _____

10. _____

11. _____

12. _____

13. _____

14. _____

15. _____

16. _____

16. $20 - 6(3 + 2) \div 15 =$

17. _____

17. $[(14 + 26) \div 8] + 3^3 - \sqrt{25 - 9} =$

18. $\dfrac{5^2 + 27}{(4 + 2)^2} =$

18. _____

19. Calculate the area of the shape.

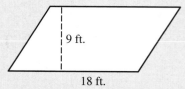
9 ft.

18 ft.

19. _____

20. An accountant is given the following spreadsheet of expenses and income for a company. What is the final balance?

20. _____

Description of expense	Amount	Description of assets/income	Amount
Payroll	$15,380	Checking account	$12,503
Utilities	$ 285	Income	$25,282
Water	$ 121		
New inventory	$ 1,387		

21. _____

21. Participants in a walk-run race are given identification cards that have a letter, A–Z paired with a digit, 0–9. How many identification cards can be made in all?

22. _____

22. A college admits 2487 new students. All new students must enroll in a college orientation course. If the maximum number of students allowed in a single class is 30, how many classes should be offered to accommodate all the new students?

23. _____

23. A rectangular playground measures 65 ft. by 80 ft. What is the perimeter of the playground? If fence material costs $6 per foot, how much would fencing cost for the playground?

24. _____

24. A walk-in refrigerator at a restaurant is a box measuring 8 ft. by 6 ft. by 7 ft. What is the volume?

25. _____

25. A wall that has a doorway is to be painted. The wall is 16 ft. by 9 ft. and the doorway is 3 ft. by 7 ft. How many square feet must be painted?

2

Integers

Learning Styles

Every person has a learning style. Some learn best through seeing (visual learner), some through touching or doing (tactile learner), and some through hearing (auditory learner). If you do not know your learning style then take a moment to complete the learning styles inventory in the *To the Student* part at the beginning of the book.

In Chapter 2, we will be exploring positive and negative numbers. The operations of arithmetic used with these numbers require that you know rules. Knowing your learning style can guide you in learning the rules and procedures.

If you are a visual learner then use the number lines where possible. When you study the rules, read them over and over to yourself. As a visual learner, you take in information best through your eyes, so reading the rules over and over will help you digest them.

If you are an auditory learner then say the rules and procedures. Before beginning a homework problem, say the rule out loud. Make up songs that have the rules and/or procedures as lyrics. Record yourself reading the rules and procedures and listen to the tape as often as possible.

If you are a tactile learner then write the rules and procedures. Do lots of practice problems. You will probably find it beneficial to do more problems than the instructor assigns. Try to simulate real situations as often as possible. Think of positive numbers as moving in one direction and negative numbers as moving in the opposite direction.

> *"He is educated who knows how to find out what he doesn't know."*
> —GEORGE SIMMEL

> *"Seeing, hearing, and feeling are miracles, and each part and tag of me is a miracle."*
> —WALT WHITMAN

> *"Let your performance do the thinking."*
> —H. JACKSON BROWN JR.

2.1 Introduction to Integers

OBJECTIVES

1 Identify integers.

2 Graph integers on a number line.

3 Use < or > to make a true statement.

4 Find the absolute value.

5 Find the additive inverse.

OBJECTIVE 1 *Identify integers.*

In Chapter 1, we studied natural and whole numbers. We are now ready to build on that foundation and look at a new set of numbers called the *integers*. One thing you should notice is as we build each new set of numbers, we merely add numbers to the sets we've already studied. Recall the natural and whole numbers:

Natural numbers: 1, 2, 3, ...

Whole numbers: 0, 1, 2, 3, ...

The natural numbers are the counting numbers beginning with 1. The whole numbers merely added 0 to the natural numbers. Well, the integers merely add negative natural numbers to the whole number set.

DEFINITION	**Integers:** ... , −3, −2, −1, 0, 1, 2, 3, ...

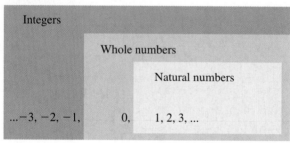

Figure 2.1

In Figure 2.1 we see that the set of integers contains all the whole numbers and natural numbers. The set of whole numbers contains all the natural numbers.

Let's relate integers to familiar situations. This will help in thinking through the arithmetic.

Historical Note

Negative numbers took some time gaining acceptance. The first known mention of negative numbers is found in the work by the Indian mathematician Brahmagupta (c. 628). He called them "negative and affirmative quantities." There have been many different ways of indicating negative numbers. The Chinese would use colors, black for negative numbers and red for positive. In sixteenth-century Europe it was popular to use the letter m for minus, to indicate negative numbers as in m:7 to indicate −7.

Positive and negative numbers can be related to money, temperature, elevation, and direction. All are familiar uses in everyday life.

Figure 2.2

	Negative Number	**Positive Number**
Money	Debt/Liability	Credit/Asset
Temperature	Below zero	Above zero
Elevation	Below sea level	Above sea level
Direction	Reverse/go down	Forward/go up

Notation

We write a positive number with a plus sign or no sign:

+5 or 5 means positive 5 and is read "positive five," "plus five," or "five."

We write a negative number with a minus sign:

−6 means negative 6 and is read "negative six" or "minus six."

The number 0 has no sign.

EXAMPLE 1 Express each amount as a positive or negative.

a. Angela owes $480 on her credit card.

Answer: −480

Explanation: Because she owes the money it is a debt and therefore a negative amount.

b. It is 29°C outside.

Answer: 29

Explanation: Because 29°C is above zero, it is a positive amount. Notice when writing a positive number, we do not have to write +29. We can merely write 29.

c. A research submarine at a depth of 245 ft.

Answer: −245

Explanation: Because the submarine is below sea level, its position is a negative amount.

Do Margin 1. ▶

OBJECTIVE 2 *Graph integers on a number line.*

Many people find a number line helpful in doing calculations with integers. A number line is like a ruler. It is a line with marks evenly spaced and with each mark representing a number.

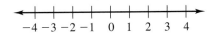

Procedure *To graph a number on a number line, we draw a dot or point on the mark for that number.*

EXAMPLE 2 Graph 3 on a number line.

Solution:

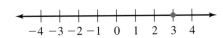

EXAMPLE 3 Graph −4 on a number line.

Solution:

Do Margin 2. ▶

OBJECTIVE 3 *Use < or > to make a true statement.*

We can use inequality symbols to express which of two given integers is larger. One way we can compare integers is by using a number line.

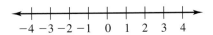

We can draw the following conclusions about integers on a number line.

Conclusion: Zero and every positive integer is to the right of every negative integer on a number line.

Conclusion: In comparing two integers, the integer that is higher or farther right on a number line is the larger integer.

◆ Margin 1

Express each amount as a positive or negative.

a. Ian has $3250 in a savings account.

b. According to the U.S. National Debt Clock, on 9/9/99 the national debt was $5,656,705,447,748.

c. The melting point of gold is 1063°C.

d. A submarine is at a depth of 250 ft.

◆ Margin 2

Graph each number on a number line.

a. −2

b. 0

c. 5

Answers to Margin 1: **a.** 3250 **b.** −5,656,705,447,748 **c.** 1063 **d.** −250

Answers to Margin 2:

a.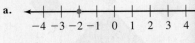

b.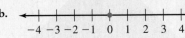

c.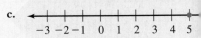

♦ **Margin 3**

Use < or > to make a true statement.

a. 15 ? −26

b. −19 ? −17

c. −14 ? −2

d. 0 ? −9

♦ **Margin 4**

Find the absolute value of each given number.

a. 14

b. −27

c. −1347

d. 0

Answers to Margin 3:
a. 15 > −26 **b.** −19 < −17
c. −14 < −2 **d.** 0 > −9

Answers for Margin 4: **a.** 14 **b.** 27
c. 1347 **d.** 0

EXAMPLE 4 Use < or > to make a true statement.

a. 9 ? −14

Answer: 9 > −14

Explanation: The larger number is the number farther right on a number line. If we graph 9 and −14 on a number line we see that 9 is farther right.

Or you could think about elevation. If zero is sea level, then positive 9 is 9 ft. above sea level and −14 is 14 ft. below sea level. Obviously 9 ft. above sea level is higher.

b. −12 ? −7

Answer: −12 < −7

Explanation: The larger number is the number farther right on a number line, in this case −7.

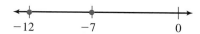

Or, think about elevation. Which is higher? −7, which means 7 ft. down, or −12, which means 12 ft. down. The −7 is closer to the surface so it is the higher number.

♦ **Conclusion:** When both numbers are negative, the integer closer to zero is the larger integer.

◀ **Do Margin 3.**

OBJECTIVE 4 *Find the absolute value.*

The word *value* calls to mind how much something is worth or the size of something. In essence, that's what *absolute value* means. It is the size of a number. The technical definition is:

DEFINITION | Absolute value: The absolute value of a given number is its distance from zero.

EXAMPLE 5 Find the absolute value of 5.

Answer: 5

Explanation: Because 5 is five steps from zero, its absolute value is 5. You could use a number line to count the steps if needed.

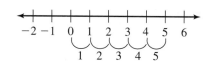

EXAMPLE 6 Find the absolute value of −6.

Answer: 6

Explanation: −6 is six steps from zero so its absolute value is 6. Notice on a number line we have to travel six steps to the left. It doesn't matter which way we have to travel, we're just counting the steps. As an analogy, imagine a size 6 shoe. It doesn't matter which way you point the shoe or where it is located, it's still a size 6 shoe.

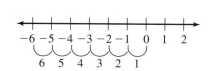

◀ **Do Margin 4.**

There are symbols for absolute value. In Example 5, we were asked to find the absolute value of 5 and we found the answer to be 5. We can translate this to symbols:

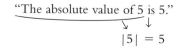

"The absolute value of 5 is 5."

$$|5| = 5$$

The vertical lines on both sides of the number indicate to find its absolute value. Notice that the word *is* translates to an equal sign. It is important to learn to read math symbols because most math books are written with such symbols as a shorthand.

EXAMPLE 7 Simplify. $|-8| =$

Answer: 8

Explanation: We are being asked to find the absolute value of -8. Because -8 is eight steps away from zero, the answer is 8.

EXAMPLE 8 Simplify. $|27| =$

Answer: 27

Explanation: We are being asked to find the absolute value of 27. Because 27 is 27 steps away from zero, the answer is 27.

Do Margin 5. ▶

OBJECTIVE 5 *Find the additive inverse.*

Notice 5 and -5 have the same absolute value because they are both the same distance from zero. However, they are on opposite sides of zero on a number line. They have opposite signs, one sign is positive and the other sign is negative. We say that 5 and -5 are **additive inverses.**

Why call them *additive inverses?* Remember inverse operations *undo* each other. Additive inverses *undo* one another in that their sum is always zero.

$$5 + (-5) = 0$$

DEFINITION **Additive inverses:** Two numbers whose sum is zero.

EXAMPLE 9 Find the additive inverse of 3.

Answer: -3

Explanation: $3 + (-3) = 0$, therefore 3 and -3 are additive inverses.

EXAMPLE 10 Find the additive inverse of -4.

Answer: 4

Explanation: $-4 + 4 = 0$, therefore -4 and 4 are additive inverses.

EXAMPLE 11 Find the additive inverse of 0.

Answer: 0

Explanation: $0 + 0 = 0$, therefore 0 is its own additive inverse.

EXAMPLE 12 Find the additive inverse of the additive inverse of -2.

Answer: -2.

Explanation: There are two additive inverses here. The additive inverse of -2 is 2, and the additive inverse of 2 is -2.

Conclusion: When we find the additive inverse of the additive inverse of a number the outcome is the original number.

Do Margin 6. ▶

The symbol for additive inverse is the minus sign. This may seem strange that the same symbol is used to mean subtraction, a negative number, and additive inverse. How do you tell the difference?

Tip

The absolute value of a number is always positive or zero.

The absolute value of a negative number is always positive.

The absolute value of a positive number is always positive.

The absolute value of zero is zero.

◆ **Margin 5**

Simplify.

a. $|-19| =$

b. $|90| =$

c. $|0| =$

d. $|42| =$

◆ **Margin 6**

a. Find the additive inverse of -6.

b. Find the additive inverse of 9.

c. Find the additive inverse of the additive inverse of 7.

Answers to Margin 5: **a.** 19 **b.** 90 **c.** 0 **d.** 42

Answers to Margin 6: **a.** 6 **b.** -9 **c.** 7

Margin 7

Simplify.

a. $-(8) =$

b. $-(-(7)) =$

c. $-(-(-(11))) =$

d. $-(-(-(-17))) =$

Margin 8

Translate to words, then simplify.

a. $-|20| =$

b. $|-(-4)| =$

c. $-(-(-(14))) =$

Answers to Margin 7: **a.** -8 **b.** 7 **c.** -11 **d.** 17

Answers to Margin 8: **a.** The additive inverse of the absolute value of 20 is -20. **b.** The absolute value of the additive inverse of -4 is 4. **c.** The additive inverse of the additive inverse of the additive inverse of 14 is -14.

Consider that in the English language, there are many cases of the same word having different meanings. For example, consider the word *can*.

If we say, "Open a can of paint," the word *can* is a noun.

If we say, "I can do this," the word *can* acts as a verb.

How do you tell the difference? The answer is by context; that is, we look at how it is used in the sentence. The same is true for a minus sign. We determine what it means by the context.

If we write: $5 - 2$ then we interpret it to mean "five subtract two."

If we write: -2 then we interpret this to mean "negative two."

If we write: $-(2)$ then we are saying "the additive inverse of two."

Subtracting two is the same as getting the additive inverse of two, which is negative two. That's why the same symbol can be used for all three.

EXAMPLE 13 Simplify. $-(3) =$

Answer: -3

> **Connection**
> Example 13 is the symbolic version of Example 9.

Explanation: $-(3) =$ means "the additive inverse of 3 is ? " So we must find the additive inverse of 3, which is -3.

EXAMPLE 14 Simplify. $-(-4) =$

Answer: 4

> **Connection**
> Example 14 is the symbolic version of Example 10.

Explanation: $-(-4) =$ means, "the additive inverse of -4 is ? " So we must find the additive inverse of -4, which is 4. Notice a pair of minus signs translates to positive.

EXAMPLE 15 Simplify. $-(0) =$

Answer: 0

> **Connection**
> Example 15 is the symbolic version of Example 11.

Explanation: $-(0) =$ means, "the additive inverse of 0 is ? " So we must find the additive inverse of 0, which is 0.

EXAMPLE 16 Simplify. $-(-(-2)) =$

Answer: -2

> **Connection**
> Example 16 is the symbolic version of Example 12.

Explanation: $-(-(-2)) =$ means, "the additive inverse of the additive inverse of -2 is ? " So we must find the additive inverse of the additive inverse of -2, which is -2.

◀ **Do Margin 7.**

EXAMPLE 17 Translate to words, then simplify. $-|-18| =$

Translation: The additive inverse of the absolute value of -18 is _____.

Answer: -18

Explanation: We must first get the absolute value of -18, then we must find the additive inverse of that result. The absolute value of -18 is 18 and the additive inverse of 18 is -18.

◀ **Do Margin 8.**

> **Connection**
> Absolute value symbols act like parentheses. Operations within absolute value symbols are to be computed before operations outside of the absolute value symbols. When we revisit the order of operations agreement in Section 2.5, we will amend the agreement to include absolute value symbols with parentheses, brackets, and braces.

2.1 Exercises

FOR EXTRA HELP

 Videotape 1

 InterAct Math
Tutorial Software

 www.carsonmath.com

 AWL Math Tutor Center

InterAct MathXL www.mathxl.com

 Student's Solutions
Manual

Trivia Bite

The English Channel Tunnel is approximately 31 miles long, 23 of which are under water. The rubble from excavating the tunnel increased the size of Britain by 90 acres.

For Exercises 1–10, express each amount as a positive or negative integer.

1. Sonja receives a paycheck of $450.

2. Adam's business made a profit of $215,000 last year.

3. The Titanic is lying on the bottom of the North Atlantic at a depth of approximately 13,200 ft.

4. The tunnel connecting Britain to France is approximately 220 m below the surface of the English Channel.

5. A college football team gains 40 yd. in one play.

6. The countdown to launch time of a space shuttle is stopped with 20 min. to go.

7. The earth accelerates objects toward its surface at a rate of 32 ft. per sec. every sec.

8. Mt. Everest is the tallest mountain in the world. Its peak is 29,028 ft. above sea level.

9. Carbon dioxide (CO_2) freezes at 78° below zero.

10. A financial planner analyzes the Smith's financial situation and finds they have a total debt of $75,243.

Trivia Bite

Mt. Everest is in the central Himalaya range on the border of Tibet and Nepal. Its summit is the highest elevation in the world and was first reached in 1953 by Edmund Hillary from New Zealand and Tenzing Norgay, a Nepalese guide.

Source: Microsoft Bookshelf '98

For Exercises 11–18, graph each integer on a number line.

11. 8

12. 7

13. −9

14. −6

15. 0

16. −4

17. −5

18. 10

For Exercises 19–26, use < or > to make a true statement.

19. −20 ? 17

20. 26 ? −18

21. −30 ? −16

22. −12 ? −19

23. 0 ? −16

24. −4 ? −5

25. −19 ? −17

26. 0 ? 5

For Exercises 27–32, complete the sentence.

27. The absolute value of 26 is _____.

28. The absolute value of 19 is _____.

29. The absolute value of −18 is _____.

30. The absolute value of −4 is _____.

31. The absolute value of 0 is _____.

32. The absolute value of −7 is _____.

For Exercises 33–42, simplify.

33. $|14| =$

34. $|242| =$

35. $|-18| =$

36. $|-9| =$

37. $|0| =$

38. $|-200| =$

39. $|47| =$

40. $|97| =$

41. $|-377| =$

42. $|-84| =$

For Exercises 43–52, complete the sentence.

43. The additive inverse of -18 is _____.

44. The additive inverse of 42 is _____.

45. The additive inverse of 61 is _____.

46. The additive inverse of -540 is _____.

47. The additive inverse of 0 is _____.

48. The additive inverse of -17 is _____.

49. The additive inverse of the additive inverse of -8 is _____.

50. The additive inverse of the additive inverse of -16 is _____.

51. The additive inverse of the additive inverse of 43 is _____.

52. The additive inverse of the additive inverse of 75 is _____.

For Exercises 53–66, simplify.

53. $-(6) =$

54. $-(9) =$

55. $-(-14) =$

56. $-(-19) =$

57. $-(0) =$

58. $-(-75) =$

59. $-(63) =$

60. $-(29) =$

61. $-(-(-4)) =$

62. $-(-(-12)) =$

63. $-(-(7)) =$

64. $-(-(-37))) =$

65. $-(-(-87)) =$

66. $-(-(-(42))) =$

For Exercises 67–76, translate the expression to words, then simplify.

67. $|-4| =$

68. $|42| =$

69. $-(-8) =$

70. $-(140) =$

71. $-(-(14)) =$

72. $-(-(-28)) =$

73. $-|-5| =$

74. $-|16|$

75. $-|-(-4)| =$

76. $|-(-(-17))|$

REVIEW EXERCISES

1. $84{,}759 + 9506 =$

2. $900{,}406 - 35{,}918 =$

3. $457 \times 609 =$

4. $85{,}314 \div 42 =$

5. Amelia begins a month with $452 in savings. Below is a list of her deposits and withdrawals for the month. What is her balance at the end of the month?

Deposits	Withdrawals
$45	$220
$98	$25
$88	$10
$54	

2.2 Adding Integers

OBJECTIVES

1 Add integers with like signs.

2 Add integers with different signs.

3 Add integers.

4 Solve applications involving addition of integers.

OBJECTIVE **1** *Add integers with like signs.*

Let's examine some realistic situations in order to discover how we add with signed numbers.

EXAMPLE 1 Suppose you have $200 in your checking account, and you deposit $40. What would be the balance?

Solution: This is a standard addition situation.

$$200 + 40 = 240$$

Explanation: Because the $200 is money in the account, it is an asset and therefore is positive. Because the $40 is a deposit, you are increasing your assets so we add the values to end up with a balance of $240.

We can also use a number line. Think of the first addend, 200, as the starting position. The absolute value of the second addend indicates the distance to move from the starting position and its sign indicates which direction. Because 40 is positive, we move to the right 40 steps. The final position is the answer.

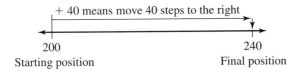

+ 40 means move 40 steps to the right

200
Starting position

240
Final position

EXAMPLE 2 Suppose you have a debt of $200 on a credit card and you purchase $40 worth of merchandize using the same credit card. What is the balance after the purchase?

Solution: Because the $200 is a debt, we write it as -200. A purchase using a credit card is a debt so the $40 is written as -40. The addition statement looks like this:

$$-200 + (-40) = -240$$

Note: In math sentences, parentheses are placed around negative numbers that *follow* operation symbols. They actually are not needed, but are there to clarify the meaning of the plus and minus signs.

Explanation: By using that credit card to purchase merchandise, you are simply increasing your debt, so we add the money you already owe to the new debt to get a total debt.

On a number line, $-200 + (-40)$ means start at -200 and move 40 steps in the negative direction, which is to the left.

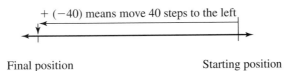

+ (−40) means move 40 steps to the left

Final position

Starting position

Notice in Examples 1 and 2 we added the absolute values because we were either increasing assets or increasing debt. On a number line we were starting positive and moving more positive or starting negative and moving more negative.

Conclusion: When we add two numbers that have the same sign, we add their absolute values and keep the same sign.

◀ **Do Margin 1.**

OBJECTIVE 2 *Add integers with different signs.*

EXAMPLE 3 Suppose you have a debt of $200 on a credit card and you make a payment of $80. What would the balance be?

Solution: The $200 is a debt, so we write it as −200, and the $80 is payment toward that debt. The addition statement looks like this:

$$-200 + 80 = -120$$

Explanation: Notice you didn't send enough to pay off the debt so you would still have a debt balance, but it would be less debt than before. Because we are decreasing the debt, we have to subtract the values and since we had more debt than payment, we end up still owing money on the credit card. Notice the negative outweighed the positive so the outcome was negative.

On a number line, −200 + 80 means start at −200 and move 80 steps to the right. Notice it would take a full 200 steps to get back to 0, therefore the result is still negative.

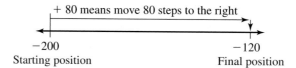

Notice we actually subtracted 80 from 200 to figure the outcome. (Addition with signed numbers does not always mean that we add.) Also notice that because of the commutative law of addition, we can write the statement as:

$$80 + (-200) = -120$$

It doesn't matter which way it is arranged, a −$200 debt with $80 in assets results in a net debt of −$120.

EXAMPLE 4 Suppose you have a debt of $200 on the credit card and you make a payment of $240. What is the balance?

Solution: Again, the $200 is a debt so we write it as −200. The payment of $240 is credit towards that debt so it is a positive amount. The addition statement looks like this:

$$-200 + 240 = 40$$

Explanation: In this case you overpaid the debt. In other words, you paid more than you owe so you now have a credit of $40 on that credit card. Notice the positive outweighed the negative so the result was positive.

♦Margin 1

Add.

a. $19 + 74 =$

b. $-42 + (-39) =$

c. $-15 + (-18) =$

d. $20 + 13 =$

Answers: **a.** 93 **b.** −81 **c.** −33 **d.** 33

On a number line, $-200 + 240$ means start at -200 and move 240 steps to the right. Notice 200 steps to the right puts us at 0, so 240 steps moves 40 steps past 0, making the result 40.

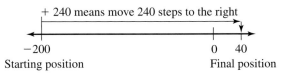

+ 240 means move 240 steps to the right

-200
Starting position

0   40
Final position

From Examples 3 and 4 we can conclude the following:

Conclusion: When we add two numbers that have different signs, we subtract their absolute values and keep the sign of the number that has the larger absolute value.

Do Margin 2. ▷

OBJECTIVE 3 *Add integers.*

In the first four examples, we discovered how to add integers. Following is a summary of the two conclusions we made. It is helpful to relate each addition situation to reconciling debts and credits.

> **Rules for Adding Integers**
> 1. When adding two numbers that have the same sign, we add their absolute values and keep the same sign. (Think: two credits or two debts)
> 2. When we add two numbers that have different signs, we subtract their absolute values and keep the sign of the number that has the larger absolute value. (Think: payment toward a debt)

EXAMPLE 5 Explain each situation in terms of debts and credits, then add.

a. $-40 + (-21) =$

Solution: We are asked to add two debts. This merely increases the debt, so we add and the result is still a debt; that is, we add the absolute values and keep the negative sign.

$$-40 + (-21) = -61$$

b. $28 + (-16) =$

Solution: We are asked to add a credit of 28 and a debt of 16. If we use the 28 to pay off the 16 debt, we'll still have a credit. We must subtract the absolute values, and because the positive integer, 28, has the larger absolute value, the result will be positive.

$$28 + (-16) = 12$$

c. $10 + (-24) =$

Solution: We are asked to add a credit of 10 and a debt of 24. If we use the 10 to pay toward the 24 debt, we will not have enough to pay off the debt. We must subtract the absolute values, and because the negative integer, -24, has the larger absolute value, the outcome will be negative.

$$10 + (-24) = -14$$

Do Margin 3. ▷

◆ **Margin 2**

Add.

a. $-18 + 6 =$

b. $-20 + 28 =$

c. $30 + (-26) =$

d. $14 + (-24) =$

e. $15 + (-15) =$

◆ **Margin 3**

Explain each situation in terms of debts and credits, then add.

a. $-40 + (-12) =$

b. $28 + 15 =$

c. $16 + (-10) =$

d. $-26 + 14 =$

e. $-22 + 22 =$

Answers to Margin 2: **a.** -12 **b.** 8 **c.** 4 **d.** -10 **e.** 0

Answers to Margin 3: **a.** Two debts, so add and the result is a debt/negative; -52. **b.** Two credits, so add and the result is a credit/positive; 43. **c.** A credit with a debt, so subtract and because the credit, 16, is more than the debt, the result is a credit/positive; 6. **d.** A debt with a credit, so subtract, and because the debt, -26, is more than the credit, the result is a debt/negative; -12. **e.** Paying 22 on a 22 debt means the balance is 0.

◆ **Margin 4**

Add.

a. $-18 + (-14) + 30 + (-2) + 15 =$

b. $42 + (-25) + 16 + (-8) + 19 =$

c. $-38 + 64 + 40 + (-100) =$

d. $28 + (-14) + 5 + (-12) + (-10) + 3 =$

EXAMPLE 6   Add. $-14 + 26 + (-18) + 15 + (-32) =$

Solution: There are two popular approaches to adding more than two integers. We can work our way from left to right.

	$-14 + 26 + (-18) + 15 + (-32)$	Add -14 and 26 to get 12.
$=$	$12 + (-18) + 15 + (-32)$	Add 12 and -18 to get -6.
$=$	$-6 + 15 + (-32)$	Add -6 to 15 to get 9.
$=$	$9 + (-32)$	Add 9 and -32 to get -23.
$=$	-23	

Or, the other popular approach is to add all the positives and negatives separately, then reconcile the result. The commutative and associative laws of addition make this method possible because they allow us to rearrange and regroup addition any way we wish. This is the method banks use to do their statements. They add all the credits, then all the debts, and then reconcile the two amounts.

Get all positives together.

Add positives: $26 + 15 = 41$.

Reconcile. $41 + (-64) = -23$.

$-14 + 26 + (-18) + 15 + (-32)$

$= 26 + 15 + (-14) + (-18) + (-32)$

$= 41 + (-64)$

$= -23$

Get all negatives together

Add negatives: $-14 + (-18) + (-32) = -64$.

Explanation: Think: The total credits are 41 and the total debts are 64. Because the debts outweigh the credits we end up with a debt of 23.

◀ **Do Margin 4.**

OBJECTIVE **4** *Solve applications involving addition of integers.*

The key to solving addition problems that involve integers is to simulate the situation.

EXAMPLE 7   The following table lists the Carter family's assets and debts. Find their net worth.

Assets	Debts
Savings = $2538	Credit card balance = $2452
Checking = $853	Mortgage = $76,483
Automobile #1 = $3500 (paid off)	Automobile #2 = $2456
Furniture = $16,500	Student loans = $6782
Jewelry = $4500	

Solution: Use the problem-solving process.

Understand: Net worth is the sum of all the assets and debts.

Working with Scientific Calculators

Using a scientific calculator to add integers.

Type the addition sentence as you read it from left to right. Remember, to enter a negative number, you must first type the value, then press the +/− button. Be sure to press the equal sign when finished.

Example: $-42 + 17 + (-28) =$

Enter: 42 +/− + 17 + 28 +/− =

Answer: -53

Plan: Write an addition statement adding all the assets and debts.

Execute: Net worth = 2538 + 853 + 3500 + 16,500 + 4500 +
  (−2452) + (−76,483) + (−2456) + (−6782)
  = 27,891 + (−88,173)
  = −60,282

Adding all the positives computes the total assets.

Adding all the negatives computes the total debts.

Answer: Because total debt −88,173 has larger absolute value than total assets 27,891, the Carter family's net worth was a debt/negative amount of −60,282.

Check: Instead of adding total assets and total debts, we could compute the sum from left to right. We will leave this to the reader.

Do Margin 5. ▶

EXAMPLE 8   A research submarine is at a depth of 458 ft. The submarine's sonar detects that the ocean floor is another 175 ft. down. What will be the depth if the submarine descends to the ocean floor? The submarine is designed to withstand the pressure from the surrounding water up to a depth of 600 ft. Is it safe to allow the submarine to descend to the ocean floor in this case?

Solution: Use the problem-solving process.

Understand: We must decide whether it is safe for a submarine to proceed from its current depth to the ocean floor.

Plan: We must first calculate the depth of the ocean floor. We can then assess whether it is beyond the submarine's depth limit of 600 ft. Because −458 ft. is its starting position and it is to proceed another 175 ft., we can add to get the depth of the ocean floor.

Execute: Ocean floor depth = −458 + (−175) = −633

Answer: The depth of the ocean floor is −633 ft. This would be beyond the submarine's 600 ft. depth limit. It would not be safe to allow the submarine to descend to the ocean floor.

Check: We can use a vertical number line.

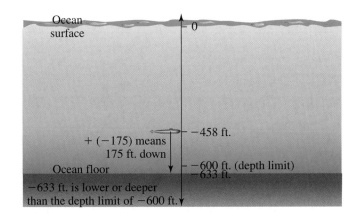

Do Margin 6. ▶

◆ **Margin 5**

The following table lists the assets and debts for the Cromwell family. Calculate their net worth.

Assets

Savings = $3528
Checking = $1242
Furniture = $21,358
Stocks = $8749

Debts

Credit card balance = $3718
Mortgage = $55,926
Automobile 1 = $4857
Automobile 2 = $3310

◆ **Margin 6**

At sunset, one cold winter evening in North Dakota, the temperature was −5°F. By midnight, the temperature had dropped another 15°F. What was the temperature at midnight?

Answer to Margin 5: −$32,934

Answer to Margin 6: −20°F

EXAMPLE 9 We can use integers to analyze the forces acting on an object. The result-ant force is the sum of the forces acting on the object. A steel beam weighs 250 lb. and has two wires attached to hold it up. Each wire has an upward force measuring 140 lb. (See the diagram below.) Find the resultant force on the beam. What does this force tell you?

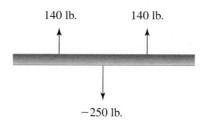

Solution: Use the problem-solving process.

Understand: Draw a diagram to get a sense of the direction of the forces. Because the weight is a force due to gravity pulling downward on objects, it has a negative value. The wires are pulling upward against gravity, so they have positive values.

Plan: To find the resultant force, we compute the sum of the forces.

Execute: Resultant force $= -250 + 140 + 140$

$$= -250 + 280$$

$$= 30$$

Answer: The resultant force is 30 lb. Because the resultant force is positive, it means the beam is traveling upward.

Check: Compute $-250 + 140 + 140$ from left to right.

$$-250 + 140 + 140$$

$$= -110 + 140$$

$$= 30$$

Trivia Bite

> In engineering and physics, the diagram above is called *a free-body diagram*. Because forces have a value as well as a direction of action, we refer to a force as a *vector*. We use vectors in situations when both quantity and direction affect the situation. Force, velocity, and acceleration are all vec-tor quantities because they have both value and direction.

◀ **Do Margin 7.**

◆ **Margin 7**

A 500 lb. concrete slab is suspended by four ropes tied to the corners. Each rope has an upward force of 110 lb. What is the re-sultant force? What does the resultant force tell you?

110 lb.   110 lb.

110 lb.   110 lb.

500 lb.

Answer: -60 lb.; Because the resultant force is negative, the slab is moving downwards.

2.2 Exercises

FOR EXTRA HELP

 Videotape 1

 InterAct Math
Tutorial Software

 www.carsonmath.com

 AWL Math Tutor Center

 InterAct
MathXL www.mathxl.com

Student's Solutions
Manual

For Exercises 1–10, add.
Key thought: Adding integers with the same sign is similar to increasing assets or increasing debts.

1. $14 + 9 =$

2. $74 + 13 =$

3. $-15 + (-8) =$

4. $-14 + (-16) =$

5. $-9 + (-6) =$

6. $-31 + (-9) =$

7. $12 + 18 + 16 =$

8. $4 + 15 + 41 =$

9. $-38 + (-17) + (-21) =$

10. $-60 + (-19) + (-32) =$

For Exercises 11–20, add.
Key thought: Adding integers with different signs is similar to reconciling assets and debts to get net worth.

11. $28 + (-18) =$

12. $17 + (-5) =$

13. $-34 + 20 =$

14. $-25 + 16 =$

15. $-21 + 35 =$

16. $-35 + 47 =$

17. $35 + (-53) =$

18. $16 + (-29) =$

19. $-24 + 80 =$

20. $75 + (-43) =$

For Exercises 21–42, solve by explaining each in terms of assets and debts.

21. $48 + 90 =$

22. $35 + 16 =$

23. $-45 + (-27) =$

24. $-68 + (-42) =$

25. $84 + (-23) =$

26. $68 + (-15) =$

27. $-15 + 42 =$

28. $-54 + 65 =$

29. $-81 + 60 =$

30. $-45 + 21 =$

31. $37 + (-58) =$

32. $62 + (-92) =$

33. $-45 + 45 =$

34. $64 + (-64) =$

35. $48 + (-18) + 16 + (-12) =$

36. $69 + (-20) + (-15) + 11 =$

37. $-36 + (-17) + 94 + (-9) =$

38. $-42 + 25 + (-61) + (-3) =$

39. $25 + (-17) + (-33) + 19 + 6 =$

40. $32 + 16 + (-51) + 3 + (-8) =$

41. $-75 + (-14) + 38 + 9 + (-17) =$

42. $-60 + 18 + (-40) + (-12) + 93 =$

For Exercises 43–52, solve.

43. The following table lists the assets and debts for the Smith family. Calculate their net worth.

Assets	Debts
Savings = $1498	Credit card balance = $1841
Checking = $2148	Mortgage = $74,614
Furniture = $18,901	Automobile #1 = $5488
Jewelry = $3845	

44. The following table lists the assets and debts for the Jones family, who just financed a new car. Calculate their net worth.

Assets	Debts
Savings = $214	Student loans = $15,988
Checking = $1242	Automobile #1 = $4857
Furniture = $21,358	Automobile #2 = $23,410 (new)

45. Jason's checking account shows a balance of $24. Unfortunately, he forgot about a check for $40 and it clears. The bank then charges $17 for insufficient funds. What is his new balance?

46. Charlene has a current balance of −$1243 on her credit card. During the month, she makes the following transactions. What will be her balance at the end of the month?

Charges:	
Truman's	$58
Dave's Diner	$13
Fuel'n Go	$15
Finance charge	$18
Payment:	$150

47. A scientist working during the summer at the South Pole station notes the temperature is about −21°F. During winter she can expect the temperature to be as much as 57°F colder. What temperature might she expect during the winter?

Trivia Bite

Because of the tilt of the Earth, the south pole experiences daylight for 6 months (summer) and then night for 6 months (winter). The sun rises in mid-September and then sets in mid-March.

48. The Inuit people in the arctic regions of North America sometimes build igloos of snow blocks. During winter months the temperatures can be as low as −58°F. A small heater can raise the internal temperature of a snow block igloo by 108°F. What would be the temperature inside the igloo?

49. A research submarine is attached to a crane on a ship. The submarine is currently at a depth of 147 ft. The crane raises the submarine 69 ft. and then must stop to be repaired. What is the elevation of the submarine?

50. A mining team is working 147 ft. below ground. During the day they dig down another 28 ft. What is their new position?

51. An elevator weighs 745 lb. When the elevator is on the first floor of a 12-story building, the steel cable connected to the top of it adds another 300 lb. Three people enter the elevator. One person weighs 145 lb., the second weighs 185 lb., and the third 168 lb. The motor exerts an upward force of 1800 lb. What is the resultant force?

52. A concrete block is suspended by two cables. The block weighs 500 lb. Each cable is exerting 250 lb. of upward force. What is the resultant force? What does this mean?

Puzzle Problem

A snail is at the bottom of a well that is 30 ft. deep. The snail climbs 3 ft. per hr. then stops to rest for an hour. During each rest period, the snail drops back down 2 ft. If the snail began climbing at 8 A.M., what time will it be when the snail gets out of the well?

REVIEW EXERCISES

1. Write 42,561,009 in expanded form.

2. Write the word name for 2,407,006.

3. Becky plans to place a fence around some property. Below is a plot plan of the property. She plans to have two 10 ft. gates. The fence company charges $7 per ft. for the fencing and $80 for each gate. What will be the total cost?

4. Subtract. 60,041 − 4596 =

5. Solve and check. 15 + n = 28

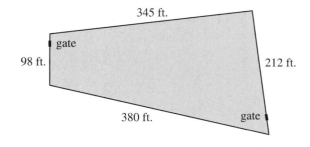

2.3 Subtracting Integers and Solving Equations

OBJECTIVES

1 Write subtraction statements as equivalent addition statements.

2 Solve equations containing a missing addend.

3 Solve applications involving subtraction of integers.

OBJECTIVE 1 *Write subtraction statements as equivalent addition statements.*

Although many people find subtraction statements with integers more difficult to think through than addition statements, we can write subtraction statements as equivalent addition statements. Let's consider some real situations in order to develop the method.

EXAMPLE 1 Marty has $30 in his checking account. He writes a check for $42. What is his balance?

Solution: Because checks are deducted from the account, we can write a subtraction statement.

$$30 - 42 = -12$$

Explanation: Because the amount being deducted was more than the amount in the account, Marty will have a negative balance.

In Section 2.1 we solved addition statements that were equivalent to this. The equivalent addition statement for this problem is

$$30 + (-42) = -12$$

Thus, $30 - 42$ and $30 + (-42)$ are equivalent because they both have the same answer.

◆ **Conclusion:** $30 - 42 = 30 + (-42)$

Although the equivalent addition statement was not needed to figure out the problem, the subtraction statements will get more complex. Therefore the equivalent addition statements will become easier to think through.

How did we write the subtraction statement as an equivalent addition statement? What changed? What stayed the same?

EXAMPLE 2 Marty has a balance of $-$12$ and the bank deducts another $17 as an insufficient funds charge. What is his new balance?

Solution: The charge is a deduction so we can write a subtraction statement.

$$-12 - 17 = -29$$

Explanation: He had a balance of $-$12$ to begin with and had another $17 deducted, so his balance continues in the negative direction. It is the same as the following addition statement:

$$-12 + (-17) = -29$$

Conclusion: $-12 - 17 = -12 + (-17)$

Do you see a pattern developing in the way we're writing the subtraction statements as equivalent addition statements? You should see that the first number stays the same. The operation symbol changes from a minus sign to a plus sign and the second number changes sign.
◆ In math language we change the subtrahend to its additive inverse.

◆ **Margin 1**

Write each as an equiva-lent addition statement, then evaluate.

a. $30 - 48 =$

b. $24 - 82 =$

c. $-16 - 25 =$

d. $-33 - 44 =$

Procedure *To write a subtraction statement as an equivalent addition statement:*

1. Change the operation symbol from minus sign to plus sign.

2. Change the subtrahend to its additive inverse.

◀ **Do Margin 1.**

You probably noticed that in each of the previous problems, we were subtracting a positive number. What happens if the subtrahend is negative and, therefore, we subtract a debt?

EXAMPLE 3 David has a balance of $28 in his checking account. The bank discovers they charged David's account a $5 service charge in error. The bank tells David they will remove the charge. What will be his new balance?

Solution: Because the $5 is a charge, it is written as -5. Because the bank is removing that amount, we must subtract. The subtraction statement looks like this:

$$28 - (-5) = 33$$

Explanation: To cancel or remove the charge, the bank must deposit that money back into David's account. Because removing a debt is the same as making a deposit, it is the same as $28 + 5$. Notice this follows the same pattern that we've already established.

Operation sign changes from minus sign to plus sign	$28 - (-5)$ $\downarrow\ \downarrow$ $= 28 + 5$ $= 33$	Subtrahend changes to its additive inverse.

Tip

$$28 - (-5)$$
$$\uparrow\quad\uparrow$$

Removing or canceling debt translates to a credit or deposit.

$$\downarrow$$
$$= 28 + 5$$

◆ **Margin 2**

Write each as an equivalent addition, then evaluate.

a. $42 - (-21) =$

b. $-48 - (-10) =$

c. $-36 - (-40) =$

EXAMPLE 4 Angela owes a friend $50. Out of kindness, her friend decides to cancel (or remove) $30 of that debt. What is Angela's balance?

Solution: Because Angela *owes* $50, we write it as -50. The $30 is debt, so it is written as -30. Because the friend is removing an amount, we can write a subtraction statement that looks like this:

$$-50 - (-30) = -20$$

Explanation: If Angela's friend cancels (removes) $30 of the $50 she owes, then Angela doesn't owe as much. She now only owes $20. Notice this is the same as $-50 + 30$. Again it follows the same pattern.

Operation sign changes from minus sign to plus sign	$-50 - (-30)$ $\downarrow\ \downarrow$ $= -50 + 30$ $= -20$	Subtrahend changes to its additive inverse.

◀ **Do Margin 2.**

Answers to Margin 1:
a. $30 + (-48) = -18$
b. $24 + (-82) = -58$
c. $-16 + (-25) = -41$
d. $-33 + (-44) = -77$

Answers to Margin 2:
a. $42 + 21 = 63$
b. $-48 + 10 = -38$
c. $-36 + 40 = 4$

OBJECTIVE 2 *Solve equations containing a missing addend.*

In Section 1.3, we learned how to solve equations involving a missing addend. Let's recall the procedure.

> **Procedure** *To find a missing addend, write a related subtraction sentence. Subtract the known addend from the sum.*

EXAMPLE 5 Solve and check.

a. $25 + x = 10$

Solution: We subtract the known addend from the sum.

$$25 + x = 10 \qquad \text{Subtract the known addend from the sum.}$$
$$x = 10 - 25 \qquad \text{We are subtracting a larger amount from a}$$
$$\qquad\qquad\qquad \text{smaller amount, so the result is negative.}$$
$$x = -15$$

Check: Verify that -15 can replace x in $25 + x = 10$, and make the equation true.

$$25 + (-15) \; ? \; 10$$
$$10 = 10$$

Explanation: Think about the original addition statement. If we start with 25 and end with 10, then a deduction has taken place. The only way this happens with addition is if a negative number is added.

b. $-14 + x = -20$

Solution: Subtract the known addend from the sum.

$$-14 + x = -20 \qquad \text{Subtract the known addend from the sum.}$$
$$x = -20 - (-14) \qquad \text{To simplify, write as an equivalent addition statement.}$$
$$x = -20 + 14$$
$$x = -6$$

> Because -20 and 14 have different signs, we subtract the absolute values and keep the sign of the number with the larger absolute value.

Check: Verify that -6 can replace x in $-14 + x = -20$, and make a true statement.

$$-14 + (-6) \; ? \; -20$$
$$-20 = -20$$

Explanation: Think through the original addition statement. Starting at -14, to end up at -20 we must continue 6 more steps in the negative direction. Therefore the second addend must be -6.

◆ **Do Margin 3.** ▶

◆ **Margin 3**

Solve and check.

a. $18 + y = 12$

b. $c + 24 = 9$

c. $-15 + d = -23$

d. $k + (-21) = -6$

e. $9 + t = -17$

Answers: **a.** -6 **b.** -15 **c.** -8 **d.** 15 **e.** -26

OBJECTIVE 3 *Solve applications involving subtraction of integers.*

Remember, the key to solving problems is to get an understanding of the situation. Let's recall the meanings of subtraction.

Subtraction means: 1. Take away. (When an amount is removed)

2. Difference. (When finding the distance between two positions)

3. Missing addend. (When looking for an amount needed to get a given total)

In the first four examples of this section, we saw take-away situations with money. Sometimes money situations involve a net amount. In business, the **net** is the money that remains after subtracting the expenses or **costs** from the **revenue**. If the net is positive, we say the net is a **profit**; if the net is negative, we say it is a **loss**.

DEFINITIONS   **Net:** Money remaining after subtracting costs from revenue.

Cost: Money spent on production, operation, labor, and debts.

Revenue: Income.

Profit: A positive net.

Loss: A negative net.

Conclusion: Net = Revenue − Cost

$$N = R - C$$

EXAMPLE 6   The financial report for a business indicates that the total revenue for 1998 was $2,453,530 and the total costs were $2,560,000. What was the net? Did the business experience a profit or loss?

Solution: Follow the problem-solving process.

Understand: We must find the net given the revenue and cost. The formula for net is $N = R - C$.

Plan: Replace R with 2,453,530 and C with 2,560,000 in $N = R - C$, then subtract.

Execute: $N = R - C$

$N = 2,453,530 - 2,560,000$

$N = -106,470$

Answer: The net for 1998 was −$106,470, which means the business had a loss of $106,470.

Check: Reverse the process. $-106,470 + 2,560,000 \ ? \ 2,453,530$

$$2,453,530 = 2,453,530$$

◀ **Do Margin 4.**

Now, let's consider difference problems. In difference problems you will be given two numbers and you must find the amount in between the two numbers. To get the difference, we subtract the smaller number from the larger number.

◆ **Margin 4**

The revenue for one month for a small business was $45,382, and the total costs were $42,295. What was the net? Was it a profit or loss?

Answer: $3087; Profit

EXAMPLE 7 On the evening news the meteorologist says the current temperature is 25°F. She then says that the evening low could get down to −6°F. How much of a change is this from the current temperature?

Solution: Follow the problem-solving process.

Understand: Because we are asked to find the amount of change, we must calculate a difference.

Plan: Find the difference between 25 and −6.

Execute: Temperature difference = 25 − (−6)   Write as an equivalent addition
$$= 25 + 6$$ statement.
$$= 31$$

Answer: From 25°F to −6°F is a change of 31°F.

Check: Use a number line or picture.

> **Connection**
> A thermometer is essentially a vertical number line.

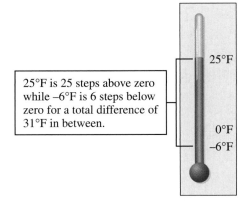

25°F is 25 steps above zero while –6°F is 6 steps below zero for a total difference of 31°F in between.

Do Margin 5. ▶

Now let's consider the missing addend situation. In this situation, you will be given a starting position and asked to find how much more is needed to achieve a new position.

EXAMPLE 8 A submarine is at −126 ft. How much must the submarine climb to reach −40 ft.?

Solution: Follow the problem-solving process.

Understand: The word *climb* indicates that we must add an amount to −126 ft. to reach a value of −40 ft. This is a missing addend situation. A picture is helpful.

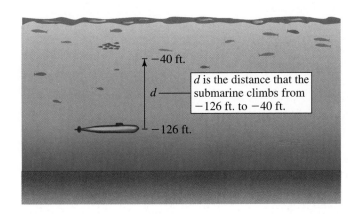

d is the distance that the submarine climbs from −126 ft. to −40 ft.

◆ **Margin 5**

Solve.

a. The surface temperature on the moon at lunar noon can reach 120°C. During the lunar night, the temperature can drop to −190°C. How much of a change in temperature occurs from day to night on the moon?

b. A submarine goes from a depth of −78 ft. to a depth of −34 ft. What is the difference in depth?

Answers: **a.** 310°C **b.** 44 ft.

Solve.

a. Barbara currently has a balance of −$58 in her checking account. To avoid any further charges she must have a minimum balance of $25. How much should she deposit to have the minimum balance?

Plan: Write a missing addend equation, then solve.

Execute: $-126 + d = -40$ To solve, subtract the known addend, -126, from the sum -40.

$$d = -40 - (-126)$$ Write as an equivalent addition statement.

$$d = -40 + 126$$

$$d = 86$$

> **Connection**
>
> In the diagram we can see that the missing addend amount is the same as the difference between -40 and -126. This explains why the missing addend statement, $-126 + d = -40$, becomes the difference statement, $d = -40 - (-126)$.

Answer: The submarine must climb 86 ft. to reach a depth of -40 ft.

Check: Verify that if the submarine climbs 86 ft. from -126 ft. it will be at -40 ft.

$$-126 + 86 \ ? \ -40$$
$$-40 = -40 \quad \text{It checks.}$$

◀ **Do Margin 6.**

2.3 Exercises

For Exercises 1–12, write as an equivalent addition statement, then evaluate.

1. $18 - 25 =$ **2.** $30 - 54 =$ **3.** $-15 - 18 =$ **4.** $-21 - 6 =$

5. $20 - (-8) =$ **6.** $33 - (-12) =$ **7.** $-14 - (-18) =$ **8.** $-20 - (-30) =$

9. $-15 - (-8) =$ **10.** $-26 - (-14) =$ **11.** $0 - (-5) =$ **12.** $0 - 16 =$

For Exercises 13–26, write as an equivalent addition statement as needed, then evaluate.

13. $-21 - 19 =$ **14.** $-15 - (-12) =$ **15.** $-4 - (-19) =$

16. $26 - (-6) =$ **17.** $31 - 44 =$ **18.** $-13 - 28 =$

19. $35 - (-10) =$ **20.** $-19 - (-24) =$ **21.** $-28 - (-16) =$

22. $9 - 15 =$ **23.** $0 - 18 =$ **24.** $-32 - 14 =$

25. $-27 - (-16) =$ **26.** $0 - (-8) =$

For Exercises 27–38, solve and check.

27. $18 + t = 12$ **28.** $21 + c = 16$ **29.** $d + (-6) = 9$

30. $a + (-14) = 15$ **31.** $-28 + u = -15$ **32.** $-22 + p = -19$

33. $-14 + m = 0$ **34.** $16 + n = 0$ **35.** $h + (-35) = -40$

36. $r + (-29) = -38$ **37.** $-13 + k = -25$ **38.** $-36 + x = -53$

For Exercises 39–52, solve.

39. Belinda has a balance of $126 in her bank account. A check written against her account for $245 arrives at the bank.

    a. What is her balance?

    b. Because Belinda has insufficient funds to cover the check amount, the bank assesses a service charge of $20. What is her balance after the service charge?

40. Brian has a credit of $86 on his account with an audio-video store. He uses his credit account to purchase some equipment at a total cost of $585. What is his new balance after the purchase?

41. The revenue for an insurance company in 1996 was $3,568,250. The total cost for the company was $1,345,680. What was the net? Was it a profit or loss?

42. As of 1998, *Titanic* was the top grossing movie of all time, grossing $600,743,440 in total box office sales. If the budget for making the film was $200,000,000, what was the profit?

43. Florence spent a total of $26,458 for her car, including the cost of financing and the cost of maintenance. She then sells the car for $4500. What is her net? Is it a profit or loss?

44. Gary bought a Fender Stratocaster guitar in 1960 for $150. In 1998, he takes the guitar to a guitar show and sells it for $12,000. What was the net? Is this a profit or loss?

▸ **Discussion** How would inflation affect the answer to Exercises 43 and 44?

45. The temperature at sunset was reported to be 19°F. By midnight it is reported to be −27°F. What is the amount of the decrease?

46. Liquid nitrogen has a temperature of −208°C. An orange is placed into the liquid nitrogen that causes the liquid to boil at −196°C. What is the amount of the increase in temperature?

47. Molecular motion is minimized at −273°C. Oxygen goes from liquid to solid at −218°C. How much colder does liquid oxygen need to get for its molecular motion to be minimized?

48. A container of liquid hydrogen fluoride is at a temperature of −90°C. The boiling point for hydrogen fluoride is 19°C. How much must the temperature rise for the hydrogen fluoride to boil?

49. Derrick has a balance of −$37 in his bank account. To avoid further charges he must have a balance of $30. What is the minimum he can deposit to avoid further charges?

50. Danielle owns a business that is in financial trouble. Her accountant informs her that her net worth is −$5267. Her net worth should be at least $2500 by the end of the month, otherwise she should close her business. What is the minimum profit she must clear in order to remain in business?

51. The lowest level of the Sears Tower in Chicago is at an elevation of −48 ft. The skydeck is at an elevation of 1353 ft. What is the distance between the lowest level and the skydeck?

52. As a result of initial expenses, a new business has a net worth of −$45,000. The goal for the first month is to make enough profit to have a net worth of $15,800. How much must the business make in that first month to achieve their goal?

REVIEW EXERCISES

1. Multiply. (145)(209)

2. Evaluate. $(3)^5$

3. Jaqueline owns 20 sq. mi. (square miles) of land that borders a national park. The government wishes to purchase an 8 mi. by 2 mi. rectangular strip of her land. How much land will she have left if she sells?

4. Find the missing factor. $3y = 24$

5. $\sqrt{225} =$

2.4 Multiplying, Dividing, and Solving Equations with Integers

OBJECTIVES

1 Multiply integers.

2 Evaluate numbers in exponential form.

3 Divide integers.

4 Solve equations containing a missing factor.

5 Evaluate square roots.

6 Solve applications involving multiplication and division.

OBJECTIVE 1 *Multiply integers.*

To discover the rules for multiplying and dividing signed numbers, consider the following pattern. We'll keep the first factor the same and gradually decrease the other factor until it becomes negative. What is happening to the product as we decrease the factor?

$2 \cdot 4 = 8$

$2 \cdot 3 = 6$ Each time we decrease the second factor by 1 we lose one of the repeatedly added 2's so that the product decreases by 2.

$2 \cdot 2 = 4$

$2 \cdot 1 = 2$ When we decrease the 1 factor to 0, the product must continue to decrease by 2. This is why $2 \cdot 0$ must equal 0. This confirms,

$2 \cdot 0 = 0$ as we already established, that any number multiplied by 0 is 0.

$2 \cdot (-1) = -2$ As we continue, the pattern must continue. Notice when we decrease the 0 factor to -1, we must continue to decrease

$2 \cdot (-2) = -4$ the product by 2. This explains why multiplying $2 \cdot -1$ must equal -2, and so on.

From this pattern, we can conclude that multiplying a positive number by a negative number equals a negative product.

The commutative property of multiplication tells us that the reverse must also be true. Multiplying a negative number times a positive number also equals a negative product.

$$\text{Both} \quad 2 \cdot (-2) \text{ and } (-2) \cdot 2 \text{ equal } -4.$$

Conclusion: When multiplying two numbers that have different signs, the product is negative.

EXAMPLE 1 Multiply.

a. $8 \cdot (-6) =$

Answer: $8 \cdot (-6) = -48$

b. $-12 \cdot 5 =$

Answer: $-12 \cdot 5 = -60$

Do Margin 1. ▶

Margin 1
Multiply.

a. $9 \cdot (-7) =$

b. $-8 \cdot 10 =$

c. $-4(5) =$

d. $6(-3) =$

Answers: **a.** -63 **b.** -80 **c.** -20 **d.** -18

What about multiplying two numbers with the same sign?

Again, let's consider a pattern and see if we can discover the rule. Let's start with a negative number times a positive number and gradually decrease the positive factor until it becomes negative.

$(-2) \cdot 4 = -8$	From the rule we already established, a negative number times a positive number is a negative product.
$(-2) \cdot 3 = -6$	As we decrease the positive factor, what happens to the product?
$(-2) \cdot 2 = -4$	Notice the product is actually increasing. From -6 to -4 is an increase of 2.
$(-2) \cdot 1 = -2$	
$(-2) \cdot 0 = 0$	
$(-2) \cdot (-1) = 2$	To continue the same pattern, when we decrease the 0 factor to -1, we must continue to increase the product by 2. This means the product must become positive.
$(-2) \cdot (-2) = 4$	

We already knew that when we multiply two positive numbers, the product is positive. Now we've discovered that when we multiply two negative numbers, the product is positive as well.

Conclusion: When multiplying two numbers that have the same sign, the product is positive.

EXAMPLE 2 Multiply.

a. $-6 \cdot (-8) =$

Answer: $-6 \cdot (-8) = 48$

b. $(-9)(-7) =$

◆ **Answer:** $(-9)(-7) = 63$

◀ **Do Margin 2.**

What if we must multiply several signed numbers?

We could simply follow the order of operations and work our way from left to right.

EXAMPLE 3 Multiply. $-2(-3)(-4) =$

Solution: $-2(-3)(-4) = -24$

Explanation: Working from left to right, $-2(-3)$ is positive 6 because we are multiplying two numbers that have the same sign. If we then multiply that positive 6 by the -4, we must get a negative number because we are now multiplying two numbers that have different signs.

EXAMPLE 4 Multiply. $-1(-1)(-1)(-5) =$

Solution: $-1(-1)(-1)(-5) = 5$

Explanation: As we work from left to right, -1 times -1 is 1. That result times the third -1 is -1. This -1 result times the -5 is positive 5!

Connection

Multiplying by -1 is like finding additive inverses. For example, finding the additive inverse of $-(-(-(-5)))$ is the same as the multiplication problem $-1(-1)(-1)(-5)$.

Just as we simply counted the total signs to work through the longer additive inverse problems, we do the same thing when multiplying.

Answers: **a.** 40 **b.** 36 **c.** 130

Rule *When multiplying signed numbers, count the total number of negative factors:*
If there are an even number of negative factors, then the product is positive.
If there are an odd number of negative factors, then the product is negative.

Do Margin 3. ▶

OBJECTIVE 2 *Evaluate numbers in exponential form.*

Because exponents mean repeated multiplication, we can now extend what we've learned to exponents.

EXAMPLE 5 Evaluate $(-5)^2$

Solution: $(-5)^2 = (-5)(-5) = 25$

Explanation: An exponent of 2 means we must multiply the base by itself. Because the base was negative, we had to multiply two negative numbers, which makes a positive product.

Conclusion: Any time the exponent is an even number, we are multiplying an even number of bases; therefore the product will be positive.

EXAMPLE 6 Evaluate $(-2)^3$

Solution: $(-2)^3 = (-2)(-2)(-2) = -8$

Explanation: An exponent of 3 means we must multiply the base by itself three times. When we multiply the first two -2's, the result is positive 4. Then the 4 times the last -2 is -8. Because we multiplied the -2 by itself an odd number of times, the result is negative!

Conclusion: Any time the exponent is an odd number with a negative base, we are multiplying an odd number of negative numbers, and therefore the result will be negative.

Rules If the base of an exponential form is a negative number and the exponent is even, the product is positive.
If the base is a negative number and the exponent is odd, the product is negative.

Do Margin 4. ▶

WARNING The parentheses are very important!

$$(-2)^4 \text{ is different from } -2^4.$$

The parentheses around the base indicate that the base number is a negative number. Without the parentheses, we must view the sign as an additive inverse. We must calculate the exponent first with a positive base, then get the opposite of the result! Look at how each group of symbols is actually read:

$(-2)^4$ is read as, "Negative two raised to the fourth power."
Solution: $(-2)^4 = (-2)(-2)(-2)(-2) = 16$

-2^4 is read as, "The additive inverse of the result of two, raised to the fourth power."
Solution: $= -[2 \cdot 2 \cdot 2 \cdot 2] = -16$

◆ **Margin 3**
Multiply.

a. $(-4)(-6)(-2) =$

b. $(-1)(-1)(-2)(-5) =$

c. $-5 \cdot (-7) \cdot (-2) \cdot (-1) =$

d. $-3(-5)(2) =$

e. $4 \cdot (-4) \cdot (3) \cdot (-1) \cdot (-1) =$

◆ **Margin 4**
Evaluate.

a. $(-2)^5 =$

b. $(-2)^6 =$

Answers to Margin 3: **a.** -48 **b.** 10 **c.** 70 **d.** 30 **e.** -48

Answers to Margin 4: **a.** -32 **b.** 64

EXAMPLE 7 Evaluate -3^4

Solution: $-3^4 = -[3 \cdot 3 \cdot 3 \cdot 3]$
$$= -81$$

Explanation: -3^4 means find "the additive inverse of the result of 3 raised to the fourth power." 3 raised to the fourth power is 81 and the additive inverse of 81 is -81.

Connection

We discovered that we can think of additive inverses as multiplying by -1. We can think of -3^4 as $-1 \cdot 3^4$. The order of operations agreement says we should compute exponents before multiplying. So we calculate 3^4 first, which is 81, then multiply by -1 to get its additive inverse, -81.

◀ **Do Margin 5.**

OBJECTIVE 3 *Divide integers.*

It turns out that the sign rules for division are the same as for multiplication. We can see this by looking at the relationship between division and multiplication. Recall that we can use division to find missing factors. Consider the following statement.

$$-2n = 6 \qquad \text{Remember } -2n = 6 \text{ means } -2 \cdot \ ? = 6.$$

We know from what we learned about multiplication that a negative number multiplied by another negative number is a positive product. This means that the missing factor must be -3.

We can write the missing factor statement as a related division statement.

$$-2n = 6 \qquad \text{can be written as:} \qquad n = 6 \div (-2) = -3$$

Conclusion: When dividing a positive number by a negative number, the quotient is negative.

What about a negative number divided by a positive number? Let's consider another missing factor statement.

$$3x = -12$$

We know that when multiplying two numbers, if the outcome is negative then the two factors have different signs. Our known factor in this case is positive 3, so the missing factor must be negative, that is, -4.

Once again, let's write the missing factor statement as a division statement.

$$3x = -12$$
$$x = -12 \div 3 = -4$$

Conclusion: When dividing a negative number by a positive number the quotient is negative.

What if the two numbers have the same sign?

Consider the following missing factor statement.

$$-4y = -20$$

We know by the rules of multiplication that a negative factor must be multiplied by a positive factor in order to get a negative product. This means that the missing factor here must be positive 5.

Let's look at the related division sentence.

$$-4y = -20$$
$$y = -20 \div (-4) = 5$$

Conclusion: When dividing two numbers that have the same sign, the quotient is positive.

In summary, we can conclude that the rules for dividing signed numbers are the same as for multiplying.

> **Rules for Multiplying and Dividing Signed Numbers**
>
> When multiplying or dividing two numbers that have the same sign, the result is positive.
> When multiplying or dividing two numbers that have different signs, the result is negative.

Do Margin 6. ▶

OBJECTIVE 4 *Solve equations containing a missing factor.*

In Section 1.4 we learned how to solve equations involving a missing factor. Let's recall the procedure.

> **Procedure** *To solve for a missing factor, write a related division sentence, dividing the product by the known factor.*

EXAMPLE 8 Solve and check.

a. $9x = -63$

Solution: Write a related division sentence.

$$9x = -63$$
$$x = -63 \div 9$$
$$x = -7$$

Check: Verify that -7 can replace x in $9x = -63$, and make the equation true.

$$9(-7) \; ? \; -63$$
$$-63 = -63 \qquad \textbf{It checks.}$$

b. $-30y = -60$

Solution: Write a related division sentence.

$$-30y = -60$$
$$y = -60 \div -30$$
$$y = 2$$

Check: Verify that 2 can replace y in $-30y = -60$, and make the equation true.

$$-30(2) \; ? \; -60$$
$$-60 = -60$$

Do Margin 7. ▶

◆ **Margin 6**

Divide.

a. $36 \div (-4) =$

b. $-40 \div 8 =$

c. $-28 \div (-7) =$

d. $\dfrac{-54}{9} =$

e. $\dfrac{-72}{-12} =$

f. $-18 \div 0 =$

◆ **Margin 7**

Solve and check.

a. $7n = -42$

b. $-12k = -36$

c. $-8h = 96$

d. $0m = -14$

Answers to Margin 6: **a.** -9 **b.** -5
c. 4 **d.** -6 **e.** 6 **f.** undefined

Answers to Margin 7: **a.** -6 **b.** 3
c. -12 **d.** undefined

OBJECTIVE 5 *Evaluate square roots.*

Recall that finding a square root is the inverse of squaring a number. To find a square root of a given number, we must find a number that can be squared to equal the given number.

EXAMPLE 9 Find all square roots of 81.

Answer: Both 9 and −9

Explanation: To find a square root of 81, we must ask, "What number squared equals 81?" We must find all numbers that can satisfy the following missing number statement:

$$x^2 = 81$$

Remember $x^2 = 81$ is the same as saying $?^2 = 81$.

Of course, 9 works because $9^2 = 81$. However, notice −9 also works because $(-9)^2 = (-9)(-9) = 81$!

Conclusion: There are two square roots for every positive number. There is a positive root and a negative root.

When we need to express both a positive and a negative of the same value, we can say "both 9 and −9" as we did above, or we can state it this way: ±9.

EXAMPLE 10 Find all possible solutions for the missing number statement.

a. $x^2 = 144$

Answer: ±12

Explanation: We must find a number that can be multiplied by itself to equal 144. Both positive and negative 12 work.

$$12^2 = 12 \cdot 12 = 144 \qquad \text{and} \qquad (-12)^2 = (-12)(-12) = 144$$

b. $m^2 = -25$

Answer: No integer solution.

Explanation: There is no way to square an integer and get anything other than a positive result. If we square a positive number we get a positive result and, because of the rules for multiplying signed numbers, if we square a negative number we get a positive result.

Conclusion: We cannot square an integer to get a negative number.

Discussion If we cannot square an integer to get a negative number, why not say the answer is undefined?

◀ **Do Margin 8.**

Now let's turn our attention to the radical symbol. In Section 1.4 we learned that the radical symbol means find only the *positive* square root of the given number. This is important to remember since we've just now discovered that there is a positive and a negative root for every positive number. The positive square root is often called the *principal* square root.

EXAMPLE 11 Evaluate the radical.

a. $\sqrt{144}$

Solution: $\sqrt{144} = 12$

Explanation: The radical sign indicates that we must find the positive square root of 144, which is 12.

◆ **Margin 8**

Find all possible solutions for the missing number statement.

a. $k^2 = 100$

b. $n^2 = 25$

c. $c^2 = -36$

Answers: **a.** ±10 **b.** ±5
c. No integer solution

b. $\sqrt{-25} =$

Solution: No integer solution.

Explanation: $\sqrt{-25}$ means find a positive number that can be squared to equal -25. Because the square of every positive integer is a positive number, there is no positive integer that can be squared to equal -25.

Rules for Square Roots

Every positive number has two square roots, a positive root and a negative root.

The radical symbol means to find only the positive, or principal, square root.

The square root of a negative number is not an integer.

Do Margin 9. ▶

OBJECTIVE 6 *Solve applications involving multiplication and division.*

EXAMPLE 12 Barbara has 4 credit cards each with a balance of $-\$245$. What is her total debt?

Solution: Use the problem-solving process.

Understand: Since she has the same debt on each card, we are repeatedly adding the same amount 4 times. Since multiplication is repeated addition, we can simply multiply to calculate the total.

Plan: Multiply.

Execute: $4 \cdot (-245) = -980$

Answer: She has a total debt of $980.

Check: We reverse the process and divide.

$$-980 \div (-245) \ ? \ 4$$
$$4 = 4$$

Do Margin 10. ▶

Many situations can be described mathematically using formulas. Force and voltage situations can be described with formulas that involve multiplication.

Isaac Newton discovered a relationship for forces. The formula for force is

$$F = ma.$$

F represents force, m represents mass, and a represents acceleration. In the English/American system of measurement, force is measured in pounds, mass in slugs, and acceleration in feet per second per second or ft./s². In the metric system, force is measured in Newtons (after Isaac), mass is measured in kilograms, and acceleration is measured in meters per second per second.

Mass is a measure of how much material or matter (atoms and molecules) is in an object. Acceleration is a measure of how quickly an object increases or decreases its speed. For example when you measure how quickly a car can go from 0 mph to 60 mph, you are measuring its acceleration.

A force that we all experience is the force of gravity. Galileo Galilei discovered that all objects fall at the same rate. This seems counter to what we'd expect; however, it is easy to demonstrate.

◆ **Margin 9**

Evaluate the radical.

a. $\sqrt{121} =$

b. $\sqrt{225} =$

c. $\sqrt{-81} =$

◆ **Margin 10**

An oceanographer, studying sunlight and temperature at various depths, programs a small submarine to take readings every -18 ft. After 7 readings, what will be the submarine's position?

Answers to Margin 9: **a.** 11 **b.** 15 **c.** No integer solution

Answer to Margin 10: -126 ft.

Try this: Hold your pencil/pen in one hand and your book in the other at the same height above the floor. It should be pretty obvious to you that the book weighs more. Now drop them at the same time. Notice they both hit the floor at the same time!

The Earth accelerates all objects downwards at the same rate, 32 ft. per second per second. Because this acceleration is downwards we write the acceleration as -32 ft./s^2, which means, in each second that an object falls toward the ground, gravity increases its speed by 32 ft./s. So when you drop objects like a book or pencil, after 1 second they have a speed of 32 ft./s, after 2 seconds they have a speed of 64 ft./s, and so on.

Historical Note

Among mathematicians, Isaac Newton is considered one of the greatest to have ever lived. He was born on Christmas day, 1642, in England.

In 1661 Newton entered Trinity College, Cambridge, earning his expenses doing menial jobs. He received a B.A. in 1664, then went home to Woolsthorpe. Over the next two years he invented calculus, discovered the laws of gravity, and experimented with optics, using a prism to show that white light is composed of all colors.

Newton returned to Cambridge in 1667, and was later appointed Lucasian Professor of Mathematics. He lectured on optics and built a telescope to make his own observations of the planets.

During the years 1684–86, Newton wrote down his discoveries in mathematics and science in *Philosophiae Naturalis Principia Mathematica* (Mathematical Principles of Natural Philosophy). It is said that while writing the book, he barely ate or slept, and upon waking would sit on the edge of his bed for hours concentrating and thinking through the problems.

In 1696 Newton was appointed Warden of the Mint. Two years later he became Master of the Mint, putting England on a gold standard that would remain until 1933. In 1705, he was knighted by Queen Anne.

Isaac Newton died on March 20, 1727, and is buried in Westminster Abbey.

♦ **Margin 11**

The blue whale is the largest animal on Earth. The mass of an average Blue whale is about 2625 slugs. Find the weight.

The weight you feel when holding an object is the force of gravity trying to accelerate the object toward the ground. Just think, you're holding back all the gravitational force that Earth can muster! Newton's law tells us that to calculate that force, we just need to multiply the mass of the object by the acceleration constant -32 ft./s^2. The metric value of the acceleration due to gravity is about -10 m/s^2 which means -10 meters per second per second.

EXAMPLE 13 A person has a mass of 5 slugs. What is his weight?

Solution: Use the problem-solving process.

Understand: Weight is a force so we can use Newton's formula $F = ma$. The acceleration due to gravity is about -32 ft./s^2.

Plan: Replace m with 5 and a with -32 in $F = ma$, then multiply.

Execute: $F = ma$

$$F = (5)(-32)$$
$$F = -160$$

Answer: The person weighs -160 lb., which means the force is directed downward.

Check: Reverse the process and divide.

$$-160 \div -32 \ ? \ 5$$
$$5 = 5$$

▶ **Discussion** What if the man were on another planet? Would his mass change? Would his weight change?

Answer: $-84{,}000$ lb.

◀ **Do Margin 11.**

EXAMPLE 14 A dumbbell is measured to have a weight of 20 N. What is its mass?

Solution: Use the problem-solving process.

Understand: Because weight is a force directed downward, we write 20 N as -20 N. We use Newton's law $F = ma$. Since N is the metric unit of force, we must remember to use the metric value for the acceleration due to gravity, which is about -10 m/s^2.

Plan: Replace F with -20 and a with -10 in $F = ma$, then solve for m.

Execute:
$$F = ma$$
$$-20 = m(-10)$$
$$-20 \div -10 = m$$
$$2 = m$$

Notice we have a missing factor equation. To find the missing factor, we must divide the product by the known factor.

Answer: The dumbbell has a mass of 2 kg.

Check: Verify that a dumbbell with a mass of 2 kg would have a weight of -20 N.

$$F = ma$$
$$-20 \; ? \; (2)(-10)$$
$$-20 = -20 \qquad \textbf{It checks.}$$

Do Margin 12. ▶

◆ **Margin 12**

Find the mass of a 17,800 N elephant.

Trivia Bite

Everything that has mass also has gravity. The amount of gravitational pull an object has depends on how massive the object is. The more massive an object the more gravitational pull it has. The moon is less massive than Earth, so it does not have as much gravitational pull as Earth. This means you would weigh less on the moon than on Earth. The moon accelerates objects towards its surface at a little less than 2 meters per second every second. The Sun accelerates objects at 900 ft. per second every second, which is roughly 30 times Earth's gravity!

Georg Simon Ohm was a German scientist who studied electricity. He is credited with the discovery of the relationship among voltage, current, and resistance. His famous formula is $V = ir$.

Historical Note

Georg Ohm is credited with the discovery of the relationship among voltage, current, and resistance, but actually a British chemist by the name of Henry Cavendish discovered the relationship 46 years before Ohm. Unfortunately, Cavendish did not publish his findings. Thus, Ohm got the credit and his name used for the unit of resistance.

The V stands for voltage. The i stands for current, which is measured in amperes or amps (A). The r stands for resistance, which is measured in ohms (named after Mr. Ohm in honor of his discovery). The symbol for ohms is the Greek letter Ω (omega).

Voltage and current can be negative numbers. A voltage source like a battery has $+$ and $-$ connections called terminals. A voltage meter measures voltage and has two probes, one for $+$ and one for $-$ terminals/connections. If we connect each probe to the appropriate terminal on the battery we measure a positive voltage. If we reverse the probes we measure a negative voltage. Voltage is positive or negative depending on the orientation of the measuring device.

Current is a measure of the flow of electricity. A current meter is essentially the same as a voltage meter. If we connect the probes, and the meter indicates negative current, the current is flowing in the opposite direction of the orientation of the probes from the meter.

Answer: 1780 kg

EXAMPLE 15 Calculate the voltage in a circuit that has a resistance of 20 Ω and a current of −9 A.

Solution: Use the problem-solving process.

Understand: We must calculate voltage given resistance and current. The formula that relates voltage, current, and resistance is $V = ir$.

Plan: Replace i with −9 and r with 20 in $V = ir$, then multiply.

$$V = ir$$
$$V = (-9)(20)$$
$$V = -180$$

Connection

The formulas $F = ma$, $V = ir$, and $A = lw$ are all mathematically the same. They all involve the product of two factors. Many different phenomena have the same mathematical relationship.

Answer: The voltage is −180 V.

Check: Reverse the process.

$$-180 \div 20 \; ? \; -9$$
$$-9 = -9 \quad \text{It checks.}$$

As mathematicians, we do not necessarily need to understand what a formula relates to (although it makes the formula more interesting) in order to work with the mathematics. In the previous example, you did not have to know that V meant voltage, i meant current, and r meant resistance. As long as you were given some indication as to where each number goes, you could have worked with the formula.

◀ **Do Margin 13.**

EXAMPLE 16 The voltage in a circuit is measured to be 120 V across a 40 Ω resistor. What is the current?

Solution: Use the problem-solving process.

Understand: We must calculate current given voltage and resistance. We use the formula $V = ir$.

Plan: Replace V with 120 and r with 40 in $V = ir$, then solve for i.

Execute:
$$V = ir$$
$$120 = i\,(40)$$
$$120 \div 40 = i$$
$$3 = i$$

Notice we have a missing factor equation. To find a missing factor, we divide the product by the known factor.

Answer: The current is 3 A.

Check: Verify that a circuit with a current of 3 A and a resistance of 40 Ω will have a voltage of 120 V.

$$120 \; ? \; (3)(40)$$
$$120 = 120 \quad \text{It checks.}$$

◀ **Do Margin 14.**

Margin 13

The current in a circuit is measured to be 12 A through a 20 Ω resistor. Find the voltage.

Margin 14

The voltage is measured to be −9 V across a 3 Ω resistor. Find the current.

Answer to Margin 13: 240 V

Answer to Margin 14: −3 A

2.4 Exercises

FOR EXTRA HELP

Videotape 1

InterAct Math
Tutorial Software

www.carsonmath.com

AWL Math Tutor Center

InterAct MathXL   www.mathxl.com

Student's Solutions
Manual

For Exercises 1–18, multiply.

1. $-7 \cdot 9 =$

2. $-12 \cdot 6 =$

3. $13 \cdot (-5) =$

4. $15 \cdot (-3) =$

5. $0 \cdot (-9) =$

6. $(-1) \cdot (-32) =$

7. $(-4) \cdot (-17) =$

8. $-13 \cdot 0 =$

9. $15 \cdot (-1) =$

10. $-21 \cdot (-8) =$

11. $19\,(-20) =$

12. $(-14)(31) =$

13. $(-1)(-5)(-7) =$

14. $(-3)(-6)(-1) =$

15. $(-1)(-1)(-6)(-9) =$

16. $(-8)(7)(-2)(-1)(-1) =$

17. $(-20)(-2)(-1)(3)(-1) =$

18. $(5)(-4)(-2)(-1)(3) =$

For Exercises 19–30, evaluate.

19. $(-7)^2 =$

20. $(-12)^2 =$

21. $(-4)^3 =$

22. $(-5)^3 =$

23. $(-3)^4 =$

24. $(-2)^4 =$

25. $(-2)^6 =$

26. $(-3)^5 =$

27. $-8^2 =$

28. $-11^2 =$

29. $-10^6 =$

30. $-10^4 =$

For Exercises 31–48, divide.

31. $36 \div (-3) =$

32. $-48 \div 8 =$

33. $-81 \div 27 =$

34. $40 \div (-10) =$

35. $-32 \div (-4) =$

36. $-100 \div (-20) =$

37. $0 \div (-2) =$

38. $0 \div 12 =$

39. $31 \div (-1) =$

40. $-14 \div (-1) =$

41. $\dfrac{65}{-13} =$

42. $\dfrac{-96}{8} =$

43. $\dfrac{-41}{0} =$

44. $\dfrac{27}{0} =$

45. $\dfrac{-124}{-4} =$

46. $\dfrac{-91}{-7} =$

47. $-\dfrac{28}{4} =$

48. $-\dfrac{42}{7} =$

For Exercises 49–70, solve and check.

49. $9x = -18$

50. $-12t = 48$

51. $-6m = -54$

52. $-8n = -32$

53. $0v = -14$

54. $0b = 28$

55. $-1c = 17$

56. $-1c = -12$

57. $-18m = 0$

58. $-25b = 0$

59. $-2\,(-5)d = -50$

60. $3\,(-8)f = 72$

61. $-1(-1)(-7)g = -63$

62. $(-9)(-1)(2)w = -90$

63. $y^2 = 25$

64. $p^2 = 49$

65. $c^2 = 81$

66. $k^2 = 225$

67. $n^2 = 196$

68. $x^2 = 289$

69. $b^2 = -64$

70. $r^2 = -1$

For Exercises 71–82, evaluate the radical.

71. $\sqrt{81} =$

72. $\sqrt{36} =$

73. $\sqrt{64} =$

74. $\sqrt{100} =$

75. $\sqrt{1} =$

76. $\sqrt{25} =$

77. $\sqrt{-169} =$

78. $\sqrt{-144} =$

79. $-\sqrt{121} =$

80. $-\sqrt{256} =$

81. $\sqrt{0} =$

82. $-\sqrt{0} =$

For Exercises 83–94, solve.

83. Rohini had a balance of −$214 on her credit card in June. By November her balance had tripled. What was her balance in November?

84. Michael has five credit accounts. Three of the accounts have a balance of −$100 each. The other two accounts each have a balance of −$258. What is Michael's total debt?

85. During one difficult year, the Morrisons had insufficient funds in their checking account on 7 occasions, and each time, they were assessed a charge appearing as −$17 on their statement. What were the total insufficient funds charges that year?

86. An oil company drill team estimated the depth to an oil pocket to be −450 ft. After drilling to 3 times that depth they finally found oil. At what depth did they find oil?

87. A car engine with a mass of 157 kg is hanging from a lift by a chain. What is the force on the chain?

88. The Eiffel Tower contains approximately 437,500 slugs of iron. What is the weight of the iron in the Eiffel tower?

89. The largest bear on Earth is the polar bear. These animals can weigh up to 8000 N, which is a downward force of −8000 N. What is the mass of one of these bears?

90. The *Titanic* weighed approximately 92,656,000 lb., which is a downward force of −92,656,000 lb. What was the mass of the *Titanic*?

91. An electrical circuit has a resistance of 7 Ω and a current of −9 A. Find the voltage.

92. An electrical circuit has a resistance of 100 Ω and a current of −12 A. Find the voltage.

93. The voltage in an electrical circuit measures −220 V. If the reistance is 5 Ω, find the current.

94. It is suspected that an incorrect resistor was put into a circuit. The correct resistor should be 12 Ω. A voltage measurement is taken and found to be 60 V. The current is measured to be 10 A. Is the resistor correct?

REVIEW EXERCISES

1. A wallpaper border is to be placed on the walls just below the ceiling around a 14 ft. by 16 ft. room. How much border paper is needed?

2. Sherry wishes to paint a wall in her house. The wall is 15 ft. by 10 ft. and has a 4 ft. by 3 ft. window. How many square feet must she be prepared to paint?

3. What is the volume of the object shown?

4. Simplify. $19 - 6 \cdot 3 + 2^3$

5. Simplify. $7\sqrt{64} - 2(5 + 3)$

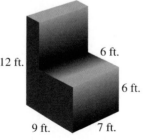

12 ft.   6 ft.

6 ft.

9 ft.   7 ft.

2.5 Order of Operations

OBJECTIVE

1 Simplify numerical expressions by following the order of operations agreement.

OBJECTIVE 1 *Simplify numerical expressions by following the order of operations agreement.*

The same order of operations agreement applies to all arithmetic. Of course, with integers it is important to remember the sign rules while following the proper order of operations. Also, in Section 2.1 we said that absolute value symbols act like parentheses, so we need to amend the order of operations to include absolute value symbols.

Order of Operations Agreement *Perform operations in the following order:*
1. Parentheses ( ), brackets [ ], braces { }, and absolute value | |
2. Exponents/Roots from left to right, in the order they occur.
3. Multiplication/Division from left to right, in the order they occur.
4. Addition/Subtraction from left to right, in the order they occur.

EXAMPLE 1   Simplify. $-8 + 6 \cdot (-2)$

Solution: $-8 + 6 \cdot (-2)$ **Multiply first.** $6(-2) = -12$
$\quad\quad\quad = -8 + (-12)$ **Add.** $-8 + (-12) = -20$
$\quad\quad\quad = -20$

> **Reminder**
> Write steps below the equal sign instead of outward to the right.

Connection
Compare Example 1 with the first example in Section 1.5, and you will see that it is essentially the same problem. The only difference is that some of the numbers are negative in Example 1.

Do Margin 1. ▶

> **Note** We divide before multiplying here because we multiply or divide from left to right in the order they occur.

EXAMPLE 2   Simplify. $-16 - 20 \div (-4) \cdot 6$

Solution: $-16 - 20 \div (-4) \cdot 6$ **Divide.** $20 \div (-4) = -5.$
$\quad\quad\quad = -16 - (-5) \cdot 6$ **Multiply.** $-5 \cdot 6 = -30.$
$\quad\quad\quad = -16 - (-30)$ **Write the subtraction as an equivalent addition.**

$\quad\quad\quad = -16 + 30$ **Add.** $-16 + 30 = 14.$
$\quad\quad\quad = 14$

Do Margin 2. ▶

◆ **Margin 1**
Simplify.
a. $-24 + (-9) \cdot 3$

b. $32 - 16 \div (-4)$

c. $-28 \div 4 + (-15)$

◆ **Margin 2**
Simplify.
a. $18 - 48 \div (-6) \cdot (-3)$

b. $20 + (-3) \cdot 4 - (-6)$

c. $-7 \cdot 5 - (-30) \div 5$

d. $12 \cdot (-4) \div (-6) + 9 - 27$

Answers to Margin 1: **a.** -51
b. 36 **c.** -22

Answers to Margin 2: **a.** -6 **b.** 14
c. -29 **d.** -10

Simplify.

a. $|-4 - (6)(-5)|$

b. $-|25 + (-34)|$

c. $-2|48 \div (-6)|$

EXAMPLE 3 Simplify. $-|-24 - 5(-2)|$ **Note** Absolute value symbols are treated like parentheses.

Solution: $-|-24 - 5(-2)|$ Multiply inside the absolute value symbols. $5(-2) = -10$

$= -|-24 - (-10)|$ Write the subtraction as an equivalent addition.

$= -|-24 + 10|$ Add. $-24 + 10 = -14$

$= -|-14|$ Find the absolute value of -14, which is 14. Then find the additive inverse of 14, which is -14. (See Example 17 in Section 2.1)

$= -14$

◀ **Do Margin 3.**

EXAMPLE 4 $8 - (-3)^2 + (-15) \div \sqrt{25}$

Solution: $8 - (-3)^2 + (-15) \div \sqrt{25}$ Evaluate the exponential and root expressions. $(-3)^2 = 9$ and $\sqrt{25} = 5$.

$= 8 - 9 + (-15) \div 5$ Divide. $-15 \div 5 = -3$

$= 8 - 9 + (-3)$ Subtract. $8 - 9 = -1$

$= -1 + (-3)$ Add. $-1 + (-3) = -4$

$= -4$

◀ **Do Margin 4.**

Simplify.

a. $14 + (-2)^3 - 8 \cdot (-3)$

b. $(-5) \cdot 6 \div (-3) - \sqrt{49}$

c. $3^3 - 4 \cdot (-8) + \sqrt{81} - |24 \div (-8)|$

EXAMPLE 5 Simplify $2(8 - 12) + (-2)^3 - [20 + (-2)] \div (-6)$

A number next to parentheses means multiply.

Solution: $2(8 - 12) + (-2)^3 - [20 + (-2)] \div (-6)$ Calculate within parentheses first. $8 - 12 = -4$

$= 2(-4) + (-2)^3 - 18 \div (-6)$ Evaluate the expression with the exponent. $(-2)^3 = -8$

$= 2(-4) + (-8) - 18 \div (-6)$ Multiply and divide, working from left to right. $2(-4) = -8$ and $18 \div (-6) = -3$

$= -8 + (-8) - (-3)$ Add. $-8 + (-8) = -16$

$= -16 - (-3)$ Write the subtraction as an equivalent addition.

$= -16 + 3$ Add. $-16 + 3 = -13$

$= -13$

Simplify.

a. $4(9 - 15) + 16 \div [2 + (-10)]$

b. $-20 + 6(3 - 8) - [8 + (-5)]^2$

c. $|19 - 29| \div 2 + [14 - 2(-6)] \div (-13)$

> **Reminder**
> To write a subtraction as an equivalent addition, change the operation from a minus sign to a plus sign and change the subtrahend to its additive inverse. In this case, think of $-16 - (-3)$ as removing a debt by making a deposit: so it becomes $-16 + 3$.

Answers to Margin 3: **a.** 26 **b.** -9 **c.** -16

Answers to Margin 4: **a.** 30 **b.** 3 **c.** 65

Answers to Margin 5: **a.** -26 **b.** -59 **c.** 3

◀ **Do Margin 5.**

EXAMPLE 6 Simplify $\{15 - 4[2 + (-5)]\} - 40 \div \sqrt{9 + 16}$

Solution: $\{15 - 4[2 + (-5)]\} - 40 \div \sqrt{9 + 16}$ Work inside the innermost parentheses first. $2 + (-5) = -3$

$= \{15 - 4[-3]\} - 40 \div \sqrt{9 + 16}$ Multiply within the bracket. $4 \cdot -3 = -12$

$= \{15 - [-12]\} - 40 \div \sqrt{9 + 16}$ Within the bracket, write subtraction as addition.

$= \{15 + 12\} - 40 \div \sqrt{9 + 16}$ Finish the bracket. $15 + 12 = 27$

$= 27 - 40 \div \sqrt{9 + 16}$ Add within the square root. $9 + 16 = 25$

$= 27 - 40 \div \sqrt{25}$ Evaluate the root. $\sqrt{25} = 5$

$= 27 - 40 \div 5$ Divide. $40 \div 5 = 8$

$= 27 - 8$ Subtract. $27 - 8 = 19$

$= 19$

Reminder

To find the square root of a sum or difference, we must add or subtract first, then get the square root of the sum or difference.

EXAMPLE 7 Simplify $\{(-4)^2 + 5[(9 - 15) \div (-2)]\} + (-3)\sqrt{16 \cdot 9}$

Solution: $\{(-4)^2 + 5[(9 - 15) \div (-2)]\} + (-3)\sqrt{16 \cdot 9}$ Work within the innermost parentheses first. $9 - 15 = -6$

$= \{(-4)^2 + 5[(-6) \div (-2)]\} + (-3)\sqrt{16 \cdot 9}$ Next, divide in the brackets. $(-6) \div (-2) = 3$

$= \{(-4)^2 + 5[3]\} + (-3)\sqrt{16 \cdot 9}$ Evaluate the exponential expression. $(-4)^2 = 16$

$= \{16 + 5[3]\} + (-3)\sqrt{16 \cdot 9}$ Multiply inside the braces. $5[3] = 15$

$= \{16 + 15\} + (-3)\sqrt{16 \cdot 9}$ Finish working in the braces by adding. $16 + 15 = 31$

$= 31 + (-3)\sqrt{16 \cdot 9}$ Multiply in the radical. $16 \cdot 9 = 144$

$= 31 + (-3)\sqrt{144}$ Evaluate the square root. $\sqrt{144} = 12$

$= 31 + (-3)12$ Multiply. $(-3)12 = -36$

$= 31 + (-36)$ Add. $31 + (-36) = -5$

$= -5$

Note We could have performed some of the calculations at the same time. For example, we could have computed $(-4)^2$, $5[3]$, and $16 \cdot 9$ all in the same step. However combining steps increases the likelihood of making careless mistakes.

Do Margin 6. ▶

◆ **Margin 6**

Simplify.

a. $\{-8 + 3[24 \div (4 + (-10))]\} + (-3)\sqrt{25 \cdot 4}$

b. $(-5)^3 - 4[(-6 + 14) \div (-7 + 6)] + \sqrt{81 \div 9}$

c. $\{[16 - 3(5 + 3)] + [26 - (3 - 9)^2]\} - 4\sqrt{25 - 9}$

d. $\{|-16(-3) - 40| + 7\} \div 3[-6 + \sqrt{121}]^2$

Answers: **a.** -50 **b.** -90 **c.** -34 **d.** 125

Simplify.

a. $\dfrac{(2-14)+4^3}{(-3)(-4)-5^2}$

b. $\dfrac{2\cdot(-4)-7^2}{9\cdot 5-3(3-7)^2}$

EXAMPLE 8 Simplify. $\dfrac{8^2-(-16)}{8-4(4-6)^3}$

> **Reminder**
> The fraction line means to divide. We must work the top (numerator) and bottom (denominator) separately in order to have numbers to divide.

Solution: $\dfrac{8^2-(-16)}{8-4(4-6)^3}$

Work the top and bottom separately. In the top: $8^2=64$. At the bottom: $4-6=-2$.

$=\dfrac{64-(-16)}{8-4(-2)^3}$

In the top: $64-(-16)=64+16$. At the bottom: $(-2)^3=-8$

$=\dfrac{64+16}{8-4(-8)}$

In the top: $64+16=80$. At the bottom: $4(-8)=-32$.

$=\dfrac{80}{8-(-32)}$

Write subtraction as equivalent addition. $8-(-32)=8+32$.

$=\dfrac{80}{8+32}$

Add. $8+32=40$.

$=\dfrac{80}{40}$

Finish by dividing: $80\div 40=2$.

$=2$

◀ **Do Margin 7.**

Answers: **a.** -4 **b.** 19

2.5 Exercises

FOR EXTRA HELP

 Videotape 1

 InterAct Math
Tutorial Software

 www.carsonmath.com

 AWL Math Tutor Center

 InterAct
MathXL www.mathxl.com

 Student's Solutions
Manual

For Exercises 1–58, simplify using order of operations.

1. $9 + 4 \cdot (-6)$

2. $15 - 5 \cdot 4$

3. $9 + 20 \div (-4) \cdot 3$

4. $-30 \div 2 \cdot (-3) + 21$

5. $9^2 - 6 \cdot (-4) \div 8$

6. $-14 - (-3)^2 \cdot 4 + 5$

7. $3 + (4 - 6^2)$

8. $3 + (4 - 6)^2$

9. $\sqrt{25 - 9}$

10. $\sqrt{25} - \sqrt{9}$

11. $(-2)^4$

12. -2^4

13. -2^5

14. $(-2)^5$

15. $-(-3)^2$

16. $[-(-3)]^2$

17. $|18 - 5(-4)|$

18. $|21 \div (-3) + 12|$

19. $-|-43 + 6 \cdot 4|$

20. $-|-16 - (-3)(-7)|$

21. $-5|26 \div 13 - 7 \cdot 2|$

22. $-4|3^2 + (-5)(6)|$

23. $28 \div (-7) + \sqrt{49} + (-3)^3$

24. $25 - 6^2 + (-52) \div \sqrt{169}$

25. $12 + 8\sqrt{81} \div (-6)$

26. $-5\sqrt{36} + 40 \div (-5) \cdot 2 - 14$

27. $4^2 - 7 \cdot 5 + \sqrt{121} - 21 \div (-7)$

28. $(-2)^3 + 3\sqrt{25} - 18 \div 3$

29. $15 + (-3)^3 + (-5)|14 + (-2)9|$

30. $-28 \div 4 + |-3(-2)(-5)| - 2^5$

31. $(-15 + 12) + 5(-6) - (9 - (-12)) \div 3$

32. $39 \div 3 + (24 - 30) - 5^2 + (-21 - (-13))$

33. $-|38 - 14 \cdot 2| + 44 \div (5 - (-6)) - 2^4$

34. $-3\sqrt{49} + |16 \div (-8)| - (28 - 40) \div (-3)$

35. $[19 - 2(4 + (-9))] - 18 \div \sqrt{25 - 16}$

36. $[12(-2 - (-5)) - 40] \div [17 + (-21)]$

37. $[(-14 + 2) + 5] \div 0 + 9^4 \cdot 7$

38. $[19(-2) - (-18)] \div [15 - 5(2 - (-1))]$

39. $\{(-3)^2 + 4[(8 - 20) \div (-2)]\} + (-5)\sqrt{25 \cdot 4}$

40. $4\sqrt{16 \cdot 9} - \{(-4)^3 + 2[18 \div (-2) + (4 - (-2))]\}$

41. $[30 - 3(-4)] \div [(15 - 6) + (-2)] + (5 - 3)^4$

42. $\{[13 - 3(1 + 6)] + [18 - (5 - 2)^2]\} - 6\sqrt{25 - 9}$

43. $\{|-12(-5) - 38| + 2\} \div 3[-9 + \sqrt{49}]^3$

44. $-3\{14 - 2|20 - 7(4)|\} + [(3)(-9) - (-21)]^2$

45. $\{2[14 - 11] + \sqrt{(4)(-9)}\} - [12 - 6 \cdot 9]$

46. $5^3 - [19 + 4\sqrt{16 - 25}] + |41 - 50|^2$

47. $\dfrac{34 - 5(2)}{7 + (9 - 4)}$

48. $\dfrac{20 + 12(-3)}{3^2 - 1}$

49. $\dfrac{(5 - 19) + 3^3}{(-2)(-6) - 5^2}$

50. $\dfrac{2^4 + 3(7 + 19)}{8^2 - (2 \cdot 8 + 1)}$

51. $\dfrac{3 \cdot (-6) - 6^2}{-2(2 - 5)^2}$

52. $\dfrac{(-2)^3 + 5(7 - 15)}{4 \cdot 11 - 2(5 - 9)^2}$

53. $\dfrac{2[5(3 - 7) + (-3)^3] - 6}{15 \cdot 5 - (12 - 7)^2}$

54. $\dfrac{[6(12 - 15) + 5(-6)] - 4^2}{10 - 2(6 - 9)^2}$

55. $\dfrac{3[14 - 2(9)] + (5 - 11)^2}{(9 - 3)^2 - 4(10 - 1)}$

56. $\dfrac{-3\{[2 - 7(3 - 4)] + 4^2\}}{5[28 + 4(2 - 9)]^3}$

57. $\dfrac{-|14 - 3^2|^2}{3(-12) + 11}$

58. $\dfrac{-84 + [4(16 - 21) + 4]}{\{41 - |12 - 15|^3\} + 6}$

59. Explain the difference between $(-2)^4$ and -2^4.

60. Explain the difference between $-(-3)^2$ and $[-(3)]^2$.

61. Why do $(-2)^5$ and -2^5 both simplify to -32?

62. Why is the answer to Exercise 37 undefined?

For Exercises 63–70, explain the mistake in each problem, then work it correctly.

63. $28 - 5(24 - 30)$

$= 28 - 5(-6)$

$= 23(-6)$

$= -138$

64. $24 \div 2 \cdot 3 + 5$

$= 24 \div 6 + 5$

$= 4 + 5$

$= 9$

65. $4 - (9 - 4)^2$

$= 4 - (9 - 16)$

$= 4 - (-7)$

$= 4 + 7$

$= 11$

66. $19 - (-2)^5$

$= 19 - 32$

$= -13$

67. $34 - [3 \cdot 5 - (14 + 8)]$

$= 34 - 3 \cdot 5 - 22$

$= 34 - 15 - 22$

$= 19 - 22$

$= -3$

68. $\dfrac{3^2 - (25 - 4^2)}{2(-7)}$

$= \dfrac{3^2 - (25 - 16)}{-14}$

$= \dfrac{3^2 - 9}{-14}$

$= \dfrac{9 - 9}{-14}$

$= \dfrac{0}{-14}$

Answer is undefined

69. $\sqrt{169 - 25}$

$= 13 - 5$

$= 8$

70. $-2[16(3 - 5) + 7]$

$= -2[16(-2) + 7]$

$= -2[32 + 7]$

$= -2[39]$

$= -78$

REVIEW EXERCISES

1. Felicia has a balance of $185 in her checking account. If she writes checks for $45, $68, and $95, what is her new balance?

2. Tina has a television that she bought for $245. She sells it at a yard sale for $32. What was the net? Was it a profit or loss?

3. A submarine at -98 ft. ascends to -25 ft. How much did the submarine ascend?

4. Jeff has 3 credit accounts each with a balance of $-$235. What is his total debt?

5. A hiker finds that his backpack has a mass of 25 kg. What is the weight of the pack?

2.6 Applications and Problem Solving

OBJECTIVES

1 Solve problems involving net.

2 Solve problems involving income taxes.

3 Solve problems involving force.

4 Solve problems involving average rate.

OBJECTIVE 1 *Solve problems involving net.*

Recall that the net profit or loss is calculated using the formula $N = R - C$, where N represents the net amount, R represents revenue, and C represents cost.

EXAMPLE 1 Cassif puts $2000 down on a car, then makes monthly payments of $275 for 5 yr. (60 mo.). After 6 yr. he sells the car for $5200. If he spent a total of $1500 in routine maintenance and minor repairs, what is his net? Is it a profit or loss?

Solution: Follow the problem-solving process.

Understand: We must calculate a net amount. To calculate the net we need total revenue and total cost. The formula for net is $N = R - C$.

Cassif's total revenue is the amount that he sold the car for, which was given to be $5200.

His total cost is the sum of all the money he spent on the car.

Plan: Calculate total cost, then subtract the cost from the total revenue to get the net.

Execute: Cost = Amount down + Total of all payments + Maintenance costs

60 payments of $275.

$$
\begin{aligned}
\text{Cost} &= 2000 + 60(275) + 1500 \\
\text{Cost} &= 2000 + 16{,}500 + 1500 \\
\text{Cost} &= \$20{,}000
\end{aligned}
$$

Now replace R with 5200 and C with 20,000 in $N = R - C$ to get the net.

$$
\begin{aligned}
N &= R - C \\
N &= 5200 - 20{,}000 \\
N &= -14{,}800
\end{aligned}
$$

Answer: Cassif's net is −$14,800, which is a loss of $14,800.

Check: Reverse the process. The net added to cost should produce the revenue.

$$
\begin{aligned}
-14{,}800 + 20{,}000 &\ ?\ 5200 \\
5200 &= 5200 \qquad \textbf{The net amount checks.}
\end{aligned}
$$

Subtracting the maintenance costs and total of all payments from the total cost should produce the down payment.

$$
\begin{aligned}
20{,}000 - 1500 - 16{,}500 &\ ?\ 2000 \\
2000 &= 2000 \qquad \textbf{The total cost checks.}
\end{aligned}
$$

Dividing the total of all the payments by 60 should produce the amount of each payment.

$$
\begin{aligned}
16{,}500 \div 60 &\ ?\ 275 \\
275 &= 275 \qquad \textbf{The total payments check.}
\end{aligned}
$$

Do Margin 1. ▶

Reminder

We developed the problem-solving process in Section 1.7.

◆ **Margin 1**

Shawna put $1200 down when she bought her car. She made 48 payments of $350 and spent $2500 in maintenance and repairs. Four years after paying off the car she sells it for $4400. What is her net? Is it a profit or loss?

Answer : −$16,100 (loss)

OBJECTIVE 2 *Solve problems involving income taxes.*

Another situation involving finances is the calculation of income taxes. The overall idea of figuring income taxes is the same as the asset and debt type problems we worked previously. The difference is that with income tax we figure total income and total **deductions.**

DEFINITION	**Deduction:** An amount that is subtracted from income.

Some common itemized deductions: Mortgage interest, medical and dental expenses, property taxes, gifts to charity, unreimbursed job related expenses, etc. If the total itemized deductions are less than the standard deduction for our filing status, then we should take the standard deduction. Table 2.1 lists the standard deductions for 1999.

Table 2.1   Standard Deduction for Most People

Single = $4250
Head of household = $6250
Married filing jointly
or Qualifying widow(er) = $7100
Married filing separately = $3550

We also get a deduction based on the number of tax exemptions we can claim. In general, we get one exemption for each of the following that applies:

1 for yourself (unless someone else claims you)

1 for your spouse

1 for each dependent (children)

A married couple with one child would get 3 exemptions. We calculate the deduction from exemptions by multiplying the total number of exemptions by $2700. So a family with 3 exemptions would get a deduction of $3 \cdot 2700 = \$8100$.

We then subtract our total deductions from the total income to get our taxable income. Once we have the taxable income we look it up in a table to see how much tax we owe.

Next, we subtract any credit amounts such as child care, child tax credit, education, and so on, from the taxes we owe. (Because tax credit calculations can get complicated, we'll only consider the simplest case of child credit whereby a family gets a credit of $400 for each qualifying child. We'll assume in our examples and problems that all children qualify.)

Finally, we compare the amount that was deducted from our pay checks throughout the year with what we actually owe, and if we owe more than we've already paid, we must pay the difference. If we paid more than we owe, we get a refund.

Process *To figure income tax:*
1. Calculate total income and total deductions.
2. Calculate the exemption amount by multiplying the number of exemptions claimed by $2700.
3. Calculate the taxable income by subtracting the total deductions from the total income.
4. Find the tax by looking up the taxable income in the provided tax tables.
5. Subtract any tax credits that apply to get the final tax amount.
6. Calculate the total tax paid throughout the year using the W-2 form.
7. Subtract the tax amount from the tax that was paid throughout the year.
8. If the result is negative, then that amount must be paid to the government. If the result is positive, then it is the amount the government must refund.

EXAMPLE 2 Emily is single, renting an apartment, and working two jobs. The following spreadsheet shows her income and deductions. What is her taxable income, tax, and amount she must pay or receive as a refund?

Annual Income	Deductions
Job 1 = $18,450	Medical = $267
Job 2 = $7600	Taxes = $1054
Interest = $34	Charitable donations = $120
Taxes paid: Job 1 W-2 = $2067	
Job 2 W-2 = $645	

Solution: Follow the problem-solving process.

Understand: We must find Emily's taxable income, tax, and amount she must pay or receive as a refund. We can follow the outline for figuring tax.

Plan: Follow the process for figuring tax.

Execute: **1.** Calculate total income and total itemized deductions.

$$\text{Total income} = \$18{,}450 + \$7600 + \$34 = \$26{,}084$$

$$\text{Total itemized deductions} = \$267 + \$1054 + \$120 = \$1441$$

From Table 2.1 we see that the standard deduction for a single person is $4250. Since Emily's total deductions are only $1441 and do not exceed the $4250 standard deduction, she will take the standard deduction.

2. Calculate the exemption amount by multiplying the number of exemptions claimed by $2700.

Because Emily is single and has no dependents, she can claim only one exemption.

$$\text{Deduction from exemptions} = 1 \cdot \$2700 = \$2700$$

3. Calculate the taxable income by subtracting the total deductions from the total income.

$$\text{Taxable Income} = \$26{,}084 - \$4250 - \$2700 = \$19{,}134$$

4. Find the tax by looking up the taxable income in Table 2.2.

In Table 2.2 we see Emily's taxable income of $19,134 is between $19,100 and $19,150, and she is single, so her tax is $2869.

Tax = $2869

Table 2.2

Taxable income:		And you are			
At least	But less than	Single	Married filing jointly	Married filing separately	Head of a household
			Your tax is		
19,050	19,100	2861	2861	2861	2861
19,100	19,150	**2869**	2869	2869	2869
19,150	19,200	2876	2876	2876	2876

5. Subtract any tax credits that apply to get the final tax amount.

She receives no tax credits.

6. Calculate the total tax paid throughout the year from the W-2.

$$\text{Total taxes paid} = \text{Job 1} + \text{Job 2}$$

$$\text{Total taxes paid} = \$2067 + \$645 = \$2712$$

7. Subtract the tax amount from the tax that was paid throughout the year. (On the W-2 form)

$$\text{Tax paid} - \text{Tax amount}$$

$$\$2712 - \$2869 = -\$157$$

8. If the result is negative, that amount must be paid to the government. If the result is positive, it is the amount the government must refund.

Because the result is negative, Emily owes the government $157.

Annual Income

Job = $27,450
Interest = $56

Deductions

Medical = $186
Taxes = $1196
Charitable donations = $450

Taxes paid (from W-2 form) = $2548

Answers: Taxable income = $19,134
Tax = $2869
Amount to pay = $157

◄ **Do Margin 2.**

EXAMPLE 3 The spreadsheet below is for the Alder family. Their filing status is married filing jointly with two children, so they get 4 exemptions. Find their taxable income, tax, and the amount they owe or receive in refund.

Solution: Follow the problem-solving process.

Understand: We must find taxable income, tax, and the amount owed or received as refund.

Plan: Follow the process outline for figuring tax.

Execute: **1.** Find the total income and total itemized deductions.

Annual Income	**Deductions**
Mr. Alder = $24,500	Mortgage Interest = $6432
Mrs. Alder = $26,200	Medical = $245
Interest = $110	Taxes = $1721
Other income = $1340	Gifts to charity = $254
Taxes paid (from W–2 forms):	Mr. Alder = $2512
	Mrs. Alder = $2845

Total income = $24,500 + $26,200 + $110 + $1340 = $52,150

Total itemized deductions = $6432 + $245 + $1721 + $254 = $8652

From Table 2.1 we see the standard deduction for married filing jointly is $7100. Because the Alder's total itemized deductions are more than the standard deduction, they should use their itemized deductions.

2. Calculate the exemption amount by multiplying the number of exemptions claimed by $2700.

Exemptions = 4 (Mr. and Mrs. Alder plus two dependents)

Deduction from exemptions = 4 · 2700 = $10,800

3. Calculate the taxable income by subtracting the total deductions from the total income.

Taxable income = $52,150 − $8652 − $10,800 = $32,698

4. Find the tax by looking up the taxable income in Table 2.3.

Table 2.3

Taxable income:		And you are			
At least	But less than	Single	Married filing jointly	Married filing separately	Head of a household
		Your tax is			
32,600	32,650	5840	4894	6382	4894
32,650	32,700	5894	**4901**	6396	4901
32,700	32,750	5868	4909	6410	4909

In Table 2.3 we see the Alder's taxable income of $32,698 is between $32,650 and $32,700. Because their filing status is married filing jointly, their tax is $4901.

Tax = $4901

5. Subtract any tax credits that apply to get the final tax amount.

Because Mr. and Mrs Alder have two children, they get a tax credit of $400 for each child.

Child tax credit = 2 · $400 = $800

Subtract this credit from the tax:

Tax amount = $4901 − $800 = $4101

6. Calculate the total tax paid throughout the year from the W-2 form.

$$\text{Tax paid} = \text{Mr. Alder} + \text{Mrs. Alder}$$
$$\text{Tax paid} = \$2512 + \$2845 = \$5357$$

7. Subtract the tax amount from the tax that was paid throughout the year.

$$\text{Tax paid} - \text{Tax amount}$$
$$\$5357 - \$4101 = \$1256$$

8. If the result is negative, that amount must be paid to the government. If the result is positive, the government must refund that amount.

Because $1256 is positive, the Alder family will receive that amount as a refund.

Answer: Taxable income = $32,698

Tax = $4101

Refund = $1256

Do Margin 3. ▶

OBJECTIVE 3 *Solve problems involving force.*

We have discussed several relationships that involve multiplying two numbers. We have considered $F = ma$, $V = ir$, and $A = lw$. Let's consider some more involved problems that use these formulas.

EXAMPLE 4 An elevator has a mass of 38 slugs and holds 4 people with the following masses: 4 slugs, 5 slugs, 6 slugs, and 3 slugs. What is the total force exerted on the cable?

Solution: Use the problem-solving process.

Understand: We must calculate the total force on the cable exerted by the elevator and its occupants. The formula for force is $F = ma$.

To calculate the downward force or weight, we need the total mass of the elevator and all its occupants. The acceleration due to gravity is -32.

Plan: Find the total mass of the elevator and all its occupants, then use $F = ma$ to calculate the force.

Execute: Total mass = Elevator + All occupants

Total mass = $38 + 4 + 5 + 6 + 3$

Total mass = 56 slugs

To find the force or weight, replace m with 56 and a with -32 in $F = ma$, then multiply.

$$F = ma$$
$$F = (56)(-32)$$
$$F = -1792 \text{ lb.}$$

Answer: The force is -1792 lb.

Check: Reverse the process. The force divided by acceleration due to gravity should equal total mass.

$$-1792 \div -32 \; ? \; 56$$
$$56 = 56 \qquad \text{The force checks.}$$

Subtracting each person's mass from the total mass should equal the mass of the elevator.

$$56 - 3 - 6 - 5 - 4 \; ? \; 38$$
$$38 = 38 \qquad \text{The total mass checks.}$$

Do Margin 4. ▶

OBJECTIVE 4 *Solve problems involving average rate.*

Another situation that has the same mathematical relationship as $F = ma$, $V = ir$, and $A = bh$ is the formula that describes the relationship among distance, rate, and time.

If we drive a car 60 mph for 2 hr. how far do we travel?

◆ **Margin 3**

The following spreadsheet is for the Jones family. Their filing status is married filing jointly with one child, so they get 3 exemptions. Find their taxable income, tax, and the amount they owe or receive in a refund. (Use Table 2.4 at the end of this section.)

Annual Income
Mr. Jones = $24,000
Mrs. Jones = $23,500
Interest = $240
Other income = $1840

Deductions
Mortgage Interest = $7254
Medical = $540
Taxes = $2300
Gifts to charity = $500
Taxes paid (from W−2 forms): Mr. Jones = $2512 Mrs. Jones = $2480

◆ **Margin 4**

A car has a mass of 1080 kg. Two passengers are in the car with masses of 68 kg and 80 kg. What is the total weight of the car with the passengers?

Answers to Margin 3: Taxable income = $30,886; tax = $4231; refund = $761

Answer to Margin 4: −12,280 N

60 mph means we travel 60 mi. for each hour we travel so in 2 hrs., we'll go 120 mi. Notice to get distance, we multiply the rate by the time. In formula form we can say:

$$\text{distance} = \text{rate} \cdot \text{time}$$
$$d = rt$$

Consider the fact that in reality we cannot drive a car at an exact speed for any lengthy period of time. Our rate will vary slightly as a result of hills, wind resistance, or even our inability to keep our foot pressure exactly the same on the accelerator. Because rates can vary we usually consider **average rate.** It is a measure of how quickly an object is able to travel a total distance in a total amount of time.

> **DEFINITION** **Average rate:** A measure of the rate at which an object travels a total distance in a total amount of time.

EXAMPLE 5 A commuter train travels at an average rate of 30 mph for 2 hr. How far does the train travel?

Solution: Follow the problem-solving process.

Understand: The rate was given to be 30 mph. The time was given to be 2 hr. The formula that relates distance, rate, and time is $d = rt$.

Plan: Replace r with 30 mph and t with 2 hr. then solve for d.

Execute: $d = rt$
$\quad\quad\quad d = (30)(2)$
$\quad\quad\quad d = 60$ mi.

Answer: The train travels 60 mi.

Check: Reverse the process. A train covering 60 mi. in 2 hr. should have an average velocity of 30 mph.

$60 \div 2 \ ? \ 30$
$30 = 30$ **It checks.**

◀ **Do Margin 5.**

EXAMPLE 6 A bus leaves at 9 A.M. and travels 40 mi. then stops for an hour. It then travels 110 mi. and arrives at its final destination at 1 P.M. Find the average rate.

Solution: Follow the problem-solving process.

Understand: To find average rate we must consider the total distance of the trip and the total time. The trip was broken into two parts. We were given the distances for both parts. We were also given a departure time and arrival time, but there was a one-hour stop that we must take into account. The formula that relates distance, rate, and time is $d = rt$.

Plan: Find the total distance of the trip and total time of the trip, then use the formula $d = rt$ to solve for r.

Execute: Total distance: $D_{total} = 40$ mi. $+ 110$ mi. $= 150$ mi.

Total time: From 9 A.M. to 1 P.M. is 4 hr. However, the bus stopped for an hour so the actual travel time was 3 hr.

Because we now have the total distance of 150 mi. and total time of 3 hr., we can use $d = rt$.

$$d = rt$$
$$150 \text{ mi.} = r(3 \text{ hr.}) \quad \text{Solve for the missing factor by dividing.}$$
$$150 \text{ mi.} \div 3 \text{ hr.} = r$$
$$50 \text{ mph} = r$$

Answer: The average rate was 50 mph.

Check: Reverse the process. If the bus traveled 50 mph for 3 hr. would it then travel 150 mi.?

$$d = rt$$
$$d = (50 \text{ mph})(3 \text{hr.})$$
$$d = 150 \text{ mi.} \quad \text{It checks.}$$

◀ **Do Margin 6.**

Margin 5

A research submarine is lowered at an average speed of 8 ft. per sec. What will be the submarine's depth after 45 sec.?

Margin 6

On a vacation trip, Candice leaves at 10 A.M. and travels 85 mi. then takes a 30-min. break. She then travels another 120 mi. and stops for another 30-min. break. Finally, she travels 70 mi. and arrives at her destination at 4 P.M. What was her average rate?

Answer to Margin 5: 360 ft.

Answer to Margin 6: 55 mph

Table 2.4

1998 Tax Table—Continued

Column Group 1

If line 39 (taxable income) is— At least	But less than	Single	Married filing jointly *	Married filing separately	Head of a household
14,000					
14,000	14,050	2,104	2,104	2,104	2,104
14,050	14,100	2,111	2,111	2,111	2,111
14,100	14,150	2,119	2,119	2,119	2,119
14,150	14,200	2,126	2,126	2,126	2,126
14,200	14,250	2,134	2,134	2,134	2,134
14,250	14,300	2,141	2,141	2,141	2,141
14,300	14,350	2,149	2,149	2,149	2,149
14,350	14,400	2,156	2,156	2,156	2,156
14,400	14,450	2,164	2,164	2,164	2,164
14,450	14,500	2,171	2,171	2,171	2,171
14,500	14,550	2,179	2,179	2,179	2,179
14,550	14,600	2,186	2,186	2,186	2,186
14,600	14,650	2,194	2,194	2,194	2,194
14,650	14,700	2,201	2,201	2,201	2,201
14,700	14,750	2,209	2,209	2,209	2,209
14,750	14,800	2,216	2,216	2,216	2,216
14,800	14,850	2,224	2,224	2,224	2,224
14,850	14,900	2,231	2,231	2,231	2,231
14,900	14,950	2,239	2,239	2,239	2,239
14,950	15,000	2,246	2,246	2,246	2,246
15,000					
15,000	15,050	2,254	2,254	2,254	2,254
15,050	15,100	2,261	2,261	2,261	2,261
15,100	15,150	2,269	2,269	2,269	2,269
15,150	15,200	2,276	2,276	2,276	2,276
15,200	15,250	2,284	2,284	2,284	2,284
15,250	15,300	2,291	2,291	2,291	2,291
15,300	15,350	2,299	2,299	2,299	2,299
15,350	15,400	2,306	2,306	2,306	2,306
15,400	15,450	2,314	2,314	2,314	2,314
15,450	15,500	2,321	2,321	2,321	2,321
15,500	15,550	2,329	2,329	2,329	2,329
15,550	15,600	2,336	2,336	2,336	2,336
15,600	15,650	2,344	2,344	2,344	2,344
15,650	15,700	2,351	2,351	2,351	2,351
15,700	15,750	2,359	2,359	2,359	2,359
15,750	15,800	2,366	2,366	2,366	2,366
15,800	15,850	2,374	2,374	2,374	2,374
15,850	15,900	2,381	2,381	2,381	2,381
15,900	15,950	2,389	2,389	2,389	2,389
15,950	16,000	2,396	2,396	2,396	2,396
16,000					
16,000	16,050	2,404	2,404	2,404	2,404
16,050	16,100	2,411	2,411	2,411	2,411
16,100	16,150	2,419	2,419	2,419	2,419
16,150	16,200	2,426	2,426	2,426	2,426
16,200	16,250	2,434	2,434	2,434	2,434
16,250	16,300	2,441	2,441	2,441	2,441
16,300	16,350	2,449	2,449	2,449	2,449
16,350	16,400	2,456	2,456	2,456	2,456
16,400	16,450	2,464	2,464	2,464	2,464
16,450	16,500	2,471	2,471	2,471	2,471
16,500	16,550	2,479	2,479	2,479	2,479
16,550	16,600	2,486	2,486	2,486	2,486
16,600	16,650	2,494	2,494	2,494	2,494
16,650	16,700	2,501	2,501	2,501	2,501
16,700	16,750	2,509	2,509	2,509	2,509
16,750	16,800	2,516	2,516	2,516	2,516
16,800	16,850	2,524	2,524	2,524	2,524
16,850	16,900	2,531	2,531	2,531	2,531
16,900	16,950	2,539	2,539	2,539	2,539
16,950	17,000	2,546	2,546	2,546	2,546

Column Group 2

If line 39 (taxable income) is— At least	But less than	Single	Married filing jointly *	Married filing separately	Head of a household
17,000					
17,000	17,050	2,554	2,554	2,554	2,554
17,050	17,100	2,561	2,561	2,561	2,561
17,100	17,150	2,569	2,569	2,569	2,569
17,150	17,200	2,576	2,576	2,576	2,576
17,200	17,250	2,584	2,584	2,584	2,584
17,250	17,300	2,591	2,591	2,591	2,591
17,300	17,350	2,599	2,599	2,599	2,599
17,350	17,400	2,606	2,606	2,606	2,606
17,400	17,450	2,614	2,614	2,614	2,614
17,450	17,500	2,621	2,621	2,621	2,621
17,500	17,550	2,629	2,629	2,629	2,629
17,550	17,600	2,636	2,636	2,636	2,636
17,600	17,650	2,644	2,644	2,644	2,644
17,650	17,700	2,651	2,651	2,651	2,651
17,700	17,750	2,659	2,659	2,659	2,659
17,750	17,800	2,666	2,666	2,666	2,666
17,800	17,850	2,674	2,674	2,674	2,674
17,850	17,900	2,681	2,681	2,681	2,681
17,900	17,950	2,689	2,689	2,689	2,689
17,950	18,000	2,696	2,696	2,696	2,696
18,000					
18,000	18,050	2,704	2,704	2,704	2,704
18,050	18,100	2,711	2,711	2,711	2,711
18,100	18,150	2,719	2,719	2,719	2,719
18,150	18,200	2,726	2,726	2,726	2,726
18,200	18,250	2,734	2,734	2,734	2,734
18,250	18,300	2,741	2,741	2,741	2,741
18,300	18,350	2,749	2,749	2,749	2,749
18,350	18,400	2,756	2,756	2,756	2,756
18,400	18,450	2,764	2,764	2,764	2,764
18,450	18,500	2,771	2,771	2,771	2,771
18,500	18,550	2,779	2,779	2,779	2,779
18,550	18,600	2,786	2,786	2,786	2,786
18,600	18,650	2,794	2,794	2,794	2,794
18,650	18,700	2,801	2,801	2,801	2,801
18,700	18,750	2,809	2,809	2,809	2,809
18,750	18,800	2,816	2,816	2,816	2,816
18,800	18,850	2,824	2,824	2,824	2,824
18,850	18,900	2,831	2,831	2,831	2,831
18,900	18,950	2,839	2,839	2,839	2,839
18,950	19,000	2,846	2,846	2,846	2,846
19,000					
19,000	19,050	2,854	2,854	2,854	2,854
19,050	19,100	2,861	2,861	2,861	2,861
19,100	19,150	2,869	2,869	2,869	2,869
19,150	19,200	2,876	2,876	2,876	2,876
19,200	19,250	2,884	2,884	2,884	2,884
19,250	19,300	2,891	2,891	2,891	2,891
19,300	19,350	2,899	2,899	2,899	2,899
19,350	19,400	2,906	2,906	2,906	2,906
19,400	19,450	2,914	2,914	2,914	2,914
19,450	19,500	2,921	2,921	2,921	2,921
19,500	19,550	2,929	2,929	2,929	2,929
19,550	19,600	2,936	2,936	2,936	2,936
19,600	19,650	2,944	2,944	2,944	2,944
19,650	19,700	2,951	2,951	2,951	2,951
19,700	19,750	2,959	2,959	2,959	2,959
19,750	19,800	2,966	2,966	2,966	2,966
19,800	19,850	2,974	2,974	2,974	2,974
19,850	19,900	2,981	2,981	2,981	2,981
19,900	19,950	2,989	2,989	2,989	2,989
19,950	20,000	2,996	2,996	2,996	2,996

Column Group 3

If line 39 (taxable income) is— At least	But less than	Single	Married filing jointly *	Married filing separately	Head of a household
20,000					
20,000	20,050	3,004	3,004	3,004	3,004
20,050	20,100	3,011	3,011	3,011	3,011
20,100	20,150	3,019	3,019	3,019	3,019
20,150	20,200	3,026	3,026	3,026	3,026
20,200	20,250	3,034	3,034	3,034	3,034
20,250	20,300	3,041	3,041	3,041	3,041
20,300	20,350	3,049	3,049	3,049	3,049
20,350	20,400	3,056	3,056	3,056	3,056
20,400	20,450	3,064	3,064	3,064	3,064
20,450	20,500	3,071	3,071	3,071	3,071
20,500	20,550	3,079	3,079	3,079	3,079
20,550	20,600	3,086	3,086	3,086	3,086
20,600	20,650	3,094	3,094	3,094	3,094
20,650	20,700	3,101	3,101	3,101	3,101
20,700	20,750	3,109	3,109	3,109	3,109
20,750	20,800	3,116	3,116	3,116	3,116
20,800	20,850	3,124	3,124	3,124	3,124
20,850	20,900	3,131	3,131	3,131	3,131
20,900	20,950	3,139	3,139	3,139	3,139
20,950	21,000	3,146	3,146	3,146	3,146
21,000					
21,000	21,050	3,154	3,154	3,154	3,154
21,050	21,100	3,161	3,161	3,161	3,161
21,100	21,150	3,169	3,169	3,169	3,169
21,150	21,200	3,176	3,176	3,176	3,176
21,200	21,250	3,184	3,184	3,190	3,184
21,250	21,300	3,191	3,191	3,204	3,191
21,300	21,350	3,199	3,199	3,218	3,199
21,350	21,400	3,206	3,206	3,232	3,206
21,400	21,450	3,214	3,214	3,246	3,214
21,450	21,500	3,221	3,221	3,260	3,221
21,500	21,550	3,229	3,229	3,274	3,229
21,550	21,600	3,236	3,236	3,288	3,236
21,600	21,650	3,244	3,244	3,302	3,244
21,650	21,700	3,251	3,251	3,316	3,251
21,700	21,750	3,259	3,259	3,330	3,259
21,750	21,800	3,266	3,266	3,344	3,266
21,800	21,850	3,274	3,274	3,358	3,274
21,850	21,900	3,281	3,281	3,372	3,281
21,900	21,950	3,289	3,289	3,386	3,289
21,950	22,000	3,296	3,296	3,400	3,296
22,000					
22,000	22,050	3,304	3,304	3,414	3,304
22,050	22,100	3,311	3,311	3,428	3,311
22,100	22,150	3,319	3,319	3,442	3,319
22,150	22,200	3,326	3,326	3,456	3,326
22,200	22,250	3,334	3,334	3,470	3,334
22,250	22,300	3,341	3,341	3,484	3,341
22,300	22,350	3,349	3,349	3,498	3,349
22,350	22,400	3,356	3,356	3,512	3,356
22,400	22,450	3,364	3,364	3,526	3,364
22,450	22,500	3,371	3,371	3,540	3,371
22,500	22,550	3,379	3,379	3,554	3,379
22,550	22,600	3,386	3,386	3,568	3,386
22,600	22,650	3,394	3,394	3,582	3,394
22,650	22,700	3,401	3,401	3,596	3,401
22,700	22,750	3,409	3,409	3,610	3,409
22,750	22,800	3,416	3,416	3,624	3,416
22,800	22,850	3,424	3,424	3,638	3,424
22,850	22,900	3,431	3,431	3,652	3,431
22,900	22,950	3,439	3,439	3,666	3,439
22,950	23,000	3,446	3,446	3,680	3,446

* This column must also be used by a qualifying widow(er).

Continued on next page

Table 2.4 continues

Table 2.4 (continued)

1998 Tax Table—Continued

23,000 / 24,000 / 25,000

At least	But less than	Single	Married filing jointly *	Married filing separately	Head of a household
23,000					
23,000	23,050	3,454	3,454	3,694	3,454
23,050	23,100	3,461	3,461	3,708	3,461
23,100	23,150	3,469	3,469	3,722	3,469
23,150	23,200	3,476	3,476	3,736	3,476
23,200	23,250	3,484	3,484	3,750	3,484
23,250	23,300	3,491	3,491	3,764	3,491
23,300	23,350	3,499	3,499	3,778	3,499
23,350	23,400	3,506	3,506	3,792	3,506
23,400	23,450	3,514	3,514	3,806	3,514
23,450	23,500	3,521	3,521	3,820	3,521
23,500	23,550	3,529	3,529	3,834	3,529
23,550	23,600	3,536	3,536	3,848	3,536
23,600	23,650	3,544	3,544	3,862	3,544
23,650	23,700	3,551	3,551	3,876	3,551
23,700	23,750	3,559	3,559	3,890	3,559
23,750	23,800	3,566	3,566	3,904	3,566
23,800	23,850	3,574	3,574	3,918	3,574
23,850	23,900	3,581	3,581	3,932	3,581
23,900	23,950	3,589	3,589	3,946	3,589
23,950	24,000	3,596	3,596	3,960	3,596
24,000					
24,000	24,050	3,604	3,604	3,974	3,604
24,050	24,100	3,611	3,611	3,988	3,611
24,100	24,150	3,619	3,619	4,002	3,619
24,150	24,200	3,626	3,626	4,016	3,626
24,200	24,250	3,634	3,634	4,030	3,634
24,250	24,300	3,641	3,641	4,044	3,641
24,300	24,350	3,649	3,649	4,058	3,649
24,350	24,400	3,656	3,656	4,072	3,656
24,400	24,450	3,664	3,664	4,086	3,664
24,450	24,500	3,671	3,671	4,100	3,671
24,500	24,550	3,679	3,679	4,114	3,679
24,550	24,600	3,686	3,686	4,128	3,686
24,600	24,650	3,694	3,694	4,142	3,694
24,650	24,700	3,701	3,701	4,156	3,701
24,700	24,750	3,709	3,709	4,170	3,709
24,750	24,800	3,716	3,716	4,184	3,716
24,800	24,850	3,724	3,724	4,198	3,724
24,850	24,900	3,731	3,731	4,212	3,731
24,900	24,950	3,739	3,739	4,226	3,739
24,950	25,000	3,746	3,746	4,240	3,746
25,000					
25,000	25,050	3,754	3,754	4,254	3,754
25,050	25,100	3,761	3,761	4,268	3,761
25,100	25,150	3,769	3,769	4,282	3,769
25,150	25,200	3,776	3,776	4,296	3,776
25,200	25,250	3,784	3,784	4,310	3,784
25,250	25,300	3,791	3,791	4,324	3,791
25,300	25,350	3,799	3,799	4,338	3,799
25,350	25,400	3,810	3,806	4,352	3,806
25,400	25,450	3,824	3,814	4,366	3,814
25,450	25,500	3,838	3,821	4,380	3,821
25,500	25,550	3,852	3,829	4,394	3,829
25,550	25,600	3,866	3,836	4,408	3,836
25,600	25,650	3,880	3,844	4,422	3,844
25,650	25,700	3,894	3,851	4,436	3,851
25,700	25,750	3,908	3,859	4,450	3,859
25,750	25,800	3,922	3,866	4,464	3,866
25,800	25,850	3,936	3,874	4,478	3,874
25,850	25,900	3,950	3,881	4,492	3,881
25,900	25,950	3,964	3,889	4,506	3,889
25,950	26,000	3,978	3,896	4,520	3,896

26,000 / 27,000 / 28,000

At least	But less than	Single	Married filing jointly *	Married filing separately	Head of a household
26,000					
26,000	26,050	3,992	3,904	4,534	3,904
26,050	26,100	4,006	3,911	4,548	3,911
26,100	26,150	4,020	3,919	4,562	3,919
26,150	26,200	4,034	3,926	4,576	3,926
26,200	26,250	4,048	3,934	4,590	3,934
26,250	26,300	4,062	3,941	4,604	3,941
26,300	26,350	4,076	3,949	4,618	3,949
26,350	26,400	4,090	3,956	4,632	3,956
26,400	26,450	4,104	3,964	4,646	3,964
26,450	26,500	4,118	3,971	4,660	3,971
26,500	26,550	4,132	3,979	4,674	3,979
26,550	26,600	4,146	3,986	4,688	3,986
26,600	26,650	4,160	3,994	4,702	3,994
26,650	26,700	4,174	4,001	4,716	4,001
26,700	26,750	4,188	4,009	4,730	4,009
26,750	26,800	4,202	4,016	4,744	4,016
26,800	26,850	4,216	4,024	4,758	4,024
26,850	26,900	4,230	4,031	4,772	4,031
26,900	26,950	4,244	4,039	4,786	4,039
26,950	27,000	4,258	4,046	4,800	4,046
27,000					
27,000	27,050	4,272	4,054	4,814	4,054
27,050	27,100	4,286	4,061	4,828	4,061
27,100	27,150	4,300	4,069	4,842	4,069
27,150	27,200	4,314	4,076	4,856	4,076
27,200	27,250	4,328	4,084	4,870	4,084
27,250	27,300	4,342	4,091	4,884	4,091
27,300	27,350	4,356	4,099	4,898	4,099
27,350	27,400	4,370	4,106	4,912	4,106
27,400	27,450	4,384	4,114	4,926	4,114
27,450	27,500	4,398	4,121	4,940	4,121
27,500	27,550	4,412	4,129	4,954	4,129
27,550	27,600	4,426	4,136	4,968	4,136
27,600	27,650	4,440	4,144	4,982	4,144
27,650	27,700	4,454	4,151	4,996	4,151
27,700	27,750	4,468	4,159	5,010	4,159
27,750	27,800	4,482	4,166	5,024	4,166
27,800	27,850	4,496	4,174	5,038	4,174
27,850	27,900	4,510	4,181	5,052	4,181
27,900	27,950	4,524	4,189	5,066	4,189
27,950	28,000	4,538	4,196	5,080	4,196
28,000					
28,000	28,050	4,552	4,204	5,094	4,204
28,050	28,100	4,566	4,211	5,108	4,211
28,100	28,150	4,580	4,219	5,122	4,219
28,150	28,200	4,594	4,226	5,136	4,226
28,200	28,250	4,608	4,234	5,150	4,234
28,250	28,300	4,622	4,241	5,164	4,241
28,300	28,350	4,636	4,249	5,178	4,249
28,350	28,400	4,650	4,256	5,192	4,256
28,400	28,450	4,664	4,264	5,206	4,264
28,450	28,500	4,678	4,271	5,220	4,271
28,500	28,550	4,692	4,279	5,234	4,279
28,550	28,600	4,706	4,286	5,248	4,286
28,600	28,650	4,720	4,294	5,262	4,294
28,650	28,700	4,734	4,301	5,276	4,301
28,700	28,750	4,748	4,309	5,290	4,309
28,750	28,800	4,762	4,316	5,304	4,316
28,800	28,850	4,776	4,324	5,318	4,324
28,850	28,900	4,790	4,331	5,332	4,331
28,900	28,950	4,804	4,339	5,346	4,339
28,950	29,000	4,818	4,346	5,360	4,346

29,000 / 30,000 / 31,000

At least	But less than	Single	Married filing jointly *	Married filing separately	Head of a household
29,000					
29,000	29,050	4,832	4,354	5,374	4,354
29,050	29,100	4,846	4,361	5,388	4,361
29,100	29,150	4,860	4,369	5,402	4,369
29,150	29,200	4,874	4,376	5,416	4,376
29,200	29,250	4,888	4,384	5,430	4,384
29,250	29,300	4,902	4,391	5,444	4,391
29,300	29,350	4,916	4,399	5,458	4,399
29,350	29,400	4,930	4,406	5,472	4,406
29,400	29,450	4,944	4,414	5,486	4,414
29,450	29,500	4,958	4,421	5,500	4,421
29,500	29,550	4,972	4,429	5,514	4,429
29,550	29,600	4,986	4,436	5,528	4,436
29,600	29,650	5,000	4,444	5,542	4,444
29,650	29,700	5,014	4,451	5,556	4,451
29,700	29,750	5,028	4,459	5,570	4,459
29,750	29,800	5,042	4,466	5,584	4,466
29,800	29,850	5,056	4,474	5,598	4,474
29,850	29,900	5,070	4,481	5,612	4,481
29,900	29,950	5,084	4,489	5,626	4,489
29,950	30,000	5,098	4,496	5,640	4,496
30,000					
30,000	30,050	5,112	4,504	5,654	4,504
30,050	30,100	5,126	4,511	5,668	4,511
30,100	30,150	5,140	4,519	5,682	4,519
30,150	30,200	5,154	4,526	5,696	4,526
30,200	30,250	5,168	4,534	5,710	4,534
30,250	30,300	5,182	4,541	5,724	4,541
30,300	30,350	5,196	4,549	5,738	4,549
30,350	30,400	5,210	4,556	5,752	4,556
30,400	30,450	5,224	4,564	5,766	4,564
30,450	30,500	5,238	4,571	5,780	4,571
30,500	30,550	5,252	4,579	5,794	4,579
30,550	30,600	5,266	4,586	5,808	4,586
30,600	30,650	5,280	4,594	5,822	4,594
30,650	30,700	5,294	4,601	5,836	4,601
30,700	30,750	5,308	4,609	5,850	4,609
30,750	30,800	5,322	4,616	5,864	4,616
30,800	30,850	5,336	4,624	5,878	4,624
30,850	30,900	5,350	4,631	5,892	4,631
30,900	30,950	5,364	4,639	5,906	4,639
30,950	31,000	5,378	4,646	5,920	4,646
31,000					
31,000	31,050	5,392	4,654	5,934	4,654
31,050	31,100	5,406	4,661	5,948	4,661
31,100	31,150	5,420	4,669	5,962	4,669
31,150	31,200	5,434	4,676	5,976	4,676
31,200	31,250	5,448	4,684	5,990	4,684
31,250	31,300	5,462	4,691	6,004	4,691
31,300	31,350	5,476	4,699	6,018	4,699
31,350	31,400	5,490	4,706	6,032	4,706
31,400	31,450	5,504	4,714	6,046	4,714
31,450	31,500	5,518	4,721	6,060	4,721
31,500	31,550	5,532	4,729	6,074	4,729
31,550	31,600	5,546	4,736	6,088	4,736
31,600	31,650	5,560	4,744	6,102	4,744
31,650	31,700	5,574	4,751	6,116	4,751
31,700	31,750	5,588	4,759	6,130	4,759
31,750	31,800	5,602	4,766	6,144	4,766
31,800	31,850	5,616	4,774	6,158	4,774
31,850	31,900	5,630	4,781	6,172	4,781
31,900	31,950	5,644	4,789	6,186	4,789
31,950	32,000	5,658	4,796	6,200	4,796

* This column must also be used by a qualifying widow(er).

Continued on next page

2.6 Exercises

FOR EXTRA HELP

 Videotape 1

 InterAct Math Tutorial Software

 www.carsonmath.com

 AWL Math Tutor Center

 InterAct MathXL www.mathxl.com

 Student's Solutions Manual

1. Darwin put $800 down when he bought a Buick Century. He made 60 payments of $288 and spent $950 in maintenance and repairs. Three years after paying off the car he sells it for $4000. What is his net? Is it a profit or loss?

2. In 1955, Malvin put $800 down on a new Chevrolet Bel-Air. He made 24 payments of $18. In 1990, he spends $2500 restoring the car for a classic car show. At the show, someone offers him $20,000 for the car. If he paid approximately $3000 for maintenance over the years, what would be his net if he accepted the offer? Is it a profit or loss?

3. Lynn takes out a loan to buy a *fixer-upper* house. She then spends $4500 in repairs and improvements. She sells the house for $80,560. If the pay-off for the loan that she took out to buy the house is $71,484, what is her net? Is it a profit or loss?

4. Scott takes out a loan to buy a *fixer-upper* house. He spends $5475 in repairs and improvements. He sells the house for $94,200 but pays the new owners $3750 in closing costs. If the pay-off for the loan that Scott took out to buy the house is $80,248, what is his net? Is it a profit or loss?

5. Valerie is a single mom with one child. Because her child lives with her and she pays more than half of the cost of the upkeep of her home, she qualifies for the head of household deduction. She rents an apartment, and works two jobs. The spreadsheet below shows her income and deductions. Find her taxable income, tax, and amount she must pay or receive as a refund.

Annual Income	Deductions
Job 1 = $21,468	Medical = $375
Job 2 = $7820	Taxes = $1278
Interest = $45	
Taxes paid: Job 1 W-2 = $1578	
Job 2 W-2 = $445	

6. Chris is single and owns a home. The spreadsheet below shows his income and deductions. Find his taxable income, tax, and amount he must pay or receive as a refund.

Annual Income	Deductions
Job 1 = $32,450	Medical = $285
Interest = $84	Taxes = $2056
Other Income = $1200	Mortgage Interest = $7248
Taxes paid: Job 1 W-2 = $2476	

7. The spreadsheet below is for the Ready family. Their filing status is married filing jointly with two children, so they get 4 exemptions. Find their taxable income, tax, and the amount they owe or receive in a refund.

Annual Income	Deductions
Mr. Ready = $23,680	Mortgage interest = $6855
Mrs. Ready = $26,250	Medical = $682
Interest = $212	Taxes = $1924
Other income = $800	Gifts to charity = $368
Taxes paid (from W-2 forms):	Mr. Ready = $2355 Mrs. Ready = $2448

8. The spreadsheet below is for the Johnson family. Their filing status is married filing jointly with three children so they get 5 exemptions. Find their taxable income, tax, and the amount they owe or receive in refund.

Annual Income	Deductions
Mr. Johnson = $35,200	Mortgage interest = $7942
Mrs. Johnson = $22,000	Medical = $684
Interest = $240	Taxes = $2384
	Gifts to charity = $1200
Taxes paid (from W-2 forms):	Mr. Johnson = $3112 Mrs. Johnson = $1856

9. A car with a mass of 88 slugs is placed on a hydraulic lift. In the trunk of the car are golf clubs with a mass of 1 slug and a tool box with a mass of 2 slugs. What force must be applied to lift the truck?

10. A ski lift has a seat suspended from a wire above. Three skiers with masses of 70 kg, 82 kg, and 20 kg are riding the lift. The seat itself has a mass of 17 kg. What is the total force exerted on the wire above?

11. Four concrete slabs each with a mass of 20 slugs are to be lifted by a crane. What force must be applied to lift the slabs?

12. A cable that can hold a maximum of 2500 lb. is to be used to lift 5 steel beams that each have a mass of 17 slugs. Will the rope hold?

For Exercises 13–18, use the chart of accelerations due to gravity for bodies in our solar system.

Accelerations Due to Gravity for Bodies in the Solar System

	Meters per second every second	Feet per second every second
Moon	−2	−5
Sun	−275	−900
Mercury	−4	−13
Venus	−9	−29
Mars	−4	−13
Jupiter	−26	−87
Saturn	−12	−39
Uranus	−10	−32
Neptune	−10	−32
Pluto	−4?	−13? (uncertain)

13. How much would an astronaut with a mass of 6 slugs weigh on Mars?

14. An astronaut has a mass of 54 kg. How much would she weigh on Jupiter?

15. How much would a 4 slug space probe weigh on Saturn?

16. How much would a 40 kg space probe weigh on Venus?

17. Neil Armstrong weighed 172 lb. at the time of the Apollo mission to the moon in 1969. His space suit and backpack added another 180 lb. How much did Neil weigh when he took that "one small step" onto the surface of the moon?

Historical Note

On July 16, 1969, Apollo 11 carrying Neil Armstrong, Edwin "Buzz" Aldrin, and Michael Collins was launched. On July 20, 1969, Armstrong and Aldrin landed the lunar module, or LM, on the moon (Collins piloted the command module in orbit around moon). Three and a half hours after landing, Neil Armstrong climbed out of the LM to become the first human to set foot on the moon. His famous words upon setting foot on the moon were: "That's one small step for man, one giant leap for mankind."

18. The Viking 1 was a probe that landed on the surface of Mars in 1976. The lander weighed about 1280 lb. on Earth. What was its weight on Mars?

Historical Note

There were two Viking spacecraft, Viking 1 and Viking 2. Viking 1 began its journey on August 20, 1975, and landed on the surface of Mars on July 20, 1976, becoming the first manmade object to touch the surface. Viking 2 followed soon after beginning on September 9, 1975, and landing on September 3, 1976. The Viking craft sent back images and valuable information and were operational for several years after landing. Viking 2 ceased functioning in April 1980. Viking 1 continued to send images and data periodically until November 1982.

Source: Milestones of flight web site, www.nasm.edu/GALLERIES/GAL100/viking.html

19. An F-16 pilot weighs 185 lb. The pilot's plane goes into a sharp turn that causes him to experience 6 times the force of Earth's gravity (6 gs). How much does the pilot weigh during the turn?

Trivia Bite

The acceleration due to Earth's gravity exerts a force of 1 g. However, if we ride a roller coaster or a jet plane or anything that accelerates more than 32 ft. per second every second, we experience more than 1 g. If we accelerate at 64 ft. per second every second we say we experience 2 gs. This means we experience a force that is twice Earth's gravity. If that force is directed downwards, then a human body weighs twice as much as normal. If the human body experiences more than about 5 gs, the blood is accelerated out of the brain causing a black out. This is why jet pilots wear g-suits. The suit reacts to the g forces and squeezes the pilot's legs and lower body, squeezing the blood back into the brain so that the pilot does not black out.

20. Madeline weighs 130 lb. During a tight turn on a roller coaster ride she experiences 3 gs. What does she weigh during the turn?

21. The elevator in the Empire State Building travels from the lobby to the 80th floor, a distance of about 968 ft., in about 44 sec. What is the average rate of the elevator?

Trivia Bite

Completed in November 1930, the Empire State Building became the tallest man-made structure at a height of 381 m with 102 floors. The building kept this distinction until 1973, when the World Trade Center Tower exceeded its height at 417 m.

22. A research submarine is lowered at an average rate of 7 ft. per sec. What will be the submarine's depth after 29 sec.?

23. The space shuttle travels at a rate of about 17,060 mph while in orbit. How far does the shuttle travel in 3 hr.?

24. A ship travels at an average rate of 20 mph. How far will the ship travel in 4 hr.?

25. On a vacation trip, Devin leaves at 11 A.M. and travels 105 mi. then takes a 30-min. break. He then travels another 140 mi. and stops for another 30-min. break. Finally, he travels 80 mi. and arrives at his destination at 5 P.M. What was his average speed?

26. On a vacation trip, Corrine leaves at 7 A.M. and travels 152 mi., then takes a 15-min. break. She then travels another 145 mi. and stops for 45 min. to eat lunch. Finally, she travels 135 mi. and arrives at her destination at 2 P.M. What was her average speed?

Puzzle Problem

A boat is in a harbor at low tide. Over the side hangs a ladder with its bottom step 6 in. below the surface of the water. The ladder steps are 12 in. apart. If the tide rises at a rate of 8 in./hr., how many steps will be under water after 5 hr.?

REVIEW EXERCISES

1. Graph $-(-3)$ on a number line.

2. $-(-(-5)) =$

3. $|15| =$

4. Solve and check. $k + 76 = -34$

5. Solve and check. $-14m = 56$

Defined Terms

Review the following terms, and for any you do not know, study its definition on the page number next to it.

Section 2.1
Integers *p. 88*
Absolute value *p. 90*
Additive inverses *p. 91*

Section 2.3
Net *p. 106*
Cost *p. 106*
Revenue *p. 106*
Profit *p. 106*
Loss *p. 106*

Section 2.6
Deduction *p. 130*
Average rate *p. 134*

Procedures, Rules, and Key Examples

Procedures/Rules	Key Example(s)
To graph a number on a number line: Draw a dot or point on the mark for that number.	Graph -3 on a number line. $-4\ -3\ -2\ -1\ \ 0\ \ 1\ \ 2\ \ 3\ \ 4$
The absolute value of a number is always positive or zero.	Find the absolute value. $\lvert -9 \rvert = 9$ $\lvert 15 \rvert = 15$ $\lvert 0 \rvert = 0$
Rules for adding integers: When adding two numbers that have the same sign, we add their absolute values and keep the same sign. When we add two numbers that have different signs, we subtract their absolute values and keep the sign of the number that has the larger absolute value.	Add. $5 + 9 = 14$ $-5 + (-9) = -14$ $-5 + 9 = 4$ $5 + (-9) = -4$
To write a subtraction statement as an equivalent addition statement: **1.** Change the operation symbol from a minus sign to a plus sign. **2.** Change the subtrahend to its additive inverse.	Subtract. $5 - 9 = 5 + (-9) = -4$ $-5 - 9 = -5 + (-9) = -14$ $5 - (-9) = 5 + 9 = 14$ $-5 - (-9) = -5 + 9 = 4$
To find a missing addend: Write a related subtraction sentence, and subtract the known addend from the sum.	Solve. $x + 30 = 18$ $\qquad x = 18 - 30$ $\qquad x = -12$

|

Rules for multiplying and dividing integers:

When multiplying or dividing two numbers that have the same sign, the result is positive.

When multiplying or dividing two numbers that have different signs, the result is negative.

Multiply.

$4 \cdot 6 = 24$

$(-4)(-6) = 24$

$4(-6) = -24$

$(-4)(6) = -24$

Divide.

$18 \div 9 = 2$

$-18 \div (-9) = 2$

$18 \div (-9) = -2$

$-18 \div 9 = -2$

Rules for evaluating expressions with exponents:

If the base of an exponential form is a negative number and the exponent is even, the product is positive.

If the base is a negative number and the exponent is odd, the product is negative.

Evaluate.

$(-2)^4 = (-2)(-2)(-2)(-2) = 16$

$(-2)^5 = (-2)(-2)(-2)(-2)(-2) = -32$

Tricky case: -2^4 means find the additive inverse of 2 raised to the 4th power.

$-2^4 = -[2 \cdot 2 \cdot 2 \cdot 2] = -16$

To solve for a missing factor:

Write a related division sentence, dividing the product by the known factor.

Solve:

$6x = -24$

$x = -24 \div 6$

$x = -4$

Rules for square:

Every positive number has two square roots, a positive root and a negative root.

The radical symbol means to find only the positive (principle) square root.

The square root of a negative number is not an integer.

Find all solutions for $n^2 = 64$

Answer: ± 8

$\sqrt{81} = 9$

$\sqrt{-25} = $ Not an integer

Order of operations agreement:

Perform operations in the following order:

1. Parentheses ( ), Brackets [ ], Braces { }, and Absolute value | |
2. Exponents and roots from left to right, in the order they occur.
3. Multiplication/Division left to right, in the order they occur.
4. Addition/Subtraction left to right, in the order they occur.

Simplify.

$\{(-4)^2 + 5[(9 - 15) \div (-2)]\} + (-3)\sqrt{16 \cdot 9}$

$= \{(-4)^2 + 5[(-6) \div (-2)]\} + (-3)\sqrt{144}$

$= \{16 + 5[3]\} + (-3) \cdot 12$

$= \{16 + 15\} + (-36)$

$= 31 + (-36)$

$= -5$

Formulas

Net: $N = R - C$

Voltage: $V = ir$

Force: $F = ma$

For Exercises 1–6, answer true or false.

1. All whole numbers are integers.

2. All integers are whole numbers.

3. The additive inverse of any number is always a negative number.

4. The absolute value of any number is always positive or zero.

5. The sum of two negative numbers is a positive number.

6. The product of two negative numbers is a positive number.

For Exercises 7–10, complete the rule.

7. When two numbers that have different signs are added, we _____
   _____

8. To write a subtraction statement as an equivalent addition statement, we _____
   _____

9. When two numbers that have different signs are multiplied, the product is _____

10. When two numbers that have different signs are divided, the quotient is _____

For Exercises 11 and 12, express each amount as an integer.

11. The *Titanic* rests at a depth of approximately 13,000 ft.

12. Water boils at 212°F.

13. Graph each integer on a number line.

    a. -8 **b.** 5

14. Use $<$ or $>$ to make a true statement.

    a. -15 ? 0   **b.** -26 ? -35 **c.** 12 ? -41

15. The absolute value of 41 is _____   16. $|16| =$

17. The additive inverse of 27 is _____   18. $-(-17) =$

19. $-(-(-26)) =$ 20. $-|-12| =$

For Exercises 21–47, calculate.

21. $-16 + 25 =$ 22. $-12 + (-14) =$

23. $18 + (-30) =$ 24. $24 - 31 =$

25. $-20 - 17 =$ 26. $-13 - (-19) =$

27. $-27 - (-22) =$ 28. $6 \cdot (-9) =$

29. $-5(-12) =$ 30. $(-1)(-3)(-6)(-2) =$

31. $(-5)(-2)(-3)$ 32. $(-3)^4 =$

33. $(-10)^3 =$ 34. $-48 \div 8 =$

35. $-63 \div (-9) =$ 36. $-12 \div 0 =$

37. $\sqrt{-36} =$ 38. $21 - 4 \cdot 7 =$

39. $-16 - 28 \div (-7) =$

40. $-|26 - 3(-2)| =$

41. $13 + 4(6 - 15) + 2^3 =$

42. $-3\sqrt{49} + 4(2 - 6)^2 =$

43. $[-12 + 4(15 - 10)] + \sqrt{100 - 36} =$

44. $\{[13 + (-7)(4)] - 2[14 - 28 \div 7 \cdot 2]\} + \sqrt{121} =$

45. $\sqrt{64 \cdot 9} + \{[20 - (5 + 8)] + (9 - 12)^3\} =$

46. $\dfrac{4 + (5 + 3)^2}{3(-13) + 5} =$

47. $\dfrac{16 - (-8)(2)^2}{4 \cdot (-5) + [16 + (-12)]} =$

For Exercises 48–52, solve and check.

48. $14 + x = -27$

49. $n + (-12) = -7$

50. $7k = -63$

51. $(-2)(-5)h = -80$

52. $y^2 = 144$

For Exercises 53–64, solve.

53. The following table lists the assets and debts for the Williams family. Calculate their net worth.

Assets	Debts
Savings = $678	Credit-card balance = $4659
Checking = $895	Mortgage = $78,956
Furniture = $7859	
Automobile = $2100	

54. A concrete block is suspended by two cables. The block weighs 600 lb. Each cable is exerting 300 lb. of upward force. What is the resultant force? What does this mean?

55. The financial report for a business indicates that the total revenue for 1998 was $1,648,200 and the total costs were $928,600. What was the net? Did the business have a profit or loss?

56. The temperature at sunset was reported to be 12°F. By midnight it was reported to be −19°F. What is the amount of the decrease?

57. Arturo has a balance of −$45 in his bank account. To avoid further charges he must have a balance of $25. What is the minimum he must deposit to avoid further charges?

58. A large air-conditioning unit with a mass of 18 slugs is placed on the roof of a building. What is the force on the roof?

59. A circuit has a resistance of 12 Ω and a current of 3 A. Calculate the voltage.

60. Branford has a debt of −$4272. He agrees to make monthly payments for 2 yr. to repay the debt. How much is each payment?

61. A boulder is estimated to have a mass of 75 slugs. A bulldozer can move up to 2500 lb. of material. Will the bulldozer be able to move the boulder?

62. Jacquelyn drives at an average rate of 65 mph for 3 hr. How far does she travel?

63. Steve begins a trip at 7 A.M. and drives 150 mi. After a 30-min. break, he drives another 110 mi. After another 30-min. break, he drives 40 mi. If he arrives at 1:00 P.M., what was his average rate of speed?

64. Sharon is single and owns a home. The following spreadsheet shows her income and deductions. Find her taxable income, tax, and amount she must pay or receive as a refund. (Use Table 2.4 at the end of section 2.6.)

Annual Income	Deductions
Job = $33,500	Medical = $396
Interest = $124	Taxes = $2142
Other income = $500	Mortgage interest = $7588
Taxes paid: Job 1 W-2 = $2516	

1. Graph -5 and 8 on a number line.

2. $|26| =$

3. $-(-(-18)) =$

4. $17 + (-29) =$

5. $-31 + (-14) =$

6. $20 - 34 =$

7. $-16 - 19 =$

8. $-30 - (-14) =$

9. Solve and check. $-19 + k = 25$

10. Below is Allison's credit card statement. What is her new balance?

Beginning balance $= -\$487$

Transactions	**Amount**
Clothing boutique	$-\$76$
Payment	$\$125$
Finance charges	$-\$14$

11. Juan takes out a loan to buy a house. He spends $6400 on repairs and improvements, then sells the house for $92,700. If the payoff for the loan is $81,348, what is his net? Is it a profit or loss?

12. $9(-12) =$

13. $(-2)(-6)(-7) =$

14. $-48 \div (-12) =$

15. Solve and check. $6n = -54$

1. _____

2. _____

3. _____

4. _____

5. _____

6. _____

7. _____

8. _____

9. _____

10. _____

11. _____

12. _____

13. _____

14. _____

15. _____

16. _____

16. Evaluate $(-4)^3$.

17. _____

17. Solve. $m^2 = 81$

18. _____

18. A block of concrete weighing 1380 N is lifted by a crane. What is the mass of the block?

19. _____

19. Lashanda drives at an average rate of 62 mph. How far will she travel in 2 hr.?

20. _____

20. $28 - 6 \cdot 5 =$

21. $19 - 4(7 + 2) + 2^4 =$

21. _____

22. _____

22. $|28 \div (2 - 6)| + \sqrt{121} =$

23. $(-3)^3 + \{[18 \div 2 + (4 - 6)] - \sqrt{49}\} =$

23. _____

24. _____

24. $\dfrac{5^2 + 3(-12)}{36 \div 2 - (3)(6)} =$

25. $\dfrac{(15 - 8)^2 + 11}{(-2)(5) + \sqrt{64}} =$

25. _____

Cumulative Review Exercises

For Exercises 1–6, answer true or false.

1. 0 is an integer.

2. −6 is a whole number.

3. $-12 \div 0 = 0$

4. $3 \cdot 4 + 3 \cdot 5 = 3(4 + 5)$

5. The sum of a negative number and a positive number is always negative.

6. $-9 - 12 = -9 + (-12)$

7. List the order in which we perform operations according to the order of operations agreement.

   1.

   2.

   3.

   4.

8. Explain the mistake in the problem.

$$-14(2) + \sqrt{16 + 9}$$
$$= -14(2) + 4 + 3$$
$$= -28 + 4 + 3$$
$$= -24 + 3$$
$$= -21$$

9. $9 \cdot (4 \cdot 3) = (9 \cdot 4) \cdot 3$ is an illustration of the _____ property of _____.

10. Explain the commutative property of addition in your own words and give an example.

11. Write 5,680,901 in expanded form.

12. Write $5 \times 10^7 + 8 \times 10^5 + 3 \times 10^4 + 6 \times 10^3 + 9 \times 1$ in standard form.

13. Write the word name for 409,254,006.

14. Graph −4 on a number line.

For Exercises 15 and 16, use , or . to make a true statement.

15. 135 ? −450

16. −930 ? −932

17. Round −23,410,512 to the nearest million.

18. Estimate 49,902 + 6,519 by rounding to the nearest thousand.

For Exercises 19–36, simplify.

19. $|16|$

20. $-(-5)$

21. $-|-8|$

22. $-(-(-(4)))$

23. $287 + 48 + (-160) + (-82)$

24. $-19 - 24$

25. $-64 - (-14)$

26. $-14(6)$

27. $-12(-8)$

28. $-1(12)(-6)(-2)$

29. $(3)^5$

30. $(-2)^6$

31. -2^6

32. $1208 \div 6$

33. $-105 \div 7$

34. $\sqrt{121}$

35. $-15 - 2[36 \div (3 + 15)]$

36. $-4(2)^3 + [18 - 12(2)] - \sqrt{25 \cdot 4}$

For Exercises 37–40, solve and check.

37. $29 + x = 54$

38. $y + 60 = 23$

39. $26x = 2652$

40. $-18x = 126$

41. Find the perimeter and area.

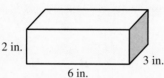

42. Find the volume.

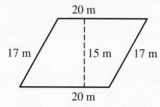

For Exercises 43–50, solve.

43. An accountant is given the following spreadsheet of expenses and income for a company. What is the final balance?

Description of expense	Amount	Description of assets/income	Amount
Payroll	$19,445	Checking account	$ 9,876
Utilities	$ 704	Income	$48,984
Water	$ 155		
Waste disposal	$ 1282		
New inventory	$21,931		

44. Sonya put $1500 down on a new car. She made 60 payments of $276 and spent $1450 on maintenance and repairs. Three years after paying off the car, she sold it for $4300. What was her net? Was it a profit or loss?

45. A landscaper needs to know the area of the yard. A plot plan showing the position of the house on the lot is shown. Find the area of the yard.

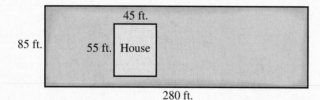

46. A computer translates everything into binary code. This means that numbers are written so that each place value either contains a 1 or a 0 digit. If a certain computer chip has 7 place values (bits), how many different numbers can be coded?

47. Jerry is to administer 300 ml of Heparin in an IV drip over the course of an hour. How many ml of Heparin should the patient receive each minute?

48. Shanisse begins a trip at 8 A.M. and drives 100 mi. She stops to rest for 15 min., then drives another 85 mi. She stops again for 45 min., then continues for another 150 mi. If she arrives at 2 P.M. what was her average rate of speed?

49. A cargo box that weighs 60,000 lb. is to be loaded onto a ship. What is the mass of the box?

50. An electrical circuit has a resistance of 8Ω. If the current is $-12A$, find the voltage.

Expressions and Polynomials

Practice

 One of the most important elements to success is practice. To get the most out of your practice or study time, develop a plan and a routine. Build into your daily routine blocks of time devoted to practice or study in an environment where you are not likely to be distracted. Some good study places are the library, a tutorial center, or an empty classroom.

To begin your practice or study sessions, the first thing you should do is tidy up your notes (see the Chapter 1 opener). If you are using the study system outlined in the *To the Student* section of the text, your next step is to update your study sheet with the new procedures and rules. Next, with your study sheet handy, proceed to work through the exercises. If you reach a problem that you are not sure about, look for examples in the text and your notes relating to that problem. If you cannot resolve the problem, mark it and move on. Above all, do not allow frustration over one problem to keep you from having a good study session.

When you've completed all that you can on your own, seek help to complete the rest. Once you've completed the assigned exercises, shift your focus to study. Study your study sheet until you know the rules and procedures. To test yourself, see if you can duplicate your study sheet from memory. Get in the frame of mind that every day is a test day. Finally, without the aid of your notes or study sheet, quiz yourself by doing additional exercises similar to those assigned. Monitor how quickly you recall the solution process for each problem. If you find yourself groping for the process, then you need to spend more time studying the procedures.

" *The formula for success is simple: Practice and concentration, then more practice and more concentration.* "
—BABE DIDRIKSON ZAHARIAS

" *It's better to look ahead and prepare, than to look back and regret.* "
—JACKIE JOYNER-KERSEE

" *No one has ever drowned in sweat.* "
—LOU HOLTZ

3.1 Expressions

OBJECTIVES

1 Differentiate between an expression and an equation.

2 Evaluate expressions.

OBJECTIVE 1 *Differentiate between an expression and an equation.*

In Chapters 1 and 2 we began developing concepts of algebra such as variables, constants, equations, and formulas. Recall the definition of an **equation.**

DEFINITION | **Equation:** A mathematical relationship that contains an equal sign.

The formulas we developed, such as $F = ma$ and $V = lwh$, are equations because they have equal signs. Notice the equal sign separates an equation into two sides, the left side and the right side. If we focus attention on the right side of $F = ma$, we see ma. Because ma by itself does not contain an equal sign, we cannot call it an equation. Instead, we call ma an **expression.**

DEFINITION | **Expression:** A mathematical relationship with constants and/or variables but no equal sign.

$F = ma$ is an equation whereas ma is merely an expression.

$4 + 7 = 11$ is an equation whereas $4 + 7$ is merely an expression.

$V = lwh$ is an equation whereas lwh is an expression.

Note Expressions or equations that contain only constants are called *numeric* whereas those that contain variables are called *algebraic*.

Think of equations as complete sentences whereas expressions are phrases or incomplete sentences. In grammar we learn that a complete sentence must have a subject and a verb. The verb in an equation is the equal sign and because expressions do not have an equal sign, they do not have a verb, and therefore are incomplete.

EXAMPLE 1 Tell whether each of the following is an expression or equation and explain why.

a. $16 \cdot 5 = 80$

Solution: $16 \cdot 5 = 80$ is an equation. More specifically, it is a numeric equation.

b. $d = rt$

Solution: $d = rt$ is an equation because it contains an equal sign. More specifically, it is an algebraic equation because it contains variables.

c. rt

Solution: rt is an expression because it does not contain an equal sign. More specifically, it is an algebraic expression because it contains variables.

◀ **Do Margin 1.**

◆ Margin 1

Tell whether each of the following is an expression or equation and explain why.

a. $E = mc^2$

b. $19 - 2\,(3 + 8)$

c. $\dfrac{\sqrt{b^2 - 4ac}}{2a}$

d. $y = mx + b$

Answers: **a.** Equation. It contains an equal sign. **b.** Expression. It does not contain an equal sign. **c.** Expression. It does not contain an equal sign. **d.** Equation. It contains an equal sign.

OBJECTIVE 2 *Evaluate expressions.*

There are two actions we can perform with an expression.

1. Evaluate

2. Rewrite

In this section, we'll focus on evaluating. The rest of this chapter will deal with rewriting.

Procedure *To evaluate an expression:*
1. Replace the variables with the corresponding given values.
2. Calculate using the order of operations agreement.

EXAMPLE 2 Evaluate $x^2 + 5xy - 7$ when $x = -3$ and $y = 2$

Solution: $\quad x^2 + 5\ x\ \ y - 7$ $\qquad$ Replace x with -3 and y with 2.

$= (-3)^2 + 5(-3)(2) - 7$ $\qquad$ Calculate. Simplify the exponential form first.

$= 9 + 5(-3)(2) - 7$ $\qquad$ Multiply.

$= 9 + (-30) - 7$ $\qquad$ Add and subtract from left to right.

$= -21 - 7$

$= -28$

Do Margin 2. ▶

EXAMPLE 3 Evaluate $\frac{3h^2}{k+5}$ when $h = 2$ and $k = -5$

Solution: $\dfrac{3h^2}{k+5}$ $\qquad$ Replace h with 2 and k with -5.

$= \dfrac{3(2)^2}{-5+5}$ $\qquad$ Calculate the top and bottom expressions separately.

$= \dfrac{3(4)}{0}$

$= \dfrac{12}{0}$, which is undefined

Explanation: When we replace k with -5, the bottom expression equals zero. Any time zero is a divisor the expression is undefined or indeterminate (0 dividend).

Recall the division properties of zero that we developed in Chapter 1.

Division Properties:
When zero is the divisor with any dividend other than zero, the quotient is undefined.

In math language $n \div 0$ is undefined where $n \neq 0$

When zero is the dividend, the quotient is zero as long as the divisor is not also zero.

In math language $0 \div n = 0$ as long as $n \neq 0$

If both dividend and divisor are zero, the quotient is indeterminate.

In math language $0 \div 0$ is indeterminate

Do Margin 3. ▶

◆ Margin 4

Evaluate.

a. $\sqrt{2x + 1} - y^2$ when
$x = 24$ and $y = -3$

b. $\sqrt{x^2 + y^2}$ when
$x = -6$ and $y = 8$

c. $2xy + \sqrt{x - y}$ when
$x = 3$ and $y = 19$

◆ Margin 5

Find the value(s) that can replace the variable and cause the expression to be undefined.

a. $\dfrac{2x + 1}{7 - x}$

b. $\dfrac{-9}{(x + 2)(x - 3)}$

EXAMPLE 4 Evaluate $\sqrt{x} + 2\sqrt{x + y}$ when $x = 9$ and $y = 16$.

Solution: $\sqrt{x} + 2\sqrt{x + y}$ Replace x with 9 and y with 16.

$\sqrt{9} + 2\sqrt{9 + 16}$ Add within the radical.

$= \sqrt{9} + 2\sqrt{25}$

$= 3 + 2(5)$ Multiply.

$= 3 + 10$ Add.

$= 13$

Reminder
When a radical contains a sum or difference we treat the sum or difference as if there were parentheses around them.

EXAMPLE 5 Evaluate $2\sqrt{m + n} + 5mn$ when $m = -6$ and $n = 2$

Solution: $2\sqrt{m + n} + 5mn$ Replace m with −6 and n with 2.

$2\sqrt{-6 + 2} + 5(-6)(2)$ Add within the radical.

$= 2\sqrt{-4} + (-30)(2)$

The answer is not an integer.

Reminder
The square root of a negative number is not an integer.

◀ **Do Margin 4.**

EXAMPLE 6 Find the value(s) that can replace the variable and cause $\dfrac{2x}{3x + 6}$ to be undefined.

Solution: In a division expression, if the divisor ever becomes 0, then the whole expression is undefined. In this case the divisor is $3x + 6$ so we must find the value for x that causes $3x + 6$ to become 0. If $x = -2$ the expression is undefined.

If we replace x with -2, then $3x + 6$ becomes 0.

$$3(-2) + 6$$
$$= -6 + 6$$
$$= 0$$

◀ **Do Margin 5.**

Answers to Margin 4: **a.** −2
b. 10 **c.** no integer solution

Answers to Margin 5: **a.** 7
b. −2 and 3

3.1 Exercises

For Exercises 1–10, indicate whether each of the following is an expression or an equation and explain why.

1. $19 + 3 \cdot 2 = 25$

2. $|14 - 29| = 15$

3. $13 + 5\sqrt{169}$

4. $\dfrac{19 + 9}{4 - 2^3}$

5. $25 - x^2$

6. $x^2 + 5xy - 2 = 7$

7. $x + y = 9$

8. $\sqrt{1 - v^2/c^2}$

9. $9 + 2(x + 3) = 14 - x$

10. $mx + b$

For Exercises 11–38, evaluate the expression using the given values.

11. $3x + 8; x = -5$

12. $5y - 7; y = -2$

13. $2x - 4y; x = 3, y = -2$

14. $-2m + n; m = 4, n = 7$

15. $3y - 5(y + 2); y = 4$

16. $-2x + 3(x - 4); \quad x = -2$

17. $m^2 + 5n - 3; m = -3, n = 1$

18. $3t^2 - 4u + 1; t = -2, u = 4$

19. $r^3 - 4rt; r = -3, t = -1$

20. $6xy - x^3; x = -2, y = 3$

21. $b^2 - 4ac; b = -4, a = -3, c = 2$

22. $b^2 - 4ac; b = 3, a = -5, c = -1$

23. $|x^2 + y|; x = 2, y -10$

24. $|r - p^3|; r = -7, p = -2$

25. $-|4a + 7b|; a = -5, b = 2$

26. $-|5x| + |y^3|; x = 6, y = -3$

27. $-6m - |n^5|$; $m = -3, n = -2$

28. $-3|c - 9cd|$; $c = 3, d = -1$

29. $\sqrt{m} + \sqrt{n}$; $m = 144, n = 25$

30. $\sqrt{m + n}$; $m = 144, n = 25$

31. $\sqrt{x^2 + y^2}$; $x = -3, y = 4$

32. $\sqrt{mn} - 5n^2$; $m = 27, n = 3$

33. $-2u + 6\sqrt{u - 3v}$; $u = 9, v = 15$

34. $3xy - 2\sqrt{5x + y}$; $x = 2, y = -1$

35. $\dfrac{3x + 5}{7 - 2y}$; $x = 10, y = 1$

36. $\dfrac{-7m + n^2}{4m + 2n}$; $m = -5, n = 1$

37. $\dfrac{a^3 - b}{5a + b^2}$; $a = -5, b = 5$

38. $\dfrac{-6uv + 14}{3u - v^2}$; $u = -1, v = 4$

For Exercises 39–46, find the value(s) that can replace the variables and cause the expression to be undefined.

39. $\dfrac{x + 3}{10 - x}$

40. $\dfrac{12}{t + 9}$

41. $\dfrac{16}{y}$

42. $\dfrac{-11}{(x + 4)(x - 5)}$

43. $\dfrac{a}{(4 + a)(7 - a)}$

44. $\dfrac{n + 2}{(n - 6)(n - 1)}$

REVIEW EXERCISES

1. What digit is in the thousands place in 472,603?

2. Write $3 \cdot 10^5$ in standard form.

3. Write 279 in expanded form.

4. What is the exponent for 7? Explain.

5. Evaluate 2^6.

3.2 Polynomials

OBJECTIVES

1 Identify monomials

2 Identify the coefficient and degree of a monomial.

3 Identify like terms.

4 Identify polynomials and their terms.

5 Identify the degree of a polynomial.

6 Write polynomials in descending order.

OBJECTIVE **1** *Identify monomials.*

As we saw in Section 3.1, there are many types of expressions. We will now focus on expressions known as *polynomials*. Polynomials are algebraic expressions that are similar to whole numbers written in expanded notation.

Consider expanded notation for the number 279. In expanded form, we see that 279 is $2 \cdot 10^2 + 7 \cdot 10 + 9$. This is essentially the same as the expression $2x^2 + 7x + 9$.

$$279 = 2 \cdot 10^2 + 7 \cdot 10 + 9 \qquad \text{Expanded form, written with base 10.}$$
$$2x^2 + 7x \qquad + 9 \qquad \text{Polynomial form, written with base } x.$$

$2x^2 + 7x + 9$ is a polynomial. In order to define polynomial, we must first define **monomial**. Think about the prefixes involved. *Poly* means *many* and *mono* means one.

DEFINITION **Monomial:** An algebraic expression that is a product of a constant and variables, each of which is raised to a power that is a whole number.

Note: Monomials are also referred to as *terms*.

EXAMPLE 1 Is the expression a monomial? Explain.

a. $5x$

Explanation: $5x$ is a monomial because it is a product of a constant, 5, and a variable, x that has an exponent of 1, which is a whole number.

From Example 1a, we can make the following conclusion:

Conclusion: Any number or variable with no apparent exponent has an understood exponent of 1.

$$5x = 5^1x^1$$

We write $5x$ instead of 5^1x^1 because it is simpler, that is, it has fewer symbols.

b. $-4xy^3$

Explanation: $-4xy^3$ is a monomial because the constant -4 and variables xy^3 are all multiplied and the variables all have whole number exponents. As we saw in Example 1a, we could write $-4xy^3$ as $(-4)^1x^1y^3$. Both of the variables' exponents, 1 and 3, are whole numbers.

c. x^6

Explanation: x^6 is a monomial. It might seem that x^6 is not a monomial because no numerical factor is visible. However, x^6 is the same as $1x^6$. There is no need to write the 1 because 1 multiplied by an amount does not affect the amount.

From Example 1c, we can make the following conclusion:

Conclusion: If a numerical factor is not apparent in a monomial, it is understood to be 1.

$$x^6 = 1x^6$$

We write x^6 instead of $1x^6$ because it is simpler.

d. 8

Explanation: It might seem that 8 is not a monomial because there is no variable visible. However, 8 is the same as $8x^0$. We do not write the x^0 because $x^0 = 1$. So, $8x^0 = 8(1)$ which is just 8. Because the exponent on the variable, 0, is a whole number, 8 is a monomial.

From Example 1d, we can make the following conclusion:

Conclusion: A numerical factor without a variable can be expressed as being multiplied by a variable with 0 as its exponent. Therefore constants are monomials.

e. $\dfrac{5}{xy}$

Explanation: $\dfrac{5}{xy}$ is not a monomial because it is not a product of a constant and variables. Rather, it is a quotient of a constant with variables. The variables make up the divisor.

f. $3x^2 + 5x - 4$

Explanation: It is not a monomial because it is not a product of a constant with variables. Rather, addition and subtraction are involved.

Note: $3x^2 + 5x - 4$ is not a monomial; however, it is a polynomial.

> **Tip**
> Monomials must be a product of constants and variables.
> Monomials cannot have variables in a divisor, nor can they involve addition or subtraction.

◀ **Do Margin 1.**

OBJECTIVE 2 *Identify the coefficient and degree of a monomial.*

Two terms are used to identify the parts of a monomial: **coefficient** and **degree**. The coefficient is the numerical factor in a monomial while degree relates to the exponents of the

> **DEFINITION** **Coefficient:** The numerical factor in a monomial.
>
> **Degree:** The sum of the exponents on all variables in a monomial.

EXAMPLE 2 variables.

Identify the coefficient and degree of each monomial.

a. $-7x^2$

Answer: Coefficient $= -7$

Degree $= 2$

Explanation: -7 is the coefficient because it is the numerical factor. 2 is the degree because it is the exponent for the variable x.

b. $-xy^3$

Answer: Coefficient $= -1$

Degree $= 1 + 3 = 4$

Explanation: From the conclusions we made in Examples 1b and 1c, we can express $-xy^3$ as $-1x^1y^3$. In this alternative form, we can see that the numerical factor is -1, which is the coefficient. We can also see that the exponents on the variables are 1 and 3. Because degree is the sum of the exponents on the variables, we add 1 and 3 which equals 4.

c. 17

Answer: Coefficient $= 17$

Degree $= 0$

Explanation: From the conclusion we made in Example 1d, we can express 17 as $17x^0$. In this alternative form it is apparent that 17 is the coefficient because it is the numerical factor, and 0 is the degree because it is the exponent on the variable.

Do Margin 2. ▷

OBJECTIVE 3 *Identify like terms.*

Remember that monomials are also referred to as terms. In Section 3.3, we will learn how to simplify polynomials that contain **like terms**.

DEFINITION **Like terms:** Monomials that have the same variables raised to the same exponents.

Notice the definition does not say anything about the coefficient. This means that the coefficients can be different. However, the variables and their exponents must match exactly.

EXAMPLE 3 Determine whether the given monomials are like terms.

a. $5x^2$ and $9x^2$

Answer: $5x^2$ and $9x^2$ are like terms because they both have x^2. They have the same variable raised to the same exponent.

b. $9xyz$ and $5xy$

Answer: $9xyz$ and $5xy$ are not like terms because the variables are different. $9xyz$ has a z while $5xy$ does not.

◆ Margin 2

Identify the coefficient and degree of each monomial.

a. m^6

b. $-7x^2y^3$

c. $14ab$

d. -10

Answers:
a. 1 **b.** -7 **c.** 14 **d.** -10
a. 6 **b.** 5 **c.** 2 **d.** 0

c. $7xy^2$ and $7x^2y$

Answer: $7xy^2$ and $7x^2y$ are not like terms because the variables are not raised to the same exponents.

d. $5a$ and $9A$

Answer: $5a$ and $9A$ are not like terms. In mathematics upper- and lowercase letters are considered different symbols, even if they are of the same letter. An uppercase A is considered a different variable than a lowercase a.

◀ **Do Margin 3.**

◆ **Margin 3**

Determine whether the given monomials are like terms.

a. $-8x^3$ and $5x^3$

b. $12mn$ and $12m$

c. $-5y^3z$ and y^3z

d. $9t$ and $12T$

e. $4ab^3c$ and $3a^3bc$

Answers: **a.** Yes **b.** No **c.** Yes **d.** No **e.** No

OBJECTIVE ◢**4** *Identify polynomials and their terms.*

An expression that is a sum of monomials is called a **polynomial**.

| **DEFINITION** | **Polynomial:** An expression that can be expressed as a sum of monomials. |

$$2x^3 + 5x + 8 \text{ is a polynomial.}$$
$$5x + 7 \text{ is a polynomial.}$$

In the expression, $9x^3 - 4x^2 + 8x - 6$, it would seem that the subtraction signs go against the word *sum* in the definition. However, $9x^3 - 4x^2 + 8x - 6$ actually is a polynomial because it can be expressed as a sum of monomials by writing the subtractions as equivalent additions.

$$9x^3 - 4x^2 + 8x - 6 = 9x^3 + (-4x^2) + 8x + (-6)$$

Mathematicians prefer to write expressions with the fewest symbols possible. Because $9x^3 - 4x^2 + 8x - 6$ has fewer symbols than $9x^3 + (-4x^2) + 8x + (-6)$, it is the preferred form. An expression written with the fewest symbols possible is said to be in **simplest form**.

| **DEFINITION** | **Simplest form:** An expression written with the fewest symbols possible. |

Remember that monomials are often referred to as *terms*. This is especially true when we talk about the individual monomials that make up a polynomial.

EXAMPLE 4 Identify the terms in the polynomial:

a. $9x^3 - 4x^2 + 8x - 6$.

Answer: First term: $9x^3$

Second term: $-4x^2$

Third term: $8x$

Fourth term: -6

Explanation: We saw that we can express $9x^3 - 4x^2 + 8x - 6$ as $9x^3 + (-4x^2) + 8x + (-6)$. Notice the dual role of the minus signs. In the original expression we interpret the minus signs to mean subtraction. In the equivalent addition form we interpret the minus signs as negative signs on the monomials.

From Example 4a, we can make the following conclusion:

Conclusion: The sign of the term is the sign to the left of the term.

b. $4y^4 - 8y^3 - 6y + 2$

Answer: First term: $4y^4$
Second term: $-8y^3$
Third term: $-6y$
Fourth term: 2

This minus sign not only means subtract $6y$, but also that the term is $-6y$.

$$4y^4 - 8y^3 - 6y + 2$$

This minus sign not only means subtract $8y^3$, but also that the term is $-8y^3$.

This plus sign not only means add 2, but also that the term is positive 2.

Do Margin 4.

Names have been developed for polynomials based on the number of terms that make up the polynomial.

Monomial: A single term.

Binomial: A polynomial that has exactly two terms.

Trinomial: A polynomial that has exactly three terms.

If polynomials have more than three terms, mathematicians just call them polynomials. We do not use prefixes to give them special names.

EXAMPLE 5

Classify each expression as a monomial, binomial, trinomial, or none of these.

a. $5x^3y$

Answer: $5x^3y$ is a monomial because it is a single term.

b. $-3x^2 + 5y$

Answer: $-3x^2 + 5y$ is a binomial because it contains two terms.

c. $4m^2 + 6m - 9$

Answer: $4m^2 + 6m - 9$ is a trinomial because it contains three terms.

d. $x^3 + 4x^2 - 9x + 2$

Answer: The answer is none of these. $x^3 + 4x^2 - 9x + 2$ is a polynomial. This expression has four terms. Mathematicians merely call it a *polynomial.*

Discussion If mathematicians had named $x^3 + 4x^2 - 9x + 2$ other than a polynomial, what would it have been called?

Do Margin 5.

OBJECTIVE 5 *Identify the degree of a polynomial.*

We can also talk about the **degree of a polynomial.** Recall that we said the degree of a monomial is the sum of the exponents on the variables. For a polynomial, the degree is the largest degree of all its terms.

DEFINITION **Degree of a polynomial:** The largest degree of all the terms that make up the polynomial.

◆ **Margin 4**

Identify the terms in the given polynomial.

a. $12x^2 - 8x + 9$

b. $4m^3 + 5m^2 - 10m - 11$

◆ **Margin 5**

Classify each expression as a monomial, binomial, trinomial, or none of these.

a. $5x + 9$

b. $4x^3 - 2x^2 + 5x - 6$

c. $8x^3y$

d. $x^2 + 2xy - y^2$

Answers to Margin 4: **a.** $12x^2$, $-8x$, 9 **b.** $4m^3$, $5m^2$, $-10m$, -11

Answers to Margin 5: **a.** binomial **b.** none of these **c.** monomial **d.** trinomial

Margin 6

Identify the degree of each polynomial.

a. $-2m^5 + 13m^7 - 8m^3 + 4m + 17$

b. $3x^{12} + 5x^6 + 8x^2 - 10x - 7$

EXAMPLE 6   Identify the degree of each polynomial.

a. $5x^3 + 9x^6 - 10x^2 + 8x - 2$

Answer: Degree $= 6$

Explanation: The highest degree of all the terms is 6.

b. $12a^7 - a^5 - 3a^9 + 4a - 16$

Answer: Degree $= 9$

◆ **Explanation:** The highest degree of all the terms is 9.

◀ **Do Margin 6.**

OBJECTIVE 6 *Write polynomials in descending order.*

Does it matter what order we write the terms of a polynomial?

$$\text{Is } -9 + 3x^3 - 6x^2 + 5x \qquad \text{the same as} \qquad 3x^3 - 6x^2 + 5x - 9 ?$$

The commutative property of addition says that changing the order of addends does not affect the sum. Because we can write $-9 + 3x^3 - 6x^2 + 5x$ as a sum in the form of $-9 + 3x^3 + (-6x^2) + 5x$, we can change the order of the terms without affecting the sum. Therefore, $-9 + 3x^3 - 6x^2 + 5x$ and $3x^3 - 6x^2 + 5x - 9$ are the same. Notice the signs move with the terms.

Conclusion: We can move a term in a polynomial as long as we keep its sign with it.

It is customary to write polynomials in what is called *descending order of degree.* Even though

---

Margin 7

Write each polynomial in descending order.

a. $9m^4 + 7m^9 - 3m^2 + 15 - 6m^7 + 2m$

b. $-14y^2 - 3y + 17 + 9y^3 - 13y^5$

c. $24 + 9x^2 - 15x + 8x^5 - 7x^3 - 4x^4$

> **Procedure** *Descending-order agreement:*
>
> To write a polynomial in descending order, place the highest degree term first, then the next highest degree, and so on.

EXAMPLE 7   changing the order does not change the expression, it is considered polite to put polynomials in descending order of degree.

Write each polynomial in descending order.

a. $5x^3 + 4x^2 - 3x^5 - 7 + 9x$

Answer: $-3x^5 + 5x^3 + 4x^2 + 9x - 7$

Explanation: The highest degree of all the terms was 5 so we placed $-3x^5$ first. Notice we kept the sign with the term. Also note that because the $5x^3$ is positive, when it was no longer the first term we had to put a plus sign to the left. Every term after the first term must have a $+$ or $-$ sign in between. When a term is first, it needs a sign only if it is negative.

b. $-16b^3 + 5b^6 + 10b^2 - 13b^4 + 19 - 12b$

◆ **Answer:** $5b^6 - 13b^4 - 16b^3 + 10b^2 - 12b + 19$

Explanation: The highest degree was 6 so we wrote $5b^6$ first. Once again, notice that we kept the signs with the terms. However, we did not write a $+$ sign to the left of $5b^6$ because it is not needed. The first term is understood to be positive if there is no sign to the left of it.

◀ **Do Margin 7.**

Answers to Margin 6: **a.** 7 **b.** 12

Answers to Margin 7: **a.** $7m^9 - 6m^7 + 9m^4 - 3m^2 + 2m + 15$ **b.** $-13y^5 + 9y^3 - 14y^2 - 3y + 17$ **c.** $8x^5 - 4x^4 - 7x^3 + 9x^2 - 15x + 24$

3.2 Exercises

FOR EXTRA HELP

▦ Videotape 3

△math InterAct Math
Tutorial Software

🌐 www.carsonmath.com

📞 AWL Math Tutor Center

InterAct MathXL www.mathxl.com

📖 Student's Solutions
Manual

For Exercises 1–12, determine whether the expression is a monomial. Explain.

1. $5x^3$

2. $7y^2$

3. $8m + 5$

4. $3t^2 + 5t - 6$

5. $\dfrac{4x^2}{5y}$

6. $\dfrac{-mn}{k}$

7. $-9x^2y$

8. 19

9. $-x$

10. $5m^3n^7p$

11. -7

12. $x^3 - 9xy$

For Exercises 13–24, identify the coefficient and degree of each monomial.

13. $3x^8$

14. $5t^2$

15. $-9x$

16. $-m$

17. 8

18. -7

19. xy^2

20. $-2t^3u^4$

21. $-a^2b^3c$

22. $15mnp$

23. -1

24. k

For Exercises 25–36, determine whether the given monomials are like terms.

25. $8x$ and $-5x$

26. $3y^2$ and y^2

27. $-5m$ and $-7n$

28. $2a$ and $2b$

29. $4mn^2$ and $7mn^2$

30. $8h^3$ and h^3

31. 14 and -6

32. T and $-T$

33. $5n$ and $9N$

34. $6xY$ and $5xy$

35. x^3y^2z and $-4y^2x^3z$

36. 1 and -5

For Exercises 37–42, identify the terms in the polynomial.

37. $5x^2 + 8x - 7$

38. $4y^3 - 2y - 9$

39. $6t - 1$

40. $9n^4 - 8n^2 + 7n - 1$

41. $-6x^3 + x^2 - 9x + 4$

42. $-9x^2 - x + 5$

For Exercises 43–54, determine whether the expression is a monomial, binomial, trinomial, or none of these.

43. $5x + 2$

44. $9y^2 + 5y - 6$

45. $18x^2$

46. 2

47. $6x^3 - 14x^2 + 9x - 8$

48. $2a - 6b$

49. $16t^2 - 8tu - 5u^2$

50. $a^5 - 6a^3 + 7a^2 - a + 2$

51. -5

52. $ab^3c^2 + a^2bc - abc$

53. $y^3 - z^5$

54. $m^4 - 5m^3 + m^2 + 6m - 9$

For Exercises 55–60, identify the degree of each polynomial.

55. $6x^3 - 9x^2 + 8x + 2$

56. $8y^4 + 2y - 5y^6 + y^3 + 4$

57. $-m^3 - 8m^5 + 3m^2 - 2m^6$

58. $5 + 6a + 3a^3 + 4a^2 - a^5$

59. $19t^9 + 4t^3 + 25t^{12} - 11t^5$

60. $-b^2 - 16b^8 + 5b^3 + 2b^4 - 22$

For Exercises 61–66, write each polynomial in descending order.

61. $14t^6 + 9t^3 + 5t^2 - 8t^4 - 1$

62. $6x + 19 - 4x^3 + 3x^2$

63. $9 - 18y^3 + 12y - 10y^2 + y^5$

64. $21m - 6m^7 + 9m^2 + 13 - 8m^5$

65. $7a^3 + 9a^5 - 6 + 2a^2 - a$

66. $2n^4 - 5n^3 - n + 2 + n^6 - 7n^8$

REVIEW EXERCISES

1. $-9 - 7 =$

2. $8 - 15 =$

3. $-15 + 8 - 21 + 2 =$

4. Is $-5 + 3$ the same as $3 + (-5)$? Explain.

5. Find the perimeter

```
        ┌─────────────────────┐
   4 m  │                     │
        └─────────────────────┘
              10 m
```

3.3 Simplifying Polynomials

OBJECTIVES

1 Simplify polynomials by combining like terms.

We saw how to evaluate expressions in Section 3.1. Now we want to consider the other action that we can perform with expressions: rewriting an expression. In Section 3.2 we defined *like terms* and *simplest form*. Expressions that contain like terms can be rewritten in a simplest form by combining those like terms, which is called *simplifying*.

One way to see how we simplify by combining like terms is to make use of the fact that repeated addition can be simplified to multiplication. Three 2's added is the same as multiplying 3 times 2.

$$2 + 2 + 2 = 3(2) = 6$$

The number 6 is simplest form because it is the simplest way to express the result of $2 + 2 + 2$.

Likewise, if we have the same variable repeatedly added, we can simplify the sum to a multiplication expression. For example,

$$x + x + x = 3x$$

The expression $3x$ is simplest form because it uses the fewest symbols possible that is still equal to $x + x + x$. We cannot simplify further because we do not know a value for x. However, if we were asked to evaluate the expression when $x = 2$, then we'd have the numeric example of three sets of 2, which simplifies to 3 times 2, which simplifies further to 6.

What if the monomials already have a coefficient? What if we had to add $3x + 2x$? Just like $3x$ means $x + x + x$ and $2x$ means $x + x$, we can expand it out this way:

$$
\begin{array}{ccc}
\overbrace{3x} & + & \overbrace{2x} \\
= x + x + x & + & x + x \\
= 5x &&
\end{array}
$$

Notice there are a total of five xs repeatedly added, so the result is $5x$. Note that the coefficient of the result is the sum of the coefficients of the original like terms.

Conclusion: We can combine like terms by adding the coefficients.

> **Procedure** *To combine like terms:*
> **1.** Add or subtract the coefficients.
> **2.** Keep the variables and their exponents the same.

Adding measurements that have the same unit is like combining like terms. Suppose we have to calculate the perimeter of the rectangle shown.

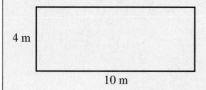

$$P = 4 \text{ m} + 10 \text{ m} + 4 \text{ m} + 10 \text{ m}$$
$$P = 28 \text{ m}$$

The above perimeter problem is similar to a problem in which we must combine like terms in a polynomial where all the variables are m.

$$4m + 10m + 4m + 10m = 28m$$

Notice if the variables were different, like $4cm + 10m + 4cm + 10m$, we would not be able to combine the terms with different variables. In a similar way, we cannot combine measurements with different units. If the width were 4 cm and the length 10 m then the perimeter is no longer 28 m because cm and m represent different distance measurements.

$$4 \text{ cm} + 10 \text{ m} + 4 \text{ cm} + 10 \text{ m} \neq 28$$

EXAMPLE 1 Combine like terms.

a. $4x + 9x$

Answer: $4x + 9x = 13x$

Explanation: We added the coefficients and kept the variable the same.

> **Reminder**
> When adding two numbers that have different signs, we subtract their absolute values and keep the sign of the number with the larger absolute value.

b. $-12y^2 + 7y^2$

Answer: $-12y^2 + 7y^2 = -5y^2$

Explanation: We subtracted the coefficients and kept the variable and its exponent the same.

c. $-4m + 4m$

Answer: $-4m + 4m = 0$

Explanation: Because $-4m$ and $4m$ are additive inverses, their sum is 0.

> **Reminder**
> When adding two numbers that have the same sign, we add their absolute values and keep the same sign.

d. $-2xy^3 - 5xy^3$

Answer: $-2xy^3 - 5xy^3 = -7xy^3$

Explanation: We added the coefficients and kept the variables and their exponents the same.

Why would we want to combine like terms? Why simplify?

Suppose you were asked to evaluate $-12y^2 + 7y^2$ for $y = 2$. You would have to replace each y with 2 and then calculate.

$$-12(2)^2 + 7(2)^2$$
$$= -12(4) + 7(4)$$
$$= -48 + 28$$
$$= -20$$

However, if we simplify the expression by combining like terms *before* evaluating, it makes the calculations a lot easier. When we combine like terms, $-12y^2 + 7y^2$ becomes $-5y^2$. Notice when we evaluate $-5y^2$ using $y = 2$, we get the same answer, but much faster.

$$= -5(2)^2$$
$$= -5(4)$$
$$= -20$$

Simplifying an expression makes the expression more compact so that when we evaluate the expression, it is much easier to do the work.

Incidentally, evaluating the original expression and its simplest form using the same variable values is a neat way to check to make sure they are equivalent. You can just choose values for the variables and evaluate both the original and simplified versions using these same values and you should get the same answer. If you don't then you're either calculating incorrectly or you've simplified incorrectly.

Do Margin 1. ▶

Sometimes there are several like terms in the same polynomial.

EXAMPLE 2 Combine like terms and write the resulting polynomial in descending order.

$$4x^3 + 9x^2 - x^3 + 5 + 7x + 2x^2 - 8 - 7x$$

Solution: $4x^3 + 9x^2 - x^3 + 5 + 7x + 2x^2 - 8 - 7x$ Collect like terms as needed.

$$= 4x^3 - x^3 + 9x^2 + 2x^2 + 7x - 7x + 5 - 8$$ Notice $7x$ and $-7x$ have a sum of 0.

$$= \quad 3x^3 \quad + \quad 11x^2 \quad + \quad 0 \quad - \quad 3$$ We can drop 0 and bring the terms together.

$$= 3x^3 + 11x^2 - 3$$

Explanation: The first thing we did was to collect the like terms. We used the commutative law of addition to rearrange the polynomial so that the like terms were next to each other. This is optional. If you're comfortable combining the like terms from just looking at the original polynomial, fine. Most people mark out the terms as they go like this:

Alternative Solution: $4x^3 + 9x^2 - x^3 + 5 + 7x + 2x^2 - 8 - 7x$

$$= 3x^3 \quad + \quad 11x^2 \quad - \quad 3$$

> **Think**
> $4x^3 + -x^3 = 3x^3$
> $9x^2 + 2x^2 = 11x^2$
> $7x + -7x = 0$
> $5 + -8 = -3$
> Mark out the terms as you combine.

> **Tips**
> 1. When combining like terms, mark through the terms that you combine so that you know what you've combined and what you have left.
> 2. Combine the terms in order of degree, combining the highest degree terms first, and so on. This will put the resulting polynomial in descending order.

◆ **Margin 1**

Combine like terms.

a. $6x + 9x$

b. $4y^2 - 20y^2$

c. $12x^3 - 12x^3$

d. $-m^2n - 12m^2n$

Answers: **a.** $15x$ **b.** $-16y^2$ **c.** 0
d. $-13m^2n$

Answers to Margin 2: **a.** $-6x^4 - 5x^3 - 3$ **b.** $9n^3 - 4n^2 + 19n - 18$ **c.** $14k^4 - 4k^2 - 6$ **d.** $-2t^3 - 18t + 6$

Answers to Margin 3: **a.** $-4x^3 - 14x^4y^2 + 9y - 14$ **b.** $12mn - 17n^2 + 29m + n - 6$ **c.** $20bc + 7a^2c - 19c - 11$

EXAMPLE 3   Combine like terms and write the resulting polynomial in descending order.

$$6x^5 + 9x^3 - 7x^6 + 15 + 2x^5 - 9x^3 + x - 18$$

Solution: Let's follow the tips and combine the terms in descending order, marking through the terms as we go.

$$6x^5 + 9x^3 - 7x^6 + 15 + 2x^5 - 9x^3 + x - 18$$
$$= -7x^6 \quad + \quad 8x^5 \quad - \quad 0 \quad + \quad x \quad - \quad 3$$
$$= -7x^6 + 8x^5 + x - 3$$

Explanation: First, we wrote $-7x^6$ because it was the highest degree term. We simply brought it down because it did not have a like term to combine with.

Next, we combined the degree-five terms.

$$6x^5 + 2x^5 = 8x^5$$

The degree-three terms equal 0.

$$9x^3 + -9x^3 = 0$$

We brought down the x term because there was no other x term.

Finally, we combined the degree-zero terms.

$$15 + -18 = -3$$

◀ **Do Margin 2.**

EXAMPLE 4   Combine like terms.

$$4xy^2 + 9x^4 - 17 - 10x^2y + 3x^2 + 8y + x^4 - 2x^2y + 5$$

Solution: If we collect the like terms and then combine, we have:

$$4xy^2 + 9x^4 - 17 - 10x^2y + 3x^2 + 8y + x^4 - 2x^2y + 5$$
$$= 4xy^2 + 9x^4 + x^4 - 10x^2y - 2x^2y + 3x^2 + 8y - 17 + 5$$
$$= 4xy^2 + 10x^4 - 12x^2y + 3x^2 + 8y - 12$$

If we follow the tip and mark through terms as we combine, we have:

$$4xy^2 + 9x^4 - 17 - 10x^2y + 3x^2 + 8y + x^4 - 2x^2y + 5$$
$$= 4xy^2 + 10x^4 - 12x^2y + 3x^2 + 8y - 12$$

Explanation: The $4xy^2$ term does not have a like term, so we bring it down.

$9x^4$ combines with x^4 to equal $10x^4$.

$-10x^2y$ combines with $-2x^2y$ to equal $-12x^2y$.

The $3x^2$ term has no like terms, so we bring it down.

The $8y$ term has no like terms, so we bring it down.

Finally, -17 and 5 combine to equal -12.

◀ **Do Margin 3.**

3.3 Exercises

FOR EXTRA HELP

 Videotape 3

 InterAct Math Tutorial Software

 www.carsonmath.com

 AWL Math Tutor Center

 InterAct MathXL www.mathxl.com

 Student's Solutions Manual

For Exercises 1–36, simplify by combining like terms.

1. $3x + 9x$

2. $5a + 2a$

3. $3y + y$

4. $b + 7b$

5. $12m - 7m$

6. $10x - 2x$

7. $n - 6n$

8. $5w - 14w$

9. $-6x + 8x$

10. $-5y + 2y$

11. $-15a - a$

12. $-9m - 3m$

13. $8a^2 + 5a^2$

14. $7x^3 - 2x^3$

15. $2y^2 - 9y^2$

16. $m^4 + 13m^4$

17. $j^3 + j^3$

18. $6n^8 - 19n^8$

19. $2b^2 - 2b^2$

20. $17c^4 - c^4$

21. $-3y^3 + 9y^3$

22. $-9x^2 + 9x^2$

23. $-4m^5 - 4m^5$

24. $-10t^2 + t^2$

25. $6xy - 7xy$

26. $-10mn + 4mn$

27. $5a^2b + 13ba^2$

28. $-8n^3m^2 - 9m^2n^3$

29. $-2t^2u^4 + 5tu^3$

30. $4ab^2c + 5abc$

31. $-19m^3n + 20m^3n$

32. $-30x^4y^2z - x^4y^2z$

33. $4a^2b + 5a^2b - 2a^2b$

34. $6xy - 9xy + 3xy$

35. $4t^3u - 9\,t^3u + 4t^3u$

36. $15x^2y^2 - 7x^2y^2 - 15x^2y^2$

For Exercises 37–46, simplify and write the resulting polynomial in descending order.

37. $9x^2 + 5x + 2x^2 - 3x + 4$

38. $7a - 8a^3 - a + 4 - 5a^3$

39. $12m^2 + 5m^3 - 11m^2 + 3 + 2m^3 - 7$

40. $4y + 15 - y^2 - 16 + 5y + 4y^3 + 6y^2$

41. $-7x + 5x^4 - 2 + 3x^4 + 7x + 10 - 3x^2$

42. $-10b^3 + 7b^4 + 1 + b^3 - 2b^2 - 12 + b^4$

43. $t^5 + 7t^2 - 20 + 6t^4 - 7t^2 + 9t^3 + 18 - 4t - 8t^5$

44. $2x^3 - x^7 + 3x + 10x^3 - 3x^7 - 3x + x^3 - 5 + 12x^2 + 4$

45. $-3y^2 + 15y - 13 - 6y^2 + 9y^4 - 8y^3 - 12y^4 + 13 + 3y - 10y^4$

46. $-8m + 17m^6 + 12m^3 - 19 - 14m - 10 - m^2 + 5m^4 - 10m^6 + 3m^2$

For Exercises 47–56, simplify.

47. $9tu^2 + 15t^3 - 17 + 4t - 3tu^2 + 3t + 5$

48. $-6x^2y - 8 + 5xy^2 + 2x^2 - x^2y + 6 + 7x^2$

49. $jk^3 + j^4 - 6jk + jk^3 - 5j^2 + 3j^4 - 2jk + 9$

50. $-a^3b - 9ab + 1 - 16a^2b^2 + 2a^3b - 8 + ab + 19a^2b^2$

51. $2m^4 + 5mn - 7m^2n^2 + 9mn - 2m^2n - 18 + 4m^2n^2 - 3 - 2m^4$

52. $-7a^2b^3 - 2ab^2 + 9a^3 + 15 - 8b^2 + 2ab^2 + 3a^2 - 10a^3 - 19 - 4a^2b^3$

53. $x^3y - 3xy^2 + 9x^2y + 5xy^2 - 1 + x^3y - 4xy - 8x^2y + 3x^3y^2 + 11 - 2xy$

54. $8tu^4 + 5tu - 20 - 3t^2u - 12tu^4 - 6tu + 14 + 9t^2u^2 - 11t^2u + 15t^2u^3 + 7t^2u^2$

55. $-4xy^2z + 18xy^2 - 20 + 5x^2y + 15xy^2 - 9xy^2z - 3 - 14y^2x - 5x^2y + 13$

56. $3a^2b^2 + 14a^2 - 19b^2 - 13abc^2 + 4 - 3ab + a^2 - 5b^2a^2 + 3ab - 2b^2 + 10ac^2b$

REVIEW EXERCISES

1. A border strip is to be glued to the edge of an office desk. How much border material is required?

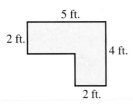

2. $-(-6) =$

3. $-285 + 173 =$

4. $-41 - (-382) =$

5. What property is illustrated by $2(3 + 4) = 2 \cdot 3 + 2 \cdot 4$?

3.4 Adding and Subtracting Polynomials

OBJECTIVES

1 Add polynomials.

2 Write an expression for the perimeter of a given shape.

3 Subtract polynomials.

OBJECTIVE 1 *Add polynomials.*

We can add and subtract expressions in the same way that we add and subtract numbers. In fact, polynomials are like whole numbers that are in expanded form.

Consider the polynomial $3x^2 + 5x + 2$. If we replace the xs with the number 10, then we have expanded form for the number 352.

$$3x^2 + 5x + 2$$
$$3 \cdot 10^2 + 5 \cdot 10 + 2$$
$$= 300 + 50 + 2$$
$$= 352$$

We say our number system is a base-10 system because each place value is a power of 10. With polynomials, the place values are variables, where x^2 is like the hundreds place, and x is like the tens place.

To add whole numbers, we add the digits in like place values. We add polynomials in the same way. However, instead of adding digits in like place values, we add like terms. Consider the following comparison:

Numeric addition:

$352 + 231 = 583$

Polynomial addition:

$(3x^2 + 5x + 2) + (2x^2 + 3x + 1) = 5x^2 + 8x + 3$

We can stack like place values to figure:

$$
\begin{array}{r}
352 \\
+\ 231 \\
\hline
583
\end{array}
$$

We can stack like terms to figure:

$$
\begin{array}{r}
3x^2 + 5x + 2 \\
+\ 2x^2 + 3x + 1 \\
\hline
5x^2 + 8x + 3
\end{array}
$$

Actually, we do not have to stack polynomials in order to add them as long as we combine all the like terms.

> **Procedure** *To add polynomials, combine like terms.*

EXAMPLE 1 Add. $(5x^3 + 9x^2 + 2x + 3) + (3x^3 + 5x + 1)$

Solution: $(5x^3 + 9x^2 + 2x + 3) + (3x^3 + 5x + 1)$

$$= 8x^3 + 9x^2 + 7x + 4$$

Explanation: We combined like terms in order of degree.

First the degree-three terms.

$$5x^3 + 3x^3 = 8x^3$$

There was no degree-two term in the second polynomial, so we brought down the $9x^2$.

Next we combined the degree-one terms.

$$2x + 5x = 7x$$

Finally we combined the degree-zero terms.

$$3 + 1 = 4$$

Note: If we had stacked the polynomials to figure the addition, we would have had to leave a blank space under the $9x^2$ because there was no degree-two term in the second polynomial. It's the same as having a zero in that place.

$$
\begin{array}{r}
5x^3 + 9x^2 + 2x + 3 \\
+3x^3 \quad\quad\quad + 5x + 1 \\
\hline
8x^3 + 9x^2 + 7x + 4
\end{array}
$$

As the polynomials get more complex, these blanks become more common, which is why stacking is not a preferred method for adding polynomials.

◀ **Do Margin 1.**

EXAMPLE 2 $(4x^5 + 7x^3 - 9x - 6) + (2x^5 - 5x^3 + 3x - 2)$

Solution: $(4x^5 + 7x^3 - 9x - 6) + (2x^5 - 5x^3 + 3x - 2)$

$$= 6x^5 + 2x^3 - 6x - 8$$

Explanation: We combined degree-five terms first.

$$4x^5 + 2x^5 = 6x^5$$

Next, we combined degree-three terms.

$$7x^3 + (-5x^3) = 2x^3$$

(Because $2x^3$ is positive, we write $+2x^3$ in the polynomial)

Next, we combined degree-one terms.

$$-9x + 3x = -6x$$

(Because $-6x$ is negative, we write $-6x$ in the polynomial)

Finally, we combined degree-zero terms.

$$-6 + (-2) = -8$$

Note: We assign the sign in between terms based on the outcome of combining.

If the outcome is positive, we place a plus sign to the left of the term.

If the outcome is negative, we place a minus sign to the left of the term.

◆ **Margin 1**

Add.

a. $(6x^3 + 5x^2 + 2x + 8) + (2x^2 + 6x + 1)$

b. $(y^4 + 3y^2 + 7y + 4) + (4y^4 + 7y^3 + y + 3)$

c. $(t^5 + 6t^3 + 4t + 2) + (3t^5 + 2t^4 + t + 1)$

Answers: **a.** $6x^3 + 7x^2 + 8x + 9$ **b.** $5y^4 + 7y^3 + 3y^2 + 8y + 7$ **c.** $4t^5 + 2t^4 + 6t^3 + 5t + 3$

EXAMPLE 3 $(4m^3 + 7mn^2 - 9mn + 13) + (-7m^3 - 9mn^2 + 4)$

Solution: $(4m^3 + 7mn^2 - 9mn + 13) + (-7m^3 - 9mn^2 + 4)$

$$= -3m^3 - 2mn^2 - 9mn + 17$$

Explanation: We combined like terms.

Beginning with the $4m^3$ and $-7m^3$ we combined to get $-3m^3$.

The next term was $7mn^2$, which combined with $-9mn^2$ to make $-2mn^2$.

Next we encountered $-9mn$. However it does not have a like term so we could not combine it with any other term.

◆ Finally, we combined 13 with 4 to get 17.

Do Margin 2. ▶

OBJECTIVE 2 *Write an expression for the perimeter of a given shape.*

EXAMPLE 4 Write an expression in simplest form for the perimeter of the rectangle shown.

Solution: To find the perimeter, we need to find the total distance around the shape. Instead of numerical measurements for the length and width, we have expressions. In this case the length is the expression $3x - 1$ and the width is the expression $2x + 5$. We can write only an expression for the perimeter; we cannot get a numerical answer until we have a value for the variable.

$2x + 5$

$3x - 1$

$$\text{Perimeter} = \text{length} + \text{width} + \text{length} + \text{width}$$
$$= (3x - 1) + (2x + 5) + (3x - 1) + (2x + 5)$$
$$= 3x + 2x + 3x + 2x - 1 + 5 - 1 + 5$$
$$= 10x + 8$$

◆ **Answer:** The expression for the perimeter is $10x + 8$.

▌ **WARNING** $10x + 8$ is simplest form. It might be tempting to try to write it as $18x$, but $10x$ and 8 are not like terms and cannot be combined.

Do Margin 3. ▶

OBJECTIVE 3 *Subtract polynomials.*

We subtract polynomials just like we subtract in arithmetic. However, polynomials are tricky because they have signs involved like the integers. When we subtracted integers we found it simpler to write the subtraction statements as equivalent addition statements. We can apply this same principle to polynomials.

◆ **Margin 2**

Add.

a. $(9x^4 - 12x^2 + 3x - 14) + (7x^4 + 2x^3 - 8x - 3)$

b. $(12x^5 + 7x^3 + 10x^2 - 13x - 15) + (8x^4 - 7x^3 + 3x^2 - 5x + 19)$

c. $(9m^3 + 4mn^2 - 8mn - 14n^2) + (2m^3 - 7mn^2 - 8mn + 3n^2)$

d. $(10a^4 - 8ab^3 + 6a^2b^2 - 15ab + c^2) + (5a^4 + 9ab^3 - 6a^2b^2 + 3ab^2 - 5c^2)$

◆ **Margin 3**

Write an expression in simplest form for the perimeter of the shape below.

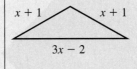

$x + 1$ $x + 1$

$3x - 2$

Answers to Margin 2: a. $16x^4 + 2x^3 - 12x^2 - 5x - 17$ ***b.*** $12x^5 + 8x^4 + 13x^2 - 18x + 4$ ***c.*** $11m^3 - 3mn^2 - 16mn - 11n^2$ ***d.*** $15a^4 + ab^3 + 3ab^2 - 15ab - 4c^2$

Answer to Margin 3: $5x$

In the following example, numeric subtraction is at the left and its related polynomial subtraction is next to it, for comparison.

Numeric subtraction:
$$896 - 254$$

Polynomial subtraction:
$$(8x^2 + 9x + 6) - (2x^2 + 5x + 4)$$

To figure, we can stack by place.

$$
\begin{array}{r}
896 \\
-254 \\
\hline
642
\end{array}
\qquad
\begin{array}{r}
8x^2 + 9x + 6 \\
-\ 2x^2 + 5x + 4 \\
\hline
6x^2 + 4x + 2
\end{array}
$$

In the numeric subtraction we subtracted digits in the same place value. In the polynomial subtraction, we subtracted like terms. We subtract 4 from 6, $5x$ from $9x$, and $2x^2$ from $8x^2$. Note that subtracting 4 from 6 is the same as combining 6 and -4, subtracting $5x$ from $9x$ is the same as combining $9x$ and $-5x$, and subtracting $2x^2$ from $8x^2$ is the same as combining $8x^2$ and $-2x^2$.

Conclusion: We can write polynomial subtraction as equivalent polynomial addition by changing the signs of each term in the subtracted polynomial.

Note: The governing principle in changing the signs is the distributive property. If we disregard the initial polynomial, we have:
$$-(2x^2 + 5x + 4) = -1(2x^2 + 5x + 4)$$
The distributive property tells us we can distribute the -1 (or minus sign) to each term inside the parentheses.
$$= -1 \cdot 2x^2 - 1 \cdot 5x - 1 \cdot 4$$
$$= -2x^2 - 5x - 4$$

We can write the polynomial subtraction as equivalent polynomial addition this way:

$$(8x^2 + 9x + 6) - (\ 2x^2 + 5x + 4)$$

Change the minus sign to a plus sign. Change all signs in the subtrahend.

$$= (8x^2 + 9x + 6) + (-2x^2 - 5x - 4)$$

$$= 6x^2 + 4x + 2$$

Just like with integers, we write subtraction as equivalent addition by figuring the additive inverse of the subtrahend. However, to get the additive inverse of an expression, we change *all* the signs in the expression.

Procedure *To subtract polynomials:*
1. Write the subtraction statement as an equivalent addition statement.
   a. Change the operation symbol from a minus sign to a plus sign.
   b. Change the subtrahend to its additive inverse. To get the additive inverse, we change *all* the signs in the polynomial.
2. Combine like terms.

EXAMPLE 5 Subtract. $(9x^3 + 7x^2 + 8x + 5) - (2x^3 + 3x^2 + 7x + 1)$

Solution: Write an equivalent addition statement, then combine like terms.

$$(9x^3 + 7x^2 + 8x + 5) - (2x^3 + 3x^2 + 7x + 1)$$

Change the minus sign to a plus sign.   Change all signs in the subtrahend.

$$= (9x^3 + 7x^2 + 8x + 5) + (-2x^3 - 3x^2 - 7x - 1)$$

$$= 7x^3 + 4x^2 + x + 4 \quad \text{Combine like terms.}$$

Explanation: We wrote the subtraction statement as an equivalent addition statement by figuring the additive inverse of the subtrahend. We changed the minus sign to a plus sign, changed all the signs in the second polynomial, and then combined the like terms.

We combined $9x^3$ and $-2x^3$ to get $7x^3$.

We combined $7x^2$ and $-3x^2$ to get $4x^2$.

We combined $8x$ and $-7x$ to get $1x$ or just x.

◆ Finally, we combined 5 and -1 to get 4.

Connection

The preceding polynomial subtraction is equivalent to the following numeric subtraction:

$$\begin{array}{r} 9785 \\ -2371 \\ \hline 7414 \end{array}$$

Notice the numeric result 7414 corresponds to the polynomial result $7x^3 + 4x^2 + x + 4$.

We could use the stacking method for subtracting polynomials; however, as we saw with adding polynomials, it is not the best method for all cases because we often have missing terms or terms that are not like terms.

Do Margin 4. ▶

EXAMPLE 6 Subtract. $(12x^3 + 5x^2 + 10x - 9) - (4x^3 + 8x^2 - 5x + 3)$

Solution: Write an equivalent addition statement, then combine like terms.

$$(12x^3 + 5x^2 + 10x - 9) - (4x^3 + 8x^2 - 5x + 3)$$

Change the minus sign to a plus sign.   Change all signs in the subtrahend.

$$= (12x^3 + 5x^2 + 10x - 9) + (-4x^3 - 8x^2 + 5x - 3)$$

$$= 8x^3 - 3x^2 + 15x - 12 \quad \text{Combine like terms.}$$

Explanation: We wrote the subtraction statement as an equivalent addition statement by figuring the additive inverse of the subtrahend.

We combined $12x^3$ with $-4x^3$ to get $8x^3$.

We combined $5x^2$ with $-8x^2$ to get $-3x^2$.

We combined $10x$ with $5x$ to get $15x$.

◆ Finally, we combined -9 with -3 to get -12.

◆ **Margin 4**

Subtract.

a. $(5x^2 + 7x + 3) - (2x^2 + 3x + 1)$

b. $(6t^3 + 8t^2 + 9t + 3) - (4t^3 + 7t^2 + t + 2)$

c. $(9y^4 + 4y^2 + 8y + 7) - (y^4 + 3y^2 + 5y + 2)$

Answers: **a.** $3x^2 + 4x + 2$ **b.** $2t^3 + t^2 + 8t + 1$ **c.** $8y^4 + y^2 + 3y + 5$

a. $(9x^2 - 7x + 5) - (2x^2 - 3x + 8)$

b. $(17x^4 + 9x^3 - 10x^2 - 12) - (13x^4 + 9x^3 - 3x^2 + 2)$

c. $(5a^3 + 7ab^2 - 12a^2b + 13b^2) - (6a^3 - 2ab^2 - 8a^2b + 10b^2)$

d. $(17m^5 - 9m^3n^2 + 14m^2n^2 - mn^2 + 13n + 5) - (20m^5 + 2m^3n^2 - mn^2 + 12n - 2)$

EXAMPLE 7 Subtract.

$$(19a^5 - 12a^3b^2 + 3ab^2 - 14ab - 7b^2) - (a^5 - 12a^3b^2 - 4ab^2 + 2a^2b - 3b^2)$$

Solution: We write an equivalent addition statement, then combine like terms.

$$(19a^5 - 12a^3b^2 + 3ab^2 - 14ab - 7b^2) - (a^5 \quad - 12a^3b^2 - 4ab^2 + 2a^2b - 3b^2)$$

Change the minus sign to a plus sign. ↓ ↓ ↓ ↓ ↓ ↓   Change all signs in the subtrahend.

$$= (19a^5 - 12a^3b^2 + 3ab^2 - 14ab - 7b^2) + (-a^5 + 12a^3b^2 + 4ab^2 - 2a^2b + 3b^2)$$

$$= 18a^5 + 7ab^2 - 14ab - 2a^2b - 4b^2$$

Explanation: We wrote the subtraction statement as an equivalent addition statement by figuring the additive inverse of the subtrahend.

We combined $19a^5$ with $-a^5$ to get $18a^5$.

Because $-12a^3b^2$ and $12a^3b^2$ are additive inverses, their sum is 0, which is not written in the final expression.

We combined $3ab^2$ with $4ab^2$ to get $7ab^2$.

$-14ab$ did not have a like term, so we wrote it in the result, unchanged.

$-2a^2b$ also did not have a like term, so we wrote it in the result, unchanged.

◆ Finally, we combined $-7b^2$ with $3b^2$ to get $-4b^2$.

◀ **Do Margin 5.**

3.4 Exercises

FOR EXTRA HELP

 Videotape 3

 InterAct Math Tutorial Software

 www.carsonmath.com

 AWL Math Tutor Center

InterAct MathXL www.mathxl.com

 Student's Solutions Manual

For Exercises 1–20, add and write the resulting polynomial in descending order.

1. $(2x + 5) + (3x + 1)$

2. $(5x + 2) + (3x + 7)$

3. $(7y - 4) + (2y - 1)$

4. $(9m - 3) + (4m - 6)$

5. $(10x + 7) + (3x - 9)$

6. $(4t - 11) + (t + 13)$

7. $(2x^2 + x + 3) + (5x + 7)$

8. $(7x^2 + 2x + 1) + (6x + 5)$

9. $(9n^2 - 14n + 7) + (6n^2 + 2n - 4)$

10. $(8t^2 + 12t - 10) + (2t^2 - 8t - 2)$

11. $(3a^3 - a^2 + 10a - 4) + (6a^2 - 9a + 2)$

12. $(9x^3 + 4x^2 + x - 11) + (3x^2 - 8x - 5)$

13. $(5x^3 + 7x^2 - 8x + 3) + (-3x^3 - 4x^2 + 8x - 5)$

14. $(7y^3 - 9y^2 + 5y - 1) + (-4y^3 + 9y^2 + y + 8)$

15. $(t^3 - 4t^2 - 9t + 5) + (8t^3 - t^2 + 10t - 1)$

16. $(m^3 + 10m^2 - 8m - 2) + (9m^3 - 11m^2 + 3m - 6)$

17. $(5m^4 + 6m^2 - 8m + 12) + (3m^3 + 9m - 15)$

18. $(7x^5 - 3x^3 + 5x^2 - 18) + (x^4 - 5x^3 - 3x^2 + 8)$

19. $(x^5 + 5x^4 - 9x^2 + 11) + (-2x^5 - 14x^4 - 6x^3 + 5)$

20. $(10a^6 + 7a^4 - 9a - 5) + (-6a^6 - 2a^4 + 11a^3 - 9)$

For Exercises 21–26, add.

21. $(12x^6 + 8x^4 + 9x^3 - 11x + 15) + (-9x^5 - 6x^4 + 3x^2 - 4x + 1)$

22. $(8u^5 - u^3 + 12u^2 - 13u - 1) + (2u^4 + u^3 - 15u^2 + 7)$

23. $(-x^5 + 9x^2y^2 - 2x^2y + 5xy - 3y^2) + (4x^5 - 10x^2y^2 + 6xy - y^2)$

24. $(-11a^6 + a^3b^2 + 2ab^3 - 6ab^2 + 9b^3) + (10a^6 - 4a^2b^3 - 3ab^3 + 10ab^2 + 2b^3)$

25. $(t^2u^3 + 9tu^2 - 18t^2u - 5tu + 6u^2 - 9) + (-2t^2u^2 - 3tu^2 + 4ut^2 - tu - 6u^2 + 8)$

26. $(5m^3n^4 - 8m^2n^2 - 11mn^2 + 12m^2n - 18n + 5) + (-6n^4m^3 - 8m^2n^2 + mn^2 + 15n - 7)$

For Exercises 27–30, write an expression for the perimeter in simplest form.

27.

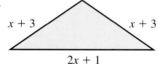

$x + 3$ $x + 3$

$2x + 1$

28.

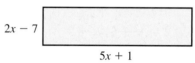

$2x - 7$

$5x + 1$

29.

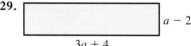

$a - 2$

$3a + 4$

30.

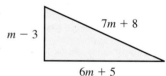

$7m + 8$

$m - 3$

$6m + 5$

For Exercises 31–34, find and explain the mistake(s) then work the problem correctly.

31. $(9y^3 + 4y^2 - 6y + 5) + (y^3 - 2y^2 + 3)$
$= 10y^3 - 2y^2 - 6y + 8$

32. $(10a^4 + 5a^3 + 3a - 11) + (5a^4 + 3a^2 + 2a - 4)$
$= 15a^4 + 3a^2 + 5a - 15$

33. $(8x^6 - 9x^4 + 15x^3 + 10x - 1) + (x^5 - 8x^3 - 9x - 3)$
$= 9x^{11} - 9x^4 + 7x^3 + x - 4$

34. $(5mn^2 + 9mn - m^2 + 2n^2 - 19) + (3mn^2 - 9mn + 2n^2 + 4)$
$= 8mn^2 + 18mn - m^2 + 4n^2 - 11$

For Exercises 35–52, subtract and write the resulting polynomial in descending order.

35. $(5x + 9) - (3x + 2)$

36. $(7m + 4) - (2m + 1)$

37. $(8t + 3) - (5t + 3)$

38. $(6a + 7) - (6a + 5)$

39. $(10n + 1) - (2n + 7)$

40. $(9x + 8) - (3x + 10)$

41. $(5x - 6) - (2x + 1)$

42. $(2m + 5) - (7m - 3)$

43. $(6x^2 + 8x - 9) - (9x + 2)$

44. $(8n^2 - 7n - 5) - (2n - 3)$

45. $(8a^2 - 10a + 2) - (-4a^2 + a + 8)$

46. $(y^2 + 3y - 9) - (-6y^2 + 8y - 1)$

47. $(7x^2 + 5x - 8) - (7x^2 + 9x + 2)$

48. $(2t^2 - t + 7) - (3t^2 - t - 6)$

49. $(-8m^3 + 9m^2 - 17m - 2) - (-8m^3 + 5m^2 + m - 6)$

50. $(5u^3 - 4u^2 + u - 10) - (-2u^3 + 4u^2 + u + 7)$

51. $(8x^5 - 6x^3 - 8x^2 + 9x + 17) - (5x^4 + 6x^3 + 10x - 2)$

52. $(t^5 - 9t^4 - 11t^3 + t + 15) - (-t^5 + 4t^4 - 11t^3 + 6t^2 + 20)$

For Exercises 53–58, subtract.

53. $(6t^3 - 5tu^2 + 8tu - 16u^2 + 9) - (6t^3 - 3tu^2 + 5u^2 + 4)$

54. $(-5x^4 + 6x^3y - 9xy^2 + 12x^3 + 13y) - (3x^4 + 6x^3y + 2xy^2 - 8x^2 + 5y + 2)$

55. $(-m^6 - m^3n^4 + 4m^2n^3 - 6mn - 10n^2 + 14) - (m^6 + 3n^4m^3 + 10m^2n^3 - 6mn + 8n^2 - 2)$

56. $(9a^3b^5 - 2ab^4 - 13a^2b^2 + 6ab^2 - ab + 7) - (-3b^5a^3 + 2ab^4 + 7a^2b^2 - ab + 12)$

57. $(11x^3y^4 + x^2y^2 + 10xy^3 - 8xy + 13y^2 - 10) - (-x^3y^4 + x^2y^2 + 3xy^3 - 2xy + 7)$

58. $(-3h^4k + 9h^2k^2 + 11hk^2 - 15hk + h^2 - 12) - (-3h^4k + 2h^2k^2 + 15k^2h - 9hk - h^2 + 4)$

For Exercises 59–62, find and explain the mistake, then work the problem correctly.

59. $(5x^2 - 6x + 10) - (3x^2 - 4x + 2)$

$= (5x^2 - 6x + 10) + (-3x^2 - 4x + 2)$

$= 2x^2 - 10x + 12$

60. $(10y^2 - 5y + 1) - (y^2 + 3y - 1)$

$= (10y^2 - 5y + 1) + (-y^2 - 3y + 1)$

$= 9y^2 + 8y + 2$

61. $(-16t^4 - 9t^2 + 4t + 11) - (3t^4 + 5t^3 + 9t^2 + 6)$

$= (-16t^4 - 9t^2 + 4t + 11) + (-3t^4 - 5t^3 - 9t^2 - 6)$

$= -19t^4 - 5t^3 + 4t + 5$

62. $(9x^5 + x^4 - 6x^2 + 3x - 1) - (12x^5 + x^4 + 2x^3 - 4x - 3)$

$= (9x^5 + x^4 - 6x^2 + 3x - 1) + (-12x^5 - x^4 - 2x^3 + 4x + 3)$

$= -3x^5 - 2x^4 - 2x^3 - 6x^2 + 7x - 2$

Puzzle Problem

A trog is a mythical creature that reproduces in a peculiar way. An adult trog will produce one newborn every hour. A newborn trog spends its first hour growing to an adult. If we place one newborn trog in a cage, how many trogs will exist after six hours? Seven hours? Explain the pattern.

REVIEW EXERCISES

1. $2 \cdot (-341) =$

2. $(-3)^4$

3. $(-2)^5 =$

4. A swimming pool measures 25 m by 15 m. What is the surface area of the water?

5. Calculate. $2^3 \cdot 2^4$

3.5 Multiplying Polynomials

OBJECTIVES

1 Multiply monomials.

2 Simplify monomials raised to a power.

3 Multiply a polynomial by a monomial.

4 Multiply a polynomial by a polynomial.

5 Write an expression for the area of a parallelogram or the volume of a box.

OBJECTIVE **1** *Multiply monomials.*

To simplify $2^3 \cdot 2^4$ we use the order of operations agreement.

$$2^3 \cdot 2^4$$
$$= 8 \cdot 16$$
$$= 128$$

However, there is an alternative. We can express the result in exponential form by expanding each of 2^3 and 2^4 into their factored forms.

$$\overbrace{2^3} \cdot \overbrace{2^4}$$

$$= \underbrace{2 \cdot 2 \cdot 2} \cdot \underbrace{2 \cdot 2 \cdot 2 \cdot 2}$$

Because there are a total of seven 2's multiplied, we can express the product as 2^7.

$$= \qquad 2^7$$

Note that $2^7 = 128$. Notice in the alternative method that the resulting exponent is the sum of the original exponents.

$$2^3 \cdot 2^4 = 2^{3+4} = 2^7$$

Conclusion: When multiplying exponential forms that have the same base, we can add the exponents and keep the same base.

> **Rule** When multiplying exponential forms that have the same base, we can add the exponents and keep the same base.
>
> **In math language:** $n^a \cdot n^b = n^{a+b}$

Discussion Why do the exponential forms have to have the same base for us to add the exponents?

We can apply this rule to any exponential forms that have the same base, even variable bases.

EXAMPLE 1 Multiply. $(x^2)(x^3)$

Solution: $(x^2)(x^3) = x^{2+3} = x^5$

Explanation: Because the bases are the same, we can add the exponents and keep the same base.

Let's break the exponential forms down.

x^2 means two xs $\quad (x^2)(x^3) \quad\quad$ x^3 means three xs

$$= (x \cdot x)(x \cdot x \cdot x) \quad \text{There are a total of five } x\text{s multiplied together.}$$
$$= x^5$$

◀ **Do Margin 1.**

Margin 1

Multiply.

a. $x^3 \cdot x^4 =$

b. $(n^6)(n^2) =$

c. $m^5(m) =$

Let's apply this to monomials that have coefficients.

EXAMPLE 2 Multiply. $(2x^4)(3x^6)$

Solution: $(2x^4)(3x^6)$
$$= 2 \cdot 3 \, x^{4+6} \quad \text{Multiply coefficients and add exponents for the like } x \text{ bases.}$$
$$= 6x^{10}$$

Explanation: This problem is one big multiplication problem. If we expanded it fully it would look like this:

$$2 \cdot x \cdot x \cdot x \cdot x \cdot 3 \cdot x \cdot x \cdot x \cdot x \cdot x \cdot x$$

We made use of the commutative property of multiplication and brought the coefficients together, then used an exponent to indicate the total number of xs that were multiplied. To get the exponent total, we simply added the exponents, $4 + 6 = 10$.

Connection

Multiplying monomials like $(2x^4)(3x^6)$ is the same as multiplying numbers in expanded form. If we replace the xs with the number 10, then we have expanded form.

$$(2x^4)(3x^6)$$
$$(2 \cdot 10^4)(3 \cdot 10^6)$$

Multiplying numbers in expanded form must work the same way as multiplying monomials. We can multiply the $2 \cdot 3$ to get 6 and add the exponents of the like bases.

$$2 \cdot 3 \cdot 10^{4+6}$$
$$= 6 \cdot 10^{10}$$

If you're not convinced, then write the expand forms in standard form then multiply.

$$(2 \cdot 10^4)(3 \cdot 10^6)$$
$$= (20,000)(3,000,000)$$
$$= 60,000,000,000$$
$$= 6 \cdot 10^{10} \quad \text{This is a 6 followed by ten 0s.}$$

Margin 2

Multiply.

a. $4x^5 \cdot 2x^8 =$

b. $(4 \cdot 10^5)(2 \cdot 10^8) =$

c. $(9a^5)(7a) =$

d. $12m^2(m^8) =$

From Examples 1 and 2, we can conclude the following:

Conclusion: To multiply monomials, we multiply the coefficients and add the exponents of the like bases.

Procedure *To multiply monomials:*
1. Multiply coefficients.
2. Add the exponents of the like bases.

◀ **Do Margin 2.**

Answers to Margin 1: **a.** x^7 **b.** n^8
c. m^6

Answers to Margin 2: **a.** $8x^{13}$
b. $8 \cdot 10^{13}$ **c.** $63a^6$ **d.** $12m^{10}$

Connection

Multiplying distance measurements in area and volume is like multiplying monomials.

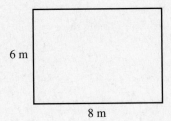

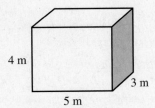

The area of the parallelogram above is:

$$A = bh$$
$$A = (8 \text{ m})(6 \text{ m})$$
$$A = 48 \text{ m}^2$$

The area calculation above is similar to multiplying two polynomials with m as the variable.

Multiply: $(8m)(6m) = 48m^2$

The volume of the box above is:

$$V = lwh$$
$$V = (5 \text{ m})(3 \text{ m})(4 \text{ m})$$
$$V = 60 \text{ m}^3$$

The volume calculation above is similar to multiplying three polynomials with m as the variable.

Multiply: $(5m)(3m)(4m) = 60m^3$

EXAMPLE 3   Multiply. $(-7a^2b^3)(9a^5bc^2)$

Solution: Multiply the coefficients and add the exponents of the like bases.

> **Reminder**
> A variable or number with no apparent exponent has an understood exponent of 1. In this case, $9a^5bc^2 = 9a^5b^1c^2$

> **Reminder**
> The product of two numbers with different signs is a negative number.

$$(-7a^2b^3)(9a^5bc^2)$$
$$= -7 \cdot 9a^{2+5} \, b^{3+1} \, c^2$$
$$= -63a^7b^4c^2$$

Do Margin 3. ▶

OBJECTIVE 2   *Simplify monomials raised to a power.*

To simplify $(2^3)^2$ we follow the order of operations agreement and evaluate the exponential form within the parentheses first.

$$(2^3)^2$$
$$= (2 \cdot 2 \cdot 2)^2$$
$$= 8^2$$
$$= 64$$

However, there is an alternative. The outside exponent, 2, indicates that we must multiply the inside exponential form by itself.

$$(2^3)^2 = 2^3 \cdot 2^3 \qquad \text{Because this is a multiplication of exponential forms}$$
$$= 2^{3+3} \qquad \qquad \text{that have the same base, we can add the exponents.}$$
$$= 2^6$$

Note that $2^6 = 64$. Notice that the exponent in the result is the product of the original exponents.

$$(2^3)^2 = 2^{3 \cdot 2} = 2^6$$

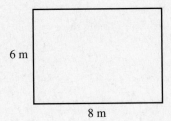

◆ **Margin 3**

Multiply.

a. $(-2xy^2)(8x^3)$

b. $-9x^2y^3(-5xy^2z^3)$

c. $(-4m^2)(3mn)(2m^3n^4)$

d. $(2a^3b)(-3a^2)(-ab^2c)$

Answers: **a.** $-16x^4y^2$ **b.** $45x^3y^5z^3$
c. $-24m^6n^5$ **d.** $6a^6b^3c$

Margin 4

Simplify.

a. $(3x^2)^4$

b. $(5a^6)^3$

c. $(-2m^4)^5$

d. $(-2t^3)^6$

Conclusion: When an exponential form is raised to a power, we can multiply the exponents and keep the same base.

> **Rule** When an exponential form is raised to a power, we can multiply the exponents and keep the same base.
>
> **In math language:** $(n^a)^b = n^{a \cdot b}$

What if we raise a monomial to a power?

EXAMPLE 4 Simplify. $(2x^3)^4$

Solution: The exponent, 4, means there are four repetitions of $2x^3$.

$$(2x^3)^4$$
$$= 2x^3 \cdot 2x^3 \cdot 2x^3 \cdot 2x^3$$
$$= 2 \cdot 2 \cdot 2 \cdot 2 \cdot x^{3+3+3+3}$$
$$= 16x^{12}$$

> **Note** $(2x^3)^4$ is read, "Two x to the third power, raised to the fourth power." Both 3 and 4 are exponents and are referred to as 3rd *power* and 4th *power*. However, in an effort to make the explanations easier to follow, we will use the word *power* to refer to the exponent outside the parentheses and the word *exponent* to refer to the exponents on the variables.

Explanation: The power outside the parentheses indicated that there were 4 factors of 2 and 4 factors of x^3. So the final coefficient is the result of evaluating 2^4, and the final exponent is the product of the power and the exponent on the variable.

Conclusion: When a monomial is raised to a power, we apply the power to each factor in the monomial; that is, we evaluate the coefficient raised to the power and multiply each variable's exponent by the power.

> **Procedure** *To simplify a monomial raised to a power:*
> 1. Evaluate the coefficient raised to that power.
> 2. Multiply each variable's exponent by the power.

◀ **Do Margin 4.**

EXAMPLE 5 Simplify. $(-3xy^5z^2)^4$

Solution: We must first evaluate the coefficient, -3, raised to the 4th power. Then we multiply each variable's exponent by 4.

$$(-3xy^5z^2)^4$$
$$= (-3)^4 x^{1 \cdot 4} y^{5 \cdot 4} z^{2 \cdot 4}$$
$$= 81x^4 y^{20} z^8$$

> **Reminder**
> $x = x^1$, so when we multiply the 1 exponent by 4 we have: $x^{1 \cdot 4} = x^4$.

◀ **Do Margin 5.**

OBJECTIVE 3 *Multiply a polynomial by a monomial.*

Suppose we have to multiply $2 \cdot 34$. To figure, we stack and multiply the 2 by each digit in the 34.

$$\begin{array}{r} 34 \\ \times 2 \\ \hline 68 \end{array}$$

Margin 5

Simplify.

a. $(2x^3y^5)^4$

b. $(-5m^6n^7)^3$

c. $(4a^3bc^8)^3$

d. $(-2r^4t^9u)^6$

Answers to Margin 4: a. $81x^8$
b. $125a^{18}$ **c.** $-32m^{20}$ **d.** $64t^{18}$

Answers to Margin 5: a. $16x^{12}y^{20}$
b. $-125m^{18}n^{21}$ **c.** $64a^9b^3c^{24}$
d. $64r^{24}t^{54}u^6$

In Chapter 1 we learned that when multiplying numbers with more than one digit we are applying the distributive property. Consider the same problem in distributive property format. We can expand the 34 into $30 + 4$ to make the problem appear like the distributive property form.

$$2(34) = 2(30 + 4)$$

If we carry out the distributive property and distribute the 2 to the 30 and the 4, we have:

$$2(30 + 4) = 2 \cdot 30 + 2 \cdot 4$$
$$= 60 + 8$$
$$= 68$$

We've established that polynomials are like numbers expressed in expanded form where the terms are like place values. We can develop a multiplication problem that is, in essence, the same as the preceding numeric problem.

$$2(34) \qquad \text{is much like} \qquad 2(3x + 4)$$

The multiplication works the same way. We must still apply the distributive property. The difference, of course, is that in the algebraic expression version we won't be able to arrive at a final numeric result because we do not know a value for x. Compare the numeric version and algebraic versions:

$$2(30 + 4) = 2 \cdot 30 + 2 \cdot 4 \qquad\qquad 2(3x + 4) = 2 \cdot 3x + 2 \cdot 4$$
$$= 60 + 8 \qquad\qquad\qquad\qquad = 6x + 8$$
$$= 68$$

Notice, if $x = 10$ in the algebraic version, we are right back to the numeric version.

Procedure *To multiply a polynomial by a monomial, use the distributive property to multiply each term in the polynomial by the monomial.*

EXAMPLE 6 Multiply. $2x(3x^2 + 4x + 1)$

Solution: We multiply each term in the polynomial by $2x$.

$2x \qquad (3x^2 + 4x + 1)$
$2x \cdot 3x^2$
$2x \cdot 4x$
$2x \cdot 1$

$= 2x \cdot 3x^2 + 2x \cdot 4x + 2x \cdot 1$
$= 6x^3 + 8x^2 + 2x$

Connection

Multiplying $2x(3x^2 + 4x + 1)$ is essentially the same as this numeric problem:

$20(300 + 40 + 1) = 20 \cdot 300 + 20 \cdot 40 + 20 \cdot 1$
$= 6000 + 800 + 20$

If you think of x^3 as thousands place, x^2 as hundreds, and x as tens, then the algebraic result is the same.

$$2x(3x^2 + 4x + 1) = 6x^3 + 8x^2 + 2x$$

Reminder
$2x = 2x^1$, and $4x = 4x^1$.
$2x \cdot 3x^2 = 2x^1 \cdot 3x^2 = 6x^{1+2} = 6x^3$
$2x \cdot 4x = 2x^1 \cdot 4x^1 = 8x^{1+1} = 8x^2$

Margin 6

Multiply.

a. $2x(4x^2 + 2x + 3)$

b. $5a^2(6a^2 - 7a + 2)$

c. $-4n^3(8n - 9)$

d. $-7m^2(4m^2 - m + 3)$

Margin 7

Multiply.

a. $3xy^2 (4x^2 + 5xy - 6y^2)$

b. $-6x^3yz (4x^2z - 9y^3z^2)$

c. $ab^4c (5ab^3 + 9a^2c^2 - 6bc + 4c)$

d. $-t^4u^2v (t^5v^2 - 9t^2uv + 2u^3v - 8tv)$

Answers to Margin 6: **a.** $8x^3 + 4x^2 + 6x$ **b.** $30a^4 - 35a^3 + 10a^2$ **c.** $-32n^4 + 36n^3$ **d.** $-28m^4 + 7m^3 - 21m^2$

Answers to Margin 7: **a.** $12x^3y^2 + 15x^2y^3 - 18xy^4$ **b.** $-24x^5yz^2 + 54x^3y^4z^3$ **c.** $5a^2b^7c + 9a^3b^4c^3 - 6ab^5c^2 + 4ab^4c^2$ **d.** $-t^9u^2v^3 + 9t^6u^3v^2 - 2t^4u^5v^2 + 8t^5u^2v^2$

EXAMPLE 7 Multiply. $-3x^2(4x^2 + 5x - 6)$

Solution: Multiply each term in the polynomial by $-3x^2$. Watch the signs. We are multiplying by a negative so the signs are affected.

$$-3x^2 \quad (4x^2 + 5x - 6)$$

$-3x^2 \cdot 4x^2$

$-3x^2 \cdot 5x$

$-3x^2 \cdot (-6)$

$$= -3x^2 \cdot 4x^2 - 3x^2 \cdot 5x - 3x^2 \cdot (-6)$$
$$= -12x^4 - 15x^3 + 18x^2$$

Reminder
The sign to the left of the term not only indicates addition or subtraction, but also indicates the sign of the term. In the polynomial $+5x$ not only means we are adding $5x$ but also means the $5x$ is positive. When we see -6 it means not only subtract 6 but also negative 6.

Explanation: When we multiplied the coefficients, the signs changed because the monomial was negative.

$$-3x^2 \cdot 4x^2 = -12x^4$$
$$-3x^2 \cdot 5x = -15x^3$$
$$-3x^2 \cdot (-6) = +18x^2$$

Tip
When we multiply a polynomial by a negative monomial, the signs of the resulting polynomial will be the opposite of the signs in the original polynomial.

$$-3x^2 (4x^2 + 5x - 6)$$
$$= -12x^4 - 15x^3 + 18x^2$$

◀ **Do Margin 6.**

EXAMPLE 8 Multiply. $-5x^2yz(8x^3y - yz^2 - 3y^3 + 4z)$

Solution: Multiply each term in the polynomial by the monomial. We must be careful with the signs because the monomial is negative.

$$-5x^2yz \quad (8x^3y - yz^2 - 3y^3 + 4z)$$

$-5x^2yz \cdot 8x^3y$

$-5x^2yz \cdot (-yz^2)$

$-5x^2yz \cdot (-3y^3)$

$-5x^2yz \cdot 4z$

$$= -5x^2yz \cdot 8x^3y - 5x^2yz \cdot (-yz^2) - 5x^2yz \cdot (-3y^3) - 5x^2yz \cdot 4z$$
$$= -40x^5y^2z + 5x^2y^2z^3 + 15x^2y^4z - 20x^2yz^2$$

Tip
Develop a consistent systematic approach. For each multiplication build each part of the resulting term one piece at a time. Multiply the coefficients first, then the variables in alphabetical order.

For $-5x^2yz \cdot 8x^3y$, think:
$$-5 \cdot 8 = -40$$
$$x^2 \cdot x^3 = x^5$$
$$y \cdot y = y^2$$
z is simply rewritten.

Result: $-40x^5y^2z$

◀ **Do Margin 7.**

OBJECTIVE **4** *Multiply a polynomial by a polynomial.*

We will see that multiplying two binomials is the same as multiplying a pair of two-digit numbers; for example, $(x + 2)(x + 3)$ is the same as $(12)(13)$.

To see how to multiply polynomials, let's once again consider how we multiply numbers. Consider the procedure for multiplying $(12)(13)$.

$$\begin{array}{r} 12 \\ \times 13 \\ \hline 36 \\ + 12 \\ \hline 156 \end{array}$$

We think to ourselves 3 times 2 is 6, then 3 times 1 is 3, which creates 36. We then move to the 1 in the tens place of the 13 and do the same thing. Because this 1 digit is in the tens place it really means 10, so when we multiply this 10 times 12, it makes 120. We usually omit writing the 0 in the ones place and write 12 in the next two places.

Behind the scenes we are still using the distributive property. We multiply each digit in one number by each digit in the other number and shift underneath as we move to each new place.

Now let's apply this same thinking to the binomials. We can stack them like above. However, we'll only stack to see how the method works. You'll find that as the polynomials get more complex, the stacking method becomes too tedious.

$$\begin{array}{r} x + 2 \\ x + 3 \\ \hline 3x + 6 \\ x^2 + 2x \\ \hline x^2 + 5x + 6 \end{array}$$

We think: 3 times 2 is 6, then 3 times x is $3x$. Now move to the x and think: x times 2 is $2x$, then x times x is x^2.

Notice how we shifted so that the $2x$ and $3x$ line up. This is because they are like terms. It is the same as lining up the tens column when we work with whole numbers.

Notice the preceding numeric result, 156, is essentially the same as the algebraic result, $x^2 + 5x + 6$.

Conclusion: We multiply each term in one polynomial by each term in the other polynomial, then combine the like terms.

Procedure *To multiply two polynomials:*
1. Multiply every term in the first polynomial by every term in the second polynomial.
2. Combine like terms.

EXAMPLE 9 Multiply. $(x + 5)(x + 1)$

Solution: We must multiply each term in $x + 1$ by each term in $x + 5$.

$$(x + 5)(x + 1)$$

$$
\begin{aligned}
&= x \cdot x + x \cdot 1 + 5 \cdot x + 5 \cdot 1 \\
&= x^2 + x + 5x + 5 \qquad\qquad \text{Combine like terms. } x + 5x = 6x \\
&= x^2 + 6x + 5 \\
&= x^2 + 6x + 5
\end{aligned}
$$

Explanation: First, we multiply x times x to get x^2.

Then, we multiply x times 1 to get x.

Next, we move to the 5 and multiply 5 times x to get $5x$.

Finally, we multiply 5 times 1 to get 5.

To finish, we combine like terms: x and $5x$ is $6x$.

Tip

The word FOIL is a popular way to remember the process of multiplying two binomials. FOIL stands for **F**irst **O**uter **I**nner **L**ast. We'll use Example 9 to demonstrate.

$(x + 1)\ (x + 5)$ **First** terms: $x \cdot x = x^2$

$(x + 1)\ (x + 5)$ **Outer** terms: $x \cdot 5 = 5x$

$(x + 1)\ (x + 5)$

   Inner terms: $1 \cdot x = x$

$(x + 1)\ (x + 5)$

   Last terms: $1 \cdot 5 = 5$

WARNING FOIL makes sense only when multiplying two binomials. If we multiply larger polynomials, there are too many inner terms. When you multiply polynomials, you must multiply *every term* in the second polynomial by *every term* in the first polynomial.

♦ Margin 8

Multiply.

a. $(x + 4)(x + 2)$

b. $(2y - 5)(3y + 1)$

c. $(n - 6)(7n - 3)$

d. $(t + 3)(4t - 1)$

Answers: **a.** $x^2 + 6x + 8$ **b.** $6y^2 - 13y - 5$ **c.** $7n^2 - 45n + 18$ **d.** $4t^2 + 11t - 3$

EXAMPLE 10   Multiply. $(2x + 4)(x - 3)$

Solution: We must multiply every term in $x - 3$ by every term in $2x + 4$. Because the problem is a multiplication of two binomials, we can think FOIL.

$$(2x\ +\ 4)\ (x\ -\ 3)$$

$2x \cdot (-3)$

$2x \cdot x$

$4 \cdot x$

$4 \cdot (-3)$

   First   **Outer**   **Inner**   **Last**

$= 2x \cdot x + 2x \cdot (-3) + 4 \cdot x + 4 \cdot (-3)$

$= 2x^2 \quad\quad -6x \quad\quad +4x \quad\quad -12$ **Combine like terms.**

$= 2x^2 \quad\quad\quad\quad -2x \quad\quad\quad -12$

$= 2x^2 - 2x - 12$

Explanation: First, we multiplied $2x \cdot x$ to get $2x^2$.

Next, we multiplied $2x \cdot (-3)$ to get $-6x$.

Next, we multiplied $4 \cdot x$ to get $4x$.

Finally, we multiplied $4 \cdot (-3)$ to get -12.

To finish, we combined like terms. The $-6x$ and $4x$ combine to get $-2x$.

◀ **Do Margin 8.**

The following sometimes occurs when multiplying two binomials.

EXAMPLE 11 Multiply. $(3x + 5)(3x - 5)$

Solution: We must multiply every term in $3x - 5$ by every term in $3x + 5$. Because they are binomials, we can think FOIL.

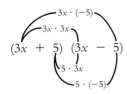

$$
\begin{array}{cccc}
\text{First} & \text{Outer} & \text{Inner} & \text{Last} \\
= 3x \cdot 3x & + 3x \cdot (-5) & + 5 \cdot 3x & + 5 \cdot (-5) \\
= 9x^2 & - 15x & + 15x & - 25 \\
= 9x^2 & & & - 25 \\
= 9x^2 - 25 & & &
\end{array}
$$

Notice that $-15x + 15x = 0$, which we do not write.

Explanation: We first multiplied $3x \cdot 3x$ to get $9x^2$.

Then, we multiplied $3x \cdot (-5)$ to get $-15x$.

Then, we multiplied $5 \cdot 3x$ to get $15x$.

Finally, we multiplied $5 \cdot (-5)$ to get -25.

To finish we combined like terms. However, in this case the like terms are additive inverses. $-15x$ and $15x$ combine to equal 0.

Notice the binomials are almost identical. The only difference is in the sign that separates the terms. We call these binomials **conjugates**.

DEFINITION **Conjugates:** Binomials that differ only in the sign separating the terms.

For example: $x + 9$ and $x - 9$ are conjugates.

$-5y - 7$ and $-5y + 7$ are conjugates.

Do Margin 9. ▶

From Example 11, we can make the following conclusion:

Conclusion: When we multiply conjugates, the result will always contain like terms that are additive inverses of each other. Because they will always be additive inverses of each other, these like terms will always add to 0.

Do Margin 10. ▶

EXAMPLE 12 Multiply. $(x + 4)(2x^2 + 5x - 3)$

Solution: We multiply every term in $2x^2 + 5x - 3$ by every term in $x + 4$.

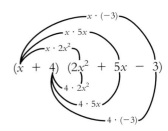

Note FOIL does not make sense here because there are too many terms. FOIL only handles the four terms from two binomials.

$$
\begin{aligned}
&= x \cdot 2x^2 + x \cdot 5x + x \cdot (-3) + 4 \cdot 2x^2 + 4 \cdot 5x + 4 \cdot (-3) \\
&= 2x^3 \quad + 5x^2 \quad - 3x \quad + 8x^2 \quad + 20x \quad - 12 \\
&= 2x^3 + 13x^2 + 17x - 12
\end{aligned}
$$

♦ **Margin 9**

State the conjugate of the given binomial.

a. $x + 7$

b. $2a - 5$

c. $-9t + 1$

d. $-6h - 5$

♦ **Margin 10**

Multiply.

a. $(x + 9)(x - 9)$

b. $(4n - 1)(4n + 1)$

c. $(-2y + 5)(-2y - 5)$

Answers to Margin 9: **a.** $x - 7$ **b.** $2a + 5$ **c.** $-9t - 1$ **d.** $-6h + 5$

Answers to Margin 10: **a.** $x^2 - 81$ **b.** $16n^2 - 1$ **c.** $4y^2 - 25$

Explanation: We first multiplied every term in the trinomial by x. We then multiplied every term in the trinomial by 4. Finally, we combined like terms. Notice we had two pairs of like terms. The $5x^2$ and $8x^2$ combined to equal $13x^2$ and $-3x$ and $20x$ combined to equal $17x$.

◀ **Do Margin 11.**

◆ **Margin 11**

Multiply.

a. $(x + 3)(4x^2 - 5x + 1)$

b. $(2n - 1)(3n^2 - 7n - 5)$

c. $(a + b)(a^2 + 2ab - 5b^2)$

d. $(y^2 + 3y - 1)(2y^2 - 5y + 4)$

OBJECTIVE 5 *Write an expression for the area of a parallelogram or the volume of a box.*

EXAMPLE 13 Write an expression for the area of the rectangle shown.

$2x - 3$

$3x + 1$

Solution: To get the area of a rectangle, we multiply length times width. In this case the length and width are expressions, so we can write only an expression for the area. We cannot get a numerical answer until we are given a value for x.

$$\text{Area} = \text{length} \cdot \text{width}$$
$$= (3x + 1)(2x - 3) \qquad \text{Multiply all terms.}$$
$$= 2x \cdot 3x + 2x \cdot 1 - 3 \cdot 3x - 3 \cdot 1$$
$$= 6x^2 + 2x - 9x - 3 \qquad \text{Combine like terms.}$$
$$= 6x^2 - 7x - 3$$

◀ **Do Margin 12.**

◆ **Margin 12**

Write an expression for the volume of the box shown.

$n + 2$

$3n$

n

Answers to Margin 11: **a.** $4x^3 + 7x^2 - 14x + 3$ **b.** $6n^3 - 17n^2 - 3n + 5$ **c.** $a^3 + 3a^2b - 3ab^2 - 5b^3$ **d.** $2y^4 + y^3 - 13y^2 + 17y - 4$

Answer to Margin 12: $3n^3 + 6n^2$

3.5 Exercises

FOR EXTRA HELP

Videotape 4

InterAct Math
Tutorial Software

www.carsonmath.com

AWL Math Tutor Center

InterAct MathXL   www.mathxl.com

Student's Solutions Manual

For Exercises 1–24, multiply.

1. $x^3 \cdot x^4$

2. $y^2 \cdot y^5$

3. $t \cdot t^6$

4. $m^4 \cdot m$

5. $7a^2 \cdot 3a$

6. $2n^3 \cdot 5n^4$

7. $-8u^4 \cdot 3u^2$

8. $9x^2(-4x)$

9. $-y^5(-8y^2)$

10. $-3n^4(-7n^6)$

11. $5xy^2 \cdot 7y^3$

12. $9a^2b^3 \cdot 8b^2$

13. $-4t^3u^2 \cdot 7t^2u$

14. $12x^5y(-3x^3y^2)$

15. $-8ab^2c(-7a^2b^3c)$

16. $(-10t^3u^2v)(-9t^2uv^5)$

17. $(9h^2j^5k)(-7hk^3)$

18. $(-13xy^6)(3xy^2z^3)$

19. $5xy(9x^2y^3)(2x^4)$

20. $-2a^2(5ab^3)(3ac^3)$

21. $3mn(-3m^2n)(4mp^5)$

22. $tu^3v(7tv^4)(-8tu^3)$

23. $3xy^4(-5xz^2)(-4xy^2z)$

24. $-6h^2k^5(h^3j^2k)(-4j^3k^2)$

For Exercises 25 and 26, find and explain the mistakes, then work the problems correctly.

25. $9x^3 \cdot 5x^4y$

$= 45x^{12}$

26. $-3m^2n \cdot 10m^5n$

$= -30m^{10}n$

For Exercises 27–38, simplify.

27. $(2x^3)^2$

28. $(3a^4)^3$

29. $(-7m^5)^2$

30. $(-4n^2)^3$

31. $(-2y^6)^5$

32. $(-3t^7)^4$

33. $(5xy^6)^2$

34. $(6a^4b^2)^3$

35. $(-5t^3uv^6)^3$

36. $(-2hj^6k^3)^6$

37. $(3xy^2)(2x^3)^4$

38. $7a^3b^2 \cdot (6ac^4)^2$

For Exercises 39 and 40, find and explain the mistakes, then work the problems correctly.

39. $(3x^4y)^4$

$= 3x^8y^4$

40. $(-2ab^4)^3$

$= -6a^3b^7$

For Exercises 41–56, multiply.

41. $5(3x - 7)$

42. $8(2t - 3)$

43. $-4(7a - 9)$

44. $-2(5m - 6)$

45. $6u(3u + 4)$

46. $3x(7x - 1)$

47. $8x(2x^2 + 3x - 4)$

48. $2n(7n^2 - 5n + 3)$

49. $-x^2(5x^2 - 6x + 9)$

50. $-t^3(3t^2 + t - 8)$

51. $4mn(m^2 + 5mn - 9n^2)$

52. $9ab^2(4a^3 - 3ab - 5b^2)$

53. $6t^2u^3(5t^3 + 7tu^2 - 9tu + 4)$

54. $3x^4y(7xy^3 + 3x^2y - 9xy - 8)$

55. $-mn^2p(4m^2p^3 + 3n^3p - mp^2 - 8m^3)$

56. $-2h^2j^3k(5h^3k^2 - 7h^2j^2 + 9hk + 2k)$

For Exercises 57–70, multiply.

57. $(x + 4)(x + 2)$

58. $(a + 3)(a + 1)$

59. $(m - 3)(m + 5)$

60. $(t + 7)(t - 2)$

61. $(y - 8)(y + 1)$

62. $(u + 4)(u - 9)$

63. $(3x + 2)(4x - 5)$

64. $(5t - 4)(3t + 1)$

65. $(4x - 1)(3x - 5)$

66. $(2a - 7)(a - 6)$

67. $(3t - 5u)(4t + u)$

68. $(h + 2k)(4h - 3k)$

69. $(a - 7b)(2a - 5b)$

70. $(2x - 5y)(3x - y)$

For Exercises 71–78, state the conjugate of the given binomial.

71. $x - 7$

72. $y + 2$

73. $2x + 5$

74. $3t - 8$

75. $-2a + 8b$

76. $-4t - u$

77. $m^2 - n$

78. $2h^3 + 5k^2$

For Exercises 79–84, multiply.

79. $(x + 3)(x - 3)$

80. $(a - 4)(a + 4)$

81. $(2m + n)(2m - n)$

82. $(5t + 6)(5t - 6)$

83. $(-6x - 1)(-6x + 1)$

84. $(-h + 2k)(-h - 2k)$

For Exercises 85–86, find and explain the mistakes, then work the problems correctly.

85. $(2x - 5)(3x + 1)$

$= 6x + 2x - 15x + 5$

$= -7x + 5$

86. $(a - b)(a - b)$

$= a^2 + b^2$

For Exercises 87–94, multiply.

87. $(6a^2 - 4b)(3a^2 + b^2)$

88. $(-3x - y^2)(4x + y)$

89. $(x + 5)(2x^2 - 3x + 1)$

90. $(a + 2)(5a^2 + 3a - 4)$

91. $(2y - 3)(4y^2 + y - 6)$

92. $(3m - 7)(m^2 - 5m - 2)$

93. $(t^2 + u)(4t^2 - 5t - 1)$

94. $(h^2 - k^2)(2h^3 - 3h + 5)$

For Exercises 95 and 96, express the area.

95.

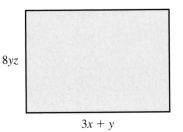

$8yz$

$3x + y$

96.

$4x - 1$

$3x + 5$

For Exercises 97 and 98, express the volume.

97.

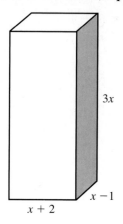

$3x$

$x - 1$

$x + 2$

98.

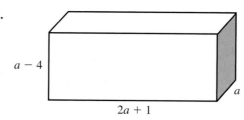

$a - 4$

a

$2a + 1$

Puzzle Problem

If there are 10^{11} stars in a galaxy and about 10^{11} galaxies in the visible universe, about how many stars are in the visible universe?

REVIEW EXERCISES

1. $846 \div 2 =$

2. Find the missing factor. $-6 \cdot (?) = 24$

3. A rectangular room has an area of 132 ft.^{2}. If one wall is 12 ft., what must the length of the other wall be?

4. The cargo box below has a volume of 210 ft.^{3}. Find the missing length.

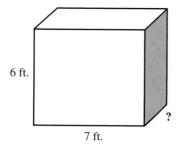

6 ft.

?

7 ft.

5. Combine like terms. $3x - 7x$

3.6 Prime Numbers and GCF

OBJECTIVES

1 Test a number to see if it is prime, composite, or neither.

2 Find the prime factorization of a given number.

3 Find all possible factors for a given number.

4 Find the greatest common factor of a given set of numbers by listing.

5 Find the greatest common factor of a given set of numbers using prime factorization.

OBJECTIVE 1 *Test a number to see if it is prime, composite, or neither.*

There are many special types of numbers, one of which is a **prime number**.

DEFINITION   **Prime number:** A number that satisfies all three of the following conditions.

1. It must be a natural number.

2. It cannot be the number 1.

3. It must be divisible only by itself and 1.

Let's develop a list of primes.

Is 0 a prime? No, because 0 is not a natural number. (0 is a whole number)

Is 1 a prime? No, because condition 2 says a prime cannot be the number 1.

Is 2 a prime? Yes, because 2 is a natural number, it is not 1, and it is divisible only by 2 (itself) and 1.

Is 3 a prime? Yes, because 3 is a natural number, it is not 1, and it is divisible only by 3 (itself) and 1.

Is 4 a prime? No, because 4 is divisible by a number other than itself and 1 (it is also divisible by 2).

If we continue this line of thinking, we get the following list of prime numbers:

$$2, 3, 5, 7, 11, 13, 17, 19, 23, 29, 31, 37, 41, 43, \ldots$$

This list of prime numbers continues without end. However, it becomes more and more difficult to tell whether a given number is prime as we get into larger and larger numbers.

Note that once we establish a prime number, such as 2, no number that is divisible by *that* prime number can be a prime number. Because every even number is divisible by 2, we can say that no even number larger than 2 is prime. Or, in the case of the prime number 3, no number that is divisible by 3 can be prime. Numbers like 4, 6, and 9 that are not prime are called **composite numbers**.

DEFINITION   **Composite number:** A natural number that is divisible by a number other than 1 and itself.

Is 0 composite? No, because it is not a natural number. Remember, 0 is a whole number.

Is 1 composite? No, because it is divisible only by itself.

Conclusion: 0 and 1 are neither prime nor composite.

The first composite number is 4. Here is a list of composites:

$$4, 6, 8, 9, 10, 12, 14, 15, 16, 18, 20, 21, 22, 24, 25, \ldots$$

Conclusion: Every composite number is divisible by at least one prime number.

This conclusion gives us a way to test a given number to see if it is prime or composite. If a given number is composite, it is divisible by a prime. All we need to do is divide a given number by each prime on the list of prime numbers, searching for a prime that divides the given number evenly. But what if we don't find a prime that divides the given number evenly? How far down the list do we need to go before we conclude a given number is itself a prime? Let's consider a couple of cases.

◆ **Discussion** Is every prime number an odd number?
Is every odd number a prime number?
Is every even number a composite?

Historical Note

Eratosthenes (*c.* 276–296 B.C.) was a Greek mathematician who resided in the ancient city of Alexandria, in what is now northern Egypt. One of his contributions was a method for finding prime numbers. He referred to the method as a *sieve* because it sifted out the composite numbers leaving only prime numbers. The idea of the method is to write odd natural numbers beginning with 3, then mark through those that are divisible by each. For instance, start with 3 then mark out all odd numbers divisible by 3. Then write 5 and mark out all odd numbers divisible by 5, and so on.

$$3, 5, 7, \cancel{9}, 11, 13, \cancel{15}, 17, 19, \cancel{21}, 23, \cancel{25}, \cancel{27}, 29, 31, \cancel{33}, \cancel{35}, 37, \cancel{39}, 41, 43, \cancel{45}, \ldots$$

Source: D. E. Smith; *History of Mathematics,* 1953

EXAMPLE 1 Is 91 prime or composite?

Solution: Divide 91 by each of the prime numbers. It is helpful to use the divisibility tricks for 2, 3, and 5 that we learned back in Section 1.4.

Does 2 divide 91 evenly? No, because 91 is odd.

Does 3 divide 91 evenly? No, because $9 + 1 = 10$ and this sum is not divisible by 3.

Does 5 divide 91 evenly? No, because 91 does not end in 0 or 5.

Does 7 divide 91 evenly? Yes.

$$
\begin{array}{r}
13 \\
7\overline{)91} \\
-7 \\
\hline
21 \\
-21 \\
\hline
0
\end{array}
$$

Answer: Because we found a number that divides 91 evenly other than 1 and 91, we conclude that 91 is a composite number.

EXAMPLE 2 Is 127 prime or composite?

Solution: Divide 127 by each of the prime numbers.

Does 2 divide 127 evenly? No, because 127 is odd.

Does 3 divide 127 evenly? No, because $1 + 2 + 7 = 10$ and this sum is not divisible by 3.

Does 5 divide 127 evenly? No, because 127 does not end in 0 or 5.

Does 7 divide 127 evenly? There isn't a good trick for 7 so we just divide.

$$\begin{array}{r} 18 \\ 7{\overline{\smash{)}\,127}} \\ -7 \\ \hline 57 \\ -56 \\ \hline 1 \end{array}$$

No, 7 doesn't divide 127 evenly.

Does 11 divide 127 evenly?

$$\begin{array}{r} 11 \\ 11{\overline{\smash{)}\,127}} \\ -11 \\ \hline 17 \\ -11 \\ \hline 6 \end{array}$$

No, 11 doesn't divide 127 evenly.

How far do we need to go? It turns out we can stop testing now and conclude that 127 is itself a prime number. The reason we can stop is that the quotient from dividing by 11 is 11. The quotient is equal to the divisor. If we get to a point where the quotient from dividing is *equal to* or *less than* the current prime divisor, then we can stop testing and conclude that the given number is itself a prime because if we progress into larger prime divisors, the quotients will continue to get smaller and actually be in the realm of numbers we've already tested. It is therefore pointless to continue.

◆ **Answer:** 127 is a prime number. It is only divisible by 1 and itself.

From what we've learned in Examples 1 and 2, we can write the following procedure for testing a given number to see if it is prime or composite.

Prime/Composite Test *To test a given number to see if it is prime or composite, divide the given number by the primes on the list of prime numbers and consider the results.*
1. If the prime number divides the given number evenly, then stop and conclude that the given number is a composite number.
2. If the prime does not divide the given number evenly, then consider the quotient.
   a. If the quotient is larger than the current prime divisor, then repeat the process with the next prime on the list of prime numbers.
   b. If the quotient is equal to or less than the current prime divisor then stop and conclude that the given number is itself a prime number.

EXAMPLE 3 Is 157 prime or composite?

Solution: Divide by the list of primes.

Does 2 divide 157 evenly? No, because 157 is odd.

Does 3 divide 157 evenly? No, because $1 + 5 + 7 = 13$ and this sum is not divisible by 3.

Does 5 divide 157 evenly? No, because 157 does not end in 5 or 0.

Does 7 divide 157 evenly?

$$\begin{array}{r} 22 \\ 7{\overline{\smash{)}\,157}} \\ -14 \\ \hline 17 \\ -14 \\ \hline 3 \end{array}$$

No, 7 does not divide 157 evenly. Notice the quotient, 22, is larger than the divisor, 7. This means we need to go on to the next prime, which is 11.

Does 11 divide 157 evenly?

$$
\begin{array}{r}
14 \\
11\overline{)157} \\
-11 \\
\hline
47 \\
-44 \\
\hline
3
\end{array}
$$

No, 11 does not divide 157 evenly. Because the quotient, 14, is larger than the divisor, 11, we must go on to the next prime, which is 13.

Does 13 divide 157 evenly?

$$
\begin{array}{r}
12 \\
13\overline{)157} \\
-13 \\
\hline
27 \\
-26 \\
\hline
1
\end{array}
$$

No, 13 does not divide 157 evenly. This time the quotient, 12, is smaller than the divisor, 13. This means that we can stop testing. We can conclude that 157 is a prime number.

◆ **Answer:** 157 is a prime number.

◀ **Do Margin 1.**

Margin 1

Categorize each given number as prime, composite, or neither.

a. 39

b. 119

c. 1

d. 107

Answers: **a.** Composite **b.** Composite **c.** Neither **d.** Prime

OBJECTIVE 2 *Find the prime factorization of a given number.*

We have learned that every composite number is divisible by at least one prime number. It turns out that we can break down composite numbers as a product of only prime factors. We call this listing a **prime factorization.**

| **DEFINITION** | **Factorization:** A multiplication equation. |

For example, here are some factorizations of 20:

$$20 = 4 \cdot 5$$
$$\text{or} \quad 20 = 2 \cdot 10$$

| **DEFINITION** | **Prime factorization:** A multiplication equation that contains only prime factors. |

A prime factorization of 20 is this:

$$20 = 2 \cdot 2 \cdot 5$$

Notice that all the factors are prime numbers.

We can write it in exponential form this way:

$$20 = 2^2 \cdot 5$$

How can we find a prime factorization? There are several popular methods. One of the most flexible methods is the use of a factor tree. First, draw two branches below the given number. For 20, it looks like this:

At the end of these branches we place two factors that multiply to equal the given number. In the case of 20, we could use 2 and 10, or 4 and 5. It is helpful to circle the prime factors as they appear.

We then keep breaking down all the composite factors until we only have prime factors.

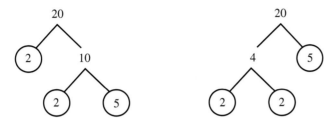

If we list all the prime numbers in a factorization, we have the prime factorization for 20. Notice it didn't matter which two factors of 20 we used to begin the factor tree. We ended up with two 2's and one 5 circled.

Procedure *To find the prime factorization of a number, use a factor tree.*

1. Draw two branches below the number.
2. Find two factors that multiply to equal the given number and place them at the end of the two branches.
3. Repeat steps 1 and 2 for every factor that is not a prime number.
4. Place all the prime factors together in a multiplication sentence.

EXAMPLE 4 Find the prime factorization of 84. Write the answer in exponential form.

Solution: Use a factor tree.

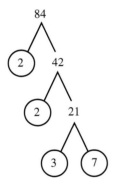

Because 84 is even, we know 2 divides evenly, so let's start with 2 and 42.

Because 42 is even, let's use 2 again. It breaks down to 2 and 21.

Finally, break down 21 to 3 and 7. Because there are no more composite numbers to break down, we're done.

Answer: $84 = 2 \cdot 2 \cdot 3 \cdot 7 = 2^2 \cdot 3 \cdot 7$

We could have started the factor tree for 84 many different ways and ended up with the same result. For example, we could have started with 4 and 21. Or we could have started with 7 and 12, or 3 and 28.

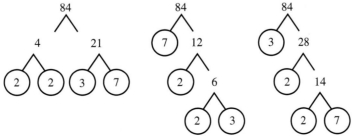

Margin 2

Find the prime factoriza-
tion of each given number.
Write your answer in
exponential form.

a. 64

b. 105

c. 5600

d. 910

No matter how we start, we end up with the same result each time. The number 84 has two 2's, a single 3, and a 7 as its prime factors.

In fact, 84 is the only number that has two 2's, a single 3, and a single 7. The product of two 2's, a single 3, and a single 7 will always equal 84.

Conclusion: No two numbers have the exact same prime factorization. In math language we say that prime factorizations are unique.

EXAMPLE 5 Find the prime factorization of 3500. Write the answer in exponential form.

Solution: Use a factor tree. Since 3500 is divisible by 10, let's start by dividing out the tens. Each division by 10 removes a zero.

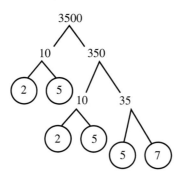

Note: We can check the prime factorization by multiplying the prime factors together. The product should be the given number.

Answer: $3500 = 2 \cdot 2 \cdot 5 \cdot 5 \cdot 5 \cdot 7 = 2^2 \cdot 5^3 \cdot 7$

◀ **Do Margin 2.**

OBJECTIVE 3 *Find all possible factors for a given number.*

Often in mathematics, we need to consider the factors or divisors of a number. Recall that factors or divisors of a given number are numbers that divide the given number with no remainder. Let's consider how to find all possible factors of a given number.

Margin 3

List all possible factors of
each given number.

a. 36

b. 60

c. 84

EXAMPLE 6 List all possible factors of 24.

Solution: To list all the factors, we'll divide by the natural numbers 1, 2, 3, and so on until we have them all. We can start with 1 and 24, then go to 2 and 12 and so forth.

$$1 \cdot 24$$
$$2 \cdot 12$$
$$3 \cdot 8$$
$$4 \cdot 6$$

Explanation: Once we established 2 and 12, all remaining factors must be between 2 and 12. When we established 3 and 8, all remaining factors must be between 3 and 8. When we established 4 and 6, any remaining factors must be between 4 and 6. The only natural number in between 4 and 6 is 5. Notice 5 doesn't divide 24 evenly. This means we have all possible factors.

◀ **Do Margin 3.**

Answers to Margin 2: **a.** 2^6 **b.** $3 \cdot 5 \cdot 7$ **c.** $2^5 \cdot 5^2 \cdot 7$ **d.** $2 \cdot 5 \cdot 7 \cdot 13$

Answers to Margin 3: **a.** 1, 2, 3, 4, 6, 9, 12, 18, 36 **b.** 1, 2, 3, 4, 5, 6, 10, 12, 15, 20, 30, 60 **c.** 1, 2, 3, 4, 6, 7, 12, 14, 21, 28, 42, 84

OBJECTIVE ◢ *Find the greatest common factor of a given set of numbers by listing.*

Suppose a fencing company has a job to enclose three sides of a yard. The left side is 32 ft. long, the center is 40 ft., and the right side is 24 ft. The company is to use prefabricated sections. What is the longest section that can be used so that no partial sections are needed?

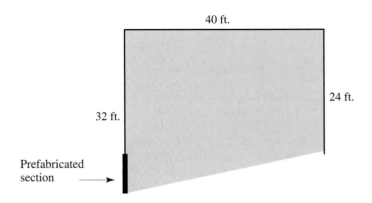

40 ft.

32 ft.

24 ft.

Prefabricated
section

Notice several sizes would work. They could use a 2-ft. section, a 4-ft. section, and an 8-ft. section because 32, 40, and 24 are all divisible by those numbers. But, the longest possible section would be the 8 ft. section. The number 8 is the largest number that divides 32, 40, and 24 evenly and is therefore called the **greatest common factor.**

DEFINITION   **Greatest common factor:** The largest number that divides all given numbers evenly.

Sometimes the greatest common factor is referred to as the *greatest common divisor.* We will abbreviate greatest common factor as GCF.

How can we get the GCF for a given set of numbers? One way is by listing factors. We list all possible factors for each given number and then search the lists for the largest factor common to all lists.

> **Procedure** *To find the greatest common factor by listing:*
> **1.** List all possible factors for each given number.
> **2.** Search the lists for the largest factor common to all lists.

EXAMPLE 7   Find the GCF of 24 and 60.

Solution: All possible factors of 24 were listed in Example 6, and you listed all possible factors of 60 in the Margin Exercise 3b.

For 24 we have: 1, 2, 3, 4, 6, 8, 12, 24

For 60 we have: 1, 2, 3, 4, 5, 6, 10, 12, 15, 20, 30, 60

Notice 12 is the largest factor common to both lists.

Answer: GCF = 12

We can check the greatest common factor by dividing it into the given numbers.

Notice these quotients share no common factors except the number 1.

$$12\overline{)24} \quad \begin{array}{r} 2 \\ \end{array} \qquad 12\overline{)60} \quad \begin{array}{r} 5 \\ \end{array}$$
$$\underline{-24} \qquad\qquad\qquad \underline{-60}$$
$$\quad 0 \qquad\qquad\qquad\quad 0$$

♦ Margin 4

Find the GCF by listing.

a. 36 and 54

b. 32 and 45

c. 28, 32, and 40

Notice that 12 not only divides the given numbers evenly, but the quotients have no common factor (except the number 1). This is always the case with GCF. If you divide the GCF into the given numbers, there should never be any factor other than 1 that is common to all the quotients.

Conclusion: If you check your answer and find there is a common factor in the quotients, it means you have found a common factor for the given numbers but not the *greatest* common factor.

In the case of 24 and 60, suppose we missed seeing the 12 and thought 6 was the GCF. Look what happens when we check 6 by dividing it into 24 and 60:

These quotients are both divisible by 2. This means 6 is not the greatest common factor.

$$6\overline{)24} \quad \begin{array}{r} 4 \\ \end{array} \qquad 6\overline{)60} \quad \begin{array}{r} 10 \\ \end{array}$$
$$\underline{-24} \qquad\qquad\qquad \underline{-60}$$
$$\quad 0 \qquad\qquad\qquad\quad 0$$

The nice thing about this check technique is that it gives a very easy way to fix a mistake. The factor that is common in the quotients is the very factor by which we are off. For instance, we merely need to multiply the answer we found, 6, by the common factor of 2 that is appearing in the quotients and we get the correct GCF, 12. The reason this works will become clearer when we look at the prime factorization method for generating GCF.

◀ **Do Margin 4.**

OBJECTIVE 5 *Find the greatest common factor of a given set of numbers using prime factorization.*

You probably found listing to be tedious. For smaller numbers it is a good method, but for larger numbers, it is not the most efficient method to use.

It turns out we can use prime factorization to generate the GCF of a given set of numbers. Consider the prime factorizations for the numbers from Example 5.

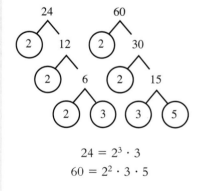

$$24 = 2^3 \cdot 3$$
$$60 = 2^2 \cdot 3 \cdot 5$$

Note: Many times, students question learning several methods for solving the same type of problem. Think of methods as if they were tools. A good analogy is a saw. For small jobs where you only need to cut through a few pieces of wood, a hand saw is a great tool and simple to use. However, for a big job involving lots of cutting or large pieces of wood, a power saw, although it takes more effort to set up, is much more efficient to use. The listing method for GCF, like the hand saw, is a good method for simpler problems involving small numbers. The prime factorization method is like the power saw. It's a good method for the bigger problems, involving large numbers or many numbers.

Answers: **a.** 18 **b.** 1 **c.** 4

We already know the answer is 12 from our listing method in Example 5. Look at the prime factorization for 12, and see if you can figure out how to use the primes from 24 and 60 to get 12.

$$\text{GCF} = 2^2 \cdot 3 = 12$$

Why are only 2 and 3 used and not 5? Why 2^2 and a single 3? Notice 2 and 3 are common to both 24 and 60, whereas 5 is not in both factorizations. We only use primes that are common to all factorizations involved. Notice 2^2 has the smallest exponent of the 2 factors and a single 3 is the smallest number of times the 3 appeared.

Conclusion: We use the smallest exponent of the common primes.

What if there were no common primes? Remember 1 is a common factor for all numbers.

Conclusion: If there are no common prime factors, then 1 is the GCF.

Procedure *To find the greatest common factor of a given set of numbers:*
1. Get the prime factorization of each given number.
2. Write the factorizations in exponential form.
3. Use the smallest exponent of the common primes.
4. Multiply or simplify.

Note: If there are no common primes, then the GCF is 1.

EXAMPLE 8 Find the GCF of 3024 and 2520.

Solution: Listing all the factors of 3024 and 2520 would be a nightmare, so let's use prime factorization.

$$3024 = 2^4 \cdot 3^3 \cdot 7$$
$$2520 = 2^3 \cdot 3^2 \cdot 5 \cdot 7$$

$$\text{GCF} = 2^3 \cdot 3^2 \cdot 7$$
$$= 8 \cdot 9 \cdot 7$$
$$= 504$$

Let's check by dividing 504 into 3024 and 2520. Remember we should find that 504 not only divides evenly, but there should not be any common factors in the quotients (except, of course, the number 1).

Notice the only factor common to both quotients is 1. This means that 504 is correct. It is the GCF for 3024 and 2520.

$$504 \overline{)3024} \qquad \begin{array}{r} 6 \\ -3024 \\ \hline 0 \end{array} \qquad\qquad 504 \overline{)2520} \qquad \begin{array}{r} 5 \\ -2520 \\ \hline 0 \end{array}$$

◆ Margin 5

Find the GCF using primes.

a. 84 and 48

b. 28 and 140

c. 42, 63, and 105

d. 60 and 77

Suppose we had left out a 3 factor when we developed the GCF from the prime factorizations. Instead of using 3^2, suppose we just used 3. We would have written this:

$$2^3 \cdot 3 \cdot 7$$
$$= 8 \cdot 3 \cdot 7$$
$$= 168$$

Let's check 168 and see what happens.

These quotients are both divisible by 3, which is the very factor that we left out of the prime factorization for the GCF.

$$\begin{array}{r} 18 \\ 168\overline{)3024} \\ -168 \\ \hline 1344 \\ -1344 \\ \hline 0 \end{array} \qquad \begin{array}{r} 15 \\ 168\overline{)2520} \\ -168 \\ \hline 840 \\ -840 \\ \hline 0 \end{array}$$

Notice the quotients are both divisible by 3, which is the very factor we left out when developing the GCF. To fix it, we merely need to multiply that 3 by the 168, which gives us the correct answer, 504.

If you leave out prime factors in developing the GCF, they will show up as common factors ◆ in the quotients when you check.

◀ **Do Margin 5.**

EXAMPLE 9   The floor of a 42 ft. by 30 ft. room is to be covered with colored squares like a checker board. The client has the following conditions:

a. All squares must have whole number dimensions.

b. The squares may not be cut or overlapped to fit inside the room.

Solution: Notice we could use 1 ft. by 1 ft. squares. But we could also use 2 ft. by 2 ft. squares. These work because 1 and 2 both divide 42 and 30 evenly. Because we want the largest possible square we must find the GCF of 42 and 30.

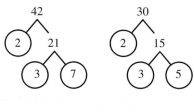

$$42 = 2 \cdot 3 \cdot 7$$
$$30 = 2 \cdot 3 \cdot 5$$
$$\text{GCF} = 2 \cdot 3 = 6$$

◆ **Answer:** The largest square that would work is a 6 ft. by 6 ft. square.

◀ **Do Margin 6.**

◆ Margin 6

A rectangular kitchen floor measures 18 ft. by 16 ft. What is the largest size square tile that can be used to cover the floor without cutting or overlapping the tiles?

Answers to Margin 5: **a.** 12 **b.** 28 **c.** 21 **d.** 1

Answer to Margin 6: 2 ft. by 2 ft. square

3.6 Exercises

FOR EXTRA HELP

 Videotape 4

 InterAct Math
Tutorial Software

 www.carsonmath.com

 AWL Math Tutor Center

InterAct
MathXL www.mathxl.com

 Student's Solutions
Manual

For Exercises 1–16, tell whether the given number is prime, composite, or neither.

1. 49	**2.** 93	**3.** 89	**4.** 71
5. 0	**6.** 153	**7.** 179	**8.** 187
9. 1	**10.** 253	**11.** 247	**12.** 311
13. 377	**14.** 389	**15.** 409	**16.** 323

17. Are all prime numbers odd? Explain.

18. Are all odd numbers prime? Explain.

For Exercises 19–34, find the prime factorization.

19. 80	**20.** 72	**21.** 156	**22.** 120
23. 268	**24.** 180	**25.** 200	**26.** 324
27. 343	**28.** 220	**29.** 975	**30.** 462
31. 378	**32.** 875	**33.** 952	**34.** 1404

For Exercises 35–40, list all possible factors of the given number.

35. 60	**36.** 40	**37.** 81
38. 80	**39.** 120	**40.** 54

For Exercises 41–46, find the GCF by listing.

41. 24 and 60	**42.** 40 and 32	**43.** 81 and 65
44. 80 and 100	**45.** 72 and 120	**46.** 54 and 63

For Exercises 47–58, find the GCF by prime factorization.

47. 140 and 196

48. 240 and 150

49. 130 and 78

50. 324 and 270

51. 336 and 504

52. 270 and 675

53. 99 and 140

54. 252 and 143

55. 60, 120 and 140

56. 120, 300, and 420

57. 64, 160, and 224

58. 315, 441, and 945

For Exercises 59–62, solve.

59. A company produces plastic storage containers. For shipping purposes, the containers must be 60 in. by 70 in. The containers are sectioned into squares. What is the largest size square that can be used without cutting or overlapping?

60. The ceiling of a 45 ft. by 36 ft. lobby of a new hotel is to be sectioned into a grid of square regions by wooden beams. What is the largest square region possible so that all regions are of equal size, and no overlapping or cutting of any region takes place?

61. A landscaper is planning a sprinkler system that will require precut PVC pipe to be connected and placed in the ground. The trenches he will dig are to be 40 ft., 32 ft., and 24 ft. in length. What is the longest length of PVC pipe he can use that will exactly fit the length of each trench and not require any further cutting? How many of the precut pipes will be used in each trench?

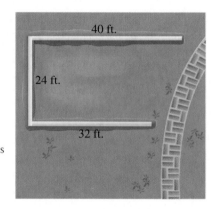

62. A professor has three classes. The first class has 30 people, the second class has 42 people, and the third class has 24 people. She wants to break the classes into project groups so that all groups are the same size no matter which class they are in. What is the largest size group she can make? How many groups will be in each class?

REVIEW EXERCISES

1. Write expanded form for 5,784,209.

2. 34,328 ÷ 17 =

3. Evaluate 2^7.

4. A college has 265 employees. The college wishes to assign an ID to each employee. The proposed ID would contain one of the digits 0–9 and one of the letters A–Z. Is the proposed ID system sufficient?

5. A company must produce square plastic panels that have an area of 36 ft.^{2}. What must be the dimensions of the squares?

3.7 Introduction to Factoring

OBJECTIVES

1 Divide monomials.

2 Find a missing factor.

3 Factor the GCF out of a polynomial.

OBJECTIVE 1 *Divide monomials.*

In Section 3.5, we learned how to multiply exponential forms. We concluded that when we multiply exponential forms that have the same base, we can add the exponents and keep the same base. For example,

$$2^3 \cdot 2^4 = 2^{3+4} = 2^7$$

Let's consider how we can reverse this process. In essence, we'll be considering how to divide exponential forms. Remember that multiplication and division are inverse operations, which means they *undo* each other. We can write a related division statement for the above multiplication statement by dividing the product by one of the factors.

If $2^3 \cdot 2^4 = 2^7$ then $2^7 \div 2^4 = 2^3$

What is happening with the exponents in the division problem? Because the exponents were added in the multiplication problem, it follows that they should be subtracted in the division problem.

> **Connection**
> We could simplify $2^7 \div 2^4$ using the order of operations.
>
> $2^7 \div 2^4 = 128 \div 16 = 8$
>
> The prime factorization of 8 is 2^3.
>
> $2^7 \div 2^4 = 128 \div 16 = 8 = 2^3$

$$2^7 \div 2^4 = 2^{7-4} = 2^3$$

Note that we subtracted the divisor's exponent from the dividend's exponent. This order is important because subtraction is not commutative.

Conclusion: When dividing exponential forms that have the same base, we subtract the divisor's exponent from the dividend's exponent and keep the same base.

> **Rule** When dividing exponential forms that have the same base, we can subtract the divisor's exponents from the dividend's exponent and keep the same base.
>
> **In math language:** $n^a \div n^b = n^{a-b}$ as long as $n \neq 0$
>
> **Note:** $n \neq 0$ because if n were replaced with 0, we would have $0^a \div 0^b$. In Section 1.4, we concluded that when the dividend and divisor are both 0, the problem is indeterminate.

EXAMPLE 1 Divide. $x^6 \div x^2$

Solution: $x^6 \div x^2 = x^{6-2} = x^4$

Explanation: Because the exponential forms have the same base, we can subtract the exponents and keep the same base. Notice we can check the result by reversing the process. We multiply the answer, x^4, by the divisor, x^2, to equal the dividend, x^6.

◆ **Check:** $x^4 \cdot x^2 = x^{4+2} = x^6$

Do Margin 1. ▶

◆ **Margin 1**

Divide.

a. $3^9 \div 3^2$

b. $a^{12} \div a^3$

c. $y^6 \div y^5$

Answers: **a.** 3^7 **b.** a^9 **c.** y

What if we have exponential forms that have the same base and the same exponent? Consider $2^3 \div 2^3$. Using the rule that we developed, we subtract the divisor's exponent from the dividend's exponent.

$$2^3 \div 2^3 = 2^{3-3} = 2^0$$

What does it mean to have 0 as an exponent? To answer this question we could evaluate $2^3 \div 2^3$ using the order of operations. The result must be equal to 2^0.

$$2^3 \div 2^3$$
$$= 8 \div 8$$
$$= 1$$

Because 2^0 and 1 are results from the same problem, they must be equal.

$$2^0 = 1$$

Conclusions: When dividing exponential forms that have the same base and the same exponent, the result is 1; and any nonzero base with an exponent of 0 simplifies to the number 1.

Note that the situation 0^0 would arise from a division situation where two exponential forms have 0 as the base and both have the same exponent.

$$0^n \div 0^n = 0^{n-n} = 0^0$$

In Section 1.4 we concluded that $0 \div 0$ is indeterminate therefore 0^0 is indeterminate.

Conclusion: 0^0 is indeterminate.

> **Rule** Any base other than 0 raised to the 0 power simplifies to the number 1.
>
> **In math language:** $n^0 = 1$, as long as $n \neq 0$;
>
> **Note:** 0^0 is indeterminate.

EXAMPLE 2 Divide. $n^3 \div n^3$ (assume $n \neq 0$)

Solution: Because we are to divide exponential forms that have the same base, we can subtract the exponents and keep the same base.

$$n^3 \div n^3 = n^{3-3} = n^0 = 1$$

To assume that $n \neq 0$ is important because if we let $n = 0$ we would have $0^3 \div 0^3$, which is indeterminate. If there is any other numeric replacement for n in $n^3 \div n^3$, the result equals 1.

◆ In all future division problems involving variables, assume that divisor variables do not equal 0.

◀ **Do Margin 2.**

Consider monomials that have coefficients, such as $20m^2 \div 4m$. When we multiplied monomials, we multiplied the coefficients and added the exponents of the like bases. Because multiplication and division are inverse operations, it follows that to divide monomials we divide the coefficients and subtract the exponents of the like bases.

◆ **Margin 2**

Simplify.

a. $x^6 \div x^6$

b. 5^0

c. $(-71)^0$

Answers: **a.** 1 **b.** 1 **c.** 1

$$20m^2 \div 4m = 5m^{2-1} = 5m$$

Note that we can check by multiplying the quotient by the divisor to equal the dividend.

$$5m \cdot 4m = 20m^{1+1} = 20m^2$$

> **Procedure** *To divide monomials:*
> 1. Divide the coefficients.
> 2. For like bases, subtract the exponent of the divisor base from the exponent of the dividend base and keep the base. If the bases have the same exponent, then they divide out, becoming 1.
> 3. Bases in the dividend that have no like base in the divisor are written unchanged in the quotient.

EXAMPLE 3 Divide.

a. $8a^7 \div (-4a)$

Solution: $8a^7 \div (-4a)$ Divide coefficients and subtract exponents of the like bases.
$$= -2a^{7-1}$$
$$= -2a^6$$

Check: $-2a^6 \cdot (-4a) = 8a^{6+1} = 8a^7$

b. $-24m^8n^5 \div (-2m^7n^2)$

Solution: $-24m^8n^5 \div (-2m^7n^2)$ Divide coefficients and subtract exponents of the like bases.
$$= 12m^{8-7}n^{5-2}$$
$$= 12mn^3$$

Check: $12mn^3 \cdot (-2m^7n^2) = -24m^{1+7}n^{3+2} = -24m^8n^5$

c. $24x^7y^2z \div 3x^3y^2$

Solution: $24x^7y^2z \div 3x^3y^2$ Divide coefficients and subtract exponents of the like bases.
$$= 8x^{7-3}y^{2-2}z$$
$\qquad\qquad$ Notice z remains the same because it doesn't have a like base in the divisor that would affect its exponent.
$$= 8x^4y^0z$$
$\qquad\qquad$ y^0 simplifies to the number 1. Because $8x^4$ is multiplying y^0, we have $8x^4(1)z$.
$$= 8x^4(1)z$$
$$= 8x^4z$$

◆ **Check:** $8x^4z \cdot 3x^3y^2 = 24x^{4+3}y^2z = 24x^7y^2z$

Note Many people say that y^2 "divides" out. Now that we recognize this result, it is not necessary to write out y^0 and then simplify to 1. From here on, if we divide two monomials that have bases with matching exponents, we will write 1, or simply leave them out of the resulting monomial.

Do Margin 3. ▶

◆ **Margin 3**

Divide.

a. $10u^7 \div 2u^3$

b. $28x^3y^8 \div (-4xy^2)$

c. $-39h^4k^9 \div (-3h^4k^8)$

d. $-24t^5uv \div 6t^2u$

Answers: **a.** $5u^4$ **b.** $-7x^2y^6$ **c.** $13k$
d. $-4t^3v$

OBJECTIVE 2 *Find a missing factor.*

Now let's use what we've learned about dividing monomials to find missing factors. Remember that we find a missing factor by dividing the product by the known factor.

EXAMPLE 4 Find the missing factor. $6x^2y \cdot (?) = 18x^5y^2$

Solution: We solve for a missing factor by dividing the product by the known factor.

$6x^2y \cdot (?) = 18x^5y^2$ **Write a related division statement.**

$(?) = 18x^5y^2 \div 6x^2y$ **Divide the coefficients and subtract the exponents of the like bases.**

$(?) = 3x^{5-2}y^{2-1}$

$(?) = 3x^3y$

Check: We can check by putting the monomial we found back into the original statement. We want to verify that the multiplication works.

$$6x^2y \cdot 3x^3y$$
$$= 18x^{2+3}y^{1+1}$$
$$= 18x^5y^2$$

EXAMPLE 5 Find the missing factor. $6a^3b^2 \cdot (?) = -30a^7b^2c^2$

Solution: We write a related division statement then divide.

$6a^3b^2 \cdot (?) = -30a^7b^2c^2$

$(?) = -30a^7b^2c^2 \div 6a^3b^2$ The bs have the same exponent, $b^2 \div b^2 = 1$, so no b appears in the result. c^2 has no like base, so it is written in the quotient unchanged.

$(?) = -5a^{7-3}c^2$

$(?) = -5a^4c^2$

Check: $6a^3b^2 \cdot -5a^4c^2 = -30a^7b^2c^2$

◀ **Do Margin 4.**

What if we had to find a missing factor that must be a polynomial? Recall that when we multiply a monomial by a polynomial, we apply the distributive property.

$$2(x + 3) = 2 \cdot x + 2 \cdot 3$$
$$= 2x + 6$$

Suppose the above problem was written as a missing factor problem.

$$2 \cdot (?) = 2x + 6.$$

Notice that the missing factor must be a polynomial. To solve for a missing factor, we must divide the product by the known factor.

$2 \cdot (?) = 2x + 6$

$(?) = (2x + 6) \div 2$ Here we apply the distributive property in reverse and *divide* each term by the 2.

$(?) = 2x \div 2 + 6 \div 2$

$(?) = x + 3$

Conclusion: To divide a polynomial by a monomial, apply the distributive property in reverse, and divide each term of the polynomial by the monomial.

EXAMPLE 6 Find the missing factor. $16x^4 - 12x^3 + 8x^2 = 4x^2 \cdot (?)$

Solution: To find the missing factor, we must divide the product by the known factor.

$$16x^4 - 12x^3 + 8x^2 = 4x^2 \cdot (?)$$
$$(16x^4 - 12x^3 + 8x^2) \div 4x^2 = (?)$$
$$16x^4 \div 4x^2 - 12x^3 \div 4x^2 + 8x^2 \div 4x^2 = (?)$$
$$4x^2 \quad - 3x \quad\quad + \quad 2 \quad = (?)$$
$$4x^2 - 3x + 2 = (?)$$

Note It doesn't matter whether the missing factor is written before or after the known factor. We can write $4x^2 \cdot (?)$ or $(?) \cdot 4x^2$. We also do not have to write the dot. We can simply write $4x^2(?)$.

Explanation: We divided each term in the polynomial by the $4x^2$.

We divided $16x^4$ by $4x^2$ to get $4x^2$.

We divided $-12x^3$ by $4x^2$ to get $-3x$.

Finally, we divided $8x^2$ by $4x^2$ to get 2.

Do Margin 5. ▷

OBJECTIVE 3 *Factor the GCF out of a polynomial.*

All of the work we've done to this point has been to develop the concept of factoring. When we rewrite a number or an expression as a product of factors, we say we are writing the number or expression in **factored form.**

DEFINITION **Factored form:** A number or expression written as a product of factors.

Factored form looks like this:

$$6x^2 (3x^2 + 2x - 1)$$

Connection
In Section 3.1, we said there were two actions that we can perform with expressions: evaluate and rewrite. Simplifying and factoring are two ways we can rewrite expressions.

If we use the distributive property and multiply it out, we get:

$$18x^4 + 12x^3 - 6x^2$$

In factoring a monomial out of a polynomial, our objective will be to start with $18x^4 + 12x^3 - 6x^2$ and find the factored form, $6x^2 (3x^2 + 2x - 1)$.

What do you notice about $6x^2$? You should notice that 6 is the largest number that divides the coefficients in $18x^4 + 12x^3 - 6x^2$. Also notice that x^2 has the smallest exponent of the xs. Does this sound familiar? As we learned in Section 3.6, $6x^2$ is the greatest common factor for the terms in $18x^4 + 12x^3 - 6x^2$.

◆ **Margin 5**

Find the missing factor.

a. $8x - 10 = 2 \cdot (?)$

b. $12h^2 + 18h = (?) \cdot 6h$

c. $9y^5 - 7y^4 + 2y^3 = (?) \, y^3$

d. $24n^5 - 30n^3 - 12n^2 = 6n^2 (?)$

Answers: **a.** $4x - 5$ **b.** $2h + 3$
c. $9y^2 - 7y + 2$ **d.** $4n^3 - 5n - 2$

You have probably noticed that the coefficients in $18x^4 + 12x^3 - 6x^2$ are also all divisible by 2 and 3. However, mathematicians have agreed that we will always take out the greatest common factor when we write factored form.

Let's recall the procedure for finding GCF and apply it to monomials.

Procedure *To find the greatest common factor of a given set of monomials or terms:*

1. Get the prime factorization of each given number.
2. Write the factorizations in exponential form. Treat variables like primes.
3. Use the smallest exponent of the common primes or variables.

Note: If there are no common primes, then the GCF is 1.

Notice the procedure calls for using the smallest exponent of the common primes. We treat variables just like primes. That's why x^2 appears in $6x^2$: because it has the smallest exponent of the common xs in the polynomial $18x^4 + 12x^3 - 6x^2$.

Let's find some GCF's, then bring this concept back into the context of writing factored form.

EXAMPLE 7 Find the GCF.

a. $10x^3y$ and $15x^5$

Solution: Get the prime factorizations and use the smallest exponent of the common primes. Remember we treat the variables like primes.

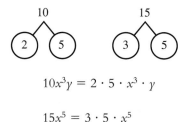

$10x^3y = 2 \cdot 5 \cdot x^3 \cdot y$

$15x^5 = 3 \cdot 5 \cdot x^5$

$GCF = 5x^3$

Note: You do not necessarily have to use primes. If you can see the GCF, fine. 5 is the biggest number that goes into both 10 and 15, and x^3 has the smallest exponent of the common xs. The y isn't used because it isn't common to both monomials.

Explanation: 5 was the only prime number common to both factorizations. Because x was common to both terms, we used the smallest exponent of the xs, which was 3. Because y was not common to both terms, we did not include it in the GCF.

b. $20x^5y$, $16x^3yz$, and $24x^2y$

Answer: $4x^2y$

Explanation: Notice 4 is the largest number that divides 20, 16, and 24. Because x was common to all the terms, we used the smallest exponent of the xs, which was x^2. Because y was also common to all three terms, we used y as well. All of the ys had an exponent of 1, so the smallest exponent of the ys is 1. Therefore we use a single y. Notice z was not common to all three terms, so we did not use z.

◀ **Do Margin 6.**

◆ **Margin 6**

Find the GCF of the given set of monomials.

a. 84 and 56

b. $16a^5$ and $24a^7b^2$

c. $14u^2t^4$, $28u^2t^5$, and $21u^2t^7v$

Answers: **a.** 28 **b.** $8a^5$ **c.** $7u^2t^4$

Now let's put what we've learned about finding GCF's back into the context of writing factored form for polynomials. Recall our objective is to write a polynomial like $18x^4 + 12x^3 - 6x^2$ in factored form, which looks like $6x^2 (3x^2 + 2x - 1)$. We now know $6x^2$ is the GCF of the terms, in the original polynomial. But how do we fill the parentheses?

Filling the parentheses involves the work we did when finding a missing factor. Once we find the GCF, we simply think of the problem as a missing factor problem.

Think: $18x^4 + 12x^3 - 6x^2 = 6x^2 \cdot (?)$

To find the missing factor, we reverse the distributive property and divide the $6x^2$ into each term of the original polynomial.

Procedure *To factor a monomial out of a given polynomial:*
1. Find the GCF of the terms that make up the polynomial.
2. Write an expression that is a product of the GCF and empty parentheses this way:
   GCF · (   )
3. Divide each term of the polynomial by the GCF and place the results inside the parentheses.

EXAMPLE 8   Factor $6x + 15$

Solution:

1. We must find the GCF of $6x$ and 15.

    The GCF of $6x$ and 15 is 3.

> **Note:**
> 3 is the largest number that divides both $6x$ and 15 evenly. We do not include the x because it is not common to both terms.

2. Write an expression that is a product of the GCF and empty parentheses.

$$6x + 15 = 3 \cdot (   )$$

3. Divide the terms in $6x + 15$ by 3 to fill the parentheses.

$$6x + 15 = 3 \cdot (6x \div 3 + 15 \div 3)$$
$$= 3 \cdot (2x + 5)$$

Answer: $6x + 15 = 3(2x + 5)$

Explanation: We had to find the GCF of $6x$ and 15 because of the agreement that we will always factor out the GCF. The GCF of $6x$ and 15 is 3 because 3 is the largest number that divides both terms. Notice we didn't include the x in the GCF because x is not common to both terms.

Check: We can check by multiplying the factored form back out using the distributive property.

$$3(2x + 5)$$
$$= 3 \cdot 2x + 3 \cdot 5$$
$$= 6x + 15$$

Keep in mind that when we factor, we are simply writing the original expression in a different form called *factored form*. It is the same as when we wrote expanded form for numbers. The factored form expression is equal to the original expression just as expanded form is equal to the original number. When we say the expressions are equal, it means that if we were given a value for the variable(s) it wouldn't matter which version of the expression we used to evaluate that value. Both factored form and the original polynomial form will give the same result.

Let's choose a value and evaluate $6x + 15$ and $3(2x + 5)$. Let's pick $x = 4$ (any number would do).

$$
\begin{array}{ll}
6x + 15 & 3(2x + 5) \\
6 \cdot 4 + 15 & 3(2 \cdot 4 + 5) \\
24 + 15 & 3(8 + 5) \\
& 3(13) \\
39 \longleftarrow \text{Same result.} \longrightarrow 39
\end{array}
$$

EXAMPLE 9 Factor. $24x^6 + 18x^4 - 30x^3$

Solution:

1. Find the GCF of $24x^6$, $18x^4$, and $-30x^3$.

$$\text{GCF} = 6x^3$$

2. Write an expression that is a product of the GCF and the empty parentheses.

$$24x^6 + 18x^4 - 30x^3 = 6x^3 \cdot (\qquad)$$

3. Divide each term in $24x^6 + 18x^4 - 30x^3$ by $6x^3$ and write the results inside the parentheses.

$$24x^6 + 18x^4 - 30x^3 = 6x^3 \cdot (24x^6 \div 6x^3 + 18x^4 \div 6x^3 - 30x^3 \div 6x^3)$$

$$= 6x^3 \cdot (4x^3 \qquad + \qquad 3x \qquad - 5)$$

Answer: $24x^6 + 18x^4 - 30x^3 = 6x^3(4x^3 + 3x - 5)$

Explanation: To find the GCF, we found the largest number that goes into 24, 18, and 30, which is 6. Because the x is common to all three terms, we included it in the GCF. The smallest exponent for the xs was 3. We then divided the GCF into all the original terms to fill the parentheses.

Check: Multiply the factored form and see if the result is the original polynomial.

$$
\begin{aligned}
&6x^3(4x^3 + 3x - 5) \\
&= 6x^3 \cdot 4x^3 + 6x^3 \cdot 3x - 6x^3 \cdot 5 \\
&= 24x^6 + 18x^4 - 30x^3
\end{aligned}
$$

◀ **Do Margin 7.**

◆ **Margin 7**

Factor.

a. $12y - 9$

b. $10x^3 + 15x$

c. $20a^4 + 30a^3 - 40a^2$

d. $42t^3u^4 - 30t^2u^3 + 18tu^2$

Answers: **a.** $3(4y - 3)$ **b.** $5x(2x^2 + 3)$ **c.** $10a^2(2a^2 + 3a - 4)$ **d.** $6tu^2(7t^2u^2 - 5tu + 3)$

3.7 Exercises

FOR EXTRA HELP

 Videotape 4

 InterAct Math Tutorial Software

 www.carsonmath.com

 AWL Math Tutor Center

 InterAct MathXL www.mathxl.com

 Student's Solutions Manual

For Exercises 1–14, divide.

1. $x^9 \div x^2$

2. $a^{11} \div a^5$

3. $m^7 \div m$

4. $y^5 \div y$

5. $u^4 \div u^4$

6. $k^6 \div k^6$

7. $-20t^6 \div 4t^2$

8. $18x^{10} \div (-2x^4)$

9. $40n^9 \div 2n$

10. $33h^{12} \div 11h$

11. $-12x^5y^2 \div (-4xy)$

12. $38m^7n^3 \div 2m^4n$

13. $14a^8b^4c \div (-a^3b^4)$

14. $-25h^7jk^2 \div h^7k$

For Exercises 15–30, find the missing factor.

15. $5a \cdot (?) = 10a^3$

16. $7m^2 \cdot (?) = 21m^5$

17. $(?) \cdot (-6x^3) = 42x^7$

18. $(?) \cdot 10y^4 = -40y^9$

19. $-36t^7u^5 = (?) \cdot (-9t^4u^4)$

20. $-48x^4y^2 = -6xy \cdot (?)$

21. $8a^9b^3c = 8a^4b^3 \ (?)$

22. $25t^7u^4v^2 = -5t^2v^2 \ (?)$

23. $8x + 12 = 4 \cdot (?)$

24. $12y - 6 = 6 \cdot (?)$

25. $5a^2 - a = a \ (?)$

26. $10m^4 + m^3 = (?) \ m^3$

27. $6t^5 - 9t^4 + 12t^2 = 3t^2 \ (?)$

28. $25k^7 - 30k^5 - 20k^2 = 5k^2 \ (?)$

29. $18x^3y^5 - 30x^4y^2z = 6x^3y^2 \ (?)$

30. $44m^7n^3p^2 + 33m^5n^6 = 11m^5n^3 \ (?)$

For Exercises 31–34, find the missing side length.

31.

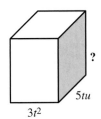

32.

33. $V = 60t^5u^2$

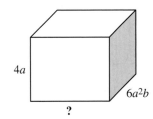

34. $V = 120a^7b$

For Exercises 35–44, find the GCF.

35. $20x$ and 30

36. 18 and $9t$

37. $8x^2$ and $14x^5$

38. $12a^5$ and $15a$

39. $36mn^3$ and $48n^2$

40. $42r^5$ and $28r^2t$

41. $16x^2y^5z$ and $36x^3y^5$

42. $21t^4uv^3$ and $35tuv^2$

43. $24m^2n^8$, $12m^3n^5p$, and $30m^2n^3$

44. $32a^4b^7c^2$, $16a^6b^2c^5$, and $40a^5c^9$

For Exercises 45–64, factor.

45. $8x - 4$

46. $10a + 20$

47. $14y + 7$

48. $8m - 20$

49. $2n^2 + 6n$

50. $5x^3 - 10x^2$

51. $7x^3 - 3x^2$

52. $12t^2 + 5t$

53. $20r^5 - 24r^3$

54. $32b^8 + 24b^4$

55. $6xy^3 + 3xy^2$

56. $5a^4b - 10a^2b$

57. $30c^6d^3f - 15c^5d^3$

58. $19h^3j^5k + 38j^3k$

59. $9a^7 - 12a^5 + 18a^3$

60. $20x^5 - 15x^3 - 25x$

61. $14m^8n + 28m^6n + 7m^5n$

62. $15a^2b^5 + 25a^2b^3 - 30a^2b^2$

63. $10x^4y^6z - 20x^5y^3z^2 - 40x^3y^5$

64. $45t^7u^9v^3 - 18t^4u^6v^2 + 27t^5u^3$

◣ REVIEW EXERCISES

1. Find the perimeter of the shape.

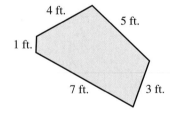

4 ft.

5 ft.

1 ft.

7 ft.   3 ft.

2. Find the area of the parallelogram.

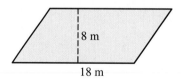

8 m

18 m

3. Find the volume of the box.

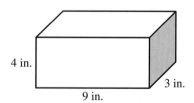

4 in.

3 in.

9 in.

4. Subtract. $(6x^3 - 9x^2 + x - 12) - (4x^3 + x + 5)$

5. Multiply. $(x + 4)(x - 6)$

3.8 Applications and Problem Solving

OBJECTIVES

1 Solve polynomial problems involving perimeter, area, and volume.

2 Solve surface area problems.

3 Solve problems involving a falling object.

4 Solve net-profit problems.

OBJECTIVE 1 *Solve polynomial problems involving perimeter, area, and volume.*

We can use mathematics to describe situations. If we can write an expression that describes the situation, then we can experiment with numbers without actually having to do real experiments. To illustrate this process, let's write expressions for perimeter, area, and volume, then evaluate those expressions using given numbers.

EXAMPLE 1 Write an expression in simplest form for the perimeter of the shape shown. What would the perimeter be if $x = 7$ ft.?

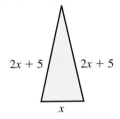

Understand: To find the perimeter or total distance around the shape, we must add the lengths of all the sides. Because those lengths are expressions, we must add the expressions. We must then find the perimeter if $x = 7$ ft.

Plan: Add the expressions to get an expression for the perimeter, then evaluate the perimeter expression using $x = 7$ ft.

Execute: Perimeter $= x + 2x + 5 + 2x + 5$

$\qquad\qquad\quad = 5x + 10$

Now evaluate $5x + 10$ using $x = 7$ ft.

$$5(7) + 10$$
$$= 35 + 10$$
$$= 45 \text{ ft.}$$

Answer: Perimeter expression $= 5x + 10$; when $x = 7$ ft., the perimeter is 45 ft.

Connection

Notice that we are rewriting the expression in simplest form and then evaluating the expression. It is much easier to evaluate the expression after simplifying. That's the whole point of simplifying. If we didn't simplify $x + 2x + 5 + 2x + 5$ and instead replaced all the xs with 7, we'd still get 45, but it is much easier to use $5x + 10$.

$$x + 2x + 5 + 2x + 5$$
$$= 7 + 2(7) + 5 + 2(7) + 5$$
$$= 7 + 14 + 5 + 14 + 5$$
$$= 45 \text{ ft.}$$

a. Write an expression in simplest form for the perimeter of the shape. What would the perimeter be if $d = 9$ m?

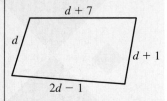

b. Write an expression in simplest form for the area of the shape. What would the area be if $w = 4$ in.?

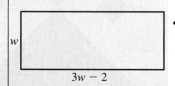

c. Write an expression in simplest form for the volume of the box. What would the volume be if $h = 2$ cm?

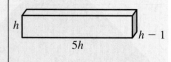

Answers: **a.** $5d + 7$; 52 m
b. $3w^2 - 2w$; 40 in.^{2} **c.** $5h^3 - 5h^2$;
20 cm^3

Check: To check the perimeter, let's look at the problem in a different way and see if we get the same result. If we replace the x with 7 ft. in the expressions for the side lengths we get numeric amounts for each side length. We should then be able to verify that the perimeter is actually 45 ft. by adding the side lengths together.

Evaluate $2x + 5$ using $x = 7$ ft.

$$2(7) + 5$$
$$= 14 + 5$$
$$= 19 \text{ ft.}$$

We now have numeric values for all the sides. We can now find the perimeter by adding all these lengths together.

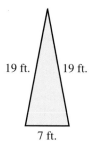

$$P = 7 \text{ ft.} + 19 \text{ ft.} + 19 \text{ ft.}$$
$$P = 45 \text{ ft. We get the same result.}$$

◄ **Do Margin 1.**

OBJECTIVE 2 *Solve surface area problems.*

We've considered the volume of a box, but there is another measurement that we can consider with boxes or any object. We can find an object's **surface area.** When we speak of surfaces, we mean the skin or shell of an object. Because area is a measure of the total number of square units, we can say a surface area is the total number of square units that completely cover the skin or shell of an object.

> **DEFINITION** **Surface area:** The total number of square units that completely cover the outer shell of an object.

In order to develop the relationship for surface area, it is helpful to break apart the object into all its surfaces. For a box, each surface is a rectangle. If we were to cut the box along the seams and fold it out flat, we would have the shape shown at the top of page 217.

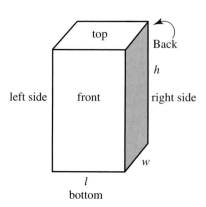

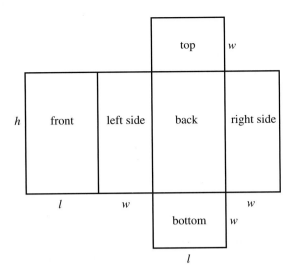

Notice there are six surfaces, top, bottom, front, back, left side, and right side. To get the total surface area, we could calculate the area of each of the six surfaces, then add it all together.

Surface area = Top + Bottom + Front + Back + Left side + Right side
of a box   area   area   area   area   area   area

$$SA = lw + lw + lh + lh + wh + wh$$

Notice we can simplify the polynomial that describes the surface area of a box by combining like terms. For example, the top and bottom are the same size, as are the front and back and the left side and right side, and therefore these represent like terms.

$$SA = 2lw + 2lh + 2wh$$

We could also make use of factoring because each term has a common 2 factor. Thus, we can write the polynomial this way:

$$SA = 2(lw + lh + wh)$$

EXAMPLE 2 A company produces metal boxes that are 2 ft. by 3 ft. by 1 ft. How many square feet of metal are needed to produce one box?

Solution: Follow the problem-solving process.

Understand: We must calculate how many square feet of metal are needed to produce a box. Because the metal will become the shell or skin of the box, we are dealing with surface area. The formula for the surface area of a box is $SA = 2(lw + lh + wh)$.

Plan: Replace l with 2 ft., w with 3 ft., and h with 1 ft. in the formula $SA = 2(lw + lh + wh)$.

Execute: $SA = 2(lw + lh + wh)$

$$SA = 2[(2 \text{ ft.})(3 \text{ ft.}) + (2 \text{ ft.})(1 \text{ ft.}) + (3 \text{ ft.})(1 \text{ ft.})]$$

$$SA = 2[6 \text{ ft.}^2 + 2 \text{ ft.}^2 + 3 \text{ ft.}^2]$$

$$SA = 2[11 \text{ ft.}^2]$$

$$SA = 22 \text{ ft.}^2$$

Answer: Each box will require 22 ft.^{2} of metal.

Check: An interesting way to check would be to use another version of the formula for surface area such as $SA = 2lw + 2lh + 2wh$ and make sure we get the same result. We will leave this to the reader.

Do Margin 2. ▶

◆ **Margin 2**

A carpenter builds wooden chests that are 2 ft. by 3 ft. by 2 ft. He finished the outside of the chests with stain. What is the surface area of the chest? What total area must he plan to cover if he intends to put three coats of stain on the box?

Answer: 32 ft.^{2}; 96 ft.^{2}

OBJECTIVE 3 *Solve problems involving a falling object.*

The polynomial $-16t^2 + s_0$ describes the height in feet of a falling object at any time during the fall. The variable s_0 represents the initial height of the object in feet from the ground. The variable t represents the number of seconds after the object is dropped from that initial height.

If we choose h to represent the height of the object after falling for t seconds, we can write the following formula:

$$h = -16t^2 + s_0$$

In Section 2.4, we mentioned that Galileo Galilei discovered that all objects fall at the same rate. Therefore, our formula for the height of an object after falling for t seconds is true for all objects, regardless of their weight. Note, however that the formula does not take into account air resistance. If we drop a feather and a lead weight at the same time, air resistance will keep the feather from reaching the ground at the same time as the lead weight. However, if we perform the same experiment in a vacuum, the feather and lead weight would reach the ground at the same time.

Historical Note: Who Was Galileo Galilei?

Born near Pisa, Italy, on February 18, 1564, Galileo Galilei was to revolutionize scientific thought. His father sent him to study medicine at the University of Pisa in 1581. However, by chance he was led to study geometry. Galileo was so successful in mathematics that he was able to persuade his father to allow him to give up medicine for mathematics and science.

In 1589, Galileo became a professor of mathematics at the University of Pisa, and in 1592, became professor of mathematics at the University of Padua. At Padua he discovered the laws of falling bodies and projectiles and conducted research on the motion of pendulums.

In 1609, Galileo constructed his own telescope, and his observations confirmed his belief in the Copernican theory that Earth and the planets revolve around the Sun. At the time, this radical theory was rejected. The prevailing view was that Earth was the center of the universe. In 1616, the Vatican arrested Galileo on charges of heresy.

In 1632, after years of silence on the subject, he repeated his belief in the Copernican system and was promptly summoned to Rome. He was brought to trial before the Inquisition in April 1633 and given a choice: either be tortured on the rack or renounce his defense of the Copernican system. He chose to renounce the Copernican system and was sentenced to house arrest for the rest of his life. He died in 1642. In 1992, the Vatican admitted its error in the Galileo case.

EXAMPLE 3 A skydiver jumps from a plane at an altitude of 2400 ft. and deploys the parachute after 8 sec. of free fall. What was the skydiver's altitude upon deploying the parachute?

Solution: Follow the problem-solving process.

Understand: We are given the initial altitude, or height, of the skydiver and the time for the free fall. To find the height after 8 sec. of free fall, we can use the formula $h = -16t^2 + s_0$.

Plan: Replace t with 8 and s_0 with 2400, then calculate.

Execute: $h = -16t^2 + s_0$

$$h = -16(8)^2 + 2400$$

$$h = -16(64) + 2400$$

$$h = -1024 + 2400$$

$$h = 1376 \text{ ft.}$$

Answer: The skydiver deployed the parachute at an altitude of 1376 ft.

Check: We can check the calculations by reversing the process.

$$1376 - 2400 \; ? \; -1024 \qquad -1024 \div 64 \; ? \; -16 \qquad \sqrt{64} = 8$$
$$-1024 = -1024 \qquad -16 = -16$$

Do Margin 3. ▶

◆ **Margin 3**

A marble is dropped from a 100-ft. tower. How far is the marble from the ground after 2 sec.?

OBJECTIVE **4** *Solve net-profit problems.*

In Section 2.3, we developed the formula for net profit or loss given revenue and cost.

$$N = R - C$$

Let's consider the situation where we are given expressions for revenue and cost. We can write an expression for net by subtracting the cost expression from the revenue expression.

EXAMPLE 4 The expression $7b + 9d + 1265$ describes the revenue for a business. The expression $8b + 3d + 742$ describes the cost. Write an expression in simplest form for the net.

Solution: Follow the problem-solving process.

Understand: We are to find net, given expressions for revenue and cost. The formula for net is $N = R - C$.

Plan: Replace R with $7b + 9d + 1265$ and C with $8b + 3d + 742$, then subtract.

Execute: $N = R - C$

$\qquad N = (7b + 9d + 1265) - (8b + 3d + 742)$

$\qquad N = (7b + 9d + 1265) + (-8b - 3d - 742)$

$\qquad N = -b + 6d + 523$

Discussion Does $-b$ in the expression indicate that the net is a total loss? Can profit still be made?

Answer: The expression that describes net is $-b + 6d + 523$.

Check: We can reverse the process. If we add the expression for net to the expression for cost, we should get the expression for revenue.

$$(-b + 6d + 523) + (8b + 3d + 742) \; ? \; 7b + 9d + 1265$$
$$7b + 9d + 1265 = 7b + 9d + 1265$$

EXAMPLE 5 Suppose the business described in Example 4 makes two different toy stuffed animals. Suppose b represents the number of toy bears and d represents the number of toy dogs. If the company sold 2345 bears and 3687 dogs in one month, what was the net?

Solution: Follow the problem-solving process.

Understand: We are to find the net, given the number of toy bears and the number of toy dogs sold. In Example 4 we found an expression for the net: $-b + 6d + 523$.

Plan: Replace b with 2345 and d with 3687 in the expression $-b + 6d + 523$, then calculate.

Execute: $-b + 6d + 523$

$\qquad -(2345) + 6(3687) + 523$

$\qquad = -2345 + 22{,}122 + 523$

$\qquad = 20{,}300$

Answer: The business made a profit of $20,300.

Check: An interesting way to check is to calculate the revenue and cost using the original expressions for revenue and cost, then calculate net. Because this procedure is the same as the check for perimeter in Example 1, we will leave this to the reader.

Answer: 36 ft.

A business sells two types of small motors. The revenue from the sales of motor A and motor B is described by $85A + 105B + 215$. The total cost of producing the two motors is described by $45A + 78B + 345$.

a. Write an expression in simplest form that describes the net.

b. If the business sells 124 of motor A and 119 of motor B in one month, what is the net for the month?

Connection

Think about the coefficients in the expressions for revenue, cost, and net. In the revenue expression, $7b + 9d + 1265$, the b represents the number of toy bears sold. The coefficient, 7, means each toy bear sells for $7. In the term, $9d$, the 9 means each toy dog sells for $9. The constant, 1265, is an income that does not depend on the number of toy bears or dogs sold. It could be interest or fees of some kind.

In the cost expression, $8b + 3d + 742$, the $8b$ term represents the cost of producing the toy bears and the $3d$ term represents the cost of producing the toy dogs. The coefficient, 8, means each bear costs $8 to produce. Similarly, 3, means each dog costs $3 to produce. The constant, 742, is a cost that does not depend on the number of toy bears or dogs produced. It could be utility and operation costs for the plant.

When we developed the expression for net by subtracting the cost expression from the revenue expression we combined like terms. In combining like terms, we subtracted the coefficients. Subtracting $8b$ from $7b$ equals $-b$. Because the $8 cost of producing each bear is more than the $7 revenue from the sale of each bear, the resulting -1 coefficient means the company has a net loss of $1 for each bear sold. Subtracting $3d$ from $9d$ equals $6d$. Because the $3 cost is less than the $9 revenue, the company has a net profit of $6 for each toy dog sold. Subtracting 742 from 1265 equals 523. Because the constant cost of $742 is less than the constant revenue of $1265, the $523 is a profit.

Putting it all together:

$$(7b + 9d + 1265) - (8b + 3d + 742) = \quad -b \quad + \quad 6d \quad + \quad 523$$

$-b$ ↑	$6d$ ↑	523 ↑
Net loss of $1 for each toy bear sold	Net profit of $6 for each toy dog sold	Net profit of $523 from constant revenue and costs

◀ **Do Margin 4.**

Answers: **a.** $40A + 27B - 130$
b. $8043

3.8 Exercises

For Exercises 1–6, write an expression in simplest form for the perimeter of the shape.

1.

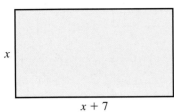

2.

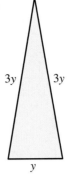

3.

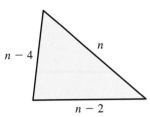

4.

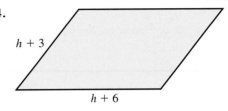

5.

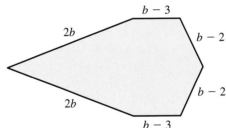

6.

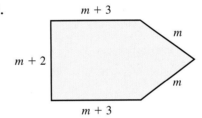

7. Calculate the perimeter of the rectangle in Exercise 1 if *x* is:

    a. 19   **b.** 27

8. Calculate the perimeter of the triangle in Exercise 2 if *y* is:

    a. 15   **b.** 22

9. Calculate the perimeter of the triangle in Exercise 3 if *n* is:

    a. 12   **b.** 8

10. Calculate the perimeter of the parallelogram in Exercise 4 if *h* is:

    a. 6   **b.** 15

11. Calculate the perimeter of the shape in Exercise 5 if *b* is:

    a. 10   **b.** 20

12. Calculate the perimeter of the shape in Exercise 6 if *m* is:

    a. 14   **b.** 32

For Exercises 13 and 14, write an expression in simplest form for the area of the shape.

13.

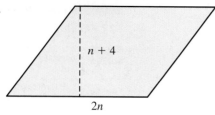

$n + 4$

$2n$

14.

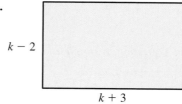

$k - 2$

$k + 3$

15. Calculate the area for the parallelogram in exercise 13 if n is:

    a. 6   **b.** 12

16. Calculate the area of the rectangle in Exercise 14 if k is:

    a. 8   **b.** 13

For Exercises 17 and 18, write an expression in simplest form for the volume of the shape.

17.

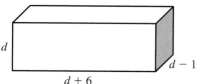

d

$d - 1$

$d + 6$

18.

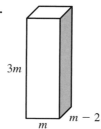

$3m$

$m - 2$

m

19. Calculate the volume of the box in Exercise 17 if d is:

    a. 3   **b.** 5

20. Calculate the volume of the box in Exercise 18 if m is:

    a. 7   **b.** 14

For Exercises 21–32, solve.

21. A company produces cardboard boxes. The boxes are 2 ft. by 2 ft. by 3 ft. each. What is the area of the cardboard that is required for each box?

22. A cube that is 5 cm along each side is to be covered with paper. What is the total area that will be covered by the paper?

23. Therese designs and builds metal sculptures. One particular sculpture is to be a large cube that will rest on one of its corners. She will cover the cube with panels that are 1 ft.^{2} in size. If the cube is to be 16 ft. on each side, how many panels will be needed to cover the cube?

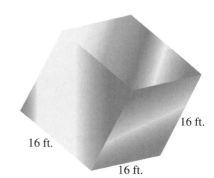

16 ft.

16 ft.

16 ft.

24. The main body of a certain satellite is a box that measures 3 ft. by 5 ft. by 6 ft. The box is covered with a skin of a thin metal alloy. What is the area covered by the metal alloy?

25. The leaning tower of Pisa is 180 ft. tall. When Galileo dropped a cannonball from the top of the leaning tower of Pisa, what was the height of the cannonball after 2 sec.? After 3 sec.?

Trivia Bite

Construction on the tower at Pisa began in 1173 and continued for more than 200 years due to continuous structural problems and soft, unstable soil. Because of the soft soil, one side of the tower began sinking before the first three stories were completed, creating a noticeable lean. The lean is estimated to be increasing at a rate of 1 mm per year.

26. Victoria Falls is a waterfall on the Zambezi river in south central Africa that is 355 ft. high. If a coin were dropped from the top of the falls, how high would the coin be after 3 sec.? 4 sec.?

27. A skydiver jumps from a plane at 12,000 ft. and deploys her parachute after 22 sec. of free fall. What is her altitude at the time she deploys her parachute?

28. Two skydivers jump from a plane at an altitude of 16,000 ft. One skydiver deploys his parachute after 27 sec. and the other 2 sec. later. What were their respective altitudes at the time each deployed his parachute?

29. A company produces two different tables, a round top and a square top. If r represents the number of round top tables and s represents the number of square top tables, then $145r + 215s + 100$ describes the revenue from the sales of the two types of table. The polynomial $110r + 140s + 345$ describes the cost of producing the two types of table.
   a. Write an expression in simplest form for net.
   b. If the company sells 120 round top tables and 106 square top tables in one month, what would be the net profit or loss?

30. Calvin installs tile. The expression $9lw + 45$ describes the revenue he receives for tiling a rectangular room where l represents the length of the room and w represents the width. The expression $3lw + t$ describes his cost for tiling a rectangular room, where t represents the number of tiles required.
   a. Write an expression in simplest form for net.
   b. Calvin gets a job to tile a 12 ft. by 14 ft. room. He uses 672 tiles. Calculate his profit.

31. Candice makes reed baskets in three sizes, small, medium, and large. She sells the small baskets for $5, medium for $9, and large for $15. The small baskets cost her $2 each to make, medium $4 each, and large $7 each.

   a. Write a polynomial that describes the revenue she receives from the sale of all three baskets.

   b. Write a polynomial that describes her cost to produce all three baskets.

   c. Write an expression in simplest form for her net.

   d. If in one day she sells 6 small, 9 medium, and 3 large baskets, what would be her net profit or loss?

32. Devon paints portraits in three sizes, small, medium, and large. He receives $50 for a small, $75 for a medium, and $100 for a large portrait. It costs him $22 for each small, $33 for each medium, and $47 for each large portrait.

   a. Write a polynomial that describes the revenue he receives from the sale of all three portrait sizes.

   b. Write a polynomial that describes his cost for all three sizes.

   c. Write an expression in simplest form for his net.

   d. In one month he sells 4 small, 6 medium, and 2 large portraits. What is his net profit or loss?

Puzzle Problem

How many cubic feet of dirt are in a hole 4 ft. wide, 6 ft. long, and 3 ft. deep?

REVIEW EXERCISES

1. Evaluate $\sqrt{2a + 7} - b^3$ for $a = 21$ and $b = -2$.

2. What is the coefficient of $-x^3$?

3. What is the degree of 5?

4. Combine like terms. $8x^3 - 9x^2 + 11 - 12x^2 + 2x - 9x^3 + 3 - 2x$

5. Evaluate 6^0.

CHAPTER 3   **Summary**

Defined Terms

Review the following terms, and for any you do not know, study its definition on the page number next to it.

Section 3.1
Equation *p. 150*
Expression *p. 150*

Section 3.2
Monomial *p. 155*
Coefficient *p. 156*
Degree *p. 156*

Like terms *p. 157*
Polynomial *p. 157*
Simplest form *p. 158*
Degree of a polynomial
   p. 159

Section 3.5
Conjugates *p. 187*

Section 3.6
Prime number *p. 193*
Composite number *p. 193*
Factorization *p. 196*
Prime factorization *p. 196*
Greatest common factor
   p. 199

Section 3.7
Factored form *p. 209*

Section 3.8
Surface area *p. 216*

Two actions that can be performed with an expression:

1. Evaluate (replace the variables with numbers and calculate)

2. Rewrite (simplify, distribute, or factor)

Procedures, Rules, and Key Examples

Procedures/Rules	Key Example(s)
Section 3.1 **To evaluate an expression:** **1.** Replace the variables with the corresponding given values. **2.** Calculate using the order of operations agreement.	Evaluate $-2x + 8$ when $x = 3$. $-2(3) + 8$ $= -6 + 8$ $= 2$
Section 3.2 **Descending order agreement:** To write a polynomial in descending order, place the highest degree term first, then the next highest degree term, and so on.	Write $6x^3 + 9x^7 + 8 - 4x^5$ in descending order. $9x^7 - 4x^5 + 6x^3 + 8$
Section 3.3 **To combine like terms:** **1.** Add or subtract the coefficients. **2.** Keep the variables and their exponents the same.	Combine like terms. $5x^3y^2 + 8x^3y^2 = 13x^3y^2$ $9a^4 - 10a^4 = -1a^4 = -a^4$
Section 3.4 To add polynomials, combine like terms.	Add. $(3x^2 - 5x + 1) + (7x^2 - 4x - 6)$ $= 3x^2 + 7x^2 - 5x - 4x + 1 - 6$ $= 10x^2 - 9x - 5$

To subtract polynomials:
1. Write the subtraction statement as an equivalent addition statement.
  a. Change the operation symbol from a minus sign to a plus sign.
  b. Change the subtrahend to its additive inverse. To get the additive inverse we change *all* the signs in the polynomial.
2. Combine like terms.
	Subtract.
	$(5a^3 - 9a + 2) - (a^3 - 6a - 7)$
	$= (5a^3 - 9a + 2) + (-a^3 + 6a + 7)$
	$= 5a^3 - a^3 - 9a + 6a + 2 + 7$
	$= 4a^3 - 3a + 9$

Procedures/Rules (cont.)	Key Example(s)

Section 3.5

When multiplying exponential forms that have the same base, we can add the exponents and keep the same base.

$$n^a \cdot n^b = n^{a+b}$$

Multiply.

$$t^5 \cdot t^4 = t^{5+4} = t^9$$

To multiply monomials:
1. Multiply coefficients.
2. Add the exponents of the like bases.

Multiply.

$$-6a^4bc^2 \cdot 5a^3b^2 = -30a^{4+3}b^{1+2}c^2$$
$$= -30a^7b^3c^2$$

When an exponential form is raised to a power, we can multiply the exponents and keep the same base.

$$(n^a)^b = n^{a \cdot b}$$

Simplify.

$$(x^2)^4 = x^{2 \cdot 4} = x^8$$

To simplify a monomial raised to a power:
1. Evaluate the coefficient raised to that power.
2. Multiply each variable's exponent by the power.

Simplify.

$$(-4hk^4)^3 = (-4)^3h^{1 \cdot 3}k^{4 \cdot 3}$$
$$= -64h^3k^{12}$$

To multiply a polynomial by a monomial:
Use the distributive law and multiply each term in the polynomial by the monomial.

Multiply.

$$2m(m^4 - 5m^2n + 3n^2)$$
$$= 2m \cdot m^4 - 2m \cdot 5m^2n + 2m \cdot 3n^2$$
$$= 2m^5 - 10m^3n + 6mn^2$$

To multiply two polynomials:
1. Multiply every term in the first polynomial by every term in the second polynomial.
2. Combine like terms.

Multiply.

$$(2x + 3)(5x - 1)$$
$$= 2x \cdot 5x + 2x \cdot (-1) + 3 \cdot 5x + 3(-1)$$
$$= 10x^2 - 2x + 15x - 3$$
$$= 10x^2 + 13x - 3$$

Section 3.6

Prime/Composite Test

To test a given number to see if it is prime or composite, divide the given number by the primes on the list of prime numbers and consider the results.
1. If the prime number divides the given number evenly, stop and conclude that the given number is a composite number.
2. If the prime does not divide the given number evenly, consider the quotient.
   a. If the quotient is larger than the current prime divisor, repeat the process with the next prime on the list of prime numbers.
   b. If the quotient is equal to or less than the current prime divisor, stop and conclude that the given number is itself a prime number.

Note: 0 and 1 are neither prime nor composite.

Test 37 to see if it is prime or composite.

Try 2:
$$\begin{array}{r} 16 \\ 2\overline{)37} \\ -2 \\ \hline 17 \\ -16 \\ \hline 1 \end{array}$$

Try 3:
$$\begin{array}{r} 12 \\ 3\overline{)37} \\ -3 \\ \hline 07 \\ -6 \\ \hline 1 \end{array}$$

2 does't work and the quotient, 16, is larger than the divisor, 2, so go on.

3 doesn't work and the quotient, 12, is larger than the divisor, 3, so go on.

Try 5:
$$\begin{array}{r} 7 \\ 5\overline{)37} \\ -35 \\ \hline 2 \end{array}$$

Try 7:
$$\begin{array}{r} 5 \\ 7\overline{)37} \\ -35 \\ \hline 2 \end{array}$$

5 doesn't work. The quotient is still larger than the divisor.

7 doesn't work. The quotient is smaller than the divisor! Stop test!

37 is a prime.

Procedures/Rules (cont.)	Key Example(s)

To find the prime factorization of a number, use a factor tree.
1. Draw two branches below the number.
2. Find two factors that multiply to make the given number and place them at the end of the two branches.
3. Repeat steps 1 and 2 for every factor that is not a prime number.
4. Place all the prime factors together in a multiplication sentence.

Find the prime factorization.

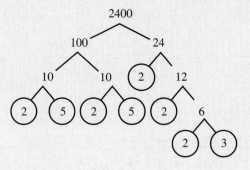

$2400 = 2 \cdot 2 \cdot 2 \cdot 2 \cdot 2 \cdot 3 \cdot 5 \cdot 5$

$\quad\quad = 2^5 \cdot 3 \cdot 5^2$

To find the greatest common factor by listing:
1. List all possible factors for each given number.
2. Search the lists for the largest factor common to all lists.

Find the GCF of 32 and 40 by listing.

Factors of 32: 1, 2, 4, 8, 16, 32
Factors of 40: 1, 2, 4, 5, 8, 10, 20, 40

GCF = 8

To find the greatest common factor of a given set of numbers:
1. Get the prime factorization of each given number.
2. Write the factorizations in exponential form.
3. Use the smallest exponent of the common primes.

Note: If there are no common primes, then the GCF is 1.

Find the GCF of 84 and 120.

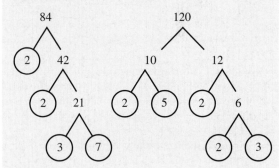

$$84 = 2^2 \cdot 3 \cdot 7$$

$$120 = 2^3 \cdot 3 \cdot 5$$

GCF $= 2^2 \cdot 3 = 2 \cdot 2 \cdot 3 = 12$

Section 3.7

When dividing exponential forms that have the same base we can subtract the divisor's exponent from the dividend's exponent and keep the same base.

$\quad n^a \div n^b = n^{a-b}$ as long as $n \neq 0$

Note: $n \neq 0$ because if n were replaced with 0, we would have $0^a \div 0^b$. In Section 1.4 we concluded that when the dividend and divisor are both 0, the problem is indeterminate.

Divide.
$t^9 \div t^2 = t^{9-2} = t^7$

Any base other than 0 raised to the 0 power simplifies to the number 1.
$\quad n^0 = 1$, as long as $n \neq 0$

Note: 0^0 is indeterminate.

Evaluate.
$245^0 = 1$
$(-5)^0 = 1$

Procedure: To divide monomials:
1. Divide the coefficients.
2. For like bases, subtract the exponent of the divisor base from the exponent of the dividend base and keep the base. If the bases have the same exponent, then they divide out, becoming 1.
3. Bases in the dividend that have no like base in the divisor are written unchanged in the quotient.

Divide.
$$-18m^3n^7p^2 \div 2m^3n^5p$$
$$= -9m^{3-3}n^{7-5}\,p^{2-1}$$
$$= -9m^0n^2p^1$$
$$= -9(1)n^2p$$
$$= -9n^2p$$

To find the greatest common factor of a given set of monomials or terms:
1. Get the prime factorization of each given number.
2. Write the factorizations in exponential form. Treat variables like primes.
3. Use the smallest exponent of the common primes or variables.
4. Multiply or simplify.

Note: If there are no common primes, the GCF is 1.

Find the GCF of $36m^3n^5p$ and $54m^2n^5$

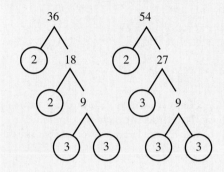

$$36 = 2^2 \cdot 3^2$$
$$54 = 2 \cdot 3^3$$
$$\text{GCF} = 2 \cdot 3^2 m^2 n^5 = 18m^2n^5$$

To factor a monomial out of a given polynomial:
1. Find the GCF of the terms that make up the polynomial.
2. Write an expression that is a product of the GCF and empty parentheses this way:
   GCF · (   )
3. Divide each term of the polynomial by the GCF and place the results inside the parentheses.

Factor.
$$9u^7 + 27u^5v - 18u^3$$
$$= 9u^3(9u^7 \div 9u^3 + 27u^5v \div 9u^3 - 18u^3 \div 9u^3)$$
$$= 9u^3(u^4 + 3u^2v - 2)$$

Formulas

Area of a parallelogram:	$A = bh$
Volume of a box:	$V = lwh$
Surface area of a box:	$SA = 2lw + 2lh + 2wh$
Height of a falling object:	$h = -16t^2 + s_0$
Net profit or loss:	$N = R - C$

For Exercises 1–6, answer true or false.

1. $4x - 8y + 9$ is an equation.

2. $3x^2 - 8x = 10$ is an equation.

3. 27 is a monomial.

4. 7 is degree 1.

5. $(-16)^0 = 0$

6. $14x^2$ and $12x$ are like terms.

For Exercises 7–10, complete the rule.

7. When combining like terms, we add or subtract the _____ and keep the _____ the same.

8. When multiplying exponential forms that have the same base, we _____ the exponents and keep the same _____.

9. When dividing exponential forms that have the same base, we _____ the _____ exponent from the _____ exponent and keep the bases the same.

10. When raising an exponential form to a power, we _____ the exponents and keep the bases the same.

For Exercises 11–14, evaluate the expression using the given values.

11. $5n^2 - 9n + 2$ when $n = -2$.

12. $\sqrt{3x + 4} - 2y$ when $x = 7$ and $y = 9$.

13. $4|t^2 - 11|; t = 3$

14. $\dfrac{-3ab + 4}{a - b^2}; a = 2, b = -4$

15. Is $9x^3yz$ a monomial? Explain.

16. Is $4x - 5$ a monomial? Explain.

For Exercises 17–20, identify the coefficient and degree of each monomial.

17. $18x$

18. y^3

19. -9

20. $-3m^5n$

21. Explain why $4xy^3$ and $9xy^3$ are like terms.

22. Explain why $7a^2b$ and $7ab^2$ are not like terms.

For Exercises 23–26, tell whether the expression is a monomial, binomial, trinomial, or none of these.

23. $9y - 5$

24. $4x^3 + 9x^2 - x + 7$

25. $-3ab$

26. $x^2 + 5x - 4$

27. What is the degree of $7a^4 - 9a + 13 + 5a^6 - a^2$?

28. Write $7a^4 - 9a + 13 + 5a^6 - a^2$ in descending order.

For Exercises 29 and 30, combine like terms and write the resulting polynomial in descending order.

29. $5m^3 + 9m - 12 + 2m^3 + 5$

30. $3x^2 - 5x + x^7 + 13 + 5x - 6x^7$

For Exercises 31 and 32, add.

31. $(y^4 + 2y^3 - 8y + 5) + (3y^4 - 2y - 9)$

32. $(4a^5 - 6a^2b^2 + 8ab - 19b) + (3a^5 - 2a^2b^3 + 6a^2b^2 - 3b)$

For Exercises 33 and 34, subtract.

33. $(19h^3 - 4h^2 + 2h - 1) - (6h^3 + h^2 + 7h + 2)$

34. $(5m^4 + 7m^3n + 2mn - 9n^2) - (6m^4 + 3m^3n + 2mn - n^2)$

For Exercises 35–37, multiply.

35. $m^3 \cdot n^4$

36. $2xy^3 \cdot (-5x^4)$

37. $-6t^3uv^2 \cdot (-t^5u^4)$

For Exercises 38 and 39, simplify.

38. $(5x^4)^3$

39. $(-2a^4bc^5)^3$

For Exercises 40–46, multiply and simplify.

40. $3n(5n^2 - n + 7)$

41. $-xy^3(2x^3 + 5xy^2 - 8y)$

42. $(a + 5)(a - 7)$

43. $(2y - 1)(5y - 8)$

44. $(x^2 + y)(x - y)$

45. $(3t + 4)(3t - 4)$

46. $(u + 2)(u^2 - 5u + 3)$

47. What is the conjugate of $-7x + 2$?

For Exercises 48 and 49, divide.

48. $r^8 \div r^2$

49. $20x^5 \div (-5x^2)$

50. Evaluate 15^0.

51. Find the missing factor. $6m^3n \cdot (?) = -54m^8n$

52. Find the missing factor. $40xy^6 - 30y^4 - 20y^2 = 10y^2 \ (?)$

53. Find an expression for the missing length in the rectangle shown. The area is $42a^3b^6$.

$3ab^5$

?

54. Determine whether 119 is prime, composite, or neither.

55. Determine whether 97 is prime, composite, or neither.

56. Find the prime factorization of 360.

57. Find the prime factorization of 4200.

58. Find the GCF of 140 and 196.

59. Find the GCF of 45 and 28.

For Exercises 60 and 61, find the GCF.

60. $48x^5$ and $36x^6y$.

61. $18a^2b^5$, $30a^4b^3c$, and $24a^3b^4c^2$

For Exercises 62 and 63, factor.

62. $9n^4 - 15n^3$

63. $18x^3 + 24x^2y - 36xz$

64. a. Write an expression in simplest form for the perimeter of the rectangle shown.

　　b. Find the perimeter if x is 9.

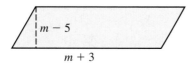

$3x - 8$

$5x + 2$

65. a. Write an expression in simplest form for the area of the parallelogram shown.

　　b. Find the area of the parallelogram if m is 8.

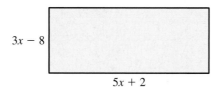

$m - 5$

$m + 3$

66. a. Write an expression for the volume of the box shown.

　　b. Find the volume of the box if n is 4.

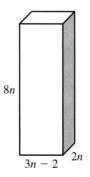

$8n$

$3n - 2$　$2n$

67. A coin is dropped from 150 ft. What is the height of the coin after 2 sec.?

68. A company produces two different tennis racquets, a normal size and an oversize version. If n represents the number of normal racquets and b represents the number of oversize racquets, then $65n + 85b + 345$ describes the revenue from the sales of the two racquets. The polynomial $22n + 42b + 450$ describes the cost of producing the two racquets.

a. Write an expression in simplest form for net.

b. If the company sells 487 normal size racquets and 246 oversize racquets in one week, what would be the net?

Practice Test

1. Evaluate $2n^3 - 3p^2$; $n = -2$, $p = 3$.

1. _____

2. Evaluate $\sqrt{x - 2} + |5y|$ when $x = 11$ and $y = -3$.

2. _____

3. Is $4x^3 - 9$ a monomial, binomial, trinomial, or none of these?

3. _____

4. What is the coefficient of $-y^4$?

4. _____

5. What is the degree of 17?

5. _____

6. What is the degree of $x^2 - 9x + 5x^4 - x^3$?

6. _____

7. Combine like terms and write the resulting polynomial in descending order.
$10a^4 + 3a^2 - 5a + 2a^2 + 5a - 11a^4$

7. _____

8. Add. $(y^4 + 2y^3 - 8y + 5) + (3y^4 - 2y - 9)$

8. _____

9. Subtract. $(12y^5 + 5y^3 - 7y - 3) - (6y^5 + y^2 - 7y + 1)$

9. _____

10. Multiply. $(-4tu^5) \cdot (-5u^2)$

10. _____

11. Simplify. $(-2a^3)^5$

11. _____

For 12 and 13, multiply.

12. $-2tu^3(t^3 - 3tu - 7)$

13. $(b - 6)(b + 6)$

12. _____

13. _____

14. What is the conjugate of $6x - 5$?

14. _____

15. Divide. $m^6 \div m^4$

15. _____

16. _____

16. Evaluate 15^0.

17. Find the missing factor. $-8x^4y \cdot (?) = -40x^7y^3$

17. _____

18. Tell whether 91 is prime, composite, or neither.

18. _____

19. Find the prime factorization of 340.

19. _____

20. Find the GCF of 180 and 396.

20. _____

21. Find the GCF of $60h^5$ and $48h^7k$.

21. _____

22. Factor $8x^7 - 16x^2$.

22. _____

23. Factor $20m^6 + 15m^4n - 30m^3$.

23. _____

24. a. Write an expression in simplest form for the area of the parallelogram.

    b. Calculate the area if n is 7.

$n - 4$

$2n + 1$

24. _____

25. A company makes two different styles of glasses frames. One style allows for rounded lenses, while the other is for rectangular lenses. If a represents the number of rounded style and b represents the number of rectangular style frames, then $145a + 176b$ describes the revenue from the sales of the frames. The polynomial, $61a + 85b$ describes the cost of producing the frames.

    a. Write an expression for the net.

    b. If the company sold 128 rounded frames and 115 rectangular frames in one month, what would be the net profit or loss?

25. _____

Cumulative Review Exercises

For Exercises 1–6, answer true or false.

1. 0 is a natural number.

2. $245 < 387$

3. $|8| = -8$.

4. $9 + (-12) = -12 + 9$

5. $15xy^2$ is a monomial.

6. The degree of 18 is 1.

7. List the order of operations agreement.

   1.

   2.

   3.

   4.

8. The sum of two negative numbers is a _____ number.

9. The product of a negative number and a positive number is a _____ number.

10. Explain how to multiply two binomials.

11. Write the word name for 2,480,045.

12. Write 7×10^8 in standard form.

13. Graph -9 on a number line.

14. Round $-105,612$ to the nearest ten thousand.

15. Estimate $682 \cdot 246$ by rounding to the nearest hundred.

16. What is the coefficient of $-x^2y$?

17. What is the degree of $5x$?

18. What is the degree of $7t^2 - 9t + 4t^5 + 12$?

For Exercises 19–29, simplify.

19. $|14 - 20|$

20. $-(-(-9))$

21. $58 - 184 - 32 + 14$

22. $-63 - (-25)$

23. $-1(-9)(-3)(-4)$

24. $(-5)^3$

25. -3^4

26. $1442 \div (-14)$

27. $\sqrt{16 \cdot 25}$

28. $2[9 + (3 - 16)] - 35 \div 5$

29. $4^3 - \{[6 + (2)8] - [12 + 9(6)]\}$

30. Evaluate $\sqrt{x - y}$ when $x = 100$ and $y = 36$.

31. Combine like terms. $5x^2 - 9x + 11x^2 - 7 - x^3 + 8x - 5$

32. Subtract. $(4y^3 - 6y^2 + 9y - 8) - (3y^2 - 12y - 2)$

33. Multiply. $(9x^3)(7xy)$

34. Multiply. $(b - 8)(2b + 3)$

35. Find the prime factorization of 360.

36. Find the GCF/GCD of $40x^2$ and $30x^3y$.

37. Divide. $18n^5 \div 3n^2$

38. Factor. $12m^4 - 18m^3 + 24m^2$

For Exercises 39 and 40, solve and check.

39. $x + 19 = 25$

40. $-16x = -128$

41. Calculate the area of the figure.

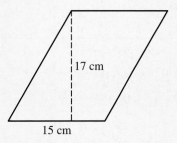

17 cm

15 cm

42. Write an expression in simplest form for the perimeter.

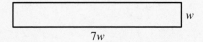

w

$7w$

43. Write an expression in simplest form for the volume.

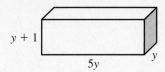

$y + 1$

$5y$

y

For Exercises 44–50, solve.

44. Andre has a monthly mortgage payment of $785. If he makes 360 payments (30 years), then what would be the total he pays over the 30 years?

45. An entrance lock requires a person to enter three things. First a digit, 0–9, then a letter of the alphabet, A–H, and finally a day of the week. How many possible combinations are there?

46. A patient is to receive 600 cc of 5% D/W (dextrose in water) solution in an IV drip over the course of an hour. How many cc should the patient receive each minute?

47. Carla has a balance of $-\$353$ in a credit card account. If she makes a payment of $150, then makes two purchases at $38 each, what is her new balance?

48. A 10 ft. by 12 ft. by 7 ft. box is to be made out of plastic. What is the surface area of the box?

49. A ride at an amusement park is a ship that rotates to an inverted position 80 ft. above the ground. While inverted, loose items such as sunglasses, coins, etc., tend to fall. How far from the ground will the falling items be after 2 sec.?

50. A company produces two different car alarm systems, basic and advanced. If b represents the number of basic alarms and a represents the number of advanced alarms, then $225b + 345a + 200$ describes the revenue from the sales of the two types of alarm systems. The polynomial $112b + 187a + 545$ describes the cost of producing the two alarm systems.
   a. Write an expression in simplest form for net.
   b. If the company sells 88 basic and 64 advanced alarm systems in one month, what would be the net profit or loss?

Equations

Preparing for a Test

If you have followed the day-by-day routine of practice and study as discussed in the *To the Student* section and the Chapter 3 opener, then most of your test preparation is already done and preparation for the actual test will be like a dress rehearsal.

Read through all your notes and the summary at the end of the chapter. Then practice duplicating your study sheet from memory. Once you know the rules and procedures, work through the review exercises that follow the chapter summary. Then take a break.

When you return, it's time for a dress rehearsal. Have ready the same materials you will use during the actual test, and set a timer for the same amount of time as you will have for the actual test. When you start the timer, first duplicate your study sheet on the scratch paper. This is often called a *memory dump*. Next, take the practice test that follows the review exercises. When you finish, calculate your score. Keep repeating this process until you get everything right every time you take the practice test.

If your instructor constructs tests that are very different from the practice tests in the text, then construct your own practice test (see the *To the Student* section of the text). Finally, be sure to take care of your body by resting and eating a healthy meal before test time. Do a quick review the morning of the test and practice the memory dump as often as you can throughout the day before test time. Above all, try to relax and trust in your preparation.

"Spectacular achievement is always preceded by spectacular preparation."
—ROBERT H. SCHULLER

"What I do is prepare myself until I know I can do what I have to do."
—JOE NAMATH

Historical Note

The modern form of the equal sign is attributed to English mathematician Robert Recorde in his book *Wetstone of Witte*, published in 1557. He wrote,

"I will sette as I doe often in woorke vse, a paire of paralleles, or Gemowe [twin] lines of one lengthe, thus: =, bicause noe .2. thynges, can be moare equalle."

Recorde's symbol was not immediately popular. The popular symbol of the times was ∝ and was commonly used into the eighteenth century.

Source: D. E. Smith, *History of Mathematics*, 1953

4.1 Equations and Their Solutions

OBJECTIVES

1 Differentiate between an expression and an equation.

2 Check a given number to see if it is a solution for a given equation.

OBJECTIVE 1 *Differentiate between an expression and an equation.*

In Section 3.1 we defined the difference between an **expression** and an **equation**. Recall these definitions.

DEFINITION **Expression:** A mathematical relationship with constants and/or variables but no equal sign.

Expressions: $3x + 5$
$7x^2 - 8x + 1$

Because expressions do not have an equal sign, think of them as phrases. If we read $3x + 5$, we say, "three x plus five." There is no verb in "three x plus five," and we must have a verb to have a complete sentence.

DEFINITION **Equation:** A mathematical relationship that contains an equal sign.

Equations: $3x + 5 = 9$
$7x^2 - 8x + 1 = 5x - 2.$

Equations have an equal sign, so think of them as complete sentences. The equal sign is a mathematical verb, like the verb *is*. When we read $3x + 5 = 9$, we can say, "Three x plus five is equal to nine." Or, we can simply say, "Three x plus five is nine." Now we have a complete sentence. We have a subject, $3x + 5$, a verb, =, and a direct object, 9.

$$3x + 5 = 9$$
$$\updownarrow$$
"Three x plus five is nine."
Subject   Verb   Direct Object

◆ **Margin 1**

Determine whether each of the following is an expression or an equation.

a. $5x - 8$

b. $7y + 9 = 15$

c. $P = 2l + 2w$

d. $4^2 + 3(9 - 1) + \sqrt{25}$

Determine whether $7x^2 - 9x + 5$ is an expression or an equation.

EXAMPLE 1 **Answer:** $7x^2 - 9x + 5$ is an expression because it does not contain an equal sign.

Determine whether $9x - 10 = 3x + 1$ is an expression or an equation.

EXAMPLE 2 **Answer:** $9x - 10 = 3x + 1$ is an equation because it contains an equal sign.

◀ **Do Margin 1.**

OBJECTIVE 2 *Check a given number to see if it is a solution for a given equation.*

In Chapter 3, we learned we can perform two actions with expressions.

1. Evaluate (replace the variables with given numbers and perform the arithmetic)

2. Rewrite (simplify, distribute, or factor)

Answers: **a.** expression **b.** equation **c.** equation **d.** expression

Our objective with an equation is to **solve** the equation. To solve an equation, we must find its **solution** or solutions.

DEFINITION   **Solve:** To find a solution or solutions to an equation.

Solution: A number that can replace the variable in an equation and make the equation true.

We have solved equations in the past. We simply called them *missing number statements* and our objective was to find the missing number. *Finding a missing number is solving an equation.* The missing number itself is the solution to the equation because it makes the equation true. Previously, we considered missing addends, missing factors, and even missing roots.

Missing addend:	Missing factor:	Missing root:
$x + 3 = 8$	$-3x = 18$	$x^2 = 81$
$x = 8 - 3$	$x = 18 \div (-3)$	$x = \pm 9$
$x = 5$	$x = -6$	

5 is the solution because it makes $x + 3 = 8$ true. $5 + 3 = 8$

-6 is the solution because it makes $-3x = 18$ true. $-3(-6) = 18$

Both 9 and -9 are solutions because they both make $x^2 = 81$ true. $(9)^2 = 81$ and $(-9)^2 = 81$

Notice what we did to verify that a number actually was a solution. When we found $x = 5$ for $x + 3 = 8$, we verified that 5 is actually the solution by putting 5 back into the equation. In more proper terms we evaluated the expression $x + 3$ with $x = 5$ and made sure it equaled 8. This is how we test or *check* a value to see if it is a solution for an equation.

Procedure   *To check a value to see if it is a solution for a given equation:*
1. Use the value to evaluate the expressions that make up the equation.
2. If the results equate, then the value is a solution.

EXAMPLE 3   Check to see if 3 is a solution for $2x - 5 = 1$.

Solution: Evaluate $2x - 5$ with $x = 3$ and see if it equals 1.

Bring down each step of $2(3) - 5$ until we get a final result. We then compare with the number on the other side.

$$2x - 5 = 1$$
$$2(3) - 5 \; ? \; 1$$
$$6 - 5 \; ? \; 1$$
$$1 = 1$$

The use of the ? here simply means we are asking ourselves, "How does $2(3) - 5$ compare with 1? Are they equal or not equal?"

Just keep bringing down 1 until we resolve whether it equates with $2(3) - 5$.

◆ **Answer:** The result $1 = 1$ is true. Therefore 3 is a solution for $2x - 5 = 1$.

EXAMPLE 4   Check to see if -2 is a solution for $y^2 + 7 = 13$.

Solution: Evaluate $y^2 + 7$ with $y = -2$ and see if the result is 13.

$$y^2 + 7 = 13$$
$$(-2)^2 + 7 \; ? \; 13$$
$$4 + 7 \; ? \; 13$$
$$11 \neq 13$$

Margin 2

Check to see if 2 is a solution for $12 = 9b - 5$.

Answer: Because 11 does not equal 13, -2 is not a solution for $y^2 + 7 = 13$.

Connection

In previous chapters we always placed results below the expression from which they came.

We wrote: $(-2)^2 + 7$

$$= 4 + 7$$
$$= 11$$

By bringing results down we were preparing for this checking technique. To check, we bring down each result from the left expression on the left side of the equation and each result from the right expression on the right side of the equation, and see if the results equate.

◀ **Do Margins 2 and 3**

Margin 3:

Check to see if -4 is a solution for $m^2 + 5 = 21$.

EXAMPLE 5 Check to see if 4 is a solution for $-3n + 8 = 2n - 17$.

Solution: Evaluate both $-3n + 8$ and $2n - 17$ with $n = 4$ and see if the results equate.

Bring down each step in evaluating $-3(4) + 8$ until we get a final result.

$$-3n + 8 = 2n - 17$$
$$-3(4) + 8 \ ? \ 2(4) - 17$$
$$-12 + 8 \ ? \ 8 - 17$$
$$-4 \neq -11$$

Bring down each step in evaluating $2(4) - 17$ until we get a final result.

The results do not equate.

◆ **Answer:** The results do not equate so 4 is not a solution.

EXAMPLE 6 Check to see if -5 is a solution for $2(a + 6) = 6a - 4(3a + 7)$.

Solution: Evaluate both $2(a + 6)$ and $6a - 4(3a + 7)$ with $a = -5$ and see if the results equate.

Margin 4

Check to see if 2 is a solution for $5m - 29 = -8m - 3$.

We keep bringing down 2 until we resolve the other side.

$$2(a + 6) = 6a - 4(3a + 7)$$
$$2(-5 + 6) \ ? \ 6(-5) - 4(3(-5) + 7)$$
$$2(1) \ ? \ -30 - 4(-15 + 7)$$
$$2 \ ? \ -30 - 4(-8)$$
$$2 \ ? \ -30 - (-32)$$
$$2 \ ? \ -30 + 32$$
$$2 = 2$$

◆ **Answer:** The results equate so -5 is a solution for $2(a + 6) = 6a - 4(3a + 7)$.

◀ **Do Margins 4 and 5.**

Margin 5

Check to see if 4 is a solution for $5x - 2(x + 3) = 7(x + 2)$.

Historical Note

Pierre de Fermat is often referred to as the *Prince of Amateurs*. He was born in Beaumont-de-Lomagne, France, in 1601. He lived a very normal, quiet, honest life as King's council in the local parliament of Toulouse, France. Fermat is probably best known for a theorem that is an extension of the Pythagorean theorem, an equation relating the side lengths of right triangles.

Fermat had the habit of writing his theorems as commentary in the margins of math books or letters to mathematicians. Consequently he did not supply proofs for his theorems. The proof of the theorem mentioned above eluded mathematicians for centuries. It became known as *Fermat's last theorem*. In 1993 Andrew Wiles of Princeton University submitted a potential proof of the theorem. Wiles' initial proof was found to have a subtle error that he was able to correct and his 150-page proof that Fermat's theorem is correct has now been accepted.

Source: E. T. Bell, *Men of Mathematics*

Answer to Margin 2: 2 is not a solution.

Answer to Margin 3: -4 is a solution.

Answer to Margin 4: 2 is a solution.

Answer to Margin 5: 4 is not a solution.

4.1 Exercises

FOR EXTRA HELP

 Videotape 5

 InterAct Math Tutorial Software

 www.carsonmath.com

 AWL Math Tutor Center

InterAct MathXL www.mathxl.com

 Student's Solutions Manual

For Exercises 1–6, determine whether each of the following is an expression or an equation.

1. $9x + 7 = 5$

2. $7x - 8$

3. $2n + 8m^3$

4. $10x^2 = 12x + 1$

5. $14 + 2^5 - 2(3 + 7)$

6. $2y = 16$

For Exercises 7–20, check to see if the given number is a solution for the given equation.

7. $6x - 7 = 41$; check $x = 8$

8. $-5a + 19 = 9$; check $a = 2$

9. $3m + 10 = -2m$; check $m = 5$

10. $18 - 7n = 2n$; check $n = -2$

11. $x^2 - 15 = 2x$; check $x = -3$

12. $x^2 - 15 = 2x$; check $x = 5$

13. $b^2 = 5b + 6$; check $b = -3$

14. $b^2 = 5b + 6$; check $b = -1$

15. $-5y + 20 = 4y + 20$; check $y = 0$

16. $10n - 14 = -n + 19$; check $n = 1$

17. $-2(x - 3) - x = 4(x + 5) - 21$; check $x = 3$

18. $3(x - 5) - 1 = 11 - 2(x + 1)$; check $x = 5$

19. $a^3 - 5a + 6 = a^2 + 3(a - 1)$; check $a = -2$

20. $2t(t + 1) = t^3 - 5t$; check $t = -3$

Puzzle Problem

A farmer has a fox, a goat, and a large head of cabbage to transport across a river. His boat is so small that it can hold only the farmer and the fox, goat, or cabbage at a time. If the farmer leaves the fox with the goat, the fox will eat the goat. If he leaves the goat with the cabbage, the goat will eat the cabbage. Somehow he transports them all across safely in the boat. Explain the process.

REVIEW EXERCISES

1. Combine like terms. $2x - 19 + 3x + 7$

2. Combine like terms. $7y + 4 - 12 - 9y$

3. Multiply. $5(m + 3)$

4. Multiply. $-(3m - 4)$

5. Solve. $x + 8 = 15$

4.2 The Addition/Subtraction Principle

OBJECTIVES

1 Determine whether a given equation is linear.

2 Solve linear equations in one variable using the addition/subtraction principle.

3 Solve equations with variables on both sides of the equal sign.

4 Solve application problems.

OBJECTIVE 1 *Determine whether a given equation is linear.*

There are many types of equations. We will focus on **linear equations.**

> **DEFINITION** **Linear equation:** An equation that is made of polynomials or monomials that are at most degree 1.

This means that linear equations can contain constants or single variable terms with an exponent of 1 on the variable. Equations that are not linear are called *nonlinear equations.*

EXAMPLE 1 Determine whether $5x + 9 = 24$ is linear.

Answer: $5x + 9 = 24$ is a linear equation because 1 is the highest degree involved.

EXAMPLE 2 Determine whether $2x - 9x^2 = 15x + 7$ is linear.

Answer: $2x - 9x^2 = 15x + 7$ is not linear because there is a term with a degree higher than 1. $-9x^2$ is a degree 2 term.

If the equation has the same variable throughout, then we say it is an equation in one variable.

$2x + 9 = 5x - 12$ is a linear equation in one variable.

$x^2 - 2x + 5 = 8$ is a nonlinear equation in one variable.

Some equations may have different variables involved in the same equation. If the equation has two different variables, we say it is an equation in two variables. The equation is still considered linear as long as all variable terms are degree 1.

$2x - 3y = 6$ is a linear equation in two variables.

$y = x^3$ is a nonlinear equation in two variables.

Do Margin 1. ▷

OBJECTIVE 2 *Solve linear equations in one variable using the addition/subtraction principle.*

In the past we have solved missing number statements like $x + 7 = 9$, which we now call *equations.* We now want to build on what we've learned about solving these equations and develop a technique for solving more complex equations. This technique for solving equations is called the **balance technique.**

Let's view equations as scales. The equal sign is like the pivot point on the scales.

◆ **Margin 1**

Determine whether the given equation is linear.

a. $2x - 8 = 14$

b. $7y = 21$

c. $x^3 = 5x + 2$

d. $y = x^2 - 5$

e. $y = 2x + 1$

Answers: **a.** Yes **b.** Yes **c.** No **d.** No **e.** Yes

When a weight is added or removed from one side of the scales, it tips the scales out of balance. To maintain balance, we must add or remove the same amount from both sides of the scale. The same principle applies to equations. When we add or remove amounts on one side of an equation, we must add or remove the same amount on the other side to keep the equation in balance. This is the essence of the balance technique.

DEFINITION **Balance technique:** If we add or remove an amount on one side of an equation, we must add or remove the same amount on the other side to keep the equation balanced.

In the language of mathematics the balance technique is divided into the use of two principles.

1. The addition/subtraction principle

2. The multiplication/division principle

In this section we will consider the addition/subtraction principle. In Section 4.3 we will consider the multiplication/division principle. Let's state the addition/subtraction principle.

The Addition/Subtraction Principle of Equality *We can add or subtract the same amount on both sides of an equation without affecting its solution(s).*
Primary purpose: To get variable terms on one side of the equal sign and constant terms on the other side.

Consider the equation $x + 2 = 5$. Notice the solution for this equation is 3 because $3 + 2 = 5$. The addition/subtraction principle tells us that we should be able to add or subtract the same amount on both sides and 3 will always be the solution. Let's arbitrarily choose the number 4 and add it on both sides to try out the principle.

$$x + 2 = 5$$
$$\underline{+4\ \ +4}$$
$$x + 6 = 9$$

Notice the new equation is $x + 6 = 9$, but the solution is still 3 because $3 + 6 = 9$. Adding 4 to both sides did not affect the solution.

How do we use the principle? Remember our goal is to solve equations. A solution statement for an equation is written like $x = 3$. So our goal is to take an equation from a form like $x + 2 = 5$ and convert it to the simpler form of $x = 3$. In the solution statement the variable is alone on one side of the equal sign and the solution appears on the other side. Our goal then is to *isolate the variable.*

Going back to our equation, $x + 2 = 5$, how can we isolate x? To isolate x, we want to clear, or undo, $+2$ from the left-hand side. To do that, we could add -2 (or subtract 2). Because $+2$ and -2 are additive inverses, their sum is 0.

To isolate x, we clear $+2$. To clear $+2$, we can add -2 (or subtract 2) to make 0.	$$\begin{aligned} x + 2 &= 5 \\ \underline{-2\ \ -2}& \\ x + 0 &= 3 \\ x &= 3 \end{aligned}$$	If we add -2 on the left side, we must add -2 on the right side also in order to keep the equation balanced.

Notice the statement $x = 3$ is the solution statement because the variable, x, is isolated on one side of the equal sign and the solution, 3, appears on the other side.

Procedure *To use the addition/subtraction principle:*
1. Decide what term you want to clear.
2. Add the additive inverse of that term to both sides of the equation. (That is, add or subtract appropriately so that the term you want to clear becomes 0.)

Connection

In the past, we solved a missing addend equation by writing a related subtraction statement. By recognizing that the inverse operation is used to solve for the missing addend, we laid the foundation for the addition/subtraction principle and the balance technique for solving equations.

Related subtraction:

$$x + 2 = 5$$

$$x = 5 - 2$$
$$x = 3$$

Balance technique:

$$x + 2 = 5$$
$$\underline{-2 \quad -2}$$
$$x + 0 = 3$$
$$x = 3$$

Note that both methods subtract 2 from 5 to arrive at the solution, 3.

EXAMPLE 3 Solve and check. $x - 8 = -15$

Solution: To isolate x, we must clear -8 from the left side, so we add $+8$. To keep the equation balanced, we must add $+8$ on both sides.

To isolate x we add $+8$ so that $-8 + 8 = 0$.

$$x - 8 = -15$$
$$\underline{+8 \qquad +8}$$
$$x + 0 = -7$$
$$x = -7$$

Because we added $+8$ on the left side we must add $+8$ on the right side as well.

Check: We learned how to check in Section 4.1. To verify that -7 is the solution for $x - 8 = -15$, we evaluate $x - 8$ using -7 in place of x and see if we get -15.

$$x - 8 = -15$$
$$-7 - 8 \; ? \; -15$$
$$-15 = -15$$

This equates, so $x = -7$ checks.

EXAMPLE 4 Solve and check. $14 = m + 5$

Solution: We need to isolate m.

$$14 = m + 5$$
$$\underline{-5 \qquad -5}$$
$$9 = m + 0$$
$$9 = m$$

It is not necessary to write the resulting 0 because when we add 0, it doesn't affect anything. However, we will continue writing this step in the text to remind you of what is happening.

Check: Evaluate $m + 5$ using $m = 9$ and see if we get 14.

$$14 = m + 5$$
$$14 \; ? \; 9 + 5$$
$$14 = 14$$

$m = 9$ makes the equation true, so it checks.

Do Margin 2. ▶

EXAMPLE 5 Solve and check. $5n - 14 - 4n = -12 + 7$

Solution: We must isolate n. However, the expressions can be simplified first by combining like terms.

◆ **Margin 2**

Solve and check.

a. $16 = x - 5$

b. $n + 13 = 20$

c. $-10 = y - 7$

d. $21 = -9 + k$

Answers: **a.** 21 **b.** 7 **c.** -3 **d.** 30

Combine $5n - 4n$ to get n. $\quad 5n - 14 - 4n = -12 + 7 \quad$ Combine $-12 + 7$ to get -5.

$$n - 14 = -5$$

Isolate n by adding $+14$.

$$\underline{+14 \quad +14}$$

$$n + 0 = \quad 9$$

$$n = 9$$

To keep balance, if we add $+14$ to the left side, we must add $+14$ to the right side as well.

Check: Evaluate $5n - 14 - 4n$ using $n = 9$ and make sure it produces the same result as $-12 + 7$.

$$5n - 14 - 4n = -12 + 7$$

$$5(9) - 14 - 4(9) \, ? \, -12 + 7$$

$$45 - 14 - 36 \, ? \, -5$$

$$31 - 36 \, ? \, -5$$

$$-5 = -5$$

Both sides produce same result, so $n = 9$ checks.

EXAMPLE 6 Solve and check. $3(y - 5) - 2y = -20 + 2$

Solution: Simplify the expressions, then isolate y.

Distribute to clear parentheses. $\quad 3(y - 5) - 2y = -20 + 2 \quad$ Combine $-20 + 2$ to get -18.

Combine $3y - 2y$ to get y. $\quad\quad 3y - 15 - 2y = -18$

$$y - 15 = -18$$

Isolate y by adding $+15$.

$$\underline{+15 \quad +15}$$

$$y + 0 = \quad -3$$

$$y = -3$$

Check: Evaluate $3(y - 5) - 2y$ using $y = -3$ and make sure it produces the same result as $-20 + 2$.

$$3(y - 5) - 2y = -20 + 2$$

$$3(-3 - 5) - 2(-3) \, ? \, -20 + 2$$

$$3(-8) - 2(-3) \, ? \, -18$$

$$-24 - (-6) \, ? \, -18$$

$$-24 + 6 \, ? \, -18$$

$$-18 = -18$$

Both sides produce the same result, so $y = -3$ checks.

◀ **Do Margin 3.**

OBJECTIVE 3 *Solve equations with variables on both sides of the equal sign.*

What if variable terms appear on both sides of the equal sign? We have to isolate the variable by getting the variable terms together on the same side of the equal sign, using the addition/subtraction principle.

EXAMPLE 7 Solve and check. $7x - 9 = 6x - 11$

Solution: Because variable terms appear on both sides of the equal sign, we must get them together on the same side of the equal sign. We can use the addition/subtraction principle to clear a term. Let's clear the $6x$ term by subtracting it. Remember we must subtract on both sides to follow the addition/subtraction principle.

◆ **Margin 3**

Solve and check.

a. $4x + 5 - 3x = 12 - 13$

b. $25 - 18 = 6c + 2 - 5c$

c. $14 - 17 = 4(u - 3) - 3u$

d. $7h - 2(3h + 4) = 15 - 12$

Answers: **a.** -6 **b.** 5 **c.** 9 **d.** 11

By subtracting $6x$ on both sides, we clear the $6x$ term from the right side and combine it with the $7x$ on the left side. We can then isolate x by adding 9 to both sides.

$$\begin{array}{rcl} 7x - 9 &=& 6x - 11 \\ \underline{-6x} & & \underline{-6x} \\ x - 9 &=& 0 - 11 \\ x - 9 &=& -11 \\ \underline{+9} & & \underline{+9} \\ x + 0 &=& -2 \\ x &=& -2 \end{array}$$

We chose to clear the $6x$ term. It actually does not matter which term you clear first (see explanation below).

Check: Evaluate $7x - 9$ and $6x - 11$ using $x = -2$ and see if the results equate.

$$\begin{array}{rcl} 7x - 9 &=& 6x - 11 \\ 7(-2) - 9 &?& 6(-2) - 11 \\ -14 - 9 &?& -12 - 11 \\ -23 &=& -23 \end{array}$$

Both sides produce the same result, so $x = -2$ is correct.

Does it matter which term you clear first? No. We could clear any term to begin with. Solving equations is much like playing chess or checkers. To begin the game you choose a piece to move. Your first move then affects how the rest of the game is played. Some first moves are considered better than other first moves. The same applies to equations. In our example, we chose to clear the $6x$ term first. Watch what happens if we choose differently. Suppose we clear the $7x$ term:

Notice what happens when we subtract $7x$ from $6x$. We get $-x$ which has a negative coefficient, whereas when we cleared the $6x$ term, we got x. Most people prefer working with positive coefficients.

$$\begin{array}{rcl} 7x - 9 &=& 6x - 11 \\ \underline{-7x} & & \underline{-7x} \\ 0 - 9 &=& -x - 11 \\ -9 &=& -x - 11 \\ \underline{+11} & & \underline{+11} \\ 2 &=& -x + 0 \\ 2 &=& -x \\ -2 &=& x \end{array}$$

What does $-x$ mean? One way to interpret it is to say it is "the additive inverse of some unknown number." The statement $2 = -x$ means we must find a number whose additive inverse is 2. Because the additive inverse of -2 is 2, x must be -2. In effect, we simply changed the signs of both sides.

In Section 4.3 we'll see another way to interpret $-x$. We'll see it as $-1x$ and divide both sides by -1 to clear the -1.

Notice the term we chose to clear in the first step did not affect the solution. However, it did change how we approached the rest of the problem. Based on what we saw in our example, how can we avoid negative coefficients?

Do Margin 4. ▶

EXAMPLE 8 Solve and check. $y - (2y + 9) = 6(y - 2) - 8y$

Solution: We must simplify the expressions, get the y terms together on the same side of the equal sign, and then isolate y.

Distribute $-(2y + 9)$
Combine like terms. $y - 2y$

Distribute $6(y - 2)$
Combine like terms. $6y - 8y$

$$\begin{array}{rcl} y - (2y + 9) &=& 6(y - 2) - 8y \\ y - 2y - 9 &=& 6y - 12 - 8y \\ -y - 9 &=& -2y - 12 \\ -y - 9 &=& -2y - 12 \\ \underline{2y} & & \underline{2y} \\ y - 9 &=& 0 - 12 \\ y - 9 &=& -12 \\ \underline{+9} & & \underline{+9} \\ y + 0 &=& -3 \\ y &=& -3 \end{array}$$

Isolate y by adding 9.

When we look at the y terms, our choices are to clear $-y$ (which means $-1y$) or $-2y$. We chose to clear the $-2y$ term because it has the smaller coefficient. Remember -2 is less than -1.

Tip
In solving an equation, when we choose a variable term to clear, we can avoid negative coefficients by clearing the term with the smaller coefficient.

◆ **Margin 4**

Solve and check.

a. $8x - 5 = 7x + 2$

b. $3b - 2 = 4b + 3$

c. $1 + 5m = 7 + 4m$

Answers: **a.** 7 **b.** -5 **c.** 6

Margin 5

Solve and check.

a. $5n - (n - 7) = 2(n + 6) + n$

b. $10 + 3(5x - 2) = 7(2x - 1) + 9$

c. $9 - 2(3c + 5) = -3 - 5(c - 1)$

Margin 6

a. A patient must receive 350 cc of a medication in three injections. He has received two injections at 110 cc each. How much should the third injection measure?

b. Daryl has a balance of $-\$568$ on a credit card. How much must he pay to bring his balance to $-\$480$?

Historical Note

The word *algebra* first appeared in a book by Arab mathematician Mohammed ibn Musa al-Khowarizmi (c. 825), entitled *al-jabr w'al-muqabalah.* The word *al-jabr*, translates to *the reunification.* Al-jabr went through many spelling changes and finally emerged as our modern word, *algebra.* When the Moors brought the word to Spain it came to mean the reunification of broken bones, an *algebrista* being one who sets broken bones. In the Middle Ages, the sign over a barber shop might read, "Algebrista y Sangrador," which meant *Bonesetter and Bloodletter.* The red-and-white striped pole that often hangs outside a barber's shop is a relic from this alternate role of the barber; red for the role of bloodletter (sangrador) and white to signify a bonesetter (algebrista).

Source: D. E. Smith, *History of Mathematics,* 1953

Answers to Margin 5: **a.** 5 **b.** -2 **c.** -3

Answers to Margin 6: **a.** 130 cc **b.** $88

Check: We must evaluate $y - (2y + 9)$ and $6(y - 2) - 8y$ using $y = -3$ and see if we get the same result.

$$y - (2y + 9) = 6(y - 2) - 8y$$
$$-3 - (2(-3) + 9) \ ? \ 6(-3 - 2) - 8(-3)$$
$$-3 - \ (-6 + 9) \ ? \quad 6(-5) \quad - 8(-3)$$
$$-3 - \qquad 3 \qquad ? \quad -30 \quad - (-24)$$
$$\qquad\qquad -6 \quad ? \quad -30 \quad + 24$$
$$\qquad\qquad -6 \quad = \quad -6$$

The results equate, so $y = -3$ is correct.

Let's put together an outline based on everything we've learned so far.

Procedure *To solve equations:*

1. Simplify both sides of the equation as needed.
   a. Distribute to clear parentheses.
   b. Combine like terms.
2. Use the addition/subtraction principle so that all variable terms are on one side of the equation and all constants are on the other side.

> **Tip**
> Clear the variable term that has the smaller coefficient. This will avoid negative coefficients.

◀ **Do Margin 5.**

OBJECTIVE 4 *Solve application problems.*

Let's put what we've learned about using the addition/subtraction principle into context of some situations, the most common being the missing addend type problem. You may recall that the way to recognize a missing addend problem is to look for key words such as "How much more is needed?"

EXAMPLE 9 Laura wants to buy a car stereo that costs $275. She currently has $142. How much more is needed?

Solution: Use the problem-solving process.

Understand: We are given the total required and the amount she currently has, and we must find how much she needs. This is a missing addend situation. We must add the needed amount to 142 and end up with 275.

Plan: Let x represent the amount Laura needs. We will translate to a missing addend equation, then solve.

Execute: current amount + needed amount = 275

$$142 \quad + \quad x \quad = 275$$
$$142 + x = 275$$
$$\underline{-142 \qquad -142}$$
$$0 + x = 133$$
$$x = 133$$

Answer: Laura needs $133 to buy the stereo.

Check: Does $142 plus the additional $133 make $275?

$$142 + 133 \ ? \ 275$$
$$275 = 275 \qquad \text{It checks.}$$

◀ **Do Margin 6.**

4.2 Exercises

For Exercises 1–12, determine whether the given equation is linear.

1. $9x - 7 = 4x + 3$

2. $7y = 14$

3. $t^2 - 5 = 20$

4. $2n - 9 = n^3 + 6n - 1$

5. $5u - 17 = u^2 + 1$

6. $14n - 9n = 5n + 3$

7. $2x + y = 5$

8. $y = x^2 - 5$

9. $t = u^3 + 2$

10. $x = -6$

11. $y = 3x + 2$

12. $(y - 5) = 3(x - 1)$

For Exercises 13–32, solve and check.

13. $n + 14 = 20$

14. $y + 19 = -6$

15. $-7 = x - 15$

16. $5 = t - 11$

17. $3r + 13 - 2r = 9$

18. $10 = 5x - 2 - 4x$

19. $9y - 8y + 11 = 12 - 2$

20. $2x - 15 - x = 9 - 17$

21. $4x - 2 = 3x + 5$

22. $6b + 7 = 5b + 3$

23. $m - 1 = 2m - 5$

24. $7t + 2 = 8t - 9$

25. $3n - 5 + 4n = 7 + 6n - 12$

26. $8u + 13 - 2u = 2u + 11 + 3u$

27. $5 + 7y + 1 - 3y = 2y - 10 + 3y - 4$

28. $-11 + x + 2 + 5x = 9x - 4 - 2x + 3$

29. $3(n + 2) = 4 + 2(n - 1)$

30. $9 - 5(b - 1) = -4(b - 2)$

31. $6t - 2(t - 5) = 9t - (6t + 2)$

32. $8 + 3(t - 5) = 2(5t + 1) - 6t$

For Exercises 33–42, translate to an equation, then solve.

33. Yolanda is to close on her new house in three weeks. The amount that she must have at closing is $4768. She currently has $3295. How much more does she need?

34. Nikki has a balance of −$457 on her credit card. What payment should she make to get the balance to −$325?

35. A patient is to receive 450 cc of a medication in three injections. The first injection is to be 200 cc and the second injection 180 cc. How much must the last injection be?

36. An entry in a chemist's notebook is smudged. The entry is the initial temperature measurement of a chemical in an experiment. In the experiment, a substance was introduced to the chemical, decreasing its temperature by 19°C. The final temperature is listed as 165°C. What was the initial temperature?

37. Jerry sells medical equipment. The company quota is set at $10,500 each month. The spreadsheet below shows Jerry's sales as of the end of the second week of July. How much more does Jerry need to sell to make the quota? Do you think Jerry will make the quota? Why?

Date	Item No.	Quantity	Price	Total Sale
7/3	45079	2	$ 800	$1600
7/8	47002	1	$4500	$4500
7/13	39077	3	$ 645	$1935

38. Marc is a waiter at a restaurant. His rent of $675 is due at the end of the week. Currently he has $487 in the bank. At the end of his shift he finds that he made $85. How much more does he need in order to pay his rent? If he has two more shifts before rent is due, do you think he will make enough to pay rent? Explain.

39. Connie is considering a new apartment. What is the length of the dining area? Is this area large enough to accommodate a 4 ft. by 4 ft. square table and four chairs?

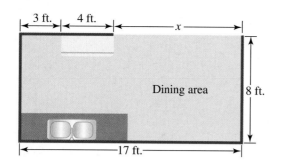

40. At an archaeological dig, the archaeologists suspect there may be a secret passage or small chamber between two chambers of a tomb. They have mapped and documented distances. What is the distance between the two chambers?

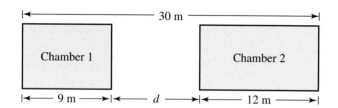

41. The perimeter of the triangle shown is 84 cm. Find the length of the missing side.

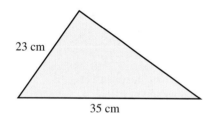

42. To take more than the standard deduction on taxes, itemized deductions must exceed the standard deduction. Because the Coleman's filing status is *married filing jointly,* their standard deduction is $7100. If the Coleman's have the following itemized deductions already, how much are they lacking to reach the level of the standard deduction?

Medical and dental expenses: $242

Taxes paid: $2447

Charitable donations: $225

Discussion What else might the Coleman's deduct?

REVIEW EXERCISES

1. $5(-8) =$

2. $24 \div (-3) =$

3. $12 + 3(-5 - 2) =$

4. $\dfrac{16 + 2(3 - 5)}{-2(5) + 4} =$

5. Solve. $5x = 15$

4.3 The Multiplication/Division Principle

OBJECTIVES

1 Solve equations using the multiplication/division principle.

2 Solve equations using both the addition/subtraction and the multiplication/division principles.

3 Solve application problems.

OBJECTIVE 1 *Solve equations using the multiplication/division principle.*

In Section 4.2 we learned to manipulate equations using the addition/subtraction principle to isolate a variable term. We solved equations like:

$$
\begin{array}{lll}
x + 3 = 7 & 3y + 5 = 2y - 1 & 4(b - 2) + 1 = 5(b - 4) - 2b \\
\underline{-3 \quad -3} & \underline{-2y \qquad -2y} & 4b - 8 + 1 = 5b - 20 - 2b \\
x + 0 = 4 & y + 5 = 0 - 1 & 4b - 7 = 3b - 20 \\
\quad x = 4 & y + 5 = -1 & \underline{-3b \qquad -3b} \\
& \underline{\quad -5 \quad -5} & b - 7 = 0 - 20 \\
& y + 0 = -6 & b - 7 = -20 \\
& \quad y = -6 & \underline{+7 \qquad +7} \\
& & b + 0 = -13 \\
& & \quad b = -13
\end{array}
$$

So far we've always ended up with a coefficient of 1 on each variable. What if we ended up with a coefficient other than 1?

Consider the equation $5n - 9 = 3n + 1$. Let's follow our plan from Section 4.2 and get the variable terms together first, then isolate n.

$$
\begin{array}{l}
5n - 9 = 3n + 1 \\
\underline{-3n \qquad -3n} \\
2n - 9 = 0 + 1 \\
2n - 9 = 1 \\
\underline{+9 \quad +9} \\
2n + 0 = 10 \\
\quad 2n = 10
\end{array}
$$

> Now what? We have a coefficient other than 1 here.

How can we solve $2n = 10$? Remember it is a missing factor statement because the 2 coefficient is multiplying the n. We learned in Chapter 1 that we can solve for a missing factor by dividing. However, we are connecting what we've learned in the past with this balance technique. If we are going to divide, we must divide on both sides to keep the equation balanced. We can write:

$$2n \div 2 = 10 \div 2 \qquad \text{or the more popular form:} \qquad \frac{2n}{2} = \frac{10}{2}$$

What happens to the 2s?

> Because $2 \div 2 = 1$, $2n \div 2 = 1n$

$$\frac{2n}{2} = \frac{10}{2}$$

> $1n = n$

$$1n = 5$$
$$n = 5$$

If we manipulate an equation using the addition/subtraction principle and we still have a coefficient other than 1, we can clear that coefficient by dividing. This is the main purpose of the multiplication/division principle.

The Multiplication/Division Principle of Equality *We can multiply or divide both sides of an equation by the same amount without affecting its solution(s).*

Primary purpose: To clear a coefficient from a term.

> **Procedure** *To use the multiplication/division principle:*
> 1. Decide what coefficient you want to clear.
> 2. Divide both sides by that coefficient.

◆ Margin 1

Solve and check.

a. $5y = 45$

b. $24 = 6n$

c. $-48 = -8a$

d. $-2x = 16$

EXAMPLE 1 Solve and check. $7x = 21$

Solution: To isolate x we must clear the coefficient 7. We divide by 7 so that we get $1x$. The multiplication/division principle tells us that we must divide on both sides to keep the equation balanced.

$$\frac{7x}{7} = \frac{21}{7}$$
$$1x = 3$$
$$x = 3$$

clear the 7 coefficent by dividing both sides by 7.

Check: We must verify that $x = 3$ makes $7x = 21$ true. We evaluate $7x$ using 3 in place of x and make sure we get 21.

$$7x = 21$$
$$7(3) \; ? \; 21$$
$$21 = 21$$

EXAMPLE 2 Solve and check. $-6m = 30$

Solution: To isolate m, we must clear the coefficient -6. We must use the multiplication/division principle and divide both sides by -6.

It is not necessary to write the resulting 1 because multiplying by 1 has no effect on the variable. However we will continue to write the 1 in the text to remind you of what is happening.

$$\frac{-6m}{-6} = \frac{30}{-6}$$
$$1m = -5$$
$$m = -5$$

Notice when we use the multiplication/division principle, we should not change the sign of the number we are dividing by.

$$\frac{-6m}{6} = \frac{30}{6}$$
$$-1m = 5$$
$$\frac{-1m}{-1} = \frac{5}{-1}$$
$$m = -5$$

Dividing by positive 6 leaves us with $-1m$, which is not completely isolated. We would need to continue and divide by -1. Dividing by -6 in the beginning avoids this extra step.

Check: Evaluate $-6m$ using $m = -5$ and make sure it equals 30.

Because $-6(-5) = 30$ we can say -5 is a solution for $-6m = 30$.

$$-6m = 30$$
$$-6(-5) \; ? \; 30$$
$$30 = 30$$

◀ **Do Margin 1.**

OBJECTIVE 2 *Solve equations using both the addition/subtraction and the multiplication/division principles.*

Now let's put the multiplication/division principle together with the addition/subtraction principle. We'll follow the same plan we developed in Section 4.2, except now we will have an extra step: to clear a coefficient at the end.

> **Procedure** *To solve equations:*
> 1. Simplify both sides of the equation as needed.
>   a. Distribute to clear parentheses.
>   b. Combine like terms.
> 2. Use the addition/subtraction principle so that all variable terms are on one side of the equation and all constants are on the other side.
> 3. Use the multiplication/division principle to clear any remaining coefficient.

Tip

Clear the variable term that has the smaller coefficient. This will avoid negative coefficients.

Answers: **a.** 9 **b.** 4 **c.** 6 **d.** -8

EXAMPLE 3 Solve and check. $-2x - 9 = 7$

Solution: We must isolate x. To isolate x we first need to clear the -9. To clear -9 we add $+9$. To clear the -2 coefficient, we'll have to divide. Our outline tells us to use the addition principle first, so let's clear the -9 by adding $+9$ to both sides.

Connection
After clearing the -9, notice the equation becomes $-2x = 16$, which is just like the equation in Margin 1d.

$$-2x - 9 = 7$$
$$\underline{+9 \quad +9}$$
$$-2x + 0 = 16$$
$$-2x = 16$$
$$\frac{-2x}{-2} = \frac{16}{-2}$$
$$1x = -8$$
$$x = -8$$

Isolate the $-2x$ term by adding 9 to both sides.

Clear the -2 coefficient by dividing both sides by -2.

Check: We evaluate $2x - 9$ using $x = -8$ and make sure the result is 7.

$$-2x - 9 = 7$$
$$-2(-8) - 9 \,?\, 7$$
$$16 - 9 \,?\, 7$$
$$7 = 7$$

$x = -8$ causes $-2x - 9 = 7$ to be true, so $x = -8$ is correct.

Do Margin 2. ▶

EXAMPLE 4 Solve and check. $7x - 1 = 3x - 21$

Solution: We must get the variable terms together on one side of the equation and constants on the other side.

Connection
After moving the $3x$ term, we have the same equation as Margin 2a.

$$7x - 1 = 3x - 21$$
$$\underline{-3x \qquad -3x}$$
$$4x - 1 = 0 - 21$$
$$4x - 1 = -21$$
$$\underline{+1 \qquad +1}$$
$$4x + 0 = -20$$
$$4x = -20$$
$$\frac{4x}{4} = \frac{-20}{4}$$
$$1x = -5$$
$$x = -5$$

The $3x$ has the smaller coefficient.
Add $-3x$ to both sides.

Isolate the $4x$ term.
Add $+1$ to both sides.

Divide both sides by 4 to clear the 4 coefficient.

Check: We must evaluate $7x - 1$ and $3x - 21$ using $x = -5$ and see if we get the same result.

$$7x - 1 = 3x - 21$$
$$7(-5) - 1 \,?\, 3(-5) - 21$$
$$-35 - 1 \,?\, -15 - 21$$
$$-36 = -36$$

Because we got the same result from both expressions, we can say $x = -5$ is correct.

Do Margin 3. ▶

EXAMPLE 5 Solve and check. $9n + 6 - 2n = 16 + 10n - 4$

Solution: Because we have like terms that are already on the same sides of the equation, we should simplify these like terms.

◆ **Margin 2**
Solve and check.

a. $4x - 1 = -21$

b. $-7b + 1 = 22$

c. $12 = 17 + 5x$

d. $-4 = 32 - 9x$

◆ **Margin 3**
Solve and check.

a. $9b + 2 = 5b + 26$

b. $n - 11 = 8n + 3$

c. $7n + 6 = 12 + 10n$

Answers to Margin 2: **a.** -5 **b.** -3 **c.** -1 **d.** 4

Answers to Margin 3: **a.** 6 **b.** -2 **c.** -2

a. $5t + 13 - t = 3 + 6t - 12$

b. $15 - 2y + 6 = 12y + 30 - 5y$

$$9n + 6 - 2n = 16 + 10n - 4 \quad \text{Combine like terms.}$$
$$\longrightarrow 7n + 6 = 12 + 10n \quad \text{7n has the smaller coefficient, so add}$$
$$\underline{-7n} \qquad\qquad \underline{-7n} \quad -7n \text{ to both sides.}$$
$$0 + 6 = 12 + 3n$$
$$6 = 12 + 3n \quad \text{Isolate 3n by subtracting 12 on both sides.}$$
$$\underline{-12} \quad \underline{-12}$$
$$-6 = 0 + 3n$$
$$-6 = 3n$$
$$\frac{-6}{3} = \frac{3n}{3} \quad \text{Divide both sides by 3 to clear the 3 coefficient.}$$
$$-2 = 1n$$
$$-2 = n$$

| **Connection**
After simplifying, we have the same equation as Margin 3c.

Check: We must evaluate $9n + 6 - 2n$ and $16 + 10n - 4$ using -2 in place of n and see if we get the same result.

$$9n + 6 - 2n = 16 + 10n - 4$$
$$9(-2) + 6 - 2(-2) \: ? \: 16 + 10(-2) - 4$$
$$-18 + 6 - (-4) \: ? \: 16 + (-20) - 4$$
$$-12 - (-4) \: ? \: -4 - 4$$
$$-12 + 4 \: ? \: -8$$
$$-8 = -8$$

◆ Because -2 produces the same result from both expressions, we can say that $-2 = n$ is correct.

◀ **Do Margin 4.**

EXAMPLE 6 Solve and check. $15 - 2(y - 3) = 6(2y + 5) - 5y$

Solution: We must simplify each side first.

a. $3(3t - 5) + 13 = 4(t - 3)$

b. $9 - 5(x + 4) = 7x - (6x + 5)$

c. $16 + 4(2c - 1) = 2 - 6(c + 3)$

| **Connection**
After distributing, we have the same equation as Margin 4b.

$$15 - 2(y - 3) = 6(2y + 5) - 5y \quad \text{Distribute to clear parentheses.}$$
$$\longrightarrow 15 - 2y + 6 = 12y + 30 - 5y \quad \text{Combine like terms.}$$
$$21 - 2y = 7y + 30$$
$$21 - 2y = 7y + 30 \quad \text{-2y has the smaller coefficient, so add 2y to both sides.}$$
$$\underline{+2y} \quad \underline{+2y}$$
$$21 + 0 = 9y + 30$$
$$21 = 9y + 30 \quad \text{Isolate 9y by subtracting 30 on both sides.}$$
$$\underline{-30} \qquad \underline{-30}$$
$$-9 = 9y + 0$$
$$\frac{-9}{9} = \frac{9y}{9} \quad \text{Divide both sides by 9 to clear the 9 coefficient.}$$
$$-1 = 1y$$
$$-1 = y$$

Check: Evaluate $15 - 2(y - 3)$ and $6(2y + 5) - 5y$ using -1 in place of y and make sure we get the same results.

$$15 - 2(y - 3) = 6(2y + 5) - 5y$$
$$15 - 2(-1 - 3) \: ? \: 6(2(-1) + 5) - 5(-1)$$
$$15 - 2(-4) \: ? \: 6(-2 + 5) - 5(-1)$$
$$15 - (-8) \: ? \: 6(3) - 5(-1)$$
$$15 + 8 \: ? \: 18 - (-5)$$
$$23 \: ? \: 18 + 5$$
$$23 = 23$$

◆ Because -1 produced the same result from both expressions, we can say that $-1 = y$ is correct.

◀ **Do Margin 5.**

OBJECTIVE 3 *Solve application problems.*

In Section 4.2, we revisited many of the applications that we solved earlier in the text and put those applications in context of using the addition principle. Now we want to do the same for the multiplication principle. The relationships that we will recall involve multiplying variables, such as:

Area of a parallelogram: $A = bh$

Volume of a box: $V = lwh$

Force: $F = ma$

Distance: $d = rt$

Voltage: $V = ir$

Notice the common thread in all of these formulas is the fact that we merely multiply numbers to achieve the result. In fact, the area of a parallelogram, force, distance, and voltage are all mathematically identical because we're simply multiplying two numbers together. The only difference is volume because we multiply three numbers. Remember the game with formulas is to replace the variables with the corresponding given numbers, then solve for the missing amount.

EXAMPLE 7 Juanita is driving at a rate of 59 ft./sec. (about 40 mph). A large piece of debris lies in the road 177 ft. ahead. How much reaction time does she have?

Solution: Use the problem-solving process.

Understand: We are given a rate and a distance and must find the time. Based on the given information we can use the formula $d = rt$.

Plan: Use $d = rt$. Replace d with 177 ft. and r with 59 ft./sec., then solve for t.

Execute:
$$d = rt$$
$$177 = 59t$$
$$\frac{177}{59} = \frac{59t}{59}$$
$$3 = 1t$$
$$3 = t$$

Answer: Juanita has 3 sec. to react to the debris in the road ahead. (The time is in units of seconds because the rate was in terms of ft./sec., which means feet per *second*.)

Check: Traveling at 59 ft./sec. for 3 seconds, will the distance be 177 ft.?

$$177 \; ? \; 59(3)$$
$$177 = 177 \qquad \textbf{It checks.}$$

Do Margin 6. ▶

We can also encounter problems that involve both the addition/subtraction and multiplication/division principles. When a relationship involves multiplication and addition or subtraction, then there is the potential for us to have to use both principles to solve for an unknown amount. Some of our past relationships where this would apply are:

Perimeter of a rectangle: $P = 2l + 2w$

Surface area of a box: $SA = 2lw + 2lh + 2wh$

◆ **Margin 6**

a. A code in a certain region requires that upstairs windows have a minimum area of 864 in.^{2}. Doris wants an upstairs window that is 18 in. wide. What must the length be?

b. A children's toy company is designing a new toy box. The box is to have a volume of 15,552 in.^{3}. The length is to be 36 in. and the width 24 in. How tall will the box be?

Answers: **a.** 48 in. **b.** 18 in.

a. The length of a rectangle is 207 cm. Find the width of the rectangle if the perimeter is 640 cm.

b. The formula $C = 8f + 50$ describes the total cost of a fence with a gate where C is the total cost and f is the number of feet to be enclosed. If Mario and Rosa have a budget of $1810, then how many feet of fencing can they afford?

Note: In the formula $C = 8f + 50$, the coefficient 8 represents the price per foot of fencing. It costs $8 for each foot of fencing. The 50 represents the cost of the gate that was mentioned. Therefore the total cost is $8 times the number of feet of fencing plus the $50 cost of the gate.

EXAMPLE 8 The total material allotted for the construction of a metal box is 3950 in.^{2}. The length is to be 25 in. and the width is to be 40 in. Find the height.

Solution: Use the problem-solving process.

Understand: We are given the surface area, length, and width of a box and must find the height using the formula $SA = 2lw + 2lh + 2wh$.

Plan: Replace SA with 3950, l with 25, w with 40, and solve for h.

Execute: $SA = 2lw + 2lh + 2wh$

$$3950 = 2(25)(40) + 2(25)h + 2(40)h$$
$$3950 = 2000 \quad + \quad 50h + \quad 80h$$
$$3950 = 2000 \quad + \quad 130h$$
$$3950 = 2000 + 130h$$
$$\underline{-2000 \quad -2000}$$
$$1950 = \quad 0 + 130h$$
$$\frac{1950}{130} = \frac{130h}{130}$$
$$15 = 1h$$
$$15 = h$$

Answer: The height must be 15 in.

Check: Is the surface area of the box 3950 in.^{2} when the length is 25 in., width is 40 in., and height is 15 in.?

$$3950 ? 2(25)(40) + 2(25)(15) + 2(40)(15)$$
$$3950 ? 2000 \quad + 750 \quad + 1200$$
$$3950 = 3950 \qquad \text{It checks.}$$

◄ **Do Margin 7.**

4.3 Exercises

For Exercises 1–26, solve and check.

1. $3x = -21$

2. $5t = 20$

3. $-4a = 36$

4. $8m = -48$

5. $-12y = -72$

6. $-15n = 45$

7. $2x + 9 = 23$

8. $5k - 4 = 31$

9. $29 = -6b - 13$

10. $45 = 15 - 10h$

11. $-7t + 11 = -73$

12. $9u - 17 = -53$

13. $7x - 16 = 5x + 6$

14. $3b + 13 = 7b + 17$

15. $9k + 19 = -3k - 17$

16. $-4y + 15 = -9y - 20$

17. $-2h + 17 - 8h = 15 - 7h - 10$

18. $5 - 8x - 24 = 7x + 11 - 5x$

19. $-4m - 7 + 13m = 12 + 2m - 5$

20. $13 - 9h - 15 = 7h + 22 - 8h$

21. $3(t - 4) = 5(t + 1) - 1$

22. $-2(x + 7) - 9 = 5x + 12$

23. $9 - 2(3x + 5) = 4x + 3(x - 9)$

24. $11x - 4(2x - 3) = 18 + 5(x + 2)$

25. $15 + 3(3x - 1) = 16 - (2x - 7)$

26. $8x - (3x + 7) = 14 - 4(x - 6)$

For Exercises 27–30, the check for each equation indicates a mistake was made. Find and correct the mistake.

27.
$$5x - 11 = 7x - 9$$
$$\underline{-5x \qquad -5x}$$
$$11 = 2x - 9$$
$$\underline{+9 \qquad +9}$$
$$20 = 2x$$
$$\frac{20}{2} = \frac{2x}{2}$$
$$10 = x$$

Check:
$$5x - 11 = 7x - 9$$
$$5(10) - 11 \; ? \; 7(10) - 9$$
$$50 - 11 \; ? \; 70 - 9$$
$$39 \neq 61$$

28. $6y + 11 = -3y + 38$

$$\underline{ -11 \qquad\qquad -11}$$

$6y + 0 = -3y + 27$

$6y = -3y + 27$

$$\underline{+3y \qquad +3y}$$

$9y = 0 + 27$

$$\dfrac{9y}{9} = \dfrac{27}{9}$$

$1y = 3$

$y = 3$

Check: $6y + 11 = -3y + 38$

$6(3) + 11 \ ? \ -3(3) + 38$

$18 + 11 \ ? \ 9 + 38$

$29 \neq 47$

29. $6 - 2(x + 5) = 3x - (x - 8)$

$6 - 2x - 10 = 3x - x - 8$

$-4 - 2x = 2x - 8$

$$\underline{+2x \quad +2x}$$

$-4 + 0 = 4x - 8$

$-4 = 4x - 8$

$$\underline{+8 \qquad\qquad +8}$$

$4 = 4x + 0$

$$\dfrac{4}{4} = \dfrac{4x}{4}$$

$1 = 1x$

$1 = x$

Check: $6 - 2(x + 5) = 3x - (x - 8)$

$6 - 2(1 + 5) \ ? \ 3(1) - (1 - 8)$

$6 - 2(6) \ ? \ 3 - (-7)$

$6 - 12 \ ? \ 3 + 7$

$-6 \neq 10$

30. $2x + 3(x - 3) = 10x - (3x - 11)$

$2x + 3x - 9 = 10x - 3x + 11$

$5x - 9 = 7x + 11$

$$\underline{-7x \qquad\quad -7x}$$

$-2x - 9 = 0 + 11$

$-2x - 9 = 11$

$$\underline{+9 \quad +9}$$

$-2x + 0 = 20$

$$\dfrac{-2x}{2} = \dfrac{20}{2}$$

$x = 10$

Check: $2x + 3(x - 3) = 10x - (3x - 11)$

$2(10) + 3(10 - 3) \ ? \ 10(10) - (3(10) - 11)$

$20 + 3(7) \ ? \ 100 - (30 - 11)$

$20 + 21 \ ? \ 100 - 19$

$41 \neq 81$

For Exercises 31–42, solve for the missing amount.

31. The area of a computer screen is to be 180 in.². Find the length if the width must be 10 in. (Use the formula $A = lw$.)

$A = 180$ in.²

32. A fish tank is to have a volume of 4320 in.^{3}. The length is to be 20 in. and the width 12 in. Find the height. (Use $V = lwh$.)

33. The Jones family is planning a 455 mi. trip. If they travel at an average speed of 65 mph, what will be their travel time? (Use $d = rt$.)

34. Jacob weighs 160 lb. If the acceleration due to gravity is about 32 feet per second per second, what is his mass? (Use $F = ma$.)

35. The surface area of a large cardboard box is to be 11,232 in.^{2}. The length is to be 36 in. and the width is to be 28 in. Find the height. (Use $SA = 2lw + 2lh + 2wh$.)

36. A stage is designed in the shape of a rectangle with a perimeter of 122 ft. If the width is 26 ft., find the length. (Use $P = 2l + 2w$.)

37. The formula $C = 28h + 40$ describes the total cost for a plumber to visit a home, where C is the total cost and h is the number of hours on site. If Li has \$152, how long can he afford for the plumber to work?

  Discussion What do you think 28 and 40 describe?

38. The formula $C = 6h + 32$ describes the total cost of renting a tiller, where C is the total cost and h is the number of hours rented. How many hours can the tiller be rented for \$50?

  Discussion What do you think 6 and 32 describe?

39. The formula $v = v_i + at$ describes the final velocity (speed) of an object after being accelerated, where v_i is the initial velocity, a is the acceleration, t is the number of seconds that the object is accelerated, and v is the final velocity. Suppose the initial velocity of a car is 30 mph, and the car accelerates at a rate of 2 mph per sec. How long will it take the car to reach a speed of 40 mph?

40. A sports car is said to be able to go from 0 mph to 60 mph in 4 sec. What acceleration is required? (Use $v = v_i + at$.)

41. The expression $54b + 1215$ describes the cost of materials for a certain computer chip, where b is the number of chips produced. The expression $25b + 4150$ describes the labor costs. How many chips can be produced with a total cost budget of $54,000?

42. Two different circuits operate within a television. The equation $V = 7i + 8 - 4i$ describes the output voltage of the first circuit. The equation $V = 3(3i - 1) - 13$ describes the output voltage of the second circuit. The variable i represents the current. The design calls for both circuits to have the same output voltage. What must be the current?

Puzzle Problem

The letters below correspond to digits. Find the digit that corresponds to each letter.

```
      F O R T Y
        T E N
  +     T E N
    ─────────────
      S I X T Y
```

REVIEW EXERCISES

1. Write the word name for 6,784,209.

2. Evaluate 7^0.

3. Evaluate $x^2 - 9x + 7$ when $x = -3$.

4. Find the prime factorization of 240.

5. Factor $24x^5 - 30x^4 + 18x^3$.

4.4 Translating Word Sentences to Equations

OBJECTIVE

1 Translate sentences to equations using key words, then solve.

In Sections 4.2 and 4.3, we developed the addition/subtraction and multiplication/division principles. We also began putting the principles in the context of application problems. In Section 4.2 we learned to expect applications that dealt with addition/subtraction because we were dealing with the addition principle. In Section 4.3 you were given formulas. However, in the real world problems generally don't come with a formula. In this section, we will further develop our problem-solving strategies by considering key words more closely.

Let's put the key words in the context of phrases and see how they translate.

Addition	Translation	Subtraction	Translation
The sum of x and 3	$x + 3$	The difference of x and 3	$x - 3$
h plus k	$h + k$	h minus k	$h - k$
7 added to t	$t + 7$	7 subtracted from t	$t - 7$
3 more than a number	$n + 3$	3 less than a number	$n - 3$
y increased by 2	$y + 2$	y decreased by 2	$y - 2$

Note: Because addition is a commutative operation, it does not matter in which order we write the translation. For "the sum of x and 3" we can write $x + 3$ or $3 + x$.	**Note:** Subtraction is not a commutative operation, therefore the way we write the translation matters. We must translate each key phrase exactly as it was presented above. Notice when we translate *less than* or *subtracted from*, the translation is in reverse order from what we read.

Notice *a number* means *an unknown number,* so we use a variable. We've usually used n, but we can select any variable we wish. Also, we will only consider sentences that translate to the addition, subtraction, and multiplication equations that we've studied so far.

Multiplication	Translation	Division	Translation
The product of x and 3	$3x$	The quotient of x and 3	$x \div 3$
h times k	hk	h divided by k	$h \div k$
twice a number	$2n$	h divided into k	$k \div h$
triple the number	$3n$	the ratio of a to b	$a \div b$

Note: Like addition, multiplication is a commutative operation. This means we can write the translation order any way we wish. h times k can be hk or kh.	**Note:** Division is like subtraction in that it is not a commutative operation, therefore we must translate division phrases exactly as presented above. Notice how *divided by* and *divided into* change the order of the translation.

Exponents	Translation	Roots	Translation
c squared	c^2	The square root of x	$\sqrt{x}$
the square of b	b^2		
k cubed	k^3		
the cube of b	b^3		
n to the fourth	n^4		
y raised to the fifth power	y^5		

The key words *sum, difference, product,* and *quotient* are the answer words for their respective operations. Notice that they all involve the word *and.* In the translation, the word *and* becomes the operation symbol indicated by the key word *sum, difference, product,* or *quotient.*

$$\text{sum of } x \text{ and } 3$$
$$\downarrow$$
$$x + 3$$

$$\text{product of } x \text{ and } 3$$
$$\downarrow$$
$$x \cdot 3$$

$$\text{difference of } x \text{ and } 3$$
$$\downarrow$$
$$x - 3$$

$$\text{quotient of } x \text{ and } 3$$
$$\downarrow$$
$$x \div 3$$

If we are going to translate sentences to equations, we must also know the key words that indicate an equal sign.

Key Words for an Equal Sign

is equal to	produces
is the same as	yields
is	results in

♦ Margin 1

Translate to an equation, then solve.

a. 15 more than a number is 40.

EXAMPLE 1 The sum of seventeen and a number is equal to fifteen. Translate to an equation, then solve for the number.

Solution: Follow the problem-solving process.

Understand: The key word *sum* indicates *addition, is equal to* indicates *an equal sign,* and *a number* indicates *a variable.*

Plan: Translate the key words to an equation, then solve the equation. We'll use n as the variable.

Execute: Translation:

$$\text{The sum of seventeen and a number is equal to fifteen.}$$
$$\qquad\qquad\downarrow\qquad\quad\downarrow\qquad\downarrow\qquad\quad\downarrow\qquad\downarrow$$
$$\qquad\qquad 17 \qquad + \qquad n \qquad = \qquad 15$$

Solve:
$$17 + n = 15$$
$$\underline{-17 \qquad\quad -17}$$
$$0 + n = -2$$
$$n = -2$$

Answer: $n = -2$

Check: Verify that our solution satisfies the sentence. In the original sentence, replace the unknown amount with -2 and make sure it is a true sentence.

b. The length increased by 9 is 25.

$$\text{The sum of seventeen and a number is equal to fifteen.}$$
$$17 \qquad + \qquad (-2) \qquad ? \qquad 15$$
$$\qquad\qquad 15 \qquad\quad = \qquad 15$$

Yes, the sum of 17 and -2 is equal to 15.

◄ Do Margin 1.

EXAMPLE 2 Twelve less than a number is ten. Translate to an equation, then solve.

Solution: Follow the problem-solving process.

Understand: The key words *less than* indicate *subtraction.* Further, we must be careful how we translate because subtraction is not commutative. When we translate *less than* the translation is in reverse order of what we read. The key word *is* means *an equal sign.*

Answers: **a.** $n + 15 = 40; n = 25$
b. $l + 9 = 25; l = 16$

Plan: Translate the key words, then solve. We'll use *n* for the variable.

Execute: Translation:

Twelve less than a number is ten.

$$n - 12 = 10$$

Solve:
$$n - 12 = 10$$
$$\underline{+12 \quad +12}$$
$$n + 0 = 22$$
$$n = 22$$

Answer: $n = 22$

Check: Verify that 22 satisfies the original sentence.

Twelve less than a number is ten.
$$22 - 12 \; ? \; 10$$
$$10 = 10$$
Yes, 12 less than 22 is 10.

Do Margin 2. ▶

EXAMPLE 3 The product of seven and *y* is negative thirty-five. Translate to an equation, then solve.

Solution: Use the problem-solving process.

Understand: Product means *multiply* and *is* means *an equal sign.*

Plan: Translate the key words, then solve.

Execute: Translation:

The product of seven and *y* is negative thirty-five.
$$\downarrow \qquad \downarrow$$
$$7 \cdot y = -35$$

Solve:
$$\frac{7y}{7} = \frac{-35}{7}$$
$$1y = -5$$
$$y = -5$$

Answer: $y = -5$

Check: Make sure that -5 satisfies the original sentence.

The product of seven and *y* is negative thirty-five.
$$7 \cdot (-5) \; ? \; -35$$
$$-35 = -35$$
Yes, the product of 7 and -5 is -35, so -5 is correct.

Do Margin 3. ▶

EXAMPLE 4 Seven more than the product of three and *r* is equal to nineteen. Translate to an equation, then solve.

Solution: Follow the problem-solving process.

Understand: more than indicates *addition, product* indicates *multiplication,* and *is equal to* indicates an *equal sign.*

Plan: Translate to an equation using the key words, then solve the equation.

◆ **Margin 2**

Translate to an equation, then solve.

a. Twelve subtracted from a number is negative eight.

b. The difference of a number and twenty-one is negative six.

◆ **Margin 3**

Translate to an equation, then solve.

a. The product of negative six and a number is negative forty-two.

b. Twelve times a number is negative seventy-two.

Answers to Margin 2: **a.** $n - 12 = -8; n = 4$ **b.** $m - 21 = -6; m = 15$

Answers to Margin 3: **a.** $-6n = -42; n = 7$ **b.** $12n = -72; n = -6$

◆ Margin 4

a. Eight less than the product of three and x is equal to thirteen.

b. Negative five times n plus eighteen is twenty-eight.

Execute: Translation:

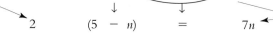

Seven more than the product of three and r is equal to nineteen.

$$7 \quad + \quad 3 \cdot r = 19$$

Solve:
$$7 + 3r = 19$$
$$\underline{ -7 \qquad -7}$$
$$0 + 3r = 12$$
$$3r = 12$$
$$\frac{3r}{3} = \frac{12}{3}$$
$$1r = 4$$
$$r = 4$$

Answer: $r = 4$

Check: Verify that 4 satisfies the original sentence. Is 7 more than the product of 3 and 4 equal to 19?

$$7 + 3 \cdot 4 \; ? \; 19$$
$$7 + 12 \; ? \; 19$$
$$19 = 19$$

Yes, 7 more than the product of 3 and 4 is equal to 19.

◀ **Do Margin 4.**

EXAMPLE 5 Twice the difference of five and n is the same as seventeen subtracted from seven times n. Translate to an equation, then solve.

Understand: *Twice* means *to multiply by 2*, *difference* is the result of *subtraction*, *is the same as* means *an equal sign*, and *subtracted from* indicates 17 is the subtrahend. Finally, *times* indicates *multiplication*.

Plan: Translate the key words to an equation, then solve.

Execute: Translation:

Twice the difference of five and n is the same as seventeen subtracted from seven times n.

$$2 \qquad (5 \; - \; n) \qquad = \qquad 7n \; - \; 17$$

Solve:
$$2(5 - n) = 7n - 17$$
$$10 - 2n = 7n - 17$$
$$\underline{ +2n \qquad +2n}$$
$$10 + 0 = 9n - 17$$
$$10 = 9n - 17$$
$$\underline{ +17 \qquad\qquad +17}$$
$$27 = 9n + 0$$
$$\frac{27}{9} = \frac{9n}{9}$$
$$3 = 1n$$
$$3 = n$$

◆ Margin 5

a. -6 times the sum of y and 3 is equal to 5 times y minus 40.

b. The difference of 8 times b and 9 is the same as triple the sum of b and 2.

Answer: $3 = n$

Check: Verify that 3 satisfies the original sentence. Twice the difference of 5 and n is the same as 17 subtracted from 7 times n.

$$2(5 - 3) \; ? \; 7(3) - 17$$
$$2(2) \; ? \; 21 - 17$$
$$4 = 4$$

Yes, twice the difference of 5 and 3 is the same as 17 subtracted from 7 times 3.

◀ **Do Margin 5.**

Answers to Margin 4:
a. $3x - 8 = 13; x = 7$
b. $-5n + 18 = 28; n = -2$

Answers to Margin 5:
a. $-6(y + 3) = 5y - 40; y = 2$
b. $8b - 9 = 3(b + 2); b = 3$

4.4 Exercises

FOR EXTRA HELP

 Videotape 5

 InterAct Math
Tutorial Software

 www.carsonmath.com

 AWL Math Tutor Center

InterAct MathXL www.mathxl.com

 Student's Solutions Manual

For Exercises 1–30, translate to an equation, then solve.

1. Five more than a number is equal to negative seven.

2. The difference of a number and nine is equal to four.

3. Six less than a number is fifteen.

4. The sum of a number and twelve is negative twenty-seven.

5. A number increased by seventeen is negative eight.

6. A number decreased by twenty-four is negative seven.

7. The product of negative three and a number is twenty-one.

8. Negative eight times a number is equal to forty.

9. A number multiplied by nine is negative thirty-six.

10. Negative seven times a number is equal to negative forty-two.

11. Four more than the product of five and x yields fourteen.

12. Eighteen subtracted from negative seven times y is equal to three.

13. The difference of negative six times m and sixteen is fourteen.

14. Twenty minus eight times b is negative four.

15. Thirty-nine minus five times x is equal to the product of eight and x.

16. Forty less than the product of three and y is equal to seven times y.

17. The sum of seventeen and four times t is the same as the difference of six times t and nine.

18. The difference of ten times n and seven is equal to twenty-five less than four times n.

19. Two times the difference of b and eight is equal to five plus nine times b.

20. Nine more than the product of eight and m is the same as three times the sum of m and thirteen.

21. Six times x plus five times the difference of x and seven is equal to nineteen minus the sum of x and six.

22. Two times r subtracted from seven times the sum of r and one is equal to three times the difference of r and five.

23. Fourteen less than negative eight times the difference of y and 3 is the same as the difference of y and 5 subtracted from the product of negative two and y.

24. The sum of n and three subtracted from twelve times n is the same as negative eleven plus the product of 2 and the difference of n and five.

For Exercises 25–30, explain the mistake in the translation. Then write the correct translation.

25. Seven less than a number is fifteen.
Translation: $7 - n = 15$

26. Nineteen subtracted from a number is eleven.
Translation: $19 - n = 11$

27. Two times the sum of x and thirteen is equal to negative nine.
Translation: $2x + 13 = -9$

28. Five times x minus twelve is the same as three times the difference of x and four.
Translation: $5x - 12 = 3x - 4$

29. Sixteen decreased by the product of six and n is the same as twice the difference of n and four.
Translation: $6n - 16 = 2(n - 4)$

30. Negative three times the sum of y and five is equal to twice y minus the difference of y and one.
Translation: $-3(y + 5) = 2y - (1 - y)$

REVIEW EXERCISES

1. Find the GCF of $30x^3y^5$ and $24xy^7z$.

2. Factor the polynomial. $18b^5 - 27b^3 + 54b^2$

3. Write an expression in simplest form for the perimeter of the rectangle.

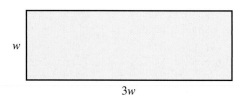

w

$3w$

4. Evaluate the perimeter expression from Review Exercise 3 when w is 6 ft.

5. Jasmine sells two sizes of holly shrub, large and small. One day, she sold 7 large holly shrubs at $7 each and some small holly shrubs at $5 each. If she sold a total of 16 holly shrubs, how many of the small shrubs did she sell? How much money did she make from the sale of the holly shrubs altogether?

4.5 Applications and Problem Solving

OBJECTIVES

1 Solve problems involving two unknown amounts.

2 Use a table in solving problems with two unknown amounts.

OBJECTIVE **1** *Solve problems involving two unknown amounts.*

Let's extend what we learned about translating sentences. Sometimes problems have two unknown amounts. Consider the following situation.

EXAMPLE 1 A carpenter is asked to make a rectangular frame out of an 8-ft. strip of wood. The length of the frame must be three times the width. What must the dimensions be?

Solution: Follow the problem-solving process.

Understand: Draw a picture.

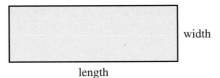

List the information:

The total frame material is 8 ft. (This means that the perimeter must be 8 ft.)

The length is three times the width.

There are two amounts missing, the length and the width. Because we only know how to solve linear equations in one variable, we must choose the length or width to be the variable and write our equation in terms of that variable.

Plan: Translate to an equation using the key words, then solve the equation.

Execute: Remember there are two amounts missing, length and width, and we'll have to select one of those amounts to be our variable.

Translation: Length must be three times the width.

$$\text{length} = 3 \cdot \text{width}$$

The translation helps us decide which amount to choose as the variable.

Because width is multiplied by 3, let's choose width to be our variable. Let w represent width. So we can now say:

$$\text{width} = w$$
$$\text{length} = 3w$$

> **Tip**
> Choose the amount that is acted on to be the variable.

To get our equation, we must use the other piece of information that we were given, which was the fact that there is a total of 8 ft. of framing material. In other words, the perimeter must be 8 ft.

$$\text{perimeter} = 8 \text{ ft.}$$

Recall that to find a perimeter we must add the lengths of all the sides.

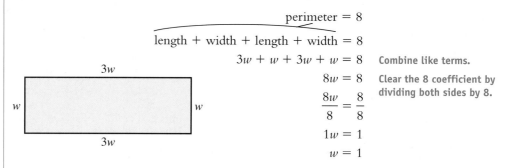

$$\overbrace{\text{perimeter} = 8}$$

$$\text{length} + \text{width} + \text{length} + \text{width} = 8$$

$$3w + w + 3w + w = 8 \qquad \text{Combine like terms.}$$

$$8w = 8 \qquad \text{Clear the 8 coefficient by}$$
$$\qquad \text{dividing both sides by 8.}$$

$$\frac{8w}{8} = \frac{8}{8}$$

$$1w = 1$$

$$w = 1$$

Answer: The width is 1 ft. Don't forget that we were asked to find both the length and width. How can we get the length? Go back to our initial translation.

The length is three times the width.

$$\text{length} = 3w$$

$$\text{length} = 3(1) = 3 \text{ ft.}$$

The carpenter must make the frame 3 ft. by 1 ft.

Check: Let's make sure everything in the original problem is satisfied. There were two restrictions in the problem. The length must be three times the width and the perimeter must be 8 ft. If the length is 3 ft. and the width is 1 ft., then the length is three times the width. Is the perimeter 8 ft.? Yes, because 3 ft. + 1 ft. + 3 ft. + 1 ft. = 8 ft.

Tips

When a problem has two unknown amounts, you must have two relationships in order to solve the problem.

1. Use one of the relationships to select the variable. Choose a variable for the unknown amount that is acted on.

"The length is three times the width." → Let width be variable w because
↓ width is multiplied by 3.
length = $3w$

2. Use the other relationship to make the equation.

"The total framing material is 8 ft." → The perimeter is 8 ft.
↓
$$\overbrace{\text{length} + \text{width} + \text{length} + \text{width} = 8}$$
$$3w \quad + \quad w \quad + \quad 3w \quad + \quad w \quad = 8$$

◀ **Do Margin 1.**

Let's consider problems involving **equilateral** and **isosceles** triangles.

DEFINITION **Equilateral triangle:** A triangle with all three sides of equal length.

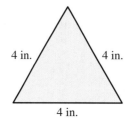

4 in.   4 in.

4 in.

The triangle to the left is an equilateral triangle because all three sides are equal in length.

◆ **Margin 1**

A TV is to have a rectangular screen with a length that is 5 in. less than twice the width and a perimeter of 44 in. What are the dimensions of the screen?

Answer: 9 in. by 13 in.

DEFINITION | **Isosceles triangle:** A triangle with two sides of equal length.

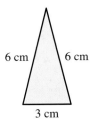

6 cm   6 cm

3 cm

The triangle to the left is an isosceles triangle because two of the sides are equal in length.

EXAMPLE 2 Suppose each of the equal length sides of an isosceles triangle is twice the length of the other side. If the perimeter is 75 cm, what are the lengths of the sides?

Solution: Follow the problem-solving process.

Understand: Isosceles triangles have two equal length sides.

These two sides are equal in length. We are told they are twice the length of the other side (bottom side).

We are also told that the perimeter is 75 cm.

Plan: Translate the relationships, write an equation, then solve.

Execute: Translation:

The equal length sides are twice the length of the other side.

equal length sides = 2 · length of the other side

Because the length of the other side is acted on (multiplied by 2), we will use the variable L to represent its length.

equal length sides = $2L$

other side = L

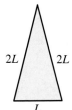

$2L$ $2L$

L

The sentence concerning the perimeter of the triangle is our second relationship. We use it to create the equation that we will solve. Perimeter is the sum of all the side lengths.

The perimeter is 75 cm.

↓

equal length sides + other side = 75

$2L + 2L \quad + \quad L \quad = 75$ Combine like terms.

$5L = 75$

$\dfrac{5L}{5} = \dfrac{75}{5}$ Clear the 5 coefficient by dividing both sides by 5.

$1L = 15$

$L = 15$

Answer: The other side is 15 cm. To get the equal length sides, use the relationship:

equal length sides = $2L = 2(15) = 30$ cm

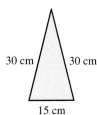

30 cm   30 cm

15 cm

Check: Because 30 is twice 15, the equal length sides are in fact twice the length of the other side. We also need to verify that the perimeter is 75 cm.

$$P = 15 + 30 + 30$$
$$P = 75 \text{ cm}$$

◀ **Do Margin 2.**

Sometimes problems do not show all the needed relationships in an obvious manner. Sometimes we must use our wits or situational knowledge to figure out the needed relationships. Let's consider problems involving angles.

Angles are formed when two lines or line segments intersect or touch. Lines segments AB and BC form an angle below. The measure of the angle is 40°.

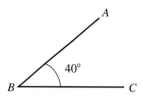

We use the symbol ∠ to indicate an angle. To indicate the angle we can write all three letters with *B* in the center like ∠*ABC* or ∠*CBA*. Or, we can simply write ∠*B*. To indicate the measure of the angle, we write ∠*B* = 40°.

Problems involving angle measurements may indicate that two unknown angles are **congruent, supplementary,** or **complementary,**. Let's state these definitions:

DEFINITION   **Congruent angles:** Angles that have the same measurement.

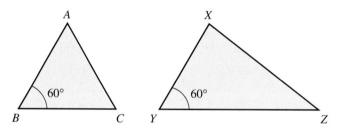

In the two triangles, ∠*B* is congruent to ∠*Y* because they have the same measurement of 60°. The symbol for congruent is ≅.

$$\angle B \cong \angle Y$$

DEFINITION   **Supplementary angles:** Two angles whose sum is 180°.

For example, a straight line forms an angle that measures 180°. Any line that intersects a straight line will cut that 180° angle into two angles that are supplementary.

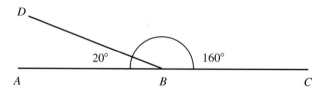

∠*ABD* and ∠*DBC* are supplementary angles because the sum of their measurements is 180°.

$$20° + 160° = 180°$$

Answer: 8 in., 11 in., and 11 in.

Complementary angles: Two angles whose sum is 90°.

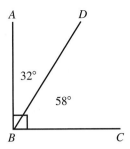

$\angle ABD$ and $\angle DBC$ are complementary because the sum of their measurements is 90°.

$$32° + 58° = 90°$$

EXAMPLE 3 In designing a roof frame, the architect wants the angled beam to meet the horizontal truss so that the outer angle measurement is 30° more than twice the inner angle measurement. What are the angle measurements?

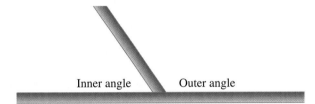

Solution: Follow the problem-solving process.

Understand: We must find the inner and outer angle measurements.

We were given a relationship about the angles: The outer angle measurement is 30° more than twice the inner angle measurement.

Because the problem involves two unknowns, we must have another relationship in order to solve it. However, this second relationship was not given in an obvious manner. We must resort to our own wits and knowledge for the second relationship.

Looking at the picture, we see that the inner and outer angles are supplementary, which means the sum of the measurements is 180°. We can say:

$$\text{inner angle} + \text{outer angle} = 180$$

Plan: Translate the relationships to an equation, then solve.

Execute: Translation:

The outer angle measurement is 30° more than twice the inner angle measurement.
$$\text{outer angle} = 30 + 2 \cdot \text{inner angle}$$

Because the inner angle measurement is multiplied by 2, we should let the variable represent the inner angle. Let's use the letter a.

$$\text{Let } a = \text{inner angle}$$
$$\text{outer angle} = 30 + 2a$$

a. Two angles are to be constructed from metal beams so that when they are joined they form a 90° angle. One of the angles is to be 6° less than 3 times the other angle. What are the angles?

b. Two electric doors are designed to swing open simultaneously when activated. The design calls for the doors to open so that the angle made between each door and the wall is the same. Because the doors are slightly different in weight, the equation $15i + 1$ describes the angle for one of the doors and $14i + 5$ describes the angle for the other door, where i represents the current supplied to the controlling motors. What current must be supplied so that the door angles are the same?

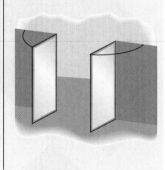

Now we tie this translation into our other relationship.

$$\text{inner angle} + \text{outer angle} = 180$$
$$a \quad + \quad 30 + 2a \quad = 180 \qquad \textbf{Combine like terms.}$$

Solve: $3a + 30 = 180$ **Clear +30 by subtracting 30 on both sides.**

$$\begin{array}{r} -30 \quad -30 \\ \hline 3a + 0 = 150 \end{array}$$

$$\frac{3a}{3} = \frac{150}{3} \qquad \textbf{Clear the 3 coefficient by dividing both sides by 3.}$$

$$1a = 50$$
$$a = 50$$

Answer: Because a represents the inner angle, we can say the inner angle measures 50°. To find the outer angle measurement, we go back to our relationships.

$$a = \text{inner angle} = 50$$
$$\text{outer angle} = 30 + 2a$$
$$= 30 + 2(50)$$
$$= 30 + 100$$
$$= 130$$

The outer angle is 130°.

Check: We must make sure that our answer satisfies all of the information in the problem. The outer angle measuring 130° is 30° more than twice the inner angle measuring 50°. The sum of
◆ the inner and outer angles is $50° + 130° = 180°$, which means they are in fact supplementary.

◀ **Do Margin 3.**

A classic situation where the relationships are not stated in an obvious way is when the unknowns are consecutive integers.

EXAMPLE 4　The sum of three consecutive even integers is 72. What are the integers?

Solution: Use the problem-solving process.

Understand: The key word *sum* means *to add.* The tricky part is to understand what is meant by *consecutive even integers. Consecutive* means *one after another. Three consecutive even integers* must mean a sequence such as 2, 4, 6 or 10, 12, 14.

What is the relationship between consecutive even integers? Notice there is a difference of 2 between all consecutive even integers. Therefore if we choose the smallest of the three unknown even integers to be x, we can say:

smallest unknown even integer: x

next even integer: $x + 2$

third even integer: $x + 4$

Plan: Write an equation, then solve.

Execute: Translation:

Connection
What if x were 10? Then we can say:
$$x = 10$$
$$x + 2 = 10 + 2 = 12$$
$$x + 4 = 10 + 4 = 14$$
10, 12, and 14 are consecutive even integers.

The sum of three consecutive even integers is 72.

$$\underset{\text{integer}}{\text{smallest even}} + \underset{\text{integer}}{\text{next even}} + \underset{\text{integer}}{\text{third even}} = 72$$

$$x \quad + \quad x + 2 \quad + \quad x + 4 \quad = 72 \qquad \textbf{Combine like terms.}$$

Answers: **a.** 24° and 66° **b.** 6A

Solve: $3x + 6 = 72$

$$3x + 6 = 72 \qquad \text{Clear 6 by subtracting 6 on both sides.}$$
$$\underline{-6 \quad -6}$$
$$3x + 0 = 66$$
$$3x = 66$$
$$\frac{3x}{3} = \frac{66}{3} \qquad \text{Clear the 3 coefficient by dividing both sides by 3.}$$
$$1x = 22$$
$$x = 22$$

Answer: Because x represents the smallest unknown even integer, we can say:

$$\text{smallest unknown even integer: } x = 22$$
$$\text{next even integer: } x + 2 = 22 + 2 = 24$$
$$\text{third even integer: } x + 4 = 22 + 4 = 26$$

Check: We must verify that the three integers are even and that their sum is 72. 22, 24, and 26 are in fact three consecutive even integers and their sum is:

$$22 + 24 + 26 = 72 \qquad \text{They check.}$$

Discussion What is the pattern for consecutive odd integers? What is the pattern for consecutive integers?

Do Margin 4. ▶

OBJECTIVE 2 *Use a table in solving problems with two unknowns.*

We have seen how drawing a picture, making a list, and using key words have helped us to understand a problem. Let's now consider problems where making a table can be helpful.

To aid in learning how to use tables, we will focus on problems that all use the same table. All tables will be four-column tables that look like this:

Categories	Value	Number	Amount

The first column, categories, will contain the two unknowns. The second column, value, will list the given values of each category. The third column, number, will contain the number of items in each category. The amount column will be found by the relationship:

$$\text{value} \cdot \text{number} = \text{amount}$$

For the number column, you will either be given a relationship about the number of items or a total number of items to split. Let's consider an example to see how it all fits together.

EXAMPLE 5 Salvador is an artist. A company produces prints of two of his paintings. The first print sells for $45 and the second print for $75. He is told that during the first day of sales, the company sold six more of the $75 print than the $45 print with a total income of $1410. How many of each print were sold?

Solution: Use the problem-solving process.

Understand: The overall relationship here is that the total income is $1410. From this we can say:

$$\begin{array}{ccc} \text{income from} & + & \text{income from} = 1410 \\ \text{first print} & & \text{second print} \end{array}$$

◆ **Margin 4**

Solve.

a. The sum of three consecutive odd integers is 63. What are the integers?

b. The sum of two consecutive integers is 91. What are the integers?

Complete a four-column table, write an equation, then solve.

a. A marketing manager wants to research the sale of two different sizes of perfume. The small bottle sells for $35 and the large bottle for $50. The report indicates that the number of large bottles sold was 24 less than the number of small bottles and the total sales was $4070. How many of each size bottle was sold?

How can we describe the income from each print? What do we know about the individual prints? We know the selling price and that there were six more $75 prints sold than $45 prints. If we knew the number of prints sold, we could multiply that number times the selling price to get the income from that particular print.

$$\text{selling price} \cdot \text{number of prints} = \text{income from the print}$$

Because of all the information we have, it is helpful to use a table. The value column is the selling price of each print. The amount column is the income from each print.

Categories	Selling price	Number of prints	Income
First print	45	n	$45n$
Second print	75	$n + 6$	$75(n + 6)$

We were given these.	We selected n to represent the number of the first print, then translated: "six more of the $75 print than the $45 print."	We multiplied straight across because:

$$\begin{array}{ccc} \text{selling} \cdot & \text{number} & = \text{income} \\ \text{price} & \text{of prints} & \text{from each print} \end{array}$$

The expressions in the last column in the table describe the income from the sale of each print.

Plan: Translate the information in the table into an equation, then solve.

Execute: Now we can use our initial relationship:

$$\begin{array}{ccc} \text{income from} & + & \text{income from} & = 1410 \\ \text{first print} & & \text{second print} \end{array}$$

$$45n \quad + \quad 75(n + 6) \quad = 1410$$
$$45n \quad + \quad 75n + 450 = 1410$$
$$120n + 450 = 1410$$
$$120n + 450 = 1410$$
$$\underline{-\,450 \quad -450}$$
$$120n + \quad 0 = \quad 960$$
$$\frac{120n}{120} = \frac{960}{120}$$
$$1n = 8$$
$$n = 8$$

b. A bottling company produces 12 oz.- and a 16 oz.-sized bottles. In one day the company produces 3 times as many 12-oz. bottles as 16-oz. bottles. If the company produces a total of 7280-oz. of the beverage, how many 12-oz. and 16-oz. bottles were filled?

Answer: We go back to our table and use the relationships about the number of each print.

$$\text{The number of \$45 prints: } n = 8$$
$$\text{The number of \$75 prints: } n + 6 = 8 + 6 = 14$$

Check: Verify that 8 of the $45 prints and 14 of the $75 prints make a total income of $1410.

$$8(45) + 14(75) \,?\, 1410$$
$$360 + 1050 \,?\, 1410$$
$$1410 = 1410 \qquad \text{It checks.}$$

◀ **Do Margin 5.**

Answers: **a.** 62 small bottles, 38 large bottles **b.** 420 12-oz. bottles, 140 16-oz. bottles

EXAMPLE 6 Jasmine sells two sizes of holly shrub. The larger size sells for $7 and the smaller size for $5. She knows she sold 16 shrubs for a total of $94, but forgot how many of each size. How many of each size did she sell?

Solution: Use the problem-solving process.

Understand: We are given the price or value of two different sized shrubs. We are also given a total amount of income from the sale of the two different shrubs and a total number of shrubs to split up. We can say:

$$\text{income from small} + \text{income from large} = \text{total income}$$

Because we have two categories of shrubs along with a value and a number of shrubs, this problem lends itself to a table setup. We are given the selling prices: $5 and $7. The tricky part is filling in the number column.

We are told that Jasmine sold a total of 16 shrubs. We can say:

$$\text{number of small} + \text{number of large} = 16$$

Or, if we knew one of the numbers, we could find the other by subtracting from 16. We can write a related subtraction this way:

$$\text{number of small} = 16 - \text{number of large}$$
$$\text{Let's choose } L = \text{number of large}$$

We could also say:

large = 16 − number of small

It works either way.

Categories	Selling price	Number	Income
Small	5	$16 - L$	$5(16 - L)$
Large	7	L	$7L$

We were given these.

Select one of the categories to be the variable. The other will be:

total number − variable

selling price · number = income

So we multiply the columns straight across.

Plan: Translate to an equation, then solve.

Execute: We can now use our initial relationship:

$$\text{income from small} + \text{income from large} = \text{total income}$$
$$5(16 - L) + \qquad 7L \qquad = 94$$
$$80 - 5L + \qquad 7L \qquad = 94$$
$$80 \qquad + 2L \qquad = 94$$
$$80 + 2L = 94$$
$$\underline{-80 \qquad \quad -80}$$
$$0 + 2L = 14$$
$$\frac{2L}{2} = \frac{14}{2}$$
$$1L = 7$$
$$L = 7$$

Answer: Number of large size: $L = 7$

Number of small size: $16 - L = 16 - 7 = 9$

Complete a four-column table, write an equation, then solve.

a. A farmer has a total of 17 pigs and chickens. The combined number of legs is 58. How many pigs and how many chickens are there?

Check: Verify that 7 large-size shrubs at $7 each and 9 small-size shrubs at $5 each is a total of 16 shrubs that sell for a total of $94.

$$7 + 9 \; ? \; 16 \qquad\qquad 7(7) + 9(5) \; ? \; 94$$
$$16 = 16 \quad \text{That checks.} \qquad 49 + 45 \; ? \; 94$$
$$94 = 94 \quad \text{That checks too.}$$

◀ **Do Margin 6.**

Tips for Table Problems

You could be given a relationship about the number of items in each category or a total number of items to split up. If you are given a relationship, use key words to translate to expressions. If you are given a total number of items to split up, choose one of the categories to be the variable and the other will be:

$$\text{total number} - \text{variable}$$

---

b. A tire factory produces two different size tires. Tire A costs $37 to produce while Tire B costs $42. At the end of the day, management gets the following report:

total tire production = 355

total cost = $13,860

How many of each tire was produced?

Answers: **a.** 12 pigs, 5 chickens
b. 210 of Tire A, 145 of Tire B

4.5 Exercises

FOR EXTRA HELP

 Videotape 5

 InterAct Math
Tutorial Software

 www.carsonmath.com

 AWL Math Tutor Center

InterAct MathXL www.mathxl.com

 Student's Solutions
Manual

For Exercises 1–22, translate to an equation, then solve.

1. An architect consultant feels that the optimal design for a new building would be a rectangular shape where the length is four times the width. Budget restrictions force the building perimeter to be 300 ft. What will be the dimensions of the building?

2. 84 ft. of border strip was used to go around a rectangular room. The width of the room is 4 ft. less than the length. What are the dimensions?

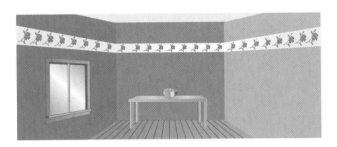

3. The Johnsons can afford 220 ft. of fencing material. They want to fence in a rectangular area with a length 20 ft. more than twice the width. What will the dimensions be?

4. After completing the trim on some kitchen cabinets, Candice has 48 in. of trim wood left. She decides to make a picture frame out of the wood. She wants to make the frame so that the length is 4 in. less than three times the width. What must the dimensions be?

5. The roof of a building is to be an isosceles triangle. The equal sides of the triangle must be 5 m less than twice the base. The perimeter is to be 70 m. What will the lengths of the base and sides be?

6. A section of a public park is in the shape of an isosceles triangle. Each equal side of the triangle is three times the length of the base. If the perimeter is 210 ft., find the lengths of the base and sides.

7. A sculpture in front of a school has an equilateral triangle sitting atop a rectangle. All sides of the equilateral triangle are the same distance as the length of the rectangle. The width of the rectangle is 10 ft. less than three times the length. The perimeter is the same for both shapes. What are the dimensions of the shapes?

8. A monument is to have the shape of an isosceles triangle sitting atop a rectangle. The base of the triangle and width of the rectangle are equal. The other two sides of the triangle are 2 ft. more than the base. The length of the rectangle is twice the width. If the perimeter of the monument is 39 ft., what are all the dimensions?

9. A wood brace is placed where a wall and floor meet. The angle between the floor and the brace is four times the angle between the brace and the wall. Find the angle measurements. (Assume the angle made by the wall and floor is 90°.)

10. A security camera is placed in the corner of a room in a museum. The camera is set so that the angle between it and one of the walls is 16° less than the angle made with the other wall. Find the angle measurements. (Assume the angle formed by the walls is 90°.)

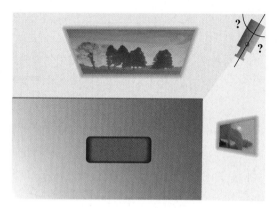

11. In an experiment a laser contacts a flat detector so that one angle between the laser beam and detector is 15° more than four times the other angle. What are the two angle measurements?

12. A suspension bridge has a steel beam that connects to the flat surface of the bridge forming two angles. The measure of the smaller of the two angles is 38° less than the measure of the other angle. What are the two angles?

13. A laser beam strikes a flat mirror forming an angle with the mirror. The beam is deflected at the same angle. The angle in between the entering and exiting beams is 10° more than 3 times the angle formed between the beam and the mirror. What are the three angles?

14. A billiard ball strikes the bumper at an angle. Because the ball has side spin, it is deflected at an angle that is 6° less than the initial angle. The angle formed between the initial and deflected lines of travel is 10° more than twice the initial angle. What are the three angles?

15. The sum of the angles in any triangle is 180°. Suppose we have a triangle with the second angle measuring 10° more than the first and the third angle 7° less than the first. What are the three angle measurements?

16. A mounting apparatus for a telescope is to be in the shape of a triangle. The second angle in the triangle is to be twice the first. The third angle is to be 12° less than the first. What are the three angle measurements?

17. The sum of two consecutive integers is 93. What are the integers?

18. The sum of two consecutive integers is 121. What are the integers?

19. The sum of two consecutive odd integers is 196. What are the integers?

20. The sum of three consecutive integers is 96. What are the integers?

21. The sum of three consecutive odd integers is 225. What are the integers?

22. The sum of three consecutive even integers is 234. What are the integers?

For Exercises 23–30, complete a four-column table, write an equation, then solve.

23. Monique sells make-up products. One of the products is a lotion that comes in two different size bottles. The smaller-size bottle sells for $8 and the larger-size for $12. She remembers that she sold 5 more smaller bottles than larger bottles that day, but doesn't remember exactly how many of each. She also remembers the total sales of the lotion was $260. How many of each size did she sell?

Categories	Value	Number	Amount

24. Byron is a wedding photographer. He develops two different size prints for a newly-wed couple. He sells the larger print for $7 each and the smaller print for $3 each. He develops twice as many smaller prints as larger prints at a total cost of $78. How many of each print did he sell?

Categories	Value	Number	Amount

25. Ian has some $5 bills and $10 bills in his wallet. If he has a total of 19 bills worth a total of $125, how many of each bill is in his wallet?

Categories	Value	Number	Amount

26. A paper mill produces two different sizes of paper stock. It costs $5 to produce a box of the smaller stock and $7 to produce a box of the larger stock. Management gets a report that 455 boxes were produced at a total cost of $2561 in one day. How many boxes of each stock were produced?

Categories	Value	Number	Amount

27. Nina has two different types of foreign bills to exchange for American currency. She is told the more valuable bill is worth three times that of the other bill. She has 12 of the more valuable bills and 18 of the other bills and receives a total of $108. What is the value of each foreign bill?

28. Cal has two different stocks. One of the stocks is more valuable than the other. He notices in the newspaper that one of his stocks is $3 more valuable than the other. He has 24 shares of the more valuable stock and 22 shares of the other stock. His total assets in the stocks is $348. How much is each stock worth?

29. Jada has her friend take a certain number of play bills. The friend says she has 13 of one bill and 17 of another. The two different bills together are worth $16. The total she has in her hand is $244. How much is each bill worth?

30. Andre sells furniture. Two different chairs go with a dining set, a chair with arms or a chair without. One armchair and one chair without arms sell together for $184. One day he sold 5 armchairs and 17 chairs without arms for a total of $1868. How much does each chair sell for?

Puzzle Problem

At 3:00:05 P.M. what are the angles made between the second hand and the minute and hour hands?

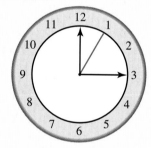

REVIEW EXERCISES

1. Check to see if $x = -3$ is a solution for $4x - 8 = 5x - 6$

2. Is $x^2 + 5 = 9$ a linear equation?

3. Solve and check. $m + 19 = 12$

4. Solve and check. $-15n = 75$

5. Solve and check. $9x - 7(x + 2) = -3x + 1$

Summary

Defined Terms

Review the following terms, and for those you do not know, study its definition on the page number next to it.

Section 4.1
Expression *p. 238*
Equation *p. 238*
Solve *p. 239*
Solution *p. 239*

Section 4.2
Linear equation *p. 243*
Balance technique *p. 244*

Section 4.5
Equilateral triangle *p. 268*
Isosceles triangle *p. 269*
Congruent angles *p. 270*

Supplementary angles *p. 270*
Complementary angles
  p. 271

Procedures, Rules, and Key Examples

Procedures/Rules	Key Example(s)
Section 4.1 **To check a value to see if it is a solution for a given equation:** 1. Use the value to evaluate the expressions that make up the equation. 2. If the results equate, the value is a solution.	Check to see if -2 is a solution for $x^2 - 5x = 14$. $$x^2 - 5x = 14$$ $$(-2)^2 - 5(-2) \ ? \ 14$$ $$4 - (-10) \ ? \ 14$$ $$4 + 10 \ ? \ 14$$ $$14 = 14$$ -2 is a solution because both sides match. Check to see if $y = 4$ is a solution for $2y - 9 = 6y - 23$. $$2y - 9 = 6y - 23$$ $$2(4) - 9 \ ? \ 6(4) - 23$$ $$8 - 9 \ ? \ 24 - 23$$ $$-1 \neq 1$$ 4 is not a solution because the results do not match.
Section 4.2 **The Addition/Subtraction Principle of Equality:** We can add or subtract the same amount on both sides of an equation without affecting its solution(s). Primary purpose: To get variable terms on one side of the equal sign and constant terms on the other side.	Solve and check. $$t + 15 = 7$$ $$t + 15 = 7$$ $$\underline{-15 \quad -15}$$ $$t + 0 = -8$$ $$t = -8$$ Check: $t + 15 = 7$ $$(-8) + 15 \ ? \ 7$$ $$7 = 7$$ $t = -8$ makes the equation true so it is correct.
Section 4.3 **The Multiplication/Division Principle of Equality:** We can multiply or divide both sides of an equation by the same amount without affecting its solution(s). Primary purpose: To clear a coefficient from a term.	Solve and check. $$-7x = 42$$ $$\frac{-7x}{-7} = \frac{42}{-7}$$ $$1x = -6$$ $$x = -6$$ Check: $-7x = 42$ $$-7(-6) \ ? \ 42$$ $$42 = 42$$ $x = -6$ makes the equation true, so it is correct.

Section 4.3

Procedure: To solve equations:
1. Simplify both sides of the equation as needed.
   a. Distribute to clear parentheses.
   b. Combine like terms.
2. Use the addition/subtraction principle so that all variable terms are on one side of the equation and all constants are on the other side.

Tip: Clear the variable term that has the smaller coefficient. This will avoid negative coefficients.

3. Use the multiplication/division principle to clear any remaining coefficient.

Solve and check.

$$12 - (2x + 7) = 13x - 5(x - 7)$$
$$12 - 2x - 7 = 13x - 5x + 35$$
$$5 - 2x = 8x + 35$$
$$5 - 2x = 8x + 35$$
$$\underline{+2x \quad +2x}$$
$$5 + 0 = 10x + 35$$
$$5 = 10x + 35$$
$$\underline{-35 \qquad -35}$$
$$-30 = 10x + 0$$
$$-30 = 10x$$
$$\frac{-30}{10} = \frac{10x}{10}$$
$$-3 = 1x$$
$$-3 = x$$

Check: $12 - (2x + 7) = 13x - 5(x - 7)$
$$12 - (2(-3) + 7) \; ? \; 13(-3) - 5(-3 - 7)$$
$$12 - (-6 + 7) \; ? \; 13(-3) - 5(-10)$$
$$12 - 1 \; ? \; -39 - (-50)$$
$$11 \; ? \; -39 + 50$$
$$11 = 11$$

Because both sides match, $-3 = x$ is correct.

Section 4.4

Translating

Addition	Translation
The sum of x and 3	$x + 3$
h plus k	$h + k$
7 added to t	$t + 7$
3 more than a number	$n + 3$
y increased by 2	$y + 2$

Note: Because addition is a commutative operation, it does not matter what order we write the translation.

For "the sum of x and 3" we can write $x + 3$ or $3 + x$.

Subtraction	Translation
The difference of x and 3	$x - 3$
h minus k	$h - k$
7 subtracted from t	$t - 7$
3 less than a number	$n - 3$
y decreased by 2	$y - 2$

Note: Subtraction is not a commutative operation, therefore the way we write the translation matters. We must translate each key phrase exactly as it was presented above. Notice when we translate *less than* or *subtracted from* the translation is in reverse order from what we read.

Multiplication	Translation
The product of x and 3	$3x$
h times k	hk
twice a number	$2n$
triple the number	$3n$

Note: Like addition, multiplication is a commutative operation. This means we can write the translation order any way we wish.

h times k can be hk or kh.

Note: Key words for an equal sign: is equal to, is the same as, is, produces, yields

Formulas

Area of a rectangle: $A = lw$

Surface area of a box: $SA = 2lw + 2lh + 2wh$

Volume of a box: $V = lwh$

Force: $F = ma$

Distance: $d = rt$

Voltage: $V = ir$

For Exercises 1–6, answer true or false.

1. $4x - 9 + 11x$ is an equation.

2. $9y - 6 = 4y - 1$ is an equation.

3. $t^2 - 9 = 16$ is a linear equation.

4. $4(m - 2) = 6m + 16$ is a linear equation.

5. -6 is a solution for $x - 9 = -15$.

6. The phrase "7 less than a number" translates to $7 - n$.

For Exercises 7–9, fill in the blank.

7. The addition/subtraction principle of equality says that we can _____ or _____ the _____ amount on both sides of an equation without affecting its solution(s).

8. The multiplication/division principle of equality says that we can _____ or _____ both sides of an equation by the _____ amount without affecting its solution(s).

9. To solve equations:

   1. _____ both sides of the equation as needed.

      a. _____ to clear parentheses.

      b. _____ like terms.

   2. Use the _____ principle so that all variable terms are on one side of the equation and all constants are on the other side.

   3. Use the _____ principle to clear any remaining coefficient.

10. Explain in your own words how to check a potential solution.

For Exercises 11–16, check to see if the given number is a solution for the given equation.

11. $-7n + 12 = 5$; check $n = 1$

12. $14 - 4y = 3y$; check $y = -2$

13. $b^2 - 15 = 2b$; check $b = -3$

14. $9r - 17 = -r + 23$; check $r = 5$

15. $5(x - 2) - 3 = 10 - 4(x - 1)$; check $x = 3$

16. $u^3 - 7 = u^2 + 6u$; check $u = -1$

For Exercises 17–28, solve and check.

17. $n + 19 = 27$

18. $y - 6 = -8$

19. $-4k = 36$

20. $-8m = -24$

21. $2x + 11 = -3$

22. $-3h - 8 = 19$

23. $9t - 14 = 3t + 4$

24. $20 - 5y = 34 + 2y$

25. $10m - 17 + m = 2 + 12m - 20$

26. $-16v - 18 + 7v = 24 - v + 6$

27. $4x - 3(x + 5) = 16 - 7(x + 1)$

28. $9y - (2y + 3) = 4(y - 5) + 2$

For Exercises 29 and 30, write a linear equation to describe the situation, then solve.

29. The temperature at 5 A.M. was reported to be -15. By 11 A.M. the temperature rose to 28. How much did the temperature rise?

30. Kari has a balance of -547 on a credit card. How much should she pay on the account to get a balance of -350?

For Exercises 31 and 32, use the formula and solve for the missing amount.

31. A swimming pool is to be designed to have a volume of 5000 ft.^{3}. The length of the pool is to be 50 ft. and the width is to be 25 ft. Find the depth (height). (Use $V = lwh$.)

32. A box company shapes flat pieces of cardboard into boxes. The flat pieces have an area of 136 ft.^{2}. If the length is 3 ft. and height is 4 ft., find the width. (Use $SA = 2lw + 2lh + 2wh$.)

For Exercises 33–44, translate to an equation, then solve.

33. Fifteen minus seven times a number is twenty-two.

34. The difference of five times x and four is the same as the product of three and x.

35. Twice the sum of n and twelve is equal to eight subtracted from negative six times n.

36. Twelve minus three times the difference of x and seven is the same as three less than the product of six and x.

37. A field is developed so that the length is 2 m less than three times the width. The perimeter is 188 m. Find the dimensions of the field.

38. A window over the entrance to a library is to be an isosceles triangle. The equal sides are to be 38 in. longer than the base. If the perimeter must be 256 in., what will be the dimensions of the triangle?

39. The supports to the basketball goals in a park are to be angled for safety. The angle made on the side facing away from the goal is to be 15° less than twice the angle made on the goal side. Find the angle measurements.

40. A security laser is placed in the corner of a room near the floor and aimed to a detector across the room. The angle made with the wall on one side of the beam is 16° more than the angle made with the other wall. What are the angles?

41. The sum of two consecutive integers is 193. Find the integers.

42. The sum of two consecutive even integers is 166. Find the integers.

43. Karen has 9 more ten dollar bills than twenty dollar bills. If the total amount of money is $330, how many of each bill does she have?

44. A store sells two different size bags of mulch. The larger bag costs $6 and the smaller bag costs $4. In one day the store sold 27 bags of mulch for a total of $146. How many of each size were sold?

1. Is $8y - 17 = 7$ an expression or equation? Why?

2. Is $9x + 7 = x^2 - 4$ linear or nonlinear? Why?

3. Check to see if -4 is a solution for $x^2 - 9 = 3x + 5$.

4. Check to see if 2 is a solution for $3x - 11 = -4(x - 2) - 5$.

For Problems 5–12, solve and check.

5. $n - 15 = -7$

6. $-9m = 54$

7. $-6y + 3 = -21$

8. $9k + 5 = 17 - 3k$

9. $-13x + 26 = 11 - 8x$

10. $4t - 13 + t = 11 + 6t - 3$

11. $7(u - 2) + 12 = 4(u - 2)$

12. $7 - 3(k + 4) = 5k - (k - 9)$

1. _____

2. _____

3. _____

4. _____

5. _____

6. _____

7. _____

8. _____

9. _____

10. _____

11. _____

12. _____

For Problems 13–20, translate to an equation and solve.

13. Daryl owes $458 in taxes. He currently has $375. How much more does he need?

14. An engineer is designing a storage box to have a volume of 20,592 in.^{3}. The length is to be 44 in. and the width is to be 26 in. Find the height. (Use $V = lwh$.)

15. Nine subtracted from four times a number is equal to twenty-three. Find the number.

16. Three times the difference of x and five is the same as nine less than four times x.

17. An entrance to a new restaurant is to be in the shape of an isosceles triangle with the equal sides 9 in. more than twice the third side. The perimeter is to be 258 in. Find the dimensions of the triangle.

18. A microphone is placed on a stand on the floor of a stage so that the angle made by the microphone and the floor on one side is 30° less than the angle made on the other side of the microphone. Find the two angle measurements.

19. The sum of two consecutive odd integers is 76. Find the integers.

20. A music store sells two models of the same guitar. Model A costs $450 while model B costs $675. In one week the store sold 12 of the guitars for a total of $6300. How many of each model did the store sell? (Use a four-column table.)

Cumulative Review Exercises

For Exercises 1–6, answer true or false.

1. $(5 + 4) + 2 = 5 + (4 + 2)$

2. $-65 < -68$

3. 91 is a prime number.

4. $x^0 = 1$ where x is any integer.

5. $4y - 9 = 3(y + 5)$ is a linear equation.

6. -6 is a solution for $5x - 8 = -24$

For Exercises 7–19, fill in the blank.

7. The quotient of two negative numbers is a _____ number.

8. When multiplying two exponential forms that have the same base, we can _____ the exponents and keep the same base.

9. When dividing two exponential forms that have the same base, we can _____ the exponents and keep the same base.

10. Explain in your own words how to evaluate an expression.

11. Write expanded form for 36,097.

12. Graph 7 on a number line.

13. Round 29,512 to the nearest thousand.

14. Estimate $5826 \div 224$ by rounding so that there is only one nonzero digit in each number.

15. What is the coefficient of $-8x^3$?

16. What is the degree of 51?

17. What is the degree of $4x^2 - 8x + 9x^5 - 6$?

For Exercises 18–24, simplify.

18. $|2 - 3(7)|$

19. $45 - (-52)$

20. $19 - 5(9)$

21. $3 + (-4)^3$

22. -2^4

23. $2(-9) - 20 \div (5)$

24. $7^2 + [8 - 5(6)][6 + 2(7)]$

25. Evaluate $2x - 5\sqrt{x + y}$ when $x = 9$ and $y = 16$.

26. Combine like terms and write your answer in descending order.

$$7t^3 - 10t + 14t^2 - 7 + t^3 - 15t$$

27. Subtract. $(10x^3 - 7x^2 - 15) - (4x^3 + x - 8)$

28. Multiply. $(-6a^4)(9a^2b)$

29. Multiply. $(3x - 5)(3x + 5)$

30. Find the prime factorization of 420.

31. Find the GCF of $36m^2n$ and $72m^5$.

32. Divide. $28k^6 \div (-4k^5)$

33. Factor. $20n^5 + 15mn^3 - 10n^2$

For Exercises 34–38, solve and check.

34. $x - 11 = 28$

35. $-7t = 42$

36. $2b - 14 = 30$

37. $9y - 13 = 4y + 7$

38. $6(n + 2) = 3n - 9$

For Exercises 39–50, solve.

39. In training for a new job, Carlos must log 500 hours of training. He has completed 278 hours. How many more hours must he log to complete the training?

40. Find the area of the shape.

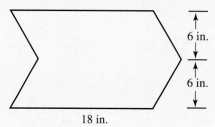

18 in.

6 in.

6 in.

41. The following table lists the assets and debts for the Goodman family. Calculate their net worth.

Assets	Debts
Savings = $985	Credit-card balance = $2345
Checking = $1862	Mortgage = $75,189
Furniture = $12,006	Automobile #1 = $4500

42. The voltage in a circuit measures -60 V. The resistance measures 15 Ω. What is the current? (Use $V = ir$.)

43. In one month, a company's total revenue was $96,408. If the profit was $45,698, find the cost.

44. A small storage building is a box that has a length of 10 ft. and a width of 9 ft. If the volume of the building is 630 ft.^{3}, what is the height?

45. Write an expression in simplest form for the area.

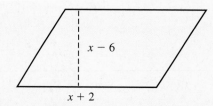

$x - 6$

$x + 2$

46. A wooden jewelry box has a lid that is 8 in. by 9 in. The surface area of the box is 314 in.^{2}. What is the height of the box?

47. Adrian drives for 45 min., stops for a break, then drives for 1 hr. and 15 min. to reach his destination. If the total trip was 124 miles, what was his average speed in mph? (Use $d = rt$.)

48. The length of a rectangle is 4 ft. more than the width. If the perimeter is 44 ft., find the length and width.

49. The sum of two consecutive odd integers is 92. Find the integers.

50. Kedra sells two sizes of hand-made blankets. The large size sells for $75 and the small size for $45. If she sold 16 blankets at a crafts fair and made a total of $930, how many of each size did she sell? (Use a four-column table.)

Fractions and Rational Expressions

Taking a Test

If you've prepared as suggested in the *To the Student* section of the text, then the key to success on the test is to relax and trust in your preparation. When taking the test, do the same things that you practiced. When you get the test, do not look at it. Instead, duplicate your study sheet on your scratch paper, just as you've practiced. This way you don't have to worry about forgetting rules as you take the test.

As you work through the test, work all the problems that you are sure about first. Show all your steps and work neatly so that your instructor can see everything you did in solving the problems. Instructors that give partial credit are much more likely to do so if they can follow the steps. After you've worked your way through the entire test, go back and work any problems that you weren't sure about. Always write something down for every problem, even if you have to guess.

Finally, use all the time you are given. If you finish early, take a break from looking at the test and then go through the entire test again. You may find it helpful to repeat your work on separate paper, then compare your solutions. Keep checking until time is up.

"You play the way you practice."
—POP WARNER

"When you're prepared you're more confident. When you have a strategy you're more comfortable."
—FRED COUPLES

5.1 Fractions, Mixed Numbers, and Rational Expressions

OBJECTIVES

1 Name the fraction represented by a shaded region.
2 Graph fractions on a number line.
3 Simplify fractions.
4 Write equivalent fractions.
5 Use $<$, $>$, or $=$ to make a true statement.
6 Write improper fractions as mixed numbers.
7 Write mixed numbers as improper fractions.

OBJECTIVE 1 *Name the fraction represented by a shaded region.*

A commercial farmer has one lot of land divided into 5 fields of equal size. Three of the fields are planted while the other 2 fields are left unplanted. In the picture, the 3 shaded regions represent the 3 planted fields and the other 2 regions represent the fields that are not planted.

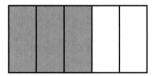

We can make the following statements about the fields:

3 out of the 5 fields are planted.

2 out of the 5 fields are not planted.

We can express the amount of all fields that are planted or not planted as parts of a whole. When we express a part of a whole, we are expressing a **fraction**.

DEFINITION	**Fraction:** A number that describes a part of a whole.

In fraction notation, the 3 fields planted out of the 5 total fields is written.

Fraction notation: $\dfrac{3}{5}$ ← Numerator
← Denominator

$\frac{3}{5}$ is read *three fifths* and it means 3 out of 5.

DEFINITION	**Numerator:** The number written in the top position in a fraction.
	Denominator: The number written in the bottom position in a fraction.

The denominator describes the total number of equal-sized divisions in the whole. The numerator describes the number of those divisions that we are interested in. If the numerator and denominator are integers, then we say the number is a **rational number.**

DEFINITION	**Rational number:** A number that can be expressed as a ratio of integers.

In math language: A number that can be expressed in the form $\frac{a}{b}$ where a and b are integers and $b \neq 0$.

EXAMPLE 1 Name the fraction represented by the shaded region.

a.

Answer: $\dfrac{7}{15}$

Explanation: There are 15 equal-size divisions in the whole region and 7 are shaded.

b.

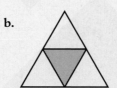

Answer: $\dfrac{3}{8}$

◆ **Explanation:** There are 8 equal-size divisions in the whole region and 3 are shaded.

Do Margin 1. ▶

EXAMPLE 2 In a group of 35 people at a conference, 17 are wearing glasses. What fraction of the people at the conference are wearing glasses? What fraction are not wearing glasses?

Answer: The fraction of the people at the conference that are wearing glasses is $\dfrac{17}{35}$. The fraction of the people at the conference that are not wearing glasses is $\dfrac{18}{35}$.

Explanation: Because a person can either wear glasses, or not wear glasses, there are only two possible categories. This means that if 17 out of the 35 people are wearing glasses, then everyone else is not. We can subtract 17 from 35 to get the number of people that are not wearing glasses. There are $35 - 17 = 18$ people who are not wearing glasses out of the 35 total people, therefore the fraction is $\dfrac{18}{35}$.

Do Margin 2. ▶

OBJECTIVE 2 *Graph fractions on a number line.*

We can graph fractions on a number line. Think of the number line as a ruler. Rulers and measuring sticks have whole inches marked off, yet also have the inches broken into halves, quarters, eighths, and even sixteenths.

EXAMPLE 3 Graph the fraction on a number line.

a. $\dfrac{3}{4}$

Solution:

Explanation: Because the denominator is 4, we draw marks to divide the distance between 0 and 1 into 4 equal spaces. Then go to the 3rd mark to the right of 0.

Discussion Where are $\dfrac{2}{4}$ and $\dfrac{4}{4}$?

Margin 1

Name the fraction represented by the shaded region.

a.

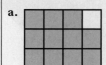

b.

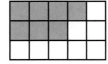

◆ **Margin 2**

a. According to a pamphlet, 1 in 5 teenagers smoke cigarettes. What fraction of teenagers smoke?

b. A football quarterback throws 28 passes during a game and 19 of those passes are completed. What fraction of the quarterback's passes were complete? What fraction were incomplete?

c. Margaret spends 8 hours each day at her office. Her company allows 1 of those hours to be a lunch break. What fraction of her time at the office is spent at lunch? What fraction is spent working?

Answers to Margin 1: a. $\dfrac{11}{12}$ b. $\dfrac{1}{4}$

Answers to Margin 2: a. $\dfrac{1}{5}$
b. $\dfrac{19}{28}$; $\dfrac{9}{28}$ c. $\dfrac{1}{8}$; $\dfrac{7}{8}$

Graph each fraction on a number line.

a. $\dfrac{1}{4}$

b. $\dfrac{5}{6}$

c. $-\dfrac{3}{8}$

d. $\dfrac{-4}{5}$

b. $\dfrac{-5}{8}$

Solution:

$-1 \quad -\frac{5}{8} \quad\quad 0$

Explanation: Because the denominator is 8, we draw marks to divide the distance between 0 and -1 into 8 equal spaces and go to the 5th mark to the left of 0.

Discussion Where are $-\dfrac{2}{8}$, $-\dfrac{4}{8}$, $-\dfrac{6}{8}$, and $-\dfrac{8}{8}$?

Note: We can write negative fractions three ways; the negative sign can be placed to the left of the fraction bar, to the left of the numerator, or to the left of the denominator.

$$-\dfrac{5}{8} = \dfrac{-5}{8} = \dfrac{5}{-8}$$

Though placing the negative sign to the left of the denominator is legal, it is considered poor form.

Connection
Recall that a fraction bar also indicates division. When dividing two numbers that have different signs, the quotient is always negative.

$$\dfrac{-6}{2} = -3 \text{ and } \dfrac{6}{-2} = -3$$

◀ **Do Margin 3.**

OBJECTIVE 3 *Simplify fractions.*

In the discussion questions in Example 3, you were asked to consider where fractions like $\dfrac{4}{4}$ and $\dfrac{2}{4}$ would be on a number line. The purpose in those questions was to get you thinking about how we might **simplify** fractions to **simplest form.**

DEFINITION **Simplify:** Write an equivalent expression with fewer symbols or smaller numbers.

Simplest form: An expression equivalent to a given expression with the fewest symbols and the smallest numbers possible.

Simplify.

a. $\dfrac{257}{1}$

b. $\dfrac{-9}{1}$

What if the denominator is 1?

EXAMPLE 4 Simplify $\dfrac{5}{1}$

Answer: $\dfrac{5}{1} = 5$

Connection
Remember the fraction line also means divide. So $\dfrac{5}{1}$ means $5 \div 1$ which is 5.

Explanation: $\dfrac{5}{1}$ is read as "five ones." Five ones can be interpreted as five wholes. On a ruler it would mean 5 inches.

Rule If the denominator of a fraction is 1, then the fraction can be simplified to the numerator.

$$\dfrac{n}{1} = n$$

Answers to Margin 3:

a.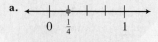

$0 \quad \frac{1}{4} \quad\quad 1$

b.

$0 \quad\quad\quad \frac{5}{6} \quad 1$

c.

$-1 \quad\quad -\frac{3}{8} \quad 0$

d.

$-1 \quad \frac{-4}{5} \quad\quad 0$

◀ **Do Margin 4.**

What if the numerator is 0?

EXAMPLE 5 Simplify $\dfrac{0}{9}$.

Answer: $\dfrac{0}{9} = 0$

Answers to Margin 4: **a.** 257 **b.** -9

Explanation: Suppose we divided a picture into 9 equal-size pieces, then didn't shade any pieces. Because the fraction represents the shaded region and there is nothing shaded, the fractional amount is $\frac{0}{9}$. This fraction is equal to 0.

> **Rule** If the numerator of a fraction is 0 and the denominator is any number other than 0, then the fraction can be simplified to 0.
>
> **In math language:** $\frac{0}{n} = 0$ as long as $n \neq 0$

> **Connection**
> This same rule appeared in Section 1.4 when we learned about dividing whole numbers. We said that 0 divided by any number other than 0 is 0. In symbolic form:
>
> **In math language:** $0 \div n = 0$ when $n \neq 0$

Do Margin 5. ▶

What if the denominator is 0?

EXAMPLE 6 Simplify $\frac{5}{0}$.

Answer: $\frac{5}{0}$ is undefined.

Explanation: Because $\frac{5}{0}$ means $5 \div 0$ we should be able to check the answer by multiplying the quotient by the divisor, in this case 0, and get the dividend, 5.

Check: quotient · divisor = dividend

$$? \cdot 0 = 5$$

Any number multiplied by 0 should make a product of 0. There's no way to get 5 by multiplying a number by 0 because there is no quotient that will satisfy $\frac{5}{0}$. Therefore $\frac{5}{0}$ is undefined.

> **Rule** If the denominator is 0 and the numerator is any number other than 0, we say the fraction is undefined.
>
> **In math language:** $\frac{n}{0}$ is undefined when $n \neq 0$

> **Connection**
> We developed this same rule in Section 1.4 as well. We said that any number other than 0 divided by 0 is undefined.
>
> **In math language:** $n \div 0$ is undefined when $n \neq 0$

Do Margin 6. ▶

What if the numerator and denominator contain the same number?

EXAMPLE 7 Simplify $\frac{4}{4}$.

Answer: $\frac{4}{4} = 1$

Explanation: This answers the discussion question from Example 3. On a number line, $\frac{4}{4}$ is the same as 1.

◆ **Margin 5**

Simplify.

a. $\frac{0}{19}$

b. $\frac{0}{-8}$

c. $-\frac{0}{11}$

◆ **Margin 6**

Simplify.

a. $\frac{97}{0}$

b. $\frac{-13}{0}$

c. $-\frac{49}{0}$

Answers to Margin 5: **a.** 0 **b.** 0 **c.** 0

Answers to Margin 6: **a.** undefined **b.** undefined **c.** undefined

Or, think about money. When we have 4 fourths, or 4 quarters, how much money do we have? Four quarters make a whole dollar. We could also draw a picture:

If we divide a whole into 4 equal parts, then shade all 4, we have shaded the whole picture.

> **Rules** A fraction with the same numerator and denominator (other than zero) can be simplified to 1.
>
> **In math language:** $\dfrac{n}{n} = 1$ as long as $n \neq 0$
>
> Zero over zero is indeterminate.
>
> **In math language:** $\dfrac{0}{0}$ is indeterminate

◀ **Do Margin 7.**

OBJECTIVE 4 *Write equivalent fractions.*

Let's return to the discussion questions from Example 3. The fractions $\frac{2}{4}$ and $\frac{4}{8}$ are both halfway between 0 and 1. In other words, they're both equivalent to the fraction $\frac{1}{2}$.

$$\frac{4}{8} = \frac{2}{4} = \frac{1}{2}$$

When two fractions name the same number, we say they are **equivalent fractions.**

| **DEFINITION** | **Equivalent fractions:** Fractions that name the same number. |

What other equivalent fractions can we see using a ruler?

The rulers reaffirm our rule that fractions with the same nonzero denominator and numerator are equal to 1.

$$1 = \frac{2}{2} = \frac{4}{4} = \frac{8}{8}$$

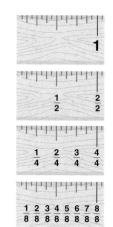

Look at how we can express $\frac{1}{2}$ in various ways.

$$\frac{1}{2} = \frac{2}{4} = \frac{4}{8}$$

We can express the quarters in terms of eighths.

$$\frac{1}{4} = \frac{2}{8} \quad \text{and} \quad \frac{3}{4} = \frac{6}{8}$$

▸ **Discussion** How many sixths would be the same as a half?

How can we get from one equivalent fraction to another? How can we get from $\frac{1}{2}$ to $\frac{2}{4}$ or $\frac{4}{8}$? Notice the numerator and denominator of $\frac{1}{2}$ can be multiplied by 2 to get the equivalent fraction $\frac{2}{4}$.

◆ **Margin 7**

Simplify.

a. $\dfrac{59}{59}$

b. $\dfrac{-7}{-7}$

Answers: **a.** 1 **b.** 1

$$\frac{1 \cdot 2}{2 \cdot 2} = \frac{2}{4}$$

We could multiply both the numerator and denominator of $\frac{1}{2}$ by 4 to get the equivalent fraction $\frac{4}{8}$.

$$\frac{1 \cdot 4}{2 \cdot 4} = \frac{4}{8}$$

Conclusion: We can multiply both the numerator and denominator by the same nonzero number to create an equivalent fraction. This process is called **upscaling** a fraction.

DEFINITION   **Upscale:** To create an equivalent fraction by multiplying both numerator and denominator by the same number.

This process is reversible as well. We can divide both the numerator and denominator by the same nonzero number to create an equivalent fraction. This process is called **reducing** a fraction.

DEFINITION   **Reduce:** To create an equivalent fraction by dividing both the numerator and denominator by the same number.

To reduce $\frac{4}{8}$ we can divide both the numerator 4 and the denominator 8 by 2 to get $\frac{2}{4}$ or divide them both by 4 to get $\frac{1}{2}$.

$$\frac{4 \div 2}{8 \div 2} = \frac{2}{4} \quad \text{and} \quad \frac{4 \div 4}{8 \div 4} = \frac{1}{2}$$

Property of Equivalent Fractions We can write an equivalent fraction by multiplying or dividing both the numerator and denominator by the same number.

Suppose we wanted to write a fraction equivalent to $\frac{1}{2}$ but with a denominator of 10. This can be written as the following missing number statement:

$$\frac{1}{2} = \frac{?}{10}$$

We could upscale by multiplying the numerator and denominator by 5 to make the equivalent fraction.

$$\frac{1 \cdot 5}{2 \cdot 5} = \frac{5}{10}$$

EXAMPLE 8   Fill in the blank so that the fractions are equivalent.

$$\frac{3}{8} \quad \frac{?}{16}$$

Solution: Upscale $\frac{3}{8}$ by multiplying the numerator and denominator by 2.

$$\frac{3 \cdot 2}{8 \cdot 2} = \frac{6}{16}$$

Do Margin 8. ▶

◆ Margin 8

Fill in the blank so that the fractions are equivalent.

a. $\dfrac{5}{9} = \dfrac{?}{36}$

b. $-\dfrac{3}{4} = \dfrac{?}{12}$

Answers: **a.** 20 **b.** −9

EXAMPLE 9 Fill in the blank so that the fractions are equivalent.

$$\frac{20}{24} = \frac{5}{?}$$

Solution: Reduce $\frac{20}{24}$ by dividing the numerator and denominator by 4.

$$\frac{20 \div 4}{24 \div 4} = \frac{5}{6}$$

◀ **Do Margin 9.**

OBJECTIVE **5** *Use $<$, $>$, or $=$ to make a true statement.*

It's easy to compare fractions that have the same denominator by simply looking at the numerators. For example, $\frac{2}{5}$ is less than $\frac{3}{5}$ because 2 is less than 3. In symbols we write

$$\frac{2}{5} < \frac{3}{5}$$

What if the fractions do not have the same denominator?

Consider $\frac{1}{2}$ and $\frac{1}{3}$. We could draw pictures to compare.

From the pictures we see that $\frac{1}{2}$ is more than $\frac{1}{3}$. In symbols we write:

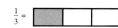

$$\frac{1}{2} > \frac{1}{3}$$

Mathematically, we could upscale each fraction to equivalent fractions with a common denominator, then compare numerators. But what would be a common denominator for 2 and 3? Because we are going to multiply both fractions to upscale them, the common denominator must have both original denominators as factors. That is, the common denominator must be a number that is divisible by both original denominators. When a number is divisible by a given number we say it is a **multiple** of the given number.

DEFINITION	**Multiple:** A number that is divisible by a given number.

Multiples of 2 would be 2, 4, 6, 8, 10, . . .

Multiples of 3 would be 3, 6, 9, 12, 15, . . .

Notice a common multiple for 2 and 3 is 6, which we can get by multiplying 2 times 3. To upscale $\frac{1}{2}$, we multiply the numerator and denominator by 3. To upscale $\frac{1}{3}$, we multiply the numerator and denominator by 2.

$$\frac{1}{2} = \frac{1 \cdot 3}{2 \cdot 3} = \frac{3}{6} \qquad \text{and} \qquad \frac{1}{3} = \frac{1 \cdot 2}{3 \cdot 2} = \frac{2}{6}$$

Notice $\frac{3}{6}$ is larger than $\frac{2}{6}$.

> **Procedure** *To compare two fractions:*
>
> 1. Upscale or reduce both fractions to equivalent fractions that have a common denominator.
> 2. Compare.

We will discuss common multiples in more detail later, but for now, a simple way to create a common multiple is to multiply the numbers as we did with 2 and 3 to get 6.

Margin 9

Fill in the blank so that the fractions are equivalent.

a. $\frac{28}{35} = \frac{4}{?}$

b. $\frac{-24}{42} = \frac{?}{7}$

Answers: **a.** 5 **b.** -4

EXAMPLE 10 Use $<$, $>$, or $=$ to write a true statement.

$$\frac{5}{8} ? \frac{4}{7}$$

Solution: Upscale both fractions to equivalent fractions that have a common denominator, then compare numerators.

A common multiple for 8 and 7 is 56.

$$\frac{5}{8} = \frac{5 \cdot 7}{8 \cdot 7} = \frac{35}{56}$$

$$\frac{4}{7} = \frac{4 \cdot 8}{7 \cdot 8} = \frac{32}{56}$$

Because 35 is larger than 32, we can say that $\frac{5}{8}$ is larger than $\frac{4}{7}$.

Answer: $\frac{5}{8} > \frac{4}{7}$

In our method of comparing fractions we find a common denominator by multiplying the denominators. As a result, when we upscale, the new numerator is the product of the original numerator and the other fraction's denominator. Because we really only need to compare the numerators, we can get those numerators quickly by doing a process called *cross multiplying*. Consider Example 10 again.

$$7 \cdot 5 = 35 \qquad 8 \cdot 4 = 32$$

$$\frac{5}{8} \bigtimes \frac{4}{7}$$

35 and 32 are the cross products for these fractions. Notice that 35 and 32 are the numerators that we get when we upscale $\frac{5}{8}$ and $\frac{4}{7}$ to fractions with 56 as the common denominator. Putting the denominators back into the picture, we can write the process this way:

$$\frac{35}{56} = \frac{5 \cdot 7}{8 \cdot 7} = \frac{5}{8} \bigtimes \frac{4}{7} = \frac{4 \cdot 8}{7 \cdot 8} = \frac{32}{56}$$

35 is the larger of the cross products. Because it is the numerator that we get when we upscale $\frac{5}{8}$, we can conclude that $\frac{5}{8}$ is the larger fraction.

Conclusion: To compare two fractions, we can compare their cross products. The larger cross product will indicate the larger fraction. Also, if the cross products are equal, so are the fractions.

EXAMPLE 11 Use $<$, $>$, or $=$ to write a true statement.

$$\frac{-7}{8} ? \frac{-9}{11}$$

Solution: Upscale both fractions to equivalent fractions that have a common denominator, then compare numerators.

A common denominator for 8 and 11 is $8 \cdot 11 = 88$.

$$\frac{-7}{8} = \frac{-7 \cdot 11}{8 \cdot 11} = \frac{-77}{88}$$

$$\frac{-9}{11} = \frac{-9 \cdot 8}{11 \cdot 8} = \frac{-72}{88}$$

> **Reminder**
> When comparing two negative numbers, the number with the smaller absolute value is the larger number.

Because -72 is closer to zero than -77, -72 is the larger of the two numerators. Therefore $\frac{-72}{88}$ or $\frac{-9}{11}$ is the larger fraction.

Answer: $\dfrac{-7}{8} < \dfrac{-9}{11}$

Using the cross products method, we get the same result:

$$\frac{-77}{88} = \frac{-7 \cdot 11}{8 \cdot 11} = \frac{-7}{8} \times \frac{-9}{11} = \frac{-9 \cdot 8}{11 \cdot 8} = \frac{-72}{88}$$

-72 is a larger number than -77. Because -72 is the numerator for $\dfrac{-9}{11}$, we conclude that $\dfrac{-9}{11}$ is the larger fraction.

◀ **Do Margin 10.**

OBJECTIVE 6 *Write improper fractions as mixed numbers.*

If the absolute value of the numerator of a fraction is greater than or equal to the absolute value of the denominator, we call the fraction an **improper fraction.**

DEFINITION **Improper fraction:** A fraction in which the absolute value of the numerator is greater than or equal to the absolute value of the denominator.

Discussion Why do we have to say "absolute value?" Why can't we just say an improper fraction is a fraction with a numerator greater than the denominator?

$\dfrac{9}{4}$ is an improper fraction because $|9| = 9$ is greater than $|4| = 4$.

$\dfrac{-7}{2}$ is an improper fraction because $|-7| = 7$ is greater than $|2| = 2$.

$\dfrac{8}{8}$ is improper because $|8| = 8$ is equal to $|8| = 8$.

What does $\dfrac{9}{4}$ mean?

Think about money. *One fourth* in money terms is a quarter. If we had 9 quarters and every 4 quarters is a dollar, then we have 2 dollars plus 1 quarter left over. Let's translate this into mathematical terms:

$$9 \text{ quarters} = 2 \text{ whole dollars} + 1 \text{ quarter}$$

$$\frac{9}{4} = 2 + \frac{1}{4}$$

The customary notation for this result is to leave out the plus sign and write $2\frac{1}{4}$. So, what do we call the $2\frac{1}{4}$? An integer combined with a fraction is called a **mixed number.**

DEFINITION **Mixed number:** An integer combined with a fraction.

When we say combined, we literally mean *added*.

$$2\frac{1}{4} = 2 + \frac{1}{4}$$ $2\frac{1}{4}$ is read as "two and one fourth" and means 2 wholes plus $\frac{1}{4}$ of another whole.

How does this apply to negative mixed numbers?

$$-3\frac{5}{8} = -\left(3\frac{5}{8}\right) = -\left(3 + \frac{5}{8}\right) = -3 - \frac{5}{8}$$

The negative sign applies to both the integer and the fraction. In terms of a number line we are saying go left 3 and then left another $\frac{5}{8}$.

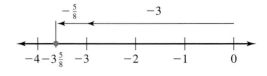

◆ **Margin 10**

Use $<$, $>$, or $=$ to write a true statement.

a. $\dfrac{3}{7}$? $\dfrac{4}{9}$

b. $\dfrac{5}{11}$? $\dfrac{4}{9}$

c. $-\dfrac{2}{7}$? $-\dfrac{12}{42}$

d. $\dfrac{-9}{10}$? $-\dfrac{8}{9}$

Answers: **a.** $\dfrac{3}{7} < \dfrac{4}{9}$ **b.** $\dfrac{5}{11} > \dfrac{4}{9}$
c. $-\dfrac{2}{7} = -\dfrac{12}{42}$ **d.** $\dfrac{-9}{10} < -\dfrac{8}{9}$

How do we express an improper fraction as a mixed number? What did we do when we expressed $\frac{9}{4}$ as a mixed number? We saw how many groups of 4 were in 9; this is a division operation.

> **Procedure** *To write an improper fraction as a mixed number:*
>
> 1. Divide the denominator into the numerator.
> 2. Write the results in the form:
>
> $$\text{quotient} \; \frac{remainder}{original \; denominator}$$

EXAMPLE 12 Write $\frac{20}{3}$ as a mixed number.

Solution: Divide 20 by 3 and write the quotient and remainder this way:

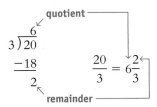

Connection

$\frac{20}{3}$ is the same as $20 \div 3$. Mixed numbers are another way to write quotients that have remainders.

Instead of 6 r2 we can now write $6\frac{2}{3}$.

EXAMPLE 13 Write $-\frac{41}{7}$ as a mixed number.

Solution: The negative sign is written with the integer in the mixed number. Divide 41 by 7 and write the quotient and remainder this way:

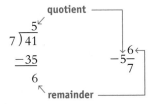

Do Margin 11. ▶

EXAMPLE 14 Divide $-2035 \div -19$ and write the quotient as a mixed number.

Solution: Because dividing two numbers with the same sign equals a positive number, we can divide and write the quotient as a positive mixed number.

$$
\begin{array}{r}
107 \\
19\overline{)2035} \\
-19 \\
\hline
13 \\
-0 \\
\hline
135 \\
-133 \\
\hline
2
\end{array}
$$

Write the remainder in the numerator.

Answer: $107\frac{2}{19}$ ◀──── Write the divisor in the denominator.

Write the quotient as the whole/integer.

Do Margin 12. ▶

◆ **Margin 11**

Write each improper fraction as a mixed number.

a. $\frac{13}{3}$

b. $\frac{32}{5}$

c. $-\frac{37}{4}$

d. $\frac{-79}{10}$

◆ **Margin 12**

Divide and write the quotient as a mixed number.

a. $49 \div 2$

b. $139 \div 13$

c. $-920 \div 9$

d. $-3249 \div -16$

Answers to Margin 11: **a.** $4\frac{1}{3}$ **b.** $6\frac{2}{5}$ **c.** $-9\frac{1}{4}$ **d.** $-7\frac{9}{10}$

Answers to Margin 12: **a.** $24\frac{1}{2}$ **b.** $10\frac{9}{13}$ **c.** $-102\frac{2}{9}$ **d.** $203\frac{1}{16}$

OBJECTIVE 7 *Write mixed numbers as improper fractions.*

In Example 12, we wrote $\frac{20}{3}$ as $6\frac{2}{3}$. How can we verify that $6\frac{2}{3}$ is correct? Because we divided 20 by 3 to get $6\frac{2}{3}$, we can reverse the process. We know that we can check division by multiplying.

$$\text{Quotient} \cdot \text{Divisor} + \text{Remainder} = \text{Dividend}$$

6	·	3	+	2	?	20	
		18	+	2	?	20	
				20	=	20	Yes, it does check.

In mixed number format, it looks like this:

1. Multiply 3 · 6 = 18   **2. Add product to numerator: 18 + 2 = 20**

$$6\frac{2}{3} = \frac{3 \cdot 6 + 2}{3} = \frac{20}{3}$$

3. Keep the same denominator.

Procedure *To write a mixed number as an improper fraction:*

1. Multiply the denominator by the integer.

2. Add the resulting product to the numerator. This resulting sum will be the numerator of the improper fraction.

3. Keep the same denominator.

EXAMPLE 15 Write $5\frac{7}{8}$ as an improper fraction.

Solution: $5\frac{7}{8} = \frac{8 \cdot 5 + 7}{8} = \frac{40 + 7}{8} = \frac{47}{8}$ **Note** We have shown the steps. However, most people perform the multiplication and addition mentally.

Explanation: We multiplied 8 by 5 to get 40, then added 7 to get 47 as the numerator of the improper fraction. We kept the denominator the same.

What if the mixed number is negative?

EXAMPLE 16 Write $-9\frac{3}{7}$ as an improper fraction.

Solution: $-9\frac{3}{7} = -\frac{7 \cdot 9 + 3}{7} = -\frac{63 + 3}{7} = -\frac{66}{7}$

Explanation: Remember the negative sign just means that the amount is to the left of 0 on a number line. We simply followed the same process and wrote the negative sign with the fraction. We multiplied 7 by 9 to get 63, then added 3 to get 66 for the numerator.

WARNING Do not make the mistake of thinking that we multiply 7 times −9 to get −63, then add 3 to get −60. Think of the negative sign as applying to all of the mixed number.

$$-9\frac{3}{7} = -\left(9\frac{3}{7}\right) = -\frac{66}{7}$$

◁ **Do Margin 13.**

◆ Margin 13

Write as an improper fraction.

a. $10\frac{5}{6}$

b. $-11\frac{2}{9}$

c. 19

Answers: **a.** $\frac{65}{6}$ **b.** $-\frac{101}{9}$ **c.** $\frac{19}{1}$

5.1 Exercises

For Exercises 1–6, name the fraction represented by a shaded region.

1.

2.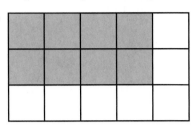

3.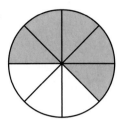

4.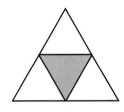

5.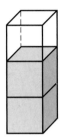

6.

For Exercises 7–12, write the fraction for each situation.

7. **a.** In an experiment, a subject is blindfolded and her nose pinned so that she cannot smell. She could accurately identify 5 out of 16 foods by taste. What fraction did she accurately identify?
   b. Still blindfolded but without her nose pinned, she could then identify 15 out of 16 foods. What fraction did she accurately identify with just the blindfold?
   c. What can we conclude from the experiment?
   d. Are there any flaws in the experiment?

8. **a.** In a memory experiment, a subject is shown 20 words in sequence. Each word is shown for 3 seconds. The subject is asked to write down all the words he can remember in any order. The subject was able to write down 11 words. What fraction of the 20 words was he able to write down?
   b. The experiment is repeated with new words and the subject is asked to write the words in the order in which they appeared. The subject was able to write 5 in order. What fraction did he get correct?
   c. What can you conclude from the experiment?
   d. Are there any flaws in this experiment?

9. A certain fish lays 800 eggs out of which only 17 survive to adulthood. What fraction describes the portion that survive to adulthood? What fraction describes the portion that do not survive to adulthood?

10. At a certain university, of the 179 entering students that declared premed as their major, only 15 completed the premed program. What fraction completed the program? What fraction did not? What can you conclude about this program?

11. Of the 258 students that have ever taken math under Mrs. Jones, 249 passed. What fraction passed? What fraction did not pass? What can you conclude about taking math with Mrs. Jones as the instructor?

12. An ad claims that a certain medicine is safe according to the FDA. You read an article that claims the medicine caused cancer in laboratory rats. As you read, you see that 9 out of 50 rats got cancer. What fraction of the rats got cancer? What can you conclude about the claims of the FDA vs. the claims of the ad?

For Exercises 13–20, graph the fraction on a number line.

13. $\dfrac{1}{4}$

14. $\dfrac{3}{4}$

15. $\dfrac{5}{6}$

16. $\dfrac{2}{3}$

17. $-\dfrac{5}{8}$

18. $-\dfrac{1}{2}$

19. $\dfrac{-3}{16}$

20. $\dfrac{9}{-10}$

For Exercises 21–28, simplify.

21. $\dfrac{23}{1}$

22. $\dfrac{19}{19}$

23. $\dfrac{0}{16}$

24. $\dfrac{12}{0}$

25. $\dfrac{-6}{-6}$

26. $\dfrac{-45}{1}$

27. $\dfrac{-2}{0}$

28. $\dfrac{0}{-18}$

For Exercises 29–36, fill in the blank so that the fractions are equivalent.

29. $\dfrac{5}{9} = \dfrac{?}{27}$

30. $\dfrac{3}{5} = \dfrac{?}{35}$

31. $\dfrac{21}{36} = \dfrac{?}{12}$

32. $\dfrac{12}{42} = \dfrac{?}{7}$

33. $-\dfrac{9}{15} = -\dfrac{18}{?}$

34. $-\dfrac{28}{36} = -\dfrac{7}{?}$

35. $\dfrac{-6}{16} = \dfrac{?}{80}$

36. $-\dfrac{24}{60} = \dfrac{2}{?}$

For Exercises 37–44, use $<$, $>$, or $=$ to make a true statement.

37. $\dfrac{4}{9}$? $\dfrac{2}{5}$

38. $\dfrac{25}{30}$? $\dfrac{5}{6}$

39. $\dfrac{12}{18}$? $\dfrac{9}{16}$

40. $\dfrac{5}{7}$? $\dfrac{8}{11}$

41. $-\dfrac{4}{15}$? $-\dfrac{6}{17}$

42. $\dfrac{-2}{9}$? $\dfrac{-5}{23}$

43. $-\dfrac{9}{12}$? $-\dfrac{15}{20}$

44. $\dfrac{-7}{8}$? $\dfrac{9}{-10}$

For Exercises 45–52, write the improper fraction as a mixed number.

45. $\dfrac{30}{7}$

46. $\dfrac{52}{9}$

47. $\dfrac{85}{4}$

48. $\dfrac{19}{2}$

49. $\dfrac{-64}{5}$

50. $-\dfrac{97}{6}$

51. $\dfrac{103}{-8}$

52. $\dfrac{-111}{20}$

For Exercises 53–60, write the mixed number as an improper fraction.

53. $5\dfrac{1}{6}$

54. $4\dfrac{3}{8}$

55. 11

56. $13\dfrac{1}{2}$

57. $-9\dfrac{7}{8}$

58. -24

59. $-1\dfrac{9}{20}$

60. $-15\dfrac{3}{4}$

REVIEW EXERCISES

1. Divide. $-48 \div (-12)$

2. Find the prime factorization of 840.

3. Find the GCF of 24 and 60.

4. Find the GCF of $40x^5y$ and $56x^2$.

5. Factor $40x^5y - 56x^2$.

5.2 Reducing Fractions and Rational Expressions

OBJECTIVES

1 Reduce fractions to lowest terms.

2 Reduce fractions within mixed numbers.

3 Reduce rational expressions.

OBJECTIVE 1 *Reduce fractions to lowest terms.*

In Section 5.1, we introduced and defined upscaling and reducing fractions. We discovered that we can make an equivalent fraction by multiplying or dividing both the numerator and denominator by the same amount. In this section we will focus on reducing fractions. Notice that we can reduce the fraction $\frac{4}{8}$ two ways.

$$\frac{4 \div 2}{8 \div 2} = \frac{2}{4} \quad \text{and} \quad \frac{4 \div 4}{8 \div 4} = \frac{1}{2}$$

We can reduce a fraction by dividing the numerator and denominator by *any* common factor. However, notice that $\frac{2}{4}$ can be reduced further by dividing both 2 and 4 by the common factor of 2, which yields $\frac{1}{2}$. Notice the 1 and 2 in $\frac{1}{2}$ have no common factors other than the number 1. The fraction $\frac{1}{2}$ is in **lowest terms**.

DEFINITION   **Lowest terms:** A fraction is in lowest terms when the greatest common factor for the numerator and denominator is 1.

> **Reminder**
> The greatest common factor is the largest number that divides all the given numbers with no remainder.

It is an unwritten rule of etiquette to always express results in lowest terms. Any time you are working with fractions, you are expected to write results in lowest terms.

$\frac{1}{2}$ is in lowest terms because the greatest common factor for 1 and 2 is 1.

$\frac{2}{4}$ is not in lowest terms because the greatest common factor for 2 and 4 is 2.

Conclusion: To write a fraction in lowest terms, we must divide both the numerator and denominator by their greatest common factor.

> **Rule** To write a fraction in lowest terms, we must divide the numerator and denominator by their greatest common factor.

EXAMPLE 1   Reduce $\frac{18}{24}$ to lowest terms.

Solution: The greatest common factor for 18 and 24 is 6, so we divide both 18 and 24 by 6.

$$\frac{18 \div 6}{24 \div 6} = \frac{3}{4}$$

Notice $\frac{3}{4}$ is in lowest terms because the greatest common factor for 3 and 4 is 1. In other words, 1 is the only number that will divide 3 and 4 evenly.

What if we did not see that the greatest common factor of 18 and 24 is 6? Suppose we just saw that 18 and 24 are both divisible by 2 and divided them both by 2.

$$\frac{18 \div 2}{24 \div 2} = \frac{9}{12}$$

But $\frac{9}{12}$ is not in lowest terms because 9 and 12 are both divisible by 3. Now we simply need to divide by the common 3.

$$\frac{9 \div 3}{12 \div 3} = \frac{3}{4}$$

Altogether, it looks like this:

$$\frac{18 \div 2}{24 \div 2} = \frac{9 \div 3}{12 \div 3} = \frac{3}{4}$$

Notice that we are still dividing by 6. We simply divided in stages. If we break down 6 to its prime factorization we have $6 = 2 \cdot 3$. We divided by the 2 factor first, then divided by the 3 factor. Whether you divide by the greatest common factor in one step or in stages you still get to lowest terms.

Recall that we can find the GCF of two numbers using their prime factorizations. Because we can use prime factorization to find the GCF, we could simply reduce by dividing out all primes that are common to the numerator and denominator. Consider $\frac{18}{24}$ once more. If we replace 18 and 24 with their prime factorizations we have:

$$\frac{18}{24} = \frac{2 \cdot 3 \cdot 3}{2 \cdot 2 \cdot 2 \cdot 3}$$

Recall that a number divided by itself simplifies to 1. We can divide a pair of 2's that are common in the numerator and denominator to equal 1. We can also divide out a pair of 3's in the same way.

$$\frac{18}{24} = \frac{\overset{1}{2} \cdot 3 \cdot \overset{1}{\cancel{3}}}{\underset{1}{2} \cdot 2 \cdot 2 \cdot \underset{1}{\cancel{3}}}$$

Connection
We are still dividing out 6 because we are dividing out the common 2 and common 3 that multiply to equal 6.

If we then multiply out the remaining factors, we have lowest terms.

$$\frac{18}{24} = \frac{\overset{1}{\cancel{2}} \cdot 3 \cdot \overset{1}{\cancel{3}}}{\underset{1}{\cancel{2}} \cdot 2 \cdot 2 \cdot \underset{1}{\cancel{3}}} = \frac{3}{4}$$

Procedure *To reduce a fraction to lowest terms:*
1. Find the prime factorization of the numerator and denominator.
2. Divide out all primes that are common to both the numerator and denominator.
3. Multiply the remaining factors.

EXAMPLE 2 Reduce $-\frac{72}{90}$ to lowest terms.

Solution: Find the prime factorization of 72 and 90.

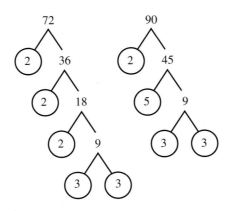

Divide out all primes that are common to both the numerator and denominator. Then multiply the remaining factors.

$$-\frac{72}{90} = -\frac{\overset{1}{\cancel{2}} \cdot 2 \cdot 2 \cdot \overset{1}{\cancel{3}} \cdot \overset{1}{\cancel{3}}}{\underset{1}{\cancel{2}} \cdot \underset{1}{\cancel{3}} \cdot \underset{1}{\cancel{3}} \cdot 5} = -\frac{1 \cdot 2 \cdot 2 \cdot 1 \cdot 1}{1 \cdot 1 \cdot 1 \cdot 5} = -\frac{4}{5}$$

Do Margin 1. ▶

EXAMPLE 3 According to a survey of 182 people, 112 responded that they had seen a particular movie. Write the lowest terms fraction of people in the survey who had not seen the movie. Interpret the results.

Solution: Follow the problem-solving process.

Understand: We must write a lowest terms fraction for the number of people who did not see the movie, then interpret the result.

Plan: Because 112 said they saw the movie, we can subtract 112 from 182 to get the number of people who had not seen the movie. We can then write the fraction and reduce.

Execute: $182 - 112 = 70$ people who did not see the movie. So the fraction of people who did not see it is $\frac{70}{182}$.

Now we need to reduce to lowest terms. We find the prime factorization of 70 and 182, divide out the common primes, then multiply the remaining factors.

$$\frac{70}{182} = \frac{\overset{1}{\cancel{2}} \cdot 5 \cdot \overset{1}{\cancel{7}}}{\underset{1}{\cancel{2}} \cdot \underset{1}{\cancel{7}} \cdot 13} = \frac{5}{13}$$

Answer: $\frac{5}{13}$ is in lowest terms because 1 is the only number that divides both 5 and 13 evenly.

Check: Because cross products are equal with equal fractions, we can verify that $\frac{5}{13}$ is the same as $\frac{70}{182}$ by cross multiplying.

$$182 \cdot 5 = 910 \qquad 13 \cdot 70 = 910$$

$$\frac{5}{13} \bowtie \frac{70}{182}$$

So $\frac{5}{13}$ is not only a lowest terms fraction, it is *the* lowest terms fraction equivalent to $\frac{70}{182}$.

♦ **Margin 1**

Reduce to lowest terms.

a. $\frac{30}{40}$

b. $\frac{48}{84}$

c. $-\frac{24}{54}$

d. $\frac{-140}{196}$

Answers: **a.** $\frac{3}{4}$ **b.** $\frac{4}{7}$ **c.** $-\frac{4}{9}$ **d.** $\frac{-5}{7}$

◆ **Margin 2**

In a survey of 300 people, 252 responded that they believed there is life elsewhere in the universe. Write in lowest terms the fraction of people in the survey who believe life exists elsewhere in the universe.

Discussion Do you think the result could be skewed?

◆ **Margin 3**

According to the NTSB (National Transportation Safety Board), in 1997 there were 2 fatalities out of 626,000,000 passengers that flew on any U.S. air carrier. Write the fraction in lowest terms.

◆ **Margin 4**

Write each fraction as a mixed number.

a. $\dfrac{28}{6}$

b. $\dfrac{80}{15}$

c. $-\dfrac{54}{24}$

d. $\dfrac{-180}{24}$

Interpreting the results: $\frac{5}{13}$ means 5 out of any 13 people in the survey did *not* see the movie. If the survey was given to people of differing ethnic groups, genders, and ages, then we should be able to ask any 13 people of those surveyed if they had seen the movie and about 5 would say no.

◆ **Note:** If a survey is not accurately conducted, we say the survey's results are *skewed*.

◀ **Do Margins 2 and 3.**

OBJECTIVE 2 *Reduce fractions within mixed numbers.*

What if a mixed number contains a fraction that can be reduced? Or, what if an improper fraction is not in lowest terms? Do we need to reduce before writing it as a mixed number? Does it matter?

EXAMPLE 4 Write $\frac{26}{12}$ as a mixed number.

Solution: We are always expected to write fractions in lowest terms. Reduce first.

$$\frac{26}{12} = \frac{\overset{1}{\cancel{2}} \cdot 13}{\underset{1}{\cancel{2}} \cdot 2 \cdot 3} = \frac{13}{6}$$

Now write $\frac{13}{6}$ as a mixed number.

$$\frac{13}{6} = 2\frac{1}{6}$$

Would the result be different if we wrote the mixed number first?

$$\frac{26}{12} = 2\frac{2}{12}$$

Notice the fraction $\frac{2}{12}$ can be reduced.

$$\frac{26}{12} = 2\frac{2}{12} = 2\frac{\overset{1}{\cancel{2}}}{\underset{1}{2 \cdot 2 \cdot 3}} = 2\frac{1}{6}$$

Conclusion: It does not affect the outcome if we reduce an improper fraction or wait to reduce the fraction part of the equivalent mixed number.

◀ **Do Margin 4.**

OBJECTIVE 3 *Reduce rational expressions.*

Let's extend what we've learned about reducing to fractions that contain variables as well as numbers. If the numerator and denominator both contain monomials or polynomials then we call the fraction a **rational expression**.

DEFINITION | **Rational expression:** An fraction that is a ratio of monomials or polynomials.

Answer to Margin 2: $\frac{21}{25}$

Answer to Margin 3: $\frac{1}{313,000,000}$

Answers to Margin 4: **a.** $4\frac{2}{3}$ **b.** $5\frac{1}{3}$
c. $-2\frac{1}{4}$ **d.** $-7\frac{1}{2}$

Connection

A rational number is a number that can be expressed as a ratio of integers, whereas a rational expression is a ratio of monomials or polynomials.

The rational expressions we will consider will only contain monomials.

EXAMPLE 5 Reduce $\frac{4x^3}{8x}$ to lowest terms.

Solution: Write the numerator and denominator in prime factored form, divide out all factors that are common to both the numerator and denominator, then multiply the remaining factors.

$$\frac{4x^3}{8x} = \frac{2 \cdot 2 \cdot x \cdot x \cdot x}{2 \cdot 2 \cdot 2 \cdot x}$$

Write the numerator and denominator in factored form.

$$= \frac{\overset{1}{\cancel{2}} \cdot \overset{1}{\cancel{2}} \cdot x \cdot x \cdot \cancel{x}}{\underset{1}{\cancel{2}} \cdot \underset{1}{\cancel{2}} \cdot 2 \cdot \underset{1}{\cancel{x}}}$$

Divide out two 2's and one x.

$$= \frac{x \cdot x}{2}$$

Multiply the remaining factors.

$$= \frac{x^2}{2}$$

Do Margin 5. ▶

Connection

x^3 means there are three x factors. If we divide out one x in the numerator and denominator, we have two x factors left. This goes back to the rule of exponents. When we divide exponential forms that have the same base, we can subtract the exponents.

Connection

We can use this idea of dividing in the numerator and denominator to better understand the rule we developed in Section 3.6 about dividing exponential forms with the same base. The rule said:

When dividing exponential forms that have the same base, we can subtract the divisor's exponent from the dividend's exponent and keep the same base.

$$\frac{2^3}{2} = 2^3 \div 2 = 2^{3-1} = 2^2$$

If we write all the factors and divide out the common 2 we have: $\dfrac{2^3}{2} = \dfrac{\overset{1}{\cancel{2}} \cdot 2 \cdot 2}{\underset{1}{\cancel{2}}} = \dfrac{2 \cdot 2}{1} = 2^2$

Conclusion: Subtracting exponents corresponds to dividing out common factors.

We can also get a better understanding of what happens when the exponents are the same.

$$\frac{3^2}{3^2} = 3^2 \div 3^2 = 3^{2-2} = 3^0$$

If we write all the factors and divide out the 3's we have: $\dfrac{3^2}{3^2} = \dfrac{\overset{1}{\cancel{3}} \cdot \overset{1}{\cancel{3}}}{\underset{1}{\cancel{3}} \cdot \underset{1}{\cancel{3}}} = \dfrac{1}{1} = 1$

Conclusion: Because $\dfrac{3^2}{3^2}$ equals both 3^0 and 1, we must conclude that $3^0 = 1$.

Discussion What if the denominator exponent is larger than the numerator exponent?

◆ **Margin 5**

Reduce to lowest terms.

a. $\dfrac{9x^4}{24x^3}$

b. $\dfrac{10a^5}{20a}$

c. $-\dfrac{4x^3y}{30x}$

d. $\dfrac{-15m^4}{20m^2n}$

Answers: a. $\dfrac{3x}{8}$ b. $\dfrac{a^4}{2}$ c. $-\dfrac{2x^2y}{15}$
d. $\dfrac{-3m^2}{4n}$

What if the variables in the denominators have larger exponents than the like variables in the numerators?

Margin 6

Reduce to lowest terms.

a. $\dfrac{14x^4y}{18x^6}$

b. $\dfrac{20m^2}{40m^5}$

c. $-\dfrac{13a^6b}{26a^4b^4}$

d. $\dfrac{-10hk^2}{5k^7}$

EXAMPLE 6 Reduce $\dfrac{-16p^3r^2s}{28p^5r^2}$ to lowest terms.

Solution: Write the numerator and denominator in factored form, divide out the common factors, then multiply the remaining factors.

$$\frac{-16p^3r^2s}{28p^5r^2} = \frac{-2\cdot 2\cdot 2\cdot 2\cdot p\cdot p\cdot p\cdot r\cdot r\cdot s}{2\cdot 2\cdot 7\cdot p\cdot p\cdot p\cdot p\cdot p\cdot r\cdot r}$$

Write the numerator and denominator in factored form.

$$= \frac{\overset{1\;\;\;1}{-2\cdot 2}\cdot 2\cdot 2\cdot \overset{1\;\;1\;\;1\;\;1\;\;1}{p\cdot p\cdot p\cdot r\cdot r}\cdot s}{\underset{1\;\;1\quad\;1\;\;1\;\;1\qquad\;\;1\;\;1}{2\cdot 2\cdot 7\cdot p\cdot p\cdot p\cdot p\cdot p\cdot r\cdot r}}$$

Divide out two 2's, three p's, and two r's.

$$= \frac{-2\cdot 2\cdot s}{7\cdot p\cdot p}$$

Multiply the remaining factors.

$$= \frac{-4s}{7p^2}$$

We ended up with p^2 in the denominator. p^5 meant that there were five p's in the denominator while p^3 meant there were only three p's in the numerator. Therefore, dividing out the three common p's leaves two p's in the denominator.

◀ **Do Margin 6.**

5.2 Exercises

FOR EXTRA HELP

 Videotape 6

 InterAct Math
Tutorial Software

 www.carsonmath.com

 AWL Math Tutor Center

 InterAct
MathXL www.mathxl.com

 Student's Solutions
Manual

For Exercises 1–12, reduce to lowest terms.

1. $\dfrac{25}{30}$

2. $\dfrac{14}{35}$

3. $\dfrac{26}{52}$

4. $\dfrac{24}{40}$

5. $\dfrac{66}{88}$

6. $\dfrac{57}{76}$

7. $-\dfrac{120}{140}$

8. $-\dfrac{196}{210}$

9. $-\dfrac{182}{234}$

10. $-\dfrac{221}{357}$

11. $\dfrac{-270}{900}$

12. $\dfrac{210}{-294}$

For Exercises 13–16, find the requested lowest terms fraction.

13. 25 minutes is what fraction of an hour?

14. 9 hours is what fraction of a day?

15. A survey is conducted in front of a department store. The person conducting the survey asks 248 women if they used a certain cosmetic product and 96 respond that they do use the product. What is the lowest terms fraction of the women surveyed that said they used the product? Interpret the results. Do you see any flaws in the survey?

16. A marketing firm develops two different television commercials for a product. To determine which is more effective, the firm conducts an experiment. Using the phone directory, they randomly select 400 people, and pay 200 to watch one of the commercials and the other 200 to watch the other. They then ask those viewers whether they would buy the product. 180 said yes in the first group, while 148 said yes in the second group. Write the lowest terms fractions for each group. Which commercial should be used? Do you see any flaws in the experiment?

Use the table below to answer Exercises 17–19. The table lists the number of miles logged by personal-use vehicles for various types of trips that occurred in 1995.

Type of trip by destination	Miles
Same state	448 million
Different state, same division	212 million
Different state, different division, but same region	133 million
Different state, different division, and different region	208 million

17. Find the lowest terms fraction of miles logged traveling to a different state within the same division.

18. Find the lowest terms fraction of miles logged traveling to a different state within the same region.

19. Find the lowest terms fraction of miles logged traveling to a different state, different region, and different division of the country.

20. A company produces 540 computer chips on a particular day. Quality-control inspectors discover 21 defective chips. What lowest terms fraction of the chips produced are defective? What lowest terms fraction is not defective?

For Exercises 21–28, write as a mixed number.

21. $\dfrac{30}{8}$

22. $\dfrac{69}{9}$

23. $\dfrac{50}{15}$

24. $\dfrac{84}{16}$

25. $-\dfrac{116}{28}$

26. $-\dfrac{168}{105}$

27. $\dfrac{-186}{36}$

28. $\dfrac{486}{-63}$

For Exercises 29–40, reduce to lowest terms.

29. $\dfrac{10x}{32}$

30. $\dfrac{15y^2}{27}$

31. $\dfrac{x^3}{xy}$

32. $\dfrac{a^4 b}{a^3 b^5}$

33. $\dfrac{6m^4 n}{15m^7}$

34. $\dfrac{30h^2 k}{18h^4 k}$

35. $\dfrac{9t^2 u}{36t^5 u^2}$

36. $\dfrac{7x^3 y}{35xy^2}$

37. $-\dfrac{42a^6 bc}{70a^5 b^2}$

38. $-\dfrac{10m^7 n}{30m^9 n^4 p}$

39. $\dfrac{-38a^4 b^5 c^2}{95a^{10} bc^7}$

40. $\dfrac{8x^2 yz^3}{-32xy^4 z^3}$

REVIEW EXERCISES

1. Multiply. $165 \cdot (-91)$

2. Multiply. $(-12)(-6)$

3. Multiply. $(5x^2)(7x^3 y)$

4. Simplify. $(-2x^3)^5$

5. Calculate the area of the figure.

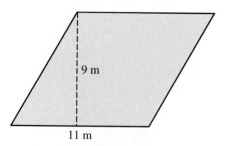

9 m

11 m

5.3 Multiplying Fractions, Mixed Numbers, and Rational Expressions

OBJECTIVES

1 Multiply fractions.

2 Multiply and reduce.

3 Multiply mixed numbers.

4 Multiply rational expressions.

5 Simplify fractions raised to a power.

6 Solve applications involving multiplying fractions.

7 Calculate the area of a triangle.

8 Calculate the radius and diameter of a circle.

9 Calculate the circumference of a circle.

OBJECTIVE **1** *Multiply fractions.*

A recipe calls for $\frac{1}{4}$ of a cup of oil. What if we only want to make half of the recipe? What is half of $\frac{1}{4}$? Consider the picture of $\frac{1}{4}$:

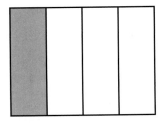

If we want half of $\frac{1}{4}$, we need to cut the single fourth into two pieces then shade one of those two pieces.

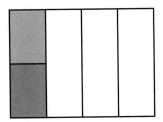

In order to name the fraction represented by half of the fourth, we need equal-size divisions. Extending the line across makes equal-size divisions.

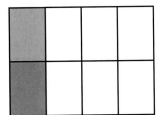

Notice the darker shaded region is half of the fourth that we originally shaded and is $\frac{1}{8}$ of the whole picture. So we can now say:

$$\text{Half of } \frac{1}{4} \text{ is } \frac{1}{8}.$$

What did we do? We can translate this statement to an equation. The word *of* is a key word for multiplication and *is* is a key word for an equal sign, so we have:

$$\text{Half of } \frac{1}{4} \text{ is } \frac{1}{8}.$$

$$\frac{1}{2} \cdot \frac{1}{4} = \frac{1}{8}$$

From this analysis, we can make the following conclusion.

Conclusion: To multiply fractions, multiply numerator by numerator and denominator by denominator.

EXAMPLE 1 Multiply.

a. $\dfrac{3}{4} \cdot \dfrac{5}{8}$

Answer: $\dfrac{3}{4} \cdot \dfrac{5}{8} = \dfrac{15}{32}$

Explanation: We multiplied numerator by numerator and denominator by denominator. Notice $\frac{15}{32}$ is in lowest terms.

b. $-\dfrac{2}{3} \cdot \dfrac{1}{5}$

Answer: $-\dfrac{2}{3} \cdot \dfrac{1}{5} = -\dfrac{2}{15}$

> **Reminder**
> The product of two numbers that have different signs is always a negative number.

Explanation: We multiplied numerator by numerator and denominator by denominator. Notice $-\frac{2}{15}$ is in lowest terms.

c. $\dfrac{-5}{7} \cdot \dfrac{9}{-11}$

Answer: $\dfrac{-5}{7} \cdot \dfrac{9}{-11} = \dfrac{45}{77}$

Explanation: We multiplied numerator by numerator and denominator by denominator.

◀ **Do Margin 1.**

OBJECTIVE 2 *Multiply and reduce.*

Consider $\dfrac{5}{6} \cdot \dfrac{3}{4}$.

$$\frac{5}{6} \cdot \frac{3}{4} = \frac{15}{24}$$

Notice $\frac{15}{24}$ can be reduced because we can divide out the common factor 3.

$$\frac{5}{6} \cdot \frac{3}{4} = \frac{15}{24} = \frac{\overset{1}{\cancel{3}} \cdot 5}{2 \cdot 2 \cdot 2 \cdot \underset{1}{\cancel{3}}} = \frac{5}{8} \qquad \text{This is now in lowest terms.}$$

> **Reminder**
> $\dfrac{-5}{7}$ can be written as $-\dfrac{5}{7}$.
>
> $\dfrac{9}{-11}$ can be written as $-\dfrac{9}{11}$.
>
> Both fractions are negative regardless of the minus sign placement. The product of two numbers that have the same sign is always a positive number.

Conclusion: We can reduce after multiplying.

However, there is an alternative. Remember that reducing is merely dividing out common factors that exist in the numerator and denominator. Because the original numerators are factors of the resulting numerator and the original denominators are factors of the resulting denominator, we can actually divide out common factors *before* multiplying.

If we write all the prime factors, we have:

$$\frac{5}{6} \cdot \frac{3}{4} = \frac{5}{2 \cdot \cancel{3}} \cdot \frac{\overset{1}{\cancel{3}}}{2 \cdot 2} = \frac{5 \cdot 1}{2 \cdot 1 \cdot 2 \cdot 2} = \frac{5}{8}$$

Or, some prefer to divide out the common factor(s) this way:

$$\frac{5}{\underset{2}{\cancel{6}}} \cdot \frac{\overset{1}{\cancel{3}}}{4} = \frac{5}{8}$$

Note For now, we will write the numerators and denominators in prime factored form so that the factors are easier to see. In Sections 5.7 and 5.8, we'll leave out the prime factored form.

Conclusion: We can reduce before or after multiplying fractions.

> **Procedure** *To multiply fractions:*
> 1. Divide out any numerator factor with any like denominator factor.
> 2. Multiply numerator by numerator and denominator by denominator.
> 3. Reduce as needed to get to the result in lowest terms. (This would be necessary if you miss common factors in the first step.)

EXAMPLE 2 Multiply $\frac{6}{8} \cdot \frac{10}{15}$. Write the product in lowest terms.

Solution: We can multiply straight across and then reduce this way:

$$\frac{6}{8} \cdot \frac{10}{15} = \frac{60}{120} = \frac{60 \div 60}{120 \div 60} = \frac{1}{2}$$

Or, we can divide out the common factors that exist in the numerators and denominators before multiplying. Using prime factorization we have:

$$\frac{6}{8} \cdot \frac{10}{15} = \frac{2 \cdot 3}{2 \cdot 2 \cdot 2} \cdot \frac{2 \cdot 5}{3 \cdot 5}$$

$$= \frac{\overset{1}{\cancel{2}} \cdot \overset{1}{\cancel{3}}}{\underset{1}{\cancel{2}} \cdot 2 \cdot 2} \cdot \frac{\overset{1}{\cancel{2}} \cdot \overset{1}{\cancel{5}}}{\underset{1}{\cancel{3}} \cdot \underset{1}{\cancel{5}}}$$

$$= \frac{1}{2}$$

No matter what technique you choose, you will end up dividing out the same common factors.

When we multiplied and then reduced, we divided out a common factor of 60.

When we divided the common primes, we divided out two 2's, a 3, and a 5 from the numerator and denominator. If you multiply out two 2's, a 3, and a 5 you get 60.

$$2 \cdot 2 \cdot 3 \cdot 5 = 60$$

Do Margin 2. ▶

Working with Scientific Calculators

On a scientific calculator, use the $\boxed{ab/c}$ key to enter fractions. We would work Example 2 this way:

$6 \boxed{ab/c} 8 \boxed{\times} 10 \boxed{ab/c} 15 \boxed{=}$

◆ **Margin 2**

Multiply and write the product in lowest terms.

a. $\dfrac{5}{9} \cdot \dfrac{3}{10}$

b. $\dfrac{6}{8} \cdot \dfrac{12}{30}$

c. $-\dfrac{10}{15} \cdot \dfrac{6}{16}$

d. $\dfrac{-18}{20} \cdot \dfrac{30}{-32}$

Answers: a. $\frac{1}{6}$ b. $\frac{3}{10}$ c. $-\frac{1}{4}$ d. $\frac{27}{32}$

OBJECTIVE 3 *Multiply mixed numbers.*

How do we multiply mixed numbers? In Section 5.1 we learned how to write mixed numbers as improper fractions and we have just learned how to multiply fractions. The logical conclusion would be to write the mixed numbers as improper fractions, then multiply.

> **Procedure** *To multiply mixed numbers:*
> 1. Write the mixed numbers as improper fractions.
> 2. Divide out any numerator factor with any like denominator factor.
> 3. Multiply numerator by numerator and denominator by denominator.
> 4. Write the product as a mixed number in simplest form.

EXAMPLE 3 Estimate the product for $10\frac{1}{2} \cdot 3\frac{1}{6}$. Then multiply and write the product as a mixed number in simplest form.

Solution: To estimate, we must round the numbers to comfortable amounts, then multiply.

$$10\frac{1}{2} \cdot 3\frac{1}{6}$$

$10\frac{1}{2}$ rounds to 11. $\qquad 3\frac{1}{6}$ rounds to 3.

$$11 \cdot 3 = 33$$

Note By *comfortable,* we mean numbers that are simple enough to work with in your head.

Remember, the purpose of the estimate is to get an idea of what the actual product should be. We now know that $10\frac{1}{2} \cdot 3\frac{1}{6}$ should be around 33 when we finish. Let's get the actual product now.

$$10\frac{1}{2} \cdot 3\frac{1}{6}$$

$$= \frac{21}{2} \cdot \frac{19}{6} \qquad \text{Write as improper fractions.}$$

$$= \frac{\overset{1}{\cancel{3}} \cdot 7}{2} \cdot \frac{19}{2 \cdot \underset{1}{\cancel{3}}} \qquad \text{Divide out a common factor of 3. Then multiply.}$$

$$= \frac{133}{4} \qquad \text{Write as a mixed number.}$$

$$= 33\frac{1}{4}$$

◆ Notice the estimate is very close to the actual product.

◀ **Do Margin 3.**

EXAMPLE 4 Multiply $-3\frac{1}{5} \cdot 2\frac{2}{9}$.

Solution: We write the mixed numbers as improper fractions, then multiply.

$$-3\frac{1}{5} \cdot 2\frac{2}{9} \qquad \text{Write as improper fractions.}$$

$$= -\frac{16}{5} \cdot \frac{20}{9}$$

$$= -\frac{16}{\underset{1}{\cancel{5}}} \cdot \frac{2 \cdot 2 \cdot \overset{1}{\cancel{5}}}{9} \qquad \begin{array}{l}\text{Divide out the}\\ \text{common factor 5.}\end{array}$$

> **Reminder**
> When multiplying two numbers with different signs, the result is negative.

$$= -\frac{64}{9} \qquad \begin{array}{l}\text{Write the improper fraction}\\ \text{as a mixed number.}\end{array}$$

$$= -7\frac{1}{9}$$

◆ Margin 3

Estimate the product, then find the actual product.

a. $3\frac{1}{3} \cdot 1\frac{4}{5}$

b. $5\frac{2}{3} \cdot 2\frac{1}{7}$

c. $2\frac{2}{5} \cdot 12$

d. $20 \cdot 3\frac{3}{4}$

Answers: **a.** 6; 6 **b.** 12; $12\frac{1}{7}$ **c.** 24; $28\frac{4}{5}$ **d.** 80; 75

Do Margin 4. ▶

OBJECTIVE 4 *Multiply rational expressions.*

EXAMPLE 5 Multiply. $-\dfrac{3a}{5b} \cdot -\dfrac{10ab^2}{9}$

Solution: If we write the monomials in prime factored form, the problem looks like this:

$$-\frac{3a}{5b} \cdot -\frac{10ab^2}{9} = -\frac{3 \cdot a}{5 \cdot b} \cdot -\frac{2 \cdot 5 \cdot a \cdot b \cdot b}{3 \cdot 3}$$

$$= -\frac{\cancel{3} \cdot a}{\cancel{5} \cdot \cancel{b}} \cdot -\frac{2 \cdot \cancel{5} \cdot a \cdot \cancel{b} \cdot b}{\cancel{3} \cdot 3} \qquad \text{Divide out a common 3, a common 5, and a common } b \text{ factor.}$$

$$= \frac{2 \cdot a \cdot a \cdot b}{3} \qquad \text{Multiply the remaining numerator and denominator factors.}$$

$$= \frac{2a^2 b}{3}$$

Reminder
When multiplying two numbers that have the same sign, the result is positive.

Do Margin 5. ▶

OBJECTIVE 5 *Simplify fractions raised to a power.*

We can have exponential forms with fraction bases. Remember an exponent indicates the number of times to use the base as a factor.

EXAMPLE 6 Evaluate $\left(\dfrac{2}{3}\right)^4$

Solution: We write the base $\frac{2}{3}$ as a factor four times.

$$\left(\frac{2}{3}\right)^4 = \frac{2}{3} \cdot \frac{2}{3} \cdot \frac{2}{3} \cdot \frac{2}{3} = \frac{16}{81}$$

There are four 2's that multiply together to make 16 in the numerator and four 3's that multiply to make 81 in the denominator. We can say:

$$\left(\frac{2}{3}\right)^4 = \frac{2^4}{3^4}$$

Rule When a fraction is raised to a power, we evaluate both the numerator and denominator raised to that power.

In math language: $\left(\dfrac{a}{b}\right)^n = \dfrac{a^n}{b^n}$ where $b \neq 0$

WARNING

$\dfrac{2^4}{3}$ and $\left(\dfrac{2}{3}\right)^4$ are not the same.

$$\frac{2^4}{3} = \frac{2 \cdot 2 \cdot 2 \cdot 2}{3} \qquad \text{while} \qquad \left(\frac{2}{3}\right)^4 = \frac{2 \cdot 2 \cdot 2 \cdot 2}{3 \cdot 3 \cdot 3 \cdot 3}$$

◆ **Margin 4**

Multiply and simplify.

a. $-4\dfrac{1}{8} \cdot 5\dfrac{1}{2}$

b. $\left(-2\dfrac{3}{5}\right)\left(-7\dfrac{1}{4}\right)$

c. $\left(2\dfrac{4}{7}\right)\left(-11\dfrac{2}{5}\right)$

d. $\left(\dfrac{-4}{9}\right)(-21)$

◆ **Margin 5**

Multiply.

a. $\dfrac{x}{6} \cdot \dfrac{9}{x^2}$

b. $\dfrac{10a^3}{3} \cdot \dfrac{12}{a}$

c. $\dfrac{18tu}{5} \cdot -\dfrac{15}{27u^4}$

d. $\dfrac{-6}{13hk^2} \cdot -\dfrac{26h}{9k}$

Working with Scientific Calculators

We would enter Example 7 this way:

| 2 | ab/c | 3 | yˣ | 4 | = |

If your calculator gives a decimal answer, then use the F◀▶D function to change the decimal back to a fraction. The F stands for fraction and the D stands for decimal. The double arrow in between indicates that pressing that key changes fractions to decimals and/or decimals to fractions.

On most scientific calculators the F◀▶D function is a secondary function for an existing key, which means you will have to first press the 2nd or shift key to use the function.

Answers to Margin 4: **a.** $-22\dfrac{11}{16}$ **b.** $18\dfrac{17}{20}$ **c.** $-29\dfrac{11}{35}$ **d.** $9\dfrac{1}{3}$

Answers to Margin 5: **a.** $\dfrac{3}{2x}$ **b.** $40a^2$ **c.** $-\dfrac{2t}{u^3}$ **d.** $\dfrac{4}{3k^3}$

◀ **Do Margin 6.**

EXAMPLE 7 Simplify the exponent. $\left(\dfrac{-2x^2y}{5}\right)^3$

Solution: We evaluate the numerator $-2x^2y$ raised to the 3rd power and also the denominator 5 raised to the 3rd power.

$$\left(\frac{-2x^2y}{5}\right)^3 = \frac{(-2x^2y)^3}{(5)^3}$$

Write $-2x^2y$ as a factor three times and 5 as a factor three times.

$$= \frac{-2x^2y \cdot -2x^2y \cdot -2x^2y}{5 \cdot 5 \cdot 5}$$

Multiply coefficients and add exponents for the like bases.

$$= \frac{-8x^6y^3}{125}$$

> **Reminder**
> When we multiply exponential forms that have the same base, we add the exponents.

We could have solved this problem using exponent rules. In Section 3.2 we learned that when we raise an exponential form to a power, we can multiply each exponent by the power as long as the expression inside the parentheses is a product or quotient.

$$\left(\frac{-2x^2y}{5}\right)^3 = \frac{(-2x^2y)^3}{(5)^3}$$

Evaluate both the numerator and denominator raised to the 3rd power.

$$= \frac{(-2)^{1\cdot3}x^{2\cdot3}y^{1\cdot3}}{5^{1\cdot3}}$$

Multiply the exponents.

$$= \frac{(-2)^3x^6y^3}{5^3}$$

Calculate $(-2)^3$ and 5^3.

$$= \frac{-8x^6y^3}{125}$$

> **Reminder**
> A number or variable with no exponent showing has an understood exponent of 1.

Margin 7

Simplify.

a. $\left(\dfrac{7x^3}{9}\right)^2$

b. $\left(\dfrac{3x^2y}{4z}\right)^3$

c. $\left(-\dfrac{2a^4b}{3c^3}\right)^4$

d. $\left(\dfrac{-m^3n^5}{2p}\right)^5$

◀ **Do Margin 7.**

OBJECTIVE 6 *Solve applications involving multiplying fractions.*

One of the difficulties in handling problems that involve fractions is identifying the arithmetic operation. Look for language that indicates we are getting a fraction of an amount to indicate multiplication with fractions. Consider the sentence:

$$\frac{3}{4} \text{ of the 28 people in the class got an A.}$$

In the above sentence we have a fraction of a given total. To calculate the actual number of people in the class who got an A, we multiply 28 by $\frac{3}{4}$. Whenever the word *of* is directly preceded by a fraction, it indicates multiplication.

The word *of* does not always mean multiply. Sometimes *of* indicates a fraction. Consider the sentence:

21 out of 28 people in the class got an A.   or   21 of 28 people in the class got an A.

Answers to Margin 6: a. $\frac{8}{125}$ **b.** $\frac{1}{64}$
c. $-\frac{27}{64}$ **d.** $\frac{25}{36}$

Answers to Margin 7: a. $\frac{49x^6}{81}$
b. $\frac{27x^6y^3}{64z^3}$ **c.** $\frac{16a^{16}b^4}{81c^{12}}$ **d.** $\frac{-m^{15}n^{25}}{32p^5}$

In *21 of 28,* the word *of* indicates the fraction $\frac{21}{28}$. Whenever the word *of* is preceded by a whole number, it indicates a fraction.

Conclusion: The word *of* indicates multiplication when preceded by a fraction and indicates a fraction when preceded by a whole number.

The two uses of the word *of* are related.

$\frac{3}{4}$ of the 28 people in the class got an A.

21 of 28 people in the class got an A.

$\frac{3}{4} \cdot 28 =$ number of people that got an A.

$\frac{21}{28} = \frac{\text{fraction of the people in the class}}{\text{that got an A.}}$

$$\frac{3}{\overset{}{\underset{1}{4}}} \cdot \frac{\overset{7}{28}}{1} = 21$$

$$\frac{21}{28} = \frac{3}{4}$$

EXAMPLE 8 An advertisement claims that 4 out of 5 dentists choose Crest toothpaste. If you were to visit a dental conference that has 345 dentists in attendance, how many would you expect to choose Crest toothpaste based on the claim of the ad?

Solution: Use the problem-solving process.

Understand: 4 out of 5 is the fraction $\frac{4}{5}$. So, according to the ad, we can say:

$$\frac{4}{5} \text{ of all dentists choose Crest.}$$

If the claim of the ad is accurate, then we can say this:

$$\frac{4}{5} \text{ of the 345 dentists at the conference should choose Crest.}$$

We substituted the 345 dentists at the conference for *all dentists* in our original statement.

Plan: We can translate our sentence to an equation and solve.

Execute: $\frac{4}{5}$ of the 345 dentists at the conference should choose Crest

$$\downarrow$$

$$\frac{4}{5} \cdot 345 = n$$

$$\frac{4}{5} \cdot \frac{345}{1} = n$$

$$\frac{4}{\overset{}{\underset{1}{5}}} \cdot \frac{3 \cdot \overset{1}{5} \cdot 23}{1} = n$$

$$276 = n$$

> **Reminder**
> An integer is written as an improper fraction by writing it over 1.

Answer: If we were to survey the dentists at the conference, we should find that out of the 345 dentists, 276 should choose Crest.

Discussion How many of the 345 dentists would you expect to choose toothpaste other than Crest? What if you conducted the survey and found an amount other than 276?

Check: 276 out of 345 dentists should be the same ratio as 4 out of 5. In other words $\frac{276}{345}$ should be the same fraction as $\frac{4}{5}$. We can verify by reducing $\frac{276}{345}$ to $\frac{4}{5}$.

Do Margin 8. ▶

◆ **Margin 8**

Solve.

a. A doctor says that 8 out of 10 patients receiving a particular treatment have no side effects. If a hospital gives 600 patients this treatment, how many can be expected to have no side effects? How many can be expected to have side effects?

b. A population study finds that there are $2\frac{1}{3}$ children for each household in a given area. If there are 6252 households in the area, how many children are there?

Discussion How can there be $2\frac{1}{3}$ children per household?

Answers: **a.** 480; 120 **b.** 14,588

EXAMPLE 9 $\frac{3}{4}$ of the students taking a particular history course passed the course. Of these, $\frac{1}{3}$ got an A. What fraction of all students taking the course got an A?

Understand: The students that got an A are a subgroup of the group that passed. We can say:

$$\frac{1}{3} \text{ of the } \frac{3}{4} \text{ that passed got an A.}$$

Plan: We can translate the sentence to an equation, then solve.

Execute: Let *a* represent the number of students that received an A.

$$\frac{1}{3} \text{ of the } \frac{3}{4} \text{ that passed got an A.}$$

$$\frac{1}{3} \cdot \frac{3}{4} = a$$

$$\frac{1}{\cancel{3}} \cdot \frac{\cancel{3}^1}{4} = a$$

$$\frac{1}{4} = a$$

Answer: $\frac{1}{4}$ of all students in the class not only passed but got an A.

Check: We can use a picture to check.

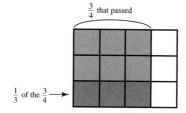

Note The fact that there are 12 divisions in the picture does not necessarily mean that there are 12 total students in the class.

♦ The darker shaded region is $\frac{3}{12}$ of the whole picture, which reduces to $\frac{1}{4}$.

◀ **Do Margin 9.**

OBJECTIVE 7 *Calculate the area of a triangle.*

We developed the formula for the area of a parallelogram in Section 1.6. We can use a parallelogram to develop the formula for calculating the area of a triangle.

Try the following experiment. You will need two pieces of paper, a pencil, a ruler, and scissors.

1. Draw a triangle on one piece of paper. It is best to place two of the corners at the corners of the paper and the top corner at the top edge of the paper like so:

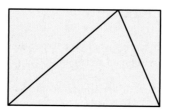

♦ **Margin 9**

Solve.

a. A study estimates that $\frac{2}{3}$ of all Americans own a car and $\frac{1}{4}$ of these cars are blue. What fraction of the population drives a blue car?

b. A survey of an area indicates that $\frac{5}{8}$ of the people responding to it said they watch more than 2 hours of television each day. Of these, $\frac{7}{10}$ were female. What fraction of all those surveyed were females who watch more than 2 hours of television each day?

Discussion Errors were made in conducting this survey. What factors might have skewed the results?

Answers: **a.** $\frac{1}{6}$ **b.** $\frac{7}{16}$

2. Label the base and height of the triangle using letters b and h like so: (Remember the height line forms a 90° angle with the base)

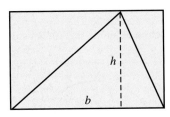

3. Now place the paper with the triangle on it on top of the blank piece of paper so that all edges match up.

4. Cut out the triangle holding both pieces together so that you create two identical triangles.

5. Rotate the blank triangle so that it is upside down, then fit it together with the labeled triangle so that two corresponding sides join. The resulting figure is a parallelogram.

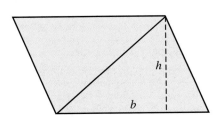

Because the triangles are identical, the angle measurements are identical. When one triangle is rotated and joined along a corresponding side, the identical angles form opposite corners of a four-sided figure. Because the opposing corner angles match, the opposing sides are parallel. Because there are two pair of parallel sides, the figure is a parallelogram.

◆ **Margin 10**

Calculate the area of each triangle.

a.

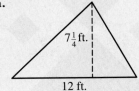

Remember that we calculate the area of a parallelogram by the formula:

$$A = bh$$

But we are interested in a single triangle. Because two identical triangles form the parallelogram, the area of one of the triangles must be half of the area of the parallelogram.

Area of a triangle = Half of the area of a parallelogram with the same base and height.

$$A = \frac{1}{2} \cdot bh$$

$$A = \frac{1}{2} bh$$

b.

EXAMPLE 10 A recycling company produces triangular stickers that are $5\frac{1}{2}$ in. along the base and $4\frac{3}{4}$ in. high. Find the area.

Solution: Use the formula $A = \frac{1}{2} bh$.

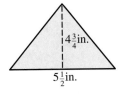

$$A = \frac{1}{2}\left(5\frac{1}{2}\right)\left(4\frac{3}{4}\right)$$

$$A = \frac{1}{2}\left(\frac{11}{2}\right)\left(\frac{19}{4}\right)$$

$$A = \frac{209}{16}$$

$$A = 13\frac{1}{16} \text{ in.}^2$$

Do Margin 10. ▶

Answers: **a.** $43\frac{1}{2}$ ft.^{2} **b.** $18\frac{2}{5}$ m^2

OBJECTIVE **8** *Calculate the radius and diameter of a circle.*

What is a **circle**?

DEFINITION **Circle:** A collection of points that are all equally distant from a central point, called the *center*.

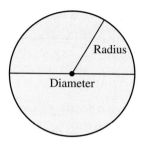

The distance from the center to any point on the circle is the same for all points on the circle. This distance is the called **radius** (plural *radii*). The distance across a circle along a straight line that passes through the center is the **diameter.**

DEFINITION **Radius:** The distance from the center to any point on the circle.

Diameter: The distance across a circle along a straight line through the center.

What are some relationships between diameter and radius?

The diameter is twice the radius.

The radius is half of the diameter.

We can use what we learned in Chapter 4 about translating key words to translate these relationships to equations.

The diameter is twice the radius.

$$d = 2 \cdot r$$

The radius is half of the diameter.

$$r = \frac{1}{2} \cdot d$$

EXAMPLE 11 Calculate the diameter of the circle shown.

Solution: Diameter is twice the radius so we multiply $3\frac{5}{8}$ cm by 2.

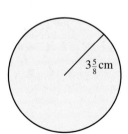

$$d = 2r$$

$$d = 2 \cdot 3\frac{5}{8}$$

$$d = \frac{\overset{1}{\cancel{2}}}{1} \cdot \frac{29}{\underset{4}{\cancel{8}}}$$

$$d = \frac{29}{4}$$

$$d = 7\frac{1}{4} \text{ cm}$$

Do Margin 11. ▷

EXAMPLE 12   Calculate the radius of the circle shown.

Solution: Radius is half of the diameter, so we need to multiply 17 ft. by $\frac{1}{2}$.

17 ft.

$$r = \frac{1}{2}d$$

$$r = \frac{1}{2} \cdot 17$$

$$r = \frac{1}{2} \cdot \frac{17}{1}$$

$$r = \frac{17}{2}$$

$$r = 8\frac{1}{2} \text{ ft.}$$

Do Margin 12. ▷

OBJECTIVE 9 *Calculate the circumference of a circle.*

DEFINITION   **Circumference:** The distance around a circle.

When ancient mathematicians began experimenting with circles they found an interesting property. When they measured the **circumference** and diameters of different-size circles and expressed the ratio of the circumference to diameter they found the fractions were very close no matter how large or small the circles. The fractions were so close that mathematicians concluded the ratio must actually be the same for all circles and the slight differences could be accounted for in their inability to accurately measure the circumference and diameter. It turns out that the value of this fraction cannot be expressed exactly, only closely approximated. We call a number like this an **irrational number.**

DEFINITION   **Irrational number:** A number that cannot be expressed exactly as a fraction.

In math language: A number that cannot be expressed in the form $\frac{a}{b}$ where a and b are integers and $b \neq 0$.

Because the ratio of the circumference to the diameter is an irrational number, mathematicians decided to use a symbol to represent the value. They chose the Greek letter **π** (pronounced *pi*).

DEFINITION   **π:** An irrational number that is the ratio of the circumference of a circle to its diameter.

In symbols we say: $\pi = \dfrac{C}{d}$, where C represents the circumference and d represents the diameter.

A circle with a diameter of 7 in. will have a circumference of about 22 in. This means the value of π is approximately $\frac{22}{7}$. Symbolically we write $\pi \approx \frac{22}{7}$. The symbol $\approx$ is read, *is approximately.*

Remember $\frac{22}{7}$ is only an approximation for the value of π. If we were to try to measure the circumference, we would find the distance to be close to 22 but not exactly. As we wrap the measuring tape around the circle, we would find the circumference mark would never line up with an exact mark on the measuring tape, no matter how precisely the tape could measure.

◆ **Margin 11**

   a. Calculate the diameter of a circle with a radius of $2\frac{3}{5}$ ft.

   b. Calculate the diameter of a circle with a radius of $\frac{7}{8}$ in.

◆ **Margin 12**

   a. Calculate the radius of a circle with a diameter of 25 ft.

   b. Calculate the radius of a circle with a diameter of $9\frac{3}{8}$ m.

Answers to Margin 11: **a.** $5\frac{1}{5}$ ft.
b. $1\frac{3}{4}$ in.

Answers to Margin 12: **a.** $12\frac{1}{2}$ ft.
b. $4\frac{11}{16}$ m

We can now use this approximate value of π to calculate circumference, given the diameter or radius of a circle. We can rearrange the relationship and say:

$$C = \pi d$$

Or, since $d = 2r$ we can say:

$$C = 2\pi r$$

EXAMPLE 13 Calculate the circumference of the circle shown.

Solution: We can use the formula $C = \pi d$ or $C = 2\pi r$. Because we were given radius, let's use $C = 2\pi r$.

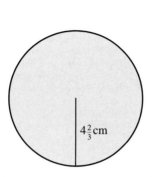

$4\frac{2}{3}$ cm

$$C = 2\pi r \qquad \text{Replace } \pi \text{ with } \frac{22}{7} \text{ and } r \text{ with } 4\frac{2}{3}.$$

$$C \approx 2 \cdot \frac{22}{7} \cdot 4\frac{2}{3}$$

$$C \approx \frac{2}{1} \cdot \frac{22}{7} \cdot \frac{14}{3} \qquad \text{Write as improper fractions.}$$

$$C \approx \frac{2}{1} \cdot \frac{22}{\overset{}{\underset{1}{7}}} \cdot \frac{\overset{2}{14}}{3} \qquad \begin{array}{l}\text{Divide out a common factor of 7}\\ \text{in 14 and 7. Then multiply.}\end{array}$$

$$C \approx \frac{88}{3} \qquad \text{Write as a mixed number.}$$

$$C \approx 29\frac{1}{3} \text{ cm}$$

Note When we replace π with $\frac{22}{7}$, the calculation becomes an approximation, so we use the ≈ symbol instead of an equal sign.

◀ **Do Margin 13.**

◆**Margin 13**

a. Calculate the circumference of a circle with a radius of $1\frac{1}{6}$ ft.

b. Calculate the circumference of a circle with a diameter of $5\frac{5}{6}$ m.

Answers: **a.** $7\frac{1}{3}$ ft. **b.** $18\frac{1}{3}$ m

5.3 Exercises

For Exercises 1–8, multiply.

1. $\dfrac{2}{5} \cdot \dfrac{4}{7}$ **2.** $\dfrac{5}{8} \cdot \dfrac{1}{4}$ **3.** $\dfrac{1}{6} \cdot \dfrac{1}{9}$ **4.** $\dfrac{1}{4} \cdot \dfrac{1}{10}$

5. $\dfrac{3}{4} \cdot \dfrac{5}{7}$ **6.** $\dfrac{9}{10} \cdot \dfrac{3}{10}$ **7.** $\dfrac{7}{100} \cdot \dfrac{7}{10}$ **8.** $\dfrac{6}{11} \cdot \dfrac{10}{19}$

For Exercises 9–22, multiply and express the product in lowest terms.

9. $\dfrac{6}{7} \cdot \dfrac{14}{15}$ **10.** $\dfrac{5}{9} \cdot \dfrac{12}{25}$ **11.** $\dfrac{8}{12} \cdot \dfrac{10}{20}$ **12.** $\dfrac{14}{18} \cdot \dfrac{9}{21}$

13. $\dfrac{24}{32} \cdot \dfrac{26}{30}$ **14.** $\dfrac{15}{33} \cdot \dfrac{22}{27}$ **15.** $\dfrac{16}{38} \cdot \dfrac{57}{80}$ **16.** $\dfrac{36}{42} \cdot \dfrac{28}{45}$

17. $-\dfrac{19}{20} \cdot \dfrac{16}{38}$ **18.** $\dfrac{25}{34} \cdot -\dfrac{17}{30}$ **19.** $-\dfrac{18}{40} \cdot -\dfrac{28}{30}$ **20.** $\dfrac{-22}{35} \cdot \dfrac{15}{-33}$

21. $\dfrac{36}{-40} \cdot \dfrac{-30}{54}$ **22.** $\dfrac{60}{-81} \cdot \dfrac{45}{72}$

For Exercises 23–34, estimate the product. Then multiply and write the product as a mixed number in simplest form.

23. $3\dfrac{1}{5} \cdot 1\dfrac{3}{4}$ **24.** $4\dfrac{1}{6} \cdot 3\dfrac{2}{5}$ **25.** $\dfrac{5}{6}(28)$

26. $12 \cdot \dfrac{7}{8}$ **27.** $4\dfrac{5}{8} \cdot 16$ **28.** $27 \cdot 2\dfrac{4}{9}$

29. $-6\dfrac{1}{8} \cdot \dfrac{4}{7}$ **30.** $2\dfrac{7}{10} \cdot -8\dfrac{1}{3}$ **31.** $-5\dfrac{2}{3} \cdot -7\dfrac{1}{5}$

32. $-3\dfrac{9}{16} \cdot -6\dfrac{1}{8}$ **33.** $\dfrac{-3}{10}(45)$ **34.** $-18 \cdot \dfrac{7}{24}$

For Exercises 35–46, multiply and express the product in lowest terms.

35. $\dfrac{x^2}{5} \cdot \dfrac{2}{3}$

36. $\dfrac{3y}{7} \cdot \dfrac{2}{5}$

37. $\dfrac{4x}{9} \cdot \dfrac{3x}{8}$

38. $\dfrac{2}{15a} \cdot \dfrac{5}{6a^2}$

39. $\dfrac{xy}{10} \cdot \dfrac{4y^3}{9}$

40. $\dfrac{3m^2}{14} \cdot \dfrac{7mn}{12}$

41. $-\dfrac{5hk^3}{9} \cdot \dfrac{3}{4h}$

42. $\dfrac{9u}{20t^2} \cdot -\dfrac{4tu^2}{15}$

43. $-\dfrac{10x^4y}{11z} \cdot -\dfrac{22z}{14x^2}$

44. $\dfrac{-a^4b^3}{5c^3} \cdot \dfrac{10ac^5}{12b}$

45. $\dfrac{9m^3}{25n^4p} \cdot \dfrac{-15n}{18m^2p}$

46. $-\dfrac{4t^6}{28tu} \cdot \dfrac{-21u^3}{18uv}$

For Exercises 47–58, simplify.

47. $\left(\dfrac{5}{6}\right)^2$

48. $\left(\dfrac{4}{9}\right)^2$

49. $\left(-\dfrac{3}{4}\right)^3$

50. $\left(-\dfrac{1}{5}\right)^3$

51. $\left(\dfrac{-1}{2}\right)^6$

52. $\left(\dfrac{-2}{3}\right)^4$

53. $\left(\dfrac{x}{2}\right)^3$

54. $\left(\dfrac{3}{y}\right)^4$

55. $\left(\dfrac{2x^2}{3}\right)^3$

56. $\left(\dfrac{t^4}{4y}\right)^3$

57. $\left(-\dfrac{m^3n}{3p^2}\right)^4$

58. $\left(\dfrac{-2xy^4}{z^3}\right)^5$

For Exercises 59–68, solve. Write all answers in lowest terms.

59. A shipping company estimates that $\frac{1}{8}$ of the produce shipped will go bad during shipment. If a single truck carries about 12,480 pieces of fruit, how many can be expected to go bad during shipment?

60. A single share of a certain computer company's stock is listed at $\$6\frac{3}{8}$ per share. What would be the total value of 500 shares of the stock?

61. $\frac{5}{6}$ of a company's employees live within a 15-mile radius of the company. Of these, $\frac{3}{4}$ live within a 10-mile radius. What fraction of all employees live within a 10-mile radius?

62. $\frac{3}{4}$ of a lot is to be landscaped. $\frac{2}{3}$ of the landscaped area is to be covered in sod. What fraction of the lot is to be covered in sod?

63. On Tanya's phone bill, 24 out of the 30 long distance calls were in state. $\frac{5}{6}$ of the in-state calls were to her parents. What fraction of all her long-distance calls were to her parents? How many calls were to her parents?

64. 16 out of the 24 people enrolled in a CPR course plan to go into a health-related field. $\frac{5}{8}$ of the people who intend to pursue a health field plan to go into nursing. What fraction of all people enrolled plan to become nurses? How many people is this?

65. A recipe calls for $1\frac{1}{3}$ cups of sugar. How much sugar should be used in making half of the recipe?

66. On a standard-size guitar, the length of the strings between the bridge and nut is $25\frac{1}{2}$ in. Guitar makers must place the 12th fret at exactly half of the length of the string. How far from the saddle or nut should one measure to place the 12th fret?

Bridge

12th Fret   Nut

Trivia Bite

Placing one's finger at the 12th fret on any string sounds out a pitch that is an octave higher than the pitch of the string. To create an octave on any stringed instrument, the length of the string must be halved.

67. A hand weight has a mass of $15\frac{1}{10}$ kg. What is its weight if the acceleration due to gravity is $-9\frac{4}{5}$ m/sec.²?

Connection

We treat units like rational expressions:

$$\text{kg} \cdot \text{m/sec.}^2 = \frac{\text{kg}}{1} \cdot \frac{\text{m}}{\text{sec.}^2} = \frac{\text{kg} \cdot \text{m}}{\text{sec.}^2}$$

$\frac{\text{kg} \cdot \text{m}}{\text{sec.}^2}$ is the unit for force. It was renamed the Newton (N) in recognition of Isaac Newton's development of the mathematics of force.

68. The Liberty Bell has a mass of $64\frac{3}{5}$ slugs. How much does it weigh if the acceleration due to gravity is $-32\frac{1}{5}$ ft./sec.²?

Reminder

The slug is the American unit for mass.

Connection

When we used $F = ma$ in the past, we said the values for the acceleration due to gravity were -10 m/sec.² for metric and -32 ft./sec.² for American. Now that we have developed fractions, we can use more accurate values.

For Exercises 69–70, find the area of each triangle.

69.

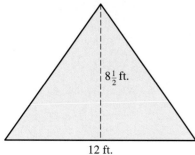

$8\frac{1}{2}$ ft.

12 ft.

70.

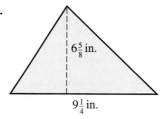

$6\frac{5}{8}$ in.

$9\frac{1}{4}$ in.

For Exercises 71–74, solve. Use $\frac{22}{7}$ for π

71. A giant sequoia tree has a diameter of 9 m.
   a. What is its radius?
   b. What is the circumference?

72. Big Ben is the name of the clock in the tower of London's Westminster Palace. The face of the clock is a circle with a diameter of 23 ft.
   a. What is the radius?
   b. What is the circumference?

73. The Fermi National Accelerator Laboratory is a circular tunnel that is used to accelerate elementary particles. The radius of the tunnel is $\frac{273}{440}$ mi.
   a. What is the diameter?
   b. What is the circumference?

74. The radius of the bottom of the Liberty Bell is $22\frac{87}{88}$ in.
   a. What is the diameter?
   b. What is the circumference?

Puzzle Problem

The Milky Way galaxy in which our solar system exists is a disc shape with spiral arms. The diameter of the galaxy is approximately 100,000 light-years. If our solar system is about $\frac{2}{3}$ of the distance from the center to the edge of the galaxy, then how far is our solar system from the center?

REVIEW EXERCISES

1. $1836 \div (-9) =$

2. $\sqrt{256} =$

3. Factor. $9x^2 - 12x$

4. Solve. $-6x = 30$

5. A 12-ft. board is to be cut into 6 pieces. How long will each piece be?

5.4 Dividing Fractions, Mixed Numbers, and Rational Expressions

OBJECTIVES

1 Divide fractions.

2 Divide mixed numbers.

3 Divide rational expressions.

4 Find the square root of a fraction.

5 Solve equations involving fractions.

6 Solve applications involving division of fractions.

OBJECTIVE **1** *Divide fractions.*

Suppose we wanted change for a $20 bill, and we want the change in $5 bills. How many $5 bills should we get? We are asking how many $5 are in $20. This is a pure division situation. It translates to the following statement:

$$20 \div 5 = 4$$

We should get four $5 bills.

What if we wanted the change in quarters? How many quarters are there in $20? Remember a quarter means $\frac{1}{4}$ of a dollar. This is still the same type of division statement that we did when we wanted to know how many $5 bills were in $20. So the statement looks like:

$$20 \div \frac{1}{4} = ?$$

Let's think it through. There are 4 quarters in every one of those $20. If each dollar has 4 quarters, then we can multiply 4 times 20 to get the total number of quarters.

$$20 \cdot 4 = 80 \text{ quarters}$$

This means our original division statement can be written as a multiplication statement.

$$20 \div \frac{1}{4} = 20 \cdot 4 = 80 \text{ quarters}$$

Remember we used an additive inverse to write a subtraction sentence as an equivalent addition sentence. To continue with this theme, mathematicians decided to say that we use a *multiplicative inverse* to write division statements as equivalent multiplication statements.

Remember inverse operations *undo* one another. Multiplicative inverses *undo* one another in that their product is always 1. In the example, $\frac{1}{4}$ and 4 are multiplicative inverses because their product is 1. Multiplicative inverses are usually called **reciprocals.**

$$\frac{1}{4} \cdot 4 = \frac{1}{\cancel{4}} \cdot \frac{\cancel{4}^{1}}{1} = 1$$

| **DEFINITION** | **Reciprocals:** Two numbers whose product is 1. |

EXAMPLE 1 Find the reciprocal.

a. $\dfrac{3}{4}$

Answer: $\dfrac{4}{3}$

Explanation: The product of $\dfrac{3}{4}$ and $\dfrac{4}{3}$ is 1.

$$\dfrac{\cancel{3}^{\,1}}{\cancel{4}_{\,1}} \cdot \dfrac{\cancel{4}^{\,1}}{\cancel{3}_{\,1}} = 1$$

b. $\dfrac{1}{5}$

Answer: 5

Explanation: The product of $\dfrac{1}{5}$ and 5 is 1.

$$\dfrac{1}{5} \cdot 5 = \dfrac{1}{\cancel{5}} \cdot \dfrac{\cancel{5}^{\,1}}{1} = 1$$

c. 6

Answer: $\dfrac{1}{6}$

Explanation: The product of 6 and $\dfrac{1}{6}$ is 1.

$$6 \cdot \dfrac{1}{6} = \dfrac{\cancel{6}^{\,1}}{1} \cdot \dfrac{1}{\cancel{6}} = 1$$

d. $-\dfrac{5}{7}$

Answer: $-\dfrac{7}{5}$

> Notice the sign stayed the same. The reciprocal of a number has the same sign as the number.

Explanation: The product of $-\dfrac{5}{7}$ and $-\dfrac{7}{5}$ is 1.

$$-\dfrac{\cancel{5}^{\,1}}{\cancel{7}_{\,1}} \cdot -\dfrac{\cancel{7}^{\,1}}{\cancel{5}_{\,1}} = 1$$

e. 0

Answer: 0 has no reciprocal.

◆ **Explanation:** The product of 0 and any number is 0, hence the product can never equal 1.

◀ **Do Margin 1.**

Now we can put the reciprocal to use.

Procedure *To divide fractions:*
1. Change the operation symbol from division to multiplication.
2. Change the divisor to its reciprocal.
3. Divide out any numerator factor with any like denominator factor.
4. Multiply numerator by numerator and denominator by denominator.
5. Write the result in simplest form.

EXAMPLE 2 Divide. $\dfrac{5}{8} \div \dfrac{3}{4}$.

Solution: Write the division statement as an equivalent multiplication statement using the reciprocal of the divisor.

$$\dfrac{5}{8} \div \dfrac{3}{4} = \dfrac{5}{8} \cdot \dfrac{4}{3} \qquad \text{Write an equivalent multiplication statement.}$$

$$= \dfrac{5}{2 \cdot 2 \cdot 2} \cdot \dfrac{\cancel{2}^{\,1} \cdot \cancel{2}^{\,1}}{3} \qquad \text{Divide out the common factors. Then multiply.}$$

$$= \dfrac{5}{6}$$

◆ **Margin 1**

Find the reciprocal.

a. $\dfrac{3}{8}$

b. $\dfrac{1}{4}$

c. -2

d. $-\dfrac{5}{11}$

Answers: **a.** $\dfrac{8}{3}$ **b.** 4 **c.** $-\dfrac{1}{2}$ **d.** $-\dfrac{11}{5}$

EXAMPLE 3 Divide. $\dfrac{-4}{9} \div -12$

Solution: Write the division statement as an equivalent multiplication statement using the reciprocal of the divisor.

$$\dfrac{-4}{9} \div -12 = \dfrac{-4}{9} \cdot -\dfrac{1}{12}$$

Write an equivalent multiplication statement.

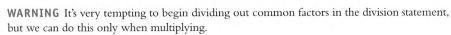

$$= \dfrac{-\overset{1}{2} \cdot \overset{1}{2}}{9} \cdot -\dfrac{1}{\underset{1}{2} \cdot \underset{1}{2} \cdot 3}$$

Now that we have a multiplication statement, we can divide out common factors.

$$= \dfrac{1}{27}$$

Reminder
The quotient is always positive when dividing two numbers that have the same sign.

WARNING It's very tempting to begin dividing out common factors in the division statement, but we can do this only when multiplying.

Recall that we sometimes expressed integer division using the fraction line. Instead of $18 \div 3 = 6$, we sometimes wrote $\dfrac{18}{3} = 6$. We can actually do the same thing with division of fractions. If we express the division of fractions in a fraction form, we call the expression a **complex fraction.**

DEFINITION **Complex fraction:** An expression that is a fraction with fractions in the numerator and/or denominator.

We can rewrite Example 2 using a complex fraction:

Instead of $\dfrac{5}{8} \div \dfrac{3}{4}$ we can write $\dfrac{\frac{5}{8}}{\frac{3}{4}}$.

Because most people prefer the regular division form, if we are given a complex fraction, we will rewrite it in that form.

Do Margin 2. ▷

OBJECTIVE 2 *Divide mixed numbers.*

We learned in Section 5.3 that when we multiply mixed numbers, we first write them as improper fractions. The same idea applies to division.

Procedure *To divide mixed numbers:*
1. Write the mixed numbers as improper fractions.
2. Write the division statement as an equivalent multiplication statement using the reciprocal of the divisor.
3. Divide out any numerator factor with any like denominator factor.
4. Multiply.
5. Write the result in simplest form.

EXAMPLE 4 Estimate the quotient of $8\dfrac{2}{5} \div 2\dfrac{1}{4}$, then find the quotient. Write the answer as a mixed number in simplest form.

Solution: To estimate, we round the mixed numbers to whole amounts.

$8\dfrac{2}{5}$ rounds to 8 because $\dfrac{2}{5}$ is below the halfway point between 8 and 9.

$2\dfrac{1}{4}$ rounds to 2 because $\dfrac{1}{4}$ is below the halfway point between 2 and 3.

◆**Margin 2**
Divide.

a. $\dfrac{3}{16} \div \dfrac{5}{12}$

b. $\dfrac{\frac{1}{14}}{\frac{1}{7}}$

c. $\dfrac{5}{18} \div -\dfrac{2}{9}$

d. $\dfrac{-1}{5} \div -4$

Answers: **a.** $\dfrac{9}{20}$ **b.** $\dfrac{1}{2}$ **c.** $-1\dfrac{1}{4}$ **d.** $\dfrac{1}{20}$

a. $5\dfrac{1}{4} \div 2\dfrac{1}{3}$

b. $\dfrac{7\dfrac{2}{7}}{7\dfrac{1}{2}}$

c. $-13\dfrac{1}{5} \div 5\dfrac{1}{10}$

d. $-4\dfrac{1}{2} \div -11\dfrac{2}{5}$

◆ Margin 4

Divide.

a. $\dfrac{8}{13m} \div \dfrac{12m}{5}$

b. $\dfrac{10x^3}{14y^2} \div \dfrac{15xy}{9}$

c. $-\dfrac{11n^3}{12p^4} \div \dfrac{33n^3}{18p}$

d. $-\dfrac{21c}{30a^3b} \div -\dfrac{14c^3b}{20a^4}$

Answers to Margin 3: **a.** $2\dfrac{1}{4}$ **b.** $\dfrac{34}{35}$
c. $-2\dfrac{10}{17}$ **d.** $\dfrac{15}{38}$

Answers to Margin 4: **a.** $\dfrac{10}{39m^2}$
b. $\dfrac{3x^2}{7y^3}$ **c.** $\dfrac{-1}{2p}$ **d.** $\dfrac{a}{b^2c^2}$

Estimate: $8 \div 2 = 4$. The actual quotient should be around 4.

To calculate the actual quotient, we write each mixed number as an improper fraction, get the reciprocal of the divisor, and then multiply.

$$8\dfrac{2}{5} \div 2\dfrac{1}{4} = \dfrac{42}{5} \div \dfrac{9}{4} \qquad \text{Write mixed numbers as improper fractions.}$$

$$= \dfrac{42}{5} \cdot \dfrac{4}{9} \qquad \text{Write as an equivalent multiplication.}$$

$$= \dfrac{2 \cdot \overset{1}{\cancel{3}} \cdot 7}{5} \cdot \dfrac{2 \cdot 2}{\underset{1}{\cancel{3}} \cdot 3} \qquad \text{Write in factored form and divide out the common factors.}$$

$$= \dfrac{56}{15} \qquad \text{Write the improper fraction as a mixed number.}$$

$$= 3\dfrac{11}{15}$$

◀ **Do Margin 3.**

OBJECTIVE 3 *Divide rational expressions.*

To divide rational expressions we follow the same procedure as numeric fractions.

EXAMPLE 5 Divide. $-\dfrac{6a^3b}{35c^4} \div \dfrac{15a}{14c^2}$

Solution: Write an equivalent multiplication statement using the reciprocal of the divisor.

$$-\dfrac{6a^3b}{35c^4} \div \dfrac{15a}{14c^2} = -\dfrac{6a^3b}{35c^4} \cdot \dfrac{14c^2}{15a} \qquad \text{Write an equivalent multiplication statement.}$$

$$= -\dfrac{2 \cdot \overset{1}{\cancel{3}} \cdot \overset{1}{\cancel{a}} \cdot a \cdot a \cdot b}{5 \cdot \underset{1}{\cancel{7}} \cdot \underset{1}{\cancel{c}} \cdot \cancel{c} \cdot c \cdot c} \cdot \dfrac{2 \cdot \overset{1}{\cancel{7}} \cdot \overset{1}{\cancel{c}} \cdot \overset{1}{\cancel{c}}}{\underset{1}{\cancel{3}} \cdot 5 \cdot \underset{1}{\cancel{a}}} \qquad \text{Write in factored form and divide out common factors. Then multiply the remaining factors.}$$

$$= -\dfrac{4a^2b}{25c^2}$$

◀ **Do Margin 4.**

OBJECTIVE 4 *Find the square root of a fraction.*

Remember a square root is a number that, when squared, equals the given number. To find a square root we must find the base that can be multiplied by itself to equal the given amount.

Let's recall what happened when we squared a fraction:

$$\left(\dfrac{3}{4}\right)^2 = \dfrac{3}{4} \cdot \dfrac{3}{4} = \dfrac{9}{16} \qquad \text{Notice that we squared the numerator and denominator to get the square of } \dfrac{3}{4}.$$

Square root reverses this process:

$$\sqrt{\frac{9}{16}} = \frac{3}{4}$$ Notice that we got the square root of the numerator and denominator to get the square root of $\frac{9}{16}$.

EXAMPLE 6   Simplify. $\sqrt{\frac{25}{36}}$

Solution: We find the square root of the numerator and denominator separately.

$$\sqrt{\frac{25}{36}} = \frac{\sqrt{25}}{\sqrt{36}} = \frac{5}{6}$$

EXAMPLE 7   Simplify. $\sqrt{\frac{45}{5}}$

Solution: Notice 45 and 5 are not perfect squares so getting the square root of the numerator and denominator separately does not simplify the problem. Notice $\frac{45}{5}$ simplifies. We can treat the fraction like a division problem. It looks like this:

$$\sqrt{\frac{45}{5}} = \sqrt{9} = 3$$

Procedure   *To get the square root of a fraction we have two options:*
1. Get the square root of the numerator and denominator separately.
or
2. Simplify the fraction, then get the square root of the quotient.

Do Margin 5. ▶

OBJECTIVE 5   *Solve equations involving fractions.*

We learned how to solve equations in Chapter 4. We learned there are two principles that we use to solve equations. The addition/subtraction principle allowed us to add or subtract the same amount on both sides of an equation without affecting its solution(s). The multiplication/division principle allowed us to multiply or divide both sides of an equation by the same nonzero amount without affecting its solution(s).

Let's apply these principles to equations that contain fractions and/or mixed numbers. Remember our goal is to isolate the variable.

EXAMPLE 8   Solve and check. $\frac{3}{4}x = 2\frac{5}{8}$

Solution: To isolate x we must clear the coefficient $\frac{3}{4}$. There are two ways that we can write the steps. In Method 1, we divide both sides of the equation by $\frac{3}{4}$, writing the division in complex fraction form. In Method 2, we multiply both sides of the equation by the reciprocal of $\frac{3}{4}$, which is $\frac{4}{3}$. The two methods are equivalent.

◆ **Margin 5**

Simplify.

a. $\sqrt{\frac{4}{9}}$

b. $\sqrt{\frac{49}{100}}$

c. $\sqrt{\frac{125}{5}}$

d. $\sqrt{\frac{392}{8}}$

Answers: **a.** $\frac{2}{3}$ **b.** $\frac{7}{10}$ **c.** 5 **d.** 7

Method 1:

$$\frac{3}{4}x = 2\frac{5}{8}$$

$$\frac{\frac{3}{4}}{\frac{3}{4}}x = \frac{\frac{21}{8}}{\frac{3}{4}}$$

> We divide both sides by $\frac{3}{4}$. Most people do not like the looks of the complex fraction form of division.

$$1x = \frac{21}{8} \cdot \frac{4}{3}$$

$$x = \frac{\overset{1}{\cancel{3}} \cdot 7}{\underset{1}{\cancel{2}} \cdot \underset{1}{\cancel{2}} \cdot 2} \cdot \frac{\overset{1}{\cancel{2}} \cdot \overset{1}{\cancel{2}}}{\underset{1}{\cancel{3}}}$$

$$x = \frac{7}{2}$$

$$x = 3\frac{1}{2}$$

Method 2:

$$\frac{3}{4}x = 2\frac{5}{8}$$

$$\frac{\overset{1}{\cancel{4}}}{\underset{1}{\cancel{3}}} \cdot \frac{\overset{1}{\cancel{3}}}{\underset{1}{\cancel{4}}}x = 2\frac{5}{8} \cdot \frac{4}{3}$$

> We multiply both sides by $\frac{4}{3}$. Most people prefer this look so this is how we will write it from here on.

$$1x = \frac{21}{8} \cdot \frac{4}{3}$$

$$x = \frac{\overset{1}{\cancel{3}} \cdot 7}{\underset{1}{\cancel{2}} \cdot \underset{1}{\cancel{2}} \cdot 2} \cdot \frac{\overset{1}{\cancel{2}} \cdot \overset{1}{\cancel{2}}}{\underset{1}{\cancel{3}}}$$

$$x = \frac{7}{2}$$

$$x = 3\frac{1}{2}$$

Check: To check we replace the x in $\frac{3}{4}x = 2\frac{5}{8}$ with $3\frac{1}{2}$, and verify that it makes the equation true.

$$\frac{3}{4}x = 2\frac{5}{8}$$

$$\frac{3}{4} \cdot 3\frac{1}{2} \; ? \; 2\frac{5}{8}$$

$$\frac{3}{4} \cdot \frac{7}{2} \; ? \; 2\frac{5}{8}$$

$$\frac{21}{8} \; ? \; 2\frac{5}{8}$$

$$2\frac{5}{8} = 2\frac{5}{8}$$

◀ **Do Margin 6.**

EXAMPLE 9 Solve and check.

$$\frac{m}{5} = -3\frac{1}{10}$$

Solution: What is the coefficient here? The expression $\frac{m}{5}$ is the same as $\frac{1m}{5}$ or $\frac{1}{5}m$ so the coefficient is $\frac{1}{5}$. To isolate m, multiply both sides of the equation by the reciprocal of the coefficient $\frac{1}{5}$, which is $\frac{5}{1}$.

$$\frac{m}{5} = -3\frac{1}{10}$$

$$\frac{\overset{1}{\cancel{5}}}{1} \cdot \frac{m}{\underset{1}{\cancel{5}}} = -\frac{31}{2 \cdot \cancel{5}} \cdot \frac{\overset{1}{\cancel{5}}}{1}$$

$$1m = -\frac{31}{2}$$

$$m = -15\frac{1}{2}$$

◆ **Margin 6**

Solve and check.

a. $\dfrac{3}{4}a = \dfrac{2}{3}$

b. $\dfrac{5}{6}y = \dfrac{3}{10}$

c. $\dfrac{4}{9}k = -\dfrac{8}{15}$

d. $\dfrac{-4}{15} = -\dfrac{1}{6}n$

Answers: **a.** $\frac{8}{9}$ **b.** $\frac{9}{25}$ **c.** $-1\frac{1}{5}$ **d.** $1\frac{3}{5}$

Check: We replace m with $-15\frac{1}{2}$ in $\frac{m}{5} = -3\frac{1}{10}$ and verify that it makes the equation true.

$$\frac{m}{5} = -3\frac{1}{10}$$

$$\frac{-15\frac{1}{2}}{5} \;?\; -3\frac{1}{10}$$

Notice we get a complex fraction when we replace m with $-15\frac{1}{2}$. Remember the complex fraction means we divide: $-15\frac{1}{2} \div 5$.

$$-15\frac{1}{2} \div 5 \;?\; -3\frac{1}{10}$$

$$-\frac{31}{2} \cdot \frac{1}{5} \;?\; -3\frac{1}{10}$$

$$-\frac{31}{10} \;?\; -3\frac{1}{10}$$

$$-3\frac{1}{10} = -3\frac{1}{10}$$

Do Margin 7. ▶

OBJECTIVE 6 *Solve applications involving division of fractions.*

Let's return to the problem about getting change for a $20 bill in quarters. You should note that the $20 is the amount that we will split up, and the size or value of each division is to be a quarter, which is the fraction $\frac{1}{4}$. We are asking how many fourths are there in $20. We can also put it into the context of a missing factor statement:

the value of the currency · the number of coins/bills = the total money

$$\frac{1}{4} \qquad\qquad \cdot \qquad\qquad n \qquad\qquad = \qquad\qquad \$20$$

$$\frac{1}{4}n = 20$$

$$\frac{\cancel{4}}{1} \cdot \frac{1}{\cancel{4}}n = \frac{20}{1} \cdot \frac{4}{1}$$

$$n = 80$$

Conclusion: Whenever we are given a total amount and either the size of the parts or the number of parts, we can write a missing factor statement and then use division to solve the problem. We can say:

size of each part · number of parts = whole amount

Of course, if you recognize the situation for what it is, then go directly to the related division statement.

size of each part = whole amount ÷ number of parts

or

number of parts = whole amount ÷ size of each part

♦ **Margin 7**

Solve and check.

a. $\frac{x}{10} = 6$

b. $\frac{y}{7} = \frac{5}{14}$

c. $\frac{n}{8} = \frac{-3}{4}$

d. $\frac{h}{-9} = -2\frac{1}{3}$

Answers: **a.** 60 **b.** $2\frac{1}{2}$ **c.** -6 **d.** 21

EXAMPLE 10   A board is $35\frac{3}{4}$ in. long. The board is to be cut into pieces that are $6\frac{1}{2}$ in. each. How many pieces can be cut?

Solution: Use the problem-solving process.

Understand: We are given the total length of a board and the size of each part, and we want to find the number of parts. This is a missing factor/division problem.

Plan: We can write a missing factor statement, then solve.

Execute: Let *n* represent the number of parts:

$$\text{size of each part} \cdot \text{number of parts} = \text{whole amount}$$

$$6\frac{1}{2} \cdot n = 35\frac{3}{4} \qquad \text{Write the mixed numbers as improper fractions.}$$

$$\frac{13}{2}n = \frac{143}{4}$$

$$\frac{2}{13} \cdot \frac{13}{2}n = \frac{143}{4} \cdot \frac{2}{13} \qquad \text{Multiply both sides by } \frac{2}{13}.$$

$$\frac{1}{\cancel{4}} \cdot \frac{\cancel{4}}{1}n = \frac{11 \cdot \cancel{13}}{2 \cdot 2} \cdot \frac{\cancel{2}}{\cancel{13}}$$

$$1n = \frac{11}{2} \qquad \text{Write as a mixed number.}$$

$$n = 5\frac{1}{2}$$

Answer: The board can be cut into $5\frac{1}{2}$ pieces that are each $6\frac{1}{2}$ in. long. Note that this means that we could cut 5 whole pieces that are each $6\frac{1}{2}$ in. long and would be left with $\frac{1}{2}$ of another $6\frac{1}{2}$ in. piece.

Check: If we multiply the length of each part by the number of parts, we should get the total length of the original board.

$$\text{size of each part} \cdot \text{number of parts} = \text{whole amount}$$

$$6\frac{1}{2} \cdot 5\frac{1}{2} \; ? \; 35\frac{3}{4}$$

$$\frac{13}{2} \cdot \frac{11}{2} \; ? \; 35\frac{3}{4}$$

$$\frac{143}{4} \; ? \; 35\frac{3}{4}$$

$$35\frac{3}{4} = 35\frac{3}{4} \qquad \text{It checks.}$$

◀ **Do Margin 8.**

◆ **Margin 8**

Solve.

a. How many $4\frac{1}{2}$ oz. servings are there in $30\frac{1}{4}$ oz. of cereal?

b. A circle has a circumference of $25\frac{41}{56}$ in. What is the diameter?

Answers: **a.** $6\frac{13}{18}$ **b.** $8\frac{3}{16}$ in.

5.4 Exercises

FOR EXTRA HELP

 Videotape 6

 InterAct Math
Tutorial Software

 www.carsonmath.com

AWL Math Tutor Center

InterAct
MathXL   www.mathxl.com

Student's Solutions
Manual

For Exercises 1–8, write the reciprocal.

1. $\dfrac{2}{3}$

2. $\dfrac{3}{8}$

3. $\dfrac{1}{6}$

4. $-\dfrac{1}{9}$

5. -15

6. 7

7. $-\dfrac{x}{4}$

8. $\dfrac{-9y}{7}$

For Exercises 9–20, divide. Express the quotient in lowest terms.

9. $\dfrac{1}{2} \div \dfrac{1}{6}$

10. $\dfrac{5}{8} \div \dfrac{2}{3}$

11. $\dfrac{3}{10} \div \dfrac{5}{6}$

12. $\dfrac{9}{16} \div \dfrac{3}{4}$

13. $\dfrac{14}{15} \div -\dfrac{7}{12}$

14. $-\dfrac{10}{13} \div \dfrac{5}{12}$

15. $\dfrac{-7}{12} \div -14$

16. $-6 \div -\dfrac{3}{4}$

17. $\dfrac{\frac{2}{5}}{\frac{4}{15}}$

18. $\dfrac{\frac{9}{10}}{\frac{21}{10}}$

19. $\dfrac{12}{\frac{-2}{3}}$

20. $\dfrac{\frac{-3}{5}}{18}$

For Exercises 21–28, estimate the quotient. Then divide and express the quotient as a mixed number in lowest terms.

21. $5\dfrac{1}{6} \div \dfrac{4}{9}$

22. $7\dfrac{1}{3} \div \dfrac{3}{8}$

23. $9\dfrac{1}{2} \div 3\dfrac{3}{4}$

24. $2\dfrac{7}{12} \div 2\dfrac{1}{6}$

25. $-2\dfrac{9}{16} \div 10\dfrac{1}{4}$

26. $4\dfrac{2}{9} \div -8\dfrac{1}{3}$

27. $\dfrac{-12\frac{1}{2}}{-6\frac{3}{4}}$

28. $\dfrac{-10\frac{2}{3}}{-2\frac{5}{6}}$

For Exercises 29–34, divide. Express the quotient in lowest terms.

29. $\dfrac{7}{15x} \div \dfrac{1}{6x^3}$

30. $\dfrac{ab}{10} \div \dfrac{4b^3}{25}$

31. $\dfrac{9m^5}{20n} \div \dfrac{12m^2n}{25}$

32. $\dfrac{t^2u^4}{18v^3} \div \dfrac{-10t^5}{12v}$

33. $\dfrac{14x^6}{28y^4z} \div \dfrac{-8x^2y}{18z}$

34. $-\dfrac{13h^6}{20hk} \div \dfrac{-65h^3k}{25k^2}$

For Exercises 35–42, simplify.

35. $\sqrt{\dfrac{64}{81}}$

36. $\sqrt{\dfrac{49}{100}}$

37. $\sqrt{\dfrac{121}{36}}$

38. $\sqrt{\dfrac{169}{100}}$

39. $\sqrt{\dfrac{180}{5}}$

40. $\sqrt{\dfrac{960}{15}}$

41. $\sqrt{\dfrac{1053}{13}}$

42. $\sqrt{\dfrac{3400}{34}}$

For Exercises 43–50, solve and check.

43. $\dfrac{3}{4}x = 12$

44. $\dfrac{2}{5}a = 10$

45. $\dfrac{5}{8} = \dfrac{3}{16}y$

46. $\dfrac{7}{10} = \dfrac{1}{5}x$

47. $\dfrac{n}{12} = -\dfrac{3}{20}$

48. $-\dfrac{1}{16} = \dfrac{k}{14}$

49. $\dfrac{-5m}{6} = \dfrac{-5}{21}$

50. $-\dfrac{9}{40} = -\dfrac{3u}{10}$

For Exercises 51–60, solve.

51. A child is to receive $1\frac{1}{3}$ tsp. of amoxicillin (an antibiotic). How many doses can be given if the bottle contains 40 tsp. (200 ml)? If the instructions are to give the dosage twice a day for 14 days, is the amount of amoxicillin in the bottle correct? If it is not correct, why would the pharmacist put more or less in the bottle than the amount prescribed?

52. A cereal box contains $13\frac{3}{10}$ oz. of cereal. The label indicates a serving size is $1\frac{4}{5}$ oz. How many servings are in a box of the cereal? The label indicates that there about 8 servings per container. Is this accurate?

53. A baker only had enough ingredients to makes $10\frac{1}{2}$ cups of batter for muffins. If he is to make a dozen muffins of equal size, how much of the batter should he place in each cup of the muffin pan?

54. Ellyn makes curtains. She wants to make 4 curtains of equal length from 78 in. of material. How long should each piece be?

55. A rectangular field has an area of $2805\frac{1}{2}$ ft.^{2} and has a width of $30\frac{1}{2}$ ft. What is the length?

56. The desired area for a triangular garden is to be $14\frac{1}{2}$ ft.^{2}. If the base is to be $6\frac{1}{2}$ ft., what must the height be?

57. The largest radio telescope in the world is the Arecibo dish located in Puerto Rico. It has a circumference of $3111\frac{2}{5}$ ft. What is the diameter?

58. A circular track has a circumference of $\frac{1}{4}$ mi. What is the radius?

59. Ron uses the odometer in his car to measure the distance from his house to where he works. He finds the distance to be $18\frac{3}{10}$ mi. It takes him $\frac{2}{5}$ hr. to get there. What was his average rate?

60. How much time does it take an F-16 fighter plane traveling at mach 2 to travel 20 mi.? Mach 2 is about $\frac{2}{5}$ mps (miles per second), which is twice the speed of sound.

REVIEW EXERCISES

1. Find the prime factorization of 378.

2. Simplify $2^2 \cdot 3 \cdot 5 \cdot x^3 \cdot y$.

3. Use $<, >,$ or $=$ to make a true statement. Explain.

$\dfrac{5}{6} \; ? \; \dfrac{10}{12}$

4. Combine like terms. $5t^3 - 8t + 9t^2 + 4t^3 - 16 - 5t$

5. Factor. $32xy + 16x$.

5.5 Least Common Multiple

OBJECTIVES

1 Find the least common multiple (LCM) by listing.

2 Find the LCM using prime factorization.

3 Find the LCM of a set of monomials.

4 Upscale fractions to equivalent fractions with the LCM as the denominator.

5 Upscale rational expressions.

OBJECTIVE 1 *Find the least common multiple (LCM) by listing.*

In Section 5.1, we defined a *multiple* to be a number that is divisible by a given number. Multiples of 6 are all divisible by 6:

6, 12, 18, 24, 30, 36, 42, 48, . . .

Multiples of 4 are all divisible by 4:

4, 8, 12, 16, 20, 24, 28, 32, 36, 40, . . .

Notice the numbers 6 and 4 have several multiples in common. In our lists we can see 12, 24, and 36 are all common multiples for 6 and 4. We will find it useful to consider the smallest common multiple **or least common multiple.**

> **DEFINITION** **Least common multiple (LCM):** The smallest natural number that is divisible by all the given numbers.

Looking again at our list of multiples for 6 and 4 we see the least common multiple is 12. One way we can find the LCM for a given set of numbers is to list multiples for each number until we find a number common to all lists. However, we can quicken the process by recognizing that we really only need to list multiples of the largest of the given numbers. We simply list its multiples until we find a multiple that is divisible by all the other given numbers.

> **Procedure** *To find the LCM by listing, list multiples of the largest given number until you find a multiple that is divisible by all the other given numbers.*

EXAMPLE 1 Find the LCM of 36 and 120 by listing.

Solution: List multiples of 120 until we find a number that is also divisible by 36. Multiples of 120 are 120, 240, 360, etc.

120	240	360
36 does not divide 120 evenly so we go to the next multiple of 120.	36 does not divide 240 evenly so we go to the next multiple of 120.	36 divides 360 an even 10 times. This is the LCM.

Because 360 is the first multiple of 120 that is also divisible by 36, it must be the LCM.

In math language, we write:

$$\text{LCM}(36,120) = 360$$

Margin 1

Find the LCM of the given numbers by listing.

a. 18 and 24

b. 21 and 35

c. 15, 18, and 30

d. 24, 45, and 120

We can check the LCM by dividing it by the given numbers.

$$36 \overline{)360} \quad \begin{array}{r} 10 \\ \hline \end{array}$$
$$\underline{-36}$$
$$00$$
$$\underline{-0}$$
$$0$$

$$120 \overline{)360} \quad \begin{array}{r} 3 \\ \hline \end{array}$$
$$\underline{-360}$$
$$0$$

Look at the quotients 10 and 3. Notice the only common factor for 10 and 3 is 1. This means that 360 is in fact the LCM for 36 and 120. If any factor other than 1 divides all the quotients evenly, it means we've got a common multiple but not the smallest common multiple; that is, we've overshot the LCM. For example, suppose we somehow ended up with 720 as our answer. Look at the check:

$$36 \overline{)720} \quad \begin{array}{r} 20 \\ \hline \end{array}$$
$$\underline{-72}$$
$$00$$
$$\underline{-0}$$
$$0$$

> Notice both quotients are divisible by 2. This means we overshot the LCM by a factor of 2.

$$120 \overline{)720} \quad \begin{array}{r} 6 \\ \hline \end{array}$$
$$\underline{-720}$$
$$0$$

To correct the problem, we simply divide our incorrect answer by the factor common to all quotients. In this case, because we overshot by a factor of 2, we divide 720 by 2, which equals the correct answer 360.

◄ **Do Margin 1.**

OBJECTIVE 2 *Find the LCM using prime factorization.*

We can use prime factorization to generate the LCM of a given set of numbers. In Example 1 we found the LCM of 36 and 120 to be 360. Let's look at the prime factorizations of 36 and 120, then compare their factorizations with the prime factorization of 360 to see what we can learn.

$$36 = 2 \cdot 2 \cdot 3 \cdot 3 = 2^2 \cdot 3^2$$
$$120 = 2 \cdot 2 \cdot 2 \cdot 3 \cdot 5 = 2^3 \cdot 3 \cdot 5$$
$$\text{LCM}(36, 120) = 360 = 2 \cdot 2 \cdot 2 \cdot 3 \cdot 3 \cdot 5 = 2^3 \cdot 3^2 \cdot 5$$

Notice each prime factor is represented in the LCM. You may remember that when we found the GCF, we included only those primes that were common to all factorizations. In this example, the 5 appears in the LCM even though it is not common to both factorizations.

Conclusion: With LCM, we must use each prime factor whether common or not.

How many of each factor do we use? Notice the factorization of 360 contains the highest exponent of each prime factor in the factorizations of 36 and 120. The 2^3 in 120 has a larger exponent than the 2^2 in 36, so we use 2^3 from 120. The 3^2 in 36 has a larger exponent than the single 3 in 120, so we use 3^2.

Conclusion: Use the largest exponent of each prime factor to generate the factorization of the LCM.

Procedure *To find the LCM using prime factorization:*
1. Find the prime factorization of each given number.
2. Write the factorizations in exponential form.
3. Use the largest exponent of each prime factor.
4. Multiply to get the LCM.

EXAMPLE 2 Find the LCM of 24, 90, and 70 using prime factorization.

Solution: We must get the prime factorization of 24, 90, and 70, then use the largest exponent of each prime factor.

$$24 = 2^3 \cdot 3$$
$$90 = 2 \cdot 3^2 \cdot 5$$
$$70 = 2 \cdot 5 \cdot 7$$
$$\text{LCM}(24, 90, 70) = 2^3 \cdot 3^2 \cdot 5 \cdot 7$$
$$= 8 \cdot 9 \cdot 5 \cdot 7$$
$$= 2520$$

The highest exponent for 2 was 2^3 in 24.

The highest exponent for 3 was 3^2 in 90.

The highest exponent for 5 was a single 5 in 90.

The highest exponent for 7 was a single 7 in 70.

The listing method is simple, direct, and great for small jobs. The prime factorization method has a more involved setup but is great for big jobs. Of course we can use either tool for the job of finding the LCM, but the idea is to select the tool that fits the job.

Do Margin 2. ▶

OBJECTIVE 3 *Find the LCM of a set of monomials.*

To find the LCM of a group of monomials, we follow the same procedure and treat the variables as prime factors. Because the variables in monomials are already in factored form, we can simply write the largest exponent of each variable in the LCM.

EXAMPLE 3 Find the LCM of $18x^3y$ and $24xz^2$.

Solution: Because the variables are already in factored form, we can simply write the highest exponent of each variable in our LCM. The LCM will contain x^3yz^2. We now just need to calculate the LCM of the coefficients by listing or prime factorization. We will use prime factorization.

$$18 = 2 \cdot 3^2$$
$$24 = 2^3 \cdot 3$$
$$\text{LCM}(18x^3y, 24xz^2) = 2^3 \cdot 3^2 x^3yz^2$$
$$= 8 \cdot 9x^3yz^2$$
$$= 72x^3yz^2$$

Do Margin 3. ▶

OBJECTIVE 4 *Upscale fractions to equivalent fractions with the LCM as the denominator.*

In Section 5.1, we learned that we can upscale a fraction to an equivalent fraction by multiplying both the numerator and denominator by the same number. Also, we showed that we can compare fractions if they have the same denominator. Then we found common denominators to be common multiples of the denominators. To keep the numbers in the fractions as small as possible, it would be in our best interest to use the least common multiple of the denominators as the common denominator.

For the fractions $\frac{3}{4}$ and $\frac{5}{6}$, there are many common multiples for the denominators. We could use 12, 24, or 36 to name a few. However, notice the LCM is 12. By upscaling the fractions to equivalent fractions with 12 as the denominator, we will keep the numbers involved as small as possible. When we use the LCM as the common denominator, we say we are using the lowest or **least common denominator.**

DEFINITION **Least common denominator (LCD):** The LCM of the denominators.

◆ **Margin 2**

Find the LCM using prime factorization.

a. 36 and 80

b. 30 and 75

c. 20, 35, and 40

d. 26, 40, and 65

◆ **Margin 3**

Find the LCM of each set of monomials.

a. $6a^3$ and $12a$

b. $9m^2n$ and $12m^4$

c. $10hk^5$ and $8h^2k$

d. $20x^3yz^2$ and $15xy^2z^5$

Answers to Margin 2: **a.** 720 **b.** 150 **c.** 280 **d.** 520

Answers to Margin 3: **a.** $12a^3$ **b.** $36m^4n$ **c.** $40h^2k^5$ **d.** $60x^3y^2z^5$

Margin 4

Upscale the fractions to equivalent fractions with the LCD.

a. $\dfrac{2}{3}$ and $\dfrac{1}{4}$

b. $\dfrac{5}{8}$ and $\dfrac{7}{12}$

c. $\dfrac{2}{5}$ and $\dfrac{3}{10}$

d. $\dfrac{9}{20}$ and $\dfrac{8}{15}$

Margin 5

Upscale the rational expressions to equivalent expressions with the LCD.

a. $\dfrac{5}{9}$ and $\dfrac{2}{3a}$

b. $\dfrac{3}{4xy}$ and $\dfrac{7}{6x^2}$

c. $\dfrac{-m}{n^4}$ and $\dfrac{3}{mn}$

d. $\dfrac{11}{12t^2u}$ and $-\dfrac{7u}{9tv}$

Answers to Margin 4:

a. $\dfrac{8}{12}$ and $\dfrac{3}{12}$

b. $\dfrac{15}{24}$ and $\dfrac{14}{24}$

c. $\dfrac{4}{10}$ and $\dfrac{3}{10}$

d. $\dfrac{27}{60}$ and $\dfrac{32}{60}$

Answers to Margin 5:

a. $\dfrac{5a}{9a}$ and $\dfrac{6}{9a}$

b. $\dfrac{9x}{12x^2y}$ and $\dfrac{14y}{12x^2y}$

c. $\dfrac{-m^2}{mn^4}$ and $\dfrac{3n^3}{mn^4}$

d. $\dfrac{33v}{36t^2uv}$ and $-\dfrac{28tu^2}{36t^2uv}$

EXAMPLE 4 Upscale $\dfrac{7}{12}$ and $\dfrac{2}{15}$ to equivalent fractions with the least common denominator.

Solution: The least common denominator (LCD) is the LCM of the denominators. We need to find the LCM of 12 and 15, then upscale the fractions. Let's use listing to find the LCM. Multiples of 15 are 15, 30, etc.

15	30	45	60
12 does not divide 15 evenly so we go to the next multiple of 15.	12 does not divide 30 evenly so we go to the next multiple of 15.	12 does not divide 45 evenly so we go to the next multiple of 15.	12 divides 60 evenly so 60 is the LCM.

So we must upscale $\dfrac{7}{12}$ and $\dfrac{2}{15}$ to equivalent fractions with 60 as the denominator. To upscale, we multiply the numerator and denominator by the same amount.

$$\frac{7 \cdot ?}{12 \cdot ?} = \frac{}{60} \qquad\qquad \frac{2 \cdot ?}{15 \cdot ?} = \frac{}{60}$$

We think, "What factor will multiply by 12 to equal 60?" Notice to find this missing factor, we can divide.

$$60 \div 12 = 5$$

So we need to multiply the numerator and denominator by 5.

We think, "What factor will multiply by 15 to equal 60?" Notice to find this missing factor, we can divide.

$$60 \div 15 = 4$$

So we need to multiply the numerator and denominator by 4.

$$\frac{7 \cdot 5}{12 \cdot 5} = \frac{35}{60} \qquad\qquad \frac{2 \cdot 4}{15 \cdot 4} = \frac{8}{60}$$

◀ **Do Margin 4.**

OBJECTIVE 5 *Upscale rational expressions.*

Rational expressions are upscaled the same way numeric fractions are. If the denominators are monomials, we must pay special attention to the variables and their exponents.

EXAMPLE 5 Upscale $\dfrac{3}{8x}$ and $\dfrac{5}{6x^2}$ to equivalent rational expressions with the least common denominator.

Solution: The least common denominator (LCD) is the LCM of the denominators. We need to find the LCM of $8x$ and $6x^2$, then upscale the rational expressions.

For the variables, x^2 has the larger exponent so the LCD will contain x^2. The LCM for 8 and 6 is 24. So the LCD is $24x^2$. This means we must upscale $\dfrac{3}{8x}$ and $\dfrac{5}{6x^2}$ to equivalent fractions with $24x^2$ as the denominator.

$$\frac{3 \cdot ?}{8x \cdot ?} = \frac{}{24x^2} \qquad\qquad \frac{5 \cdot ?}{6x^2 \cdot ?} = \frac{}{24x^2}$$

We think, "What factor will multiply by $8x$ to equal $24x^2$?" Notice to find this missing factor, we can divide.

$$24x^2 \div 8x = 3x$$

So we need to multiply the numerator and denominator by $3x$.

We think, "What factor will multiply by $6x^2$ to equal $24x^2$?" Notice to find this missing factor, we can divide.

$$24x^2 \div 6x^2 = 4$$

So we need to multiply the numerator and denominator by 4.

$$\frac{3 \cdot 3x}{8x \cdot 3x} = \frac{9x}{24x^2} \qquad\qquad \frac{5 \cdot 4}{6x^2 \cdot 4} = \frac{20}{24x^2}$$

◀ **Do Margin 5.**

5.5 Exercises

For Exercises 1–8, find the LCM by listing.

1. 10 and 6   **2.** 15 and 9   **3.** 12 and 36   **4.** 8 and 32

5. 20 and 30   **6.** 16 and 20   **7.** 4, 6, and 9   **8.** 5, 6, and 15

For Exercises 9–20, find the LCM using prime factorization.

9. 18 and 24   **10.** 24 and 50   **11.** 63 and 28   **12.** 42 and 56

13. 52 and 28   **14.** 68 and 56   **15.** 180 and 200   **16.** 210 and 420

17. 28, 32, and 60   **18.** 26, 30, and 39   **19.** 42, 56, and 80   **20.** 36, 49, and 72

For Exercises 21–28, find the LCM.

21. $12x$ and $8y$ **22.** $9ab$ and $3b$ **23.** $16mn$ and $8m$ **24.** $14t^2$ and $5tu$

25. $10y^3z$ and $6y$ **26.** $20h^5k$ and $15h^2k^3$ **27.** $18x^2y$ and $12xy^3z$ **28.** $24a^4bc$ and $30ab^2$

For Exercises 29–36, upscale the fractions to equivalent fractions with the LCD.

29. $\dfrac{3}{10}$ and $\dfrac{5}{6}$ **30.** $\dfrac{2}{15}$ and $\dfrac{4}{9}$ **31.** $\dfrac{7}{12}$ and $\dfrac{11}{36}$ **32.** $\dfrac{1}{8}$ and $\dfrac{15}{32}$

33. $\dfrac{1}{20}$ and $\dfrac{17}{30}$ **34.** $\dfrac{3}{16}$ and $\dfrac{9}{20}$ **35.** $\dfrac{3}{4}, \dfrac{1}{6},$ and $\dfrac{7}{9}$ **36.** $\dfrac{2}{5}, \dfrac{5}{6},$ and $\dfrac{13}{15}$

For Exercises 37–44, upscale the rational expressions to equivalent expressions with the LCD.

37. $\dfrac{7}{12x}$ and $\dfrac{3}{8y}$ **38.** $\dfrac{5}{9ab}$ and $\dfrac{2}{3b}$ **39.** $\dfrac{9}{16mn}$ and $\dfrac{3n}{8m}$ **40.** $\dfrac{1}{14t^2}$ and $\dfrac{2}{5tu}$

41. $\dfrac{7}{10y^3z}$ and $\dfrac{5z}{6y}$ **42.** $\dfrac{-13}{20h^5k}$ and $\dfrac{4}{15h^2k^3}$ **43.** $\dfrac{z}{18x^2y}$ and $\dfrac{-5}{12xy^3z}$ **44.** $\dfrac{5}{24a^4bc}$ and $\dfrac{c}{30ab^2}$

Puzzle Problem

Planetary conjunctions occur when two or more planets form a rough line with the sun. Using rough estimates of the orbital periods of the following planets, calculate the time that will elapse between alignments of these planets.

	Orbit Period
Earth	1 year
Mars	2 years
Jupiter	12 years
Saturn	30 years
Uranus	84 years

Trivia Bite

Because the planets do not all orbit in the same plane it is virtually impossible for more than two planets to form a straight line with the sun. However, if we were to view the solar system from above, the planets can form an apparent line (conjunction) with the sun. Theoretically, this type of alignment for all 9 planets could occur every 334,689,064 years. When conjunction dates are calculated we relax the criterion that the planets be in a straight line and choose an acceptable angle of dispersion.

The conjunction that occurred in May of 2000 included Mercury, Venus, Earth, the moon, Mars, Jupiter, and Saturn with an angle of dispersion of about 15 degrees in the sky.

REVIEW EXERCISES

1. $16 + (-28) =$

2. $-8 - (-19) =$

3. Combine like terms. $14x^2 - 16x + 25x^2 - 12 + 4x^3 + 9$

4. Solve and check. $m + 14 = 6$

5. Solve and check. $x - 37 = 16$

5.6 Adding and Subtracting Fractions, Mixed Numbers, and Rational Expressions

OBJECTIVES

1. Add and subtract fractions with the same denominator.
2. Add and subtract rational expressions with the same denominator.
3. Add and subtract fractions with different denominators.
4. Add and subtract rational expressions.
5. Add mixed numbers.
6. Subtract mixed numbers.
7. Add and subtract signed mixed numbers.
8. Solve equations.
9. Solve applications.

OBJECTIVE 1 *Add and subtract fractions with the same denominator.*

What happens when we add two half cups of milk? We should get a whole cup. Two halves make a whole.

$$\frac{1}{2} + \frac{1}{2} = 1$$

How does this work? The first thought that we might pursue is to treat addition like multiplication and add numerator plus numerator and denominator plus denominator. But, watch what happens:

$$\frac{1}{2} + \frac{1}{2} = \frac{1+1}{2+2} = \frac{2}{4} = \frac{1}{2} \qquad \text{Can't be true.}$$

When we follow that approach to its conclusion, a half plus a half equals a half. This doesn't make sense. The only situation where we can add and end up with the same amount as what we started with is $0 + 0 = 0$.

So what do we do? Notice if we add the numerators and keep the same denominator we'll have $\frac{2}{2}$ which simplifies to 1.

$$\frac{1}{2} + \frac{1}{2} = \frac{1+1}{2} = \frac{2}{2} = 1$$

Conclusion: If the denominators are the same, we add (or subtract) numerators and keep the same denominator.

> **Procedure** *To add or subtract fractions that have the same denominator:*
> 1. Add or subtract the numerators.
> 2. Keep the same denominator.
> 3. Simplify.

EXAMPLE 1 Add. $\dfrac{1}{8} + \dfrac{3}{8}$

Solution: Because the denominators are the same, add the numerators, then simplify.

$$\frac{1}{8} + \frac{3}{8} = \frac{1+3}{8} = \frac{4}{8} = \frac{1}{2}$$

Margin 1

Add or subtract.

a. $\dfrac{5}{6} + \dfrac{1}{6}$

b. $\dfrac{3}{16} + \dfrac{1}{16}$

c. $\dfrac{9}{10} - \dfrac{3}{10}$

d. $\dfrac{7}{8} - \dfrac{5}{8}$

EXAMPLE 2 Subtract. $\dfrac{8}{9} - \dfrac{2}{9}$

Solution: Because the denominators are the same, subtract the numerators, then simplify.

$$\frac{8}{9} - \frac{2}{9} = \frac{8-2}{9} = \frac{6}{9} = \frac{2}{3}$$

◀ **Do Margin 1.**

EXAMPLE 3 Add. $-\dfrac{3}{10} + -\dfrac{1}{10}$

Solution: Because the denominators are the same add the numerators, then simplify.

$$-\frac{3}{10} + \left(-\frac{1}{10}\right) = \frac{-3 + (-1)}{10} = -\frac{4}{10} = -\frac{2}{5}$$

> **Reminder**
> To add two numbers that have the same sign, add their absolute values and keep the same sign.

EXAMPLE 4 Add. $\dfrac{-7}{12} + \dfrac{5}{12}$

Solution: Because the denominators are the same, add the numerators, then simplify.

$$\frac{-7}{12} + \frac{5}{12} = \frac{-7 + 5}{12} = \frac{-2}{12} = \frac{-1}{6}$$

> **Reminder**
> To add two numbers that have different signs, subtract their absolute values and keep the sign of the number with the larger absolute value.

Margin 2

Add.

a. $\dfrac{-1}{8} + \dfrac{-3}{8}$

b. $-\dfrac{5}{12} + -\dfrac{5}{12}$

c. $\dfrac{11}{24} + \left(-\dfrac{5}{24}\right)$

d. $\dfrac{-7}{16} + \dfrac{1}{16}$

◀ **Do Margin 2.**

EXAMPLE 5 Subtract. $\dfrac{5}{9} - \dfrac{7}{9}$

Solution: Because the denominators are the same, subtract the numerators, then simplify.

$$\frac{5}{9} - \frac{7}{9} = \frac{5-7}{9} \qquad \text{Write an equivalent addition statement.}$$

$$= \frac{5 + (-7)}{9}$$

$$= -\frac{2}{9}$$

> **Reminder**
> To write a subtraction statement as an equivalent addition statement, change the subtraction sign to an addition sign and change the subtrahend to its additive inverse.

Answers to Margin 1: a. 1 b. $\frac{1}{4}$ c. $\frac{3}{5}$
d. $\frac{1}{4}$

Answers to Margin 2: a. $-\frac{1}{2}$ b. $-\frac{5}{6}$
c. $\frac{1}{4}$ d. $-\frac{3}{8}$

EXAMPLE 6 Subtract. $\dfrac{-3}{16} - \left(\dfrac{-5}{16}\right)$

Solution: Because the denominators are the same, subtract the numerators, then simplify.

$$\frac{-3}{16} - \left(\frac{-5}{16}\right) = \frac{-3 - (-5)}{16}$$

Write an equivalent addition statement.

$$= \frac{-3 + 5}{16}$$

$$= \frac{2}{16}$$

$$= \frac{1}{8}$$

Do Margin 3. ▶

OBJECTIVE 2 *Add and subtract rational expressions with the same denominator.*

To add or subtract rational expressions, we follow the same procedure. Add or subtract the numerators and keep the same denominator. Keep in mind that we can add or subtract only like terms. If the terms are not alike, we must indicate the addition or subtraction in a polynomial expression.

EXAMPLE 7 Add. $\frac{3x}{10} + \frac{x}{10}$

> **Reminder**
> Like terms have the same variable(s) raised to the same exponent(s).

Solution: Because the rational expressions have the same denominator, we add numerators and keep the same denominator.

$$\frac{3x}{10} + \frac{x}{10} = \frac{3x + x}{10} = \frac{4x}{10} = \frac{2x}{5}$$

Because $3x$ and x are like terms, they can be combined to equal $4x$.

◆ We could add the numerators because they are like terms.

EXAMPLE 8 Add. $\frac{x}{3} + \frac{2}{3}$

Solution: Because the denominators are the same, we can add the numerators and keep the same denominator.

$$\frac{x}{3} + \frac{2}{3} = \frac{x + 2}{3}$$

We cannot combine x and 2 because they are not like terms. We must express the addition as a polynomial.

EXAMPLE 9 Add. $\frac{x^2 - 3x + 1}{7} + \frac{x + 4}{7}$

Solution: The denominators are the same, so we add the numerators and keep the same denominator. Notice the numerators are polynomials. To add polynomials, we combine like terms.

$$\frac{x^2 - 3x + 1}{7} + \frac{x + 4}{7} = \frac{(x^2 - 3x + 1) + (x + 4)}{7}$$

$$= \frac{x^2 - 2x + 5}{7}$$

x^2 had no like term to combine with. We combined $-3x$ with x to get $-2x$. We combined 1 with 4 to get 5.

Do Margin 4. ▶

Let's consider subtraction of rational expressions that have the same denominator. We can still write subtraction as an equivalent addition. Remember, if we do not have like terms, we must express the subtraction as a polynomial.

◆ **Margin 3**

Subtract.

a. $\frac{1}{9} - \frac{5}{9}$

b. $-\frac{3}{15} - \frac{2}{15}$

c. $\frac{1}{7} - \left(-\frac{4}{7}\right)$

d. $-\frac{9}{10} - \left(-\frac{1}{10}\right)$

◆ **Margin 4**

Add.

a. $\frac{3}{8m} + \frac{7}{8m}$

b. $\frac{5t}{u} + \frac{2}{u}$

c. $\frac{x + 7}{9} + \frac{2 - x}{9}$

d. $\frac{3t^2 + 8}{5} + \frac{4t^2 - t - 10}{5}$

Answers to Margin 3: **a.** $-\frac{4}{9}$ **b.** $-\frac{1}{3}$ **c.** $\frac{5}{7}$ **d.** $-\frac{4}{5}$

Answers to Margin 4: **a.** $\frac{5}{4m}$ **b.** $\frac{5t + 2}{u}$ **c.** 1 **d.** $\frac{7t^2 - t - 2}{5}$

EXAMPLE 10 Subtract. $\dfrac{4n}{9} - \left(\dfrac{-5}{9}\right)$

Solution: Because the denominators are the same, we subtract numerators and keep the same denominator.

$$\dfrac{4n}{9} - \left(\dfrac{-5}{9}\right) = \dfrac{4n - (-5)}{9} \qquad \text{Write as an equivalent addition statement.}$$

$$= \dfrac{4n + 5}{9}$$

EXAMPLE 11 Subtract. $\dfrac{8x^2 - 5x + 3}{y} - \dfrac{2x - 7}{y}$

Solution: Because the denominators are the same, we subtract numerators and keep the same denominator. To subtract the polynomial numerators, we write an equivalent addition, then combine like terms.

$$\dfrac{8x^2 - 5x + 3}{y} - \dfrac{2x - 7}{y} = \dfrac{(8x^2 - 5x + 3) - (2x - 7)}{y} \qquad \text{Write as an equivalent addition statement.}$$

$$= \dfrac{(8x^2 - 5x + 3) + (-2x + 7)}{y}$$

$$= \dfrac{8x^2 - 7x + 10}{y} \qquad \begin{array}{l} 8x^2 \text{ had no like terms. } -5x \text{ and } -2x \\ \text{combine to equal } -7x. \text{ 3 and 7 combine} \\ \text{to equal 10.} \end{array}$$

> **Reminder**
> To write the equivalent addition, we must find the additive inverse of the subtrahend, $2x - 7$.
> To write the additive inverse of a polynomial, we change all the signs in the polynomial.

◀ **Do Margin 5.**

OBJECTIVE 3 *Add and subtract fractions with different denominators.*

To add or subtract fractions that have different denominators, we need to find a common denominator. We will apply what we learned in Section 5.5 about getting the LCD and upscaling.

> **Procedure** *To add or subtract fractions with different denominators:*
> **1.** Find the LCD.
>
> > **Note:** Any common multiple of the denominators will do, but using the LCD makes the numbers more manageable.
>
> **2.** Upscale the fractions to equivalent fractions that have the common denominator.
> **3.** Add or subtract the numerators and keep the common denominator.
> **4.** Simplify.

EXAMPLE 12 Add. $\dfrac{7}{12} + \dfrac{2}{15}$ **Note:** These are the same fractions we used in Example 4 in Section 5.5.

Solution: We must find a common denominator, upscale, then add the numerators and keep the common denominator. The LCD for 12 and 15 is 60. (See Example 4, Section 5.5.)

◆ **Margin 5**

Subtract.

a. $\dfrac{7}{15x} - \dfrac{2}{15x}$

b. $\dfrac{9n}{5} - \dfrac{-n}{5}$

c. $\dfrac{y}{z} - \dfrac{4}{z}$

d. $\dfrac{3 + b}{7a} - \dfrac{9 - b}{7a}$

Answers: **a.** $\dfrac{1}{3x}$ **b.** $2n$ **c.** $\dfrac{y-4}{z}$
d. $\dfrac{-6 + 2b}{7a}$

$$\frac{7}{12} + \frac{2}{15}$$

To write an equivalent fraction with 60 as the denominator we upscaled $\frac{7}{12}$ by multiplying its numerator and denominator by 5.

$$= \frac{7(5)}{12(5)} + \frac{2(4)}{15(4)}$$

$$= \frac{35}{60} + \frac{8}{60}$$

$$= \frac{43}{60}$$

To write an equivalent fraction with 60 as the denominator, we upscaled $\frac{2}{15}$ by multiplying its numerator and denominator by 4.

Do Margin 6. ▶

OBJECTIVE **4** *Add and subtract rational expressions.*

The same procedure that applies to adding and subtracting numeric fractions applies to rational expressions.

EXAMPLE 13 Add. $\dfrac{3}{8x} + \dfrac{5}{6x^2}$ **Note:** These are the same expressions in Example 5 of Section 5.5.

Solution: We must find a common denominator, upscale the fractions, then add numerators and keep the common denominator. Keep in mind that if the numerators are not like terms, we will have to express the sum as a polynomial.

The LCM for 8 and 6 is 24. For the x's, because x^2 has the larger exponent, the LCM contains x^2. The LCD is $24x^2$. (See Example 5, Section 5.5.)

To write an equivalent fraction with $24x^2$ as the denominator, we upscaled $\frac{3}{8x}$ by multiplying its numerator and denominator by $3x$.

$$\frac{3}{8x} + \frac{5}{6x^2}$$

$$= \frac{3(3x)}{8x(3x)} + \frac{5(4)}{6x^2(4)}$$

$$= \frac{9x}{24x^2} + \frac{20}{24x^2}$$

$$= \frac{9x + 20}{24x^2}$$

To write an equivalent fraction with $24x^2$ as the denominator, we upscaled $\frac{5}{6x^2}$ by multiplying its numerator and denominator by 4.

Because $9x$ and 20 are not like terms, we had to express the sum as a polynomial. Because we cannot factor $9x + 20$, we cannot reduce the rational expression.

Do Margin 7. ▶

OBJECTIVE **5** *Add mixed numbers.*

There are two ways we can add mixed numbers. Because we know how to add fractions, the logical thing would be to write the mixed numbers as improper fractions, then follow the procedure for adding fractions.

EXAMPLE 14 Add. $5\dfrac{3}{4} + 4\dfrac{2}{3}$

Solution: We will write the mixed numbers as improper fractions, then follow our procedure for adding fractions.

$$5\frac{3}{4} + 4\frac{2}{3}$$

To write an equivalent fraction with 12 as the denominator, we upscaled $\frac{23}{4}$ by multiplying the numerator and denominator by 3.

$$= \frac{23}{4} + \frac{14}{3}$$ The LCD for 4 and 3 is 12.

$$= \frac{23(3)}{4(3)} + \frac{14(4)}{3(4)}$$ To write an equivalent fraction with 12 as the denominator, we upscaled $\frac{14}{3}$ by multiplying the numerator and denominator by 4.

$$= \frac{69}{12} + \frac{56}{12}$$

$$= \frac{125}{12}$$ Write as a mixed number.

$$= 10\frac{5}{12}$$

Many people do not like writing the mixed numbers as improper fractions because the numbers get quite large during the process. Because mixed numbers are the result of addition by definition, we can make use of the commutative property of addition and add the integer parts and fraction parts separately.

$$2\frac{1}{5} = 2 + \frac{1}{5} \quad \text{and} \quad 3\frac{2}{5} = 3 + \frac{2}{5}$$

So we can say:

$$2\frac{1}{5} + 3\frac{2}{5} = 2 + \frac{1}{5} + 3 + \frac{2}{5}$$

$$= 2 + 3 + \frac{1}{5} + \frac{2}{5}$$ From here on, we will not write all the steps you see here. We will leave the mixed numbers intact but add the integers and fractions separately.

$$= 5 + \frac{3}{5}$$

$$= 5\frac{3}{5}$$

Let's repeat Example 14 using this method.

EXAMPLE 15 Add. $5\frac{3}{4} + 4\frac{2}{3}$

Solution: We will add the integer parts and fraction parts separately.

$$5\frac{3}{4} + 4\frac{2}{3}$$

$$= 5\frac{3(3)}{4(3)} + 4\frac{2(4)}{3(4)}$$

$$= 5\frac{9}{12} + 4\frac{8}{12}$$

$$= 9\frac{17}{12}$$

Why did we get $10\frac{5}{12}$ using the first method? Actually, both $10\frac{5}{12}$ and $9\frac{17}{12}$ are the same result. Notice the $\frac{17}{12}$ in $9\frac{17}{12}$ is an improper fraction. If we write $\frac{17}{12}$ as a mixed number, we get $1\frac{5}{12}$. If we combine the 1 with the 9 we get $10\frac{5}{12}$. It looks like this:

$$9\frac{17}{12} = 9 + 1\frac{5}{12} = 10\frac{5}{12}$$

Procedure *To add mixed numbers:*

Method 1: Write as improper fractions then follow the procedure for adding fractions.

Method 2: Add the integer parts and fraction parts separately.

Do Margin 8. ▶

OBJECTIVE **6** *Subtract mixed numbers.*

We can subtract mixed numbers in the same way that we add. Let's first consider an example where we write the mixed numbers as improper fractions then subtract.

EXAMPLE 16   Subtract. $7\frac{1}{4} - 2\frac{5}{6}$

Solution: We will write the mixed numbers as improper fractions then follow the procedure for adding and subtracting fractions.

$$7\frac{1}{4} - 2\frac{5}{6}$$
$$= \frac{29}{4} - \frac{17}{6}$$
$$= \frac{29(3)}{4(3)} - \frac{17(2)}{6(2)}$$
$$= \frac{87}{12} - \frac{34}{12}$$
$$= \frac{53}{12}$$
$$= 4\frac{5}{12}$$

As with addition of mixed numbers, we can subtract the integer parts and fraction parts separately. Here's why:

Consider $6\frac{4}{9} - 2\frac{1}{9}$. When we expand the mixed numbers, we get:

$$\left(6 + \frac{4}{9}\right) - \left(2 + \frac{1}{9}\right)$$

Notice this is like subtracting polynomials. We can write an equivalent addition statement by getting the additive inverse of the subtrahend. To get the additive inverse of $(2 + \frac{1}{9})$, we change the signs of the terms inside the parentheses.

$$\left(6 + \frac{4}{9}\right) + \left(-2 - \frac{1}{9}\right)$$

Now that it is an addition statement, the commutative property of addition allows us to combine the integer and fraction parts separately.

Conclusion: We can subtract the integer and fraction parts separately.

$$\left(6 + \frac{4}{9}\right) + \left(-2 - \frac{1}{9}\right) = 6 - 2 + \frac{4}{9} - \frac{1}{9}$$
$$= 4 + \frac{3}{9}$$
$$= 4 + \frac{1}{3}$$
$$= 4\frac{1}{3}$$

Let's repeat Example 16 using this method.

EXAMPLE 17   Subtract. $7\frac{1}{4} - 2\frac{5}{6}$

Solution: Leave the mixed numbers intact and subtract the integers and fractions separately.

◆ **Margin 8**

Add.

a. $4 + 9\frac{2}{7}$

b. $6\frac{1}{8} + 3\frac{5}{8}$

c. $5\frac{5}{6} + 7\frac{1}{3}$

d. $3\frac{3}{4} + 8\frac{4}{5}$

Answers: **a.** $13\frac{2}{7}$ **b.** $9\frac{3}{4}$ **c.** $13\frac{1}{6}$
d. $12\frac{11}{20}$

Margin 9

Subtract.

a. $9\frac{7}{8} - 4\frac{5}{8}$

b. $7\frac{9}{10} - 1\frac{1}{5}$

c. $10\frac{1}{5} - 3\frac{3}{4}$

d. $8 - 5\frac{1}{4}$

$$7\frac{1}{4} - 2\frac{5}{6}$$

$$= 7\frac{1(3)}{4(3)} - 2\frac{5(2)}{6(2)}$$

$$= 7\frac{3}{12} - 2\frac{10}{12}$$

We have a problem now. Notice we can subtract the 2 from 7 but we cannot subtract $\frac{10}{12}$ from $\frac{3}{12}$. We must get the $\frac{3}{12}$ larger than the $\frac{10}{12}$ in order to subtract. We take 1 away from the 7 integer, write that 1 as $\frac{12}{12}$, then add it to the $\frac{3}{12}$. The steps look like this:

$$7\frac{3}{12} = 6 + 1 + \frac{3}{12} = 6 + \frac{12}{12} + \frac{3}{12} = 6\frac{15}{12}$$

Now we can complete the subtraction.

$$7\frac{3}{12} - 2\frac{10}{12}$$

$$= 6\frac{15}{12} - 2\frac{10}{12}$$

$$= 4\frac{5}{12} \qquad \text{This is the same result as Example 16.}$$

◀ **Do Margin 9.**

OBJECTIVE 7 *Add and subtract signed mixed numbers.*

We have seen that when we add mixed numbers, the mixed-number sum often includes an improper fraction within the mixed number. We have also seen that we often have to rewrite mixed numbers in order to be able to subtract, which can be confusing. It is usually safer to use our original method of writing the mixed numbers as improper fractions, then follow the procedure for adding and subtracting fractions.

EXAMPLE 18 Subtract. $-9\frac{1}{6} - \left(-2\frac{1}{3}\right)$

Solution: First we should write an equivalent addition statement by getting the additive inverse of the subtrahend. Because signs are involved, we will be safe and write the mixed numbers as improper fractions. This will avoid borrowing.

$$-9\frac{1}{6} - \left(-2\frac{1}{3}\right) \qquad \text{Write an equivalent addition using the additive inverse of the subtrahend.}$$

$$= -9\frac{1}{6} + 2\frac{1}{3} \qquad \text{Write the mixed numbers as improper fractions.}$$

$$= -\frac{55}{6} + \frac{7}{3}$$

Upscale $-\frac{55}{6}$ to an equivalent fraction with 6 as the denominator by multiplying the numerator and denominator by 1. It already has 6 as its denominator.

$$= -\frac{55(1)}{6(1)} + \frac{7(2)}{3(2)}$$

Upscale $\frac{7}{3}$ to an equivalent fraction with 6 as the denominator by multiplying the numerator and denominator by 2.

$$= -\frac{55}{6} + \frac{14}{6} \qquad \text{Add.}$$

$$= -\frac{41}{6} \qquad \text{Write the sum as a mixed number.}$$

$$= -6\frac{5}{6}$$

◀ **Do Margin 10.**

Margin 10

Add or subtract.

a. $5\frac{3}{4} + \left(-5\frac{1}{8}\right)$

b. $7\frac{1}{2} - 10\frac{1}{3}$

c. $-5\frac{7}{10} - 2\frac{1}{4}$

d. $-8\frac{2}{9} - \left(-\frac{5}{6}\right)$

Answers to Margin 9: a. $5\frac{1}{4}$ b. $6\frac{7}{10}$
c. $6\frac{9}{20}$ d. $2\frac{3}{4}$

Answers to Margin 10: a. $\frac{5}{8}$ b. $-2\frac{5}{6}$
c. $-7\frac{19}{20}$ d. $-7\frac{7}{18}$

OBJECTIVE 8 *Solve equations.*

We learned how to solve equations in Chapter 4. We now want to put fractions in the context of what we learned. We learned there are two principles that we use to solve equations: we can add or subtract the same amount on both sides of an equation without affecting its solution(s), and we can multiply or divide both sides of an equation by the same nonzero amount without affecting its solution(s).

Let's recall how we used the first of these principles, the addition/subtraction principle.

> **Procedure** *To use the addition/subtraction principle:*
> 1. Decide what term you want to clear.
> 2. Add the additive inverse of that term to both sides of the equation. (That is, add or subtract appropriately so that the term you want to clear becomes 0.)

EXAMPLE 19 Solve. $x + \dfrac{3}{8} = \dfrac{5}{6}$

Solution: We want to isolate x, so we subtract $\dfrac{3}{8}$ from both sides.

$$x + \frac{3}{8} = \frac{5}{6}$$

$+\dfrac{3}{8} - \dfrac{3}{8} = 0$ which isolates x on the left side.

$$x + \frac{3}{8} - \frac{3}{8} = \frac{5}{6} - \frac{3}{8}$$

To complete the subtraction on the right side, we must find a common denominator. The common denominator for 6 and 8 is 24.

$$x + 0 = \frac{5(4)}{6(4)} - \frac{3(3)}{8(3)}$$

$$x = \frac{20}{24} - \frac{9}{24}$$

$$x = \frac{11}{24}$$

Check: To check we replace the x with $\dfrac{11}{24}$ and verify that it makes the equation true.

$$x + \frac{3}{8} = \frac{5}{6}$$

$$\frac{11}{24} + \frac{3}{8} \overset{?}{=} \frac{5}{6}$$

$$\frac{11(1)}{24(1)} + \frac{3(3)}{8(3)} \overset{?}{=} \frac{5}{6}$$

$$\frac{11}{24} + \frac{9}{24} \overset{?}{=} \frac{5}{6}$$

$$\frac{20}{24} \overset{?}{=} \frac{5}{6}$$

$$\frac{5}{6} = \frac{5}{6} \qquad \textbf{It checks.}$$

Do Margin 11. ▶

OBJECTIVE 9 *Solve applications.*

What clues can we look for that indicate that we need to add or subtract? With addition we look for words that indicate we must find a total. For subtraction we look for clues that indicate we must find a difference. Another clue for addition/subtraction situations is that the fractions represent distinct categories.

◆ **Margin 11**
Solve and check.

a. $\dfrac{1}{3} + y = \dfrac{3}{4}$

b. $a - \dfrac{1}{5} = \dfrac{5}{8}$

c. $m + 3\dfrac{1}{2} = -6\dfrac{1}{4}$

d. $-4\dfrac{1}{7} = t - \dfrac{2}{5}$

Answers: **a.** $\dfrac{5}{12}$ **b.** $\dfrac{33}{40}$ **c.** $-9\dfrac{3}{4}$ **d.** $-3\dfrac{26}{35}$

Margin 12

A report claims that $\frac{3}{8}$ of all cars are blue, $\frac{1}{4}$ are red, and the rest are all other colors. What fraction of all cars are other colors?

Margin 13

At the opening of the stock market a company's stock is worth $\$5\frac{3}{8}$ per share. At the close of the stock market the same stock is worth $\$6\frac{1}{4}$. How much did the stock increase in value?

Tip

If you are uncertain how to approach a problem containing fractions, try replacing the fractions with whole numbers and read the problem again. Suppose Example 20 read like this:

In a poll, 25 of the respondents said they agreed with the passing of a certain bill, 20 said they disagreed, and the rest said they had no opinion. What fraction of the respondents had no opinion?

If we knew the total number of people in the poll, to solve we could add the number that agreed plus the number that disagreed, then subtract from the total number of people polled. With fractions we do not need the total number because it is always represented by 1. Therefore, to solve our equation, we added the fraction that agreed plus the fraction that disagreed, then subtracted the sum from all respondents, which is represented by 1.

EXAMPLE 20 In a poll, $\frac{1}{4}$ of the respondents said they agreed with the passing of a certain bill, $\frac{1}{5}$ said they disagreed, and the rest said they had no opinion. What fraction of the respondents had no opinion?

Solution: Follow the problem-solving process.

Understand: There are three distinct ways to respond to the poll: one can agree, can disagree, or have no opinion. Because the three categories together make up the whole group that responded, we can say this:

$$\begin{array}{c} \text{respondents} \\ \text{who agreed} \end{array} + \begin{array}{c} \text{respondents} \\ \text{who disagreed} \end{array} + \begin{array}{c} \text{respondents who} \\ \text{had no opinion} \end{array} = \text{all respondents}$$

Because we are dealing with fractions, *all respondents* means the number 1. The number 1 always represents the whole amount.

Plan: Use the preceding formula to write an equation, then solve.

Execute: Let n represent the number of respondents that had no opinion.

$$\begin{array}{c} \text{respondents} \\ \text{who agreed} \end{array} + \begin{array}{c} \text{respondents} \\ \text{who disagreed} \end{array} + \begin{array}{c} \text{respondents who} \\ \text{had no opinion} \end{array} = \text{all respondents}$$

$$\frac{1}{4} \quad + \quad \frac{1}{5} \quad + \quad n \quad = \quad 1$$

$$\frac{1(5)}{4(5)} + \frac{1(4)}{5(4)} + n = \frac{1(20)}{1(20)}$$

$$\frac{5}{20} + \frac{4}{20} + n = \frac{20}{20}$$

$$\frac{9}{20} + n = \frac{20}{20}$$

$$\frac{9}{20} - \frac{9}{20} + n = \frac{20}{20} - \frac{9}{20}$$

$$n = \frac{11}{20}$$

Answer: $\frac{11}{20}$ had no opinion.

Check: Add the three fractions together. The whole group should be the sum, which is represented by the number 1.

$$\frac{1}{4} + \frac{1}{5} + n = 1$$

$$\frac{1}{4} + \frac{1}{5} + \frac{11}{20} \stackrel{?}{=} 1$$

$$\frac{1(5)}{4(5)} + \frac{1(4)}{5(4)} + \frac{11(1)}{20(1)} \stackrel{?}{=} 1$$

$$\frac{5}{20} + \frac{4}{20} + \frac{11}{20} \stackrel{?}{=} 1$$

$$\frac{20}{20} \stackrel{?}{=} 1$$

$$1 = 1 \qquad \text{It checks.}$$

◀ **Do Margins 12 and 13.**

Answer to Margin 12: $\frac{3}{8}$

Answer to Margin 13: $\frac{7}{8}$

5.6 Exercises

FOR EXTRA HELP

 Videotape 7

 InterAct Math
Tutorial Software

www.carsonmath.com

AWL Math Tutor Center

InterAct
MathXL www.mathxl.com

Student's Solutions
Manual

For Exercises 1–18, add or subtract.

1. $\dfrac{2}{7} + \dfrac{3}{7}$

2. $\dfrac{1}{13} + \dfrac{6}{13}$

3. $\dfrac{-4}{9} + \dfrac{-2}{9}$

4. $-\dfrac{2}{15} + -\dfrac{8}{15}$

5. $\dfrac{10}{17} - \dfrac{4}{17}$

6. $\dfrac{9}{11} - \dfrac{1}{11}$

7. $\dfrac{6}{35} - \dfrac{13}{35}$

8. $\dfrac{1}{30} - \dfrac{4}{30}$

9. $\dfrac{9}{x} + \dfrac{3}{x}$

10. $\dfrac{4}{ab} + \dfrac{16}{ab}$

11. $\dfrac{8x^2}{9} - \dfrac{2x^2}{9}$

12. $\dfrac{4m}{n} - \dfrac{6m}{n}$

13. $\dfrac{x}{6} + \dfrac{1}{6}$

14. $\dfrac{4}{5h} - \dfrac{k}{5h}$

15. $\dfrac{3x^2 + 4x}{7y} + \dfrac{x^2 - 7x + 1}{7y}$

16. $\dfrac{9t - 12}{11u^2} + \dfrac{6t + 5}{11u^2}$

17. $\dfrac{7n^2}{5m} - \dfrac{2n^2 + 3}{5m}$

18. $\dfrac{h^2 - 3h + 5}{k} - \dfrac{h^2 + 4h - 1}{k}$

For Exercises 19–34, add or subtract.

19. $\dfrac{3}{10} + \dfrac{5}{6}$

20. $\dfrac{2}{15} - \dfrac{4}{9}$

21. $\dfrac{7}{12} - \dfrac{11}{36}$

22. $\dfrac{1}{8} + \dfrac{15}{32}$

23. $\dfrac{1}{20} + \dfrac{17}{30}$

24. $\dfrac{3}{16} - \dfrac{9}{20}$

25. $\dfrac{3}{4} + \dfrac{1}{6} + \dfrac{7}{9}$

26. $\dfrac{2}{5} + \dfrac{5}{6} + \dfrac{13}{15}$

27. $\dfrac{7x}{12} + \dfrac{3x}{8}$

28. $\dfrac{a}{9} + \dfrac{2a}{3}$

29. $\dfrac{9}{16m} - \dfrac{3}{8m}$

30. $\dfrac{7}{10x} - \dfrac{5}{6x}$

31. $\dfrac{13}{20} + \dfrac{4}{5h}$

32. $\dfrac{7}{4x} + \dfrac{5}{12}$

33. $\dfrac{2}{3n^2} - \dfrac{7}{9n}$

34. $\dfrac{1}{2t} - \dfrac{3}{5t^3}$

For Exercises 35–42, add.

35. $3\dfrac{4}{9} + 7$

36. $4\dfrac{2}{5} + \dfrac{1}{5}$

37. $2\dfrac{1}{4} + 5\dfrac{1}{4}$

38. $6\dfrac{1}{7} + 1\dfrac{5}{7}$

39. $5\dfrac{5}{6} + 1\dfrac{2}{3}$

40. $3\dfrac{4}{5} + 7\dfrac{1}{4}$

41. $6\dfrac{5}{8} + 3\dfrac{7}{12}$

42. $9\dfrac{1}{2} + 10\dfrac{4}{9}$

For Exercises 43–50, subtract.

43. $9\dfrac{7}{8} - \dfrac{1}{8}$

44. $4\dfrac{5}{9} - 3$

45. $11\dfrac{7}{12} - 2\dfrac{5}{12}$

46. $8\dfrac{3}{7} - 2\dfrac{1}{7}$

47. $5\dfrac{1}{8} - 1\dfrac{5}{6}$

48. $6\dfrac{2}{5} - 2\dfrac{3}{4}$

49. $8\dfrac{1}{3} - 7\dfrac{5}{6}$

50. $10\dfrac{5}{12} - 6\dfrac{5}{8}$

For Exercises 51–58, add or subtract.

51. $6\dfrac{3}{4} + \left(-2\dfrac{1}{3}\right)$

52. $-5\dfrac{1}{2} + 2\dfrac{1}{8}$

53. $-7\dfrac{5}{6} - 2\dfrac{2}{3}$

54. $-\dfrac{4}{5} - 3\dfrac{1}{2}$

55. $\dfrac{5}{8} - 4\dfrac{1}{4}$

56. $6\dfrac{7}{8} - \left(-3\dfrac{1}{4}\right)$

57. $-7\dfrac{3}{4} - \left(-1\dfrac{1}{8}\right)$

58. $-2\dfrac{1}{6} - \left(-3\dfrac{1}{2}\right)$

For Exercises 59–66, solve and check.

59. $x + \dfrac{3}{5} = \dfrac{7}{10}$

60. $\dfrac{1}{4} + y = \dfrac{5}{8}$

61. $n - \dfrac{4}{5} = \dfrac{1}{4}$

62. $\dfrac{5}{8} = k - \dfrac{1}{6}$

63. $-\dfrac{1}{6} = b + \dfrac{3}{4}$

64. $-\dfrac{2}{3} = h - \dfrac{1}{8}$

65. $-3\dfrac{1}{2} + t = -4\dfrac{1}{5}$

66. $-2\dfrac{1}{6} = m - \dfrac{3}{4}$

For Exercises 67–74, solve.

67. A table top is $\dfrac{3}{4}$ in. thick. During sanding $\dfrac{1}{16}$ in. is removed off the top. How thick is the tabletop after sanding?

68. Amanda wants to hang a picture frame so that the bottom of the frame is $54\dfrac{1}{2}$ in. from the floor. The hanger on the back of the picture is $11\dfrac{5}{8}$ in. from the bottom of the frame. Where should she place the nail?

69. A poll is taken to assess the President's approval rating. Respondents can answer four ways: excellent, good, fair, or poor. $\dfrac{1}{8}$ said "excellent," $\dfrac{3}{5}$ said "good," and $\dfrac{1}{6}$ said "fair."

a. What fraction of the respondents said "excellent" or "good?"

b. What fraction said "poor?"

70. In Mrs. Robinson's English class, students can earn a grade of A, B, C, or F (no Ds are given). $\dfrac{1}{3}$ of the students got an A, $\dfrac{3}{8}$ got a B, and $\dfrac{1}{6}$ got a C.

a. What fraction of the students received an A or B?

b. What fraction of the students received an F?

71. A wooden frame is to be $20\dfrac{3}{4}$ in. by $13\dfrac{1}{2}$ in. What total length of wood is needed to make the frame?

72. A game preserve is in the shape of a triangle. What is the total distance around the park?

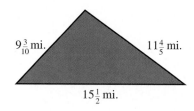

$9\dfrac{3}{10}$ mi. $11\dfrac{4}{5}$ mi.

$15\dfrac{1}{2}$ mi.

73. What is the width, w, of the hallway in the following floor plan?

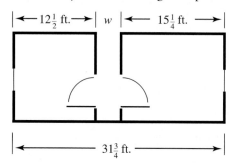

$\leftarrow 12\dfrac{1}{2}$ ft. $\rightarrow$ | w | $\leftarrow 15\dfrac{1}{4}$ ft. $\rightarrow$ |

$\leftarrow 31\dfrac{3}{4}$ ft. $\rightarrow$

74. What is the diameter, d, of the hubcap on the tire shown?

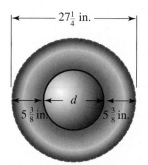

$\leftarrow 27\dfrac{1}{4}$ in. $\rightarrow$

d

$5\dfrac{3}{8}$ in. $5\dfrac{3}{8}$ in.

REVIEW EXERCISES

1. Simplify. $3^2 - 5[2 + (18 \div (-2))] + \sqrt{49}$

2. Add. $(4x^2 + 5x - 6) + (6x^2 - 2x + 1)$

3. Subtract. $(8y^3 - 4y + 2) - (9y^3 + y^2 + 7)$

4. Multiply. $(-6ab^3)(9a^2bc)$

5. Multiply. $(x + 3)(2x - 5)$

5.7 Order of Operations and Evaluating and Simplifying Expressions

OBJECTIVES

1 Use the order of operations agreement to simplify expressions containing fractions and mixed numbers.

2 Evaluate expressions.

3 Calculate the area of a trapezoid.

4 Calculate the area of a circle.

5 Simplify polynomials containing fractions.

6 Evaluate expressions with negative exponents.

OBJECTIVE **1** *Use the order of operations agreement to simplify expressions containing fractions and mixed numbers.*

EXAMPLE 1 Simplify. $\dfrac{3}{4} + 2\dfrac{4}{5} \cdot \left(-\dfrac{5}{6}\right)$

Solution: Follow the order of operations agreement. We multiply first.

$$\dfrac{3}{4} + 2\dfrac{4}{5} \cdot \left(-\dfrac{5}{6}\right) \qquad \text{Write } 2\dfrac{4}{5} \text{ as an improper fraction so that we can multiply.}$$

$$= \dfrac{3}{4} + \dfrac{14}{5} \cdot \left(-\dfrac{5}{6}\right)$$

$$= \dfrac{3}{4} + \dfrac{\overset{7}{\cancel{14}}}{\cancel{5}} \cdot \left(-\dfrac{\overset{1}{\cancel{5}}}{\cancel{6}_{3}}\right) \qquad \text{Divide out a common factor of 2 in 14 and 6, then divide out the common 5's. Multiply the remaining factors.}$$

$$= \dfrac{3}{4} + \left(-\dfrac{7}{3}\right)$$

$$= \dfrac{3(3)}{4(3)} + \left(-\dfrac{7(4)}{3(4)}\right) \qquad \text{Upscale to equivalent fractions with the LCD 12.}$$

$$= \dfrac{9}{12} + \left(-\dfrac{28}{12}\right) \qquad \text{Add.}$$

$$= -\dfrac{19}{12} \qquad \text{Write as a mixed number.}$$

$$= -1\dfrac{7}{12}$$

Do Margin 1. ▶

EXAMPLE 2 Simplify. $\left(\dfrac{1}{2}\right)^{3} - 6\left(\dfrac{3}{5} + \dfrac{2}{3}\right)$

Solution: We will follow the order of operations agreement.

◆ **Margin 1**

Simplify.

a. $\dfrac{5}{8} - \dfrac{2}{9} \cdot \dfrac{3}{4}$

b. $3\dfrac{1}{2} + \dfrac{4}{5} \div \dfrac{3}{10}$

c. $\dfrac{1}{6} + \left(-4\dfrac{1}{3}\right) \cdot 2\dfrac{1}{2}$

d. $-5\dfrac{3}{8} - \dfrac{7}{10} \div \left(-\dfrac{7}{30}\right)$

Answers: **a.** $\dfrac{11}{24}$ **b.** $6\dfrac{1}{6}$ **c.** $-10\dfrac{2}{3}$
d. $-2\dfrac{3}{8}$

◆ Margin 2

Simplify.

a. $\left(\dfrac{1}{3}\right)^2 + 16\left(\dfrac{1}{4} + \dfrac{3}{8}\right)$

b. $\left(\dfrac{1}{25} - \dfrac{4}{5}\right) - \left(\dfrac{2}{5}\right)^2$

c. $\left(1\dfrac{2}{3}\right)^2 + 2\dfrac{1}{2} \div \dfrac{5}{8}$

d. $\left(\dfrac{1}{2} + \dfrac{1}{3}\right) \div \left(\dfrac{3}{4} - \dfrac{1}{8}\right)$

Answers: **a.** $10\frac{1}{9}$ **b.** $-\frac{23}{25}$ **c.** $6\frac{7}{9}$ **d.** $1\frac{1}{3}$

$\left(\dfrac{1}{2}\right)^3 - 6\left(\dfrac{3}{5} + \dfrac{2}{3}\right)$ — Calculate the addition inside the parentheses first. The LCD for 5 and 3 is 15.

$= \left(\dfrac{1}{2}\right)^3 - 6\left(\dfrac{3(3)}{5(3)} + \dfrac{2(5)}{3(5)}\right)$ — Upscale the fractions.

$= \left(\dfrac{1}{2}\right)^3 - 6\left(\dfrac{9}{15} + \dfrac{10}{15}\right)$ — Add.

$= \left(\dfrac{1}{2}\right)^3 - 6 \cdot \dfrac{19}{15}$ — Evaluate the expression with the exponent. $\dfrac{1}{2} \cdot \dfrac{1}{2} \cdot \dfrac{1}{2} = \dfrac{1}{8}.$

$= \dfrac{1}{8} - 6 \cdot \dfrac{19}{15}$

$= \dfrac{1}{8} - \dfrac{\overset{2}{\cancel{6}}}{1} \cdot \dfrac{19}{\underset{5}{\cancel{15}}}$ — Multiply.

$= \dfrac{1}{8} - \dfrac{38}{5}$ — The common denominator for 8 and 5 is 40.

$= \dfrac{1(5)}{8(5)} - \dfrac{38(8)}{5(8)}$ — Upscale the fractions.

$= \dfrac{5}{40} - \dfrac{304}{40}$ — Subtract.

$= -\dfrac{299}{40}$

$= -7\dfrac{19}{40}$

◀ **Do Margin 2.**

OBJECTIVE 2 *Evaluate expressions.*

We learned how to evaluate expressions in Section 3.1. Let's extend those ideas to fractions. Recall, to evaluate an expression, we replace the variables with the corresponding given values and calculate using the order of operations agreement.

EXAMPLE 3 Evaluate $4xy^2 + z$ when $x = \dfrac{3}{4}$, $y = \dfrac{2}{3}$, and $z = 3\dfrac{1}{2}$.

Solution: Replace x with $\dfrac{3}{4}$, y with $\dfrac{2}{3}$, and z with $3\dfrac{1}{2}$, then follow the order of operations agreement.

$4xy^2 + z$ — Replace the variables with the corresponding given values.

$= 4\left(\dfrac{3}{4}\right)\left(\dfrac{2}{3}\right)^2 + 3\dfrac{1}{2}$ — Evaluate the expression with the exponent. We also wrote $\dfrac{4}{1}$ because we will multiply.

$= \dfrac{4}{1}\left(\dfrac{3}{4}\right)\dfrac{4}{9} + 3\dfrac{1}{2}$

$= \dfrac{\overset{1}{\cancel{4}}}{1}\left(\dfrac{\overset{1}{\cancel{3}}}{\cancel{4}}\right)\dfrac{4}{\underset{3}{\cancel{9}}} + 3\dfrac{1}{2}$ — Divide out a common 4 and then a common 3 in the 3 and 9. Then multiply.

$= \dfrac{4}{3} + 3\dfrac{1}{2}$ — We can add either by making both addends improper fractions or making both mixed numbers. We will change the $\dfrac{4}{3}$ to a mixed number.

$= 1\dfrac{1}{3} + 3\dfrac{1}{2}$

$= 1\dfrac{1(2)}{3(2)} + 3\dfrac{1(3)}{2(3)}$ — Upscale the fractions.

$= 1\dfrac{2}{6} + 3\dfrac{3}{6}$ — Add.

$= 4\dfrac{5}{6}$

Do Margin 3. ▶

OBJECTIVE 3 *Calculate the area of a trapezoid.*

Evaluating expressions is the foundation for using formulas. We now want to develop a formula for calculating the area of a **trapezoid**.

DEFINITION **Trapezoid:** A four-sided figure with one pair of parallel sides.

The following figures are examples of trapezoids:

In each figure, notice the top and bottom sides are parallel but the other sides are not. Because the top and bottom sides are different lengths, we will label them *a* and *b*.

When figuring the area of trapezoids, we have the same problem as with triangles: the square units that we use to measure the area do not fit against the slanted edges. This makes it difficult to find the area. To develop the area formula we can duplicate the trapezoid like we did for the triangle. By inverting one of the trapezoids, then joining the corresponding sides, we can make a parallelogram. It looks like this:

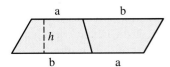

The area of this parallelogram is found by multiplying the base and height. Notice the base is $a + b$ and the height is h. So the area of the parallelogram is $h(a + b)$. However, we really want the area of one of the trapezoids. The area of a single trapezoid is half the area of the parallelogram.

Conclusion: The formula for the area of a trapezoid is: $A = \dfrac{1}{2}h(a + b)$

EXAMPLE 4 Calculate the area of the trapezoid shown.

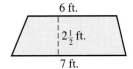

Solution: Use the formula $A = \dfrac{1}{2}h(a + b)$

$A = \dfrac{1}{2}h(a + b)$

$A = \dfrac{1}{2}\left(2\dfrac{1}{2}\right)(6 + 7)$ **Replace *h* with $2\frac{1}{2}$, *a* with 6, and *b* with 7.**

$A = \dfrac{1}{2}\left(2\dfrac{1}{2}\right)(13)$ **Write as improper fractions.**

$A = \dfrac{1}{2}\left(\dfrac{5}{2}\right)\left(\dfrac{13}{1}\right)$ **Multiply.**

$A = \dfrac{65}{4}$ ft.^{2}

$A = 16\dfrac{1}{4}$ ft.^{2}

Reminder
Area is always in square units. Because the distance units are in terms of ft., the area unit is in terms of ft.^{2}.

◆ Margin 3

Evaluate each expression using the given values.

a. mv^2; $m = 3\dfrac{1}{4}$, $v = 10$

b. $\dfrac{1}{2}at^2$; $a = -9\dfrac{4}{5}$, $t = 2$

c. $a^2 - 5b$; $a = \dfrac{3}{4}$, $b = \dfrac{7}{20}$

d. $2m(n + 3)$; $m = \dfrac{5}{8}$, $n = 1\dfrac{1}{3}$

Answers: **a.** 325 **b.** $-19\dfrac{3}{5}$ **c.** $-1\dfrac{3}{16}$ **d.** $5\dfrac{5}{12}$

Margin 4

Calculate the area of the trapezoid shown.

a.

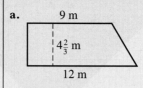

b.

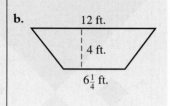

◀ **Do Margin 4.**

OBJECTIVE **4** *Calculate the area of a circle.*

In Section 5.3, we developed the circumference formula. Now let's develop a formula for area of a circle. Because area is measured in square units and circles have a curved edge, we cannot fit the square units exactly against that curved edge. If we could make a circle look like a rectangle or parallelogram, then we could calculate the area. To make a circle look like a parallelogram, we cut it up like a pizza.

First, cut the top half of the circle into six pizza slices and fan those slices out. Then do the same with the bottom half of the circle.

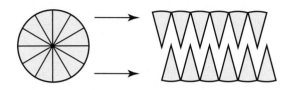

Now fit the slices together like teeth. The resulting figure closely resembles a parallelogram. This parallelogram has the same area as the circle.

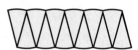

Note If we were to cut more slices, the *slight curves* along the top and bottom would be less prominent. In fact, if we were to use extremely slender slices, the curves would almost be imperceptible.

Because the area of a parallelogram is found by multiplying base times height, we need to relate the base and height of the parallelogram back to the circle. If our circle is a pizza then the height goes from point to crust, so the height corresponds to the radius of the circle. The base distance corresponds to the crust of the bottom half of the pizza, so the base is half of the circumference of the circle.

The height corresponds to the radius.

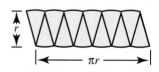

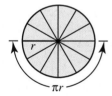

The base corresponds to half of the circumference. Full circumference is found by $2\pi r$ so half circumference is just πr.

Because the area of a parallelogram is found by multiplying base times height, we can say:

$$A = \text{base} \cdot \text{height}$$
$$A = \pi r \cdot r$$
$$A = \pi r^2$$

Conclusion: The formula for the area of a circle with radius r is: $A = \pi r^2$.

EXAMPLE 5 Calculate the area of a circle with a diameter of 5 m.

Solution: Use the formula $A = \pi r^2$. Because we were given the diameter we must first find radius.

Answers: **a.** 49 m² **b.** $36\frac{1}{2}$ ft.²

$$r = \frac{1}{2}d$$

$$r = \frac{1}{2} \cdot 5$$

$$r = \frac{1}{2} \cdot \frac{5}{1}$$

$$r = \frac{5}{2}\text{m}$$

Note: Because we'll square the radius to find area, we will leave the fraction as an improper fraction.

Now we can find the area using $A = \pi r^2$.

Reminder
$\pi \approx \frac{22}{7}$. When we replace π with $\frac{22}{7}$ the calculation becomes an approximation, so we use $\approx$ instead of an equal sign.

$$A = \pi r^2$$

$$A \approx \frac{22}{7} \cdot \left(\frac{5}{2}\right)^2$$

Replace π with $\frac{22}{7}$ and r with $\frac{5}{2}$.

$$A \approx \frac{22}{7} \cdot \frac{25}{4}$$

Square $\frac{5}{2}$ to get $\frac{24}{4}$.

$$A \approx \frac{\overset{11}{22}}{7} \cdot \frac{25}{\underset{2}{4}}$$

Divide out a common factor of 2 in 22 and 64. Then multiply.

$$A \approx \frac{275}{14}\text{ m}^2$$

Write as a mixed number.

$$A \approx 19\frac{9}{14}\text{ m}^2$$

Reminder
Area is always in square units. Because the distance unit for the radius is m, the area unit is m^2.

Do Margin 5. ▷

OBJECTIVE 5 *Simplify polynomials containing fractions.*

Let's apply fractions to simplifying polynomials. We learned in Section 3.2 that we can simplify expressions by combining like terms. Like terms have the same variables raised to the same exponents. We combine like terms by adding or subtracting the coefficients.

EXAMPLE 6 Combine like terms. $\frac{1}{4}a^3 - \frac{3}{5}a + \frac{1}{6}a^3 + \frac{1}{2}a$

Solution: Add or subtract coefficients for the like terms. To add or subtract the coefficients, we must get a common denominator and then upscale. Because we cannot combine an a^3 term with an a term, we only need to get common denominators for the fractions that we will actually combine.

Reminder
The commutative property allows us to rearrange the expression to group the like terms together. This is optional.

$$\frac{1}{4}a^3 - \frac{3}{5}a + \frac{1}{6}a^3 + \frac{1}{2}a$$

$$= \frac{1}{4}a^3 + \frac{1}{6}a^3 - \frac{3}{5}a + \frac{1}{2}a$$

The LCD for the a^3 coefficients is 12.
The LCD for the a coefficients is 10.

$$= \frac{1(3)}{4(3)}a^3 + \frac{1(2)}{6(2)}a^3 - \frac{3(2)}{5(2)}a + \frac{1(5)}{2(5)}a$$

Upscale.

$$= \frac{3}{12}a^3 + \frac{2}{12}a^3 - \frac{6}{10}a + \frac{5}{10}a$$

Add/subtract coefficients.

$$= \frac{5}{12}a^3 - \frac{1}{10}a$$

a^3 and a are not like terms, so we cannot combine further. The expression is in simplest form.

Do Margin 6. ▷

◆ **Margin 5**

a. Calculate the area of a circle with a radius of 6 yd.

b. Calculate the area of a circle with a diameter of 7 in.

◆ **Margin 6**

Combine like terms.

a. $\frac{5}{8}x + \frac{1}{2} - \frac{3}{4}x^2 + \frac{2}{3} - \frac{1}{4}x$

b. $\frac{1}{6}m - 3mn + \frac{2}{3}m + \frac{1}{7} - \frac{1}{2}mn$

Answers to Margin 5: **a.** $113\frac{1}{7}$ yd.^{2}
b. $38\frac{1}{2}$ in.^{2}

Answers to Margin 6: **a.** $-\frac{3}{4}x^2 + \frac{3}{8}x + \frac{7}{6}$ **b.** $\frac{5}{6}m - \frac{7}{2}mn + \frac{1}{7}$

◆ **Margin 7**

Simplify.

a. $\left(\frac{3}{8}n - \frac{2}{3}\right) + \left(\frac{1}{4}n - \frac{1}{6}\right)$

b. $\left(\frac{2}{5}x^2 + \frac{5}{16}\right) -$
$\left(\frac{1}{10}x^2 + \frac{3}{4}\right)$

We can also add or subtract polynomials. We learned in Section 3.4 that to add polynomials we simply combine like terms. To subtract we write an equivalent addition statement using the additive inverse of the subtrahend.

EXAMPLE 7 Simplify. $\left(\frac{3}{4}h - \frac{5}{6}\right) - \left(\frac{1}{3}h - \frac{2}{5}\right)$

Solution: When we subtract polynomials, we write an equivalent addition. We change the operation to addition and change the subtrahend to its additive inverse. To write the additive inverse of a polynomial, we change the sign of each term in the polynomial.

$$\left(\frac{3}{4}h - \frac{5}{6}\right) - \left(\frac{1}{3}h - \frac{2}{5}\right)$$ Write an equivalent addition statement.

$$= \left(\frac{3}{4}h - \frac{5}{6}\right) + \left(-\frac{1}{3}h + \frac{2}{5}\right)$$

$$= \frac{3}{4}h - \frac{1}{3}h - \frac{5}{6} + \frac{2}{5}$$ Collect like terms. (Optional)

$$= \frac{3(3)}{4(3)}h - \frac{1(4)}{3(4)}h2 - \frac{5(5)}{6(5)} + \frac{2(6)}{5(6)}$$ Upscale. Keep in mind that we only need to get common denominators for each set of like terms.

$$= \frac{9}{12}h - \frac{4}{12}h - \frac{25}{30} + \frac{12}{30}$$ Combine the coefficients of the like terms.

$$= \frac{5}{12}h - \frac{13}{30}$$

◀ **Do Margin 7.**

We can also multiply monomials and polynomials with fractional coefficients. Consider multiplying monomials first. We learned in Section 3.5 that when we multiply monomials, we multiply the coefficients and add the exponents of the like bases.

EXAMPLE 8 Multiply. $\left(-\frac{3}{10}m^4n\right)\left(\frac{2}{9}m^2np\right)$

Solution: We multiply the coefficients. When we multiply the variables, we add the exponents of the like bases.

$$\left(-\frac{3}{10}m^4n\right)\left(\frac{2}{9}m^2np\right) = -\frac{\overset{1}{\cancel{3}}}{\underset{5}{\cancel{10}}} \cdot \frac{\overset{1}{\cancel{2}}}{\underset{3}{\cancel{9}}}m^{4+2}n^{1+1}p$$ Divide out a common factor of 3 in the 3 and 9 and a common factor of 2 in the 2 and 10. Then multiply.

$$= -\frac{1}{15}m^6n^2p$$

◀ **Do Margin 8.**

Next, as we learned in Section 3.5, when we multiply a polynomial by a monomial, we use the distributive property and multiply each term inside the polynomial by the monomial.

◆ **Margin 8**

Multiply.

a. $\left(\frac{1}{4}a^3b\right)\left(\frac{2}{7}a^5\right)$

b. $\left(\frac{-5}{12}h^4k\right)\left(\frac{6}{-11}h^3k^2\right)$

EXAMPLE 9 Multiply. $\frac{3}{5}\left(\frac{1}{4}x^2 - \frac{2}{3}x + 10\right)$

Solution: We use the distributive property and multiply each term in $\frac{1}{4}x^2 - \frac{2}{3}x + 10$ by $\frac{3}{5}$.

$$\frac{3}{5}\left(\frac{1}{4}x^2 - \frac{2}{3}x + 10\right)$$ Distribute $\frac{3}{5}$.

$$= \frac{3}{5} \cdot \frac{1}{4}x^2 - \frac{3}{5} \cdot \frac{2}{3}x + \frac{3}{5} \cdot \frac{10}{1}$$

$$= \frac{3}{5} \cdot \frac{1}{4}x^2 - \frac{3}{5} \cdot \frac{\overset{1}{\cancel{2}}}{\underset{1}{\cancel{3}}}x + \frac{3}{\cancel{5}} \cdot \frac{\overset{2}{\cancel{10}}}{1}$$ Divide out common factors. Then multiply.

$$= \frac{3}{20}x^2 - \frac{2}{5}x + 6$$

Answers to Margin 7: **a.** $\frac{5}{8}n - \frac{5}{6}$
b. $\frac{3}{10}x^2 - \frac{7}{16}$

Answers to Margin 8: **a.** $\frac{1}{14}a^8b$
b. $\frac{5}{22}h^7k^3$

Do Margin 9. ▶

Finally, we can multiply polynomials. We learned in Section 3.5 that when we multiply polynomials, we multiply every term in the second polynomial by every term in the first polynomial.

EXAMPLE 10 Multiply. $\left(\dfrac{3}{4}x - 5\right)\left(\dfrac{1}{2}x + 8\right)$

Solution: We multiply every term in the second polynomial by every term in the first polynomial.

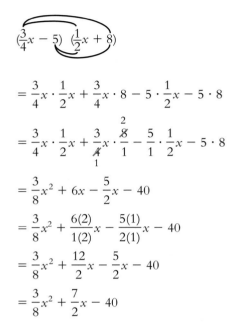

We distribute by multiplying every term in the second polynomial by every term in the first polynomial.

$$= \frac{3}{4}x \cdot \frac{1}{2}x + \frac{3}{4}x \cdot 8 - 5 \cdot \frac{1}{2}x - 5 \cdot 8$$

$$= \frac{3}{4}x \cdot \frac{1}{2}x + \frac{3}{\overset{1}{4}}x \cdot \frac{\overset{2}{8}}{1} - \frac{5}{1} \cdot \frac{1}{2}x - 5 \cdot 8$$

Divide out a common 4 in 4 and 8. Then multiply.

$$= \frac{3}{8}x^2 + 6x - \frac{5}{2}x - 40$$

Combine like terms. We must get a common denominator for the coefficients. The LCD is 2.

$$= \frac{3}{8}x^2 + \frac{6(2)}{1(2)}x - \frac{5(1)}{2(1)}x - 40$$

Upscale coefficients of the like terms.

$$= \frac{3}{8}x^2 + \frac{12}{2}x - \frac{5}{2}x - 40$$

Subtract coefficients of the like terms.

$$= \frac{3}{8}x^2 + \frac{7}{2}x - 40$$

Do Margin 10. ▶

OBJECTIVE 6 *Evaluate expressions with negative exponents.*

It may seem strange, but exponents can be negative. We can use the rule that we developed for dividing exponential forms to discover how to interpret negative exponents. Let's recall the rule:

Rule When dividing exponential forms that have the same base, we can subtract the divisor's exponent from the dividend's exponent and keep the same base.

$$n^a \div n^b = n^{a-b} \qquad \text{as long as} \qquad n \neq 0$$

or in fraction form: $\dfrac{n^a}{n^b} = n^{a-b}$

We can develop the rule by dividing out common factors.

$$\frac{2^5}{2^2} = \frac{\overset{1}{\cancel{2}} \cdot \overset{1}{\cancel{2}} \cdot 2 \cdot 2 \cdot 2}{\underset{1}{\cancel{2}} \cdot \underset{1}{\cancel{2}}} = \frac{2^3}{1} = 2^3$$

Because we divided out two 2's out of the five 2's in the numerator/dividend, we are left with three 2's in the numerator/dividend and the number 1 in the denominator/divisor. We must have the same base, or we would have no common factors to divide out.

Conclusion: Dividing out common factors corresponds to subtracting exponents.

We can say: $\dfrac{2^5}{2^2} = 2^{5-2} = 2^3$

What would happen if the numerator and denominator were interchanged?

◆ **Margin 9**
Multiply.

a. $\dfrac{2}{3}\left(\dfrac{3}{7}a^3 - 9a^2 - \dfrac{1}{2}a\right)$

b. $-\dfrac{5}{12}\left(6m^2 + \dfrac{2}{5}mn - \dfrac{4}{15}n^2\right)$

◆ **Margin 10**
Multiply.

a. $\left(3a - \dfrac{1}{8}\right)\left(2a - \dfrac{1}{4}\right)$

b. $\left(\dfrac{1}{5}m - \dfrac{2}{3}\right)\left(\dfrac{1}{5}m + \dfrac{2}{3}\right)$

Answers to Margin 9: **a.** $\dfrac{2}{7}a^3 - 6a^2 - \dfrac{1}{3}a$ **b.** $-\dfrac{5}{2}m^2 - \dfrac{1}{6}mn + \dfrac{1}{9}n^2$

Answers to Margin 10: **a.** $6a^2 - a + \dfrac{1}{32}$ **b.** $\dfrac{1}{25}m^2 - \dfrac{4}{9}$

$$\frac{2^2}{2^5} = \frac{\overset{1}{\cancel{2}} \cdot \overset{1}{\cancel{2}}}{\underset{1}{\cancel{2}} \cdot \underset{1}{\cancel{2}} \cdot 2 \cdot 2 \cdot 2} = \frac{1}{2^3}$$

Notice we still divide out two 2's but this time three 2's are left in the denominator and 1 is in the numerator. This should make sense because when we interchanged the numerator and denominator we were getting the reciprocal of our original problem, so the outcome should be the reciprocal of our original outcome, and it is.

Now consider the rule. If we follow the rule and subtract the divisor/denominator exponent from the dividend/numerator exponent, we get a negative exponent.

$$\frac{2^2}{2^5} = 2^{2-5} = 2^{-3}$$

Because 2^{-3} and $\frac{1}{2^3}$ are correct results from the same problem, we can say: $2^{-3} = \frac{1}{2^3}$

Recall that $\frac{1}{2^3}$ can also be expressed as $\left(\frac{1}{2}\right)^3$. So we can say: $2^{-3} = \frac{1}{2^3} = \left(\frac{1}{2}\right)^3$

Conclusion: An exponential form with a negative exponent is equivalent to the reciprocal of the base with the exponent made positive.

Exponential forms with negative exponents are not considered to be in simplest form.

> **Rule** An exponential form with a negative exponent is equivalent to the reciprocal of the base with the exponent made positive.
>
> $$x^{-a} = \frac{1}{x^a} \qquad \text{and} \qquad \left(\frac{a}{b}\right)^{-n} = \left(\frac{b}{a}\right)^n$$

> **Procedure** *To evaluate an expression with a negative exponent:*
> **1.** Write the reciprocal of the base.
> **2.** Change the sign of the exponent to positive.
> **3.** Calculate.

EXAMPLE 11 Evaluate.

a. 3^{-4}

Solution: To evaluate we need to write an equivalent expression with a positive exponent. We write the reciprocal of the base and change the sign of the exponent.

$$3^{-4} = \frac{1}{3^4} = \frac{1}{3 \cdot 3 \cdot 3 \cdot 3} = \frac{1}{81}$$

b. $(-2)^{-6}$

Solution: To evaluate we need to write an equivalent expression with a positive exponent. We write the reciprocal of the base and change the sign of the exponent.

$$(-2)^{-6} = \frac{1}{(-2)^6} = \frac{1}{-2 \cdot -2 \cdot -2 \cdot -2 \cdot -2 \cdot -2} = \frac{1}{64}$$

c. $\left(-\frac{3}{4}\right)^{-2}$

Solution: To evaluate we need to write an equivalent expression with a positive exponent. We write the reciprocal of the base and change the sign of the exponent.

$$\left(-\frac{3}{4}\right)^{-2} = \left(-\frac{4}{3}\right)^2 = -\frac{4}{3} \cdot -\frac{4}{3} = \frac{16}{9}$$

◀ **Do Margin 11.**

◆ **Margin 11**

Evaluate.

a. 5^{-2}

b. $(-4)^{-3}$

c. $\left(-\frac{4}{9}\right)^{-2}$

Answers: **a.** $\frac{1}{25}$ **b.** $-\frac{1}{64}$ **c.** $\frac{81}{16}$

5.7 Exercises

FOR EXTRA HELP

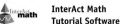 Videotape 7

InterAct Math
Tutorial Software

www.carsonmath.com

AWL Math Tutor Center

InterAct MathXL   www.mathxl.com

Student's Solutions
Manual

For Exercises 1–14, simplify.

1. $\dfrac{1}{4} + 2 \cdot \dfrac{5}{8}$

2. $\dfrac{1}{6} - 5 \cdot \dfrac{7}{10}$

3. $4\dfrac{3}{5} - \dfrac{1}{2} \div \dfrac{5}{6}$

4. $5\dfrac{1}{3} + \dfrac{3}{4} \div \dfrac{5}{12}$

5. $2\dfrac{1}{2} - \left(\dfrac{3}{4}\right)^2$

6. $\left(\dfrac{4}{5}\right)^2 + 5\dfrac{2}{3}$

7. $\dfrac{1}{4} - \dfrac{2}{3}\left(6 - 1\dfrac{1}{2}\right)$

8. $\dfrac{2}{5} + 9\left(\dfrac{1}{3} + \dfrac{1}{6}\right)$

9. $5\dfrac{3}{4} - 2\sqrt{\dfrac{80}{5}}$

10. $7\dfrac{1}{4} + \dfrac{3}{4}\sqrt{\dfrac{4}{9}}$

11. $\left(\dfrac{1}{2}\right)^3 + 2\dfrac{1}{4} - 5\left(\dfrac{3}{10} + \dfrac{1}{5}\right)$

12. $\left(\dfrac{2}{3}\right)^4 \div \dfrac{1}{3}\left(1\dfrac{1}{3} + \dfrac{2}{9}\right)$

13. $5\left(\dfrac{1}{2} - 3\dfrac{4}{5}\right) - 2\left(\dfrac{1}{6} + 4\right)$

14. $1\dfrac{1}{4} - \left(2\dfrac{1}{4}\right)\left(\dfrac{1}{3} + \dfrac{1}{2}\right)^2$

For Exercises 15–22, evaluate the expression using the given values.

15. $x + vt; \; x = 2\dfrac{1}{2}, v = 30, t = \dfrac{1}{4}$

16. $x + vt; \; x = 6\dfrac{4}{5}, v = 40, t = \dfrac{1}{6}$

17. $mv^2; \; m = 1\dfrac{1}{4}, v = \dfrac{3}{4}$

18. $mv^2; \; m = 6\dfrac{1}{2}, v = -\dfrac{2}{5}$

19. $\dfrac{1}{2}at^2; \; a = -9\dfrac{4}{5}, t = \dfrac{1}{2}$

20. $\dfrac{1}{2}at^2; \; a = -32\dfrac{1}{5}, t = 2\dfrac{1}{2}$

21. $\dfrac{Mm}{d^2}; \; M = 200, m = 5\dfrac{1}{2}, d = \dfrac{1}{2}$

22. $\dfrac{Mm}{d^2}; \; M = 10, m = 2\dfrac{1}{3}, d = \dfrac{2}{3}$

For Exercises 23–26, find the area.

23.

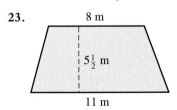

8 m / $5\frac{1}{2}$ m / 11 m

24.

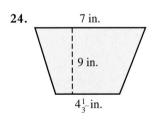

7 in. / 9 in. / $4\frac{1}{3}$ in.

25.

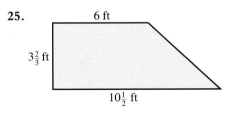

6 ft / $3\frac{2}{3}$ ft / $10\frac{1}{2}$ ft

26.

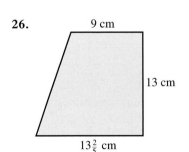

9 cm / 13 cm / $13\frac{2}{5}$ cm

For Exercises 27–30, use $\dfrac{22}{7}$ for π.

27. Find the area of a circle with a radius of 14 in.

28. Find the area of a circle with a radius of 3 ft.

29. Find the area of a circle with a diameter of $2\dfrac{1}{2}$ m.

30. Find the area of a circle with a diameter of $10\dfrac{1}{3}$ cm.

31. Find the area of the shape shown.

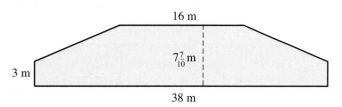

16 m

$7\frac{7}{10}$ m

3 m

38 m

32. Find the area of the shaded region. The diameter of the circle is $4\frac{1}{2}$ in.

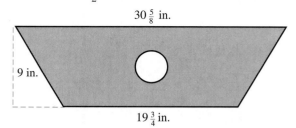

$30\frac{5}{8}$ in.

9 in.

$19\frac{3}{4}$ in.

For Exercises 33 and 34, combine like terms.

33. $\frac{1}{2}x^2 - \frac{3}{4}x + x^3 - \frac{1}{2}x^2 - 6x + 5$

34. $\frac{3}{8}n^3 + \frac{2}{5}n - 3n^3 - \frac{1}{4} + n - 2$

For Exercises 35–38, add or subtract.

35. $\left(5y^3 - \frac{4}{5}y^2 + y - \frac{1}{6}\right) + \left(y^3 + 3y^2 - \frac{2}{3}\right)$

36. $\left(a^4 + \frac{1}{4}a^2 - a - \frac{1}{6}\right) + \left(\frac{3}{10}a^3 + a^2 - \frac{2}{3}a + 4\right)$

37. $\left(\frac{4}{5}t^3 + \frac{2}{3}t^2 - \frac{1}{6}\right) - \left(\frac{7}{10}t^3 + \frac{1}{4}t^2 + \frac{1}{2}\right)$

38. $\left(2x^4 + \frac{3}{5}x^2 - \frac{1}{8}x + 1\right) - \left(6x^4 + \frac{1}{4}x^2 - \frac{1}{2}x - \frac{1}{3}\right)$

For Exercises 39–44, multiply.

39. $\left(-\frac{1}{6}mn\right)\left(\frac{3}{5}m^3n^2\right)$

40. $\left(\frac{-7}{15}x^4y\right)\left(\frac{5}{-21}x^2y\right)$

41. $\frac{5}{8}\left(\frac{4}{5}t^2 - \frac{2}{3}t - \frac{1}{10}\right)$

42. $-\frac{1}{15}\left(3h^2 + \frac{5}{6}hk - \frac{3}{4}k^2\right)$

43. $\left(\frac{2}{3}x - 6\right)\left(3x - \frac{1}{4}\right)$

44. $\left(\frac{1}{4}u - \frac{1}{3}\right)\left(\frac{1}{2}u + \frac{2}{3}\right)$

For Exercises 45–52, evaluate.

45. 3^{-2}

46. 5^{-3}

47. $(-2)^{-5}$

48. $(-6)^{-2}$

49. $\left(\frac{3}{10}\right)^{-4}$

50. $\left(\frac{2}{11}\right)^{-2}$

51. $\left(-\frac{3}{4}\right)^{-3}$

52. $\left(-\frac{2}{5}\right)^{-4}$

Puzzle Problem

The original size of Khufu's pyramid in Egypt was 754 ft. along the base with a height of 481 ft. It was originally covered with blocks cut so that the faces were all smooth. (Most of these blocks are now gone.) What was the total surface area of the four faces of Khufu's pyramid?

REVIEW EXERCISES

1. Solve and check. $4x - 9 = 11$

2. Solve and check. $5(n - 3) = 3n + 7$

3. Six times the sum of n and two is equal to three less than n. Translate to an equation, then solve.

4. Florence consults a map and finds that her trip will be 240 mi. If she averages 60 mph, how long will it take her to get to her destination?

5. David has $105 all in tens and fives. If he has 15 bills how many tens and how many fives does he have?

5.8 Solving Equations

OBJECTIVES

1 Use the LCD to simplify equations.

2 Translate sentences to equations, then solve.

3 Solve applications involving one unknown.

4 Solve applications involving two unknowns.

OBJECTIVE 1 *Use the LCD to simplify equations.*

In Section 5.4, we reviewed solving equations using the multiplication/division principle. In Section 5.6 we reviewed solving equations using the addition/subtraction principle. Now let's consider equations that require both principles. We will also develop a way to use the multiplication/division principle to simplify equations that contain fractions.

EXAMPLE 1 Solve and check. $\dfrac{3}{4}x - \dfrac{2}{3} = \dfrac{1}{6}$

Solution: To isolate x, first add $\dfrac{2}{3}$ to both sides. Then clear the $\dfrac{3}{4}$ coefficient by multiplying both sides by its reciprocal $\dfrac{4}{3}$.

$$\frac{3}{4}x - \frac{2}{3} = \frac{1}{6}$$

> **Reminder**
> $-\dfrac{2}{3}$ and $\dfrac{2}{3}$ and are additive inverses so their sum is 0.

$$\frac{3}{4}x - \frac{2}{3} + \frac{2}{3} = \frac{1}{6} + \frac{2}{3} \qquad \text{Add } \tfrac{2}{3} \text{ to both sides.}$$

$$\frac{3}{4}x + 0 = \frac{1(1)}{6(1)} + \frac{2(2)}{3(2)} \qquad \text{Upscale.}$$

$$\frac{3}{4}x = \frac{1}{6} + \frac{4}{6} \qquad \text{Add.}$$

$$\frac{3}{4}x = \frac{5}{6}$$

> **Reminder**
> $\dfrac{3}{4}$ and $\dfrac{4}{3}$ are multiplicative inverses/reciprocals, so their product is 1.

$$\overset{1}{\underset{3}{\cancel{4}}} \cdot \overset{1}{\underset{\cancel{4}}{\cancel{3}}} x = \frac{5}{\underset{3}{\cancel{6}}} \cdot \frac{\overset{2}{\cancel{4}}}{3} \qquad \text{Multiply both sides by } \tfrac{4}{3}.$$
$$\qquad\qquad\qquad\qquad\qquad \text{Then simplify.}$$

$$1x = \frac{10}{9}$$

$$x = 1\frac{1}{9}$$

Check: To check, we replace x in $\frac{3}{4}x - \frac{2}{3} = \frac{1}{6}$ with $1\frac{1}{9}$ and verify that it makes the equation true.

$$\frac{3}{4}x - \frac{2}{3} = \frac{1}{6}$$

$$\frac{3}{4}\left(1\frac{1}{9}\right) - \frac{2}{3} \; ? \; \frac{1}{6}$$

$$\frac{3}{4} \cdot \frac{10}{9} - \frac{2}{3} \; ? \; \frac{1}{6}$$

$$\overset{1}{\underset{2}{\cancel{3}}} \cdot \overset{5}{\underset{3}{\cancel{10}}} - \frac{2}{3} \; ? \; \frac{1}{6}$$

$$\frac{5}{6} - \frac{2}{3} \; ? \; \frac{1}{6}$$

$$\frac{5(1)}{6(1)} - \frac{2(2)}{3(2)} \; ? \; \frac{1}{6}$$

$$\frac{5}{6} - \frac{4}{6} \; ? \; \frac{1}{6}$$

$$\frac{1}{6} = \frac{1}{6} \qquad \text{It checks.}$$

Example 1 illustrates the approach to solving equations that we learned in Chapter 4. But working with fractions can be tedious. However, we can use the multiplication/division principle to clear the fractions, leaving us with integers, which simplifies the process considerably.

Remember, the multiplication/division principle simply says that we can multiply or divide both sides of an equation by the same amount without affecting its solution(s). The principle does not say that we have to wait until the last step of the process to use it. If we multiply both sides by a number that is divisible by all the denominators, we can divide out all the denominators. To keep the numbers as small as possible, it is best to use the least common multiple of the denominators, which is the LCD (lowest common denominator).

Consider Example 1 again using this technique.

EXAMPLE 2 Solve and check. $\dfrac{3}{4}x - \dfrac{2}{3} = \dfrac{1}{6}$

Solution: We will clear the fractions by multiplying both sides by the LCD, then solve. The LCD of 4, 3, and 6 is 12.

$$\frac{3}{4}x - \frac{2}{3} = \frac{1}{6}$$

$$12\left(\frac{3}{4}x - \frac{2}{3}\right) = \left(\frac{1}{6}\right)12 \qquad \text{Multiply both sides by 12.}$$

$$\frac{12}{1} \cdot \frac{3}{4}x - \frac{2}{3} \cdot \frac{12}{1} = \frac{1}{6} \cdot \frac{12}{1} \qquad \text{Multiply each term by 12.}$$

$$\overset{3}{\underset{1}{\cancel{12}}} \cdot \frac{3}{\underset{1}{\cancel{4}}}x - \frac{2}{\underset{1}{\cancel{3}}} \cdot \overset{4}{\underset{1}{\cancel{12}}} = \frac{1}{\underset{1}{\cancel{6}}} \cdot \overset{2}{\underset{1}{\cancel{12}}}$$

After dividing out the common factors, we are left with an equation that contains only integers, which is more comfortable to work with for most people.

$$9x - 8 = 2$$

$$9x - 8 = 2$$

$$\underline{+8 \quad +8} \qquad \text{Add 8 to both sides to isolate } 9x.$$

$$9x + 0 = 10$$

$$\frac{9x}{9} = \frac{10}{9} \qquad \text{Clear the 9 coefficient by dividing both sides by 9.}$$

$$1x = \frac{10}{9}$$

$$x = 1\frac{1}{9}$$

Notice we end up with the same result as that in Example 1. Also, the check will be the same as the check used for Example 1.

> **WARNING** Always use the original equation in the check. It is very tempting to check using a later step like $9x - 8 = 2$. If you make a mistake in simplifying the equation and continue solving correctly, and then use one of the equations after the mistake to check, you will mistakenly check the incorrect answer. The wrong answer will check because you solved correctly *after* the mistake so the answer you will get is a solution for the incorrect equation, not for the original equation. To guard against this error, use the original equation to check.

Do Margin 1. ▶

We can modify the process outline for solving equations that we developed in Chapter 4 to include this procedure for clearing fractions.

Procedure *To solve equations:*
1. Simplify both sides of the equation as needed.
   a. Distribute to clear parentheses.
   b. Clear fractions by multiplying both sides by the LCD of all denominators. (optional)
   c. Combine like terms.

Tip: Clear the variable term that has the smaller coefficient. This will avoid negative coefficients.

2. Use the addition/subtraction principle so that all variable terms are on one side of the equation and all constants are on the other side.
3. Use the multiplication/division principle to clear any remaining coefficients.

EXAMPLE 3 Solve and check. $\frac{2}{3}(x - 3) = \frac{1}{2}x - \frac{2}{5}$

Solution: We will start by distributing the $\frac{2}{3}$ to clear the parentheses. We will then clear the fractions by multiplying by their LCD. The LCD for 3, 2, and 5 is 30.

Note We could actually start the solution by multiplying by the LCD. But many times people get confused in handling the parentheses, so that's why we are choosing to distribute first. Remember, you can clear the fractions at any time during the process, or not at all if you so choose.

$$\frac{2}{3}(x - 3) = \frac{1}{2}x - \frac{2}{5}$$

Distribute $\frac{2}{3}$ to clear the parentheses.

$$\frac{2}{3}x - \frac{2}{\cancel{3}} \cdot \frac{\cancel{3}}{1} = \frac{1}{2}x - \frac{2}{5}$$

$$\frac{2}{3}x - 2 = \frac{1}{2}x - \frac{2}{5}$$

$$30\left(\frac{2}{3}x - 2\right) = \left(\frac{1}{2}x - \frac{2}{5}\right)30$$

Clear the fractions by multiplying both sides by the LCD, 30. Distribute 30 to every term in the equation.

$$\frac{\cancel{30}^{10}}{1} \cdot \frac{2}{\cancel{3}} x - 30 \cdot 2 = \frac{\cancel{30}^{15}}{1} \cdot \frac{1}{\cancel{2}} x - \frac{2}{\cancel{5}} \cdot \frac{\cancel{30}^{6}}{1}$$

Simplify by dividing out all the common factors.

From here on, the problem is exactly the same as the one we did in Chapter 4. See Example 4 in Section 4.3.

$$20x - 60 = 15x - 12$$

Get the *x* terms together by subtracting 15x on both sides.

$$20x - 60 = 15x - 12$$

Simplify.

$$\underline{-15x \qquad\qquad -15x}$$

$$5x - 60 = 0 - 12$$

$$5x - 60 = -12$$

Isolate 5x by adding 60 to both sides.

$$\underline{+60 \qquad +60}$$

$$5x - 0 = 48$$

$$\frac{5x}{5} = \frac{48}{5}$$

Isolate x by dividing both sides by 5.

$$1x = \frac{48}{5}$$

$$x = 9\frac{3}{5}$$

◆ **Margin 1**

Solve and check.

a. $\frac{5}{8}y - 3 = \frac{1}{2}$

b. $\frac{1}{4} + \frac{1}{8}m = \frac{5}{16}$

c. $-\frac{7}{9} = \frac{2}{3}n + \frac{1}{6}$

d. $\frac{-1}{7} = -2 - \frac{3}{5}k$

Answers: a. $5\frac{3}{5}$ b. $\frac{1}{2}$ c. $-1\frac{5}{12}$
d. $-3\frac{2}{21}$

Check: We replace x with $9\frac{3}{5}$ in $\frac{2}{3}(x-3) = \frac{1}{2}x - \frac{2}{5}$ and verify that it makes the equation true. We will leave this check to the reader.

◀ **Do Margin 2.**

OBJECTIVE 2 *Translate sentences to equations, then solve.*

There are two new key words that are associated with fractions. In Section 5.3 we learned that the key word *of* indicates multiplication when preceded by a fraction.

$$\frac{2}{5} \textbf{ of a number} \qquad \text{translates to} \qquad \frac{2}{5}n$$

The second key word is reciprocal.

$$\textbf{The reciprocal of } n \qquad \text{translates to} \qquad \frac{1}{n}$$

The key words we've learned in the past, such as sum, difference, less than, product, and quotient translate the same regardless of the types of numbers involved.

EXAMPLE 4 The sum of $2\frac{2}{3}$ and $\frac{3}{5}$ of n is the same as 2 less than $\frac{1}{3}$ of n. Solve for n.

Understand: We must solve for n given key words. The word *sum* indicates addition. *Less than* indicates subtraction in reverse order of the sentence. $\frac{3}{5}$ *of n* and $\frac{1}{3}$ *of n* indicate multiplication because *of* is preceded by the fraction in both cases.

> **Reminder**
> When used with the key words sum, difference, product, and quotient, the word *and* becomes the operation symbol.

Plan: Translate to an equation, then solve.

Execute: The sum of $2\frac{2}{3}$ and $\frac{3}{5}$ of n is the same as 2 less than $\frac{1}{3}$ of n.

$$2\frac{2}{3} + \frac{3}{5}n = \frac{1}{3}n - 2$$

$$\frac{8}{3} + \frac{3}{5}n = \frac{1}{3}n - 2$$

$$15\left(\frac{8}{3} + \frac{3}{5}n\right) = \left(\frac{1}{3}n - 2\right)15$$

Clear the fractions by multiplying all terms on both sides by the LCD, which is 15 in this case. Distribute 15 to every term in the equation.

$$\frac{\overset{5}{\cancel{15}}}{1} \cdot \frac{8}{\underset{1}{\cancel{3}}} + \frac{\overset{3}{\cancel{15}}}{1} \cdot \frac{3}{\underset{1}{\cancel{5}}}n = \frac{\overset{5}{\cancel{15}}}{1} \cdot \frac{1}{\underset{1}{\cancel{3}}}n - 2 \cdot \frac{15}{1}$$

$$40 + 9n = 5n - 30$$

$$40 + 9n = 5n - 30$$

$$\underline{-40 \qquad\qquad -40}$$

$$0 + 9n = 5n - 70$$

$$9n = 5n - 70$$

$$\underline{-5n \quad -5n}$$

$$4n = 0 - 70$$

$$\frac{4n}{4} = \frac{-70}{4}$$

$$1n = -\frac{35}{2}$$

$$n = -17\frac{1}{2}$$

◆ **Margin 2**

Solve and check.

a. $\frac{1}{4}y + 5 = \frac{1}{6}y + 2$

b. $\frac{2}{3} - \frac{4}{5}a = \frac{1}{5} + a$

c. $\frac{1}{4}(m + 2) = \frac{3}{4}m - \frac{2}{3}$

d. $\frac{3}{5}(x - 10) = \frac{1}{2}(x + 6) - 4$

Answers: **a.** -36 **b.** $\frac{7}{27}$ **c.** $2\frac{1}{3}$ **d.** 50

Answer: $n = -17\frac{1}{2}$

Check: Replace n with $-17\frac{1}{2}$ to show that the answer checks, the way that we did in Example 3. We will leave this check to the reader.

Do Margin 3. ▶

OBJECTIVE 3 *Solve applications involving one unknown.*

EXAMPLE 5 Find the missing base in the trapezoid if the area is 9 ft.^{2}.

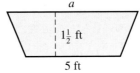

Understand: We must solve for a. Because we are given the area of the trapezoid, we can use the formula $A = \frac{1}{2}h(a + b)$.

Plan: Replace the variables in the formula with the known amounts and solve for a.

Execute:

$$A = \frac{1}{2}h(a + b)$$

$$9 = \frac{1}{2}\left(1\frac{1}{2}\right)(a + 5)$$
We prepare to multiply $\frac{1}{2}\left(1\frac{1}{2}\right)$ by writing improper fractions.

$$9 = \frac{1}{2}\left(\frac{3}{2}\right)(a + 5)$$

$$9 = \frac{3}{4}(a + 5)$$

$$9 = \frac{3}{4}a + \frac{15}{4}$$

$$4(9) = \left(\frac{3}{4}a + \frac{15}{4}\right)4$$

$$36 = \frac{\cancel{4}}{1} \cdot \frac{3}{\cancel{4}}a + \frac{\cancel{4}}{1} \cdot \frac{15}{\cancel{4}}$$

$$36 = 3a + 15$$

$$36 = 3a + 15$$
$$\underline{-15 \qquad\quad -15}$$
$$21 = 3a + 0$$

$$\frac{21}{3} = \frac{3a}{3}$$

$$7 = 1a$$

$$7 = a$$

Answer: The missing side is 7 ft.

Check: Verify that the area is 9 ft.^{2} when $a = 7$ ft., $b = 5$ ft., and $h = 1\frac{1}{2}$ ft. We will leave this check to the reader.

Do Margin 4. ▶

OBJECTIVE 4 *Solve applications involving two unknowns.*

In Chapter 4 we developed techniques for solving problems that have two unknowns. We discussed that it is necessary to have two relationships in order to solve problems with two unknowns. We broke our discussion into types of problems: problems where we are given the relationships in key word form and problems where the relationships weren't stated so clearly. As the information became more ambiguous, we began using tables to help. We will follow the same format here. First consider problems where the relationships are stated in clear, distinct key words.

◆ **Margin 3**

Translate to an equation, then solve.

a. $\frac{2}{3}$ of a number is $3\frac{5}{6}$.

b. The difference of x and $\frac{4}{9}$ is equal to $-3\frac{1}{6}$.

c. The sum of 3 and $\frac{3}{4}$ of k is the same as the reciprocal of 8.

d. $\frac{1}{2}$ of the difference of n and 4 is equal to $\frac{2}{5}$ of n.

◆ **Margin 4**

Find the missing side if the area of the trapezoid is $4\frac{1}{2}$ m^2.

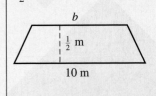

Answers to Margin 3:
a. $\frac{2}{3}n = 3\frac{5}{6};\ n = 5\frac{3}{4}$
b. $x - \frac{4}{9} = -3\frac{1}{6};\ x = -2\frac{13}{18}$
c. $3 + \frac{3}{4}k = \frac{1}{8};\ k = -3\frac{5}{6}$
d. $\frac{1}{2}(n - 4) = \frac{2}{5}n;\ n = 20$

Answer to Margin 4: $b = 8$ m

Margin 5

Two boards are joined. One board is $\frac{2}{3}$ the length of the other. The two boards combine to be $7\frac{1}{2}$ ft. What are the lengths of both boards?

EXAMPLE 6 A small sales company has two salespersons. In one particular week, Benjamin's total sales are half of the amount sold by Florence. If the total amount of sales for the week is $12,000, how much did each person sell?

Solution: Use the problem-solving process.

Understand: There are two sentences that describe relationships.

Benjamin's total sales are half of the amount sold by Florence.

And   . . . the total amount of sales for the week is $12,000. . . .

We are to find the amount that Benjamin sold and the amount that Florence sold.

Plan: Translate the key words. Combine the relationships into a single equation and solve.

Execute: Let f be the amount sold by Florence.

Benjamin's total sales are half of the amount sold by Florence.

$$\text{Benjamin's sales} = \frac{1}{2} \cdot f$$

. . . the total amount of sales for the week is $12,000. . . .

$$\text{Benjamin's sales} + \text{Florence's sales} = \$12,000$$

$$\frac{1}{2}f + f = 12{,}000$$ Multiply both sides throughout by 2 in order to clear the fraction.

$$2\left(\frac{1}{2}f + f\right) = (12{,}000)2$$

$$f + 2f = 24{,}000$$ Combine like terms.

$$3f = 24{,}000$$

$$\frac{3f}{3} = \frac{24{,}000}{3}$$ Divide both sides by 3 to clear the 3 coefficient.

$$f = 8000$$

Answer: Because f represents Florence's sales, we can say her sales amounted to $8000. If Benjamin's sales were $\frac{1}{2}$ of Florence's then his totaled only $4000.

Check: Notice the sum of their sales is in fact $12,000.

$$\$4000 + \$8000 = \$12{,}000$$

◀ **Do Margins 5 and 6.**

Now let's examine problems involving geometry terms, such as perimeter, supplementary, and complementary where the two relationships aren't as clearly stated. Let's recall the definitions of these geometry terms.

DEFINITION   **Perimeter:** The total distance along the edge of a shape.

Supplementary angles: Angles whose sum is 180°.

Complementary angles: Angles whose sum is 90°.

Margin 6

The length of a rectangular frame is $3\frac{1}{2}$ in. longer than the width. If the perimeter is 42 in., what are the dimensions of the frame?

EXAMPLE 7 Figure $ABCD$ is a rectangle. If $\angle DBC$ is $14\frac{1}{3}°$ less than $\angle ABD$ then what are the angle measurements?

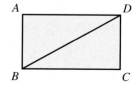

Solution: Use the problem-solving process.

Understand: We must find the angle measurements. It seems we are only given one relationship:

$$\angle DBC \text{ is } 14\frac{1}{3}° \text{ less than } \angle ABD$$

The other relationship is in the shape. Because the shape is a rectangle, the corner angles measure 90°. Therefore $\angle DBC$ and $\angle ABD$ are complementary, which means the sum of their measurements is 90°.

Answer to Margin 5: $4\frac{1}{2}$ ft.; 3 ft.

Answer to Margin 6: $8\frac{3}{4}$ in. by $12\frac{1}{4}$ in.

Plan: Translate the relationships to an equation, then solve.

Execute: Let a represent the measurement of $\angle ABD$.

$\angle DBC$ is $14\frac{1}{3}°$ less than $\angle ABD$

$$\angle DBC = a - 14\frac{1}{3}$$

The sum of the angle measurements is 90.

$$\angle DBC + \angle ABD = 90$$

$$a - 14\frac{1}{3} + a = 90$$

$$2a - 14\frac{1}{3} = 90$$

$$2a - 14\frac{1}{3} + 14\frac{1}{3} = 90 + 14\frac{1}{3}$$

$$2a + 0 = 104\frac{1}{3}$$

$$\frac{1}{2} \cdot 2a = \frac{313}{3} \cdot \frac{1}{2}$$

$$1a = \frac{313}{6}$$

$$a = 52\frac{1}{6}$$

Answer: Because $\angle ABD$ is represented by the variable a, we can say $\angle ABD = 52\frac{1}{6}°$. Because $\angle DBC = a - 14\frac{1}{3}$, we can find its value replacing a with $52\frac{1}{6}$ and then subtracting:

$$\angle DBC = 52\frac{1}{6} - 14\frac{1}{3}$$

$$\angle DBC = 52\frac{1(1)}{6(1)} - 14\frac{1(2)}{3(2)}$$

$$\angle DBC = 52\frac{1}{6} - 14\frac{2}{6}$$

$$\angle DBC = 51\frac{7}{6} - 14\frac{2}{6}$$

$$\angle DBC = 37\frac{5}{6}°$$

Check: We should verify that the sum of the angle measurements is 90°.

$$\angle DBC + \angle ABD = 90$$

$$37\frac{5}{6} + 52\frac{1}{6} \text{ ? } 90$$

$$89\frac{6}{6} \text{ ? } 90$$

$$90 = 90$$

Do Margins 7 and 8. ▶

In Section 4.5 we developed the use of tables with currency problems finding the number of bills of two different types of currency. In this final group of similar problems, we will use a table to help organize information involving distance, rate, and time.

We previously developed the formula for the relationship among distance, rate, and time to be:

$$d = rt$$

Next, we will find the distance between two people or objects that are traveling in opposite directions. It won't matter whether the two people or objects are headed toward one another or away from one another, the math will be the same. The tables we will use will look like this:

Categories	Rate	Time	Distance

◆ **Margin 7**

A piece of steel is welded to a horizontal truss forming two angles with the truss. The larger of the two angles is $2\frac{1}{2}°$ more than twice the smaller angle. Find the angle measurements.

◆ **Margin 8**

A rectangle has a width that is half of the length. The perimeter is $37\frac{1}{2}$ m. What are the dimensions of the rectangle?

Answer to Margin 7: $59\frac{1}{6}°$, $120\frac{5}{6}°$

Answer to Margin 8: $12\frac{1}{2}$ m by $6\frac{1}{4}$ m

◆ **Margin 9**

Andrea is bicycling east at 8 mph along a trail. Carla is running west on the same trail at $4\frac{1}{2}$ mph. If they are $\frac{1}{4}$ mi. apart, how long will it be until they meet?

◆ **Margin 10**

Jennifer and Phil are traveling in cars going toward one another on an interstate highway. Jennifer is traveling at 70 mph and Phil at 65 mph. From the time they eventually meet and pass each other, how long until they are 27 mi. apart?

Note: When two objects travel toward one another or away from one another, we can describe their closing or separation speed. When we combined like terms, we combined the numbers representing the speeds of the cars.

20 mph + 40 mph = 60 mph

60 mph is the closing speed for the two cars. It is the same as if we kept one car still and drove the other car at 60 mph for 1 mi. It would take the same amount of time.

The reasoning is the same if the two cars are going away from each other. If they are going 40 mph and 20 mph, then they are separating at a rate of 60 mph. This is why the math is the same, whether the cars are going toward each other or away from each other.

Answer to Margin 9: $\frac{1}{50}$ hr.

Answer to Margin 10: $\frac{1}{5}$ hr.

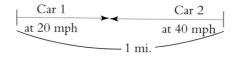 **EXAMPLE 8** In a crash test, two remote-guided cars are 1 mi. apart and driven toward each other to collide somewhere in between. One car is driven at a speed of 20 mph, while the other is driven at 40 mph. How long will it take them to collide?

Solution: Follow the problem-solving process.

Understand: Let's draw a picture of the situation:

```
     Car 1              Car 2
   at 20 mph          at 40 mph
   |————————►◄————————|
            1 mi.
```

Notice car 2 will make up more of the distance that separates the cars because it is going faster.

From our picture we can say that the sum of the individual distances traveled will be 1 mi.

distance traveled by car 1 + distance traveled by car 2 = total distance apart

Use a table to develop the individual distances.

Categories	Rate	Time	Distance
Car 1	20 mph	t	$20t$
Car 2	40 mph	t	$40t$

If we used a stopwatch to time the crash test, we would start the watch when the cars are separated by 1 mi., and stop the watch when they collide. Both cars travel for the same amount of time, so the time is t for both.

We calculate distance by multiplying the rate column and time column to get the distance column in the table.

We now have expressions for the individual distances and know that the sum of the individual distances will be 1 mi.

Plan: Use the information to write an equation, then solve.

Execute: distance traveled by car 1 + distance traveled by car 2 = total distance apart

$$20t \quad + \quad 40t \quad = 1$$
$$60t = 1$$
$$\frac{60t}{60} = \frac{1}{60}$$
$$1t = \frac{1}{60}$$
$$t = \frac{1}{60}$$

Discussion How many minutes is $\frac{1}{60}$ of an hour?

Answer: The cars will collide in $\frac{1}{60}$ of an hour.

Check: We can verify that in $\frac{1}{60}$ of an hour the cars will travel a combined distance of 1 mi.

Car 1 individual distance: $d = 20\left(\frac{1}{60}\right) = \frac{20}{60} = \frac{1}{3}$ mi.

Car 2 individual distance: $d = 40\left(\frac{1}{60}\right) = \frac{40}{60} = \frac{2}{3}$ mi.

Notice car 2 goes twice as far because it is going twice as fast.

Combined distance: car 1 + car 2 = total distance

$$\frac{1}{3} + \frac{2}{3} \;?\; 1$$
$$\frac{3}{3} \;?\; 1$$
$$1 = 1$$

◄ **Do Margins 9 and 10.**

5.8 Exercises

For Exercises 1–8, use the LCD to simplify the equation, then solve and check.

1. $3x - \dfrac{1}{4} = \dfrac{1}{2}$

2. $\dfrac{2}{3}y - 8 = \dfrac{1}{6}$

3. $\dfrac{3}{4}x + \dfrac{1}{6} = \dfrac{1}{2}$

4. $\dfrac{1}{5}x + \dfrac{2}{3} = 2$

5. $\dfrac{4}{5} - \dfrac{a}{2} = \dfrac{3}{4} - 1$

6. $\dfrac{1}{8} + n = \dfrac{5}{6}n - \dfrac{2}{3}$

7. $\dfrac{1}{2}(x - 6) = \dfrac{1}{4}x - \dfrac{2}{5}$

8. $\dfrac{5}{6}(x - 8) = \dfrac{1}{5}x - \dfrac{1}{3}$

For Exercises 9–16, translate to an equation, then solve.

9. $\dfrac{3}{8}$ of a number is $4\dfrac{5}{6}$. Find the number.

10. The product of $2\dfrac{1}{3}$ and n is $-4\dfrac{1}{2}$.

11. $8\dfrac{7}{10}$ less than y is $-2\dfrac{1}{2}$.

12. $6\dfrac{2}{3}$ more than m is $5\dfrac{1}{12}$.

13. $3\dfrac{1}{4}$ more than twice n is $-\dfrac{1}{6}$.

14. $\dfrac{5}{8}$ less than $3\dfrac{4}{5}$ times k is $\dfrac{9}{16}$.

15. $\dfrac{3}{4}$ of the sum of b and 10 is equal to $1\dfrac{1}{6}$ added to b.

16. $\dfrac{2}{5}$ of the difference of 1 and h is the same as $1\dfrac{1}{4}$ times h.

For Exercises 17–30, solve.

17. Find the length of the missing side if the area is $7\dfrac{41}{50}$ cm^2.

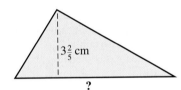

$3\dfrac{2}{5}$ cm

?

18. Find the length of the missing side if the area is $122\dfrac{1}{2}$ in.^{2}.

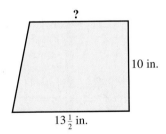

?

10 in.

$13\dfrac{1}{2}$ in.

19. Mary is part of a team of social workers trying to raise funds to build an outreach center for a neighborhood. Mary alone raises $2\dfrac{1}{2}$ times what the rest of the team does. The total amount raised was \$31,542. How much did Mary raise?

20. Rosa discovers that her salary is $\dfrac{5}{6}$ of Rick's salary, yet they have the same job. The difference in their salaries is \$4700. What are each of their salaries?

21. Two chemical storage tanks hold the same amount. Tank A is $\dfrac{3}{4}$ full while tank B is $\dfrac{2}{3}$ full. The two tanks currently have a combined total of 1700 gallons of CO_2. How much do the tanks hold?

22. A computer has two equal-size hard drives. Drive C is $\dfrac{4}{5}$ full while drive D is only $\dfrac{1}{4}$ full. If the two drives have a combined total of $6\dfrac{3}{10}$ gigabytes stored, how much does each drive hold?

23. In a survey, $\frac{3}{4}$ of the people contacted agreed to answer a single yes or no question. Of these, $\frac{1}{5}$ answered yes and the rest answered no. What fraction of all people contacted said no?

24. In a political poll, $\frac{2}{5}$ of the respondents said they would vote for a certain candidate. Of these, $\frac{1}{3}$ were men. What fraction of all respondents that voted for the candidate were women?

25. Two boards come together as shown below. The smaller angle measurement is $\frac{1}{3}$ the measurement of the larger angle. Find the angle measurements.

26. At exactly 3:00:10 find the angles made between the second hand and the minute and hour hands of a clock.

27. An F-16 and MiG-29 are flying directly toward one another. The F-16 is flying at Mach 2, which is $1483\frac{3}{5}$ mph. The MiG-29 is flying at Mach 1 which is $741\frac{4}{5}$ mph. If the two planes are 20 mi. apart, how long until they meet?

28. A jury is listening to a testimony in a case of a head-on collision. As the defendant swerved to avoid a construction crew, he moved slightly into the lane of oncoming traffic. After passing the construction crew, he hit an on-coming car. He claims he did not have time to react to the on-coming car because of a bend in the road ahead. It was determined that the plaintiff was at the bend in the road at the same time that the defendant was passing the construction crew. From the site where the construction crew was working to the bend was measured to be 200 ft. The defendant was traveling at 44 ft./sec. The plaintiff was traveling at $51\frac{1}{3}$ ft./sec. How much time did the defendant have to react? Do you think it was sufficient time?

29. Greg and Debbie start back to back and walk away from each other until they are $\frac{1}{2}$ mi. apart. If Greg walks at 2 mph and Debbie at $2\frac{1}{2}$ mph, how long will this take?

30. Laura and Thomas are swimming in opposite directions. Laura is swimming 2 m/sec. and Thomas is swimming $2\frac{3}{5}$ m/sec. If we begin a stopwatch at the time they pass one another, how long will it take them to get 40 m apart?

REVIEW EXERCISES

1. Use $<, >,$ or $=$ to make a true sentence.

$$\frac{6}{17} \quad ? \quad \frac{7}{19}$$

2. Reduce to lowest terms. $\dfrac{9x}{24x^2}$

3. Multiply. $\dfrac{-4}{9} \cdot \dfrac{21}{30}$

4. Divide. $\dfrac{14a}{15ab^4} \div \dfrac{21}{25ab}$

5. Add. $\dfrac{7}{8} + 5\dfrac{2}{3}$

Summary

Defined Terms

Review the following terms and for those you do not know, study its definition on the page number next to it.

Section 5.1
Fraction *(p. 290)*
Numerator *(p. 290)*
Denominator *(p. 290)*
Rational number *(p. 290)*
Simplify *(p. 292)*
Simplest form *(p. 292)*
Equivalent fractions *(p. 294)*
Upscale *(p. 295)*
Reduce *(p. 295)*

Muliple *(p. 296)*
Improper fraction *(p. 298)*
Mixed number *(p. 298)*

Section 5.2
Lowest terms *(p. 303)*
Rational expression *(p. 306)*

Section 5.3
Circle *(p. 320)*
Radius *(p. 320)*

Diameter *(p. 320)*
Circumference *(p. 321)*
Irrational number *(p. 321)*
Pi (π) *(p. 321)*

Section 5.4
Reciprocals *(p. 327)*
Complex fraction *(p.329)*

Section 5.5
Least common muliple
(LCM) *(p. 337)*
Least common denominator
(LCD) *(p. 339)*

Section 5.7
Trapezoid *(p. 357)*

Procedures, Rules, and Key Examples

Procedures/Rules	Key Example(s)
Section 5.1	
If the denominator of a fraction is 1, the fraction can be simplified to the numerator. $$\frac{n}{1} = n$$	$$\frac{8}{1} = 8$$
If the numerator of a fraction is 0 and the denominator is any number other than 0, the fraction can be simplified to 0. $$\frac{0}{n} = 0 \text{ as long as } n \neq 0$$	$$\frac{0}{11} = 0$$
If the denominator is 0 with any number other than 0 in the numerator, we say the fraction is undefined. $$\frac{n}{0} \text{ is undefined when } n \neq 0$$	$$\frac{14}{0} \text{ is undefined}$$
A fraction with the same numerator and denominator (other than zero) can be simplified to 1. $$\frac{n}{n} = 1 \text{ as long as } n \neq 0$$	$$\frac{60}{60} = 1$$
Zero over zero is indeterminate. $$\frac{0}{0} \text{ is indeterminate}$$	$$\frac{0}{0} \text{ is indeterminate}$$

| | Key Example(s) |

Property of equivalent fractions:
We can write an equivalent fraction by multiplying or dividing both the numerator and denominator by the same number.

Upscale:
$$\frac{4}{5} = \frac{4 \cdot 3}{5 \cdot 3} = \frac{12}{15}$$

Reduce:
$$\frac{12}{15} = \frac{12 \div 3}{15 \div 3} = \frac{4}{5}$$

To compare two fractions:
1. Upscale or reduce both fractions to equivalent fractions that have a common denominator.
2. Compare.

Tip: We can upscale quickly using cross products. If the cross products are equal, so are the fractions. The larger cross product will indicate the larger fraction.

Use $<$, $>$, or $=$ to make a true statement.

$$\frac{6}{9} \, ? \, \frac{12}{18}$$

Reduce each:
$$\frac{6}{9} = \frac{6 \div 3}{9 \div 3} = \frac{2}{3} \qquad \frac{12}{18} = \frac{12 \div 6}{18 \div 6} = \frac{2}{3}$$

They are equal: $\dfrac{6}{9} = \dfrac{12}{18}$

$$\frac{5}{8} \, ? \, \frac{4}{7}$$

Upscale each (using cross products here):
$$\frac{35}{56} = \frac{5 \cdot 7}{8 \cdot 7} = \frac{5}{8} \quad \frac{4}{7} = \frac{4 \cdot 8}{7 \cdot 8} = \frac{32}{56}$$

$$\frac{35}{56} > \frac{32}{56} \text{ therefore } \frac{5}{8} > \frac{4}{7}$$

To write an improper fraction as a mixed number:
1. Divide the denominator into the numerator.
2. Write the results in the form:

$$\text{quotient} \, \frac{\text{remainder}}{\text{original denominator}}$$

Write $\dfrac{19}{8}$ as an improper fraction.

$$\begin{array}{r} 2 \\ 8 \overline{)19} \\ -16 \\ \hline 3 \end{array} \qquad 2\frac{3}{8}$$

To write a mixed number as an improper fraction:
1. Multiply the denominator by the integer.
2. Add the resulting product to the numerator. This resulting sum will be the numerator of the improper fraction.
3. Keep the same denominator.

$$5\frac{7}{8} = \frac{8 \cdot 5 + 7}{8} = \frac{40 + 7}{8} = \frac{47}{8}$$

Most people do not write these steps, performing the calculations mentally.

Section 5.2

To reduce a fraction to lowest terms:
1. Find the prime factorization of the numerator and denominator.
2. Divide out all primes that are common to both the numerator and denominator.
3. Multiply the remaining factors.

$$\frac{70}{182} = \frac{\overset{1}{\cancel{2}} \cdot 5 \cdot \overset{1}{\cancel{7}}}{\underset{1}{\cancel{2}} \cdot \underset{1}{\cancel{7}} \cdot 13} = \frac{5}{13}$$

Reducing a rational expression:

$$-\frac{18x^3y}{24x^2y^3} = -\frac{\overset{1}{\cancel{2}} \cdot \overset{1}{\cancel{3}} \cdot 3 \cdot \overset{1}{\cancel{x}} \cdot \overset{1}{\cancel{x}} \cdot x \cdot \overset{1}{\cancel{y}}}{\underset{1}{\cancel{2}} \cdot 2 \cdot 2 \cdot \underset{1}{\cancel{3}} \cdot \underset{1}{\cancel{x}} \cdot \underset{1}{\cancel{x}} \cdot \underset{1}{\cancel{y}} \cdot y \cdot y} = -\frac{3x}{4y^2}$$

Section 5.3

To multiply fractions:
1. Divide out any numerator factor with any like denominator factor.
2. Multiply numerator by numerator and denominator by denominator.
3. Reduce as needed to get to lowest terms. (This would be necessary if you miss common factors in the first step.)

$$-\frac{14}{15} \cdot \frac{9}{16} = -\frac{\overset{1}{\cancel{2}} \cdot 7}{\underset{1}{\cancel{8}} \cdot 5} \cdot \frac{\overset{1}{\cancel{3}} \cdot 3}{\underset{1}{\cancel{2}} \cdot 2 \cdot 2 \cdot 2} = -\frac{21}{40}$$

Multiplying rational expressions:

$$\frac{9a}{10b} \cdot \frac{-16}{21a^2b} = \frac{\overset{1}{\cancel{3}} \cdot 3 \cdot \overset{1}{\cancel{a}}}{2 \cdot 5 \cdot b} \cdot \frac{-2 \cdot 2 \cdot 2 \cdot 2}{\underset{1}{\cancel{3}} \cdot 7 \cdot \underset{1}{\cancel{a}} \cdot a \cdot b}$$

$$= -\frac{24}{35ab^2}$$

To multiply mixed numbers:
1. Write the mixed numbers as improper fractions.
2. Divide out any numerator factor with any like denominator factor.
3. Multiply numerator by numerator and denominator by denominator.
4. Write the product as a mixed number in simplest form.

$$-6\frac{3}{8} \cdot -1\frac{1}{3} = -\frac{51}{8} \cdot -\frac{4}{3}$$

$$= -\frac{\cancel{3} \cdot 17}{2 \cdot 2 \cdot 2} \cdot -\frac{\cancel{2} \cdot \cancel{2}}{\cancel{3}}$$

$$= \frac{17}{2}$$

$$= 8\frac{1}{2}$$

When a fraction is raised to a power, we evaluate both the numerator and denominator raised to that power. In symbols:

$$\left(\frac{a}{b}\right)^n = \frac{a^n}{b^n} \qquad \text{where } b \neq 0$$

$$\left(\frac{2}{3}\right)^4 = \frac{2}{3} \cdot \frac{2}{3} \cdot \frac{2}{3} \cdot \frac{2}{3} = \frac{16}{81}$$

$$\left(\frac{-2x^2y}{5}\right)^3 = \frac{-2x^2y}{5} \cdot \frac{-2x^2y}{5} \cdot \frac{-2x^2y}{5}$$

$$= \frac{-8x^6y^3}{125}$$

Section 5.4

To divide fractions:
Write the division statement as an equivalent multiplication statement:
1. Change the operation symbol from division to multiplication.
2. Change the divisor to its reciprocal.
3. Divide out any numerator factor with any like denominator factor.
4. Multiply numerator by numerator and denominator by denominator.
5. Write the result in simplest form.

$$\frac{9}{16} \div \frac{3}{10} = \frac{9}{16} \cdot \frac{10}{3}$$

$$= \frac{\cancel{3} \cdot 3}{\cancel{2} \cdot 2 \cdot 2 \cdot 2} \cdot \frac{\cancel{2} \cdot 5}{\cancel{3}}$$

$$= \frac{15}{8}$$

$$= 1\frac{7}{8}$$

Dividing rational expressions:

$$-\frac{5k}{16m^2n} \div \frac{25n}{2m} = -\frac{5k}{16m^2n} \cdot \frac{2m}{25n}$$

$$= -\frac{\cancel{5} \cdot k}{\cancel{2} \cdot 2 \cdot 2 \cdot 2 \cdot \cancel{m} \cdot m \cdot n} \cdot \frac{\cancel{2} \cdot \cancel{m}}{\cancel{5} \cdot 5 \cdot n}$$

$$= -\frac{k}{40mn^2}$$

To divide mixed numbers:
1. Write the mixed numbers as improper fractions.
2. Write the division statement as an equivalent multiplication statement using the reciprocal of the divisor.
3. Divide out any numerator factor with any like denominator factor.
4. Multiply.
5. Write the result in simplest form.

$$-2\frac{2}{5} \div 5\frac{1}{4} = -\frac{12}{5} \div \frac{21}{4}$$

$$= -\frac{12}{5} \cdot \frac{4}{21}$$

$$= -\frac{2 \cdot 2 \cdot \cancel{3}}{5} \cdot \frac{2 \cdot 2}{\cancel{3} \cdot 7}$$

$$= -\frac{16}{35}$$

To get the square root of a fraction, we have two options:
1. Get the square root of the numerator and denominator separately.
or
2. Simplify the fraction, then get the square root of the quotient.

$$\sqrt{\frac{49}{100}} = \frac{\sqrt{49}}{\sqrt{100}} = \frac{7}{10}$$

$$\sqrt{\frac{75}{3}} = \sqrt{25} = 5$$

Section 5.5

To find the LCM by listing:
List multiples of the largest given number until you find a multiple that is divisible by all the other given numbers.

Find the LCM of 18 and 24.
Listing method:

24	48	72
is not divisible by 18.	is not divisible by 18.	is divisible by 18.

$$\text{LCM}(18, 24) = 72$$

To find the LCM using prime factorization
1. Find the prime factorization of each given number.
2. Write the factorizations in exponential form.
3. Use the largest exponent of each prime factor.
4. Multiply to get the LCM.

Prime factorization method:
$$18 = 2 \cdot 3^2$$
$$24 = 2^3 \cdot 3$$
$$\text{LCM}(18, 24) = 2^3 \cdot 3^2$$
$$= 2 \cdot 2 \cdot 2 \cdot 3 \cdot 3$$
$$= 72$$
Find the LCM of $18x^3y$ and $24xz^2$.
$$\text{LCM}(18x^3y, 24xz^2) = 2^3 \cdot 3^2 x^3 y z^2$$
$$= 8 \cdot 9x^3yz^2$$
$$= 72x^3yz^2$$

Section 5.6

To add or subtract fractions or rational expressions that have the same denominator:
1. Add or subtract numerators.
2. Keep the same denominator.
3. Simplify.

Note: With rational expressions we can only combine numerators that have like terms. For unlike terms, we must express the sum or difference as a polynomial.

$$\frac{2}{9} + \frac{1}{9} = \frac{3}{9} = \frac{1}{3}$$

Adding/subtracting rational expressions:

$$\frac{5h}{2} - \frac{1}{2} = \frac{5h - 1}{2}$$

$$\frac{2x - 9}{5y} - \frac{x + 1}{5y} = \frac{(2x - 9) - (x + 1)}{5y}$$
$$= \frac{(2x - 9) + (-x - 1)}{5y}$$
$$= \frac{x - 10}{5y}$$

To add or subtract fractions with different denominators:
1. Find the LCD.

Note: Any common multiple of the denominators will do, but using the LCD will make the numbers more manageable.

2. Upscale the fractions to equivalent fractions that have the common denominator.
3. Add or subtract the numerators and keep the common denominator.
4. Simplify.

$$\frac{1}{6} + \frac{3}{4} = \frac{1(2)}{6(2)} + \frac{3(3)}{4(3)}$$
$$= \frac{2}{12} + \frac{9}{12}$$
$$= \frac{11}{12}$$

$$\frac{1}{2} - \frac{5}{8} = \frac{1(4)}{2(4)} - \frac{5(1)}{8(1)}$$
$$= \frac{4}{8} - \frac{5}{8}$$
$$= -\frac{1}{8}$$

$$\frac{3}{5x^2} - \frac{9}{10xy} = \frac{3(2y)}{5x^2(2y)} - \frac{9(x)}{10xy(x)}$$
$$= \frac{6y}{10x^2y} - \frac{9x}{10x^2y}$$
$$= \frac{6y - 9x}{10x^2y}$$

Procedures/Rules (cont.)	Key Example(s)

To add mixed numbers:

Method 1: Write as improper fractions, then follow the procedure for adding fractions.

$$2\frac{3}{5} + 5\frac{3}{4} = \frac{13}{5} + \frac{23}{4} = \frac{13(4)}{5(4)} + \frac{23(5)}{4(5)}$$

$$= \frac{52}{20} + \frac{115}{20}$$

$$= \frac{167}{20}$$

$$= 8\frac{7}{20}$$

Method 2: Add the integer parts and fraction parts separately.

$$2\frac{3}{5} + 5\frac{3}{4} = 2\frac{3(4)}{5(4)} + 5\frac{3(5)}{4(5)}$$

$$= 2\frac{12}{20} + 5\frac{15}{20}$$

$$= 7\frac{27}{20}$$

$$= 8\frac{7}{20}$$

To subtract mixed numbers:

Method 1: Write as improper fractions, then follow the procedure for adding/subtracting fractions.

$$8\frac{1}{4} - 5\frac{2}{3} = \frac{33}{4} - \frac{17}{3} = \frac{33(3)}{4(3)} - \frac{17(4)}{3(4)}$$

$$= \frac{99}{12} - \frac{68}{12}$$

$$= \frac{31}{12}$$

$$= 2\frac{7}{12}$$

Method 2: Subtract the integer parts and fraction parts separately.

$$8\frac{1}{4} - 5\frac{2}{3} = 8\frac{1(3)}{4(3)} - 5\frac{2(4)}{3(4)}$$

$$= 8\frac{3}{12} - 5\frac{8}{12}$$

$$= 7\frac{15}{12} - 5\frac{8}{12}$$

$$= 2\frac{7}{12}$$

Section 5.7

An exponential form with a negative exponent is equivalent to the reciprocal of the base with the exponent made positive.

$$x^{-a} = \frac{1}{x^a} \text{ and } \left(\frac{a}{b}\right)^{-n} = \left(\frac{b}{a}\right)^n$$

Evaluate the exponent:

$$3^{-3} = \frac{1}{3^3} = \frac{1}{3 \cdot 3 \cdot 3} = \frac{1}{27}$$

To evaluate an expression with a negative exponent:
1. Get the reciprocal of the base.
2. Change the sign of the exponent to positive.
3. Calculate.

$$\left(\frac{2}{3}\right)^{-4} = \left(\frac{3}{2}\right)^4 = \frac{3}{2} \cdot \frac{3}{2} \cdot \frac{3}{2} \cdot \frac{3}{2} = \frac{81}{16}$$

Section 5.8

To solve equations:

1. Simplify both sides of the equation as needed.
   a. Distribute to clear parentheses.
   b. Clear fractions by multiplying all denominators on both sides by the LCD (optional).
   c. Combine like terms.

2. Use the addition/subtraction principle so that all variable terms are on one side of the equation and all constants are on the other side.

Tip: Clear the variable term that has the smaller coefficient. This will avoid negative coefficients.

3. Use the multiplication/division principle to clear any remaining coefficients.

$$\frac{2}{3}(x - 1) = \frac{1}{2}x - \frac{2}{5}$$

$$\frac{2}{3}x - \frac{2}{3} = \frac{1}{2}x - \frac{2}{5}$$

$$30\left(\frac{2}{3}x - \frac{2}{3}\right) = \left(\frac{1}{2}x - \frac{2}{5}\right)30$$

$$\overset{10}{\underset{1}{\frac{\cancel{30}}{1}}} \cdot \frac{2}{\underset{1}{\cancel{3}}}x - \frac{2}{\underset{1}{\cancel{3}}} \cdot \overset{10}{\frac{\cancel{30}}{1}} = \overset{15}{\frac{\cancel{30}}{1}} \cdot \frac{1}{\underset{1}{\cancel{2}}}x - \frac{2}{\underset{1}{\cancel{5}}} \cdot \overset{6}{\frac{\cancel{30}}{1}}$$

$$20x - 20 = 15x - 12$$

$$20x - 20 = 15x - 12$$

$$\underline{-15x \qquad\quad -15x}$$

$$5x - 20 = \ \ 0 \ - 12$$

$$5x - 20 = -12$$

$$\underline{+20 \qquad +20}$$

$$5x - \ 0 = \quad 8$$

$$\frac{5x}{5} = \frac{8}{5}$$

$$1x = \frac{8}{5}$$

$$x = 1\frac{3}{5}$$

Formulas

Area of a parallelogram: $\qquad A = bh$

Area of a triangle: $\qquad A = \frac{1}{2}bh$

Area of a trapezoid: $\qquad A = \frac{1}{2}h(a + b)$

Circles:

Diameter: $\qquad d = 2r$

Radius: $\qquad r = \frac{1}{2}d$

Circumference of a circle: $\qquad C = \pi d$ or $C = 2\pi r$

Area of a circle: $\qquad A = \pi r^2$

For Exercises 1–6, answer true or false.

1. $\frac{3}{5}$ is in lowest terms.

2. Every integer can be expressed in fraction form.

3. $\frac{54}{0} = 0$.

4. A fraction with the same numerator and denominator (other than 0) can be simplified to the number 1.

5. When finding the LCM using primes, the rule is to use only those prime factors that are common to all the factorizations.

6. When multiplying mixed numbers, we can multiply the integer parts and fraction parts separately.

7. We can write an equivalent fraction by _____ or _____ both the numerator and denominator by the same number.

8. Explain in your own words how to reduce a fraction to lowest terms.

9. Explain in your own words how to divide fractions.

10. Explain in your own words how to add fractions.

11. Name the lowest-terms fraction represented by the shaded region.

  a. 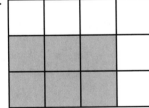   b.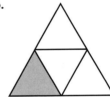

12. Graph each number on a number line.

  a. $4\frac{1}{5}$ b. $-2\frac{5}{8}$

13. Write each improper fraction as a mixed number.

  a. $\frac{40}{9}$ b. $-\frac{29}{4}$

14. Write each mixed number as an improper fraction.

  a. $6\frac{2}{3}$ b. $-5\frac{1}{2}$

15. Simplify.

  a. $\frac{18}{1}$ b. $\frac{-14}{0}$ c. $\frac{0}{19}$ d. $\frac{-2}{-2}$

16. Reduce to lowest terms.

a. $\dfrac{15}{35}$

b. $-\dfrac{84}{105}$

17. Reduce to lowest terms.

a. $\dfrac{-8m^3n}{26n^2}$

b. $\dfrac{6xy^2}{20y^5}$

18. Use $<$, $>$, or $=$ to write a true statement.

a. $\dfrac{5}{9}$ **?** $\dfrac{7}{13}$

b. $\dfrac{3}{16}$ **?** $\dfrac{5}{24}$

c. $\dfrac{10}{16}$ **?** $\dfrac{25}{40}$

19. Multiply. Write your answer as a mixed number where appropriate.

a. $\dfrac{4}{9} \cdot \dfrac{12}{20}$

b. $-4\dfrac{1}{2} \cdot 2\dfrac{2}{3}$

20. Multiply.

a. $\dfrac{10n}{6p^3} \cdot \dfrac{14np}{8n}$

b. $\dfrac{h^3}{4k} \cdot \dfrac{-12k}{15h}$

21. Divide. Write your answer as a mixed number where appropriate.

a. $-\dfrac{5}{6} \div \dfrac{-15}{28}$

b. $4\dfrac{1}{5} \div \dfrac{3}{5}$

22. Divide.

a. $\dfrac{20a}{12ab^3} \div \dfrac{10}{8b}$

b. $\dfrac{-8x^6y}{25x^3} \div \dfrac{2xy}{5z^2}$

23. Evaluate the square root.

a. $\sqrt{\dfrac{100}{36}}$

b. $\sqrt{\dfrac{50}{2}}$

24. Find the LCM.

a. 28 and 24

b. $15x$ and $20x^2y$

25. Add or subtract.

a. $\dfrac{5}{6} + \dfrac{1}{5}$

b. $4\dfrac{5}{8} + 6\dfrac{2}{3}$

c. $\dfrac{9}{15} - \dfrac{4}{5}$

d. $-5\dfrac{1}{6} - \left(-2\dfrac{1}{3}\right)$

26. Add or subtract.

a. $\dfrac{3n}{8} + \dfrac{n}{8}$

b. $-\dfrac{2}{9x} + \left(-\dfrac{4}{9x}\right)$

c. $\dfrac{7}{6h} - \dfrac{3}{4h}$

d. $\dfrac{5}{8a} - \dfrac{7}{12}$

27. Simplify.

    a. $3\dfrac{1}{2} + \dfrac{1}{4}\left(\dfrac{2}{3} - \dfrac{1}{6}\right)$

    b. $\left(4 + \dfrac{1}{12}\right) - 8 \div \dfrac{3}{4}$

28. Evaluate the expressions.

    a. $\left(\dfrac{1}{2}\right)^{6}$

    b. $\left(-\dfrac{3}{5}xy^{3}\right)^{2}$

    c. 4^{-2}

    d. $\left(-\dfrac{2}{3}\right)^{-3}$

29. Simplify.

    a. $\dfrac{3}{5}x^{2} + 9x - \dfrac{1}{2}x^{2} - 2 - 11x$

    b. $\left(\dfrac{1}{4}n^{2} - \dfrac{2}{3}n - 3\right) + \left(\dfrac{3}{8}n^{2} + 1\right)$

    c. $\left(12y^{3} - \dfrac{5}{6}y + \dfrac{2}{5}\right) - \left(\dfrac{1}{2}y^{3} + 2y^{2} + \dfrac{1}{3}\right)$

    d. $\left(\dfrac{2}{7}ab^{3}\right)\left(-\dfrac{7}{8}abc^{2}\right)$

    e. $\left(\dfrac{1}{4}x + 2\right)\left(6x - \dfrac{3}{8}\right)$

30. Solve and check.

    a. $y - 3\dfrac{4}{5} = \dfrac{1}{3}$

    b. $-\dfrac{3}{4}n = \dfrac{2}{3}$

    c. $\dfrac{1}{2}m - 5 = \dfrac{3}{4}$

    d. $\dfrac{1}{2}\left(n - \dfrac{2}{3}\right) = \dfrac{3}{4}n + 2$

31. Find the area of the triangle.

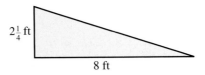

32. Find the area of the trapezoid.

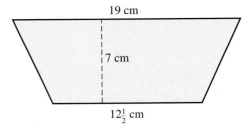

33. The circular top of a can of beans has a diameter of $3\dfrac{1}{2}$ in. Find the radius.

34. A wheel on a piece of exercise equipment has a radius of $14\dfrac{1}{5}$ cm. Find the diameter.

35. A tree has a diameter of $2\dfrac{1}{2}$ ft. What is the circumference? Use $\dfrac{22}{7}$ for π.

36. Find the surface area of a circular tabletop with a diameter of 4 ft. Use $\dfrac{22}{7}$ for π.

37. Find the area of the shaded region.

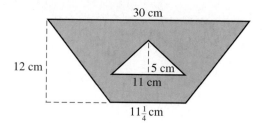

38. At a clothing store, $\frac{5}{9}$ of the merchandise is discounted. $\frac{3}{10}$ of these discounted items are shirts. What fraction of all items in the store are discounted shirts?

39. A box of rice contains $4\frac{1}{2}$ cups of uncooked rice. The label indicates that a single serving is $\frac{1}{4}$ cup of uncooked rice. How many servings are in the box?

40. Respondents to a survey can answer in three ways: agree, disagree, or no opinion. $\frac{2}{5}$ said they agreed, while $\frac{1}{4}$ said they disagreed. What fraction had no opinion?

41. $\frac{1}{4}$ of the sum of n and 2 is the same as $\frac{3}{5}$ less than $\frac{3}{4}$ of n. Translate to an equation, then solve.

42. The area of the trapezoid shown is 19 m². Find the length of the missing side.

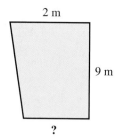

43. A section of PVC pipe is $2\frac{1}{3}$ times the length of a second section. The two pipes together have a length of 35 ft. What are the lengths of the two sections?

For Exercises 44 and 45, use a four-column table.

44. Emma and Bill pass each other going in opposite directions. Emma is walking at $2\frac{1}{2}$ mph and Bill is jogging at 5 mph. How long until they are $\frac{7}{10}$ mi. apart?

45. Two asteroids are separated by 2000 mi. and headed directly at one another. Asteroid A is traveling at 50 mi./sec. and asteroid B is traveling at 120 mi./sec. How long until they collide?

Practice Test

1. Name the fraction represented by the shaded region.

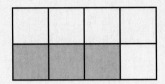

2. Graph $-3\frac{1}{4}$ on a number line.

3. Use $<$, $>$, or $=$ to write a true statement.

   a. $\frac{3}{4}$? $\frac{9}{12}$

   b. $\frac{14}{15}$? $\frac{5}{6}$

4. Write $\frac{37}{6}$ as a mixed number.

5. Write $-4\frac{5}{8}$ as an improper fraction.

6. Simplify.

   a. $\frac{-16}{1}$

   b. $\frac{3}{0}$

   c. $\frac{0}{14}$

   d. $\frac{27}{27}$

7. Reduce to lowest terms.

   a. $\frac{24}{40}$

   b. $-\frac{9x^5y}{30x^3}$

8. Multiply.

   a. $-2\frac{1}{4} \cdot 5\frac{1}{6}$

   b. $\frac{2a}{7b^3} \cdot \frac{14b}{3a}$

9. Divide.

   a. $3\frac{1}{8} \div \frac{5}{4}$

   b. $\frac{-2m^4n}{9n^2} \div \frac{5m}{12n^2}$

10. Find the LCM of $18t^3$ and $12t^2u$.

11. Add or subtract.

   a. $\frac{3}{10} + \frac{1}{4}$

   b. $-3\frac{1}{8} - \left(-1\frac{1}{2}\right)$

12. Add or subtract.

   a. $-\frac{4x}{15} + \left(-\frac{x}{15}\right)$

   b. $\frac{7}{4a} - \frac{5}{6a}$

13. Evaluate.

   a. $\left(\frac{2}{5}\right)^2$

   b. 4^{-3}

1. _____

2. _____

3. _____

4. _____

5. _____

6. _____

7. _____

8. _____

9. _____

10. _____

11. _____

12. _____

13. _____

14. _____

14. Calculate. $2\frac{1}{4} + \frac{5}{8}\left(\frac{4}{5} - \frac{3}{10}\right)$

15. Simplify.

15. _____

a. $\left(\frac{1}{4}n^2 - \frac{1}{2}n + 2\right) - \left(\frac{1}{4}n^2 + n + 5\right)$ **b.** $\left(m + \frac{3}{4}\right)\left(2m - \frac{1}{2}\right)$

16. _____

16. Solve and check. $\frac{1}{4}m - \frac{2}{3} = \frac{5}{6}$

17. Find the area of the triangle.

17. _____

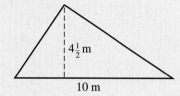
4½ m

10 m

18. _____

18. Find the area of the trapezoid.

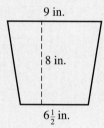
9 in.

8 in.

$6\frac{1}{2}$ in.

19. _____

19. A compact disc has a diameter of $4\frac{3}{4}$ in. What is the circumference? Use $\frac{22}{7}$ for π.

20. _____

20. Calculate the area of a circle with a diameter of 5 in. Use $\frac{22}{7}$ for π.

21. _____

21. In a new housing development, $\frac{3}{4}$ of the lots were sold in the first year. $\frac{5}{6}$ of these lots had finished houses on them by the end of the first year. What fraction of all lots in the development had finished houses on them by the end of the first year?

22. _____

22. Jason needs pieces of wire that are $4\frac{3}{8}$ in. long. How many pieces can he make out of 35 in. of wire?

23. _____

23. A testing company examines the results of a certain multiple-choice test question that has four choices, A, B, C, or D. $\frac{5}{8}$ of those who took the test chose A, while $\frac{1}{6}$ chose B, and nobody chose D. What fraction chose C?

24. _____

24. $\frac{2}{3}$ of the sum of n and 5 is the same as $\frac{3}{4}$ less than $\frac{1}{2}$ of n. Translate to an equation, then solve.

25. _____

25. Ken and Keisha are traveling toward each other on the same path. Ken is bicycling at 8 mph and Keisha is jogging at $5\frac{1}{2}$ mph. If they are $\frac{3}{4}$ of a mile apart, how long until they meet?

Cumulative Review Exercises

For Exercises 1–6, answer true or false.

1. -1 is a whole number.

2. The sum of two numbers that have the same sign is always positive.

3. $2x = 5x^2 + 8$ is a linear equation.

4. -7 is a solution for $3a + 9 = -30$.

5. $\left(\dfrac{3}{4}\right)^2 = \dfrac{3^2}{4^2}$

6. $\dfrac{3}{8} > \dfrac{5}{16}$

7. When raising an exponential form to a power, we can _____ the exponents and keep the same base.

8. Explain in your own words how to find the GCF/GCD.

9. Explain the mistake. $4x - 9 = 10x + 3$

$$\begin{array}{r} 4x - 9 = 10x + 3 \\ \underline{-4x \qquad\quad -4x} \\ -9 = 6x + 3 \\ \underline{-9 = 6x + 3} \\ \underline{-3 \qquad -3} \\ -6 = 6x \\ \dfrac{-6}{6} = \dfrac{6x}{6} \\ -1 = x \end{array}$$

10. Explain in your own words how to add or subtract fractions.

11. Write the word name for 4,582,601.

12. Graph -9 on a number line.

13. Round 851,412 to the nearest ten thousand.

14. Estimate $461 \cdot 72$ by rounding so that there is only one nonzero digit in each number.

15. What is the degree of $-x$?

16. Find the prime factorization of 3600.

17. Find the GCF/GCD of $54x^3y$ and $45x$.

For Exercises 18–27, simplify.

18. $-(-(-15))$

19. $5 - 12 + (-17) + 9$

20. $2^3 - 7(-4)$

21. $-3^4 - 5\sqrt{100 - 64}$

22. $18 - 5(6) \div [9 + 3(-7)]$

23. $\dfrac{36}{40}$

24. $4\dfrac{1}{3} \div 5\dfrac{1}{9}$

25. $-\dfrac{5x^3}{9} \cdot \dfrac{27}{40x}$

26. $7\dfrac{1}{6} - 3\dfrac{1}{2}$

27. $\dfrac{4}{5x} + \dfrac{2}{3}$

28. Evaluate $mn - 2n^3$ when $m = \dfrac{5}{6}$ and $n = -4$.

29. Combine like terms and write your answer in descending order.

$$12b^2 - \frac{2}{3}b + 3b^2 - b^3 - \frac{1}{4}b$$

30. Subtract. $(x^4 - 5x^3 - 10x + 18) - (x^4 + 2x^3 - 3x - 8)$

31. Multiply. $(-8x^3)(-4x^2)$

32. Multiply. $(x - 7)(x + 7)$

33. Divide. $t^9 \div t^4$

34. Factor. $18m^3 + 24m^2 - 30m$

For Exercises 35–38, solve and check.

35. $x + 24 = 8$

36. $\frac{2}{3}a = -18$

37. $\frac{3}{4}n - 1 = -\frac{2}{5}$

38. $3(y - 4) - 8 = 7y - 16$

For Exercises 39–50, solve.

39. A combination lock consists of 4 rollers with the numerals 0–9 on each roller. How many different combinations are possible?

40. The following table lists the assets and debts for the Krueger family. Calculate their net worth.

Assets	Debts
Savings = $1260	Credit-card balance = $872
Checking = $945	Mortgage = $57,189
Furniture = $13,190	Automobile 1 = $3782
	Automobile 2 = $12,498

41. An acorn falls from a height of 92 ft. What is the height of the acorn after 2 sec.?
(Use the formula $h = -16t^2 + s_0$.)

42. a. Write an expression in simplest form for the perimeter of the rectangle shown.

   b. Find the perimeter if y is 8.

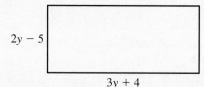

$2y - 5$

$3y + 4$

43. Write an expression in simplest form for the area.

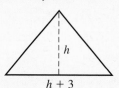

h

$h + 3$

44. A tree has a circumference of 44 in. Calculate the diameter.

45. A storage chest has a volume of 24 ft.^{3}. If the length is 4 ft. and the width is 3 ft., find the height?

46. A support beam is to be angled so that the angle made between the beam and the ground on one side of the beam is 35° less than twice the adjacent angle on the other side of the beam. What are the angle measurements?

47. The length of a rectangle is 3 times the width. If the perimeter is 48 ft., find the length and width.

48. The sum of two consecutive even integers is 130. Find the integers.

49. A movie theater sells two different-size boxes of popcorn. The large box costs $3 and the small box costs $2. In one day the theater sold 657 boxes of popcorn for a total of $1612. How many of each size were sold? (Use a four-column table.)

50. Shelly and Ryan pass each other going in opposite directions. Shelly is running at $5\frac{1}{2}$ mph and Ryan is jogging at 3 mph. How long until they are $3\frac{2}{5}$ mi. apart?

Decimals

Getting Help

The learning process is not a comfortable process, because in learning we must allow ourselves to be vulnerable. Remember that to learn, you must be challenged. There will be times in your education that you will encounter difficulties. The key to success is learning to respond to those difficulties in a positive way.

When you encounter difficulties that you cannot resolve, seek help immediately. Even if your question or problem seems small, it can often make a difference in understanding future topics. As soon as possible, talk with your instructor. Most instructors have office hours or are available to discuss your problems.

In addition, most colleges have tutorial centers with flexible hours that offer a variety of services. The centers will likely have tutors as well as the videotapes and computer software that are keyed to the text. Some schools may have these resources available in their library as well. Another tutorial source you may use is Addison Wesley Longman's tutorial center. If you received a PIN with your text you can call, email, or fax your question, and our staff of experienced math teachers will answer your questions. (Phone numbers, email address, and fax numbers are listed in the *To the Student* section.)

Another great source for help is your classmates. Form a study group and exchange phone numbers and email addresses so that you can discuss problems and encourage and support each other. Above all, do not let questions go unanswered. Take advantage of all the resources available to you to get the help you need to succeed.

"In the middle of difficulty lies opportunity"
—ALBERT EINSTEIN

"No matter what accomplishments you make, someone helped you."
—ALTHEA GIBSON

6.1 Decimals and Rational and Irrational Numbers

OBJECTIVES

1 Write decimals as fractions or mixed numbers.
2 Write a word name for a decimal number.
3 Graph decimals on a number line.
4 Use < or > to make a true statement.
5 Round decimal numbers to a specified place.

OBJECTIVE **1** *Write decimals as fractions or mixed numbers.*

In Chapter 5 we developed fractions and rational expressions. We now want to explore a different notation for writing fractions, **decimal notation.**

| **DEFINITION** | **Decimal notation:** A base-ten notation for expressing fractions. |

By *base-ten notation* we mean that each place value is a power of 10, an exponential expression with a base of 10. Whole numbers are expressed to the left of the decimal point and have familiar place values such as ones, tens, hundreds, etc. To the right of the decimal point are the fractional place values. For each fractional place, the power of 10 is in the denominator of a fraction.

Decimal Point
↓

Whole numbers			**Fractions**							
Hundreds	Tens	Ones	Tenths	Hundredths	Thousandths	Ten-thousandths	Hundred-thousandths	Millionths	Ten-millionths	Hundred-millionths

Notice that the names follow the same pattern as the whole number side of the decimal point. We simply place *th* in each whole place name. *Tens* becomes *tenths, hundreds* becomes *hundredths,* and so on.

◆ **Discussion** Why is there no oneths place?

In Chapter 5, we defined **rational numbers.** Let's recall that definition:

| **DEFINITION** | **Rational number:** Any number that can be expressed as a ratio of integers. |

In math language, a rational number is a number that can be expressed in the form $\frac{a}{b}$, where a and b are integers and $b \neq 0$.

Because all integers can be expressed as fractions by placing the integer over 1, all integers are rational numbers. All mixed numbers are rational numbers because we can write them as fractions.

Decimal notation is another way to express fractional amounts, but are all decimal numbers rational numbers? We need to explore further to answer this question.

To see the connection between decimals and fractions, let's have a look at a number written in expanded form, which is written as the sum of the digits multiplied by their respective place values. Consider the number 23.791.

Decimal Point
↓

Whole numbers			Fractions							
Hundreds	Tens	Ones	Tenths	Hundredths	Thousandths	Ten-thousandths	Hundred-thousandths	Millionths	Ten-millionths	Hundred-millionths
	2	3 .	7	9	1					

Expanded form: $23.791 = 2 \cdot 10 + 3 \cdot 1 + 7 \cdot \dfrac{1}{10} + 9 \cdot \dfrac{1}{100} + 1 \cdot \dfrac{1}{1000}$

We can develop the fraction equivalent by multiplying and adding.

$$23.791 = 2 \cdot 10 + 3 \cdot 1 + 7 \cdot \dfrac{1}{10} + 9 \cdot \dfrac{1}{100} + 1 \cdot \dfrac{1}{1000}$$

$$= 20 + 3 + \dfrac{7}{1} \cdot \dfrac{1}{10} + \dfrac{9}{1} \cdot \dfrac{1}{100} + \dfrac{1}{1} \cdot \dfrac{1}{1000}$$

$$= 23 + \dfrac{7}{10} + \dfrac{9}{100} + \dfrac{1}{1000}$$

$$= 23 + \dfrac{7(100)}{10(100)} + \dfrac{9(10)}{100(10)} + \dfrac{1(1)}{1000(1)}$$

$$= 23 + \dfrac{700}{1000} + \dfrac{90}{1000} + \dfrac{1}{1000}$$

$$= 23 + \dfrac{791}{1000}$$

$$= 23\dfrac{791}{1000}$$

To add the fractions we need a common denominator. Notice the last place value determines the common denominator. In this case the LCD is 1000.

Notice the whole number part becomes the whole/integer part of a mixed number.

The numerator of the fraction is a combination of all the decimal digits and the denominator is simply the last place value.

From the expanded form we can see a simple way to write a decimal number as a fraction or mixed number:

Procedure *To write a decimal number as a fraction or mixed number:*
1. Write all digits to the left of the decimal (whole digits) as the whole/integer in a mixed number.
2. Write all digits to the right of the decimal (fraction digits) in the numerator of a fraction.
3. Write the last place value in the denominator.
4. Reduce/Simplify.

EXAMPLE 1 Write 0.84 as a reduced fraction.

Solution: Write all the decimal digits in the numerator and write the last place value in the denominator.

$$0.84 = \dfrac{84}{100} = \dfrac{84 \div 4}{100 \div 4} = \dfrac{21}{25}$$
$$\uparrow$$
hundredths

Because $0.84 = \dfrac{84}{100}$, we read 0.84 as "eighty-four hundredths."

Tip
Because the denominators will always be powers of ten, and the primes that divide powers of ten are 2 and 5, these fractions will only reduce if the numerators are divisible by 2 or 5.

Margin 1

Write as a fraction or mixed number in lowest terms.

a. 0.79

b. −0.08

c. 2.6

d. −14.675

Margin 2

Write the word name.

a. 0.91

b. −0.602

c. 2.7

d. 124.90017

Answers to Margin 1: **a.** $\frac{79}{100}$ **b.** $-\frac{2}{25}$ **c.** $2\frac{3}{5}$ **d.** $-14\frac{27}{40}$

Answers to Margin 2: **a.** ninety-one hundredths **b.** negative six hundred two thousandths **c.** two and seven tenths **d.** one hundred twenty-four and ninety thousand seventeen hundred-thousandths.

EXAMPLE 2 Write −217.5 as a mixed number in lowest terms.

Solution: We write the whole/integer part of the decimal number as the whole/integer part of a mixed number, then write the decimal digits in the numerator and the last place value in the denominator.

$$-217.5 = -217\frac{5}{10} = -217\frac{5 \div 5}{10 \div 5} = -217\frac{1}{2}$$

−217.5 can be read several ways. It is common to say, "negative two hundred seventeen point five." From the mixed number form, we can say "negative two hundred seventeen and five tenths." Or in simplest form, "negative two hundred seventeen and one half."

◄ **Do Margin 1.**

OBJECTIVE **2** *Write a word name for a decimal number.*

Because decimal numbers can be expressed as fractions, when we write a word name for a decimal number we write the name of the fraction or mixed number equivalent.

$0.39 = \frac{39}{100}$ and is written: Thirty-nine hundredths

$19.8571 = 19\frac{8571}{10,000}$ and is written: Nineteen and eight thousand five hundred seventy-one ten-thousandths.

In a mixed number the word *and* separates the whole/integer and decimal/fraction parts of both mixed numbers and decimal numbers. The word *and* takes the place of the decimal point in the word name for a decimal number that has a whole/integer amount.

> *To write the word name for decimal digits (fraction part of a decimal number):*
> **1.** Write the word name for digits as if they represented a whole number.
> **2.** Write the name of the last place value.
>
> *To write a word name for a decimal number with both whole integer and fraction parts:*
> **1.** Write the name of the whole/integer number part.
> **2.** Write the word *and* for the decimal point.
> **3.** Write the name of the fractional part.

EXAMPLE 3 Write the word name.

a. 0.0000489.

b. 98.10479.

Answers: a. Four hundred eighty-nine ten-millionths **b.** Ninety-eight and ten thousand four hundred seventy-nine hundred-thousandths

◄ **Do Margin 2.**

OBJECTIVE **3** *Graph decimals on a number line.*

Decimal numbers can be graphed on a number line. The denominator of the fraction equivalent describes the number of divisions in between integers. The numerator of the fraction equivalent describes how many of those divisions we are interested in.

EXAMPLE 4 Graph 3.6 on a number line.

Solution: Because 3.6 means $3\frac{6}{10}$ we divide the distance between 3 and 4 into 10 equal-size divisions and draw a dot on the 6th division mark.

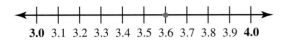

EXAMPLE 5 Graph 14.517 on a number line.

Solution: Because 14.517 means $14\frac{517}{1000}$ we should divide the space in between 14 and 15 into 1000 divisions and count to the 517th mark. Of course, this is rather tedious so we'll gradually zoom in on smaller and smaller sections of the number line:

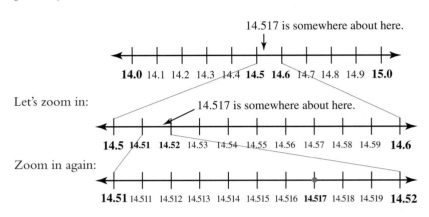

Let's zoom in:

Zoom in again:

Note: 14.0 and 14 are the same just as 14.5 and 14.50 are the same. Zero digits that go on forever beyond the last nonzero digit in a decimal number are understood. Writing those zeros is optional, but it is simplest not to write them.

Rather than go through all the *zoom* stages, we could just go directly to the last number line. Notice that the endpoints on the last number line are determined by the next to last place, in this case the hundredths place. We then divide the space in between into ten divisions and count to the 7th mark.

EXAMPLE 6 Graph -2.75 on a number line.

Solution: Let's zoom in on tenths in order to graph the hundredths. Because the number is negative, the values of the numbers decrease as you move left. However, the absolute value of the numbers will increase toward the left.

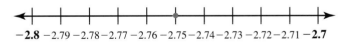

Do Margin 3. ▷

OBJECTIVE 4 *Use < or > to make a true statement.*

Now that we have a sense of what decimal numbers mean and where they are on a number line, we can compare two decimal numbers.

◆ **Margin 3**

Graph on a number line.

a. 0.7

b. 12.45

c. -6.21

d. -19.106

Answers:

a.

b.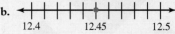

c.

d.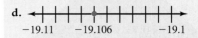

EXAMPLE 7 Use $<$ or $>$ to make a true statement.

$$13.7 \; ? \; 13.2$$

Solution: We could use a number line. The number 0.7 is farther to the right on a number line than 0.2 so 0.7 is the larger number.

13.7 is farther right than 13.2

Answer: $13.7 > 13.2$

Look at the digits in 13.7 and 13.2 from left to right. They match until we get to the tenths place. Notice it is this place that has different digits that tells us which number is larger. The larger digit is in the larger number.

> **Procedure** *To determine which of two decimal numbers is larger:*
> 1. Compare the digits in the numbers from left to right until you reach two different digits in the same place value.
> 2. The larger of these digits is in the larger number.
>
> **Note:** If both numbers are negative, the opposite applies, the smaller digit is in the larger number because the number is closer to zero.

EXAMPLE 8 Use $<$ or $>$ to make a true statement.

$$35.619 \; ? \; 35.625$$

Answer: $35.619 < 35.625$

Explanation: Compare the digits in the numbers from left to right.

35.619
$\updownarrow$
35.625

Each digit matches until we get to the hundredths place.

In the hundredths places, because 2 is larger than 1, we can say 35.625 is larger than 35.619. On a number line, 35.625 would be farther to the right.

EXAMPLE 9 Use $<$ or $>$ to make a true statement.

$$-0.00981 \; ? \; -0.009806$$

Answer: $-0.00981 < -0.009806$

Explanation: When we compare the digits from left to right, we see the hundred-thousandths place is the first place to have different digits.

−0.00981
$\updownarrow$
−0.009806

Hundred-thousandths digits do not match.

Because both numbers are negative, the number closest to zero will be the larger number. This means the smaller of the two different digits is actually in the larger number. Because 0 is smaller than 1, −0.009806 is the larger number.

◀ **Do Margin 4.**

◆ Margin 4

Use $<$ or $>$ to write a true statement.

a. 0.61 ? 0.65

b. 45.192 ? 45.092

c. −5.71 ? −5.8

d. −0.004701 ? −0.00471

Answers: **a.** $0.61 < 0.65$
b. $45.192 > 45.092$
c. $-5.71 > -5.8$
d. $-0.004701 > -0.00471$

OBJECTIVE 5 *Round decimal numbers to a specified place.*

Now that we have talked about graphing and comparing decimal numbers, we can study rounding. The idea in rounding is to round a given number to the nearest number in a specified place. When we graph on a number line, the endpoints are determined by the place prior to the last nonzero digit. This place will therefore determine how we round. Suppose we want to round 14.517 to the nearest hundredth. We graphed this number in Example 5. Let's recall that graph:

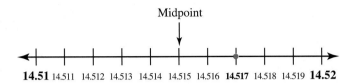

Notice 14.517 is between 14.51 and 14.52. Because 14.517 is above the midpoint on this scale, it is closer to 14.52 and we say it rounds to 14.52. If we wanted to round to a different place value, we would use a different scale. Suppose we wanted to round 14.517 to the nearest tenth. We would use a number line that had nearest tenths as the endpoints.

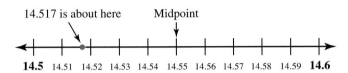

Because 14.517 is below the halfway point on this scale, it rounds down to 14.5. If we had to round to the nearest whole (ones place), we would use 14 and 15 as our endpoints and the tenths place would determine which of the two nearest wholes 14.517 is closest to.

Because 14.517 is slightly higher than the midpoint on this scale, it rounds up to 15. Even if it were exactly halfway, we agree to round up.

Conclusion: We round decimal numbers the same way we round whole numbers.

> **Procedure** *To round a number to a given place value, consider the digit to the right of the desired place value.*
> If this digit is 5 or higher, round up.
> If this digit is 4 or less, round down.

EXAMPLE 10 Round 917.28053 to the specified place.

a. Tenths

Answer: 917.3

Explanation: The nearest tenths are 917.2 and 917.3. The halfway point is 917.25. The 8 in the hundredths place means that 917.28053 is above the halfway point therefore we round up.

◆ **Margin 5**

Round 59.61538 to the specified place.

a. Tenths

b. Hundredths

c. Ten-thousandths

d. Whole number

b. Thousandths

Answer: 917.281

Explanation: The nearest thousandths are 917.280 and 917.281. The halfway point is 917.2805. The 5 in the ten-thousandths place means that 917.28053 is at the halfway point, therefore we round up to 917.281.

c. Whole number

Answer: 917

Explanation: The nearest whole number means to round to the ones place. The nearest wholes are 917 and 918 with the midpoint being 917.5. The 2 in the tenths place means that 917.28053 is below this midpoint so we round down to 917.

◀ **Do Margin 5.**

6.1 Exercises

FOR EXTRA HELP

  Videotape 8

InterAct Math
Tutorial Software

www.carsonmath.com

AWL Math Tutor Center

InterAct MathXL www.mathxl.com

Student's Solutions
Manual

For Exercises 1–20, write as a fraction or mixed number in lowest terms.

1. 0.2

2. 0.6

3. 0.25

4. 0.75

5. 0.375

6. 0.125

7. 0.24

8. 0.48

9. 1.5

10. 4.2

11. 18.75

12. 6.4

13. 9.625

14. 14.375

15. 7.36

16. 9.54

17. −0.008

18. −0.0005

19. −13.012

20. −219.105

For Exercises 21–32, write the word name.

21. 0.097

22. 0.415

23. 0.002015

24. 0.030915

25. 4159.6

26. 31.98

27. 521.608

28. 2109.003

29. −107.99

30. −2106.1

31. −0.50092

32. −0.000982

For Exercises 33–44, graph on a number line.

33. 0.8

34. 0.4

35. 1.3

36. 6.9

37. 4.25

38. 13.75

39. 8.06

40. 11.05

41. −3.21

42. −6.45

43. −19.017

44. −15.002

For Exercises 45–54, use < or > to make a true statement.

45. 0.81 ? 0.8

46. 0.15 ? 0.5

47. 2.891 ? 2.8909

48. 153.77 ? 153.707

49. 0.001983 ? 0.001985

50. 0.090194 ? 0.091094

51. -1.01981 ? -1.10981

52. -18.1095 ? -18.1085

53. -145.7183 ? -14.57183

54. -26.71803 ? -267.1803

For Exercises 55–60, round 610.28315 to the specified place.

55. Tenths

56. Hundredths

57. Ten-thousandths

58. Whole number

59. Tens

60. Hundreds

For Exercises 61–66, round 0.95106 to the specified place.

61. Whole number

62. Tens

63. Tenths

64. Hundredths

65. Thousandths

66. Ten-thousandths

For Exercises 67–72, round 408.06259 to the specified place.

67. Tenths

68. Hundredths

69. Ten-thousandths

70. Thousandths

71. Tens

72. Whole number

REVIEW EXERCISES

1. Add. $\dfrac{37}{100} + \dfrac{4}{100}$

2. Subtract. $-29 - (-16)$

3. Simplify. $7a^2 - a + 15 - 10a^2 + a - 7a^3$

4. Simplify. $(5x^2 + 9x - 18) - (x^2 - 12x - 19)$

5. Solve. $x - 19 = -36$

6. Find the perimeter of a $12\frac{1}{2}$ ft. by 16 ft. room.

6.2 Adding and Subtracting with Decimals and Solving Equations

OBJECTIVES

1 Add decimal numbers.

2 Subtract decimal numbers.

3 Add and subtract signed decimals.

4 Combine like terms.

5 Add polynomials.

6 Subtract polynomials.

7 Solve equations using the addition/subtraction principle.

8 Solve applications.

OBJECTIVE 1 *Add decimal numbers.*

In Section 6.1 we saw how decimal numbers use the base-ten place value system. That means adding and subtracting decimal numbers should be similar to adding and subtracting whole numbers and integers. Let's have a look:

$$2.82 + 43.01$$

Recall that the digits to the right of the decimal point mean fractional amounts. We can write the addends as mixed numbers to see how addition works.

$$2.82 + 43.01$$
$$= 2\frac{82}{100} + 43\frac{1}{100}$$
$$= 45\frac{83}{100}$$
$$= 45.83$$

Compare the sum with the original addends. Notice that we could have simply added the digits in the corresponding place values in the addends just as we did with whole numbers and integers.

$$\begin{array}{r} 2.82 \\ + 43.01 \\ \hline 45.83 \end{array}$$

Connection

When we added mixed numbers, we discovered we could add the whole/integer parts and fraction parts separately. Because the decimal point separates the whole/integer part from the fractional part in a decimal number, we can do the same thing with decimals. We can line up the decimal points, add all the whole/integer digits and fraction/decimal digits separately, then bring down the decimal point into the sum.

Procedure *To add decimal numbers:*

1. Stack the numbers by place value. (Line up the decimal points.)
2. Add the digits in the corresponding place values.
3. Place the decimal point in the sum so that it is aligned with the decimal points in the addends.

EXAMPLE 1 Add. 51.092 + 23.64

Solution: We stack the numbers by place value then add. We can align the corresponding place values by aligning the decimal points.

$$\begin{array}{r} 1 \\ 51.092 \\ + 23.640 \\ \hline 74.732 \end{array}$$

There are understood 0's beyond the last nonzero digit in every decimal number. Some people choose to write these instead of having blank spaces.

EXAMPLE 2 Add. 32.09 + 4.103 + 19.6423

Solution: We stack the numbers by the corresponding place value then add.

$$\begin{array}{r} 1 \quad 1 \\ 32.0900 \\ 4.1030 \\ + 19.6423 \\ \hline 55.8353 \end{array}$$

EXAMPLE 3 Add. 76 + 0.0189 + 3.99

Solution: We stack the numbers by the corresponding place value then add.

$$\begin{array}{r} 11\ 1 \\ 76.0000 \\ 0.0189 \\ + 3.9900 \\ \hline 80.0089 \end{array}$$

Whole numbers/integers have an understood decimal point to the right of the ones place. Once again, we can write the understood 0's as desired.

◀ **Do Margin 1.**

OBJECTIVE 2 *Subtract decimal numbers.*

We subtract decimal numbers the same way we add, except that we subtract the digits in the corresponding place values.

To subtract decimal numbers

1. Stack the number with the larger absolute value on top so that the place values align. (Line up the decimal points.)
2. Subtract the digits in the corresponding place values.
3. Place the decimal point in the difference so that it aligns with the decimal points in the minuend and subtrahend.

EXAMPLE 4 Subtract. 58.941 − 2.54

Solution: Stack the number with the larger absolute value on top.

$$\begin{array}{r} 58.941 \\ - 2.540 \\ \hline 56.401 \end{array}$$

Reminder
There are understood 0 digits beyond the last nonzero digit in every decimal number.

◆ **Margin 1**

Add.

a. 14.103 + 7.035

b. 0.1183 + 0.094

c. 6.981 + 0.57 + 23.1

d. 0.46 + 781 + 34

Answers: **a.** 21.138 **b.** 0.2123
c. 30.651 **d.** 815.46

EXAMPLE 5 Subtract. $179 - 48.165$

Solution: Stack the number with the larger absolute value on top.

$$
\begin{array}{r}
{}^{8\ 9\ 9\ 10} \\
17\cancel{9}.\cancel{000} \\
-\ 48.165 \\
\hline
130.835
\end{array}
$$

> **Reminder**
> Because 179 is a whole number, the decimal point is understood to be to the right of the 9 digit. We can write as many 0's to the right of the decimal point as desired.

EXAMPLE 6 Subtract. $98.036 - 9.482$

Solution: Stack the number with the larger absolute value on top.

$$
\begin{array}{r}
{}^{17} \\
{}^{8\ 7\ 9\ 13} \\
\cancel{9}\cancel{8}.\cancel{0}\cancel{3}6 \\
-\ 9.482 \\
\hline
88.554
\end{array}
$$

Do Margin 2. ▷

OBJECTIVE 3 *Add and subtract signed decimals.*

To add or subtract signed decimal numbers, we follow the same rules that we developed in Chapter 2 for integers.

EXAMPLE 7 Add. $-15.79 + 8.4$

Solution: Because we must add two numbers that have different signs, we subtract their absolute values and keep the sign of the number with the larger absolute value. Because -15.79 is the number with the larger absolute value, our sum will be negative.

> **Reminder**
> To subtract, stack the number with the larger absolute value on top and align the decimal points.

$$
\begin{array}{r}
{}^{0\ 15} \\
1\cancel{5}.79 \\
-\ 8.40 \\
\hline
7.39
\end{array}
$$

We subtracted the absolute values and then put the negative sign in the answer sentence.

Answer: $-15.79 + 8.4 = -7.39$.

Think: We have a debt of $15.79 and make a payment of $8.40. Because our payment is not enough to pay off the debt, we still have a debt of $7.39.

EXAMPLE 8 Add. $-68.95 + (-14.50)$

Solution: Because we must add two numbers that have the same sign, we add and keep the same sign.

$$
\begin{array}{r}
{}^{1\ 1} \\
68.95 \\
+\ 14.50 \\
\hline
83.45
\end{array}
$$

Notice that we took the negative signs out to figure, then put them back in the answer statement.

Answer: $-68.95 + (-14.50) = -83.45$.

Think: We have a debt of $68.95 and add another debt of $14.50, so we are increasing our debt to $83.45.

Do Margin 3. ▷

◆ **Margin 2**

Subtract.

a. $146.79 - 35.14$

b. $809 - 162.648$

c. $0.80075 - 0.06791$

d. $10.302 - 9.8457$

◆ **Margin 3**

Add.

a. $90.56 + (-59.8)$

b. $-104 + (-54.88)$

c. $-5.14 + 0.75 + (-4.50) + 2.95$

d. $54.6 + (-14.50) + (-6.78) + (-1.3)$

Answers to Margin 2: **a.** 111.65 **b.** 646.352 **c.** 0.73284 **d.** 0.4563

Answers to Margin 3: **a.** 30.76 **b.** -158.88 **c.** -5.94 **d.** 32.02

Margin 4

Subtract.

a. $1.987 - 10.002$

b. $4.1 - (-0.951)$

c. $-213.4 - 90.9$

d. $-0.0178 - (-0.05021)$

Margin 5

Simplify.

a. $5.2x^2 - 9.87x + 15x^2 - 2.35x - 10x^3 + 1.6$

b. $4.2a^4 - 2.67ab + b + 15.7a^2b + a^4 - 5.1b - 3.9ab$

EXAMPLE 9 To subtract signed decimal numbers, we can write equivalent addition statements.

Subtract. $0.08 - 5$

Solution: If we write an equivalent addition statement, we see that we are adding two numbers with different signs.

> **Reminder**
> To write a subtraction as an equivalent addition, we change the subtraction sign to an addition sign and change the subtrahend to its additive inverse.

$$0.08 - 5$$
$$= 0.08 + (-5)$$

Because we are adding numbers with different signs, we subtract and keep the sign of the number with the larger absolute value. The number with the larger absolute value is -5, so our result will be negative.

$$\begin{array}{r} 5.00 \\ -\ 0.08 \\ \hline 4.92 \end{array}$$ Subtract absolute values.

Answer: $0.08 - 5 = -4.92$

Think: We have $0.08 in a bank account (asset) and write a check for $5. This means we are overdrawn by $4.92.

EXAMPLE 10 Subtract. $-16.2 - (-9.07)$

Solution: If we write an equivalent addition statement, we see that we are adding two numbers with different signs.

$$-16.2 - (-9.07)$$
$$= -16.2 + 9.07$$

$$\begin{array}{r} 16.20 \\ -9.07 \\ \hline 7.13 \end{array}$$ Subtract absolute values.

Answer: $-16.2 - (-9.07) = -7.13$

◀ **Do Margin 4.**

OBJECTIVE 4 *Combine like terms.*

In Section 3.3 we learned that we can simplify expressions by combining like terms. Recall that like terms have the same variables raised to the same exponents. To combine like terms, add or subtract the coefficients and keep the variables the same.

EXAMPLE 11 Simplify. $4.2n^2 - 6.1n + 9 + 2.35n - 10n^2 + 2.09$

Solution: Combine like terms.

$$4.2n^2 - 6.1n + 9 + 2.35n - 10n^2 + 2.09$$
$$= \underline{4.2n^2 - 10n^2}\ \underline{-6.1n + 2.35n}\ \underline{+9 + 2.09}$$
$$\qquad\qquad \downarrow \qquad\qquad \downarrow \qquad\qquad \downarrow$$
$$= -5.8n^2 \qquad - 3.75n \qquad + 11.09$$

Some people find it helpful to write the like terms together.

◀ **Do Margin 5.**

Answers to Margin 4: **a.** -8.015 **b.** 5.051 **c.** -304.3 **d.** 0.03241

Answers to Margin 5: **a.** $-10x^3 + 20.2x^2 - 12.22x + 1.6$ **b.** $5.2a^4 + 15.7a^2b - 6.57ab - 4.1b$

OBJECTIVE 5 *Add polynomials.*

We learned in Section 3.4 that we can add polynomials by combining like terms.

EXAMPLE 12 Add. $(8.2y^3 - 6.1y^2 + 10.65) + (y^3 - y - 12)$

Solution: Combine like terms.

$$(8.2y^3 - 6.1y^2 + 10.65) + (y^3 - y - 12)$$
$$= \underbrace{8.2y^3 + y^3}_{} - 6.1y^2 - y + \underbrace{10.65 - 12}_{} \qquad \text{Collect and combine like terms.}$$
$$= \qquad 9.2y^3 \qquad - 6.1y^2 - y \qquad - 1.35$$

> Notice we simply brought down terms that have no like terms to combine.

Do Margin 6. ▷

OBJECTIVE 6 *Subtract polynomials.*

We also learned how to subtract polynomials in Section 3.4. To subtract polynomials, write an equivalent addition statement using the additive inverse of the subtrahend. Then combine like terms.

EXAMPLE 13 Subtract. $(12.6x^3 - 19.15x^2 + 8x - 7) - (2.9x^3 + 10.1x^2 + x - 9.1)$

Solution: Write the subtraction as an equivalent addition. Then combine like terms.

$$(12.6x^3 - 19.15x^2 + 8x - 7) - (2.9x^3 + 10.1x^2 + x - 9.1) \qquad \text{Write an equivalent addition.}$$
$$= (12.6x^3 - 19.15x^2 + 8x - 7) + (-2.9x^3 - 10.1x^2 - x + 9.1) \qquad \text{Combine like terms.}$$
$$= \underbrace{12.6x^3 - 2.9x^3}_{} \quad \underbrace{- 19.15x^2 - 10.1x^2}_{} \quad \underbrace{+ 8x - x}_{} \quad \underbrace{- 7 + 9.1}_{}$$
$$= \qquad 9.7x^3 \qquad\qquad - 29.25x^2 \qquad\quad + 7x \qquad + 2.1$$

Do Margin 7. ▷

OBJECTIVE 7 *Solve equations using the addition/subtraction principle.*

In Section 4.2 we learned to solve equations using the addition/subtraction principle of equality. The addition/subtraction principle states that the same number may be added or subtracted on both sides of an equation without affecting its solution(s).

EXAMPLE 14 Solve and check. $n + 4.78 = -5.62$

Solution: To isolate n we subtract 4.78 from both sides.

> **Reminder**
> Because $+4.78$ and -4.78 are additive inverses, their sum is 0. This isolates n.

$$n + 4.78 = -5.62$$
$$\begin{array}{r} n + 4.78 = -5.62 \\ \underline{-4.78 \qquad -4.78} \\ n + \quad 0 = -10.40 \\ n = -10.4 \end{array}$$
Subtract 4.78 from both sides.

Check: Replace n in the original equation with -10.4, and verify that it makes the equation true.

$$n + 4.78 = -5.62$$
$$-10.4 + 4.78 \overset{?}{=} -5.62$$
$$-5.62 = -5.62$$

Do Margin 8. ▷

Reminder
y^3 has a coefficient of 1.
$$8.2y^3 + 1y^3 = 9.2y^3$$

◆ **Margin 6**

Add.

a. $(t^3 - 4.15t^2 + 7.3t - 16) + (0.4t^3 + 3.9t^2 + 1.73)$

b. $(8.1m^4 + 2.78m^2 - 5.15mn + 10.3n) + (1.7m^4 - 5.2m^2 - 3.1mn - 15n)$

Reminder
We write an equivalent addition using the additive inverse of the subtrahend. To get the additive inverse of a polynomial, change the sign of each term in the polynomial.

◆ **Margin 7**

Subtract.

a. $(n^3 + 15.7n^2 + 0.2n - 1.8) - (3.5n^3 + 2.17n^2 + 0.97)$

b. $(14.91x^3 - 0.14x^2y + 2.9xy + 4.8y^2) - (2.98x^3 - x^2y - 9.35y^2 + 12.8)$

◆ **Margin 8**

Solve and check.

a. $-10.15 + x = 2.8$

b. $-14.6 = y - 2.99$

Answers to Margin 6: **a.** $1.4t^3 - 0.25t^2 + 7.3t - 14.27$ **b.** $9.8m^4 - 2.42m^2 - 8.25mn - 4.7n$

Answers to Margin 7: **a.** $-2.5n^3 + 13.53n^2 + 0.2n - 2.77$ **b.** $11.93x^3 + 0.86x^2y + 2.9xy + 14.15y^2 - 12.8$

Answers to Margin 8: **a.** 12.95 **b.** -11.61

OBJECTIVE 8 Solve applications.

Many applications of adding and subtracting decimals can be found in the field of accounting. We developed some principles of accounting in Chapters 1 and 2. Let's extend those principles to decimal numbers.

EXAMPLE 15 Following is Reed's register from his checkbook. Find his final balance.

Date	No.	Transaction Description	Subtractions	Additions	Balance
3 - 30		Deposit		1254 88	1380 20
3 - 30	2341	National Mortgage	756 48		
3 - 30	2342	Electric & Gas	125 56		
3 - 30	2343	Bell South	43 70		
3 - 30	2344	Visa	150 00		
3 - 30	2345	Domino's	12 50		
4 - 01		Transfer from savings		200 00	

Understand: We must find Reed's balance after he writes some checks and transfers $200 from his savings account to his checking account. Checks are deductions and a transfer into the checking account adds to the account.

Plan: Because all the checks are deductions, we deduct/subtract their sum from the initial balance then add the $200 transfer.

Execute: balance = initial balance − sum of all deductions + transfer amount

$$B = 1380.20 - (756.48 + 125.56 + 43.70 + 150.00 + 12.50) + 200.00$$
$$B = 1380.20 - 1088.24 + 200.00$$
$$B = 291.96 + 200.00$$
$$B = 491.96$$

Answer: $491.96

Check: We can check by reversing the process. We start with $491.96, deduct the transfer, and add in all the check amounts. The result should be the initial balance of $1380.20.

$$491.96 - 200 + 756.48 + 125.56 + 43.70 + 150 + 12.50$$
$$= 1380.20$$

◀ **Do Margin 9.**

Date	No.	Transaction Description	Subtractions	Additions	Balance
12-28	1320	Publix	45 28		580 20
12-30		Deposit		640 50	
12-30	1321	Electric & Gas	155 48		
12-30	1322	Bell South	89 45		
12-30	1323	Mastercard	100 00		
12-30	1324	Riverside Apartments	565 00		

Meyer, Lewis, Callahan, Garrity and Pederson, Esq.
One Financial Center
Boston, Massachusetts 02111

700 Philadelphia Avenue
Washington, D.C. 20004
Stephen L. Meyer

Direct Dial Number
617.555.0000

October 23, 1997

Gross amount from borrower: (Debts)

Amounts paid by the borrower: (Credits)

		Earnest Money	1000.00
Contract Sales Price	89,980.00	Principal Amount of	
Settlement Charges	1025.12	New Loan(s)	85,300.00
		Portion of Appraisal	125.10
		County Taxes	188.36

◆ **Margin 9**

Solve.

a. To the right is Carol's checkbook register. Find her final balance.

b. To the right is a settlement sheet for the Jacksons, who will be closing on their new house in a few days. How much should they bring to closing?

Answers: **a.** $310.77 **b.** $4391.66

6.2 Exercises

For Exercises 1–6, add.

1. $34.51 + 125.2$

2. $9.167 + 12.32$

3. $58.915 + 0.8361$

4. $0.5082 + 1.946$

5. $312.98 + 6.337 + 14$

6. $0.183 + 75 + 245.9$

For Exercises 7–12, subtract.

7. $864.55 - 23.1$

8. $1964.86 - 213.5$

9. $809.06 - 24.78$

10. $6.002 - 4.135$

11. $0.1005 - 0.08261$

12. $5.0002 - 0.19873$

For Exercises 13–24, add or subtract.

13. $25 + (-14.89)$

14. $-20 + 8.007$

15. $-0.16 + (-4.157)$

16. $-7.09 + (-10.218)$

17. $0.0015 - 1$

18. $0.0209 - 1$

19. $-30.75 - 159.27$

20. $-980.4 - 19.62$

21. $-90 - (-16.75)$

22. $-100 - (-63.912)$

23. $-0.008 - (-1.0032)$

24. $-0.005 - (-2.0031)$

For Exercises 25–28, simplify.

25. $5y - 2.81y^2 + 7.1 - 10y^2 - 8.03$

26. $0.8x^3 - 7x + 2.99 - 1.2x - 9 + 0.91x^3$

27. $a^3 - 2.5ab + 9a^2 + ab - 12 + 3.5a^3 + 1.4$

28. $4.2m^2 + m^3 - 9mn + 0.6 - 1.02mn - m^3 + 0.5$

For Exercises 29–32, add.

29. $(x^2 + 5.2x + 3.4) + (9.2x^2 - 6.1x + 5)$

30. $(0.4y^3 - 8.2y + 2.1) + (3.4y^3 - y^2 - 3)$

31. $(3.1a^3 - a^2 + 7.5a - 0.01) + (6.91a^3 - 3.91a^2 + 6)$

32. $(20.5m^4 + 16mn^2 - 18.3mn - 0.77n^2) + (0.9m^4 - 2.08mn^2 + n^2)$

For Exercises 33–36, subtract.

33. $(1.4n^2 + 8.3n - 0.6) - (n^2 - 2.5n + 3)$

34. $(5k^3 + 2.5k^2 - 6.2k - 0.44) - (2.2k^3 - 6.2k - 0.5)$

35. $(0.08t^4 + 1.9t^2 - 0.2t + 0.45) - (1.5t^4 + 2.4t^3 + 0.95t^2 - 1)$

36. $(12.91x^3 - 16.2x^2y^2 - xy + 7) - (10x^3 - 5.1x^2y - 3.6xy + 11.45)$

For Exercises 37–42, solve and check

37. $m + 3.67 = 14.5$

38. $x - 5.8 = 4.11$

39. $2.19 = 8 + y$

40. $6.09 = n + 7.1$

41. $k - 4.8 = -8.02$

42. $-16.7 = h - 7.59$

For Exercises 43–52, solve.

43. Below is Adam's checkbook register. Find his final balance.

Date	No.	Transaction Description	Subtractions		Additions		Balance	
2-24		Deposit			428	45	682	20
2-25	952	Food Lion	34	58				
2-25	953	Cash	30	00				
2-25	954	Green Hills Apartments	560	00				
2-25	955	Electric & Gas	58	85				
2-26	956	American Express	45	60				

44. Below is Jada's checkbook register. Find her balance.

Date	No.	Transaction Description	Subtractions		Additions		Balance	
10-26	1011	Sears	36	70			250	20
10-27	1012	Frank's Costumes	25	00				
10-30	1013	Credit Union Mortgage	858	48				
10-30	1014	Electric Co.	119	45				
10-30	1015	Water Co.	17	50				
10-30	1016	Food Mart	84	95				
10-31		Deposit			1355	98		

45. Tim runs a landscaping company. Below is a balance sheet that shows the charges for a client. What is the client's final balance?

Charges		Payments Received	#200
Trencher rental	#75.00		
PVC pipe	#15.50		
Sprinkler heads	#485.65		
Back flow prevention	#45.79		

46. Below is a settlement sheet for the Smiths, who will be closing on their new house in a few days. How much should they bring to closing?

Meyer, Lewis, Callahan, Garrity and Pederson, Esq.
One Financial Center
Boston, Massachusetts 02111

700 Philadelphia Avenue
Washington, D.C. 20004
Stephen L. Meyer

Direct Dial Number
617.555.0000

October 23, 1997

Gross amount from borrower: (Debts)		Amounts paid by the borrower: (Credits)	
		Earnest Money	1500.00
Contract Sales Price	124,480.00	Principal Amount of	
Settlement Charges	1265.08	New Loan(s)	118,300.00
Construction Overage	345.00	Portion of Appraisal	145.05
		County Taxes	218.24

47. Below is Catherine's receipt from the grocery store. If she gave the cashier $20, how much would she get back in change?

Milk	$1.59
Yogurt	$2.49
Bread	$1.79
Soup	$0.79
Cereal	$3.45
Tax	$0.51

48. Below is Selina's receipt from a clothing store. She has a $50 bill; is this enough? If not, how much more does she need?

Tank T	$12.95
Jeans	$35.95
Tax	$2.45

49. Maria gets paid once a month. Following is a list of deductions from her paycheck. Find Maria's net pay.

CURRENT GROSS:	2904.17		
Taxes		Other Deductions	
FICA	174.25	MED SPEND	45.00
MEDICARE	39.64	DENT	12.74
FED W/H	235.62	BLUE CROSS	138.15
STATE W/H	131.56	401K	200.00
		STD LIFE	7.32

50. Tony gets paid twice each month. Following is a list of deductions from his paycheck. Find Tony's net pay.

CURRENT GROSS:		1067.08	
Taxes		Other Deductions	
FICA	62.27	HMO BLUE	210.96
MEDICARE	14.57	DENT	7.54
FED W/H	128.05	STATE RET	64.02
STATE W/H	53.35	STATE OP LIFE	6.45

51. Find the missing side length if the perimeter is 96.1 m.

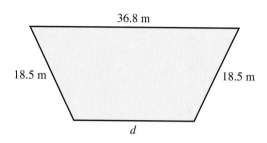

36.8 m

18.5 m   18.5 m

d

52. Find the missing side length if the perimeter is 5.9 mi.

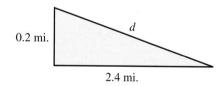

0.2 mi.   d

2.4 mi.

Puzzle Problem

At a bookstore a calculator and pen cost a total of $15.50. The calculator costs $15 more than the pen. What is the cost of each item?

REVIEW EXERCISES

1. Multiply. $-12 \cdot 25$

2. Multiply. $\dfrac{3}{10} \cdot \dfrac{7}{10}$

3. Divide. $10404 \div 34$

4. Divide. $9\dfrac{1}{6} \div 3\dfrac{1}{4}$

5. $\sqrt{121}$

6. Multiply. $-2x^2 \cdot 9x^5$

7. Multiply. $(y + 3)(y - 5)$

8. Solve and check. $-8x = 72$

6.3 Multiplying Decimals; Exponents with Decimal Bases

OBJECTIVES

1 Multiply decimal numbers.

2 Multiply signed decimal numbers.

3 Evaluate exponential forms with decimal bases.

4 Write a number in scientific notation in standard form.

5 Write standard form numbers in scientific notation.

6 Multiply monomials.

7 Multiply polynomials.

8 Solve applications.

OBJECTIVE 1 *Multiply decimal numbers.*

Because decimal numbers are base-ten numbers, we should be able to multiply them in the same way that we multiplied whole numbers. But, we must be careful in placing the decimal point in the product. Consider the following:

$$(0.3)(0.7)$$

Because we know how to multiply fractions, let's write these decimals as equivalent fractions to see what we can discover about the decimal point placement in the product.

$$(0.3)(0.7)$$

$$= \frac{3}{10} \cdot \frac{7}{10}$$ 0.3 is equivalent to $\frac{3}{10}$ and 0.7 is equivalent to $\frac{7}{10}$.

$$= \frac{21}{100}$$ Notice there are two decimal places in the product because we multiplied tenths times tenths to create hundredths.

$$= 0.21$$

What if the problem were $0.03 \cdot 0.7$?

$$0.03 \cdot 0.7$$

$$= \frac{3}{100} \cdot \frac{7}{10}$$ 0.03 is equivalent to $\frac{3}{100}$.

$$= \frac{21}{1000}$$ Notice there are now three decimal places in the product because we multiplied hundredths times tenths to create thousandths.

$$= 0.021$$

What can we say about multiplying decimal numbers?

Notice that the numerators of the fraction equivalents were whole numbers.

Conclusion: When we multiply decimal numbers, we can multiply the decimal digits just as we would multiply whole numbers.

Notice that the fraction equivalents will always have denominators that are powers of ten.

Conclusion: Because we multiply the denominators, we can say that the number of decimal places in the product will be equal to the total number of decimal places in the factors.

Margin 1

Multiply.

a. $6.8 \cdot 12.5$

b. $2.04 \cdot 0.09$

c. $1.001 \cdot 34.6$

d. $15 \cdot 0.007$

Procedure *To multiply decimal numbers:*

1. Stack the numbers. It is not necessary to align the decimal points in the factors.
2. Multiply as if the numbers were whole numbers.
3. Place the decimal in the product so that it has the same number of decimal digits as the total number of decimal digits in the factors.

EXAMPLE 1 Multiply. $(2.31)(1.7)$

Solution: We multiply the numbers as if they were whole numbers. Because 2.31 has two decimal places and 1.7 has one decimal place, the product will have three decimal places. Count the places in the product starting from the right.

Reminder
We do not need to align the decimal points because the process does not depend on decimal alignment.

$$
\begin{array}{r}
2 \\
2.31 \leftarrow \textbf{2 decimal places} \\
\times\ 1.7 \leftarrow \textbf{+1 decimal place} \\
\hline
1617 \\
+231\ \ \ \\
\hline
3.927 \leftarrow \textbf{3 decimal places}
\end{array}
$$

Connection

$2.31 \cdot 1.7$ is the same as:

$$2\tfrac{31}{100} \cdot 1\tfrac{7}{10}$$
$$= \tfrac{231}{100} \cdot \tfrac{17}{10}$$
$$= \tfrac{3927}{1000}$$
$$= 3\tfrac{927}{1000}$$
$$= 3.927$$

Connection

Adding the total number of places in the factors corresponds to adding exponents on the denominator powers of ten. We add the exponents because we are multiplying exponential forms that have the same base.

$$\frac{231}{100} \cdot \frac{17}{10} = \frac{231}{10^2} \cdot \frac{17}{10^1} = \frac{3927}{10^{2+1}} = \frac{3927}{10^3} = 3.927$$

Number of decimal places in 2.31 | Number of decimal places in 1.7 | Number of decimal places in product.

◀ **Do Margin 1.**

OBJECTIVE 2 *Multiply signed decimal numbers.*

Recall that when multiplying two numbers that have the same sign, the result is positive. When multiplying two numbers that have different signs, the result is negative.

EXAMPLE 2 Multiply $(0.09)(-0.051)$

Solution: We multiply as if they were whole numbers. Because 0.09 has two decimal places and -0.051 has three decimal places, our product will have five decimal places. Because we are multiplying two numbers that have different signs, the product will be negative.

Notice that we took out the sign while calculating but put the sign back in the final answer.

$$
\begin{array}{r}
0.051 \leftarrow \textbf{3 decimal places} \\
\times\ 0.09 \leftarrow \textbf{+2 decimal places} \\
\hline
0459 \\
0000\ \ \\
+\ 0000\ \ \ \\
\hline
0.00459 \leftarrow \textbf{5 decimal places}
\end{array}
$$

◆ **Answer:** $(0.09)(-0.051) = -0.00459$

◀ **Do Margin 2.**

Answers to Margin 1: **a.** 85 **b.** 0.1836 **c.** 34.6346 **d.** 0.105

Answer to Margin 2: **a.** -4.365 **b.** 51.6

OBJECTIVE 3 *Evaluate exponential forms with decimal bases.*

Because an exponent signifies to repeatedly multiply the base, we must be careful how we place the decimal in the result.

EXAMPLE 3 Evaluate. $(-0.04)^3$

Solution: Use -0.04 as a factor three times.

$$(-0.04)^3 = (-0.04)(-0.04)(-0.04)$$
$$= -0.000064$$

Tip
We can quickly tell how many places should be in the product by multiplying the number of decimal places in the base by the exponent.

number of · exponent = number of
places in   places in
base   product.

$$2 \quad · \quad 3 \quad = \quad 6$$

Reminder
The product is negative because we had an odd number of negative factors.

Do Margin 3. ▶

OBJECTIVE 4 *Write a number in scientific notation in standard form.*

Scientists, mathematicians, and engineers often work with very large or very small numbers. When these very large or very small numbers have lots of 0 digits, it is convenient to express these numbers in **scientific notation.**

DEFINITION **Scientific notation:** A number whose absolute value is greater than or equal to 1 but less than 10, multiplied by 10 raised to an integer exponent.

2.37×10^4 is in scientific notation because the absolute value of 2.37 is 2.37, which is greater than or equal to 1 but less than 10, and the exponent 4 on the power of 10 is an integer. 23.7×10^3 is not in scientific notation because the absolute value of 23.7, which is 23.7, is greater than 10.

What does 2.37×10^4 mean?

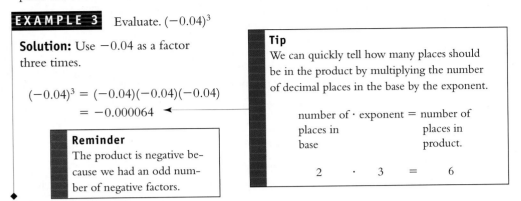

$$2.37 \times 10^4 = 2.37 \times 10 \times 10 \times 10 \times 10$$

Notice each multiplication by 10 moved the decimal point one more place to the right. Because we multiplied by four factors of 10, the decimal point moved a total of four places to the right from where it started in 2.37.

Conclusion: The number of 10's is determined by the exponent, so the exponent actually determines the number of places that the decimal point moves.

Procedure *To change from scientific notation with a positive integer exponent to standard notation for a given number, move the decimal point to the right the number of places indicated by the exponent.*

Margin 3
Evaluate.

a. $(0.008)^2$

b. $(0.32)^0$

c. $(-0.12)^2$

d. $(-0.6)^3$

Answers: **a.** 0.000064 **b.** 1
c. 0.0144 **d.** −0.216

EXAMPLE 4 Write 2.37×10^6 in standard form, then write the word name.

Solution: The exponent 6 means multiply 2.37 by 6 tens. Each ten will move the decimal point one place to the right, so move the decimal point 6 places to the right from its original position in 2.37. Moving two of those 6 places puts the decimal after the 37 so we must place four 0's after the 7 digit to account for all 6 places.

$$2.37 \times 10^6 = 2,370,000$$

Connection

Because 10^6 is another way to represent 1,000,000, it is popular to read 2.37×10^6 as 2.37 million.

◆ Word name: Two million, three hundred seventy thousand

EXAMPLE 5 Write -5.1782×10^9 in standard form, then write the word name.

Solution: The 9 exponent means that we are multiplying -5.1782 by 9 tens. Each of those tens will move the decimal point to the right one place, so the decimal point moves to the right a total of 9 places from its original position in -5.1782. Moving four of the 9 places puts the decimal point after the 2 in 1782 so we must place five 0's after the 2 to account for all 9 places.

$$-5.1782 \times 10^9 = -5,178,200,000$$

Connection

Because 10^9 is another way to represent 1,000,000,000, it is popular to read -5.1782×10^9 as -5.1782 billion.

◆ Word name: Negative five billion, one hundred seventy-eight million, two hundred thousand

◀ **Do Margin 4.**

OBJECTIVE 5 *Write standard form numbers in scientific notation.*

Suppose we want to express 62,910,000 in scientific notation. Because scientific notation must have a number whose absolute value is greater than or equal to 1 but less than 10, our first step will be to establish the decimal position. Notice that the only place the decimal point could go is between the 6 and 2 digits.

62,910,000

Decimal goes here to express a number whose absolute value is greater than or equal to 1 but less than 10.

Notice that there are 7 place values to the right of the decimal point in its new position between the 6 and 2. To make the scientific notation equal to the original number, we must account for the seven place values with the power of 10. Each place value is a power of 10. So the power of 10 is 7 in this case.

$$62,910,000 = 6.2910000 \times 10^7$$

Decimal goes here to express a number whose absolute value is greater than or equal to 1 but less than 10.

There are 7 places to the right of the new decimal position.

$$62,910,000 = 6.2910000 \times 10^7 = 6.291 \times 10^7$$

These 0's are no longer significant now that they are on the other side of the decimal point. 0's to the right of the last nonzero decimal digit are not significant; that means we do not need them.

◆ Margin 4

Write each number in standard form, then write the word name.

a. 7.8×10^3

b. 9.102×10^9

c. 4.1518×10^{12}

d. 5.4×10^7

Answers: a. 7800; Seven thousand eight hundred. **b.** 9,102,000,000; Nine billion, one hundred two million **c.** 4,151,800,000,000; Four trillion, one hundred fifty-one billion, eight hundred million. **d.** 54,000,000; Fifty-four million

Procedure *To write a standard form number greater than 1 in scientific notation:*

1. Place a decimal point in the number so that it expresses a number whose absolute value is greater than or equal to 1 but less than 10. (Tip: Place the decimal point to the right of the first nonzero digit.)
2. Count the number of places to the right of the new decimal position. This is the power of 10.
3. Delete 0's to the right of the last nonzero digit.

EXAMPLE 6 Write 417,000,000,000 in scientific notation.

Solution: Place the decimal to the right of the first digit. We then count the places to the right of this decimal position to establish the exponent on the power of 10. Finally we delete all the zeros to the right of the last nonzero digit, which is the 7 in this case.

$$417,000,000,000 = 4.17 \times 10^{11}$$

Decimal goes here to express a number whose absolute value is greater than or equal to 1 but less than 10.

There are 11 places to the right of the new decimal position.

EXAMPLE 7 Write −2,917,300,000 in scientific notation.

Solution: Place the decimal to the right of the first digit. Then count the places to the right of this decimal position to establish the power of 10. Finally, delete all the 0's to the right of the last nonzero digit, which is 3 in this case, and write the negative sign in the answer.

$$-2,917,300,000 = -2.9173 \times 10^{9}$$

Decimal goes here to express a number whose absolute value is greater than or equal to 1 but less than 10.

There are 9 places to the right of the new decimal position.

Do Margin 5. ▷

OBJECTIVE 6 *Multiply monomials.*

We learned how to multiply monomials in Section 3.5. Recall that the procedure is to multiply the coefficients, add exponents for the like bases, and write any unlike bases as they are.

EXAMPLE 8 Multiply. $(2.1y^7)(3.43y^8)$

Solution: We multiply the coefficients, add exponents for the like bases, and write the unlike bases as they are.

$(2.1y^7)(3.43y^8)$

$= 2.1 \cdot 3.43 \cdot y^7 \cdot y^8$

$= 7.203y^{7+8}$

$= 7.203y^{15}$

Connection

Multiplying monomials is essentially the same as multiplying numbers expressed in scientific notation.

$(2.1y^7)(3.43y^8)$ corresponds to $(2.1 \times 10^7)(3.43 \times 10^8)$

$(2.1y^7)(3.43y^8)$	$(2.1 \times 10^7)(3.43 \times 10^8)$
$= 2.1 \cdot 3.43 \cdot y^7 \cdot y^8$	$= 2.1 \times 3.43 \times 10^7 \times 10^8$
$= 7.203y^{7+8}$	$= 7.203 \times 10^{7+8}$
$= 7.203y^{15}$	$= 7.203 \times 10^{15}$

Margin 6

Multiply.

a. $(1.7n^3)(6n^2)$

b. $(-4.03ab^2)(-2.1a^3c)$

Margin 7

Evaluate the expression.

a. $(0.3xy^4)^3$

b. $(-2.5m^2n^3)^2$

Margin 8

Multiply.

a. $5.3(1.4a^2 + 2.5a - 1)$

b. $-2.8xy^2(3.5y^3 - 4xy + 5.1x - 3)$

Margin 9

Multiply.

a. $(7x - 4.1)(3.2x - 2)$

b. $(1.9y + 5)(1.9y - 5)$

Answers to Margin 6: **a.** $10.2n^5$
b. $8.463a^4b^2c$

Answers to Margin 7: **a.** $0.027x^3y^{12}$
b. $6.25m^4n^6$

Answers to Margin 8: **a.** $7.42a^2 + 13.25a - 5.3$ **b.** $-9.8xy^5 + 11.2x^2y^3 - 14.28x^2y^2 + 8.4xy^2$

Answers to Margin 9: **a.** $22.4x^2 - 27.12x + 8.2$ **b.** $3.61y^2 - 25$

EXAMPLE 9 Multiply. $(6.13x^2y)(-4.5x^4y^3z)$

Solution: We multiply the coefficients, add exponents for the like bases, and write the unlike bases as they are.

$$(6.13x^2y)(-4.5x^4y^3z)$$
$$= 6.13 \cdot -4.5 \cdot x^2 \cdot x^4 \cdot y \cdot y^3 \cdot z$$
$$= -27.585 \cdot x^{2+4} \cdot y^{1+3} \cdot z$$
$$= -27.585x^6y^4z$$

◀ **Do Margin 6.**

We learned in Section 3.5 that we can raise monomials to a power. It works the same way as raising a number in scientific notation to a power. We can raise the coefficient and all the variables to the power.

EXAMPLE 10 Evaluate the expression. $(0.2a^2bc^4)^3$

Solution: We evaluate 0.2 to the 3rd then multiply each variable's exponent by 3.

$$(0.2a^2bc^4)^3 = 0.008a^{2\cdot3}b^{1\cdot3}c^{4\cdot3}$$
$$= 0.008a^6b^3c^{12}$$

◀ **Do Margin 7.**

OBJECTIVE 7 *Multiply polynomials.*

In Section 3.5, we learned that the key to multiplying polynomials was the distributive property. Let's first consider multiplying a polynomial by a monomial. To multiply a polynomial by a monomial, use the distributive property and multiply every term in the polynomial by the monomial.

EXAMPLE 11 Multiply. $4.2x^2(5x^3 - 6xy + 9x - 7)$

Solution: We multiply every term in the polynomial by the monomial.

$$4.2x^2(5x^3 - 6xy + 9x - 7)$$

$$= 4.2x^2 \cdot 5x^3 - 4.2x^2 \cdot 6xy + 4.2x^2 \cdot 9x - 4.2x^2 \cdot 7$$
$$= 21x^5 - 25.2x^3y + 37.8x^3 - 29.4x^2$$

◀ **Do Margin 8.**

Consider multiplying two polynomials. To multiply two polynomials, multiply each term in the second polynomial by every term in the first polynomial then simplify.

EXAMPLE 12 Multiply. $(3.1n - 2)(1.7n + 6.5)$

Solution: We multiply every term in the second polynomial by every term in the first polynomial, then simplify.

$$(3.1x - 2) \quad (1.7n + 6.5)$$

$$= 3.1n \cdot 1.7n + 3.1n \cdot 6.5 - 2 \cdot 1.7n - 2 \cdot 6.5$$
$$= 5.27n^2 + 20.15n - 3.4n - 13$$
$$= 5.27n^2 + 16.75n - 13$$

◀ **Do Margin 9.**

OBJECTIVE 8 *Solve applications.*

One very common multiplication situation occurs when a total amount of money is calculated based on a **unit price.**

DEFINITION | **Unit price:** The price for each unit of an item.

For example, suppose gasoline is sold at $1.09 per gallon. The unit price is $1.09 per gallon, which means we pay $1.09 for every 1 gallon of gasoline that we purchase. The unit price can be expressed as a fraction.

$$\$1.09 \text{ per gallon} = \$1.09/\text{gallon} = \frac{\$1.09}{1 \text{ gallon}}$$

Note: The word *per* translates to the symbol /, or a fraction line, and indicates division.

If we purchase 10 gallons of gasoline, because the unit price is $1.09 for every 1 gallon, we can multiply the unit price by the quantity we buy to get the total price. As a formula we can say:

$$\text{total} = \text{quantity} \cdot \text{unit price}$$
$$T = qp$$

In our gasoline example, it looks like this:

$$T = 10 \text{ gal} \cdot \frac{\$1.09}{1 \text{ gal}}$$
$$T = \frac{10 \cancel{\text{ gal}}}{1} \cdot \frac{\$1.09}{1 \cancel{\text{gal}}}$$
$$T = \$10.90$$

Connection

We left the units in to illustrate how the units of measurement can be *divided out* just like numbers and variables. Because *gal* appears in the numerator and denominator, we can *divide out* this common factor. Notice that the only unit of measurement left is $. This unit or dimensional analysis, as it is called, gives some insight as to how unit pricing works. We will discuss unit analysis in more detail when we develop rates more fully in Chapter 7.

EXAMPLE 13 For county property taxes, a vehicle is assessed to have a value of $1010. The property taxes are $0.2954 for each dollar of the assessed value of the vehicle. How much are the taxes?

Solution: Use the problem-solving process.

Understand: We must find the property taxes for the vehicle. We know the assessed value of the vehicle is $1010 and the tax rate is $0.2954 for each dollar of the assessed value. Because $0.2954 is a unit price used to calculate total tax based on the $1010 assessment, we will multiply.

Plan: Multiply the value of the vehicle by the rate to find t, the tax.

Execute: $t = 1010 \cdot 0.2954$

$t = 298.354$

Answer: Because taxes are monetary charges and we can only actually pay to the nearest hundredth of a dollar (cent), we must round our answer to the nearest hundredth.

$$\text{tax} = \$298.35$$

Check: The best check would be to reverse the process and divide. Because we have not covered division of decimals yet, we will check by estimating the product and verifying that the estimate is reasonably close to the actual product.

◆ Margin 10

Solve.

a. The long distance phone company that Janice uses charges $0.09 per minute. If she talks for 76 minutes, how much will the call cost?

b. A broker purchases 150 shares of a stock that costs $4.25 per share. What is the total cost of the purchase?

To estimate $1010 \cdot 0.2954$ we will round the numbers, then multiply the rounded numbers. When we estimate we want to round so that we can perform the calculation mentally. In this case, we'll round 1010 to the nearest thousand and 0.2954 to the nearest tenth.

$$1010 \cdot 0.2954$$
$$\downarrow \qquad \downarrow$$
$$1000 \cdot 0.3 = 300$$

Our estimate of 300 is very close to the actual product of 298.354.

◀ **Do Margin 10.**

In earlier chapters we discussed force. Newton's formula for calculating force is:

$$F = ma$$

Because we were working with whole numbers in Chapter 2 we rounded the measurements for the acceleration due to gravity to integers. In English/American units we said the acceleration is -32 ft./sec.^{2} and in metric we said -10 m/sec.^{2}. The acceleration is negative because gravity accelerates objects downwards. In Chapter 5 we developed fractions and refined these values a bit. We said the English/American measurement of the acceleration is about $-32\frac{1}{5}$ ft./sec.^{2} and metric is about $-9\frac{4}{5}$ m/sec.^{2}. We can now use decimal values for the acceleration due to gravity. In English/American units, the decimal value is about -32.2 ft./sec.^{2} and in metric it is about -9.8 m/sec.^{2}.

Connection

If we write -32.2 and -9.8 as equivalent mixed numbers we have:

$$-32.2 = -32\frac{2}{10} = -32\frac{1}{5}$$

$$-9.8 = -9\frac{8}{10} = -9\frac{4}{5}$$

These are the same mixed number values we used in Chapter 5. In Section 6.4 we will see how to write fractions and mixed numbers as equivalent decimal numbers.

In English/American units, the mass unit is the *slug* while in metric it is the *kilogram* (kg). In the English/American system, the unit of force is the *pound* (lb.) while in the metric system the unit of force is the *Newton* (N).

EXAMPLE 14   Find the weight of a person with a mass of 68.1 kg.

Understand: We must calculate the weight given the mass. Weight is force due to gravity and force is calculated using Newton's formula, $F = ma$. The metric value for the acceleration due to gravity on Earth is -9.8 m/sec.^{2}.

Plan: Use the formula $F = ma$ with $m = 68.1$ kg and $a = -9.8$ m/sec.^{2}.

Execute: $F = ma$

$$F = (68.1)(-9.8)$$
$$F = -667.38 \text{ N}$$

Answer: The weight of the person is -667.38 N. The negative sign indicates that the force is downward.

Check: We will check by estimating the product. If we round the mass and acceleration to the tens place, we have:

$$(68.1)(-9.8)$$
$$\downarrow \qquad \downarrow$$
$$(70)\ (-10) = -700 \text{ N}$$

The -700 N estimate is reasonably close to the actual weight of -667.38 N.

◆ Keep in mind that estimates are not as strong of a check as reversing the process.

◀ **Do Margin 11.**

Answers to Margin 10: **a.** $6.84 **b.** $637.50

Answers to Margin 11:
a. -0.0245 N **b.** -241.5 lb.

6.3 Exercises

For Exercises 1–16, multiply.

1. $(0.6)(0.9)$
2. $(0.7)(0.4)$
3. $(2.9)(3.65)$
4. $(6.81)(4.2)$

5. $(9.81)(71.62)$
6. $(10.4)(12.31)$
7. $(0.01)(619.45)$
8. $145(0.1)$

9. $19.6(10)$
10. $(1000)5.413$
11. $(0.1508)(-100)$

12. $(-10,000)(0.052)$
13. $152(-0.001)$
14. $-0.01(6)$

15. $(-0.029)(-0.15)$
16. $(-0.18)(-0.234)$

For Exercises 17–24, evaluate the expression.

17. $(0.9)^2$
18. $(0.13)^2$
19. $(2.1)^3$
20. $(3.5)^3$

21. $(-0.03)^4$
22. $(-0.2)^5$
23. $(-0.4)^3$
24. $(-1.43)^2$

For Exercises 25–30, write the scientific notation in standard form. Then write the word name.

25. The speed of light is approximately 3×10^8 m/sec.

26. A light year (lt-yr) is the distance light travels in one year and is approximately 5.76×10^{12} mi.

27. The approximate distance to the Andromeda galaxy is 2.14×10^6 lt-yr.

28. The national debt is about $\$1.568 \times 10^{12}$.

29. There are approximately 2.73×10^8 people in the United States.

30. Earth is about 9.292×10^7 mi. from the sun.

For Exercises 31–36, write each number in scientific notation.

31. There are 2,300,000 blocks of stone in Khufu's pyramid.

32. There are approximately 191,000 words in this textbook.

33. The distance from the Sun to Pluto is approximately 3,540,000,000 mi.

34. The nearest star to ours (the Sun) is Proxima Centauri at a distance of about 24,700,000,000,000 mi.

35. Human DNA consists of 5,300,000,000 nucleotide pairs.

36. An FM radio station broadcasts at a frequency of 102,300,000 Hz.

For Exercises 37–42, multiply.

37. $(3.2x^5)(1.8x^3)$

38. $(9.15y^2)(7.2y^4)$

39. $(0.2a^2b)(-0.08abc)$

40. $(-1.67m^2n^3)(2.4mp^2)$

41. $(0.5t^3u^2)^2$

42. $(-0.2xy^2z^5)^3$

For Exercises 43–52, multiply.

43. $0.5(1.8y^2 + 2.6y + 38)$

44. $0.2(6.8m^3 + 0.44m - 7)$

45. $-6.8(0.02a^3 - ab + 1.9b)$

46. $-0.14(0.9n^2 - 40n - 8.7)$

47. $(0.4x - 1.22)(16x + 3)$

48. $(6.1y + 2)(0.8y - 5)$

49. $(5.8a + 9b)(5.8a - 9b)$

50. $(0.1m - 0.41n)(0.1m + 0.41n)$

51. $(k - 0.1)(14.7k - 2.8)$

52. $(9t - 0.5)(2t - 6.01)$

For Exercises 53–70, solve.

53. A gas company charges $0.9674 per therm used. If Ruth uses 76 therms during a particular month, what will be the cost?

54. An electric company charges $0.0801 per kwh. Margaret uses 698 kwh one particular month. What will be the cost?

Time	Rate
8 A.M.—5 P.M.	$0.12/minute
5 P.M.—10 P.M.	$0.10/minute
10 P.M.—8 A.M.	$0.09/minute

55. Joanne makes a long distance call at 4:12 P.M. and talks for 32 minutes. How much does the call cost?

56. Joanne makes another long distance call at 10:31 P.M. and talks for 21 minutes. How much does the call cost?

57. A van is assessed to have a value of $6400. Property tax on vans in the owner's county is $0.2855 per dollar of the assessed value. What will be the tax on the van?

58. One year after the purchase of a new sedan, it is assessed to have a value of $9850. Property tax for the sedan in the county in which the owner lives is $0.0301 per dollar of assessed value. What is the property tax on the sedan?

59. Stan buys 4.75 lb. of onions at $0.79/lb. What is the total cost of the onions?

60. Nadine buys 5.31 lb. of chicken at $1.19/lb. What is the total cost of the chicken?

61. Find the weight of 0.003 kg of aspirin.

62. Find the weight of a 16.5 kg backpack.

63. Find the weight of a piece of road maintenance equipment with a mass of 21.8 slugs.

64. In a test, the speed of a car is measured to be 12.5 ft./sec. How far will the car travel in 4.5 sec.?

65. A square top covers the components in an electronic device. The top is 4.8 cm on each side. What is the area of the top?

66. A computer processor is 3.4 cm by 3.6 cm. What is the area of the processor?

67. Find the area of the lot shown.

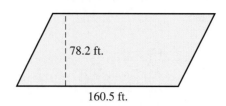

78.2 ft.

160.5 ft.

68. Find the area of the brace shown.

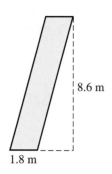

8.6 m

1.8 m

69. Find the volume.

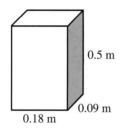

0.5 m

0.09 m

0.18 m

70. Find the volume.

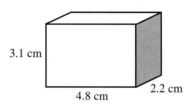

3.1 cm

4.8 cm

2.2 cm

REVIEW EXERCISES

1. $3280 \div 16 =$

2. $\sqrt{169} =$

3. Find the radius and circumference of a circle with a 14 in. diameter.

4. Solve. $9t = -81$

5. Divide. $12x^7 \div 2x^3$

6.4 Dividing and Roots with Decimals

OBJECTIVES

1 Divide decimal numbers.

2 Write fractions and mixed numbers as decimals.

3 Evaluate square roots.

4 Write a number expressed in scientific notation with a negative exponent in standard form.

5 Write standard form numbers less than 1 in scientific notation.

6 Divide monomials with decimal coefficients.

7 Solve equations using the multiplication/division principle.

8 Solve applications.

OBJECTIVE **1** *Divide decimal numbers.*

Suppose we must divide $4.8 \div 2$. To gain some insight on how to approach the problem, we will write the decimal number as an equivalent fraction or mixed number.

$$4.8 \div 2 = 4\frac{8}{10} \div 2$$

$$= \frac{48}{10} \div 2$$

$$= \frac{48}{10} \div \frac{2}{1}$$

$$= \frac{\overset{24}{\cancel{48}}}{10} \cdot \frac{1}{\underset{1}{\cancel{2}}}$$ We divide 2 into 48. This suggests that we can divide the original digits as if they were whole numbers.

$$= \frac{24}{10}$$ The 10 denominator indicates that the quotient is still in terms of tenths.

$$= 2\frac{4}{10}$$ We could have reduced the fraction, but we left it in terms of tenths so that writing the equivalent decimal number would be simple.

$$= 2.4$$

By using equivalent fractions and dividing them, we can see that $4.8 \div 2 = 2.4$.

Conclusion: We can set up the problem in long division form and write the decimal in the quotient directly above its position in the dividend.

$$
\begin{array}{r}
2.4 \\
2\overline{)4.8} \\
\underline{-4} \\
08 \\
\underline{-8} \\
0
\end{array}
$$

Procedure *To divide a decimal number by a whole number:*

1. Set up a long division problem.
2. Divide the divisor into the dividend as if both numbers were whole numbers. Be sure to align the digits in the quotient properly.
3. Write the decimal in the quotient directly above its position in the dividend.
4. Write extra 0 digits to the right of the last nonzero digit as needed until you get a remainder of 0, or until a digit or group of digits repeats without end in the quotient.

To indicate repeated digits, write a repeat bar over the digit or group of digits that repeat.

◆ Margin 1

Divide.

a. $29 \div 5$

b. $211 \div 4$

c. $-12.8 \div 8$

d. $0.591 \div -2$

EXAMPLE 1 Divide. $9 \div 2 =$

Solution: Set up a long division problem and divide. Write the decimal in the quotient directly above its position in 9.0.

> **Reminder**
> 9 can be expressed as 9.0000. We can write as many 0's as needed to the right of the decimal point.

$$\begin{array}{r} 4.5 \\ 2\overline{)9.0} \\ -8 \\ \hline 10 \\ -10 \\ \hline 0 \end{array}$$

> **Connection**
> We saw that we could express division using mixed numbers in Chapter 5. For $9 \div 2$ it looks like:
>
> $$\begin{array}{r} 4 \\ 2\overline{)9} \\ -8 \\ \hline 1 \end{array}$$
>
> Answer: $4\frac{1}{2}$
>
> Our decimal answer 4.5 is the same as $4\frac{1}{2}$.
>
> $$4.5 = 4\frac{5}{10} = 4\frac{1}{2}$$

◀ **Do Margin 1.**

EXAMPLE 2 Divide. $3.2 \div 3$

Solution: Set up a long division problem and divide. Write the decimal in the quotient directly above its position in 3.2.

$$\begin{array}{r} 1.066... \\ 3\overline{)3.200} \\ -3 \\ \hline 02 \\ -0 \\ \hline 20 \\ -18 \\ \hline 20 \end{array}$$

> We keep getting the same digit in the quotient. In fact, the 6 digit will continue in an unending pattern.

To indicate that the 6 digit repeats in an unending pattern, we use a repeat bar over the 6.

Answer: $1.0\overline{6}$

Note: The 0 digit is not part of the repeated pattern, so we do not write the repeat bar over the 0. Writing the bar over the 0 as well as the 6 indicates a different number.

$$1.\overline{06} = 1.060606...$$

> This is a different pattern than we see in our quotient.

> **WARNING** Be careful when writing the repeat bar that you only place it over decimal digits that repeat in an unending pattern.

◆ Margin 2

Divide.

a. $10 \div 6$

b. $14 \div 9$

c. $-12.5 \div 3$

d. $-0.16 \div 15$

◀ **Do Margin 2.**

What if the divisor is a decimal number? Consider $1.68 \div 0.5$. Because we now know how to divide a decimal number by a whole number, might it be possible to write an equivalent division statement with a whole number divisor?

Recall that division can be expressed in fraction form this way:

$$1.68 \div 0.5 = \frac{1.68}{0.5}$$

In Chapter 5 we learned that we can upscale fractions to equivalent fractions by multiplying both the numerator and denominator by the same amount. Notice if we multiply the

Answers to Margin 1: **a.** 5.8 **b.** 52.75 **c.** -1.6 **d.** -0.2955

Answers to Margin 2: **a.** $1.\overline{6}$ **b.** $1.\overline{5}$ **c.** $-4.1\overline{6}$ **d.** $-0.010\overline{6}$

denominator by 10, the decimal moves to the right one place so that 0.5 becomes the whole number 5. This achieves our goal of writing the divisor as a whole number. But to keep the fractions equivalent, we must multiply the numerator by 10 as well so that 1.68 becomes 16.8. So we have:

$$1.68 \div 0.5 = \frac{1.68}{0.5} = \frac{1.68 \cdot 10}{0.5 \cdot 10} = \frac{16.8}{5}$$

Conclusion: We can write equivalent division problems by moving the decimal in the dividend and divisor the same number of places.

Now we can divide.

$$
\begin{array}{r}
3.36 \\
5\overline{)16.80} \\
-15 \\
\hline
18 \\
-15 \\
\hline
30 \\
-30 \\
\hline
0
\end{array}
$$

We can now amend our procedure for dividing decimal numbers.

Procedure *To divide when the divisor is a decimal number:*

1. Set up a long division problem.
2. Move the decimal point in the divisor to the right enough places to make the divisor an integer.
3. Move the decimal point in the dividend the same number of places.
4. Divide the divisor into the dividend as if both numbers were whole numbers. Be sure to align the digits in the quotient properly.
5. Write the decimal point in the quotient directly above its position in the dividend.

EXAMPLE 3 Divide. $24.1 \div 0.04$

Solution: Because the divisor is a decimal number, we will write an equivalent division statement with a whole number divisor. To do this we multiply both the divisor and dividend by an appropriate power of 10. In this case, it's 100. This moves the decimal to the right two places in both divisor and dividend.

$$24.1 \div 0.04 = 2410 \div 4$$

> When we move the decimal point two places to the right in 2410, we move past an understood 0 digit that is to the right of 1.

Now divide:

$$
\begin{array}{r}
602.5 \\
4\overline{)2410.0} \\
-24 \\
\hline
01 \\
-0 \\
\hline
10 \\
-8 \\
\hline
20 \\
-20 \\
\hline
0
\end{array}
$$

Do Margin 3. ▷

◆ **Margin 3**

Divide.

a. $0.768 \div 0.6$

b. $13.49 \div 1.42$

c. $-605.3 \div 0.02$

d. $16 \div -0.03$

Answers: **a.** 1.28 **b.** 9.5 **c.** 30,265 **d.** $533.\overline{3}$

OBJECTIVE 2 *Write fractions and mixed numbers as decimals.*

Because a fraction is a notation for division, we can write fractions as decimals by dividing the denominator into the numerator. We follow the same procedure as for dividing decimal numbers when the divisor is a whole number.

> **Procedure** *To write a fraction as a decimal number, divide the denominator into the numerator.*

EXAMPLE 4 Write $\frac{5}{8}$ as a decimal number.

Solution: Divide the denominator into the numerator.

$$
\begin{array}{r}
0.625 \\
8\overline{)5.000} \\
-48 \\
\hline
20 \\
-16 \\
\hline
40 \\
-40 \\
\hline
0
\end{array}
$$

Connection

We can verify that $0.625 = \frac{5}{8}$ by writing 0.625 as a fraction in lowest terms.

$$0.625 = \frac{625}{1000} = \frac{625 \div 125}{1000 \div 125} = \frac{5}{8}$$

Answer: $\dfrac{5}{8} = 0.625$

How can we write mixed numbers as decimals? We saw in Section 6.1 that $23.791 = 23\frac{791}{1000}$. In reverse, if we want to write a mixed number as a decimal, the whole/integer part of the mixed number, in this case 23, is the whole number part of the decimal number, and the fraction part, $\frac{791}{1000}$ becomes the decimal digits.

> **Procedure** *To write a mixed number as a decimal:*
> 1. Write the whole/integer part of the mixed number to the left of a decimal point.
> 2. Divide the denominator into the numerator to get the decimal digits.

EXAMPLE 5 Write $-17\frac{2}{11}$ as a decimal number.

Solution: The -17 part of the mixed number will be written to the left of the decimal. The fraction part will become the decimal digits.

$$
\begin{array}{c}
-17 \quad \dfrac{2}{11} \\
\downarrow \ \downarrow \ \downarrow \\
= -17 \, . \ ?
\end{array}
$$

To get the decimal digits, we divide 11 into 2.

$$
\begin{array}{r}
0.1818... \\
11\overline{)2.0000} \\
-11 \\
\hline
90 \\
-88 \\
\hline
20 \\
-11 \\
\hline
90
\end{array}
$$

$\frac{2}{11}$ becomes a decimal number with digits that repeat in an unending pattern, so we use a repeat bar over those digits that repeat.

Answer: $-17\dfrac{2}{11} = -17.\overline{18}$

Notice $-17.\overline{18}$ is a rational number because it is equivalent to $-17\frac{2}{11}$, which we can express as the fraction $\frac{-189}{11}$. It turns out that all nonterminating decimal numbers with repeating digits can be expressed as fractions.

♦ **Conclusion:** All nonterminating decimal numbers with repeating digits are rational numbers.

Are there decimal numbers that are nonterminating without repeating digits? We'll have to explore further.

Do Margin 4. ▶

OBJECTIVE 3 *Evaluate square roots.*

In the past we have discussed square roots of perfect squares. We have considered problems like:

$$\sqrt{64} = 8 \quad \text{because} \quad (8)^2 = 64$$

$$\sqrt{\frac{100}{121}} = \frac{10}{11} \quad \text{because} \quad \left(\frac{10}{11}\right)^2 = \frac{100}{121}$$

The numbers 64, 100, and 121 are all perfect squares because their square roots are integers. (For a list of perfect squares, see Table 1.1.)

We now consider square roots of numbers that are not perfect squares. Consider the following:

$$\sqrt{14} = ?$$

We must find a positive number that can be squared to equal 14. (The radical sign means we only want the positive square root.)

$$(?)^2 = 14$$

Notice that 14 is between the two perfect squares: 9 and 16. The positive square root of 9 is 3 and the positive square root of 16 is 4. Because 14 is between 9 and 16, we can conclude that $\sqrt{14}$ must be between 3 and 4.

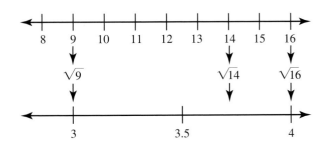

Notice that $\sqrt{14}$ is not an integer. Based on our number line, we can make a rough estimate that $\sqrt{14}$ is about 3.7 and can test our guess by squaring it.

$$(3.7)^2 = 13.69$$

Because 13.69 is less than 14, it means that 3.7 is too small. If we try 3.8, we get:

$$(3.8)^2 = 14.44$$

So $\sqrt{14}$ is between 3.7 and 3.8. We need to go to hundredths to refine further.

We could continue guessing and refining indefinitely because $\sqrt{14}$ is a nonterminating decimal number that does not have a pattern of unending repeating digits. And because we cannot express its exact value as a decimal or fraction, $\sqrt{14}$ is an **irrational number.** In fact the square root of any whole number that is not a perfect square is irrational. We defined irrational numbers in Section 5.3 when we discussed the irrational number π.

DEFINITION   **Irrational number:** A number that cannot be expressed in the form of $\frac{a}{b}$ where a and b are integers and $b \neq 0$.

The rational numbers and irrational numbers together form the set of numbers known as the **real numbers.** If we were to travel the entire number line from end to end, every number we encounter would either be rational or irrational.

DEFINITION   **Real numbers:** The set of all rational and irrational numbers.

The only way to express the exact value of the square root of a whole number that is not a perfect square is to use the radical sign.

Connection
The symbol π is the exact representation of the ratio of the circumference of a circle to its diameter.

$\sqrt{14}$ is the exact expression for the square root of fourteen.

Because we cannot write down the entire decimal equivalent, we must approximate the value. To approximate we can guess, verify, and refine as we did earlier, or use a calculator.

Using a calculator, we find:

$$\sqrt{14} \approx 3.741657387$$

Reminder
This symbol means *is approximately.*

If we round the result to the nearest hundredth, we have:

$$\sqrt{14} \approx 3.74$$

EXAMPLE 6   Approximate $\sqrt{29}$ to the nearest hundredth.

♦ **Answer:** $\sqrt{29} \approx 5.39$

◀ **Do Margin 5.**

What if we must find the square root of a decimal number? Some decimal numbers have exact decimal square roots; others are irrational.

$$\sqrt{0.09} = 0.3 \quad \text{because} \quad (0.3)^2 = 0.09$$
$$\sqrt{1.21} = 1.1 \quad \text{because} \quad (1.1)^2 = 1.21$$
$$\sqrt{0.0064} = 0.08 \quad \text{because} \quad (0.08)^2 = 0.0064$$

Notice that each of the radicands is a perfect square if we take out the decimal point.

0.09 is similar to 9

1.21 is similar to 121

and 0.0064 is similar to 64

Also notice that the square roots have the same digits as the square roots of the corresponding perfect squares. The tricky part is the decimal point placement. If we square a decimal number, the product will always have twice as many places as the original decimal number.

Working with Scientific Calculators

To find the decimal value of a square root on a scientific calculator, enter the radicand and then press the $\boxed{\sqrt{x}}$ key.

Answer: 3.741657387

Note that this is still only an approximation. The digits continue without terminating or repeating in a pattern that would allow us to use the repeat bar.

♦ **Margin 5**

Approximate the square root to the nearest hundredth.

a. $\sqrt{50}$

b. $\sqrt{94}$

c. $\sqrt{128}$

d. $\sqrt{212}$

Answer: **a.** 7.07 **b.** 9.70 **c.** 11.31 **d.** 14.56

$$(0.3)^2 = 0.09 \qquad (0.03)^2 = 0.0009 \qquad (0.003)^2 = 0.000009$$

$$\underset{\substack{\uparrow \\ \text{1 place}}}{} \quad \underset{\substack{\uparrow \\ \text{2 places}}}{} \qquad \underset{\substack{\uparrow \\ \text{2 places}}}{} \quad \underset{\substack{\uparrow \\ \text{4 places}}}{} \qquad \underset{\substack{\uparrow \\ \text{3 places}}}{} \quad \underset{\substack{\uparrow \\ \text{6 places}}}{}$$

This means that when we find the square root of a decimal number that corresponds to a perfect square, the decimal number must have an *even number of decimal places* if it is to have an exact decimal root. Otherwise, the root is irrational.

Conclusion: If the number has an even number of decimal places and corresponds to a perfect square, the square root will have half the number of decimal places and the same digits as the square root of the corresponding perfect square.

$$\sqrt{0.0081} = 0.09 \qquad \text{whereas} \qquad \sqrt{0.081} \approx 0.284604989$$

$$\underset{\substack{\uparrow \\ \text{4 places}}}{} \quad \underset{\substack{\uparrow \\ \text{2 places}}}{} \qquad\qquad \underset{\substack{\uparrow \\ \text{irrational}}}{}$$

Rule The square root of a decimal number that corresponds to a perfect square and has an even number of decimal places will be a decimal number with half the number of places and the same digits as the square root of the corresponding perfect square.

EXAMPLE 7 Find the square root. $\sqrt{0.000196}$

Solution: Because 196 is a perfect square and there are an even number of decimal places in 0.000196, the square root will have half the number of places and the same digits as the square root of 196.

$$\sqrt{0.000196} = 0.014 \qquad \text{Because 0.000196 has six places, its square root will have three places. Because } \sqrt{196} = 14, \text{ we can say that } \sqrt{0.000196} = 0.014.$$

Do Margin 6. ▷

OBJECTIVE 4 *Write a number expressed in scientific notation with a negative exponent in standard form.*

What happens when we divide a decimal number by a power of 10?

$$2.5 \div 10 = 0.25 \qquad 2.5 \div 100 = 0.025 \qquad 2.5 \div 1000 = 0.0025$$

$$10\overline{)2.50} \atop \begin{array}{r} 0.25 \\ -20 \\ \hline 50 \\ -50 \\ \hline 0 \end{array} \qquad 100\overline{)2.500} \atop \begin{array}{r} 0.025 \\ -200 \\ \hline 500 \\ -500 \\ \hline 0 \end{array} \qquad 1000\overline{)2.5000} \atop \begin{array}{r} 0.0025 \\ -2000 \\ \hline 5000 \\ -5000 \\ \hline 0 \end{array}$$

Notice the decimal point moves from its original position in 2.5 to the left the same number of places as the 0 digits in the power of ten. Writing these divisions in exponential form and fractional form we have:

$$2.5 \div 10 = 2.5 \div 10^1 = \frac{2.5}{10^1} = 0.25$$

$$2.5 \div 100 = 2.5 \div 10^2 = \frac{2.5}{10^2} = 0.025$$

$$2.5 \div 1000 = 2.5 \div 10^3 = \frac{2.5}{10^3} = 0.0025$$

Dividing by a power of 10 causes the decimal point to move to the left the same number of places as the exponent (or 0 digits) in the power of 10. We can connect this to scientific notation with negative exponents.

$$2.51 \times 10^{-4} = 2.51 \times \frac{1}{10^4}$$

We can write 10^{-4} with a positive exponent as $\frac{1}{10^4}$.

$$= \frac{2.51}{1} \times \frac{1}{10^4}$$

To multiply a decimal number by a fraction, we can write the decimal number over 1 then multiply straight across.

$$= \frac{2.51}{10^4}$$

$$= 0.000251$$

This is division by a power of 10, so the decimal point will move left the same number of places as the exponent.

Conclusion: Multiplying by a power of 10 with a negative exponent is the same as dividing by the power of 10 with the exponent made positive.

> **Procedure** *To write a number expressed in scientific notation with a negative exponent in standard form, move the decimal point to the left the same number of places as the absolute value of the exponent.*

A negative exponent in scientific notation indicates that the number is less than 1. Such small numbers occur in chemistry and physics in measuring the mass or size of atomic particles or measuring the wavelengths of certain radiation. We would also encounter such small numbers in measuring the time it takes a computer to do a single calculation.

EXAMPLE 8 Write 8.45×10^{-9} in standard form.

Solution: Multiplying by 10^{-9} is the same as dividing by 10^9, so we move the decimal point to the left 9 places.

$$8.45 \times 10^{-9} = 0.00000000845$$

> **Tip**
> When we write a number expressed in scientific notation with a negative exponent in standard form, we can write the same number of 0's as the exponent to the left of the leftmost digit and place the decimal point in between the first two 0's.

 WARNING A negative exponent does not necessarily mean that the base number or evaluated outcome is negative.

◀ **Do Margin 7.**

OBJECTIVE 5 *Write standard form numbers whose absolute value is less than 1 and greater than 0 in scientific notation.*

We can write decimal numbers whose absolute value is less than 1 and greater than 0 in scientific notation using negative exponents on the power of 10.

We have seen that $8.45 \times 10^{-9} = 0.00000000845$ How can we express a standard form number in scientific notation? In scientific notation we write a decimal number whose absolute value is greater than or equal to 1 but less than 10, multiplied by a power of 10.

To write 0.00000000845 in scientific notation, we must place the decimal between the 8 and 4 digits so that the absolute value of 8.45 is a number greater than or equal to 1 but less than 10. Moving the decimal point to this position means that we have 9 places to account for with the power of 10.

◆ **Margin 7**

Write in standard form.

a. 4.51×10^{-6}

b. 2.78×10^{-8}

c. -6×10^{-4}

d. -1.993×10^{-7}

Answers: **a.** 0.00000451
b. 0.0000000278 **c.** −0.0006
d. −0.0000001993

This suggests the following procedure:

Procedure *To write a decimal number whose absolute value is less than 1 and greater than 0 in scientific notation:*

1. Place the decimal point in the number so that it expresses a number whose absolute value is between 1 and 10.
2. Count the number of decimal places between the original decimal position and the new decimal position.
3. Write this number of places as a negative exponent with a base of 10.

EXAMPLE 9 Write 0.0000139 in scientific notation.

Solution: We must move the decimal point between the 1 and 3 digits so that 1.39 is a number between 1 and 10. There are 5 decimal places in between the original decimal position and the new position, so the exponent will be -5.

$$0.0000139 = 1.39 \times 10^{-5}$$

There are 5 decimal places in between the original position and the new position.	Decimal point moves here to form a number whose absolute value is greater than or equal to 1 but less than 10.

Do Margin 8. ▶

OBJECTIVE 6 *Divide monomials with decimal coefficients.*

We learned how to divide monomials in Section 3.7. Recall that the procedure is to divide the coefficients, subtract exponents for the like bases, and write any unlike bases as they are.

EXAMPLE 10 Divide. $8.2n^6 \div 2.5n^2$

Solution: Divide the coefficients, subtract exponents for the like bases, and write the unlike bases as they are.

$$8.2n^6 \div 2.5n^2$$
$$= (8.2 \div 2.5)\, n^{6-2}$$
$$= 3.28n^4$$

Connection

Dividing monomials is like reducing a rational expression that has monomials in the numerator and denominator. Example 10 can be written:

$$8.2n^6 \div 2.5n^2 = \frac{8.2n^6}{2.5n^2}$$

We can divide out the variables and divide the coefficients.

$$\frac{8.2n^6}{2.5n^2} = \frac{8.2 \cdot \overset{1}{\cancel{n}} \cdot \overset{1}{\cancel{n}} \cdot n \cdot n \cdot n \cdot n}{2.5 \cdot \underset{1}{\cancel{n}} \cdot \underset{1}{\cancel{n}}} = 3.28n^4$$

EXAMPLE 11 Divide. $-3.85x^5yz \div 0.5x^4y$

Solution: Divide the coefficients, subtract exponents for the like bases, and write the unlike bases as they are.

$$-3.85x^5yz \div 0.5x^4y$$
$$= -3.85 \div 0.5\, x^{5-4}\, y^{1-1}\, z$$
$$= -7.7x^1y^0z$$
$$= -7.7xz$$

Reminder
y^0 simplifies to 1.

♦ **Margin 8**

Write in scientific notation.

a. 0.000591

b. 0.0000004

c. −0.003194

d. −0.000002051

Answers: **a.** 5.91×10^{-4}
b. 4×10^{-7} **c.** -3.194×10^{-3}
d. -2.051×10^{-6}

Connection

Example 11 written in rational expression form:

$$-3.85x^5yz \div 0.5x^4y = \frac{-3.85x^5yz}{0.5x^4y}$$

We can divide out the variables and divide the coefficients.

$$\frac{-3.85x^5yz}{0.5x^4y} = \frac{-3.85 \cdot \overset{1}{\cancel{x}} \cdot \overset{1}{\cancel{x}} \cdot \overset{1}{\cancel{x}} \cdot \overset{1}{\cancel{x}} \cdot x \cdot \overset{1}{\cancel{y}} \cdot z}{0.5 \cdot \underset{1}{\cancel{x}} \cdot \underset{1}{\cancel{x}} \cdot \underset{1}{\cancel{x}} \cdot \underset{1}{\cancel{x}} \cdot \underset{1}{\cancel{y}}} = -7.7xz$$

◄ **Do Margin 9.**

OBJECTIVE 7 *Solve equations using the multiplication/division principle.*

In Section 4.3, we learned to solve equations using the multiplication/division principle of equality. The principle states: We can multiply or divide both sides of an equation by the same amount without affecting its solution(s). Let's apply the principle to equations that contain decimal numbers.

EXAMPLE 12   Solve and check. $-14.2x = 39.05$

Solution: To isolate x, we must divide out the -14.2 coefficient. According to the multiplication/division principle, we must divide by -14.2 on both sides of the equation so that we do not affect the solution.

$$-14.2x = 39.05$$
$$\frac{-14.2x}{-14.2} = \frac{39.05}{-14.2}$$
$$1x = -2.75$$
$$x = -2.75$$

Check: We replace x in $-14.2x = 39.05$ with -2.75, and verify that it makes the equation true.

$$-14.2x = 39.05$$
$$-14.2(-2.75) \; ? \; 39.05$$
$$39.05 = 39.05 \qquad \text{It checks.}$$

◄ **Do Margin 10.**

OBJECTIVE 8 *Solve applications.*

EXAMPLE 13   Laura owes a hospital $1875.84. She arranges to pay the debt in monthly installments over 1 year. How much is each monthly payment?

Understand: Laura is to make monthly payments to pay off an $1875.84 debt in 1 year. Because there are 12 months in a year, she will make 12 payments.

Plan: To calculate the amount of each payment we divide $1875.84 by 12.

Execute: $P = 1875.84 \div 12$
$\qquad\quad P = 156.32$

Answer: Each payment should be $156.32.

Check: We can check by reversing the process. We want to verify that 12 payments of $156.32 is a total of $1875.84, so we multiply $156.32 by 12.

$$156.32 \cdot 12 = 1875.84$$

◆ **Margin 9**

Divide.

a. $1.28y^5 \div 0.4y$

b. $-0.8tu^7 \div (-1.6u^3)$

◆ **Margin 10**

Solve and check.

a. $37.994 = 3.14d$

b. $-9.8m = 158.76$

Answers to Margin 9: **a.** $3.2y^4$
b. $0.5tu^4$

Answers to Margin 10: **a.** 12.1
b. -16.2

Do Margin 11. ▶

Let's have another look at circles. In Chapter 5 we used the fraction $\frac{22}{7}$ as an approximate value for π. Now that we have discussed how to write fractions as equivalent decimals, we can find a decimal approximation for π by dividing 22 by 7.

$$
\begin{array}{r}
3.14... \\
7\overline{)22.00} \\
-21 \\
\hline
10 \\
-7 \\
\hline
30 \\
-28 \\
\hline
2
\end{array}
$$

Remember, π is an irrational number so we cannot express its value exactly as a fraction or decimal. The numbers $\frac{22}{7}$ and 3.14 are merely approximations of π, although accurate enough for most calculations. The π key on a scientific calculator gives an even more accurate approximation:

$$\pi \approx 3.141592654$$

In Chapter 5 we developed formulas for the circumference and area of a circle.

$$C = \pi d \quad \text{or} \quad C = 2\pi r$$
$$A = \pi r^2$$

EXAMPLE 14 The diameter of Earth is about 12,756 km at the equator. Find the circumference of the Earth at the equator. Use 3.14 for π.

Understand: We must find the circumference of a circle with a diameter of 12,756 km.

Plan: Use the formula $C = \pi d$.

Execute: $C = \pi d$

$\qquad C = \pi(12,756)$

$\qquad C = 12,756\pi$

$\qquad C \approx 12,756(3.14)$

$\qquad C \approx 40,053.84$ km

Connection

Earlier in this section, we encountered other irrational numbers. We said that for square roots that are irrational, we express the exact irrational number by leaving the radical sign in the expression. Similarly, the only way we can express the *exact* value of a calculation involving π is to leave the symbol π in the expression. Calculating with 3.14 or $\frac{22}{7}$ is an approximation.

Exact	Approximation
$\sqrt{17}$	≈ 4.123105626
$12,756\pi$	$\approx 40,053.84$

Answer: The circumference of Earth is about 40,053.84 km at the equator.

◆ **Margin 11**

Solve.

a. To avoid interest charges, Mark takes a 90-day same-as-cash option to pay off a VCR purchase. The cash amount would be $256.86. If he designs his budget to make 3 payments of equal amount, how much should each payment be?

b. A patient is to receive 1.5 L of saline over 60 minutes. How much should the patient receive every minute?

Answers: **a.** $85.62 **b.** 0.025 L

Solve.

a. The radius of an engine cylinder is 1.78 cm. What is the circumference? Use 3.14 for π.

b. The circumference of a crater is 16.8 km. What is the diameter? Use 3.14 for π.

Check: We can check by reversing the process and dividing. If we divide the product by one of the factors, we should get the other factor. Let's divide 40,053.84 by 3.14, and we should get the diameter.

$$40{,}053.84 \div 3.14 = 12{,}756 \qquad \text{It checks.}$$

◀ **Do Margin 12.**

Answers: **a.** 11.1784 cm
b. $\approx$ 5.35 km

6.4 Exercises

FOR EXTRA HELP

Videotape 8

InterAct Math
Tutorial Software

www.carsonmath.com

AWL Math Tutor Center

InterAct
MathXL   www.mathxl.com

Student's Solutions
Manual

For Exercises 1–20, divide.

1. $286.2 \div 12$ **2.** $194.16 \div 24$ **3.** $0.48 \div 20$ **4.** $3.392 \div 32$

5. $5640 \div 9.4$ **6.** $93.96 \div 10.8$ **7.** $0.288 \div 0.06$ **8.** $0.7208 \div 0.08$

9. $-6.4 \div 0.16$ **10.** $28.5 \div -0.15$ **11.** $1 \div 0.008$ **12.** $0.1 \div 0.002$

13. $20.6 \div -1000$ **14.** $-8.93 \div -10,000$ **15.** $-8.145 \div -0.01$ **16.** $21.7 \div 0.001$

17. $19.6 \div 0.11$ **18.** $16.3908 \div 0.18$ **19.** $-0.0901 \div -0.085$ **20.** $339.2 \div 0.106$

For Exercises 21–36, write as a decimal.

21. $\dfrac{3}{5}$ **22.** $\dfrac{1}{8}$ **23.** $\dfrac{9}{20}$ **24.** $\dfrac{7}{25}$

25. $-\dfrac{7}{16}$ **26.** $-\dfrac{5}{32}$ **27.** $\dfrac{13}{30}$ **28.** $\dfrac{1}{6}$

29. $13\dfrac{1}{4}$ **30.** $25\dfrac{1}{2}$ **31.** $-17\dfrac{5}{8}$ **32.** $-5\dfrac{4}{5}$

33. $104\dfrac{2}{3}$ **34.** $76\dfrac{5}{6}$ **35.** $-216\dfrac{4}{7}$ **36.** $-99\dfrac{5}{9}$

For Exercises 37–48, evaluate the square root. If the root is irrational, approximate the square root to the nearest hundredth.

37. $\sqrt{0.0016}$ **38.** $\sqrt{2.25}$ **39.** $\sqrt{0.25}$ **40.** $\sqrt{0.0081}$

41. $\sqrt{24}$ **42.** $\sqrt{108}$ **43.** $\sqrt{200}$ **44.** $\sqrt{78}$

45. $\sqrt{1.69}$ **46.** $\sqrt{0.0256}$ **47.** $\sqrt{0.009}$ **48.** $\sqrt{2.5}$

For Exercises 49–54, write each number in standard form.

49. A speck of dust has a mass of about 6.5×10^{-10} kg.

50. A single streptococcus bacterium is approximately 7.5×10^{-7} m in diameter.

51. At the center of the visible light spectrum is yellow-green light with a wavelength of about 5.5×10^{-7} m.

52. Radiation with a wavelength of 4.7×10^{-9} m would be considered x-rays.

53. The radius of a hydrogen atom is approximately 5.3×10^{-11} m.

54. The smallest viruses are about 2×10^{-8} m in diameter.

For Exercises 55–60, write each number in scientific notation.

55. A grape has a mass of about 0.0028 kg.

56. Radiation with a wavelength of 0.000000027 m is in the ultraviolet portion of the spectrum.

57. The radius of the nucleus of a hydrogen atom is about 0.000000000000001 m.

58. The laser in a CD-ROM drive has a wavelength of about 0.00000078 m.

59. A single byte of information is downloaded to a computer in 0.0000177 sec.

60. A satellite receives a signal from a transmitter and returns the signal to receiving dishes in about 0.0048 sec.

For Exercises 61–66, divide.

61. $10.2x^6 \div 0.4x^4$

62. $5.44y^4 \div 1.7y$

63. $-0.96m^3n \div 0.15m$

64. $3.42hk^7 \div (-3.8k^2)$

65. $-3.03a^5bc \div (-20.2a^4b)$

66. $-2.7t^4u^3v^2 \div (-0.75uv^2)$

For Exercises 67–76, solve and check.

67. $2.1b = 12.642$

68. $81.5y = 21.19$

69. $-0.88h = 1.408$

70. $-4.066 = 0.38k$

71. $-2.28 = -3.8n$

72. $-2.89 = -8.5p$

73. $-28.7 = 8.2x$

74. $-10.8t = 9.072$

For Exercises 75–90, solve.

75. Carla purchases a computer on a 90-day same-as-cash option. The purchase price is $1839.96. If she plans to make three equal payments to pay off the debt, how much should each payment be?

76. Brad has a student loan balance of $1875.96. If he agrees to make equal monthly payments over a three-year period, how much will each payment be?

77. A patient is to receive radiation treatment for cancer. The treatment calls for a total of 324 rads administered in 8 bursts of focused radiation. How many rads should each burst be?

78. The FDA recommends that for a healthy diet a person should limit fat intake to about 40 g of fat per day. If a person splits the recommended fat intake equally among three meals, how much fat is allowed in each meal?

79. The voltage in a circuit is 12 V. The current is measured to be 0.03 A. What is the resistance? (Use $V = ir$.)

80. A pallet of merchandise weighs 485 lb. If the acceleration due to gravity is 32.2 ft./sec.^{2}, then what is the mass? (Use $F = ma$.)

81. A plane is flying at 350 mph. How long will it take the plane to reach a city that is 600 mi. away? (Use $d = rt$.)

82. The area of a rectangular painting is 358.875 in.^{2}. If the length is 16.5 in., what is the width? (Use $A = bh$.)

For Exercises 83–86, use 3.14 to approximate π.

83. The radius of the moon is about 1087.5 mi. Calculate the circumference of the moon at its equator.

84. The solid rocket boosters used to propel the space shuttle into orbit have O-rings that fit around the boosters at connections. If the diameter of one rocket booster is 12.17 ft., what is the circumference of the O-ring?

Historical Note

On January 28, 1986, the space shuttle *Challenger* exploded 73 seconds after launch, killing all 7 crew members. The cause of the disaster was a weakness in an attach ring on one of the rocket boosters. At the time, the rings did not completely circle the cases of the rocket boosters. After the disaster, the rings were redesigned to completely circle the rocket boosters.

85. Haleakala crater on Maui island, Hawaii, is the largest inactive volcanic crater in the world and has a circumference of about 21 mi. at its rim. What is the diameter at the rim?

86. To manufacture a light fixture, a company bends 6.5 ft. of metal tubing into a circle. What is the radius of this circle?

REVIEW EXERCISES

1. Simplify. $16 - 9[12 + (8 - 13)] \div 3$

2. Simplify. $\dfrac{3}{5} \cdot 30$

3. Evaluate. $\dfrac{1}{2}at^2$ when $a = -10$ and $t = 6$

4. Find the area.

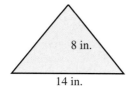

8 in.

14 in.

5. Find the area.

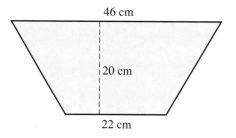

46 cm

20 cm

22 cm

6.5 Order of Operations

OBJECTIVES

1 Simplify numerical expressions using the order of operations agreement.
2 Simplify expressions containing fractions and decimals.
3 Evaluate expressions.
4 Find the area of a triangle.
5 Find the area of a trapezoid.
6 Find the area of a circle.
7 Find the volume of a cylinder.
8 Find the volume of a pyramid.
9 Find the volume of a cone.
10 Find the volume of a sphere.
11 Find the area or volume of a composite form.

OBJECTIVE 1 *Simplify numerical expressions using the order of operations agreement.*

EXAMPLE 1 Simplify. $2.75 - 8.2(3.4 + 4)$

Solution: Follow the order of operations agreement.

$2.75 - 8.2(3.4 + 4)$ Add inside parentheses. **3.4 + 4 = 7.4**
$= 2.75 - 8.2(7.4)$ Multiply 8.2(7.4) to get 60.68.
$= 2.75 - 60.68$ Subtract. **2.75 − 60.68 = −57.93**
$= -57.93$

Do Margin 1. ▷

EXAMPLE 2 Simplify. $12.6 \div 0.2 - 6.5(0.4)^2$

Solution: Follow the order of operations agreement.

$12.6 \div 0.2 - 6.5(0.4)^2$ Evaluate the expression with the exponent. **(0.4)² = 0.16** Multiply and divide in order from left to right.
$= 12.6 \div 0.2 - 6.5(0.16)$ **12.6 ÷ 0.2 = 63 and 6.5(0.16) = 1.04**
$= 63 - 1.04$ Subtract. **63 − 1.04 = 61.96**
$= 61.96$

Do Margin 2. ▷

EXAMPLE 3 Simplify. $[(1.2 + 0.3) \div 5] \div 2\sqrt{0.09}$

Solution: Follow the order of operations agreement.

$[(1.2 + 0.3) \div 5] \div 2\sqrt{0.09}$ Calculate within the innermost parentheses. **1.2 + 0.3 = 1.5**
$= [1.5 \div 5] \div 2\sqrt{0.09}$ Calculate in the brackets. **1.5 ÷ 5 = 0.3**
$= 0.3 \div 2\sqrt{0.09}$ Calculate the square root. **√0.09 = 0.3**
$= 0.3 \div 2(0.3)$ Multiply and divide from left to right. **0.3 ÷ 2 = 0.15**
$= 0.15(0.3)$ Multiply. **0.15(0.3) = 0.045**
$= 0.045$

Do Margin 3. ▷

◆ **Margin 1**

Simplify.

a. $6.5 - 1.2(14.4 + 0.6)$

b. $1.8 + 0.4(9.02 - 10.52)$

◆ **Margin 2**

Simplify.

a. $-0.2 \div 1.6 - 20(0.2)^3$

b. $(1.5)^2 + 2.52 \div 0.4(-0.6)$

◆ **Margin 3**

Simplify.

a. $[(12.8 - 15.3) \div 10] - 0.6\sqrt{0.16}$

b. $\sqrt{1.69} + 0.2[(0.011 + 0.013) \div (0.5 - 0.8)]$

Answers to Margin 1: **a.** -11.5
b. 1.2

Answers to Margin 2: **a.** -0.285
b. -1.53

Answers to Margin 3: **a.** -0.49
b. 1.284

OBJECTIVE 2 *Simplify expressions containing fractions and decimals.*

How do we handle problems that contain both decimals and fractions? Here are a few ways:

> **Procedure** *To simplify numerical expressions that contain both fractions and decimals:*
> 1. Write all decimals as fractions. (Method 1)   or
>   Write all fractions as decimals. (Method 2)   or
>   Write decimals over the number 1. (Method 3)
> 2. Follow the order of operations agreement.

Consider an example using all three methods.

EXAMPLE 4 $\dfrac{2}{3}(6.09) - \dfrac{3}{8}$

Method 1: Write all decimals as fractions.

$$\frac{2}{3}(6.09) - \frac{3}{8} = \frac{2}{3} \cdot \frac{609}{100} - \frac{3}{8}$$

Write all decimals as fractions. $6.09 = \dfrac{609}{100}$

$$= \frac{\overset{1}{\cancel{2}}}{\underset{1}{\cancel{3}}} \cdot \frac{\overset{203}{\cancel{609}}}{\underset{50}{\cancel{100}}} - \frac{3}{8}$$

Divide out common factors and multiply.

$$= \frac{203}{50} - \frac{3}{8}$$

The LCD for 50 and 8 is 200.

$$= \frac{203(4)}{50(4)} - \frac{3(25)}{8(25)}$$

Upscale each fraction to equivalent fractions with 200 as the denominator.

$$= \frac{812}{200} - \frac{75}{200}$$

Subtract numerators and keep the common denominator.

$$= \frac{737}{200} \quad \text{or} \quad 3\frac{137}{200} \quad \text{or} \quad 3.685$$

Working with fractions can be tedious, so let's explore two other methods.

Method 2: Write all fractions as decimals.

$$\frac{2}{3}(6.09) - \frac{3}{8} = 0.\overline{6}(6.09) - 0.375$$

Write the fractions as decimals.
$\dfrac{2}{3} = 0.\overline{6}$ and $\dfrac{3}{8} = 0.375$

The repeated digits of $0.\overline{6}$ create a problem. We must round $0.\overline{6}$ in order to calculate further. We'll round to thousandths. This means our answer will not be exact.

$$\approx 0.667(6.09) - 0.375$$

$$\approx 4.06203 - 0.375$$

$$\approx 3.68703$$

Because we used fractions in Method 1, we got an exact answer of 3.685. We can see from Method 2 that if a fraction becomes a decimal number with an unending pattern of repeated digits, the answer will not be exact.

Method 3: Write decimals over 1 in fraction form. We can write the decimal number over 1 without affecting the problem because any number divided by 1 is equal to that same number.

$$\frac{2}{3}(6.09) - \frac{3}{8} = \frac{2}{3} \cdot \frac{6.09}{1} - \frac{3}{8}$$

Write 6.09 over 1 and multiply.
$2 \cdot 6.09 = 12.18$ and $3 \cdot 1 = 3$

$$= \frac{12.18}{3} - \frac{3}{8}$$

Divide denominators into numerators to get decimals. $12.18 \div 3 = 4.06$ and $3 \div 8 = 0.375$

$$= 4.06 - 0.375$$

$$= 3.685$$

The answer from Method 3 is exact because dividing 12.18 by 3 makes a terminating decimal. Nonterminating decimal numbers result from division, so if we can legally avoid a division that creates a nonterminating decimal, we can get an exact answer. The only way to completely avoid nonterminating decimal numbers is to stick with fractions.

Do Margin 4. ▶

OBJECTIVE 3 *Evaluate expressions.*

EXAMPLE 5 Evaluate $vt + \frac{1}{2}at^2$ when $v = 30$, $t = 0.4$, and $a = 12.8$

Solution: Replace the variables in the expression with the corresponding numbers and calculate following the order of operations agreement.

$vt + at^2$	Replace the variables with the given numbers.
$(30)(0.4) + \frac{1}{2}(12.8)(0.4)^2$	Write $\frac{1}{2}$ as a 0.5.
$= (30)(0.4) + 0.5(12.8)(0.4)^2$	Calculate $(0.4)^2 = 0.16$
$= (30)(0.4) + 0.5(12.8)(0.16)$	Multiply. $(30)(0.4) = 12$ and $0.5(12.8)(0.16) = 1.024$
$= 12 + 1.024$	Add. $12 + 1.024 = 13.024$
$= 13.024$	

Do Margin 5. ▶

OBJECTIVE 4 *Find the area of a triangle.*

Now that we have discussed how to handle fractions and decimals in the same problem, we can consider the area of triangles with decimal base and/or height. In Section 5.3 we learned the formula for the area of a triangle, $A = \frac{1}{2}bh$.

EXAMPLE 6 Find the area.

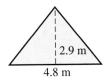

2.9 m
4.8 m

Solution: Use the formula $A = \frac{1}{2}bh$.

$$A = \frac{1}{2}bh$$

$$A = \frac{1}{2}(4.8)(2.9)$$

$$A = \frac{1}{2}(13.92)$$

$$A = \frac{1}{2} \cdot \frac{13.92}{1}$$

$$A = \frac{13.92}{2}$$

$$A = 6.96 \text{ m}^2$$

Do Margin 6. ▶

OBJECTIVE 5 *Find the area of a trapezoid.*

We can now calculate the area of a trapezoid with decimal lengths and/or height. In Section 5.6 we learned the formula, $A = \frac{1}{2}h(a + b)$.

EXAMPLE 7 Find the area.

Solution: Use the formula $A = \frac{1}{2}h(a + b)$.

◆ **Margin 4**

Multiply.

a. $\frac{3}{4}(24.6)$

b. $(-5.052)\frac{5}{6}$

c. $\frac{1}{2}(7.9) - \frac{2}{5}$

d. $\frac{1}{4} + (-2.7) \div \frac{9}{2}$

◆ **Margin 5**

Evaluate.

a. mv^2 when $m = 10$ and $v = 0.5$

b. $\frac{Mm}{d^2}$ when $M = 3.5 \times 10^5$, $m = 4.8$, and $d = 2$

c. $\frac{2}{3}h - k^3$ when $h = 0.45$ and $k = -0.2$

◆ **Margin 6**

Find the area.

a.

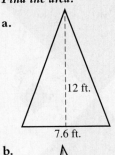

12 ft.
7.6 ft.

b.

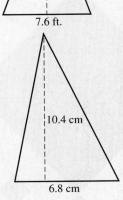

10.4 cm
6.8 cm

Answers to Margin 4: **a.** 18.45 **b.** −4.21 **c.** 3.55 **d.** −0.35

Answers to Margin 5: **a.** 2.5 **b.** 420,000 **c.** 0.308

Answers to Margin 6: **a.** 45.6 ft.^{2} **b.** 35.36 cm^2

Find the area.

a.

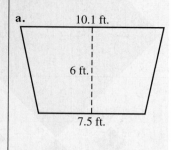

b.

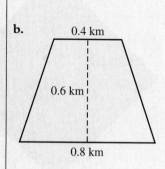

$$A = \frac{1}{2}h(a + b)$$

$$A = \frac{1}{2}(9)(4.86 + 5.34)$$

$$A = \frac{1}{2}(9)(10.2)$$

$$A = \frac{1}{2}(91.8)$$

$$A = \frac{1}{2} \cdot \frac{91.8}{1}$$

$$A = \frac{91.8}{2}$$

$$A = 45.9 \text{ m}^2$$

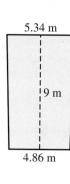

◀ **Do Margin 7.**

OBJECTIVE 6 *Find the area of a circle.*

We saw in Section 6.4 that the value of π can be approximated as the decimal number 3.14. Let's use this approximation to calculate the area of a circle. The formula for the area of a circle is $A = \pi r^2$.

EXAMPLE 8 Find the area. Use 3.14 for π.

Solution: Use the formula $A = \pi r^2$. We are given the diameter of 6.8 cm and we need the radius. Because radius is half of the diameter, the radius is 3.4 cm.

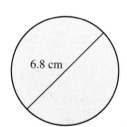

$$A = \pi r^2$$

$$A = \pi(3.4)^2$$

$$A = \pi(11.56)$$

$$A = 11.56\pi \text{ cm}^2$$

$$A \approx 11.56(3.14) \text{ cm}^2$$

$$A \approx 36.2984 \text{ cm}^2$$

Note: Because π is irrational, 11.56π is the exact representation of the area. When we use an approximate value for π, the calculation becomes an approximation. This is why we use $\approx$.

◀ **Do Margin 8.**

OBJECTIVE 7 *Find the volume of a cylinder.*

Previously we discussed how to calculate the volume of a box. In Section 1.7 we developed the formula for the volume of a box to be $V = lwh$. Notice that the bottom (or top) of the box is a rectangle and lw calculates the area of that rectangle. To get the volume of the box we multiply the area of the base, which is lw, by the height of the box, h.

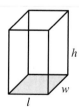

We can find the volume of any three-dimensional object that is formed by repeatedly stacking the same shape by multiplying the area of the base shape by the height of the object.

To understand what we mean by repeatedly stacking a shape, imagine that the box is a stack of papers. A single sheet of paper is very thin when viewed along its edge, but is a rectangle when viewed from the top or bottom. When we stack lots of these thin rectangles on top of each other, we create a box shape.

Margin 8

Find the area.

a.

b.

Answers to Margin 7: **a.** 52.8 ft.^{2}
b. 0.36 km^2

Answers to Margin 8: **a.** 0.785 m^2
b. 13.8474 ft.^{2}

440 ■ Chapter 6 Decimals

Let's now apply this idea to a cylinder. A cylinder has a circular base that is repeatedly stacked. Imagine coins stacked. Because a cylinder is formed by repeatedly stacking thin circles, we can find its volume by multiplying the area of the base by the height of the cylinder.

V = area of base · height of cylinder
(circle)

$V = \pi r^2 \cdot h$

$V = \pi r^2 h$

EXAMPLE 9 Find the volume of the cylinder. Use 3.14 for π.

Solution: Use the formula $V = \pi r^2 h$.

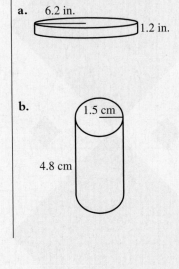

4.5 in.

12 in.

$$V = \pi r^2 h$$
$$V = \pi (4.5)^2 (12)$$
$$V = \pi (20.25)(12)$$

Exact answer → $V = 243\pi$ in.^{3}

$V \approx 243(3.14)$ in.^{3}

Approximation → $V \approx 763.02$ in.^{3}

> **Reminder**
> Volume is always expressed in cubic units. We write cubic inches as in.^{3}.

Do Margin 9. ▷

OBJECTIVE 8 *Find the volume of a pyramid.*

A pyramid is an object formed when lines drawn from the four corners of a square or rectangle meet at a single point. Notice a pyramid would fit inside a box that has a base of the same length and width.

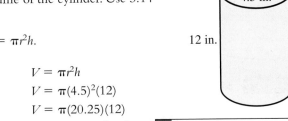

It turns out that a pyramid has $\frac{1}{3}$ of the volume of a box with the same size base and same height.

$$V = \frac{1}{3} \cdot \text{volume of a box with the same length, width, and height.}$$
$$V = \frac{1}{3} lwh$$

EXAMPLE 10 Find the volume of the pyramid.

Solution: Use the formula $V = \frac{1}{3} lwh$.

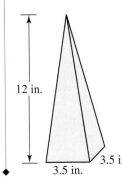

12 in.

3.5 in.

3.5 in.

$$V = \frac{1}{3}(3.5)(3.5)(12)$$

$$V = \frac{1}{3}(147)$$

$$V = \frac{1}{3} \cdot \frac{147}{1}$$

$$V = \frac{147}{3}$$

$$V = 49 \text{ in.}^3$$

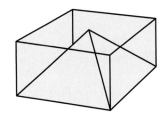

Multiply (3.5)(3.5)(12) first, then write the product over 1 and multiply by $\frac{1}{3}$. We don't want to use the decimal equivalent of $\frac{1}{3}$ because it is a nonterminating decimal and its use would lead to an approximate answer.

Do Margin 10. ▷

◆ **Margin 9**

Find the volume.

a. 6.2 in.

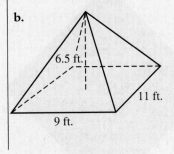

1.2 in.

b.

1.5 cm

4.8 cm

◆ **Margin 10**

Find the volume.

a.

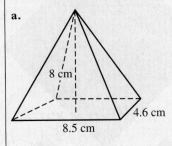

8 cm

8.5 cm

4.6 cm

b.

6.5 ft.

9 ft.

11 ft.

Answers to Margin 9:
a. 144.84 in.^{3} b. 33.912 cm^3

Answers to Margin 10:
a. 104.2$\overline{6}$ cm^3 b. 214.5 ft.^{3}

◆ Margin 11

Find the volume. Use 3.14 for π.

a.

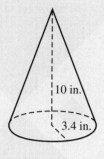

10 in.

3.4 in.

b.

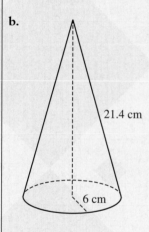

21.4 cm

6 cm

OBJECTIVE 9 *Find the volume of a cone.*

A cone is much like a pyramid except the base is a circle. Also, like a pyramid, which is $\frac{1}{3}$ the volume of a box with the same-size base and height, a cone is $\frac{1}{3}$ the volume of a cylinder with the same-size base and height.

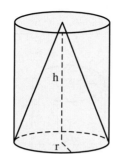

$$V = \frac{1}{3} \cdot \text{volume of a cylinder with the same-size base and height}$$

$$V = \frac{1}{3} \pi r^2 h$$

EXAMPLE 11 Find the volume of the cone. Use 3.14 for π.

Solution: Use the formula $V = \frac{1}{3} \pi r^2 h$.

$$V = \frac{1}{3} \pi r^2 h$$

$$V = \frac{1}{3} \pi (3.5)^2 (6)$$

$$V = \frac{1}{3} \pi (12.25)(6)$$

$$V = \frac{1}{3} (73.5) \pi$$

$$V = \frac{1}{3} \cdot \frac{73.5}{1} \pi$$

$$V = \frac{73.5}{3} \pi$$

$$V = 24.5 \pi \text{ cm}^3$$

$$V \approx 24.5(3.14) \text{ cm}^3$$

$$V \approx 76.93 \text{ cm}^3$$

6 cm

3.5 cm

◀ **Do Margin 11.**

OBJECTIVE 10 *Find the volume of a sphere.*

A ball is a sphere. Every point on the surface of a sphere is equally distant from the center. It turns out the volume of a sphere is equal to 4 cones that have the same radius as the sphere and height equal to the radius.

$$\text{the volume of one such cone} = \frac{1}{3} \pi r^2 r = \frac{1}{3} \pi r^{2+1} = \frac{1}{3} \pi r^3$$

Because the height of the cone is equal to the radius of the sphere, we use r in place of h.

Because the volume of a sphere is equal to 4 of these cones, we must multiply the expression we found above by 4.

$$\text{volume of a sphere} = 4 \cdot \frac{1}{3} \pi r^3 = \frac{4}{1} \cdot \frac{1}{3} \pi r^3 = \frac{4}{3} \pi r^3$$

Conclusion: The volume of a sphere can be found using the formula $V = \frac{4}{3} \pi r^3$.

EXAMPLE 12   Find the volume of the sphere. Use 3.14 for π.

Solution: Use the formula $V = \frac{4}{3}\pi r^3$.

$$V = \frac{4}{3}\pi r^3$$

$$V = \frac{4}{3}\pi (0.2)^3$$

$$V = \frac{4}{3}\pi (0.008)$$

$$V = \frac{4}{3}\left(\frac{0.008}{1}\right)\pi$$

$$V = \frac{0.032}{3}\pi$$

$$V = 0.010\overline{6}\,\pi$$

$$V \approx 0.010\overline{6}(3.14)$$

$$V \approx 0.033493\,\text{m}^3$$

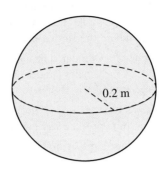

0.2 m

Discussion What is a reasonable number of decimal places in an answer?

Do Margin 12. ▶

OBJECTIVE 11 *Find the area or volume of a composite form.*

Composite forms are geometrical forms that combine fundamental forms. The key to finding the area of a composite form is to determine whether two or more forms have been put together or whether a form or forms have been removed from a larger form.

EXAMPLE 13   Find the area. Use 3.14 for π.

Understand: This shape is a combination of a half-circle and a triangle. The area of a full circle is found by $A = \pi r^2$, so for a half-circle, the area formula is $A = \frac{1}{2}\pi r^2$. Note that 1.6 m is the diameter, so the radius is 0.8 m. Also, the formula for the area of a triangle is $A = \frac{1}{2}bh$.

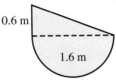

0.6 m

1.6 m

Plan: Add the area of the half-circle and triangle.

Execute: $A =$ area of half-circle + area of triangle

$$A = \quad \frac{1}{2}\pi r^2 \quad + \quad \frac{1}{2}bh$$

$$A \approx \frac{1}{2}(3.14)(0.8)^2 \quad + \quad \frac{1}{2}(1.6)(0.6)$$

$$A \approx \frac{1}{2}(3.14)(0.64) \quad + \quad \frac{1}{2}(1.6)(0.6)$$

$$A \approx \frac{1}{2}(2.0096) \quad + \quad \frac{1}{2}(0.96)$$

$$A \approx \frac{1}{2}\cdot\frac{2.0096}{1} \quad + \quad \frac{1}{2}\cdot\frac{0.96}{1}$$

$$A \approx \frac{2.0096}{2} \quad + \quad \frac{0.96}{2}$$

$$A \approx 1.0048 \quad + \quad 0.48$$

$$A \approx 1.4848\,\text{m}^2$$

We use $\approx$ when we replace π with an approximate value.

Answer: The area of the shape is 1.4848 m².

Check: We can check by reversing each calculation. We will leave the check to the reader.

Do Margin 13. ▶

◆ Margin 12

a. Find the volume of a sphere with a radius of 3 ft.

b. Find the volume of a sphere with a radius of 8 cm.

◆ Margin 13

Find the area.

145 ft.

42.5 ft.

40.2 ft.   40.2 ft.

Answers to Margin 12:
a. 113.04 ft.³ **b.** 2143.57$\overline{3}$ cm³

Answer to Margin 13: 7800.47 ft.²

EXAMPLE 14 Find the actual space for clothes inside the washing machine shown.

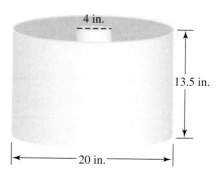

Understand: A washing machine is a small cylinder within a larger cylinder. The space in between the cylinders is where the clothes are placed.

The volume of a cylinder is found by $V = \pi r^2 h$. The radius of the smaller cylinder is 2 in. The radius of the larger cylinder is 10 in.

Plan: Subtract the volume of the inner cylinder from the volume of the outer cylinder.

Execute: V = volume of outer cylinder − volume of inner cylinder

$$V = \quad \pi r^2 h \quad - \quad \pi r^2 h$$

$$V = \pi(10)^2(13.5) - \pi(2)^2(13.5)$$

$$V = \pi(100)(13.5) - \pi(4)(13.5)$$

$$V = 1350\pi - 54\pi$$

$$V = 1296\pi \text{ in.}^3$$

$$V \approx 1296(3.14) \text{ in.}^3$$

$$V \approx 4069.44 \text{ in.}^3$$

> **Connection**
>
> Because the volume for both shapes involves π, we chose to leave π in symbolic form.
>
> Because 1350π and -54π are both in terms of π, they are like terms. We can combine by subtracting the coefficients and keeping the result in terms of π.

◆ *Answer:* The space for clothes inside the washing machine is about 4069.44 in.^{3}.

◀ **Do Margin 14.**

◆ **Margin 14**

Find the amount of storage space under the countertop.

(sink) 17 in.   20.25 in.

25.5 in.

48 in.

Historical Note

In any list of the greatest mathematicians to have ever lived, the name Archimedes is usually at the top. Most scholars place Archimedes, Isaac Newton, and Carl Friedrich Gauss as the top three mathematicians to have ever lived.

Archimedes was born in 287 B.C. in the city of Syracuse in what is now Sicily. There is some indication that he was related to Heiron II, king of Syracuse. In his youth he studied in Alexandria, Egypt, where he met and befriended Eratosthenes (famous for his sieve method for finding primes).

Though he preferred pure mathematics, Archimedes made great contributions to astronomy and mechanics. One popular invention was the Archimedian pump. This pump was an auger-like device within a cylinder. The pump worked by tilting the apparatus at an angle with one end in water. At the other end, the operator would turn the auger which would force water out the top of the cylinder. This pump remained in use for centuries.

Archimedes is also known for his discovery of principles of hydrostatics. The story goes that the king had given a smith a lump of pure gold to make a crown. After the crown was completed, the king became suspicious that the smith had substituted the gold for a lessor metal within the crown, keeping most of the gold for himself. The king asked Archimedes to figure a way to determine if the crown was pure gold without destroying the crown. Sometime thereafter Archimedes was taking a bath and as he lowered himself into the water, he realized that his body weighed less as a result of the water pushing upwards, and this force is equal to the volume displacement of the water. He realized that this principle could be applied to the problem of the crown. It is said that in his excitement over the discovery he jumped from the water and ran naked through the streets of ancient Syracuse in what is now Italy shouting, "Eureka, Eureka!" (I have found it, I have found it!)

In pure mathematics, Archimedes developed methods for finding the areas of many shapes and objects including the cylinder, cone, and sphere. He developed a method for calculating π. He placed the value as being between $3\frac{1}{7}$ and $3\frac{10}{71}$. He even developed some of the ideas involved in calculus, preceding Isaac Newton (the accepted inventor of calculus) by 1800 years.

Answer: 23,500.43 in.^{3}

6.5 Exercises

FOR EXTRA HELP

 Videotape 8

 InterAct Math Tutorial Software

 www.carsonmath.com

 AWL Math Tutor Center

 InterAct MathXL www.mathxl.com

 Student's Solutions Manual

For Exercises 1–12, simplify.

1. $0.64 + 2.5(0.8)$

2. $9.28 - 0.56(12)$

3. $88.2 - 2.2(0.45 + 20.1)$

4. $9.5(2.6 - 0.99) - 4.33$

5. $(0.4)^2 - 2.8 \div 0.2(1.6)$

6. $16.1 \div 0.4(0.3) + (1.3)^2$

7. $10.7 - 18\sqrt{1.96} + (74.6 - 88.1)$

8. $7.5 + 2.2\sqrt{0.25} - 36.8 \div 8$

9. $\sqrt{0.0081} + 120.8 \div 4(2.5 - 6.4)$

10. $40.1 - 6.9 \div 4.6(1.4)^2 + \sqrt{0.36}$

11. $[-5.53 \div (0.68 + 0.9)] \div (0.6)^2$

12. $[(6.2 - 10) \div 8] - 2\sqrt{1.21}$

For Exercises 13–20, simplify.

13. $\dfrac{3}{5}(-0.85)$

14. $\dfrac{7}{10}(12.88 - 4.38)$

15. $\dfrac{1}{4} + \dfrac{2}{3}(0.06)$

16. $(-12.9)\dfrac{5}{6} + \dfrac{3}{4}$

17. $-2\dfrac{4}{9}(1.8) - \dfrac{3}{8}$

18. $3\dfrac{1}{8} - (0.45)\left(5\dfrac{1}{3}\right)$

19. $\dfrac{2}{5} \div (-0.8) + \left(\dfrac{1}{3}\right)^2$

20. $\dfrac{1}{6} - \dfrac{1}{24}(0.4 + 1.2)$

For Exercises 21–28, evaluate the expression using the given numbers.

21. $\dfrac{1}{2}mv^2$ when $m = 1.25$ and $v = -8$

Trivia Bite

$\dfrac{1}{2}mv^2$ is an expression used to calculate the energy of a moving object where m is the mass of the object and v is the velocity. As a formula, we would write $E = \dfrac{1}{2}mv^2$. Energy is measured in

22. $\dfrac{1}{2}mv^2$ when $m = 84.6$ and $v = 0.2$

23. $vt + \dfrac{1}{2}at^2$ when $v = 20$, $t = 0.6$, and $a = -12.5$

Trivia Bite

$vt + \dfrac{1}{2}at^2$ is an expression used to calculate the distance an object travels that has an initial velocity, v, and acceleration or deceleration, a, over a period of time, t. As a formula, we would write $d = vt + \dfrac{1}{2}at^2$.

24. $vt + \dfrac{1}{2}at^2$ when $v = 35$, $t = 1.2$, and $a = 10$

25. mc^2 when $m = 2.5 \times 10^{-6}$ and $c = 3 \times 10^8$

Trivia Bite

mc^2 is an expression developed by Albert Einstein to calculate the energy of a particle where m is the mass of the particle and c is the speed of light. It appears in possibly the most famous formula of physics: $E = mc^2$.

26. mc^2 when $m = 3.6 \times 10^{-4}$ and $c = 3 \times 10^8$

27. $\dfrac{w}{h^2} \cdot 705$ when $w = 150$ and $h = 67$

Trivia Bite

> $\dfrac{w}{h^2} \cdot 705$ is used to calculate a person's body mass index (BMI) where w is the person's weight in pounds and h is the person's height in inches. According to the National Institute of Health, people with a BMI of 25 or more have an increased risk for cardiovascular and other diseases.

28. $\dfrac{w}{h^2} \cdot 705$ when $w = 185$ and $h = 70$

For Exercises 29–34, find the area. Use 3.14 for π.

29.

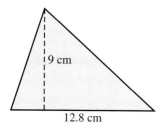

9 cm

12.8 cm

30.

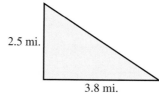

2.5 mi.

3.8 mi.

31.

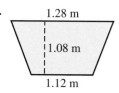

1.28 m

1.08 m

1.12 m

32.

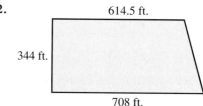

614.5 ft.

344 ft.

708 ft.

33.

10.5 in.

34.

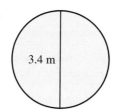

3.4 m

For Exercises 35–42, find the volume. Use 3.14 for π.

35. A quarter.

0.2 cm

2.4 cm

36. Canned drink.

2.5 in.

4.2 in.

37. The Castillo at Chichen Itza in the Yucatan forest in Mexico is 90 ft. tall.

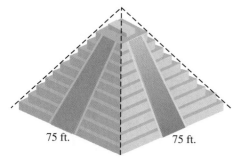

75 ft.   75 ft.

Trivia Bite

> The Castillo at Chichen Itza is a pyramid temple built by the Mayans around A.D. 500. The pyramid is a calendar. There are 4 sets of steps, one set on each side of the pyramid with 91 steps each. At the top is a small temple with 1 step on all four sides for a total of 365 steps, one for each day of the year. At the bottom of each set of steps are serpent heads that stare out from either side of the steps. At the equinoxes the sunlight hits the ridged sides of the pyramid at the perfect angle to cast a shadow that connects with the serpent head, giving the appearance of a serpent snaking its way down the pyramid.

38. Menkaure's pyramid at Giza, Egypt, is 203 ft. tall.

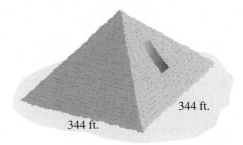

344 ft.

344 ft.

39. Onyx decorative cone.

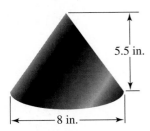

5.5 in.

8 in.

40. Afterburner fire cone from a jet engine.

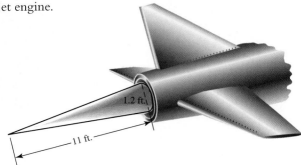

1.2 ft.

11 ft.

41. If Earth is a sphere with a radius of 3884.3 mi., find the volume of Earth.

42. If the Sun is a sphere with a radius of 423,858 mi., find the volume of the Sun.

◆ **Discussion** Approximately how many Earths could fit inside the Sun?

For Exercises 43 and 44, find the area of the composite shape.

43.

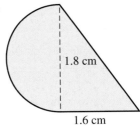

1.8 cm

1.6 cm

44.

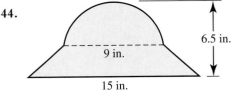

9 in.

6.5 in.

15 in.

For Exercises 45 and 46, find the area of the shaded region.

45.

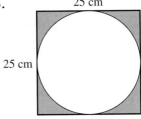

25 cm

25 cm

46.

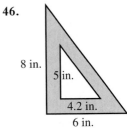

8 in.

5 in.

4.2 in.

6 in.

For Exercises 47–48, find the volume of the composite object.

47. Water tower.

|← 64 ft. →|

34.5 ft.

►|20 ft.|◄

48. Mini-sub.

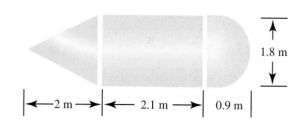

1.8 m

|←—2 m—→|←— 2.1 m —→| 0.9 m |

49. Pyramid with inner chamber. The chamber is 5.8 m by 5.2 m by 10.8 m. The height of the pyramid is 146.5 m.

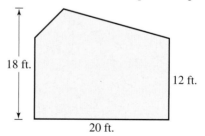

230 m

230 m

50. Bowling ball. The diameter of the ball is 10 in. Each hole is a cylinder 3 in. deep with a 1 in. diameter.

51. a. A grain of salt is a cube that is 0.03 in. along each side. What is the volume of a grain of salt?

b. About how many grains of salt would be in 1 in.^{3}?

52. A grain of salt is a cube that is 0.03 in. along each side. Calculate the surface area of a grain of salt.

Puzzle Problem

Estimate the volume of our galaxy. It is a spiral galaxy 100,000 lt-yr (light years) across and about 1000 lt-yr thick at the center.

REVIEW EXERCISES ───────────────────────

1. Solve and check. $4(x - 6) + 1 = 9x - 2 - 8x$

2. Solve and check. $\frac{1}{3}y - \frac{3}{4} = 6$

3. Find the area of the composite shape.

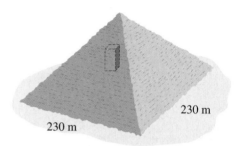

18 ft.

12 ft.

20 ft.

4. Find the area of the shaded region.

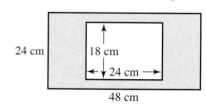

24 cm

18 cm

↕ 24 cm →

48 cm

5. Translate to an equation, then solve. $\frac{1}{2}$ of the sum of x and 12 is equal to 9 less than $\frac{3}{4}$ of x.

6. Marla has $160 in tens and fives. If she has 4 more tens than fives, how many of each bill does she have?

6.6 Solving Equations and Problem Solving

OBJECTIVES

1 Solve equations using the addition/subtraction and multiplication/division principles of equality.

2 Simplify decimal equations using the multiplication principle.

3 Solve application problems.

4 Solve problems involving one unknown.

5 Solve problems using the Pythagorean theorem.

6 Solve problems involving two unknowns.

OBJECTIVE **1** *Solve equations using the addition/subtraction and multiplication/division principles of equality.*

EXAMPLE 1   Solve and check. $6 - 4.8t = 3.552$

Solution: To isolate t, we can start by using the addition/subtraction principle. Subtract 6 on both sides to isolate $-4.8t$. Then clear the -4.8 coefficient by using the multiplication/division principle and dividing by -4.8 on both sides.

$$6 - 4.8t = 3.552$$

$$
\begin{array}{ll}
6 - 4.8t = 3.552 & \\
\underline{-6 \qquad\quad -6} & \text{Subtract 6 on both sides to isolate } -4.8t. \\
0 - 4.8t = -2.448 & \\
\dfrac{-4.8t}{-4.8} = \dfrac{-2.448}{-4.8} & \text{Divide both sides by } -4.8 \text{ to isolate } t. \\
1t = 0.51 & \\
t = 0.51 &
\end{array}
$$

Check: Replace t in $6 - 4.8t = 3.552$ with 0.51, and verify that it makes the equation true.

$$
\begin{array}{ll}
6 - 4.8t = 3.552 & \\
6 - 4.8(0.51) \ ? \ 3.552 & \\
6 - 2.448 \ ? \ 3.552 & \\
3.552 = 3.552 & \text{It checks.}
\end{array}
$$

Do Margin 1. ▶

OBJECTIVE **2** *Simplify decimal equations using the multiplication principle.*

In Chapter 5, we learned to use the multiplication principle of equality to clear fractions in an equation. The idea was to multiply both sides throughout by the LCD. Because decimal numbers are equivalent to fractions, we can use the same principle with equations that contain decimal numbers. When we write decimals as fractions, the denominators are always powers of 10. This means we can clear decimals by multiplying both sides by an appropriate power of 10.

We saw in Section 6.3 that when we multiply a decimal number by a power of 10, the decimal point moves to the right. Multiplying a decimal number by a large enough power of 10 can yield a whole number. We saw this again in Section 6.4 when we wanted to have whole numbers as divisors.

◆ **Margin 1.**

Solve and check.

a. $4.5y - 21.99 = 6$

b. $0.08n + 0.4 = 0.244$

Answers: **a.** 6.22 **b.** -1.95

How do we decide what power of 10 to use?

$$2.5x - 0.64 = 8$$

In the above equation, if we multiplied every term by 10, the decimal point would move to the right 1 place in each number.

$$10 \cdot 2.5x - 10 \cdot 0.64 = 10 \cdot 8$$
$$25x - 6.4 = 80$$

10 is not large enough because 0.64 becomes 6.4, which still contains a decimal point. Because 0.64 has two places we should use 100. Notice the decimal number with the most places determines what power of 10 to use.

$$2.5x - 0.64 = 8$$
$$100(2.5x - 0.64) = 100 \cdot 8$$
$$100 \cdot 2.5x - 100 \cdot 0.64 = 100 \cdot 8$$
$$250x - 64 = 800$$

Multiplying every term on both sides by 100 transforms the equation from a decimal equation to an equation that contains only integers.

Notice if we write the original equation in terms of equivalent fractions, 100 is the LCD.

$$\frac{25}{10}x - \frac{64}{100} = 8$$

100 is the LCD of 10 and 100.

So we are still multiplying by the LCD in the same way that we did with fractions, although the LCD will be a power of 10 with the same number of 0 digits as there are decimal places in the number with the most decimal places.

We can amend our procedure for solving equations to include this technique.

Procedure *To solve equations:*
1. Simplify both sides of the equation as needed.
   a. Distribute to clear parentheses.
   b. Clear fractions and/or decimals by multiplying both sides by the LCD of all denominators. In the case of decimals, the LCD is the power of 10 with the same number of 0 digits as there are decimal places in the number with the most decimal places.
   c. Combine like terms.
2. Use the addition/subtraction principle so that all variable terms are on one side of the equation and all constants are on the other side.
3. Use the multiplication/division principle to clear any remaining coefficients.

Let's look at Example 1 again using this technique.

EXAMPLE 2 Solve and check. $6 - 4.8t = 3.552$

Solution: We can clear the decimals by multiplying both sides throughout by an appropriate power of 10. Use the power of 10 with the same number of 0's as there are decimal places in the number with the most decimal places. In this case 3.552 has the most decimal places so we multiply both sides by 1000.

$$6 - 4.8t = 3.552$$

$$1000 \cdot 6 - 1000 \cdot 4.8t = 1000 \cdot 3.552$$ **Multiply both sides by 1000 to clear the decimal numbers.**

$$6000 - 4800t = 3552$$

$$6000 - 4800t = 3552$$

$$\underline{-6000 \qquad\qquad -6000}$$ **Subtract 6000 from both to isolate** $-4800t$.

$$0 - 4800t = -2448$$

$$\frac{-4800t}{-4800} = \frac{-2448}{-4800}$$ **Divide both sides by** -4800 **to isolate** t.

$$1t = 0.51$$

$$t = 0.51$$

Check: Replace t in $6 - 4.8t = 3.552$ with 0.51, and verify that it makes the equation true.

$$6 - 4.8t = 3.552$$

$$6 - 4.8(0.51)\ ?\ 3.552$$

$$6 - 2.448\ ?\ 3.552$$

$$3.552 = 3.552$$ **It checks.**

Do Margin 2. ▶

EXAMPLE 3   Solve and check. $0.3(x - 1.2) = 1.46 - (2.1x + 0.22)$

Solution: First, we will distribute to clear the parentheses. Then we will use the multiplication/division principle to clear the decimal numbers. Because 1.46 and 0.22 have the most decimal places we will multiply both sides throughout by 100 to clear the decimals.

$$0.3(x - 1.2) = 1.46 - (2.1x + 0.22)$$ **Distribute to clear parentheses.**

$$0.3x - 0.36 = 1.46 - 2.1x - 0.22$$

$$100 \cdot 0.3x - 100 \cdot 0.36 = 100 \cdot 1.46 - 100 \cdot 2.1x - 100 \cdot 0.22$$ **Multiply both sides by 100 to clear the decimal numbers.**

$$30x - 36 = 146 - 210x - 22$$ **Combine like terms.**

$$30x - 36 = 124 - 210x$$

$$30x - 36 = 124 - 210x$$

$$\underline{+210x \qquad\qquad +210x}$$ **Add** $210x$ **to both sides so that the** x **terms are on the same side of the equation.**

$$240x - 36 = 124 + 0$$

$$240x - 36 = 124$$ **Add 36 to both sides to isolate** $240x$.

$$\underline{+36 \qquad +36}$$

$$240x + 0 = 160$$

$$\frac{240x}{240} = \frac{160}{240}$$ **Divide both sides by 240 to isolate** x.

$$1x = 0.\overline{6}$$

$$x = 0.\overline{6}$$

Check: Because $0.\overline{6}$ is a nonterminating decimal number, we will have to round if we intend to check with it. If we round, we must be aware that both sides will not match exactly. We should expect that they will be close, though.

The only way to get the exact same result on both sides is to use the fraction equivalent of $0.\overline{6}$, which is $\frac{2}{3}$. But the method for writing nonterminating decimals as fractions is beyond the scope of our discussion. Therefore, we will round the decimal and keep in mind that both sides will not match exactly, but should be reasonably close.

We will round $0.\overline{6}$ to the nearest thousandth, 0.667.

$$0.3(x - 1.2) = 1.46 - (2.1x + 0.22)$$
$$0.3(0.667 - 1.2) \ ? \ 1.46 - [2.1(0.667) + 0.22]$$
$$0.3(-0.533) \ ? \ 1.46 - [1.4007 + 0.22]$$
$$-0.1599 \ ? \ 1.46 - 1.6207$$
$$-0.1599 \ ? \ -0.1607$$

As predicted, the values are not equal, but they are reasonably close. If we had rounded to ten-thousandths, or hundred-thousandths, the outcome would be even closer.

◀ Do Margin 3.

OBJECTIVE 3 *Solve application problems.*

Let's first consider more advanced problems similar to those we've discussed in previous sections.

EXAMPLE 4   The table shows the pricing schedule for a phone company.

Time	Rate
8 A.M.–5 P.M.	$0.12/minute
5 P.M.–10 P.M.	$0.10/minute
10 P.M.–8 A.M.	$0.09/minute

Below is a portion of Muffy's long distance call summary. Find the total cost of the calls.

Time of call	City	Number of minutes
10:05 P.M.	Richmond, VA	13
9:52 A.M.	Columbus, OH	21
4:47 P.M.	Atlanta, GA	24

Understand: Because the unit cost depends on the time of the call, we must consider the time of each call.

The 13 minutes to Richmond will cost $0.09/minute because the call was made at 10:05 P.M.. The 21 minutes to Columbus will cost $0.12/minute because the call was made at 9:52 A.M.. Notice the 24–minute call to Atlanta started at 4:47 P.M. and lasted until 5:11 P.M. This means the first 13 minutes of the call will cost $0.12/minute and the last 11 minutes will cost $0.10/minute because the rate changes at 5 P.M..

Plan: Multiply the number of minutes by their respective rates, then add to get the total cost.

Execute: total cost = cost of call to Richmond + cost of call to Columbus + cost of call to Atlanta

$$C = 0.09(13) \quad + \quad 0.12(21) \quad + \quad 0.12(13) + 0.10(11)$$
$$C = 1.17 \quad + \quad 2.52 \quad + \quad 1.56 + 1.10$$
$$C = 6.35$$

Answer: The total cost of the long distance calls is $6.35.

Check: We can reverse the calculations. Start with 6.35 and subtract all the last three addends and see if we get the first addend.

$$6.35 - 1.10 - 1.56 - 2.52 \ ? \ 1.17$$
$$1.17 = 1.17$$

◆ Margin 3

Solve and check.

a. $18.6k - 1.25 = 9.4k + 1.326$

b. $11 - 2(5.6 + x) = 0.7x + 1.3$

Answers: **a.** 0.28 **b.** $-0.\overline{5}$

We can check the multiplication by dividing each product by one of the factors to see if we get the other factor.

$1.17 \div 13 \; ? \; 0.09$	$2.52 \div 21 \; ? \; 0.12$	$1.56 \div 13 \; ? \; 0.12$	$1.10 \div 11 \; ? \; 0.10$
$0.09 = 0.09$	$0.12 = 0.12$	$0.12 = 0.12$	$0.10 = 0.10$

Do Margin 4. ▷

OBJECTIVE 4 *Solve problems involving one unknown.*

Let's first consider some problems where we translate key words directly to an equation.

EXAMPLE 5 5.9 more than 0.2 times *n* is equal to 6.7 less than the product of 1.4 and *n*. Translate to an equation, then solve for *n*.

Understand: Focus on the key words.

5.9 <u>more than</u> 0.2 <u>times</u> *n* <u>is equal to</u> 6.7 <u>less than</u> the <u>product</u> of 1.4 and *n*.
addition   multiplication   equal sign   subtraction   multiplication

Plan: Translate the key words to an equation, then solve.

Execute: 5.9 <u>more than</u> 0.2 <u>times</u> *n* <u>is equal to</u> 6.7 <u>less than</u> the <u>product</u> of 1.4 and *n*.

$$5.9 \;+\; 0.2n \;=\; 1.4n - 6.7$$
$$5.9 + 0.2n = 1.4n - 6.7$$
$$10(5.9 + 0.2n) = 10(1.4n - 6.7)$$
$$10 \cdot 5.9 + 10 \cdot 0.2n = 10 \cdot 1.4n - 10 \cdot 6.7$$
$$59 + 2n = 14n - 67$$
$$\underline{\; -2n \qquad -2n \;}$$
$$59 + 0 = 12n - 67$$
$$59 = 12n - 67$$
$$\underline{\; +67 \qquad\qquad +67 \;}$$
$$126 = 12n + 0$$
$$\frac{126}{12} = \frac{12n}{12}$$
$$10.5 = 1n$$
$$10.5 = n$$

> **Reminder**
> When translating *less than* to subtraction, the subtrahend appears before the words *less than* and the minuend appears after.

Answer: $n = 10.5$

Check: Verify that 10.5 satisfies the original statement.

$$5.9 + 0.2n = 1.4n - 6.7$$
$$5.9 + 0.2(10.5) \; ? \; 1.4(10.5) - 6.7$$
$$5.9 + 2.1 \; ? \; 14.7 - 6.7$$
$$8 = 8 \qquad \textbf{It checks.}$$

Do Margin 5. ▷

EXAMPLE 6 Karen uses an Internet provider that charges $14.95 per month; this allows for 200 minutes of on-line time. After 200 minutes, it costs $0.10 for each additional minute. If Karen's bill is $20.75, how many minutes did she spend on-line in all?

Understand: We must calculate Karen's total time on-line. The total charges are $20.75. To calculate the total charges, the company charged a flat fee of 14.95 for the first 200 minutes, then $0.10 for each additional minute.

flat fee + cost of additional minutes = total charges

♦ **Margin 4**

Solve.

a. Use the rate schedule in Example 4 to find the total cost of the long distance calls.

Time of call	City	Number of minutes
2:35 P.M.	Los Angeles, CA	19
9:55 P.M.	Tulsa, OK	29
11:02 A.M.	Houston, TX	31

b. Angie has the following medical charges. She arranges with the hospital to make three equal payments to pay off the debt. How much is each payment?

Service	Cost
Lab	$125
Radiology	$287.25
Treatment	$482.56

♦ **Margin 5**

a. 18.6 less than *y* is the same as 1.25 times *y* plus 2.18. Translate to an equation, then solve for *y*.

b. 0.4 times the difference of *m* and 1.8 is equal to -0.8 times *m* minus 9.36. Translate to an equation, then solve for *m*.

Answers to Margin 4: **a.** $8.66
b. $298.27

Answers to Margin 5:
a. $y - 18.6 = 1.25y + 2.18$; $y = -83.12$
b. $0.4(m - 1.8) = -0.8m - 9.36$; $m = -7.2$

Because $0.10 is a unit price, to get the cost of additional minutes, we multiply 0.10 times the number of additional minutes. If we let m be the number of additional minutes, then $0.10m$ describes the cost of those additional minutes.

Plan: We will write an equation and then solve.

Execute: flat fee + cost of additional minutes = total charges

$$14.95 \ + \qquad\qquad 0.10m \qquad\qquad = 20.75$$
$$14.95 + 0.10m = 20.75$$
$$100 \cdot 14.95 + 100 \cdot 0.10m = 100 \cdot 20.75 \qquad \text{**Clear decimals by multiplying throughout by 100.**}$$
$$1495 + 10m = 2075$$
$$\underline{-1495 \qquad\qquad -1495}$$
$$0 + 10m = 580$$
$$\frac{10m}{10} = \frac{580}{10}$$
$$1m = 58$$
$$m = 58$$

Answer: Karen spent 58 additional minutes on-line. This is in addition to the 200 minutes that she paid for with $14.95, so she spent a total of:

$$200 + 58 = 258 \text{ minutes on-line}$$

Check: Verify that 258 minutes at $14.95 for the first 200 minutes and $0.10 per minute thereafter comes to a total of $20.75.

$$14.95 + 0.10(58) \; ? \; 20.75$$
$$14.95 + 5.80 \; ? \; 20.75$$
$$20.75 = 20.75 \qquad \text{**It checks.**}$$

◀ **Do Margin 6.**

OBJECTIVE 5 *Solve problems using the Pythagorean theorem.*

One of the most popular theorems in mathematics is the Pythagorean theorem. The theorem is named after the Greek mathematician, Pythagoras. The theorem describes a relationship that exists among the sides of all **right triangles.**

DEFINITION | **Right triangle:** A triangle that has one right angle.

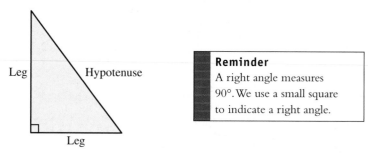

Reminder
A right angle measures 90°. We use a small square to indicate a right angle.

In a right triangle, the two sides that form the 90° angle are called **legs.** The side directly across from the 90° angle is called the **hypotenuse.**

DEFINITION | **Legs:** The sides that form the 90° angle in a right triangle.

Hypotenuse: The side directly across from the 90° angle in a right triangle.

♦ **Margin 6**

Solve.

a. A plumber charges a flat fee of $45 plus $7.25 for every quarter of an hour spent working. Dina suspects that the $95.75 that he charged is too much. How long should he have worked?

b. A cell-phone company charges $19.95 for up to 30 minutes of use plus $0.50 for each additional minute. Ron's bill comes to a total of $33.45. How many minutes did he spend using his cell phone?

Answers: **a.** 7 quarter hours or 1.75 hr. **b.** 57 min.

The Pythagorean Theorem The sum of the areas of the squares on the legs is the same as the area of the square on the hypotenuse.

Consider a right triangle with side lengths of 3 ft., 4 ft., and 5 ft.

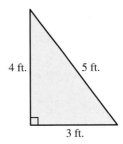

The theorem tells us that adding the areas of squares on the legs gives the same area as a square on the hypotenuse.

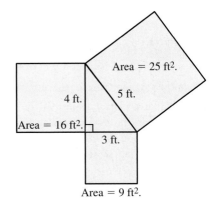

If we add the areas of the squares on the legs, $9 + 16$, we get the same area as the square on the hypotenuse, 25.

$$16 \text{ ft.}^2 + 9 \text{ ft.}^2 = 25 \text{ ft.}^2$$

Conclusion: The sum of the squares of the lengths of the legs is equal to the square of the length of the hypotenuse.

$$(\text{leg})^2 + (\text{leg})^2 = (\text{hypotenuse})^2$$

Using variables a and b for the lengths of the legs and c for the length of the hypotenuse, we can express the theorem as:

$$a^2 + b^2 = c^2$$

The Pythagorean theorem can be used to find a missing length in a right triangle if the other two lengths are known.

EXAMPLE 7 To steady a telephone pole, a wire is to be attached 21 ft. up on the pole and then tied to a stake 20 ft. out from the base of the pole. What length of wire will be needed between the pole and the stake?

Understand: Draw a picture. We assume the pole makes a 90° angle with the ground. The wire connects to the pole to form a right triangle. This means we can use the Pythagorean theorem.

Plan: Use the Pythagorean theorem. The missing length is the hypotenuse, which is c in the formula.

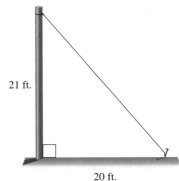

Solve.

a. In constructing a roof, three boards are used to form a right triangle frame. An 8 ft. board and a 6 ft. board are brought together to form a 90° angle. How long must the third board be?

Note: We have kept showing the units in the equation as a reminder that 441 ft.^{2} and 400 ft.^{2} are the areas of the squares on the legs. Their sum, 841 ft.^{2}, is the area of the square on the hypotenuse. To get the length from the area, we must use a square root.

$$a^2 + b^2 = c^2$$
$$(21 \text{ ft.})^2 + (20 \text{ ft.})^2 = c^2$$
$$441 \text{ ft.}^2 + 400 \text{ ft.}^2 = c^2$$
$$841 \text{ ft.}^2 = c^2$$
$$\sqrt{841} \text{ ft.} = c$$
$$29 \text{ ft.} = c$$

Answer: The wire must be at least 29 ft. to connect between the pole and the stake.

Check: Verify that $a^2 + b^2 = c^2$ is true when a is 21, b is 20, and c is 29.

$$(21)^2 + (20)^2 \,?\, (29)^2$$
$$441 + 400 \,?\, 841$$
$$841 = 841$$

EXAMPLE 8 How high up on the side of a building will a 20 ft. ladder reach if the base is placed 8 ft. from the base of the building?

Understand: Draw a picture of the situation.

When the ladder is placed against the building, it forms a right triangle. The 20 ft. ladder is the hypotenuse and the 8 ft. distance from the base of the building is one of the legs. We must find the vertical distance, which is the other leg, by using the Pythagorean theorem.

Plan: Use the Pythagorean theorem to find the missing length of the leg, given the hypotenuse and other leg.

Execute:
$$a^2 + b^2 = c^2$$
$$(8)^2 + b^2 = (20)^2$$
$$64 + b^2 = 400$$
$$64 + b^2 = 400$$
$$\underline{-64 \qquad\quad -64}$$
$$0 + b^2 = 336$$
$$b^2 = 336$$
$$\sqrt{b^2} = \sqrt{336}$$
$$b = \sqrt{336}$$
$$b \approx 18.33 \text{ ft.}$$

To undo the squaring of b, we get the square root. To keep the equation balanced we must get the square root on both sides.

Because $\sqrt{336}$ is irrational, this is the exact answer.

This is the approximate answer rounded to the nearest hundredth.

b. A wire is attached to a telephone pole then pulled taut and attached to a stake in the ground 4 m from the base of the pole. If the length of wire is 10 m, how high up the pole is the wire attached?

Answer: The ladder would reach approximately 18.33 ft. up the side of the building.

Check: Verify that the sum of the squares of the legs is the same as the square of the hypotenuse.

$$(8)^2 + (18.33)^2 \,?\, (20)^2$$
$$64 + 335.9889 \,?\, 400$$
$$399.9889 \,?\, 400$$

Because 18.33 is an approximation, we cannot expect the results to match exactly, but they should be close. 399.9889 is close enough to 400 for us to accept that 18.33 is a good approximation.

◀ **Do Margin 7.**

Answer: **a.** 10 ft. **b.** ≈ 9.17 m

Historical Note

Pythagoras was a Greek mathematician who lived in the town of Croton in what is now Italy. He is believed to have lived from 569 B.C. to 500 B.C. The Pythagorean theorem is named after Pythagoras not because he discovered the relationship, as commonly thought, but because he and his followers are the first to have proved that the relationship is true for all right triangles.

The Pythagoreans also believed in the mystical power of numbers. Pythagoras believed that everything in the universe was built on whole numbers. The very theorem that he and his followers proved wreaked havoc in their beliefs, however. It quickly became obvious that the legs could be measured by whole numbers as desired, but

the hypotenuse would have to be a value that *couldn't* be expressed exactly. For example, if the legs are both 1 unit in length, we have:

$$1^2 + 1^2 = c^2$$
$$1 + 1 = c^2$$
$$2 = c^2$$
$$\sqrt{2} = c$$

$\sqrt{2}$ is definitely not a whole amount and, it turns out, cannot be expressed exactly as a decimal or fraction. It is said that this distressed Pythagoras tremendously because he was so sure that the universe was based on whole numbers.

OBJECTIVE 6 *Solve problems involving two unknowns.*

In Section 4.5 we developed the use of a table to help organize information in problems involving two unknowns. The table we used had four columns:

Categories	Value	Number	Amount

The two unknowns are written in the categories column. The value column contains the given values of each category. The number column will describe the number of items in each category. The amount column will be found by the relationship:

$$\text{value} \cdot \text{number} = \text{amount}$$

EXAMPLE 9 Betty tells her son that she has 15 coins in her change purse that total $2.85. If she only keeps quarters and dimes in her change purse, how many of each coin does she have?

Understand: There are two unknowns in the problem. We must find the number of quarters and the number of dimes. Since we only know how to solve equations involving one unknown, we must select a variable for either quarters or dimes.

Let's begin filling in the chart. Let n represent the number of quarters. How can we represent the number of dimes? Suppose she has 1 quarter. Then if there are 15 coins there must be 14 dimes. If she has 2 quarters, then there must be 13 dimes and so on. We can find the number of dimes by subtracting the number of quarters from 15.

> **Note:** We could let the variable represent either quarters or dimes. We will see that by letting n be quarters, we avoid negatives.

Conclusion: If she has n quarters, she must have $15 - n$ dimes.

Categories	Value	Number	Amount
Quarters	0.25	n	$0.25n$
Dimes	0.10	$15 - n$	$0.10(15 - n)$

> Because value · number = amount, we multiply the expression in the Value column and in the Number column to get the expression in the Amount column.

We are also given the total amount of money, which is $2.85. This total is calculated by adding the amount in quarters plus the amount in dimes.

$$\text{amount in quarters} + \text{amount in dimes} = \text{total amount of money}$$

Plan: Write an equation, then solve.

Execute: amount in quarters + amount in dimes = total money

$$0.25n \qquad + \qquad 0.10(15 - n) = 2.85 \qquad \text{Distribute 0.10 to clear parentheses.}$$
$$0.25n \qquad + \qquad 1.5 - 0.10n = 2.85$$
$$100 \cdot 0.25n + 100 \cdot 1.5 - 100 \cdot 0.10n = 100 \cdot 2.85 \qquad \text{Multiply each term by 100 to clear decimals.}$$
$$25n + 150 - 10n = 285$$
$$15n + 150 = 285$$

Notice when we combine $25n$ and $-10n$ we get $15n$, which has a positive coeffi-cient. Choosing to let n rep-resent the larger valued coin, quarters, leads to this positive coefficient. We'll see that if we let n be the number of dimes, we get a negative coefficient.

$$15n + 150 = 285$$
$$\underline{ -150 \quad -150}$$
$$15n + 0 = 135$$
$$15n = 135$$
$$\frac{15n}{15} = \frac{135}{15}$$
$$1n = 9$$
$$n = 9$$

Answer: Because n represents the number of quarters, there are 9 quarters. We get the number of dimes by subtracting the number of quarters from 15.

$$9 \text{ quarters}$$
$$15 - 9 = 6 \text{ dimes}$$

Check: Verify that 9 quarters and 6 dimes are 15 coins that total $2.85.

$$9 + 6 = 15 \text{ coins} \qquad\qquad 0.25(9) + 0.10(6) \; ? \; 2.85$$
$$2.25 + 0.60 \; ? \; 2.85$$
$$2.85 = 2.85 \qquad \text{Everything checks.}$$

We could have let n be the number of dimes. If so then $15 - n$ would describe the number of quarters. The problem would have proceeded this way:

Categories	Value	Number	Amount
Quarters	0.25	$15 - n$	$0.25(15 - n)$
Dimes	0.10	$0.10n$	$0.10n$

$$\text{amount in quarters} + \text{amount in dimes} = \text{total}$$
$$0.25(15 - n) \qquad + \qquad 0.10n \qquad = 2.85$$
$$3.75 - 0.25n \qquad + \qquad 0.10n \qquad = 2.85$$
$$3.75 - 0.15n = 2.85$$
$$3.75 - 0.15n = 2.85$$
$$\underline{-3.75 \qquad\qquad -3.75}$$
$$0 - 0.15n = -0.90$$
$$\frac{-0.15n}{-0.15} = \frac{-0.90}{-0.15}$$
$$1n = 6$$
$$n = 6$$

Note: We chose to work this version of the problem with-out clearing the decimals.

Now when we combine $-0.25n$ and $0.10n$ the result, $-0.15n$ has a negative coefficient. Choosing to let n represent the number of the smaller valued coin leads to this negative coefficient.

◆ Notice we get the same answer regardless of which number of coins we let n represent.

◀ **Do Margin 8.**

6.6 Exercises

For Exercises 1–12, solve and check.

1. $4.5n + 7 = 7.9$

2. $3x - 0.8 = 1.6$

3. $15.5y + 11.8 = 21.1$

4. $12.1x + 5.6 = 10.924$

5. $16.7 - 3.5t = 17.12$

6. $0.15 = 0.6 - 1.2n$

7. $0.62k - 12.01 = 0.17k - 14.8$

8. $0.8n + 1.22 = 0.408 - 0.6n$

9. $4.1 - 1.96x = 4.2 - 1.99x$

10. $5.1m + 7.5 = 4.05 - 4.9m$

11. $0.4(8 + t) = 5t + 0.9$

12. $8k - 2.98 = 2.6(k - 0.8)$

13. $20(0.2n + 0.28) = 3.98 - (0.3 - 3.92n)$

14. $4(2.55 - x) - 5.8x = 12.2 - (8 + 11.4x)$

For Exercises 15–22, solve.

Use the table below for Exercises 15 and 16. The table shows the pricing schedule for a phone company.

Time	Rate
8 A.M.–5 P.M.	$0.12/minute
5 P.M.–10 P.M.	$0.10/minute
10 P.M.–8 A.M.	$0.09/minute

15. Below is a portion of Vicki's long distance call summary. Find the total cost of the calls.

Time of call	City	Number of minutes
1:09 P.M.	Denver, CO	15
12:32 A.M.	Austin, TX	38
7:57 A.M.	Phoenix, AZ	12

16. Below is a portion of Teshieka's long distance call summary. Find the total cost of the calls.

Time of call	City	Number of minutes
5:01 P.M.	Boston, MA	22
9:07 P.M.	New York, NY	64
7:32 A.M.	Washington, DC	24

17. Below is a summary of charges from an electric company. Find the total charges if 1433 kWh were used.

Transaction Summary	
Previous balance:	$50.94
Payment on 4/10:	$40.00
Late fee:	$ 0.55
Electric charge:	First 800 kWh @ $0.07107 per kWh
	Remaining kWh @ $0.05675 per kWh

18. Below is a summary of gas charges from an electric and gas company. Find the total charges if 38 therms were used in one month.

Transaction Summary	
Basic facilities charge:	$ 3.00
First 25 therms	@ $ 0.71171 per therm
Remaining therms	@ $ 0.64647 per therm

For Exercises 19 and 20, use the following table of a company's reimbursements for travel on business.

Reimbursements for Travel on Business
$0.35 per mile when using own vehicle
$0.15 per mile when using company vehicle
$10 breakfast when traveling between 12 A.M. and 11 A.M.
$12 lunch when traveling between 11A.M. and 3 P.M.
$15 dinner when traveling between 3 P.M. and 12 A.M.

19. Aimee uses her own vehicle on a business trip. She leaves on a Tuesday at 11:30 A.M. and notes the odometer reads 75618.4. She returns Thursday and arrives at 6:45 P.M.. At the conclusion of her trip, she finds the odometer reads 76264.1. How much should she be reimbursed for mileage and food from her company?

20. Miguel uses a company vehicle and leaves at 7:00 A.M. on a Monday. He notes the odometer reads 45981.6. He returns Friday, arriving at 2:15 P.M.. At the conclusion of the trip he notes the odometer reads 46610.8. How much should he be reimbursed for mileage and food?

21. Chan receives the following printout of charges that he plans to pay on a 90-day same-as-cash option. He wants to split the charges into three equal payments. How much is each payment?

Description	Amount
1 EPS keyboard	$1645.95
2 cables	$14.95 each
Sales tax	$83.79

22. Shannon makes the following purchases for her new apartment using a special credit agreement with the store. The credit line is interest-free as long as she pays off the balance in 6 months. After 6 months, all 6 months' worth of interest will be added onto the remaining balance. To avoid the interest, she decides to split the charges into six equal payments. How much is each payment?

Description	Amount
1 table	$354.95
4 chairs	$74.95 each
Sales tax	$32.74

For Exercises 23–34, solve.

23. 18.75 more than 3.5 times t is equal to 1.5 minus t.

24. The product of 3.2 and n is the same as the sum of n and 1.43.

25. 0.48 less than 0.2 times y is the same as 0.1 times the difference of 2.76 and y.

26. 0.6 times the sum of k and 1.5 is equal to 0.42 plus the product of 1.2 and k.

27. There is a huge crater on Mimas, one of Saturn's moons. The crater is circular and has an area of 7850 km². What is the diameter of the crater?

Trivia Bite

Grooves on the opposite side of the moon suggest that the object that impacted with Mimas to form the crater nearly tore the moon apart.

28. Crater Lake in southern Oregon is actually the top of an inactive volcano (Mount Mazama). The lake is circular and has an area of 20 mi². What is the diameter of the lake?

Trivia Bite

Crater Lake is known for its beautiful blue color. The lake was formed by snow and rainwater after the volcano became inactive. It is the second deepest lake in North America at 1932 ft. Only Great Slave Lake in northwestern Canada is deeper at 2015 ft.

29. Rebekah is a scientist studying plant life in the rain forests of South America. From her base camp, she hikes 4 mi straight south then heads east 3 mi. Because it is getting late in the day, she decides to head straight back to camp from her present location rather than backtrack the way she came. What will be the distance back to camp?

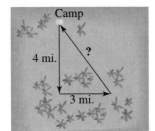

30. Jennifer, a long-distance swimmer wants to estimate the distance between two docks on a lake. She measures from the north dock directly south along the shoreline, then makes a 90° turn and measures east until she reaches the southern dock. The measurements are shown in the diagram. How far apart are the docks? If there are 5280 ft. in a mile, how many times must she swim the distance between the docks to swim 1 mi.? (Ignore the lengths of the docks.)

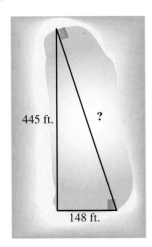

31. A 14.5 ft. beam is to be attached to a vertical beam 3.5 ft. above a horizontal support beam. How long must the horizontal support beam be to attach to the bottom of the 14.5 ft. beam?

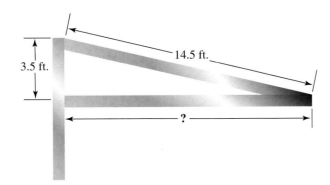

32. A steel support beam on a tower needs to be replaced. It is known that the connecting joint is 24 ft. above the ground and the base of the support beam is 8 ft. out from a point directly below the connecting joint. How long is the beam?

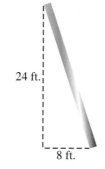

24 ft.

8 ft.

33. A fire truck has a ladder that can extend up to 60 ft. The bottom of the ladder sits atop the truck so that the bottom of the ladder is 8 ft. above the ground. If the truck parks 30 ft. away from the bottom of a building, what is the highest point the ladder can reach? If a person is in a sixth-story window awaiting rescue, can the ladder reach them if each story is 10.5 ft.?

34. A screen on a small notebook computer is 6 in. by 8 in. What is the distance along the diagonal?

35. A softball diamond is a square with bases at each corner. The distance between bases is 65 ft. What is the distance from home plate to second base?

36. An 8 ft. by 15 ft. steel frame is to be fitted with a diagonal beam welded to the corners of the frame. How long must the diagonal beam be?

For Exercises 37–40, use a four-column table to solve.

37. Latisha has a change purse in which she only keeps quarters and half dollars. She has 12 more quarters than half dollars. If the total she has in the purse is $6.75, how many of each coin does she have?

38. Jose has 22 coins totaling $4.10. If all of the 22 coins are either nickels or quarters, how many of each coin does he have?

39. Bernice sells drinks at college football games and gets paid by the number of each size drink she sells. There are two drink sizes, 12 oz. and 16 oz. The 12 oz. sells for $1.50 and the 16 oz. for $2.00. She knows she sold 65 drinks but cannot remember how many of each size. If her total sales is $109, how many of each size did she sell?

40. Arturo sells two different home security systems. He receives $225 commission for each A system and $275 commission for each B system that he sells. In one month he sells 14 systems. If he receives a total commission of $3400, how many of each system did he sell?

Puzzle Problem

A cube has a surface area of 10.14 ft.². What are the dimensions of the cube?

REVIEW EXERCISES

1. Write the word name for 24,915,000,204.

2. Round 46,256,019 to the nearest hundred-thousand.

3. Add. $4\frac{5}{8} + 8\frac{5}{6}$

4. Subtract. $12\frac{1}{4} - 7\frac{4}{5}$

5. Divide. $28x^9 \div 7x^3$

6. Explain the mistake, then work the problem correctly.

$$-3x - 18 = -24$$
$$\underline{+ 18 \quad +18}$$
$$-3x + 0 = -6$$
$$\frac{-3x}{3} = \frac{-6}{3}$$
$$x = -2$$

Defined Terms

Review the following terms and for those you do not know, study its definition on the page number next to it.

Section 6.1
Decimal notation *p. 390*
Rational number *p. 390*

Section 6.3
Scientific notation *p. 411*
Unit price *p. 415*

Section 6.4
Irrational number *p. 426*
Real numbers *p. 426*

The Real Number System

Real Numbers	
Irrational Numbers: Numbers that cannot be expressed in the form $\frac{a}{b}$ where a and b are integers and $b \neq 0$, such as: $-\sqrt{2}, -\sqrt{3}, \sqrt{0.8}, \pi$.	**Rational Numbers:** Numbers that can be expressed in the form $\frac{a}{b}$ where a and b are integers and $b \neq 0$, such as: $-4\frac{3}{4}, -\frac{2}{3}, 0.018, 0.\overline{3}, \frac{5}{8}$.

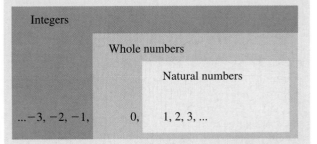

Integers
Whole numbers
Natural numbers

$\dots -3, -2, -1,$ $0,$ $1, 2, 3, \dots$

Our Place Value System

Decimal Point
↓

Whole numbers			Fractions							
Hundreds	Tens	Ones	Tenths	Hundredths	Thousandths	Ten-thousandths	Hundred-thousandths	Millionths	Ten-millionths	Hundred-millionths

Procedures, Rules, and Key Examples

Procedures/Rules	Key Example(s)
Section 6.1 **To write a decimal number as a fraction or mixed number:** 1. Write all digits to the left of the decimal (whole digits) as the whole/integer in a mixed number. 2. Write all digits to the right of the decimal (fraction digits) in the numerator of a fraction. 3. Write the last place value in the denominator. 4. Reduce/Simplify.	$0.24 = \dfrac{24}{100} = \dfrac{6}{25}$ $16.082 = 16\dfrac{82}{1000} = 16\dfrac{41}{500}$

To write the word name for decimal digits (fraction part of a decimal number):
1. Write the word name for digits as if they represented a whole number.
2. Write the name of the last place value.

0.0351
= three hundred fifty-one ten-thousandths.

To write a word name for a decimal number with both whole/integer and fraction parts:
1. Write the name of the whole/integer number part.
2. Write the word *and* for the decimal point.
3. Write the name of the fractional part.

147.29
= one hundred forty-seven and twenty-nine hundredths.

To determine which of two decimal numbers is larger:
1. Compare the digits in the numbers from left to right until you reach two different digits in the same place value.
2. The larger of these digits is in the larger number.

Note: If both numbers are negative, the opposite applies; the smaller digit is in the larger number because the number is closer to zero.

Use $<$, $>$, or $=$ to make a true sentence.
$$0.517 > 0.516$$
$$-0.4912 < -0.4812$$

To round a number to a given place value, consider the digit to the right of the desired place value.

If this digit is 5 or higher, round up.
If this digit is 4 or less, round down.

Round 4.619 to the nearest tenth.
Answer: 4.6

Round 18.5141 to the nearest whole.
Answer: 19

Round 0.00127 to the nearest ten-thousandth.
Answer: 0.0013

Section 6.2

To add decimal numbers:
1. Stack the numbers by place value. (Line up the decimal points.)
2. Add the digits in the corresponding place values.
3. Place the decimal point in the sum so that it is aligned with the decimal points in the addends.

$2.035 + 15.98$

$$\begin{array}{r} \overset{1\ 1}{2.035} \\ +\ 15.980 \\ \hline 18.015 \end{array}$$

To subtract decimal numbers:
1. Stack the number with the larger absolute value on top so that the place values align. (Tip: Line up the decimal points.)
2. Subtract the digits in the corresponding place values.
3. Place the decimal point in the difference so that it is aligned with the decimal points in the minuend and subtrahend.

$60.043 - 4.798$

$$\begin{array}{r} \overset{\ \ \ 13}{\overset{5\,9\ 9\cancel{3}1\cancel{3}}{60.043}} \\ -\ 4.798 \\ \hline 55.245 \end{array}$$

To combine like terms, add or subtract the coefficients and keep the variables the same.

$5.1x^3 + 2.91x - 7x^2 + 9.8 - 3.4x - 3.8x^2$
$= 5.1x^3 - 10.8x^2 - 0.49x + 9.8$

Section 6.3

To multiply decimal numbers:
1. Stack the numbers. It is not necessary to align the decimal points in the factors.
2. Multiply as if the numbers were whole numbers.
3. Place the decimal in the product so that it has the same number of decimal digits as the total number of decimal digits in the factors.

6.45×2.8

$$\begin{array}{r} \overset{1\ 1}{6.45} \\ \times\ 2.8 \\ \hline 5\ 160 \\ 12\ 90 \\ \hline 18.060 \end{array}$$

To change from scientific notation with a positive exponent to standard notation for a given number, move the decimal point to the right the number of places indicated by the exponent.

$3.59 \times 10^7 = 35{,}900{,}000$
$-4.2 \times 10^5 = -420{,}000$

Procedures/Rules (cont.)	Key Example(s)

To write a standard form number greater than or equal to 1 in scientific notation:
1. Place a decimal in the number so that it expresses a number whose absolute value is greater than or equal to 1 but less than 10. (Tip: Place the decimal point to the right of the first nonzero digit.)
2. Count the number of places to the right of the new decimal position. This number is the power of 10.
3. Delete 0's to the right of the last nonzero digit.

$$4{,}791{,}000{,}000 = 4.791 \times 10^9$$

$$-87{,}050{,}000{,}000{,}000 = -8.705 \times 10^{13}$$

Section 6.4

To divide decimal numbers by a whole number:
1. Set up a long-division problem.
2. Divide the divisor into the dividend as if both numbers were whole numbers. Be sure to align the digits in the quotient properly.
3. Write the decimal in the quotient directly above its position in the dividend.
4. Write extra 0 digits to the right of the last nonzero digit as needed until you get a remainder of 0, or until a digit or group of digits repeats without end in the quotient.

To indicate repeated digits, write a repeat bar over the digit or group of digits that repeat.

$20.4 \div 6$

$$
\begin{array}{r}
3.4 \\
6\,\overline{)20.4} \\
-18 \\
\hline
24 \\
-24 \\
\hline
0
\end{array}
$$
Answer: 3.4

$3.68 \div 11$

$$
\begin{array}{r}
0.33454545\ldots \\
11\,\overline{)3.6800000} \\
-33 \\
\hline
38 \\
-33 \\
\hline
50 \\
-44 \\
\hline
60 \\
-55 \\
\hline
50
\end{array}
$$
Answer $= 0.33\overline{45}$

To divide when the divisor is a decimal number:
1. Set up a long-division problem.
2. Move the decimal point in the divisor to the right enough places to make the divisor an integer.
3. Move the decimal point in the dividend the same number of places.
4. Divide the divisor into the dividend as if both numbers were whole numbers. Be sure to align the digits in the quotient properly.
5. Write the decimal point in the quotient directly above its position in the dividend.

$50.8 \div 0.02 = 5080 \div 2$

$$
\begin{array}{r}
2540 \\
2\,\overline{)5080} \\
-4 \\
\hline
10 \\
-10 \\
\hline
08 \\
-8 \\
\hline
0
\end{array}
$$
Answer $= 2540$

$0.694 \div 0.22 = 69.4 \div 22$

$$
\begin{array}{r}
3.154545\ldots \\
22\,\overline{)69.400000} \\
-66 \\
\hline
34 \\
-22 \\
\hline
120 \\
-110 \\
\hline
100 \\
-88 \\
\hline
12
\end{array}
$$
Answer $= 3.1\overline{54}$

To write a fraction as a decimal:
1. Divide the denominator into the numerator.

Note: Continue dividing until there is no remainder or a digit or group of digits repeats without end. If a digit or group of decimal digits repeats forever, write a repeat bar over the digit or group of digits that repeat.

Write $\dfrac{5}{8}$ as a decimal.

Answer: $\dfrac{5}{8} = 0.625$

$$
\begin{array}{r}
0.625 \\
8\,\overline{)5.000} \\
-48 \\
\hline
20 \\
-16 \\
\hline
40 \\
-40 \\
\hline
0
\end{array}
$$

Procedures/Rules (cont.)	Key Example(s)

To write a mixed number as a decimal:
1. Write the whole/integer as the whole part of a decimal number (to the left of a decimal point).
2. Divide the denominator into the numerator to get the decimal digits.

Note: Continue dividing until there is no remainder or a digit or group of digits repeats without end. If a digit or group of decimal digits repeat forever, write a repeat bar over the digit or group of digits that repeat.

Write $14\frac{2}{3}$ as a decimal.

Answer: $14.\overline{6}$

$$\begin{array}{r} 0.66... \\ 3\overline{)2.00} \\ -18 \\ \hline 20 \\ -18 \\ \hline 20 \end{array}$$

To write a number expressed in scientific notation with a negative exponent in standard form, move the decimal point to the left the same number of places as the absolute value of the negative exponent.

$5.671 \times 10^{-6} = 0.000005671$

$-2.6 \times 10^{-9} = -0.0000000026$

To write a decimal number whose absolute value is less than 1 and greater than 0 in scientific notation:
1. Place a decimal point in the number so that it expresses a number whose absolute value is between 1 and 10.
2. Count the number of decimal places between the original decimal position and the new decimal position.
3. Write this number of places as a negative exponent with a base of 10.

$0.00000902 = 9.02 \times 10^{-6}$

$-0.0002194 = -2.194 \times 10^{-4}$

Section 6.5

To simplify numerical expressions that contain both fractions and decimals:
1. Write all decimals as fractions. (Method 1) or
   Write all fractions as decimals. (Method 2) or
   Write decimals over 1 in fraction form. (Method 3)
2. Follow the order of operations agreement.

(Method 3 illustrated here)
$$\frac{2}{3}(6.09) - \frac{3}{8} = \frac{2}{3} \cdot \frac{6.09}{1} - \frac{3}{8}$$
$$= \frac{12.18}{3} - \frac{3}{8}$$
$$= 4.06 - 0.375$$
$$= 3.685$$

Section 6.6

Procedure: To solve equations:
1. Simplify both sides of the equation as needed.
   a. Distribute to clear parentheses.
   b. Clear fractions and/or decimals by multiplying both sides by the LCD of all denominators. In the case of decimals the LCD is the power of 10 with the same number of 0 digits as decimal places in the number with the most decimal places.
   c. Combine like terms.
2. Use the addition/subtraction principle so that all variable terms are on one side of the equation and all constants are on the other side.
3. Use the multiplication/division principle to clear any remaining coefficient.

$$8.3 - (2.4x + 0.7) = 14.6x - 5(x + 6.16)$$
$$8.3 - 2.4x - 0.7 = 14.6x - 5x - 30.8$$
$$10(8.3 - 2.4x - 0.7) = (14.6x - 5x - 30.8)10$$
$$83 - 24x - 7 = 146x - 50x - 308$$
$$76 - 24x = 96x - 308$$
$$76 - 24x = 96x - 308$$
$$\underline{+24x \quad +24x}$$
$$76 + 0 = 120x - 308$$
$$\underline{76 = 120x - 308}$$
$$\underline{+308 \qquad +308}$$
$$384 = 120x + 0$$
$$384 = 120x$$
$$\frac{384}{120} = \frac{120x}{120}$$
$$3.2 = 1x$$
$$3.2 = x$$

Check:
$$8.3 - (2.4x + 0.7) = 14.6x - 5(x + 6.16)$$
$$8.3 - [2.4(3.2) + 0.7] \ ? \ 14.6(3.2) - 5(3.2 + 6.16)$$
$$8.3 - [7.68 + 0.7] \ ? \ 46.72 - 5(9.36)$$
$$8.3 - [7.68 + 0.7] \ ? \ 46.72 - 46.8$$
$$8.3 - 8.38 \ ? \ 46.72 - 46.8$$
$$-0.08 = -0.08$$
Because both sides match, 3.2 is correct.

Formulas

Total cost given a unit price: $T = qp$

Volume of a pyramid: $V = \frac{1}{3}lwh$

Volume of sphere: $V = \frac{4}{3}\pi r^3$

Volume of a cylinder: $V = \pi r^2 h$

Volume of a cone: $V = \frac{1}{3}\pi r^2 h$

Review Exercises

For Exercises 1–6, answer true or false.

1. Every rational number can be expressed as a decimal number.

2. Every decimal number is a rational number.

3. $5.\overline{16}$ means $5.16666\ldots$

4. $0.11 - 0.095$ equals a positive number.

5. $\sqrt{90} \approx 7.487$

6. 0.0851 is between 0.085 and 0.086.

7. Explain the mistake.

$$\begin{array}{r} 0.24 \\ \times\ 0.13 \\ \hline 72 \\ 24 \\ \hline 3.12 \end{array}$$

8. Explain in your own words how to add decimal numbers.

9. To write a fraction as a decimal number, divide the _____ into the _____.

10. Explain in your own words how to write a decimal number as a fraction.

11. Write the word name.

   a. 24.39   **b.** 581.459   **c.** 0.2917

12. Write as a fraction in lowest terms.

   a. 0.8   **b.** 0.024   **c.** -2.65

13. Use $<$, $>$, or $=$ to make a true statement.

   a. 2.001 ? 2.0009 **b.** -0.016 ? -0.008

14. Round 31.8056 to the nearest:

   a. whole **b.** tenth **c.** hundredth **d.** thousandth

15. Add or subtract.

   a. $24.81 + 54.2 + 9.005 =$ **b.** $9.005 - 1.089 =$

   c. $6.7 + (-12.9) =$ **d.** $(-20) - (-13.88) =$

   e. $-1.082 + 29.1 - 4.9 =$ **f.** $14.1 - (-6.39) =$

16. Simplify.

   a. $9y^4 + 8xy - 12.2y^2 + 2.1y^4 - 16xy + 3.8$

   b. $(0.3a^2 + 8.1a + 6) + (5.1a^2 - 2.9a + 4.33)$

   c. $(2.6x^3 - 98.1x^2 + 4) - (5.3x^3 - 2.1x + 8.2)$

17. Multiply.

   a. $(3.44)(5.1) =$ **b.** $(1.5)(-9.67) =$

   c. $(1.2)^2 =$ **d.** $(-0.2)^3 =$

   e. $(2.5h)(0.01hk^3)$ **f.** $(-0.9x^3y)(4.2x^5yc)$

   g. $0.2(0.6n^3 - 1.4n^2 + n - 18)$ **h.** $-1.1(2.8a - 4.9b)$

   i. $(4.5a - 6.1)(2a - 0.5)$ **j.** $(1.3x + 4.2)(1.3x - 4.2)$

18. Divide.

   a. $21 \div 4 =$

   b. $3.6 \div 9 =$

   c. $17.304 \div 2.06 =$

   d. $-54 \div (-0.09) =$

   e. $3.18x^7 \div 0.4x^5$

   f. $-2.38a^4bc \div 6.8ab$

19. Find the square root.

   a. $\sqrt{0.64}$

   b. $\sqrt{1.21}$

20. Approximate the square root to the nearest hundredth.

   a. $\sqrt{148}$

   b. $\sqrt{1.6}$

21. Write as a decimal.

   a. $\dfrac{3}{5}$

   b. $\dfrac{1}{6}$

   c. $6\dfrac{1}{4}$

   d. $-4\dfrac{2}{11}$

22. Write in standard form.

   a. 3.58×10^9

   b. -4.2×10^5

   c. 6.51×10^{-7}

   d. -2.19×10^{-6}

23. Write in scientific notation.

   a. $9,200,000,000$

   b. $-103,000$

   c. 0.000019

   d. -0.000000406

24. Simplify:

   a. $4.59 - 2.8 \cdot 6.7 + \sqrt{1.44}$

   b. $3.6 + (2.5)^2 - 1.5(9 - 0.4)$

   c. $\dfrac{3}{4}(0.86) - \dfrac{3}{8}$

   d. $\dfrac{1}{6}(0.24) + \dfrac{2}{3}(-3.3)$

25. Find the radius, diameter, circumference, and area of the circle.

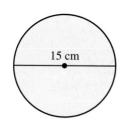

15 cm

26. Find the area.

   a.

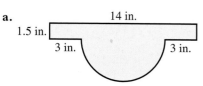

14 in.

1.5 in.

3 in.   3 in.

   b.

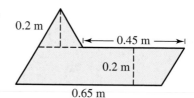

0.2 m

0.45 m

0.2 m

0.65 m

27. Find the area of the shaded region.

   a. The inner circle has a diameter of 3 in.

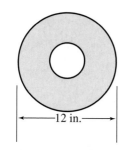

12 in.

   b.

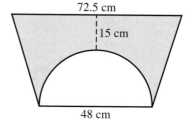

72.5 cm

15 cm

48 cm

28. Find the volume.

a.
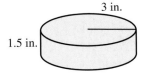

b.
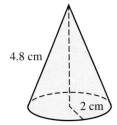

c.
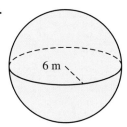

d.
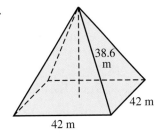

e.
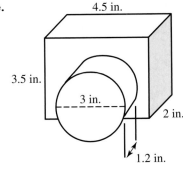

29. Find the volume of the shaded region.

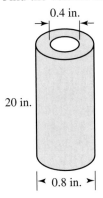

30. Solve and check.

  a. $x + 0.56 = 2.1$

  b. $y - 0.58 = -1.22$

  c. $0.8y = 2.4$

  d. $-1.6n = 0.032$

  e. $4.5k - 2.61 = 0.99$

  f. $51.2 = 0.4a + 51.62$

  g. $2.1x - 12.6 = 1.9x - 10.98$

  h. $3.6n - (n - 13.2) = 10(0.4n + 0.62)$

For Exercises 31–38, solve.

31. Below is Sonja's checkbook register. Find her balance.

Date	No.	Transaction Description	Subtractions		Additions		Balance	
3 - 21	1201	Papa John's	16	70			150	20
3 - 22	1202	Food Lion	84	61				
3 - 25	1203	Ford Motor Credit	245	16				
3 - 25	1204	Electric & Gas	67	54				
3 - 25	1205	Fleet Mortgage	584	95				
3 - 26		Deposit			1545	24		

32. Naja has a student loan balance of $2854.82. If he agrees to make equal monthly payments over a five-year period, how much will each payment be? How much is the last payment?

33. Use BMI $= \frac{w}{h^2} \cdot 705$ to calculate a person's body mass index if the person weighs 210 lb. and is 72 in. tall.

34. Below is a portion of Ngyen's long distance call summary. Find the total cost of the calls using the rate table provided.

Time of call		City	Number of minutes
2:49	P.M.	Sacramento, CA	45
7:42	A.M.	Miami, FL	28
8:57	P.M.	Miami, FL	16

Time	Rate
8 A.M.−5 P.M.	$0.12/minute
5 P.M.−10 P.M.	$0.10/minute
10 P.M.−8 A.M.	$0.09/minute

35. A 12 ft. board is resting against a tree. The bottom of the board is 3 ft. from the base of the tree. How far is the top of the board from the base of the tree? (Assume the tree forms a right angle with the ground.)

36. An electrician charges $45 for all service calls then $4.50 for every 15 minutes spent working. If Brittany has $72, how long can she afford for the electrician to work?

37. Elissa has a change purse that contains only nickels and dimes. She has 9 more dimes than nickels. If she has a total of $2.40 in the purse, how many of each coin is there?

38. Conrad makes two different sizes of glass figurines. He sells the larger size for $14.50, and the smaller size sells for $9.50. If he sells a total of 24 of the glass crafts at a fair and makes a total of $273, how many of each size did he sell?

1. Write the word name for 56.789.

2. Write 0.68 as a fraction in lowest terms.

3. Use $<$, $>$, or $=$ to make a true statement.

$$-0.0059 \ ? \ -0.0058$$

4. Round 2.0915 to the nearest:

  a. whole   **b.** tenth   **c.** hundredth   **d.** thousandth

5. $4.591 + 34.6 + 2.8 =$ **6.** $9.005 - 1.089 =$

7. $(4.5)(-8.61) =$ **8.** $4.8 \div 16 =$

9. $(-55.384) \div (-9.2) =$ **10.** $\sqrt{0.81} =$

11. Write $3\frac{5}{6}$ as a decimal.

12. a. Write 2.97×10^{-6} in standard form.

    b. Write $-35{,}600{,}000$ in scientific notation.

13. Simplify: $3.2 + (0.12)^2 - 0.6(2.4 - 5.6)$

14. Evaluate $\frac{1}{2} mv^2$ when $m = 12.8$ and $v = 0.2$

15. Simplify. $(2.6x^3 - 98.1x^2 + 4) - (5.3x^3 - 2.1x + 8.2)$

16. Multiply.

  a. $2.5x^3 \cdot 0.6xy$ **b.** $(4.5a - 6.1)(2a - 0.5)$

17. Solve and check. $6.5t - 12.8 = 13.85$

18. Solve and check. $1.8(k - 4) = -3.5k - 4.02$

19. A circle has a radius of 12 ft. Use 3.14 to approximate π.

  a. Find the circumference

  b. Find the area.

20. Find the volume. Use 3.14 to approximate π.

  a.
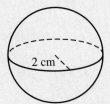

  b.
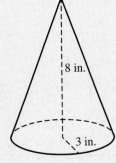

1. _____

2. _____

3. _____

4. _____

5. _____

6. _____

7. _____

8. _____

9. _____

10. _____

11. _____

12. _____

13. _____

14. _____

15. _____

16. _____

17. _____

18. _____

19. _____

20. _____

21. _____

21. Find the area.

22. _____

22. Below is a copy of Dedra's bill at a restaurant. She pays with a $20 bill. How much change should she receive?

Guest Check	
Club sandwich	7.95
Salad	2.95
Bev.	1.50
Tax	0.57

23. _____

23. Bill receives the following printout of charges that he plans to pay on a 90-day same-as-cash option. He wants to split the charges into three equal payments. How much is each payment?

Description	Amount
1 TV	$385.95
2 speakers	$59.95 each
Sales tax	$30.35

24. _____

24. Carlton has a 20-ft. ladder that he places 5 ft. from the base of his house. How high up the side of the house will the ladder reach?

25. _____

25. Yolanda sells two different kinds of cakes, round-layered cakes and sheet cakes. The round-layered cakes sell for $14 and the sheet cakes sell for $12.50. In one month, she sold 12 cakes and made a total of $162. How many of each cake did she sell?

Cumulative Review Exercises

For Exercises 1–6, answer true or false.

1. $4x^3 - 7$ is a binomial.

2. 221 is a composite number.

3. π is a rational number.

4. The conjugate of $x - 9$ is $-x + 9$.

5. The addition/subtraction principle of equality states that adding or subtracting the same amount on both sides of an equation will not affect its solution(s).

6. $-0.118 < -0.119$

7. Any number other than 0 divided by 0 is _____.

8. When combining like terms, we add or subtract the _____ and keep the _____ the same.

9. Explain the mistake. $(x + 2)(x + 3) = x^2 + 6$

10. Explain in your own words how to divide fractions.

11. Write the word name for 29.6081.

12. Graph $-2\frac{3}{4}$ on a number line.

13. Round 2.0185 to the nearest hundredth.

14. Estimate $2439 \div 49$ by rounding so that there is only one nonzero digit in each number.

15. What is the degree of $3b^2 - b^4 + 9b + 4b^6 - 17$?

16. Find the prime factorization of 840.

17. Find the LCD of 56 and 42.

For Exercises 18–27, simplify.

18. $\{9 - 4[6 + (-16)]\} \div 7$

19. $(-2)^5 - 4\sqrt{16 + 9}$

20. $-\dfrac{48}{60}$

21. $\dfrac{18x^2y}{24xy^4}$

22. $-\dfrac{1}{9} \cdot 5\dfrac{1}{6}$

23. $(5)^{-2}$

24. $-\dfrac{12n^2}{13m} \div \dfrac{-15p}{26m}$

25. $4\dfrac{5}{8} + 7\dfrac{1}{2}$

26. $\dfrac{x}{6} + \dfrac{3}{4}$

27. $0.48 + 0.6(4\frac{1}{3}) + \sqrt{0.81}$

28. Evaluate $4h^2 - k^3$ when $h = \frac{3}{4}$ and $k = -0.2$.

29. Combine like terms and write your answer in descending order.

$2.6x^3 - \dfrac{1}{4}x^2 + 3.8 - x^3 + x^2 - \dfrac{1}{4}$

30. Subtract. $(14m^3 - 5.7m^2 + m + 7) - (9.1m^3 + m^2 - 11.6)$

31. Find the missing factor. $4x^3 \cdot \boxed{?} = -48x^5y$

32. Multiply. $(4y - 3)(y + 2)$

33. Write 7.68×10^8 in standard form. Then write the word name.

34. Factor. $30mn^5 + 15mn^2 - 25mn$

For Exercises 35–38, solve and check.

35. $y - 15.2 = 9.5$

36. $-0.06k = 0.408$

37. $\dfrac{2}{3}x + \dfrac{1}{4} = \dfrac{5}{6}$

38. $2n + 8 = 5(n - 3) + 2$

For Exercises 39–50, solve.

39. The temperature at sunset was reported to be 24°F. By midnight it is reported to be $-8°$F. What is the amount of the decrease?

40. The financial report for a business indicates that the total revenue for 1998 was $2,609,400 and the total costs were $1,208,500. What was the net? Did the business have a profit or loss?

41. Sherry has an annual salary of $23,272. What is her gross monthly salary?

42. a. Write an expression for the volume of the box shown.

    b. Find the volume of the box if w is 6 cm.

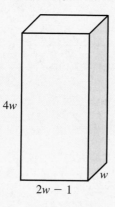

43. A box company shapes flat pieces of cardboard into boxes. The flat pieces have an area of 21.4 ft.^{2}. If the length is 2 ft. and the width is 1.5 ft., find the height. (Use the formula $SA = 2lw + 2lh + 2wh$.)

44. Find the area of the shape.

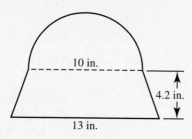

45. Five minus twice the difference of x and nine is the same as seven more than the product of six and x.

46. A study at a business indicates that $\frac{3}{4}$ of its employees have taken business supplies for use in other than the business. Of those employees that have taken supplies, $\frac{2}{3}$ have been with the company more than 5 years. What fraction of all employees have taken supplies and been with the company for more than 5 years?

47. A question on a survey allows for a response of yes, no, or no opinion. $\frac{1}{6}$ of the respondents said yes and $\frac{3}{5}$ said no. What fraction had no opinion?

48. An entrance to an amusement park is to be an isosceles triangle. The equal sides are to be 12.5 ft. longer than the base. If the perimeter must be 70 ft., then what will be the dimensions of the triangle?

49. A 20 ft. ladder is resting against the side of a building. The bottom of the ladder is 6 ft. from the base of the building. How far is the top of the ladder from the base of the building?

50. Cedrick makes two different sizes of clay bowls. He sells the larger size for $12.50, and the smaller size sells for $8.50. If he sells a total of 20 of the bowls at a fair and makes a total of $206, how many of each size did he sell? (Use a four-column table.)

Ratios, Proportions, and Measurement

Working with a Group

One of the most rewarding ways to grow as a student is to form a study group. When you form a group, keep in mind that smaller is better, usually four people or fewer. By forming a group, you have another source for information. Group members can answer questions when you are unable to reach your instructor, or they can offer notes if you miss a class.

If you meet with your group to go over homework, it is best to have completed as much of the homework as possible on your own. Then you can go over problems together that you or other group members did not understand. You will find that explaining things to each other will strengthen your understanding of the material.

If you meet with your group to prepare for tests, again, it is best to have prepared on your own ahead of time. Then using your study sheets, go from person to person quizzing each on procedures and rules. When quizzing someone or watching someone work a problem, be careful that you don't offer tips or hints while they speak or work. Let them say or do the whole procedure, even if you note a mistake. Only correct after they've completed the procedure. Remember, nobody will be there to hint or give clues that something is wrong during a test.

Keep in mind that your study group cannot be expected to carry you through a course. The responsibility for your success remains with you. However, the support, encouragement, and accountability you gain from a study group can be a valuable part of achieving that success.

"*Alone we can do so little, together we can do so much.*"
—Helen Keller

"*Never doubt that a small group of thoughtful, committed citizens can change the world. Indeed, it is the only thing that ever has.*"
—Margaret Mead

7.1 Ratios, Probability, and Rates

OBJECTIVES

1 Write ratios in simplest form.
2 Calculate probabilities.
3 Calculate unit ratios.
4 Write rates in simplest form.
5 Use unit price to determine the better buy.

OBJECTIVE **1** *Write ratios in simplest form.*

Suppose 18 females and 12 males are on a committee, a total of 30 people. We can use a fraction to compare the number of males to the total number of people:

$$\text{The number of males out of the total number of people} = \frac{12}{30} = \frac{2}{5}$$

Or we can use a fraction to compare the number of females to the total number of people:

$$\text{The number of females out of the total number of people} = \frac{18}{30} = \frac{3}{5}$$

A fraction or quotient comparison of two quantities is called a **ratio.**

> **DEFINITION** **Ratio:** A quotient comparison of two quantities.

In the above example, we could write other ratios:

$$\text{The ratio of males to females} = \frac{12}{18} = \frac{2}{3}$$

$$\text{The ratio of females to males} = \frac{18}{12} = \frac{3}{2}$$

Note: It is common to see the colon used to indicate ratios.

$$2:3 \text{ means } \frac{2}{3}$$

$$3:2 \text{ means } \frac{3}{2}$$

In mathematics we use only the fractional notation.

Notice a key word indicating ratio is the word *to*. In translating, think of the word *to* as the fraction line. The amount described to the left of the word *to* is the numerator of the ratio and the amount to the right of the word *to* is the denominator.

> **Procedure** *To translate a phrase involving the word* ratio:
> **1.** Find the word *to*
> **2.** Write the amount expressed to the left of the word *to* in the numerator.
> **3.** Write the amount expressed to the right of the word *to* in the denominator.

EXAMPLE 1 The pitch, or degree of slant, of a roof is to have a 24 in. vertical rise for every 40 in. of horizontal length. Write the ratio of vertical rise to horizontal length in simplest form.

Solution: Use the word *to* to guide the translation. Then simplify.

Translation: "vertical rise to horizontal length"

$$\downarrow \qquad \qquad \downarrow$$
$$24 \text{ in. to } 40 \text{ in.}$$
$$\hookrightarrow \frac{24 \text{ in.}}{40 \text{ in.}} \longleftarrow$$

Now simplify. $\dfrac{24 \text{ in.}}{40 \text{ in.}} = \dfrac{3}{5}$

Note: We can treat the units as if they were factors. Because the units are the same in the numerator and denominator, we can eliminate them as if we divided out common factors. This idea will be used when we discuss dimensional analysis later in the chapter.

This means that measuring out 40 in. then up 24 in. is equivalent to measuring out 5 in. then up 3 in. The pitch of the roof is the same using either ratio.

Connection
The ratio of the vertical rise to the horizontal length is known as the *slope of an incline*. Slope is often referred to as the ratio of *rise to run* or *rise over run*.

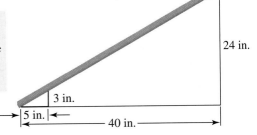

24 in.

3 in.

5 in.

40 in.

Do Margin 1. ▶

◆ **Margin 1**

Stacie measures 40 in. from her feet to her navel and 24 in. from her navel to the top of her head. Write the ratio of the smaller length to the larger length in simplest form.

Sometimes ratios can have fractions or decimals in the numerators and/or denominators. This can make simplifying more challenging.

EXAMPLE 2   For every $2\frac{1}{2}$ rotations of the pedals on a bicycle, the back wheel rotates $3\frac{1}{4}$ times. Write the ratio of the number of pedal rotations to the number of back wheel rotations.

Solution: Use the word *to* to guide the translation. Then simplify.

Translation: "the number of pedal rotations to number of back wheel rotations"

$$2\frac{1}{2} \text{ to } 3\frac{1}{4} = \frac{2\frac{1}{2}}{3\frac{1}{4}}$$

Trivia Bite

Gears in machines operate on the principle of ratios. In an automobile, the number of revolutions of the drive shaft is converted by a gear to a number of tire revolutions. In a lower gear, the engine is turning at many more revolutions than the tire, while in a higher gear the opposite is true.

This is a complex fraction. We learned how to simplify complex fractions in Chapter 5. To simplify we can simply divide the mixed numbers.

$$\frac{2\frac{1}{2}}{3\frac{1}{4}} = 2\frac{1}{2} \div 3\frac{1}{4}$$
$$= \frac{5}{2} \div \frac{13}{4}$$
$$= \frac{5}{\cancel{2}} \cdot \frac{\cancel{4}^{\,2}}{13}$$
$$= \frac{10}{13}$$

This means that every 10 rotations of the pedals result in 13 rotations of the back wheel.

Conclusion: To simplify a ratio of two fractions or mixed numbers, divide the fractions/mixed numbers.

◆ **Margin 2**

A recipe calls for $2\frac{1}{2}$ cups of milk and $\frac{3}{4}$ of a cup of sugar. Write the ratio of milk to sugar in simplest form.

Procedure   *To simplify a ratio of two fractions or mixed numbers, divide the fractions or mixed numbers.*

Do Margin 2. ▶

Answer to Margin 1: $\frac{3}{5}$

Answer to Margin 2: $\frac{10}{3}$

Consider simplifying ratios of decimal numbers.

EXAMPLE 3 The diameter of a piece of pipe is 0.42 m. The length is 3.5 m. Write the ratio of the diameter to the length in simplest form.

Solution: Use the word *to* to guide the translation. Then simplify.

Translation: "diameter to length"

$$0.42 \text{ to } 3.5 = \frac{0.42}{3.5}$$

To write simplest form, we must first write the fraction in terms of integers and clear the decimals. We learned in Chapter 5 that we can write an equivalent fraction by upscaling. In Chapter 6 we learned that we can clear decimals by multiplying by an appropriate power of 10. Because 0.42 has the most decimal places, it determines the power of 10 that we use. Therefore, in this case, we multiply both the numerator and denominator by 100.

$$\frac{0.42}{3.5} = \frac{0.42(100)}{3.5(100)} = \frac{42}{350} = \frac{3}{25}$$

Conclusion: We can clear decimals in a ratio by multiplying the numerator and denominator by an appropriate power of 10.

> **Procedure** *To clear decimals in a ratio, multiply the numerator and denominator by an appropriate power of 10 as determined by the decimal number with the most decimal places.*

◀ **Do Margin 3.**

Tip

Because we multiply the numerator and denominator by the same power of 10, the decimal will move to the right the same number of places in the numerator and denominator.

◆ **Margin 3**

An ant weighing 0.004 g can carry an object weighing up to 0.02 g. Write the ratio of the object's weight to the ant's weight in simplest form.

Trivia Bite

Various parts of the human body are in a fixed ratio to each other. This ratio is called the Golden ratio. The Golden ratio is an irrational number with an approximate value of 0.618 to 1. The following ratios of measurements on the human body are in Golden ratio:

$$\frac{\text{Navel to top of head}}{\text{Floor to navel}} \qquad \frac{\text{Center of eyes to top of head}}{\text{Bottom of chin to center of eyes}} \qquad \frac{\text{Shoulders to top of head}}{\text{Navel to shoulders}} \qquad \frac{\text{Bottom to tip of middle finger}}{\text{Wrist to bottom of middle finger}}$$

Of course, everyone's body is slightly different so when measuring, not everyone will get exactly 0.618. This is quite normal.

The Golden ratio also appears in plant growth, art, architecture, and musical instruments. The front of the Greek building known as the *Parthenon* fits inside a rectangle with a ratio of width to length that is in the Golden ratio. The ratio of the length of the neck of a violin to the length of its body is in the Golden ratio.

OBJECTIVE 2 *Calculate probabilities.*

Ratios are used in calculating probabilities. Suppose a friend approached you with a coin and suggested the following game. If it lands heads up you win; if it lands tails up your friend wins. What is the likelihood that you will win? Because there are two possible outcomes, heads or tails, and you win with one of those two possibilities, we say you have a 1 in 2 chance of winning.

Note that we can write "1 in 2" as the fraction $\frac{1}{2}$. The fraction $\frac{1}{2}$ is the **theoretical probability** that you will win.

DEFINITION **Theoretical probability:** The ratio of the number of favorable outcomes to the total number of possible outcomes.

$$\text{The probability of a coin landing heads up} = \frac{1}{2} \begin{matrix} \leftarrow \textbf{1 side is heads} \\ \textbf{out of} \\ \leftarrow \textbf{2 possible outcomes} \end{matrix}$$

Notice that your friend has an equal chance of winning because there is 1 tail out of 2 possible outcomes.

Answer: $\frac{5}{1}$

The probability of a coin landing tails up $= \dfrac{1}{2}$

Procedure *To write a probability:*

1. Write the number of favorable outcomes in the numerator.
2. Write the total number of possible outcomes in the denominator.
3. Simplify.

Note that this does not mean that if your friend wins one time, you are guaranteed to win the next time. In fact, your friend could have a streak of several wins before you win, if at all. On any given day, the number of times you win may be more or less than $\dfrac{1}{2}$ of the time. However, over a lifetime, if you were to look over the record of how many times you won, it would be close to $\dfrac{1}{2}$ of the time. Taken to extreme, if you could play the game forever, it would be exactly $\dfrac{1}{2}$ of the time.

Some of the problems we will explore involve things like dice and cards. A six-sided game die is a cube with dots numbering each of the six sides.

A standard deck of cards contains 52 cards with four suits of 13 cards. There are two red suits, diamonds and hearts, and 2 black suits, spades and clubs. The 13 cards are as follows: Ace, 2, 3, 4, 5, 6, 7, 8, 9, 10, Jack, Queen, King. The Jack, Queen, and King are referred to as *face cards*.

EXAMPLE 4 Write each probability in simplest form.

a. What is the probability of rolling a 2 on a six-sided die?

Solution: Write the ratio of the number of 2's to the total number of outcomes on a six-sided die. There is only 1 two on a six-sided die.

$$P = \dfrac{1}{6} \quad \begin{array}{l} \leftarrow \text{ Number of 2's, on a six-sided die.} \\ \leftarrow \text{ Number of possible outcomes.} \end{array}$$

b. What is the probability of tossing a 7 on a six-sided die?

Solution: Write the ratio of the number of 7's to the total number of outcomes on a six-sided die. There are no 7's on a normal six-sided die.

$$P = \dfrac{0}{6} = 0$$

c. What is the probability of tossing a 1, 2, 3, 4, 5, or 6 on a six-sided die?

Solution: There are six favorable outcomes out of six possible outcomes.

$$P = \dfrac{6}{6} = 1$$

d. What is the probability of drawing a red King out of a standard deck of 52 cards?

Solution: Write the ratio of the number of red Kings to the total number of cards in the deck. There are two red Kings, the King of diamonds and King of hearts, in a standard deck of 52 cards.

$$P = \dfrac{2}{52} = \dfrac{1}{26}$$

Margin 4

Write each probability in simplest form.

a. What is the probability of selecting a 2 from a standard deck of cards?

b. Suppose 28 males and 30 females put their names in a hat for a drawing. If the winner's name is selected from the hat at random, what is the probability that the winner will be male?

Margin 5

A family has a total debt of $7842 and a gross income of $34,240. Write a unit ratio of debt to income and interpret the results.

Reminder

0.1298 has an understood 1 in the denominator so that the ratio is 0.1298 to 1.

Margin 6

A certain college has a total of 14,280 students and 788 faculty. Write a unit ratio of students to faculty.

Answers to Margin 4: a. $\frac{1}{13}$ b. $\frac{14}{29}$

Answer to Margin 5: 0.229; This ratio means that the family owes about $0.23 for every $1 of income.

Answer to Margin 6: 18.1

◀ **Do Margin 4.**

OBJECTIVE 3 *Calculate unit ratios.*

Sometimes it is more useful to write a ratio as a **unit ratio.** A unit ratio is a ratio that has a denominator of 1.

DEFINITION **Unit ratio:** A ratio in which the denominator is 1.

For example, debt-to-income ratios are more informative when expressed as unit ratios. If a person has a debt to income ratio of $20 to $79, we can write the ratio in fraction form.

$$\text{Debt-to-income ratio} = \frac{20}{79}$$

Note that this ratio is in simplest form. The ratio indicates that the person has $20 of debt for every $79 of income.

It might be more informative or useful to write the ratio as a unit ratio. Because a unit ratio has a denominator of 1, we divide the denominator into the numerator.

$$\text{Debt-to-income unit ratio} = \frac{20}{79} \approx 0.2532$$

Interpreting this result, we can conclude that the person has a debt of about $0.25 for every $1 of income.

Conclusion: Dividing the denominator into the numerator creates a unit ratio.

Procedure *To calculate a unit ratio, divide the denominator into the numerator.*

EXAMPLE 5 A family has total debt of $5865 and a gross annual income of $45,200. Calculate their debt-to-income ratio as a unit ratio and interpret the results.

Understand: We are to calculate a unit ratio of debt to income.

Plan: Write the ratio of debt to income, then divide to calculate the unit ratio.

Execute: Debt to income $= \dfrac{5865}{45,200} \approx 0.1298$

Answer: The family has a debt-to-income ratio of approximately 0.13. This means that they owe about $0.13 for every $1 of income.

Check: Verify the calculation by multiplying 0.1298 by 45,200 to get approximately 5865. You will not get exactly 5865 because 0.1298 was an approximation.

◀ **Do Margin 5.**

EXAMPLE 6 A certain college has 9845 students and 468 faculty members. Write the student-to-faculty ratio as a unit ratio.

Understand: We are to calculate a unit ratio of students to faculty.

Plan: Write the ratio of students to faculty, then divide to calculate the unit ratio.

Execute: Students to faculty $= \dfrac{9845}{468} \approx 21.04$

Answer: The student to faculty ratio is approximately 21.04. This means that there are about 21 students for every faculty member at the college.

Check: Verify the calculation by multiplying 21.04 by 468 to get approximately 9845.

◀ **Do Margin 6.**

OBJECTIVE 4 *Write rates in simplest form.*

We can now build on what we've learned about unit ratios to discuss rates. We have informally discussed rates in the past. Velocity and unit price are examples of rates. Let's now formalize **rate.**

DEFINITION	**Rate:** A unit ratio comparing two different measurements.

Let's explore some rates. Velocity or speed is the rate at which an object changes its position with respect to time. It is the ratio of distance to time. An average velocity is the ratio of the total distance traveled to the total time of the trip.

$$v = \frac{d}{t}$$

Units for velocity always have a distance unit over a time unit. For example, miles per hour is written mi./hr. and feet per second is written ft./sec.

EXAMPLE 7 An athlete is clocked running the 40-yard dash in 4.4 seconds. What is the athlete's average rate in yards per second?

Understand: We are to find the rate in yards per second. The unit yards per second indicates writing a ratio of the number of yards to the number of seconds.

Plan: We will use r to represent the rate. We write a ratio of the number of yards to the number of seconds. Then divide.

Execute: $r = \dfrac{40 \text{ yd.}}{4.4 \text{ sec.}}$

$r = 9.\overline{09} \text{ yd./sec.}$

$r \approx 9.1 \text{ yd./sec.}$

Note: The / is read *per* so that the unit yd./sec. is read, *yards per second.*

Answer: The athlete ran at an average rate of about 9.1 yd./sec.

Check: Verify that running at a rate of 9.1 yd./sec. would cover 40 yd. in 4.4 sec. We can use the formula $d = rt$.

$d = (9.1)(4.4)$

$d = 40.04 \text{ yd.}$

Reminder
Our answer is not exactly 40 because we used the approximate value 9.1 for the rate. 40.04 is reasonably close considering our rounding of $9.\overline{09}$.

Do Margin 7. ▶

Another common rate is unit price. Unit price is a rate that compares price to quantity.

$$U = \frac{p}{q}$$

Units written for unit price always have a money unit over a quantity unit. For example, $/lb. or ¢/oz.

EXAMPLE 8 A 15-oz. box of cereal costs $2.39. What is the unit price in cents per ounce?

Understand: We are asked to find the unit price in cents per ounce. Because $2.39 is in terms of dollars, we must express the amount in cents. Because there are 100¢ in every dollar, we multiply 2.39 by 100. In Chapter 6 we learned that multiplying a decimal number by 100 moves the decimal point to the right 2 places.

$$\$2.39 = 239¢$$

Plan: Let U represent the unit price. We will write the ratio of cents to ounce, then divide.

Execute: $U = \dfrac{239¢}{15 \text{ oz.}}$ $U = 15.9\overline{3} \text{ ¢/oz.}$

Answer: The unit price is $15.9\overline{3}$ ¢/oz.

Check: Verify that purchasing 15 oz. at a unit price of $15.9\overline{3}$ ¢/oz. equals a total price of 239¢. Use the formula $P = uq$.

$P = (15.93)(15)$

$P = 238.95$

Reminder
Our answer is not exactly 239 because we used the approximate value 15.93 for the rate. 238.95 is reasonably close to 239 given our rounding of $15.9\overline{3}$.

◆ **Margin 7**

Solve.

a. A cyclist rides 20 mi. in 48 min. What was the cyclist's average rate in mi./min?

b. Sasha drives 45 mi. in $\frac{3}{4}$ hr. After a break she drives another 60 mi. in $\frac{4}{5}$ hr. What was her average rate for the entire trip in mi./hr?

Answers: **a.** $0.41\overline{6}$ mi./min.
b. 67.7 mi./hr.

◀ **Do Margin 8.**

Margin 8

Solve.

a. A 50-oz. bottle of detergent costs $3.58. What is the unit price in ¢/oz?

b. 4.5 lb. of chicken cost $5.22. What is the unit price in $/lb?

Margin 9

Solve.

a. Which is the better buy?

50-oz. bottle of detergent for $3.58
or
100-oz. bottle of detergent for $6.19

b. Which is the better buy?

10.5-oz. can of soup for 79¢
or
16-oz. can of soup for $1.29

Margin 10

Solve.

a. Which is the better buy?

16-oz. can of beans, 2 for $1.59
or
a 28-oz. can of beans for $1.19

b. Which is the better buy?

Three 2-lb. bags of store brand fries for $4.00
or
a 2-lb. bag of name brand fries for $1.30

Answers to Margin 8: **a.** 7.16 ¢/oz. **b.** $1.16/lb.

Answers to Margin 9: **a.** 100-oz. bottle **b.** 10.5-oz. can

Answers to Margin 10: **a.** 28-oz. can **b.** 2-lb. bag of brand name fries

---

OBJECTIVE 5 *Use unit price to determine the better buy.*

As consumers, we are often faced with a choice between two different sizes of the same item. Unit prices can be used to determine which of two different-size quantities is a better buy.

EXAMPLE 9 Which is the better buy?

15-oz. box of raisin bran for $2.45 or 20-oz. box of raisin bran for $3.05

Understand: We must determine which of the two different sizes is the better buy. We are given the price for each box and the quantity in each box.

Plan: Let F represent the unit price of the 15-oz. box and T represent the unit price of the 20-oz. box. Calculate the unit price for each box, then compare unit prices. The box with the smaller unit price is the better buy.

Execute: 15-oz. box   20-oz. box

$$F = \frac{\$2.45}{15 \text{ oz.}} \qquad T = \frac{\$3.05}{20 \text{ oz.}}$$

$$F = \$0.16\overline{3}/\text{oz.} \qquad T = \$0.1525/\text{oz.}$$

Answer: Because $0.1525/oz. is the smaller unit price, the 20-oz. box is the better buy. In other words, we would pay less for each ounce of raisin bran by buying the 20-oz. box than we would pay if we bought the 15-oz. box of the same raisin bran.

♦ *Check:* Verify the calculations by reversing the process. We will leave this to the reader.

◆ **Discussion** Under what circumstances might it actually be better to purchase the smaller quantity even if it is not the better buy?

◀ **Do Margin 9.**

EXAMPLE 10 Which is the better buy?

10.5-oz. of name brand yogurt, 2 for $1.19 or 32-oz. of store brand yogurt for $1.95

Understand: We must determine which is the better buy. Note that buying 2 of the 10.5-oz. yogurt means that we are buying $2(10.5) = 21$ oz. of yogurt for $1.19.

Plan: Let N represent the unit price of the name brand yogurt and S represent the unit price of the store brand yogurt. Calculate the unit prices, then compare.

Execute: Name brand: Store Brand:

$$N = \frac{\$1.19}{21 \text{ oz.}} \qquad S = \frac{\$1.95}{32 \text{ oz.}}$$

$$N = \$0.05\overline{6}/\text{oz.} \qquad S \approx \$0.061/\text{oz.}$$

Answer: Because $0.05\overline{6}$/oz. is the smaller unit price, buying 2 of the 10.5-oz. name brand yogurt is a better buy than buying 32 oz. of the store brand.

♦ *Check:* Verify the calculations by reversing the process. We will leave this to the reader.

◀ **Do Margin 10.**

7.1 Exercises

FOR EXTRA HELP

 Videotape 9

 InterAct Math
Tutorial Software

 www.carsonmath.com

 AWL Math Tutor Center

InterAct
MathXL www.mathxl.com

 Student's Solutions
Manual

For Exercises 1–10, write each ratio in simplest form.

1. There are 62 males and 70 females attending a conference.
   a. What is the ratio of males to total attendance?
   b. What is the ratio of females to total attendance?
   c. What is the ratio of males to females?
   d. What is the ratio of females to males?

2. The table below shows the number of mature trees in a region of forest.

Tree type	Number of mature trees
Pine	488
Maple	264
Oak	114
Other	295

   a. What is the ratio of pine trees to maple trees?
   b. What is the ratio of maple trees to oak trees?
   c. What is the ratio of pine trees to total trees?
   d. What is the ratio of oak trees to total trees?

3. Calcium nitride is a chemical compound that is composed of 3 calcium atoms and 2 nitrogen atoms.
   a. What is the ratio of calcium to nitrogen?
   b. What is the ratio of nitrogen to calcium?
   c. What is the ratio of calcium to total atoms in the compound?
   d. What is the ratio of nitrogen to total atoms in the compound?

Trivia Bite

Chemical compounds are written with subscripts that indicate the number of atoms of each element in the compound. Calcium nitride is written Ca_3N_2.

4. Hydrogen phosphate is a chemical compound composed of 3 atoms of hydrogen, 1 atom of phosphorous, and 4 atoms of oxygen.
   a. What is the ratio of hydrogen to oxygen?
   b. What is the ratio of phosphorous to oxygen?
   c. What is the ratio of hydrogen to total atoms in the compound?
   d. What is the ratio of oxygen to total atoms in the compound?

Trivia Bite

Hydrogen phosphate is written H_3PO_4. Note that there is no subscript written for P. The subscript is an understood 1 just as when there is no exponent written for a particular base in exponential form.

5. A roof is to have a pitch, or slant, of 14 in. vertically to 16 in. horizontally. Write the ratio of vertical distance to horizontal distance in simplest form.

6. The steps on a staircase rise 8 in. and are 10 in. wide. Write the ratio of the rise to the width of each step in simplest form.

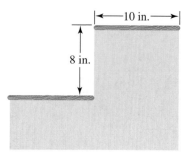

7. The back wheel of a bicycle rotates $3\frac{1}{2}$ times with $2\frac{1}{4}$ rotations of the pedals. Write the ratio of back wheel rotations to pedal rotations in simplest form.

8. At 20 mph in 2nd gear, a car engine is running about 3000 revolutions per minute (rpm). The drive tire is rotating about 260 rpm. What is the ratio of the engine rpm to the tire rpm?

9. In one week, Nina receives 7 bills, 3 letters, 17 advertisements, and 3 credit card offers in her mail.
   a. What is the ratio of bills to total mail?
   b. What is the ratio of advertisements to total mail?
   c. What is the ratio of letters to credit card offers?
   d. What is the ratio of bills and letters to advertisements and credit card offers?

10. The bridge and nut on a standard guitar are set so that the length of each string in between is $25\frac{1}{2}$ in. The seventh fret is placed on the neck of the guitar exactly 17 in. from the bridge.
   a. What is the ratio of the distance between the bridge and the seventh fret to the total length of the strings?
   b. What is the ratio of the distance between the bridge and the seventh fret to the distance between the seventh fret and the nut?
   c. What is the ratio of the distance between the seventh fret and the nut to the total length of the string?

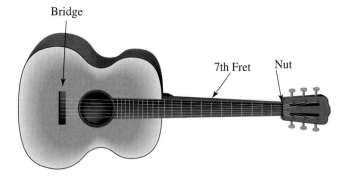

Bridge

7th Fret   Nut

Trivia Bite

Musical tones are created on stringed instruments by changing the string length. Shortening the string length causes the pitch to get higher, which with guitars, violins, and basses, is achieved by placing a finger on a string. By placing a finger at the 7th fret on a guitar, the string length is shortened from 25.5 in. to 17 in., which creates a musical tone that is higher in pitch than the tone created by plucking the string with no finger placement (open). In fact, the musical tone created at the 7th fret is a musical 5th from the open string tone. Musical notes are named after the letters of the alphabet, so if the tone of an open string is named A then the note at the 7th fret is named E, which is the 5th letter from A.

For Exercises 11–22, calculate the probability.

11. What is the probability of selecting a King in a standard deck of cards?

12. What is the probability of selecting a red Ace in a standard deck of cards?

13. What is the probability of selecting an Ace or a King in a standard deck of cards?

14. What is the probability of selecting a Queen or a Jack?

15. What is the probability of tossing a 1 or 6 on a six-sided game die?

16. What is the probability of tossing a 1, 2, or 3, on a six-sided game die?

17. What is the probability of tossing a 7 on a six-sided game die?

18. What is the probability of selecting a 1 in a standard deck of cards?

19. Suppose 12 males and 14 females put their names in a hat for a drawing. What is the probability that the winner will be a female?

20. Miwa sends 14 entries in a random drawing contest. If the company receives 894,264 entries, what is the probability that Miwa will win?

21. The table below shows the colors of candies in a bag of M & M candies.

Green	42
Red	28
Brown	30
Blue	25

   a. What is the probability that a candy chosen at random will be blue?
   b. What is the probability that a candy chosen at random will be red?
   c. What is the probability that a candy chosen at random will be green or brown?
   d. What is the probability that a candy chosen at random will be a color other than red?

22. The table below shows the home states for those attending a conference.

Georgia	46
South Carolina	42
North Carolina	38
Tennessee	18
Florida	14

   a. What is the probability that a person chosen at random is from North Carolina?
   b. What is the probability that a person chosen at random is from Florida?
   c. What is the probability that a person chosen at random is from Georgia or South Carolina?
   d. What is the probability that a person chosen at random is from any state other than Tennessee?

For Exercises 23–28, calculate each ratio as a unit ratio and interpret the result.

23. A family has a total debt of $4798 and a gross income of $29,850. What is their debt-to-income ratio?

Trivia Bite

Debt-to-income ratios are one of the determining factors in qualifying for a loan. More specifically, lenders consider the ratio of monthly debt to gross monthly income. This ratio is called the *back-end ratio*. We'll say more about this in Section 7.6.

24. A family has a total debt of $9452 and a gross income of $42,325. What is their debt-to-income ratio?

Trivia Bite

The ratio of mortgage payment to gross monthly income is another determining factor in qualifying for a loan. This ratio is called the *front-end* ratio.

25. A family is seeking a loan to purchase a new home. If they are approved for the loan, their new payment will be $945. If their gross monthly income is $3650, what is their payment-to-income ratio?

26. A family pays $550 in rent. If their gross monthly income is $2240, what is their rent-to-income ratio?

27. At a certain college campus there are 12,480 students and 740 faculty. What is the student-to-faculty ratio?

28. A certain college has 4212 students and 310 faculty. What is the student-to-faculty ratio?

29. At the close of the stock market on a particular day, each share of a computer company's stock was selling for $54\frac{1}{8}$. The annual earnings per share were $4.48. What is the ratio of selling price to annual earnings?

Trivia Bite

The ratio of selling price to annual earnings for a stock is call its *price-to-earnings ratio,* or *P/E ratio* for short. Companies with higher (15+) P/E ratios generally show a trend of increasing profits. Companies with lower (less than 5) P/E ratios are considered high-risk investments.

30. At the close of the stock market on a particular day, each share of a textile company's stock was selling for $12\frac{5}{16}$. The annual earnings per share were $3.52. What is the ratio of selling price to annual earnings?

For Exercises 31–36, calculate the rate.

31. Tara drives 306.9 mi. in 4.5 hr. What was her average rate in miles per hour?

32. Li drives 130.2 km in $2\frac{1}{3}$ hr. What was Li's average rate in kilometers per hour?

33. A long distance phone call lasting 23 minutes costs $2.76. What was the rate in dollars per minute?

34. 16.4 gal. of gas cost exactly $18. What was the rate in dollars per gallon?

35. Gina paid $3.75 for $\frac{1}{2}$ lb. of shrimp. What was the unit price in dollars per pound?

36. A 32-oz. bag of frozen french fries cost $1.50. What was the unit price in dollars per ounce?

For Exercises 37–42, determine which is the better buy.

37. a 10.5-oz. can of soup for $0.89   or   a 16-oz. can of soup for $1.19

38. a 15-oz. can of fruit cocktail for $1.09   or   a 20-oz. can of fruit cocktail for $1.49

39. a bag containing 16 diapers for $4.89   or $7.95 for a bag containing 24 diapers

40. a 28-oz. box of rice for $1.59   or $1.79 for a 32-oz. box of rice

41. Two 15.5-oz. boxes of store brand raisin bran $5.00   or
a 20-oz. box of name brand raisin bran for $3.45

42. Three 15-oz. cans of mixed vegetables for $1.98   or
Two 12.5-oz. cans of mixed vegetables for $1.19

REVIEW EXERCISES

1. $(9.6)(12.5) =$

2. $5\frac{1}{4} \cdot \frac{6}{7} =$

3. $4.5\left(2\frac{3}{4}\right) =$

4. Solve. $10.4x = 89.44$

5. Solve. $\frac{2}{3}y = 5$

7.2 Proportions

OBJECTIVES

1 Determine whether two ratios are proportional.

2 Solve for a missing number in a proportion.

3 Solve proportion problems.

4 Use proportions to solve for missing lengths in figures that are similar.

OBJECTIVE **1** *Determine whether two ratios are proportional.*

In Chapter 5 we discussed equivalent fractions. We saw that fractions like $\frac{5}{8}$ and $\frac{10}{16}$ are equivalent because $\frac{5}{8}$ can be upscaled to $\frac{10}{16}$ or $\frac{10}{16}$ can be reduced to $\frac{5}{8}$. We also saw that the cross products are equal with equivalent fractions.

$$16 \cdot 5 = 80 \qquad 8 \cdot 10 = 80$$

$$\frac{5}{8} \bowtie \frac{10}{16}$$

This is true for all **equivalent ratios**.

DEFINITION   **Equivalent ratios:** Ratios that name the same number.

Suppose we have two equivalent ratios $\frac{a}{b} = \frac{c}{d}$ where $b \neq 0$ and $d \neq 0$. We can use the multiplication/division principle of equality to clear the fractions by multiplying both sides by the LCD. In this case the LCD is bd.

$$\frac{\overset{1}{\cancel{b}d}}{1} \cdot \frac{a}{\cancel{b}} = \frac{c}{\cancel{d}} \cdot \frac{b\overset{1}{\cancel{d}}}{1}$$

$$ad = bc$$

Note that ad and bc are the cross products.

$$ad \qquad\qquad bc$$

$$\frac{a}{b} \bowtie \frac{c}{d}$$

Conclusion: Cross products of equivalent ratios are equal.

> **Rule** *If ratios are equivalent, then the cross products are equal.*
>
> ***In math language:*** Given two ratios: $\frac{a}{b}$ and $\frac{c}{d}$ where $b \neq 0$ and $d \neq 0$, if $\frac{a}{b} = \frac{c}{d}$ then

When two ratios are equivalent, we say they are *proportional*. Subsequently, we call the equation a **proportion.**

DEFINITION   **Proportion:** An equation with two ratios set equal to one another.

We can use cross products to determine whether two ratios are proportional. If the cross products are equal, then the ratios must be proportional.

EXAMPLE 1 Determine whether the ratios are proportional.

a. $\dfrac{3.5}{8}$? $\dfrac{1.4}{3.2}$

Solution: Compare the cross products. If the cross products are equal, then the ratios are proportional.

$$(3.2)(3.5) = 11.2 \qquad (8)(1.4) = 11.2$$

$$\dfrac{3.5}{8} \times \dfrac{1.4}{3.2}$$

Because the cross products are equal, these ratios are proportional.

b. $\dfrac{\frac{3}{4}}{\frac{3}{10}}$? $\dfrac{\frac{5}{12}}{\frac{1}{2}}$

Solution: Compare the cross products. If the cross products are equal, then the ratios are proportional.

$$\dfrac{1}{2} \cdot \dfrac{3}{4} = \dfrac{3}{8} \qquad\qquad \dfrac{\overset{1}{\cancel{3}}}{\cancel{10}_{2}} \cdot \dfrac{\overset{1}{\cancel{5}}}{\cancel{12}_{4}} = \dfrac{1}{8}$$

$$\dfrac{\frac{3}{4}}{\frac{3}{10}} \times \dfrac{\frac{5}{12}}{\frac{1}{2}}$$

◆ Because the cross products are not equal, these ratios are not proportional.

◀ **Do Margin 1.**

OBJECTIVE 2 *Solve for a missing number in a proportion.*

Cross products can be used to solve for a missing number in a proportion.

> **Procedure** *To solve a proportion using cross products:*
> **1.** Calculate the cross products.
> **2.** Set the cross products equal to one another.
> **3.** Use the multiplication/division principle to isolate the variable.

EXAMPLE 2 Solve. $\dfrac{5}{12} = \dfrac{x}{8}$

Solution: Because this is a proportion, the cross products must be equal.

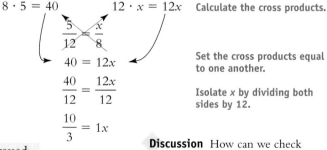

$$8 \cdot 5 = 40 \qquad\qquad 12 \cdot x = 12x \qquad \text{Calculate the cross products.}$$

$$\dfrac{5}{12} = \dfrac{x}{8}$$

$$40 = 12x \qquad\qquad \text{Set the cross products equal to one another.}$$

$$\dfrac{40}{12} = \dfrac{12x}{12} \qquad\qquad \text{Isolate } x \text{ by dividing both sides by 12.}$$

$$\dfrac{10}{3} = 1x$$

$$3\dfrac{1}{3} = x$$

Note: We could have expressed the answer as $3.\overline{3}$.

Discussion How can we check the solution?

♦ Margin 1

Determine whether the ratios are proportional.

a. $\dfrac{14}{16}$? $\dfrac{18}{20}$

b. $\dfrac{3.6}{5}$? $\dfrac{18}{25}$

c. $\dfrac{3\frac{1}{4}}{16}$? $\dfrac{6}{29\frac{7}{13}}$

Answers: **a.** Not proportional
b. Proportional **c.** Proportional

Do Margin 2.

EXAMPLE 3 Solve. $\dfrac{2\frac{1}{4}}{8} = \dfrac{-9}{b}$

Solution: Because this is a proportion, the cross products must be equal.

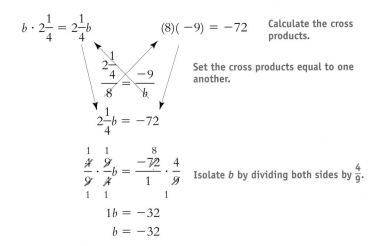

$b \cdot 2\frac{1}{4} = 2\frac{1}{4}b$ $(8)(-9) = -72$ Calculate the cross products.

$\dfrac{2\frac{1}{4}}{8} = \dfrac{-9}{b}$ Set the cross products equal to one another.

$2\frac{1}{4}b = -72$

$\dfrac{\cancel{4}}{\cancel{9}} \cdot \dfrac{\cancel{9}}{\cancel{4}}b = \dfrac{-\cancel{72}}{1} \cdot \dfrac{4}{\cancel{9}}$ Isolate b by dividing both sides by $\frac{4}{9}$.

$1b = -32$

$b = -32$

Do Margin 3. ▶

OBJECTIVE 3 *Solve proportion problems.*

In the typical proportion problem, you will be given a ratio. You will then be asked to upsize or downsize from that ratio. The key to solving proportion problems lies in identifying the two equivalent ratios and then equating them.

EXAMPLE 4 Bobbie can read 40 pages in 50 minutes. How long will it take her to read 220 pages?

Understand: We must calculate the time it takes Bobbie to read 220 pages. We are given a rate of 40 pages in 50 minutes.

$$\text{“40 pages in 50 minutes” translates to: } \dfrac{40 \text{ pages}}{50 \text{ min.}}$$

If we translate the rate with the unknown amount in the same way, we have:

$$220 \text{ pages in ? minutes translates to: } \dfrac{220 \text{ pages}}{x \text{ min.}}$$

Plan: Because the rates are equivalent, we can set up a proportion by equating them. Then solve.

Execute: $\dfrac{40 \text{ pages}}{50 \text{ min.}} = \dfrac{220 \text{ pages}}{x \text{ min.}}$

$x \cdot 40 = 40x$ $50 \cdot 220 = 11{,}000$

$\dfrac{40}{50} = \dfrac{220}{x}$

$40x = 11{,}000$

$\dfrac{40x}{40} = \dfrac{11{,}000}{40}$

$1x = 275$

$x = 275 \text{ min.}$

> **Tip**
> When setting up the proportion, write the given rate any way you wish. Then write the rate with the unknown in logical correspondence to the given rate. One way to ensure logical correspondence is to match the units straight across.
>
> Notice that the units in the numerators match and the units in the denominators match.

Margin 2

a. $\dfrac{5}{9} = \dfrac{n}{27}$

b. $\dfrac{y}{16} = \dfrac{7}{10}$

Margin 3

Solve.

a. $\dfrac{4\frac{2}{3}}{m} = \dfrac{\frac{3}{4}}{-9}$

b. $\dfrac{-6.5}{y} = \dfrac{-13}{14}$

Answers to Margin 2: **a.** 15 **b.** 11.2

Answers to Margin 3: **a.** -56 **b.** 7

Answer: It will take Bobbie 275 minutes to read 220 pages.

Check: Do a quick estimate. Bobbie reads 40 pages in 50 minutes. 220 pages is a little over 5 times the number of pages, therefore it ought to take her a little over 5 times the amount of time. $5 \cdot 50 = 250$ minutes, so a little more than that would make 275 minutes a very reasonable answer.

◆ **Margin 4**

Solve.

a. Erica drove 303.8 mi. using 12.4 gal. of gasoline. At this rate, how much gasoline would it take for her to drive 1000 mi?

Discussion How did she measure the 303.8 mi. and the 12.4 gal?

If you set the proportion up incorrectly you will get an answer that is quite unreasonable. Doing a quick estimate check is a powerful way to recognize when you've set up a proportion incorrectly.

For example, suppose we had set up Example 3 this way:

$$\frac{40 \text{ pages}}{50 \text{ min.}} = \frac{x \text{ min.}}{220 \text{ pages}}$$

Note that this is not a logical correspondence.

Solving we get:

$$40 \cdot 220 = 8800 \qquad 50 \cdot x = 50x$$

$$\frac{40}{50} \overset{\times}{=} \frac{x}{220}$$

$$8800 = 50x$$

$$\frac{8800}{50} = \frac{50x}{50}$$

$$176 = 1x$$

$$176 \text{ min.} = x$$

This answer is unreasonable because it is a little over 3 times the amount of time ($3 \cdot 50 = 150$ minutes), while she is reading more than 5 times the number of pages ($5 \cdot 40 = 200$ pages). Below are a few other correct ways we could have set up the proportion for Example 3. Note that they all have logical correspondence in the rates.

$$\frac{50 \text{ min.}}{40 \text{ pages}} = \frac{x \text{ min.}}{220 \text{ pages}} \qquad \frac{40 \text{ pages}}{220 \text{ pages}} = \frac{50 \text{ min.}}{x \text{ min.}} \qquad \frac{50 \text{ min.}}{x \text{ min.}} = \frac{40 \text{ pages}}{220 \text{ pages}}$$

Think:		Think of this version as an analogy:	This version is a rearrangement of the previous version:
In 50 min. she reads 40 pages	so	In *x* min. she reads 220 pages	If 40 pages in 50 min. then 220 pages in *x* min.	If 50 min. yields 40 pages then *x* min. yields 220 pages

b. Xion can mow a 2500-ft.² lawn in 20 min. At this rate how long will it take him to mow a 12,000-ft.² lawn?

Procedure *To solve proportion problems:*
1. Set up the given ratio any way you wish.
2. Set up the ratio with the unknown so that it logically corresponds to the way you set up the given ratio.
3. Solve.

◀ **Do Margin 4.**

OBJECTIVE 4 *Use proportions to solve for missing lengths in figures that are similar.*

Though the two triangles to the right have different measurements for the lengths of the sides, their angles are congruent.

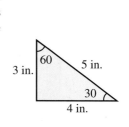

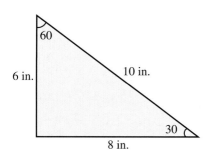

Answers: **a.** 40.8 gal. **b.** 96 min.

Two figures with the same number of sides and all angles congruent are said to be **similar figures.**

DEFINITION   **Similar figures:** Figures that have the same number of sides and all angles congruent.

We can think of two similar figures as upscaled or reduced versions of the same figure. Notice that the sides in the larger triangle are twice as long as the corresponding sides of the smaller triangle. The ratios of the lengths in the larger triangle to the corresponding lengths in the smaller triangle are all equivalent to 2 to 1, which we can write as 2.

$$\text{Larger triangle} \longrightarrow \quad \frac{6}{3} = \frac{10}{5} = \frac{8}{4} = 2$$
$$\text{to}$$
$$\text{Smaller triangle} \longrightarrow$$

The sides in the smaller triangle are all half as long as the corresponding sides in the larger triangle. Therefore the ratios of the lengths in the smaller triangle to the corresponding lengths in the larger triangle are all equivalent to $\frac{1}{2}$.

$$\text{Smaller triangle} \longrightarrow \quad \frac{3}{6} = \frac{5}{10} = \frac{4}{8} = \frac{1}{2}$$
$$\text{to}$$
$$\text{Larger triangle} \longrightarrow$$

Conclusion: The ratios of the lengths of the sides in similar figures are proportional.

EXAMPLE 5   The triangles below are similar. Find the missing length.

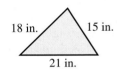

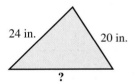

Understand: We must find the missing length in the larger triangle. Because the triangles are similar, we know that the ratios of the lengths are proportional. It is helpful to correlate the sides this way:

Smaller triangle		Larger triangle
18 in.	◄――――►	24 in.
15 in.	◄――――►	20 in.
21 in.	◄――――►	?

We can write ratios of the lengths in the smaller triangle to the corresponding lengths in the larger triangle this way:

$$\text{Smaller triangle} \longrightarrow \quad \frac{18}{24} = \frac{15}{20} = \frac{21}{?}$$
$$\text{to}$$
$$\text{Larger triangle} \longrightarrow$$

Plan: Let *n* represent the unknown length. Write a proportion, then solve. We can use either $\frac{18}{24}$ or $\frac{15}{20}$ as the given ratio in our proportion.

Execute:
$$\frac{18}{24} = \frac{21}{n} \qquad \text{or} \qquad \frac{15}{20} = \frac{21}{n}$$

$$18n = 504 \qquad\qquad 15n = 420$$

$$\frac{18n}{18} = \frac{504}{18} \qquad\qquad \frac{15n}{15} = \frac{420}{15}$$

$$1n = 28 \qquad\qquad 1n = 28$$

$$n = 28 \qquad\qquad n = 28$$

◆ **Margin 5**

The figures below are similar. Solve for the length of the missing side.

a.

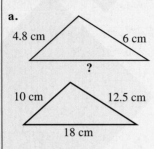

4.8 cm   6 cm   ?

10 cm   12.5 cm   18 cm

> **Reminder**
>
> There are many correct ways to write the proportions. Here is another correct way:
>
> | Smaller Triangle | | Larger Triangle |
> |---|---|---|
> | 18 | corresponds to | 24 |
> | 21 | corresponds to | *n* |
>
> $$\frac{18}{21} = \frac{24}{n}$$
>
> Here we relate the corresponding sides across instead of in the same ratio.

Answer: The missing length is 28 in.

Check: Verify that 28 is in the same ratio using a different pair of corresponding sides. In other words, if you used $\frac{18}{24}$ as the given ratio to solve for *n*, then use $\frac{15}{20}$ as a check.

◀ **Do Margin 5.**

Sometimes more than one length is missing in problems involving similar figures. In order to solve these problems we must be given the length of at least one pair of corresponding sides.

b.

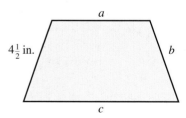

20 ft.

$14\frac{1}{2}$ ft.

32 ft.

?

EXAMPLE 6 The figures below are similar. Find the missing lengths.

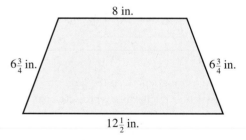

8 in.

$6\frac{3}{4}$ in. $6\frac{3}{4}$ in.

$12\frac{1}{2}$ in.

a

$4\frac{1}{2}$ in.   *b*

c

Answers: **a.** 8.64 cm **b.** 23.2 ft.

Understand: We are to find the missing lengths. Because the figures are similar, the ratios of the lengths of the corresponding sides are proportional. We can correlate the corresponding sides this way:

Larger trapezoid		Smaller trapezoid
$6\frac{3}{4}$ in.	⟷	$4\frac{1}{2}$ in.
8 in.	⟷	a
$6\frac{3}{4}$ in.	⟷	b
$12\frac{1}{2}$ in.	⟷	c

$6\frac{3}{4}$ in. corresponds to $4\frac{1}{2}$ in. We'll write proportions using this correspondence to find the missing lengths.

Plan: Write proportions relating the corresponding sides, then solve.

Execute: It actually isn't necessary to write a proportion to find b. Notice that two of the sides in the larger trapezoid measure $6\frac{3}{4}$ in. This means that the corresponding sides must match in the smaller trapezoid. Because $6\frac{3}{4}$ in. corresponds to $4\frac{1}{2}$ in. we can conclude that $b = 4\frac{1}{2}$ in.

To find a:

$$\frac{6\frac{3}{4}}{4\frac{1}{2}} = \frac{8}{a}$$

$$a \cdot \frac{27}{4} = \frac{27}{4}a \qquad \frac{9}{2} \cdot \frac{\overset{4}{\cancel{8}}}{1} = 36$$

$$\frac{\dfrac{27}{4}}{\dfrac{9}{2}} \diagup\!\!\!\!\!\diagdown \frac{8}{a}$$

$$\frac{27}{4}a = 36$$

$$\frac{1}{\cancel{4}} \cdot \frac{\overset{1}{\cancel{27}}}{\cancel{\cancel{27}}}a = \frac{\overset{4}{\cancel{36}}}{1} \cdot \frac{4}{\cancel{27}}_{3}$$

$$a = \frac{16}{3}$$

$$a = 5\frac{1}{3} \text{ in.}$$

To find c:

$$\frac{6\frac{3}{4}}{4\frac{1}{2}} = \frac{12\frac{1}{2}}{c}$$

$$c \cdot \frac{27}{4} = \frac{27}{4}c \qquad \frac{9}{2} \cdot \frac{25}{2} = \frac{225}{4}$$

$$\frac{\dfrac{27}{4}}{\dfrac{9}{2}} \diagup\!\!\!\!\!\diagdown \frac{\dfrac{25}{2}}{c}$$

$$\frac{27}{4}c = \frac{225}{4}$$

$$\frac{1}{\cancel{4}} \cdot \frac{\overset{1}{\cancel{27}}}{\cancel{\cancel{27}}}c = \frac{\overset{25}{\cancel{225}}}{\cancel{4}} \cdot \frac{4}{\cancel{27}}_{3}$$

$$c = \frac{25}{3}$$

$$c = 8\frac{1}{3} \text{ in.}$$

Answer: The missing lengths are: $a = 5\frac{1}{3}$ in., $b = 4\frac{1}{2}$ in., and $c = 8\frac{1}{3}$ in.

◆ **Margin 6**

Find the missing lengths in the similar shapes.

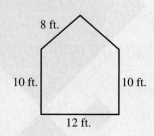

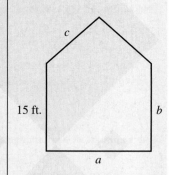

Check: We can do a quick-estimate check. Notice that in the larger trapezoid, the $12\frac{1}{2}$ in. side is about twice the length of the $6\frac{3}{4}$ in. side. The same relationship should hold with the corresponding sides in the smaller trapezoid. Side c should be about twice the length of the $4\frac{1}{2}$ in. side in the smaller trapezoid. We found side c to be $8\frac{1}{3}$ in., which is about twice $4\frac{1}{2}$ in. Further, the 8 in. side is closer in length to $6\frac{3}{4}$ in. than $12\frac{1}{2}$ in. Relating this to the smaller trapezoid we should find that side a is closer in length to $4\frac{1}{2}$ in. than to $8\frac{1}{3}$ in. We found side

◆ a to be $5\frac{1}{3}$ in. These estimates are quite reasonable.

◀ **Do Margin 6.**

7.2 Exercises

For Exercises 1–8, determine whether the ratios are proportional.

1. $\dfrac{3}{4}$? $\dfrac{9}{12}$

2. $\dfrac{15}{4}$? $\dfrac{7}{2.4}$

3. $\dfrac{15}{24}$? $\dfrac{2.5}{3.2}$

4. $\dfrac{21}{30}$? $\dfrac{49}{70}$

5. $\dfrac{16.8}{40.2}$? $\dfrac{17.5}{42.5}$

6. $\dfrac{9.5}{14}$? $\dfrac{28.5}{42}$

7. $\dfrac{\frac{4}{5}}{\frac{7}{10}}$? $\dfrac{12}{10\frac{1}{2}}$

8. $\dfrac{2\frac{1}{3}}{3\frac{1}{2}}$? $\dfrac{5\frac{1}{2}}{7\frac{1}{3}}$

For Exercises 9–20, solve for the missing number.

9. $\dfrac{x}{8} = \dfrac{20}{32}$

10. $\dfrac{28}{42} = \dfrac{y}{6}$

11. $\dfrac{-3}{8} = \dfrac{n}{20}$

12. $\dfrac{m}{12} = \dfrac{2}{-5}$

13. $\dfrac{18}{h} = \dfrac{21.6}{30}$

14. $\dfrac{20}{32.8} = \dfrac{8}{k}$

15. $\dfrac{-14}{b} = \dfrac{3.5}{6.25}$

16. $\dfrac{4.9}{6.4} = \dfrac{-24.5}{c}$

17. $\dfrac{4\frac{3}{4}}{5\frac{1}{2}} = \dfrac{d}{16\frac{1}{2}}$

18. $\dfrac{\frac{3}{5}}{6\frac{1}{2}} = \dfrac{1\frac{2}{3}}{j}$

19. $\dfrac{-9.5}{22} = \dfrac{-6\frac{1}{3}}{t}$

20. $\dfrac{u}{-\frac{3}{8}} = \dfrac{-4.2}{5}$

For Exercises 21–32, solve.

21. William drove 358.4 mi. using 16.4 gal. of gasoline. At this rate, how much gasoline would it take for him to drive 750 miles?

22. A carpet cleaning company claims they will clean 400 ft.^{2} for \$54.95. At this rate how much should they charge to clean 600 ft.^{2}?

23. Gunther estimates that a 12.5-lb. turkey will have 30 servings. At this rate, how many servings would a 15-lb. turkey have?

24. A 20-lb. bag of fertilizer covers 5000 ft.^{2}. How many pounds of fertilizer should be used for a lawn that is 12,000 ft.^{2}? If the fertilizer only comes in 20-lb. bags, how many bags must be purchased?

25. Nathan notes that he consumes 16 oz. of yogurt in five days. At this rate how many ounces will he consume in a year? (1 year $= 365\frac{1}{4}$ days.)

26. In taking a patient's pulse, a nurse counts 19 heartbeats in 15 sec. How many heartbeats is this in 1 min? (1 min. $=$ 60 sec.)

27. A recipe for bran muffins calls for $1\frac{1}{2}$ cups of bran flakes. The recipe yields 8 muffins. How much bran flakes should be used to make a dozen muffins?

28. The same recipe for bran muffins calls for $2\frac{1}{2}$ teaspoons of baking powder. How much baking powder should be used to make a dozen muffins?

29. A $42\frac{1}{2}$-ft. wall measures $8\frac{1}{2}$ in. on a scale drawing. How long should a $16\frac{1}{4}$-ft. wall measure on the drawing?

30. Corrine is drafting the blueprints for a house. The scale for the drawing is $\frac{1}{4}$ in. = 1 ft. How long should she draw the line representing a wall that is $13\frac{1}{4}$ ft. long?

31. A building company estimates that construction on a high-rise building will move at a pace of 18 vertical feet every 30 days. At this rate, how long will it take them to complete a building that is 980 ft. tall?

32. A water treatment plant can treat 40,000 gal. of water in 24 hr. How many gallons does the plant treat in a week? (1 wk. = 168 hr.)

For Exercises 33–40, find the missing lengths in the similar shapes.

33.

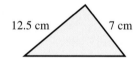
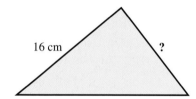

34.

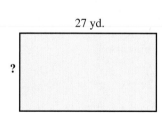

35.

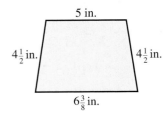

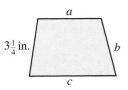

36.

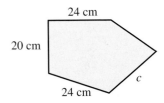

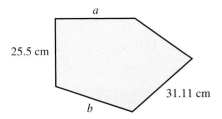

37. To estimate the height of a tree, a forester uses the concept of similar triangles. A particular tree is found to have a shadow measuring 84 ft. in length. The forester has her own shadow measured at the same time. Her shadow measures $7\frac{1}{2}$ ft. If she is $5\frac{1}{2}$ ft. tall, how tall is the tree?

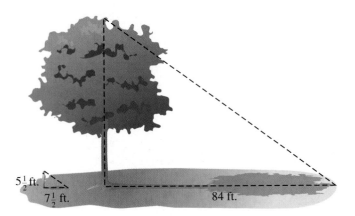

38. To estimate the height of the great pyramid in Egypt, the Greek mathematician Thales performed a procedure similar to that in Exercise 37, except that he supposedly used a staff instead of his body. Suppose the shadow from the staff measured to be 6 ft. and, at the same time, the shadow of the pyramid measured to be $524\frac{2}{3}$ ft. If the staff was $5\frac{1}{2}$ ft. tall, what was the height of the pyramid?

Historical Note

Thales was a Greek mathematician who lived from 640 B.C. to 546 B.C. He was born in Miletus, a city on the coast of what is now Turkey. Thales is considered to be the father of science and mathematics in the Greek culture. He is the first to have proposed the importance of proof. It is said that he asserted that water is the origin of all things and that matter is infinitely divisible.

39. To estimate the distance across a river an engineer creates similar triangles. The idea is to create two similar right triangles as shown below. Calculate the width of the river.

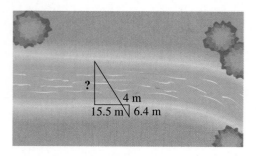

4 m
15.5 m
6.4 m
?

Discussion How would one create the similar right triangles?

40. The method described in Exercise 39 could be used to determine the width of the Grand Canyon at a given point. Consider the measurements below. Calculate the width of the canyon.

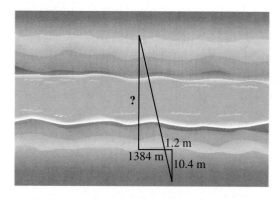

?
1.2 m
1384 m
10.4 m

Trivia Bite

Located in northwest Arizona, the Grand Canyon is a gorge created by the erosion of the Colorado River. The canyon is about 277 miles (446 km) long and ranges in width from 4 to 18 miles (6.4 to 29 km). The canyon reaches depths of up to 1 mile (1.6 km).

Puzzle Problem

It takes a clock 5 sec. to strike 6 o'clock. How many seconds will it take for the clock to strike 12 o'clock?

REVIEW EXERCISES

1. $\dfrac{4}{9} \cdot \dfrac{5}{12}$

2. $59.2 \cdot \dfrac{1}{2} \cdot \dfrac{1}{2} \cdot \dfrac{1}{4}$

3. $9\dfrac{1}{3}(144)$

4. $70.5 \div 12 =$

5. $60\dfrac{3}{4} \div 3$

7.3 American Measurement

OBJECTIVES

1 Convert American units of length.

2 Convert American units of area.

3 Convert American units of capacity.

4 Convert American units of weight.

5 Convert units of time.

6 Convert American units of speed.

OBJECTIVE 1 *Convert American units of length.*

To convert from one unit of measurement to another, we will use a method known as *dimensional analysis* or *unit analysis*. First, consider units of length. The table below lists facts about the American units of length.

$$1 \text{ foot} = 12 \text{ inches}$$
$$1 \text{ yard} = 3 \text{ feet}$$
$$1 \text{ mile} = 5280 \text{ feet}$$

We can express these facts as ratios.

$$\frac{1 \text{ ft.}}{12 \text{ in.}} \quad \text{or} \quad \frac{12 \text{ in.}}{1 \text{ ft.}} \qquad \frac{1 \text{ yd.}}{3 \text{ ft.}} \quad \text{or} \quad \frac{3 \text{ ft.}}{1 \text{ yd.}} \qquad \frac{1 \text{ mi.}}{5280 \text{ ft.}} \quad \text{or} \quad \frac{5280 \text{ ft.}}{1 \text{ mi.}}$$

Because each numerator and denominator of each of these ratios represents the same amount, these ratios are all equivalent to the number 1 and are called **unit fractions.**

DEFINITION   **Unit fraction:** A fraction with a ratio equivalent to 1.

Because the ratio of any given unit fraction is 1, multiplying by unit fractions is a way of finding an equivalent amount. To use dimensional analysis, we multiply the given measurement by unit fractions so that the undesired units divide out leaving the desired unit.

Suppose we want to convert 4 ft. to units of inches. Because 4 ft. has the undesired unit feet in a numerator, we must multiply by a unit fraction that has the unit feet appearing in the denominator. We can then divide out the undesired feet units, leaving us with the desired unit, inches.

$$4 \text{ ft.} = \frac{4 \cancel{\text{ ft.}}}{1} \cdot \frac{12 \text{ in.}}{1 \cancel{\text{ ft.}}} = 48 \text{ in.}$$

Procedure *To convert units using dimensional analysis, multiply the given measurement by unit fractions so that the undesired units divide out leaving the desired units.*

Tip

If an undesired unit is in a numerator, that unit should appear in the denominator of a unit fraction.

If an undesired unit is in the denominator, that unit should appear in the numerator of a unit fraction.

EXAMPLE 1 Convert using dimensional analysis.

a. 3.5 mi. to feet.

Solution: Because 3.5 mi. can be expressed in a numerator over 1, we will use a unit fraction with miles in the denominator. Because the desired unit is feet, we will use the fact that there are 5280 ft. to 1 mi.

$$3.5 \text{ mi.} = \frac{3.5 \text{ mi.}}{1} \cdot \frac{5280 \text{ ft.}}{1 \text{ mi.}} = 18{,}480 \text{ ft.}$$

b. $8\frac{2}{3}$ yd. to feet.

Solution: Because $8\frac{2}{3}$ yd. can be expressed in a numerator over 1, we will use a unit fraction with yard in the denominator. Because the desired unit is feet, we will use the fact that there are 3 ft. to 1 yd.

$$8\frac{2}{3} \text{ yd.} = \frac{26 \text{ yd.}}{3} \cdot \frac{\overset{1}{3} \text{ ft.}}{1 \text{ yd.}} = 26 \text{ ft.}$$

c. 78 in. to feet.

Solution: Because 78 in. can be expressed in a numerator over 1, we will use a unit fraction with inch in the denominator. Because the desired unit is feet, we will use the fact that there is 1 ft. to 12 in.

$$78 \text{ in.} = \frac{\overset{13}{78} \text{ in.}}{1} \cdot \frac{1 \text{ ft.}}{\underset{2}{12} \text{ in.}} = \frac{13}{2} \text{ ft.} = 6\frac{1}{2} \text{ ft. or } 6.5 \text{ ft.}$$

◀ **Do Margin 1.**

Sometimes we may not know a direct fact between the given unit and the desired unit. In such cases we must convert to one or more intermediate units to get to the desired unit. This means we will use more than one unit fraction to convert to the desired unit.

EXAMPLE 2 Convert using dimensional analysis.

a. 10 yd. to inches

Solution: If we do not know a direct fact relating yards to inches, we can first convert yards to feet, then feet to inches. To convert from yards to feet, we use the fact that there are 3 ft. to 1 yd. Then to convert from feet to inches, we use the fact that there are 12 in. to 1 ft.

$$10 \text{ yd.} = \frac{10 \text{ yd.}}{1} \cdot \frac{3 \text{ ft.}}{1 \text{ yd.}} \cdot \frac{12 \text{ in.}}{1 \text{ ft.}} = 360 \text{ in.}$$

Note: Multiplying these two ratios gives the direct fact that there are 36 in. to 1 yd.

b. 4400 yd. to miles

Solution: Because we do not have a direct fact relating yards to miles, we can convert yards to feet, then convert feet to miles. We use the fact that there are 3 ft. to 1 yd. and then use 1 mi. to 5280 ft.

$$4400 \text{ yd.} = \frac{4400 \text{ yd.}}{1} \cdot \frac{3 \text{ ft.}}{1 \text{ yd.}} \cdot \frac{1 \text{ mi.}}{5280 \text{ ft.}} = \frac{13{,}200}{5280} \text{ mi.} = 2\frac{1}{2} \text{ mi. or } 2.5 \text{ mi.}$$

Discussion What is the direct fact relating yards and miles?

◆ **Margin 1**

Convert using dimensional analysis.

a. 20 ft. to inches

b. $4\frac{3}{10}$ mi. to feet

c. 62 ft. to yards

Answers: **a.** 240 in. **b.** 22,704 ft.
c. $20\frac{2}{3}$ ft. (or $20.\overline{6}$ ft.)

Notice that if we had known the direct facts in Example 2, we would not have had to use several unit fractions. The more facts you know, the fewer unit fractions you have to use. However, this means more to remember. So there's a tradeoff: remember more facts and have simpler calculations, or remember fewer facts and be prepared to use more unit fractions. Each has its merits, but in the end it's up to the individual to choose which is better.

Do Margin 2. ▶

◆ **Margin 2**

Convert using dimensional analysis.

a. 0.8 yd. to inches

b. $\frac{7}{10}$ mi. to yards

OBJECTIVE **2** *Convert American units of area.*

Now that we have discussed dimensional analysis, we can apply this method to other types of measurement. Consider units of area. Recall that area is always represented in square units like square inches (in.^{2}), square feet (ft.^{2}), square yards (yd.^{2}), or square miles (mi.^{2}).

A square foot is a 1 ft. by 1 ft. square. If we sub-divide that square into inches, we have a 12 in. by 12 in. square.

Calculating the area of 1 ft. by 1 ft. square as a 12 in. by 12 in. square, we find that it contains 144 in.^{2} Note that if we square both sides of the fact for length, we get the fact for area.

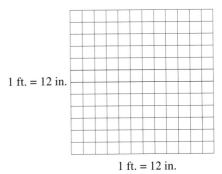

1 ft. = 12 in.

1 ft. = 12 in.

$$1 \text{ ft.} = 12 \text{ in.}$$
$$(1 \text{ ft.})^2 = (12 \text{ in.})^2$$
$$1 \text{ ft.}^2 = 144 \text{ in.}^2$$

Conclusion: We can derive the facts for area by squaring the facts for length.

Because 1 yd. = 3 ft., we can conclude 1 yd.^{2} = 9 ft.^{2}.

Because 1 mi. = 5280 ft., we can conclude that 1 mi.^{2} = 27,878,400 ft.^{2}.

EXAMPLE 3 Convert using dimensional analysis.

a. 8 ft.^{2} to square inches

Solution: Because the given measurement is in square feet, we will use a unit fraction with ft.^{2} in the denominator. Because we are to convert to square inches, we will use the ratio of 144 in.^{2} to 1 ft.^{2}.

$$8 \text{ ft.}^2 = \frac{8 \cancel{\text{ft.}^2}}{1} \cdot \frac{144 \text{ in.}^2}{1 \cancel{\text{ft.}^2}} = 1152 \text{ in.}^2$$

b. 150 ft.^{2} to square yards

Solution: Because the given measurement is in square feet, we will use a unit fraction that has ft.^{2} in the denominator. Because we are to convert to square yards we use the ratio of 1 yd.^{2} to 9 ft.^{2}.

$$150 \text{ ft.}^2 = \frac{\overset{50}{\cancel{150 \text{ ft.}^2}}}{1} \cdot \frac{1 \text{ yd.}^2}{\underset{3}{\cancel{9 \text{ ft.}^2}}} = \frac{50}{3} \text{ yd.}^2 = 16\frac{2}{3} \text{ yd.}^2 \quad \text{or} \quad 16.\overline{6} \text{ yd.}^2$$

◆ **Margin 3**

Convert using dimensional analysis.

a. 12 yd.^{2} to square feet

b. 6000 in.^{2} to square feet

c. 8 yd.^{2} to square inches

Do Margin 3. ▶

EXAMPLE 4 A 12.5 ft. by 20 ft. room is to be carpeted. Calculate the area of the room in square yards.

Understand: We must calculate the area of a rectangular room. The formula for the area of a rectangle is $A = bh$.

Answers to Margin 2: **a.** 28.8 in.
b. 1232 yd.

Answers to Margin 3: **a.** 108 ft.^{2}
b. $41\frac{2}{3}$ ft.^{2} (or $41.\overline{6}$ ft.^{2})
c. 10,368 in.^{2}

The given measurements are in terms of feet. If we calculate the area with these measurements, the area unit will be square feet. The desired unit is square yards. There are two approaches to this problem:

1. Calculate the area in terms of square feet, then convert square feet to square yards.

2. Convert feet to yards and calculate the area with the measurements in terms of yards so that the outcome is in square yards.

Plan: Let's use the first approach. We will calculate the area using the given measurements so that the area will be in square feet. We will then convert square feet to square yards.

Execute: $A = bh$

$$A = (12.5)(20)$$
$$A = 250 \text{ ft.}^2$$

Now convert to square yards.

$$250 \text{ ft.}^2 = \frac{250 \text{ ft.}^2}{1} \cdot \frac{1 \text{ yd.}^2}{9 \text{ ft.}^2} = \frac{250}{9} \text{ yd.}^2 = 27.\overline{7} \text{ yd.}^2$$

Answer: The room will require $27.\overline{7}\text{yd.}^2$ of carpet.

Check: Let's use the second method we described as a check. We will convert the given measurements to yards, then calculate the area.

$$12.5 \text{ ft.} = \frac{12.5 \text{ ft.}}{1} \cdot \frac{1 \text{ yd.}}{3 \text{ ft.}} = \frac{12.5}{3} \text{ yd.} = 4.1\overline{6} \text{ yd.}$$

$$20 \text{ ft.} = \frac{20 \text{ ft.}}{1} \cdot \frac{1 \text{ yd.}}{3 \text{ ft.}} = \frac{20}{3} \text{ yd.} = 6.\overline{6} \text{ yd.}$$

$$A = (4.1\overline{6})(6.\overline{6})$$
$$A \approx (4.17)(6.67) \approx 27.81 \text{ yd.}^2$$

Note: We chose to express the measurements as decimal numbers with repeating digits. Because we must round these values to calculate the area, our answer is an approximation. The approximation is reasonably close to the area we found using the first method.

OBJECTIVE 3 *Convert American units of capacity.*

Now consider **capacity.**

DEFINITION   **Capacity:** A measure of the amount of liquid a container holds.

In the American system, we use units like gallons (gal.), quarts (qt.), pints (pt.), cups (c.), and ounces (oz.) to measure the capacity of a container. Below are the facts for capacity units.

1 cup = 8 ounces
1 pint = 2 cups
1 quart = 2 pints
1 gallon = 4 quarts

1 cup   1 pt.   1 qt.   1 gal.

EXAMPLE 5   Convert using dimensional analysis.

a. 5 pt. to ounces

Solution: Because we do not have a direct fact relating pints to ounces, we first convert pints to cups. Because the given measurement is in pints, the first unit fraction should have pints in the denominator. We use the ratio of 2 c. to 1 pt. The next unit fraction will have to have cups in the denominator so we use the ratio of 8 oz. to 1 c.

$$5 \text{ pt.} = \frac{5 \text{ pt.}}{1} \cdot \frac{2 \text{ c.}}{1 \text{ pt.}} \cdot \frac{8 \text{ oz.}}{1 \text{ c.}} = 80 \text{ oz.}$$

b. 240 oz. to gallons

Solution: Because we do not have a direct fact relating ounces to gallons, we first convert ounces to cups. Because the given measurement is in ounces, the first ratio will have ounces in the denominator. We use the ratio 1 c. to 8 oz. We then use 1 pt. to 2 c. to convert from cups to pints. Next we use the ratio 1 qt. to 2 pt. to convert from pints to quarts. Finally, to convert from quarts to gallons we use the ratio 1 gal. to 4 qt.

$$240 \text{ oz.} = \frac{240 \text{ oz.}}{1} \cdot \frac{1 \text{ c.}}{8 \text{ oz.}} \cdot \frac{1 \text{ pt.}}{2 \text{ c.}} \cdot \frac{1 \text{ qt.}}{2 \text{ pt.}} \cdot \frac{1 \text{ gal.}}{4 \text{ qt.}} = \frac{240}{128} \text{ gal.} = 1\frac{7}{8} \text{ gal.} \quad \text{or} \quad 1.875 \text{ gal.}$$

Do Margin 4. ▷

OBJECTIVE 4 *Convert American units of weight.*

We have discussed weight as being the force due to gravity. Now let's discuss converting units of weight. In the American system, weight can be expressed in terms of ounces (oz.), pounds (lb.), and tons (T). Below is a list of the facts for weight.

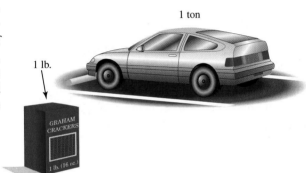

1 ton

1 lb.

    1 pound = 16 ounces
    1 ton = 2000 pounds

EXAMPLE 6 Convert using dimensional analysis.

a. 12 oz. to pounds

Solution: Because the given measurement is in ounces, we multiply by the ratio 1 lb. to 16 oz.

$$12 \text{ oz.} = \frac{\overset{3}{\cancel{12}} \text{ oz.}}{1} \cdot \frac{1 \text{ lb.}}{\underset{4}{\cancel{16}} \text{ oz.}} = \frac{3}{4} \text{ lb.} \quad \text{or} \quad 0.75 \text{ lb.}$$

b. 0.2 T to ounces

Solution: Because we do not have a direct fact relating tons to ounces, we convert tons to pounds, then convert pounds to ounces.

$$0.2 \text{ T} = \frac{0.2 \text{ T}}{1} \cdot \frac{2000 \text{ lb.}}{1 \text{ T}} \cdot \frac{16 \text{ oz.}}{1 \text{ lb.}} = 6400 \text{ oz.}$$

Do Margin 5. ▷

OBJECTIVE 5 *Convert units of time.*

Now let's consider units of time. Time is measured in units of years (yr.), months (mo.), weeks (wk.), days (d.), hours (hr.), minutes (min.), and seconds (sec.). We will focus our attention on converting between years, days, hours, minutes, and seconds.

$$1 \text{ year} = 365\frac{1}{4} \text{ or } 365.25 \text{ days}$$

    1 day = 24 hours
    1 hour = 60 minutes
    1 minute = 60 seconds

Note: Unit abbreviations for time vary. Years can be y. or yr. Days can be d. or da. Hours can be h. or hr. Seconds can be s. or sec.

Margin 4

Convert using dimensional analysis.

a. 6 qt. to cups

b. 20 oz. to pints

c. 3 gal. to cups

Margin 5

Convert using dimensional analysis.

a. 12 lb. to ounces

b. 5000 lb. to tons

Answers to Margin 4: **a.** 24 c. **b.** 1.25 pt. **c.** 48 c.

Answers to Margin 5: **a.** 192 oz. **b.** 2.5 T

Margin 6

Convert using dimensional analysis.

a. 8.5 min. to seconds

b. 5 d. to minutes

c. 1 yr. to hours

EXAMPLE 7 Convert using dimensional analysis.

a. $4\frac{1}{2}$ hr. to seconds

Solution: Because we do not have a direct fact relating hours to seconds, we will convert hours to minutes, then minutes to seconds.

$$4\frac{1}{2} \text{ hr.} = \frac{9 \text{ hr.}}{\overset{}{\underset{1}{\cancel{2}}}} \cdot \frac{\overset{30}{\cancel{60} \text{ min.}}}{1 \text{ hr.}} \cdot \frac{60 \text{ sec.}}{1 \text{ min.}} = 16{,}200 \text{ sec.}$$

b. 15 min. to day

Solution: Because we do not have a direct fact relating minutes to days, we will convert minutes to hours, then hours to days.

$$15 \text{ min.} = \frac{\overset{1}{\cancel{15} \text{ min.}}}{1} \cdot \frac{1 \text{ hr.}}{\underset{4}{\cancel{60} \text{ min.}}} \cdot \frac{1 \text{ d.}}{24 \text{ hr.}} = \frac{1}{96} \text{ d.} \quad \text{or} \quad 0.010416\overline{6} \text{ d.}$$

◀ **Do Margin 6.**

OBJECTIVE 6 *Convert American units of speed.*

The final type of measurement that we will consider is speed. Speed is the ratio of distance to time like feet per second (ft./sec.) or miles per hour (mi./hr.). We can convert either the distance unit or the time or both within the speed unit.

EXAMPLE 8 Convert using dimensional analysis.

a. 90 ft./sec. to feet per minute

Solution: Note that the distance unit is the same in the given speed unit and the desired speed unit. This means that we only need to convert the time part of the speed unit from seconds to minutes. Because seconds is in the denominator of the given unit, we must place seconds in the numerator of the unit fraction. This means we must use the ratio of 60 sec. to 1 min.

$$90 \text{ ft./sec.} = \frac{90 \text{ ft.}}{1 \text{ sec.}} \cdot \frac{60 \text{ sec.}}{1 \text{ min.}} = \frac{5400 \text{ ft.}}{1 \text{ min.}} = 5400 \text{ ft./min.}$$

b. 60 mi./hr. to feet per second

Solution: Both the distance and time units must be converted. We are to convert the distance units from miles to feet. Because the given speed has miles in the numerator, we use a unit fraction with miles in the denominator. Because the given speed has hours in the denominator, we use a unit fraction with hours in the numerator.

We can carry out the two conversions in two stages:

$$60 \text{ mi./hr.} = \frac{60 \text{ mi.}}{1 \text{ hr.}} \cdot \frac{5280 \text{ ft.}}{1 \text{ mi.}} = \frac{316{,}800 \text{ ft.}}{1 \text{ hr.}} = 316{,}800 \text{ ft./hr.}$$

$$316{,}800 \text{ ft./hr.} = \frac{316{,}800 \text{ ft.}}{1 \text{ hr.}} \cdot \frac{1 \text{ hr.}}{60 \text{ min.}} \cdot \frac{1 \text{ min.}}{60 \text{ sec.}} = \frac{316{,}800 \text{ ft.}}{3600 \text{ sec.}} = 88 \text{ ft./sec.}$$

Or we can perform both conversions in one long line.

$$60 \text{ mi./hr.} = \frac{60 \text{ mi.}}{1 \text{ hr.}} \cdot \frac{5280 \text{ ft.}}{1 \text{ mi.}} \cdot \frac{1 \text{ hr.}}{60 \text{ min.}} \cdot \frac{1 \text{ min.}}{60 \text{ sec.}} = \frac{316{,}800 \text{ ft.}}{3600 \text{ sec.}} = 88 \text{ ft./sec.}$$

◀ **Do Margin 7.**

Margin 7

Convert using dimensional analysis.

a. 500 ft./min. to feet per second

b. 28 ft./sec. to miles per hour

c. 9 yd./sec. to feet per minute

Answers to Margin 6: **a.** 510 sec. **b.** 7200 min. **c.** 8766 hr.

Answers to Margin 7: **a.** $8\frac{1}{3}$ ft./sec. or 8.3 ft./sec. **b.** $19\frac{1}{11}$ mi./hr. or $19.\overline{09}$ mi./hr. **c.** 1620 ft./min.

7.3 Exercises

FOR EXTRA HELP

 Videotape 9

 InterAct Math Tutorial Software

 www.carsonmath.com

 AWL Math Tutor Center

InterAct MathXL www.mathxl.com

 Student's Solutions Manual

For Exercises 1–54, convert using dimensional analysis.

1. 52 yd. to feet

2. 6 ft. to inches

3. 90 in. to feet.

4. 45 ft. to yards

5. 6.5 mi. to feet

6. 7392 ft. to miles

7. $5\frac{1}{2}$ ft. to inches

8. $15\frac{2}{3}$ yd. to feet

9. 20.8 mi. to yards

10. 0.2 yd. to inches

11. 99 in. to yards

12. 3168 yd. to miles

13. 30 yd.^{2} to square feet

14. 648 in.^{2} to square feet

15. $10\frac{1}{2}$ yd.^{2} to square inches

16. 40 ft.^{2} to square yards

17. A 13.5 ft. by 12.5 ft. room is to be carpeted. Calculate the area in square yards.

18. A 16.5 ft. by 14 ft. room is to be carpeted. Calculate the area in square yards.

19. 9 pt. to cups

20. 6 qt. to pints

21. 2.5 c. to ounces

22. 16.4 gal. to quarts

23. 240 oz. to pints

24. 12 c. to pints

25. $8\frac{1}{2}$ pt. to quarts

26. 60 oz. to quarts

27. 0.4 gal. to ounces

28. 5.2 gal. to pints

29. 40.8 pt. to gallons

30. $6\frac{1}{3}$ c. to pints

31. $12\frac{1}{4}$ lb. to ounces

32. $9\frac{1}{2}$ lb. to ounces

33. 180 oz. to pounds

34. 124 oz. to pounds

35. 1200 lb. to tons

36. 6200 lb. to tons

37. 4.2 T to pounds

38. 6.4 T to pounds

39. 150 min. to hours

40. 200 min. to hours

41. 90 sec. to minutes

42. 20 sec. to minutes

43. $10\frac{1}{2}$ min. to seconds

44. 12.2 min. to seconds

45. $2\frac{1}{4}$ hr. to minutes

46. $8\frac{1}{2}$ hr. to minutes

47. 0.2 hr. to seconds

48. 1.5 hr. to seconds

49. 30 d. to years

50. 90 d. to years

51. 1 d. to minutes

52. 7 d. to seconds

53. 72 yr. to minutes

54. 85 yr. to seconds

For Exercises 55–60, solve using dimensional analysis.

▦ **55.** The speed of sound in dry air at 0°C is 1088 ft./sec. Calculate the speed of sound in miles per minute.

56. Calculate the speed of sound in miles per hour.

57. An investigating officer examining skid marks at the scene of an accident estimates that the speed of the vehicle was 80 ft./sec. The driver of the vehicle claims to have been going 40 mi./hr. Are they in agreement?

58. A defendant in a collision case claims to have been traveling at 35 mi./hr. around a blind curve and did not have time to swerve to miss rear-ending the plaintiff, who had just pulled into the road. The investigating officer determines that any speed less than 60 ft./sec. would have allowed time to brake or swerve to avoid the collision. How would you judge the defendant?

▦ **59.** The escape velocity for Earth is 36,687.6 ft./sec. Calculate this speed in miles per hour.

Trivia Bite

Escape velocity is the speed that an object must achieve to escape the gravitational pull of a planet. Any object traveling slower than the escape velocity will either return to the planet or orbit it.

▦ **60.** The minimum speed required to achieve orbit around Earth is 25,957 ft./sec. Calculate this speed in miles per hour.

Trivia Bite

An object traveling less than the minimum speed to achieve orbit will return to the planet. Objects traveling at speeds greater than the minimum speed to achieve orbit but less than the escape velocity will orbit the planet. For Earth, objects traveling at exactly 25,957 ft./sec. will orbit in a circle. As speed is increased, the path becomes more elliptical.

Puzzle Problem

A light year (lt-yr) is the distance that light travels in one year. If light travels at a rate of 186,282 mi./sec., how far is 1 lt-yr in miles? *Proxima Centauri* is the next nearest star to our star (the Sun) at a distance of about 4.2 lt-yr. What is this distance in miles?

REVIEW EXERCISES

1. 8.4(100) =

2. 0.95(1000)

3. 4800 ÷ 1000 =

4. 45.6 ÷ 10 =

5. 9 ÷ 1000 =

7.4 Metric Measurement

OBJECTIVES

1 Convert units of metric length.
2 Convert units of metric capacity.
3 Convert units of metric mass.
4 Convert units of metric area.

OBJECTIVE **1** *Convert units of metric length.*

The problem with the American system is the use of strange numbers that are not easy to remember. The metric system was designed to be easy to use and remember because it is a base-10 system. This means that different units within a type of measurement are larger or smaller by a power of 10.

Because our decimal notation is a base-10 system, when we increase or decrease the size of the measurement in the metric system, we will be multiplying or dividing by powers of 10 which, as we have seen, merely moves the decimal point either right or left. Thus, converting within the metric system will involve moving the decimal point either right or left an appropriate number of places.

Each type of measurement has a **base unit.**

| **DEFINITION** | **Base unit:** A basic unit; other units are named relative to it. |

To the right is a list of the base units for common metric measurements:

Type of measurement	Base unit
Length	meter (m)
Capacity	liter (l)
Mass	gram (g)

To further simplify the system, prefixes indicate the size of the unit within each type of measurement relative to the base unit. Below is a unit chart showing the most common prefixes relative to the base unit. In the chart below, the units to the left of the base unit are larger than the base unit while the units to right of the base unit are smaller than the base unit.

kilo	hecto	deka	Base unit	deci	centi	milli
(k)	(h)	(da)		(d)	(c)	(m)

A great feature of the system is that conversions work the same whether converting length units, capacity units, or mass units. Let's first consider metric length. As we've mentioned, the base unit for length is the meter (m), which is a little larger than a yard. Each of the other units for length is a power of 10 larger or smaller than the base unit. Putting the base unit with the prefixes we have:

kilometer	hectometer	dekameter	meter	decimeter	centimeter	millimeter
(km)	(hm)	(dam)		(dm)	(cm)	(mm)

1 km = 1000 m   1 hm = 100 m   1 dam = 10 m   1 m   1 m = 10 dm   1 m = 100 cm   1 m = 1000 mm

Here are some comparisons to help you visualize the size of these units.

A kilometer is about 0.6 of a mile.

A hectometer is about 110 yd., which is the length of a football field including one end zone.

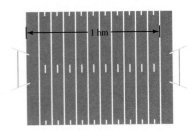

A dekameter is about the length of a school bus.

A decimeter is about the width of your hand across your palm (include your thumb)

A centimeter is about the width of your pinky finger at the tip.

A millimeter is about the thickness of your fingernail.

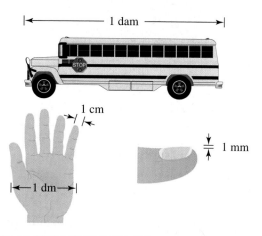

Trivia Bite

> Originally, a meter was defined to be 1/40,000,000 of a circle on Earth's surface passing through the poles. Now, we define a meter to be the distance traveled by light in empty space in 1/299,792,458 of a second.

We can use these facts to convert metric length using dimensional analysis. Suppose we want to convert 2.8 m to centimeters. We use the fact that there are 100 cm to 1 m.

$$2.8 \text{ m} = \frac{2.8 \ \cancel{m}}{1} \cdot \frac{100 \text{ cm}}{1 \ \cancel{m}} = 280 \text{ cm}$$

Note that we multiplied by 100, which causes the decimal point to move 2 places to the right of its original position in 2.8 m. Also notice that from the base unit to centi is two places to the right on the unit chart.

kilo	hecto	deka	Base unit	deci	centi	milli
(k)	(h)	(da)		(d)	(c)	(m)

Suppose we want to convert 2.8 m to km. We use the fact that there is 1 km to 1000 m. Based on what we saw above, what do you expect to happen?

$$2.8 \text{ m} = \frac{2.8 \ \cancel{m}}{1} \cdot \frac{1 \text{ km}}{1000 \ \cancel{m}} = \frac{2.8}{1000} \text{ km} = 0.0028 \text{ km}$$

Note that we divided by 1000, which caused the decimal point to move to the left 3 places. Also notice that from the base unit to kilo on the prefix chart is 3 places to the left.

kilo	hecto	deka	Base unit	deci	centi	milli
(k)	(h)	(da)		(d)	(c)	(m)

Conclusion: The decimal point moves the same number of places and in the same direction as the number of units and direction taken to get from the given unit to the desired unit on the unit chart.

EXAMPLE 1 Convert.

a. 0.15 m to decimeters

Solution: Let's first use dimensional analysis. Because the given measurement is in meters, we will use a ratio with m in the denominator. We use the ratio 10 dm to 1 m.

$$0.15 \text{ m} = \frac{0.15 \ \cancel{m}}{1} \cdot \frac{10 \text{ dm}}{1 \ \cancel{m}} = 1.5 \text{ dm}$$

Note that we multiplied 0.15 by 10, which causes the decimal point to move one place to the right. Using the unit chart, decimeter is the 1st unit to the right of meter. To convert meters to decimeters we move the decimal point 1 place to the right of its position in 0.15 m.

$$0.1\,5 \text{ m} = 1.5 \text{ dm}$$

b. 56.1 dam to centimeters

Solution: Using dimensional analysis, because we do not know a direct fact from dekameters to centimeters, we first convert dekameters to meters. We then convert meters to centimeters.

$$56.1 \text{ dam} = \frac{56.1 \text{ dam}}{1} \cdot \frac{10 \text{ m}}{1 \text{ dam}} \cdot \frac{100 \text{ cm}}{1 \text{ m}} = 56{,}100 \text{ cm}$$

Note that we are multiplying 56.1 by 1000, which causes the decimal point to move three places to the right from its original position in 56.1. Using the unit chart, centimeter is the 3rd unit to the right of dekameter (including the base unit). To convert from dekameters to centimeters, we move the decimal point 3 places to the right of its original position in 56.1 dam.

$$56.1\,0\,0 \text{ dam} = 56{,}100 \text{ cm}$$

Do Margin 1. ▶

EXAMPLE 2 Convert.

a. 2500 mm to meters

Solution: Using dimensional analysis, because the initial unit is millimeters, we multiply by a ratio that has mm in the denominator. We use the ratio of 1 m to 1000 mm.

$$2500 \text{ mm} = \frac{2500 \text{ mm}}{1} \cdot \frac{1 \text{ m}}{1000 \text{ mm}} = \frac{2500}{1000} \text{ m} = 2.5 \text{ m}$$

Note that we divided 2500 by 1000, which causes the decimal point to move to the left 3 places from its original position in 2500. Using the unit chart, we see that meters is the 3rd unit to the left of millimeters so we move the decimal point 3 places to the left of its position in 2500 mm.

$$2\,5\,0\,0 \text{ mm} = 2.5 \text{ m}$$

> **Reminder**
> The decimal point is understood to be to the right of the last digit in a whole number.

> **Reminder**
> 2.5 is the same as 2.500. However the 0 digits are no longer significant because they are to the right of the last nonzero decimal digit.

b. 970 cm to hectometers

Solution: Using dimensional analysis, because we do not know a direct fact from centimeters to hectometers, we first convert centimeters to meters. We then convert meters to centimeters.

$$970 \text{ cm} = \frac{970 \text{ cm}}{1} \cdot \frac{1 \text{ m}}{100 \text{ cm}} \cdot \frac{1 \text{ hm}}{100 \text{ cm}} = \frac{970}{10{,}000} \text{ hm} = 0.097 \text{ hm}$$

Note that we divided 970 by 10,000, which causes the decimal point to move four places to the left from its original position in 970. On the unit chart, we see that hectometers is four units to the left of centimeters, so we move the decimal point four places to the left of its position in 970 cm.

$$0\,9\,7\,0 \text{ cm} = 0.097 \text{ hm}$$

◆ **Margin 1**

Convert.

a. 9.74 m to millimeters

b. 0.075 hm to meters

c. 0.61 dam to decimeters

Answers: **a.** 9740 mm **b.** 7.5 m **c.** 61 dm

Convert.

a. 6800 mm to meters

b. 456 m to hectometers

c. 20,800 cm to kilometers

◀ **Do Margin 2.**

From Examples 1 and 2 we see that the beauty of the metric system is that the movement from one unit to another on the unit chart mirrors the movement of the decimal point in the number. From here on we'll simply use the chart to convert within the metric system.

Procedure *To convert metric units of length, capacity, or mass, use the following chart:*

kilo (k)	hecto (h)	deka (da)	Base unit	deci (d)	centi (c)	milli (m)

1. Locate the given unit on the chart.
2. Count the number of units and note the direction (right or left) to get to the desired unit.
3. Move the decimal point in the given measurement the same number of places and in the same direction.

> **Tip**
> Many people remember the order of units by making up a sentence with the initial letters of each unit. Here's an example:
>
> King Hector's Data Base Decidedly Centralized Millions

OBJECTIVE 2 *Convert units of metric capacity.*

Consider metric capacity. The base unit for metric capacity is the liter (l). A liter is the amount of liquid that would fill a 10 cm by 10 cm by 10 cm box. In other words, a liter of liquid occupies the same space as 1000 cm³.

We need to make an important connection. There are 1000 ml to 1 l. If there are 1000 cm³ and 1000 ml to 1 l, then we can conclude that 1 ml must occupy 1 cm³ of space. The common abbreviation for cm³ is cc (cubic centimeter).

Conclusion: 1 ml = 1 cm³   or   1 cc

This is an extremely important connection for people entering health-related fields to make. Medications are often prescribed in units of ml, yet the packaging for the medication is expressed in ccs or vice versa. The person administering the medication needs to recognize that ml and cc are used interchangeably. We can alter the chart for capacity slightly by including the cc with ml.

kiloliter (kl)	hectoliter (hl)	dekaliter (dal)	liter	deciliter (dl)	centiliter (cl)	milliliter or cubic centimeter (ml or cm³ or cc)
1 kl = 1000 l	1 hl = 100 l	1 dal = 10 l	1 l	1 l = 10 dl	1 l = 100 cl	1 l = 1000 ml 1 l = 1000 cm³ or cc

A liter is slightly more than a quart.

A milliliter (or cubic centimeter) is a little less than a quarter of a teaspoon.

 WARNING Be careful that you see the difference between cl and cc.

EXAMPLE 3 Convert.

a. 680 dl to hectoliters

Solution: Because hl is the 3rd unit to the left of dl, we move the decimal point 3 places to the left in 680.

$$680 \text{ dl} = 0.68 \text{ hl}$$

Answers: **a.** 6.8 m **b.** 4.56 hm **c.** 0.208 km

b. 0.0184 dal to milliliters

Solution: Because ml is 4 units to the right of dal, we move the decimal point 4 places to the right in 0.0184.

$$0.0184 \text{ dal} = 184 \text{ ml}$$

c. 0.0085 l to cubic centimeters

Solution: Because cubic centimeters and milliliters are interchangeable, we consider the conversion as moving from liters to milliliters on the chart. Because the milliliter is the 3rd unit to the right from liters, we move the decimal point 3 places to the right in 0.0085.

$$0.0085 \text{ l} = 8.5 \text{ cc}$$

Do Margin 3. ▶

OBJECTIVE 3 *Convert units of metric mass.*

Recall that mass and weight are different measurements. Weight is the force due to gravity and is dependent on the mass of an object and the gravitational pull on that object by a planet or other object. Mass on the other hand is a measure of the amount of matter that makes up an object.

DEFINITION	**Mass:** A measure of the amount of matter.

As we have mentioned, the base unit for metric mass is the gram (g).

kilogram (kg)	hectogram (hg)	dekagram (dag)	gram (g)	decigram (dg)	centigram (cg)	milligram (mg)
1 kg = 1000 g	1 hg = 100 g	1 dag = 10 g	1 g	1g = 10 dg	1 g = 100 cg	1 g = 1000 mg

A large paper clip has a mass of about 1 g. Because this is quite small, most objects are measured in terms of kg. 1 l of water has a mass of 1 kg. So, a 2 l bottle of soda has a mass of 2 kg. You are probably familiar with the mg as well. Medicines in tablet form are usually measured in mg. Common over-the-counter pain relief medicines are usually in doses of around 200 mg.

EXAMPLE 4 Convert.

a. 250 mg to grams

Solution: Because g is the 3rd unit to the left of mg, we move the decimal point 3 places to the left in 250.

$$250 \text{ mg} = 0.25 \text{ g}$$

b. 0.062 hg to grams

Solution: Because g is the 2nd unit to the right of hg, we move the decimal point 2 places to the right in 0.062.

$$0.062 \text{ hg} = 6.2 \text{ g}$$

Do Margin 4. ▶

For very large measurements of mass we use the *metric ton* (t).

$$1 \text{ t} = 1000 \text{ kg}$$

How can we convert to metric tons? If we can get to kilograms on the unit chart, then we can use the above fact to convert kilograms to metric tons.

◆ **Margin 3**

Convert.

a. 1.2 kl to dekaliters

b. 7800 cl to liters

c. 0.17 cl to cubic centimeters

Connection
Because 1 l is the amount of liquid contained in a 10 cm by 10 cm by 10 cm cube, we can say that that same cube filled with water has a mass of 1 kg.

◆ **Margin 4**

Convert.

a. 6500 mg to dekagrams

b. 0.15 kg to grams

c. 2.6 cg to decigrams

Answers to Margin 3: **a.** 120 dal **b.** 78 l **c.** 1.7 cc

Answers to Margin 4: **a.** 0.65 dag **b.** 150 g **c.** 0.26 dg

Connection

We have said that a 10 cm by 10 cm by 10 cm cube full of water has a capacity of 1 l and a mass of 1 kg. A metric ton would be the mass of a cube of water 1000 times the volume of the 10 cm by 10 cm by 10 cm cube. This cube would be 100 cm by 100 cm by 100 cm, which is actually 1 m by 1 m by 1 m. In other words, 1 t is the mass of a 1 m by 1 m by 1 m cube full of water. The volume of this cube is 1 m³ and the capacity is 1000 l or 1 kl.

Suppose we want to convert 4500 kg to tons. We could use the above fact in a dimensional analysis setup.

$$4500 \text{ kg} = \frac{4500 \text{ kg}}{1} \cdot \frac{1 \text{ t}}{1000 \text{ kg}} = \frac{4500}{1000} \text{ t} = 4.5 \text{ t}$$

Note that we divided by 1000 to get from kilograms to tons, which caused the decimal point to move left 3 places in 4500. Notice that this is the same as moving three steps to the left on the chart.

Conclusion: We can think of the metric ton as being three units to the left of kilograms.

EXAMPLE 5 Convert.

a. 64,000 kg to tons

Solution: We can think of tons as being the 3rd unit to the left of kilograms, so we move the decimal point 3 places to the left in 64,000.

$$64,000 \text{ kg} = 64 \text{ t}$$

b. 0.0058 t to kilograms

Solution: If tons is the 3rd unit to the left of kilograms, then kilograms is the 3rd unit to the right of tons, so we move the decimal point 3 places to the right in 0.0058.

$$0.0058 \text{ t} = 5.8 \text{ kg}$$

c. 240,000 dag to tons

Solution: Kilograms is 2 units to the left of dekagrams and tons is 3 units to the left of kilograms. Therefore tons is a total of 5 units to the left of dekagrams, so we move the decimal point 5 places to the left in 240,000.

$$240,000 \text{ dag} = 2.4 \text{ t}$$

◀ **Do Margin 5.**

OBJECTIVE **4** *Convert units of metric area.*

To discover the facts for area in the American system, we figured how many square inches would fit in 1 ft.². We can perform the same analysis with the metric system. Let's figure the number of square decimeters that fit in 1 m². Because there are 10 dm to 1 m, a 1 m by 1 m square is a 10 dm by 10 dm square.

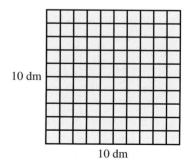

If we calculate the area of this square meter in terms of decimeters, we have 10 · 10 = 100 dm².

$$1 \text{ m}^2 = 100 \text{ dm}^2$$

We concluded in our discussion of area facts for the American system that we could simply square both sides of the length facts. Notice that the same applies with the metric facts.

$$1 \text{ m} = 10 \text{ dm}$$
$$(1 \text{ m})^2 = (10 \text{ dm})^2$$
$$1 \text{ m}^2 = 100 \text{ dm}^2$$

Deriving the other facts in this way we have:

square kilometer (km²)	square hectometer (hm²)	square dekameter (dam²)	square meter (m²)	square decimeter (dm²)	square centimeter (cm²)	square millimeter (mm²)
1 km² = 1,000,000 m²	1 hm² = 10,000 m²	1 dam² = 100 m²	1 m²	1m² = 100 dm²	1 m² = 10,000 cm²	1 m² = 1,000,000 mm²

We can still use the chart to convert units of metric area, but we have to alter our thinking a bit. Consider the following conversions and look for a pattern.

Suppose we want to convert 2.4 m² to square decimeters.

$$2.4 \text{ m}^2 = \frac{2.4 \text{ m}^2}{1} \cdot \frac{100 \text{ dm}^2}{1 \text{ m}^2} = 240 \text{ dm}^2$$

When we converted length measurements of m to dm, we multiplied by 10 causing the decimal point to move 1 place to the right. When converting m² to dm² we multiply by 100, which moves the decimal point 2 places to the right.

Now let's convert 2.4 m² to square centimeters.

$$2.4 \text{ m}^2 = \frac{2.4 \text{ m}^2}{1} \cdot \frac{10,000 \text{ cm}^2}{1 \text{ m}^2} = 24,000 \text{ cm}^2$$

When we converted length measurements of m to cm, we multiplied by 100 causing the decimal point to move 2 places to the right. When converting m² to cm² we multiply by 10,000, which moves the decimal point 4 places to the right.

Do you see a pattern? Consider 2.4 m² to square millimeters.

$$2.4 \text{ m}^2 = \frac{2.4 \text{ m}^2}{1} \cdot \frac{1,000,000 \text{ mm}^2}{1 \text{ m}^2} = 2,400,000 \text{ mm}^2$$

When we converted length measurements of m to mm, we multiplied by 1000 causing the decimal point to move 3 places to the right. When converting m² to mm² we multiply by 1,000,000, which moves the decimal point 6 places to the right.

Conclusion: The decimal point moves *twice* the number of places as units from the initial area unit to the desired area unit on the chart.

EXAMPLE 6   Convert.

a. 0.0081 m² to square millimeters

Solution: Square millimeter is the 3rd unit to the right of square meters. With area units we move the decimal point twice the number of places; so that means the decimal point moves to the right 6 places.

$$0.0081 \text{ m}^2 = 8100 \text{ mm}^2$$

Tip
Use the exponent of 2 in the area units as a reminder to move the decimal point twice the number of units from the initial unit to the desired unit on the chart.

Convert.

a. 0.25 m^2 to square centimeters

b. 4750 m^2 to square decimeters

c. $95{,}100{,}000 \text{ cm}^2$ to square hectometers

b. $45{,}100 \text{ m}^2$ to square hectometers

Solution: Square hectometer is the 2nd unit to the left of square meters. With area units we move the decimal point twice the number of places; so that means the decimal point moves to the left 4 places.

$$45{,}100 \text{ m}^2 = 4.51 \text{ hm}^2$$

c. 0.000075 km^2 to square decimeters

Solution: dm^2 is the 4th unit to the right of km^2. With area units we move the decimal point twice the number of places; so that means the decimal point moves to the right 8 places.

$$0.000075 \text{ km}^2 = 7500 \text{ dm}^2$$

◀ **Do Margin 6.**

Connection

We can connect metric prefixes with powers of ten and word names. There are more unit prefixes in the metric system than we've presented here. These unit prefixes are used in science and computers. You may have heard of computer hard drives with storage capacity in gigabytes. The prefix *giga* is a metric prefix. We can use powers of ten to indicate the size of a unit with a particular prefix relative to the base unit. For example, because a km is 1000 times the size of 1 m, we can say 1 km = 1000 m $= 10^3$ m. Or, because there are 1000 mm in 1 m we can say that 1 mm is 1/1000 of 1 m. To express 1/1000, we use a negative exponent in the power of ten notation. We say 1 mm = 1/1000 m = 10^{-3} m. Below is a list of the prefixes, their size relative to the base unit, and the common word name.

Prefix	Factor	Word name
Tera (T)	$= 10^{12}$	Trillion
Giga (G)	$= 10^{9}$	Billion
Mega (M)	$= 10^{6}$	Million
Kilo (k)	$= 10^{3}$	Thousand
Hecto (h)	$= 10^{2}$	Hundred
Deka (da)	$= 10^{1}$	Ten
Base unit	$= 10^{0} = 1$	
Deci (d)	$= 10^{-1}$	Tenth
Centi (c)	$= 10^{-2}$	Hundredth
Milli (m)	$= 10^{-3}$	Thousandth
Micro (μ)	$= 10^{-6}$	Millionth
Nano (n)	$= 10^{-9}$	Billionth
Pico (p)	$= 10^{-12}$	Trillionth

We can connect these powers of ten with scientific notation. For example 6.5 gigabytes can be expressed thus:

$$6.5 \text{ GB} = 6.5 \times 10^9 \text{ bytes} = 6.5 \text{ billion bytes} = 6{,}500{,}000{,}000 \text{ bytes}$$

On the smaller side, x-rays have a wavelength of about 3×10^{-9} m.

$$3 \times 10^{-9} \text{ m} = 3 \text{ nanometers} = 3 \text{ billionths of a meter} = 0.000000003 \text{ m}$$

Answers: **a.** 2500 cm^2
b. $475{,}000 \text{ dm}^2$ **c.** 0.951 hm^2

7.4 Exercises

FOR EXTRA HELP

 Videotape 9

 InterAct Math
Tutorial Software

 www.carsonmath.com

 AWL Math Tutor Center

InterAct MathXL www.mathxl.com

 Student's Solutions
Manual

For Exercises 1–44, convert.

1. 4.5 m to centimeters

2. 0.81 m to millimeters

3. 0.07 km to meters

4. 1.3 hm to meters

5. 3800 m to kilometers

6. 125 cm to meters

7. 9500 mm to meters

8. 1540 dm to dekameters

9. 3.48 dam to meters

10. 0.45 cm to millimeters

11. 112 m to hectometers

12. 79 mm to decimeters

13. 0.12 l to milliliters

14. 145 dl to liters

15. 0.005 kl to liters

16. 0.05 l to milliliters

17. 9.5 cl to dekaliters

18. 0.248 hl to liters

19. 0.08 l to cubic centimeters

20. 6 cc to liters

21. 12 cc to milliliters

22. 0.085 dal to cubic centimeters

23. 0.4 l to cubic centimeters

24. 145 cl to cubic centimeters

25. 12 g to milligrams

26. 0.009 kg to grams

27. 4500 g to kilograms

28. 900 mg to grams

29. 0.05 dag to centigrams

30. 0.045 hg to milligrams

31. 3600 kg to tons

32. 850 kg to tons

33. 0.106 t to kilograms

34. 5.07 t to kilograms

35. 65,000 hg to tons

36. 0.008 t to dekagrams

37. 1.8 m² to square centimeters

38. 0.05 cm² to square millimeters

39. 2200 cm² to square meters

40. 19,000 m² to square hectometers

41. 580,000 m² to square kilometers

42. 2.4 dm² to square centimeters

43. 44,300 dm² to square hectometers

44. 0.068 km² to square meters

Puzzle Problem

A light year (lt-yr) is the distance that light travels in 1 year. The speed of light is 3×10^8 m/sec. Calculate the distance of 1 lt-yr in kilometers. The Magellanic Cloud is the nearest galaxy to our own Milky Way galaxy at a distance of 150,000 lt-yr. Calculate this distance in kilometers.

REVIEW EXERCISES

1. $8.4(12) \div 100 =$

2. $6\dfrac{1}{2} \div 3 \cdot 10 =$

3. Combine like terms. $6.5x^2 - 8x + 14.4 - 10.2x - 20$

4. Distribute. $100(4.81y^2 - 2.8y + 7)$

5. Solve. $7x - 3.98 = 2.4x + 8.9$

7.5 Converting Between the American and Metric Systems

OBJECTIVES

1 Convert units of length.

2 Convert units of capacity.

3 Convert units of mass/weight.

4 Convert units of temperature.

OBJECTIVE 1 *Convert units of length.*

When converting between systems of measurement, we use dimensional analysis. The key is to have at least one fact that acts as a bridge from one system to the other. However, the more facts we know, the more likely we'll be able to convert directly to the desired unit.

Let's consider units of length. The following table lists common facts for converting metric and American units of length. Note that the conversions are approximations.

American to Metric	Metric to American
1 in. ≈ 2.54 cm	1 m ≈ 3.3 ft.
1 yd. ≈ 0.914 m	1 m ≈ 1.094 yd.
1 mi. ≈ 1.609 km	1 km ≈ 0.621 mi.

EXAMPLE 1 Convert.

a. Convert 100 yd. to meters

Solution: Because the initial unit is yards, we will multiply by a unit fraction that has yards in the denominator. Because the desired unit is meters, we need to use a fact that relates yards to meters. Note that we can choose to use 0.914 m to 1 yd. or 1 m to 1.094 yd.

$$100 \text{ yd.} \approx \frac{100 \text{ yd.}}{1} \cdot \frac{0.914 \text{ m}}{1 \text{ yd.}} \approx 91.4 \text{ m} \qquad \text{or}$$

$$100 \text{ yd.} \approx \frac{100 \text{ yd.}}{1} \cdot \frac{1 \text{ m}}{1.094 \text{ yd.}} \approx \frac{100}{1.094} \text{ m} \approx 91.4 \text{ m}$$

b. Convert 5 km to miles

Solution: Because the initial unit is kilometers we will multiply by a unit fraction that has kilometers in the denominator. Because the desired unit is miles, we need to use a fact that relates kilometers to miles. Note that we can choose to use 1 mi. to 1.609 km or 0.621 mi. to 1 km.

$$5 \text{ km} \approx \frac{5 \text{ km}}{1} \cdot \frac{1 \text{ mi.}}{1.609 \text{ km}} \approx \frac{5}{1.609} \text{ mi.} \approx 3.1 \text{ mi.} \qquad \text{or}$$

$$5 \text{ km} \approx \frac{5 \text{ km}}{1} \cdot \frac{0.621 \text{ mi.}}{1 \text{ km}} \approx 3.1 \text{ mi.}$$

Do Margin 1. ▷

Sometimes, we may not have a fact that directly relates the initial unit to the desired unit. In these situations, we simply select a fact that relates to a unit that is close to the desired unit in the target system, then convert from that unit to the desired unit.

EXAMPLE 2 Convert.

a. Convert 0.3 mi. to meters.

Solution: The facts that we have relate miles to kilometers. We will use the fact of 1.609 km to 1 mi. Once we have an answer in kilometers, we will then use the unit chart and move the decimal point 3 places to the right to get to meters.

◆ **Margin 1**

Convert.

a. Convert 60 mi. to kilometers

b. Convert 42 m to feet

c. Convert $9\frac{1}{2}$ in. to centimeters

Answers: **a.** 96.54 km **b.** 138.6 ft.
c. 24.13 cm

$$0.3 \text{ mi.} \approx \frac{0.3 \text{ mi.}}{1} \cdot \frac{1.609 \text{ km}}{1 \text{ mi.}} \approx 0.4827 \text{ km}$$

Because m is the 3rd unit to the right of km, we move the decimal point 3 places to the right in 0.4827.

$$0.4827 \text{ km} = 482.7 \text{ m}$$

b. Convert 64 mm to inches

Solution: The only fact that we have relating to inches is 1 in. to 2.54 cm. To use this fact, we need the initial unit to be in centimeters. Therefore, we first convert 64 mm to centimeters.

Because centimeters is 1 unit to the left of millimeters on the metric chart, we move the decimal point 1 place to the left in 64.

$$64 \text{ mm} = 6.4 \text{ cm}$$

Now we can use the fact of 1 in. to 2.54 cm.

$$6.4 \text{ cm} \approx \frac{6.4 \text{ cm}}{1} \cdot \frac{1 \text{ in.}}{2.54 \text{ cm}} \approx \frac{6.4}{2.54} \text{ in.} \approx 2.52 \text{ in.}$$

◄ **Do Margin 2.**

OBJECTIVE 2 *Convert units of capacity.*

Though the units and numbers change, the method for converting units of capacity does not change.

American to Metric	Metric to American
1 qt. ≈ 0.946 l	1 l ≈ 1.057 qt.

Margin 3

Convert.

a. 6 qt. to liters

b. 30 gal. to liters

c. 60 ml to ounces

d. 20 cc to ounces

EXAMPLE 3   Convert.

a. Convert 2 gal. to liters

Solution: The facts that we have relate quarts to liters. Because the initial unit is gallons, we must first convert gallons to quarts. We can then use the fact that 1 qt. ≈ 0.946 l.

$$2 \text{ gal.} \approx \frac{2 \text{ gal.}}{1} \cdot \frac{4 \text{ qt.}}{1 \text{ gal.}} \cdot \frac{0.946 \text{ l}}{1 \text{ qt.}} \approx 7.568 \text{ l}$$

> **Reminder**
> There are 4 quarts (qt.) to 1 gallon (gal.)

b. Convert 40 ml to ounces

Solution: The facts that we have relate quarts to liters. Because the initial unit is milliliters, we must first convert milliliters to liters. Because liters is 3 units to the left of milliliters, we move the decimal point 3 places to the left.

$$40 \text{ ml} = 0.04 \text{ l}$$

> **Reminder**
> Because ml and cc are the same size, we would follow the exact same procedure to convert cubic centimeters to ounces.

We can now use the fact that 1 l ≈ 1.057 qt. However, our target unit is ounces, so we'll need to convert quarts to ounces.

$$0.04 \text{ l} \approx \frac{0.04 \text{ l}}{1} \cdot \frac{1.057 \text{ qt.}}{1 \text{ l}} \cdot \frac{2 \text{ pt.}}{1 \text{ qt.}} \cdot \frac{2 \text{ c.}}{1 \text{ pt.}} \cdot \frac{8 \text{ oz.}}{1 \text{ c.}} \approx 1.35 \text{ oz.}$$

◄ **Do Margin 3.**

Answers to Margin 2: **a.** 1248.$\overline{48}$ dm
b. 19.8 in. **c.** 2188 yd.

Answers to Margin 3: **a.** 5.676 l
b. 113.52 l **c.** 2.03 oz. **d.** 0.676 oz.

OBJECTIVE 3 *Convert units of mass/weight.*

The primary conversion facts that link American weight to metric mass are as follows:

American to Metric	Metric to American
1 lb. ≈ 0.454 kg	1 kg ≈ 2.2 lb
1 T ≈ 0.907 t	1 t ≈ 1.1 T

EXAMPLE 4 Convert.

a. 180 lb. to kilograms

Solution: We have a direct fact relating pounds to kilograms. We can then use the fact that 1 lb. ≈ 0.454 kg.

$$180 \text{ lb.} \approx \frac{180 \text{ lb.}}{1} \cdot \frac{0.454 \text{ kg}}{1 \text{ lb.}} \approx 81.72 \text{ kg}$$

b. $6\frac{1}{2}$ T to metric tons.

Solution: We have a direct fact relating American tons to metric tons. We can then use the fact that 1 T ≈ 0.907 t.

$$6\frac{1}{2} \text{ T} \approx \frac{6\frac{1}{2} \text{ T}}{1} \cdot \frac{0.907 \text{ t}}{1 \text{ T}} \approx 5.9 \text{ t}$$

c. 250 g to ounces

Solution: The facts that we have relate kilograms to pounds. Because the initial unit is grams, we must first convert grams to kilograms. Because kilograms is 3 units to the left of grams, we move the decimal point 3 places to the left.

$$250 \text{ g} = 0.25 \text{ kg}$$

We can now use the fact that 1 kg ≈ 2.2 lb. However, our target unit is ounces, so we will need to convert pounds to ounces.

$$0.25 \text{ kg} \approx \frac{0.25 \text{ kg}}{1} \cdot \frac{2.2 \text{ lb.}}{1 \text{ kg}} \cdot \frac{16 \text{ oz.}}{1 \text{ lb.}} \approx 8.8 \text{ oz.}$$

Do Margin 4. ▷

◆ **Margin 4**

Convert.

a. 50 lb. to kilograms

b. 4.2 t to American tons

c. 40 oz. to grams

> **Reminder**
> There are 16 ounces (oz.) in 1 pound (lb.)

OBJECTIVE 4 *Convert units of temperature.*

Temperature is a measure of the heat of a substance. In the American system, temperature is measured in degrees Fahrenheit (°F). In the metric system, temperature is measured in degrees Celsius (°C). To relate the two systems, we need two reference points. For simplicity, we will use the freezing point and the boiling point of water.

On the Fahrenheit scale the freezing point of water is 32°F and the boiling point of water is 212°F. Note that the difference between these two points is 212°F − 32°F = 180°F.

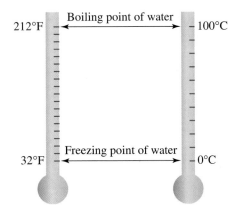

On the Celsius scale the freezing point of water is 0°C and the boiling point of water is 100°C. Note that the difference between these two points is 100°C.

Answers: **a.** 22.7 kg **b.** 4.62 T **c.** 1135 g

Margin 5

Convert.

a. 60°C to degrees Fahrenheit

b. 120.5°C to degrees Fahrenheit

c. 20°C to degrees Fahrenheit

d. −5°C to degrees Fahrenheit

Note that the differences between the two reference points can be related in a ratio. A change of 180°F is the same as a change of 100°C. Writing this in ratio form, we can simplify the ratio by reducing.

$$\frac{180°F}{100°C} = \frac{9°F}{5°C}$$

We should be able to use this ratio to convert between the two temperature scales. Let's try it with one of the reference points. If we convert 100°C to degrees Fahrenheit, we should get 212°F.

$$100°C = \frac{100°C}{1} \cdot \frac{9°F}{5°C} = 180°F$$

But 180°F isn't correct. We should get 212°F. The problem is the fact that the two scales do not have the same reference point for 0. The Fahrenheit scale is shifted by 32°F. This means we need to add 32°F when converting from degrees Celsius to degrees Fahrenheit.

$$100°C = \frac{100°C}{1} \cdot \frac{9°F}{5°C} + 32$$
$$= 180 + 32$$
$$= 212°F$$

We can write a formula to perform this conversion. To convert from degrees Celsius to degrees Fahrenheit, multiply degrees Celsius by $\frac{9}{5}$ then add 32.

Conclusion: $F = \frac{9}{5} C + 32$

EXAMPLE 5 Convert 45°C to degrees Fahrenheit.

Solution: Because the desired unit is degrees Fahrenheit we use the formula $F = \frac{9}{5} C + 32$.

$$F = \frac{9}{5}(45) + 32$$

$$F = \frac{9}{\overset{}{5}}\left(\frac{\overset{9}{\cancel{45}}}{1}\right) + 32$$

$$F = 81 + 32$$
$$F = 113°F$$

◀ **Do Margin 5.**

What if we need to convert degrees Fahrenheit to degrees Celsius? For example, suppose we wanted to convert 86°F to degrees Celsius? We could use the same formula. We could replace F with 86, then solve for C. Watch the order of operations as we solve. We'll use this order to develop the formula for converting degrees Fahrenheit to degrees Celsius.

$$F = \frac{9}{5}C + 32$$
$$86 = \frac{9}{5}C + 32$$
$$86 = \frac{9}{5}C + 32 \quad \text{Subtract 32 from}$$
$$\underline{-32 \qquad\qquad -32} \quad \text{both sides.}$$
$$54 = \frac{9}{5}C + 0$$

So 86°F = 30°C. Note the order in which we solved the equation. We first subtracted 32 from the given degrees Fahrenheit, then multiplied this result by $\frac{5}{9}$.

$$54 = \frac{9}{5}C$$

Conclusion: $C = \frac{5}{9}(F - 32)$

The parentheses are needed so that the subtraction occurs before the multiplication.

$$\frac{5}{\cancel{9}} \cdot \frac{\overset{6}{\cancel{54}}}{1} = \frac{\cancel{5}}{\cancel{9}} \cdot \frac{\cancel{9}}{\cancel{5}}C \quad \text{Multiply both sides by } \frac{5}{9}.$$

$$30 = C$$

EXAMPLE 6 Convert 80°F to degrees Celsius.

Solution: Because the desired unit is degrees Celsius, we use the formula $C = \frac{5}{9}(F - 32)$.

$$C = \frac{5}{9}(80 - 32)$$

$$C = \frac{5}{9}(48)$$

$$C = \frac{5}{\overset{}{9}}\left(\frac{\overset{16}{\cancel{48}}}{1}\right)$$

$$C = 26\frac{2}{3}°C \text{ or } 26.\overline{6}°C$$

◀ **Do Margin 6.**

Margin 6

Convert.

a. 86°F to degrees Celsius

b. 14°F to degrees Celsius

c. −28°F to degrees Celsius

d. 0°F to degrees Celsius

Answers to Margin 5: **a.** 140°F **b.** 248.9°F **c.** 68°F **d.** 23°F

Answers to Margin 6: **a.** 30°C **b.** −10°C **c.** −33.3̄°C **d.** −17.7̄°C

7.5 Exercises

FOR EXTRA HELP

 Videotape 9

 InterAct Math Tutorial Software

 www.carsonmath.com

 AWL Math Tutor Center

InterAct MathXL www.mathxl.com

 Student's Solutions Manual

For Exercises 1–38, convert.

1. A football field measures 100 yd. Convert the length of the field to meters.

2. A circular table has a diameter of 44 in. Convert the diameter to centimeters.

3. The Boston marathon is 26.2 mi. Convert to kilometers.

4. A mechanic uses a $\frac{3}{4}$ in. wrench. Convert to millimeters.

5. A carpenter uses a $\frac{3}{8}$ in. drill bit. Convert to millimeters.

6. A family builds a 4 ft. fence. Convert to centimeters.

7. An athlete prepares for the 400 m dash. Convert to yards.

8. A swimmer prepares for the 100 m freestyle. Convert to feet.

9. A sign in Germany indicates that it is 40 km to Berlin. Convert to miles.

10. A sign in France indicates that road construction is 0.4 km ahead. Convert to feet.

11. An oceanographer measures a fish to be 45 cm. What is this in inches?

12. The width of a pencil is about 7 mm. What is this in inches?

13. How many liters are in a 16 oz. bottle of soda?

14. How many milliliters are in an 8 oz. can of soda?

15. A recipe calls for $2\frac{1}{2}$ c. of milk. Convert to milliliters.

16. How many liters are in a half-gallon of milk?

17. How many gallons are in a 3 l cola?

18. How many ounces are in a 2 l cola?

19. The instructions on a pain reliever for children indicate to give the child 20 ml. How many ounces is this?

20. A child is to receive 40 ml of liquid medicine. How many ounces is this?

21. A woman weighs 135 lb. How many kilograms is this?

22. A man weighs 168 lb. How many kilograms is this?

23. How many grams is a 12 oz. steak?

24. How many grams is a 2.5 oz. serving of rice?

25. A pill contains 500 mg of medicine. How many ounces is this?

26. A pill contains 300 mg of aspirin. How many ounces is this?

27. A person from Russia claims that her mass is 55 kg. What is her weight in pounds?

28. A person from Japan claims to have a mass of 68 kg. What is his weight in pounds?

29. Normal body temperature is 98.6°F. What is normal body temperature in degrees Celsius?

30. A recipe calls for the oven to be preheated to 350°F. What is this temperature in degrees Celsius?

31. On a hot summer day, the temperature is reported to be 102°F. What is this in degrees Celsius?

32. It is suggested to keep the thermostat at 78°F. What is this in degrees Celsius?

33. During January, the temperature is reported to be −5°F. What is this in degrees Celsius ?

34. At the South Pole, the temperatures can get down to −40°F. What is this in degrees Celsius?

35. Temperatures in the Sahara desert can reach $54\frac{4}{9}$°C. What is this in degrees Fahrenheit?

36. Temperatures on the moon can reach 127°C at noon. What is this in degrees Fahrenheit?

37. On a cold day in December, the temperature is reported to be 6°C. What is this in degrees Fahrenheit?

38. The lowest temperatures on the moon can reach −173°C just before dawn. What is this in degrees Fahrenheit?

39. Absolute zero is the temperature at which molecular energy is at a minimum. This temperature is −273.15°C. What is this in degrees Fahrenheit?

Trivia Bite

Absolute zero refers to 0 on the Kelvin scale. The Kelvin scale is what most scientists use to measure temperatures. Increments on the Kelvin scale are equivalent to increments on the Celsius scale so an increase or decrease of 1 K is the same as an increase or decrease of 1°C.

40. Liquid oxygen boils at a temperature of −182.96°C. What is this in degrees Fahrenheit?

Discussion What is the boiling point of oxygen on the Kelvin scale?

Puzzle Problem

A common unit of antiquity was the cubit, which was the length of the forearm from the tip of the middle finger to the elbow. In the Biblical story of Noah's ark, the ark was to be built 300 cubits long, 50 cubits wide, and 30 cubits high. If the cubit was about 1.5 ft. in length, what were the dimensions of the ark in feet? yards? meters?

REVIEW EXERCISES

1. 896 ÷ 3200 =

2. (90 + 110 + 84) ÷ 2400 =

3. Combine like terms. $-\frac{1}{3}x + \frac{2}{5}y - x - \frac{3}{4}y$

4. Solve. $2(t + 1) = 5t$

5. Solve. $\dfrac{m}{20.4} = \dfrac{8}{15}$

7.6 Applications and Problem Solving

OBJECTIVES

1 Use debt-to-income ratios to decide loan qualification.

2 Calculate the maximum monthly PITI payment that meets the front-end ratio qualification for a loan.

3 Calculate the maximum debt that meets the back-end ratio qualification for a loan.

4 Solve problems involving two objects moving in the same direction.

5 Calculate medical dosages.

OBJECTIVE **1** *Use debt-to-income ratios to decide loan qualification.*

In Section 7.1 we considered ratios of debt-to-income. Lenders consider debt-to-income ratios to determine whether an applicant qualifies for a loan. The two ratios commonly used are **front-end** and **back-end ratios.**

DEFINITION	**Front-end ratio:** The ratio of the total monthly house payment to gross monthly income.
	Back-end ratio: The ratio of the total monthly debt payments to gross monthly income.

The total monthly house payment includes the principal and interest for the mortgage, property taxes, and home owners insurance premiums. It is usually referred to as the PITI payment.

Suppose a family with a gross monthly income of $4000 is trying to qualify for a loan. If they were to get the loan, their monthly PITI payment would be approximately $1000.

$$\text{Front-end ratio} = \frac{1000}{4000} = 0.25$$

Because the general guideline for lenders on a conventional loan is that the front-end ratio should not be greater than 0.28, this family would meet the qualification for the front-end ratio.

Now let's consider their back-end ratio. The total monthly debt includes the monthly PITI payment, minimum payments for all credit cards, all auto loan payments, and student loan payments. Sometimes student loans and car loans are not considered if they are to be paid off within a certain time frame. Suppose the family has the following monthly debts:

$$
\begin{aligned}
\text{Monthly PITI payment} &= \$1000 \\
\text{Credit card 1 minimum payment} &= \$50 \\
\text{Credit card 2 minimum payment} &= \$40 \\
\text{Car loan} &= \$253 \\
\underline{\text{Student loan}} &= \underline{\$104} \\
\text{Total monthly debt} &= \$1447
\end{aligned}
$$

Their back-end ratio is:

$$\text{Back-end ratio} = \frac{1447}{4000} = 0.36175$$

The general guideline for lenders on a conventional loan is that the back-end ratio should not be greater than 0.36. This family's back-end ratio is just slightly greater than the maximum allowed value for the back-end ratio.

In summary, the front-end ratio for this family tells us that they have enough income to make the monthly PITI payment, but their back-end ratio indicates that they have a bit too much total debt.

Note that these two ratios are only part of what lenders consider in deciding whether a person qualifies for a loan/mortgage. Two other factors that are considered are credit score and the type of loan (FHA and VA have different qualifying values for the front- and back-end ratios).

Credit scores are calculated by national credit bureaus. Generally lenders will request credit reports from several different bureaus. For example, it is common to request three reports and use the middle of the three scores in the decision process. Credit scores range from about 450 up to 900. A score of 650 is generally considered acceptable, greater than 650 is good, and less than 650 is considered risky.

Trivia Bite

Some factors that impact a credit score in order of importance are:

1. Payment history/unresolved collections (Nonpayment and late payment has a negative impact.)

2. Excessive lines of credit (more than 3 different lines of credit has a negative impact.)

3. Balances (A balance at or near the maximum has a negative impact.)

The type of loan affects the factors considered as well. For example, with VA (Veteran Affairs) loans, the front-end ratio and credit score are not considered. The back-end ratio can be up to 0.41. Of course, the primary qualification with a VA loan is that one must be a veteran. With an FHA (Federal Housing Authority) loan the credit score is not considered. The front-end ratio can be up to 0.29 and back-end ratio up to 0.41.

Table 7.1 Factors for loan qualification

	Conventional	VA	FHA
1. Front-end ratio should not exceed	0.28	N/A	0.29
2. Back-end ratio should not exceed	0.36	0.41	0.41
3. Credit score should be	650 or higher	N/A	N/A

EXAMPLE 1 The Jones family is trying to qualify for a conventional loan. Their gross monthly income is $4200. If they get the loan, their monthly PITI payment will be approximately $985. Their total monthly debt excluding the PITI payment is $535. They have a credit score of 720. Do they qualify?

Understand: We are to decide whether the Jones family qualifies for the loan. Because they are applying for a conventional loan, we must consider their front-end ratio, back-end ratio, and credit score.

Plan: **1.** Calculate their front-end ratio. Their front-end ratio should not exceed 0.28.

$$\text{Front-end ratio} = \frac{\text{Monthly PITI payment}}{\text{Gross monthly income}}$$

2. Calculate their back-end ratio. Their back-end ratio should not exceed 0.36.

$$\text{Back-end ratio} = \frac{\text{Total monthly debt payments}}{\text{Gross monthly income}}$$

3. Consider their credit score.

Execute: Front-end ratio $= \dfrac{985}{4200} \approx 0.235$

Back-end ratio $= \dfrac{1520}{4200} \approx 0.362$

Credit score $= 720$

Note: The total monthly debt payment is found by adding the monthly PITI payment to the current monthly debt.

$$985 + 535 = 1520$$

Answer: The back-end ratio is just slightly greater than 0.36. However, the Jones family has a good credit score at 720. Their good credit score should be taken into consideration to offset their 0.362 back-end ratio. They would most likely qualify for the loan.

◆ *Check:* Verify the ratio calculations by reversing the process. This will be left to the reader.

Do Margin 1. ▶

OBJECTIVE 2 *Calculate the maximum monthly PITI payment that meets the front-end ratio qualification for a loan.*

One useful piece of information in looking to buy or build a home is the maximum monthly PITI payment for which you can qualify.

EXAMPLE 2 Suppose a family has a gross monthly income of $3000. What is the maximum monthly PITI payment that would meet the front-end ratio qualification for a conventional loan?

Understand: We are to calculate the maximum monthly PITI payment that would meet the front-end ratio qualification for a conventional loan. The general guideline is that the front-end ratio should not exceed 0.28. The front-end ratio is calculated as follows:

$$\text{Front-end ratio} = \frac{\text{Monthly PITI payment}}{\text{Gross monthly income}}$$

Plan: Let p represent the monthly PITI payment. Replace each part of the front-end ratio formula with appropriate values and solve for p.

Execute:

$$0.28 = \frac{p}{3000}$$

$$\frac{3000}{1} \cdot \frac{0.28}{1} = \frac{p}{3000} \cdot \frac{3000}{1}$$

$$840 = p$$

Answer: The maximum PITI payment that would meet the front-end ratio qualification for a conventional loan with a gross monthly income of $3000 is $840.

Check: Verify that the front-end ratio with a monthly PITI payment of $840 and gross monthly income of $3000 is in fact 0.28.

$$\text{Front-end ratio} = \frac{840}{3000} = 0.28$$

Do Margin 2. ▶

OBJECTIVE 3 *Calculate the maximum debt that meets the back-end ratio qualification for a loan.*

In addition to figuring the maximum monthly PITI payment to qualify with the front-end ratio, it is helpful to also calculate the maximum debt one can have yet still satisfy the back-end ratio qualification.

◆ **Margin 1**

Use Table 7.1 to solve.

a. The Cicone family is trying to qualify for a conventional loan. Their gross monthly income is $2870. If they get the loan, their monthly PITI payment will be approximately $897. Their total monthly debt excluding the monthly PITI payment is $1223. They have a credit score of 700. Do they qualify?

b. The Mathias family is trying to qualify for an FHA loan. Their gross monthly income is $2580. If they get the loan, their monthly PITI payment will be approximately $710. They have the following monthly debt payments:

Credit card 1 = $25
Car loan 1 = $265
Student loan = $78

Their credit score is 750. Do they qualify?

◆ **Margin 2**

Suppose a person has a gross monthly income of $1680. What is the maximum monthly PITI payment that would meet the front-end ratio qualification for a FHA loan?

Answers to Margin 1: **a.** Front-end = 0.313 Back-end = 0.39 They do not qualify. **b.** Front-end = 0.275 Back-end = 0.418 Though the back-end ratio is just slightly greater than the required 0.41, the other factors warrant qualification.

Answer to Margin 2: $487.20

EXAMPLE 3 Suppose a family has a gross monthly income of $3200. What is the maximum debt that would meet the back-end ratio qualification for an FHA loan?

Understand: We are to calculate the maximum debt that would meet the back-end ratio qualification for an FHA loan. The general guideline is that the back-end ratio should not exceed 0.41. The back-end ratio is calculated as follows:

$$\text{Back-end ratio} = \frac{\text{Total monthly debt payments}}{\text{Gross monthly income}}$$

Plan: Let d represent the total monthly debt. Replace each part of the back-end ratio formula with appropriate values and solve for d.

Execute:

$$0.41 = \frac{d}{3200}$$

$$\frac{3200}{1} \cdot \frac{0.41}{1} = \frac{d}{3200} \cdot \frac{3200}{1}$$

$$1312 = d$$

Answer: The maximum monthly debt that would meet the back-end ratio qualification for an FHA loan with a gross monthly income of $3200 is $1312.

Check: Verify that the back-end ratio with a total monthly debt of $1312 and gross monthly income of $3200 is in fact 0.41.

$$\text{Back-end ratio} = \frac{1312}{3200} = 0.41$$

◀ **Do Margin 3.**

OBJECTIVE 4 *Solve problems involving two objects moving in the same direction.*

In Section 5.8, we considered distance/rate/time problems in which objects were moving in opposite directions. What if the objects are traveling in the same direction? The usual objective is to calculate the time for one object to catch up to another object ahead. We will use the same table.

Categories	Rate	Time	Distance

EXAMPLE 4 John and Karen are traveling south in separate cars on the same interstate. Karen is traveling at 65 mph and John at 70 mph. Karen passes Exit 38 at 1:15 P.M. John passes the same exit at 1:30 P.M. At what time will John catch up to Karen?

Understand: To find the time that John catches up to Karen, we must calculate the amount of time it will take him to catch up to her. We can then add that time to 1:30 P.M.

He passed the same exit 15 minutes after she did. In order to use $d = rt$, the time units must match the time units within the rate. Because 65 and 70 were in mph (miles per hour), the time units must be in terms of hours. We must convert minutes to hours.

$$15 \text{ min.} = \frac{\overset{1}{\cancel{15} \text{ min.}}}{1} \cdot \frac{1 \text{ hr.}}{\underset{4}{\cancel{60} \text{ min.}}} = \frac{1 \text{ hr.}}{4} \quad \text{or} \quad 0.25 \text{ hr.}$$

Let's begin filling in the chart:

◆ **Margin 3**

Suppose a family has a gross monthly income of $2350. What is the maximum debt that would meet the back-end ratio qualification for a VA loan?

Answer: $963.50

Because we want to find John's time to catch up we let t represent the amount of time it takes him to catch up. But, Karen's travel time is a bit tricky. Suppose John catches up at 2:15 P.M. Notice this is 1 hr. after Karen passed Exit 38. For John, it is only 45 min. after he passed Exit 38.

Conclusion: At the time John catches up, Karen's travel time from Exit 38 is 15 min. (0.25 hr.) more than John's travel time from the same exit. This will be the case no matter what time he catches up, so Karen's travel time is $t + 0.25$.

Categories	Rate	Time	Distance
Karen	65	$t + 0.25$	$65(t + 0.25)$
John	70	t	$70t$

Because $d = rt$, we multiply the rate and time columns to get the distance column.

What can we conclude about the distances they each will have traveled when John catches up? Notice we used Exit 38 as the reference point from which we described the time of travel. When John catches up, he and Karen will be exactly the same distance from Exit 38!

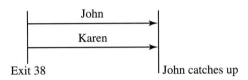

Conclusion: At the time John catches up, he and Karen will have traveled the exact same distance from Exit 38.

Plan: Because they will have gone the same distance past Exit 38 when John catches up, we set the expressions of their individual distances equal and solve for t.

Execute: John's distance = Karen's distance (at the time he catches up)

When we subtract $65t$ from both sides, we subtract the coefficients of the like terms, which are the individual rates. $70 - 65 = 5$ means that John's rate relative to Karen's is 5 mph. This means he must make up the 16.25 mi. that separates them at a relative rate of 5 mph.

$$70t = 65(t + 0.25)$$
$$70t = 65t + 16.25$$
$$70t = 65t + 16.25$$
$$\underline{-65t \quad -65t}$$
$$5t = 0 + 16.25$$
$$\frac{5t}{5} = \frac{16.25}{5}$$
$$1t = 3.25$$
$$t = 3.25$$

When we distribute 65 and multiply $65 \cdot 0.25$, we are multiplying rate by time, which makes distance. $65 \cdot 0.25 = 16.25$ mi., which is the distance that Karen is ahead of John. At the end of the 15 min. (0.25 hr.) John is at Exit 38 and she is 16.25 miles ahead. This is the distance John must make up to catch her.

When we divide the 16.25 mi. by the 5 mph, we get the time it takes John to make up the distance between him and Karen.

Answer: It will take John 3.25 hr., which is 3 hr. and 15 min., after passing Exit 38 to catch up to Karen. If it was 1:30 P.M. when he passed Exit 38, then it will be 4:45 P.M. when he catches up.

Check: If John travels for 3.25 hr. after Exit 38, then Karen travels for $3.25 + 0.25 = 3.5$ hr. after Exit 38. We can check by verifying that they indeed are the exact same distance after traveling for their respective amounts of time.

John: $d = rt$ Karen: $d = rt$
$d = (70)(3.25)$ $d = (65)(3.5)$
$d = 227.5$ mi. $d = 227.5$ mi.

Do Margin 4. ▶

◆ **Margin 4**

Carlos and Rita are running in the Boston marathon. Carlos passes Faneuil hall at 10:30 and Rita at 11:00. Carlos is pacing himself at 4 mph while Rita is pushing at 5 mph. How much time will it take Rita to catch up with Carlos?

Answer: 2 hr.

OBJECTIVE 5 *Calculate medical dosages.*

We can use the method of dimensional analysis to determine appropriate medical dosages. Recall that the main idea is to note the initial unit and the desired unit. Then multiply by appropriate unit fractions so that the undesired units divide out leaving the desired units.

EXAMPLE 5 A patient with a mass of 42 kg is to receive an antibiotic. The order is to administer 20 mg per kg. How much of the antibiotic should be given?

Understand: The order of 20 mg per kg means that the patient should receive 20 mg of the antibiotic for every 1 kg of the patient's mass.

Plan: Multiplying the patients mass by the dosage per unit mass should yield the total dose.

Execute: patients mass · dosage per unit of mass = total dose

$$\frac{42 \text{ kg}}{1} \cdot \frac{20 \text{ mg}}{1 \text{kg}} = 840 \text{ mg}$$

Answer: The patient should receive 840 mg of the antibiotic.

Check: Verify that if a 42 kg patient is receiving 840 mg of antibiotic, then he or she is receiving 20 mg of antibiotic per kg of mass.

$$\frac{840 \text{ mg}}{42 \text{ kg}} = 20 \text{ mg/kg}$$

◀ **Do Margin 5.**

Many medications are administered in intravenous (IV) solutions. A regulator is used to regulate the number of drops of the solution the patient receives over a period of time (usually 1 minute). The medical orders are usually not written in terms of the number of drops per minute so, in order that the regulator may be properly set, they must be converted to number of drops per minute.

EXAMPLE 6 A patient is to receive 500 ml of 5% D/W solution IV over 4 hr. The label on the box of the IV indicates that 10 drops dissipate 1 ml of the solution. How many drops should the patient receive each minute?

Understand: The order is for 500 ml over 4 hr. With this particular IV, 10 drops of the solution is equal to 1 ml. We can write these rates as:

$$\frac{500 \text{ ml}}{4 \text{ hr}} \quad \text{and} \quad \frac{10 \text{ drops}}{1 \text{ ml}}$$

We want to end up with a description of the number of drops per minute. This means that milliliters and hours are undesired units. We need to convert hours to minutes and divide out milliliters completely.

Plan: Multiply the above rates to eliminate milliliters. Use the fact that 1 hr. is 60 min. to convert hours to minutes.

Execute: $\dfrac{500 \text{ ml}}{4 \text{ hr.}} \cdot \dfrac{10 \text{ drops}}{1 \text{ ml}} \cdot \dfrac{1 \text{ hr.}}{60 \text{ min.}} = \dfrac{5000 \text{ drops}}{240 \text{ min.}} = 20.8\overline{3}$ drops/min.

Answer: The patient should receive about 20 to 21 drops per minute.

Check: Verify that a rate of 21 drops/min. with every 10 drops equal to 1 ml is a total of 500 ml after 4 hr.

$$\frac{4 \text{ hr}}{1} \cdot \frac{60 \text{ min.}}{1 \text{ hr.}} \cdot \frac{21 \text{ drops}}{1 \text{ min.}} = 5040 \text{ drops}$$

If the patient recieves 5040 drops in 4 hr. and every 10 drops is 1 ml, we can calculate the total number of ml that the patient receives:

$$\frac{5040 \text{ drops}}{1} \cdot \frac{1 \text{ ml}}{10 \text{ drops}} = \frac{5040}{10} \text{ ml} = 504 \text{ ml}$$

Because we rounded the number of drops to whole amounts, this result does not match 500 ml exactly. However, it is reasonable.

◀ **Do Margin 6.**

◆ Margin 5

A patient with a mass of 68 kg is to receive isoniazid, an antitubercular drug. The order is to administer 10 mg per kg. How much isoniazid should be given?

◆ Margin 6

A patient is to receive 250 cc of 5% D/W solution IV over 4 hr. The label on the box of the IV indicates that 15 drops dissipates 1 cc of the solution. How many drops should the patient receive each minute?

Answer to Margin 5: 680 mg

Answer to Margin 6: 15 to 16 drops per minute.

7.6 Exercises

For Exercises 1–16, use the table to solve.

Factors for loan qualification

	Conventional	VA	FHA
1. Front-end ratio should not exceed	0.28	N/A	0.29
2. Back-end ratio should not exceed	0.36	0.41	0.41
3. Credit score should be	650 or higher	N/A	N/A

1. The Wu family is trying to qualify for a conventional loan. Their gross monthly income is $3850. If they get the loan, their monthly PITI payment will be approximately $845. Their monthly debt excluding the monthly PITI payment is $435. They have a credit score of 700. Do they qualify?

2. The Deas family is trying to qualify for a conventional loan. Their gross monthly income is $3240. If they get the loan, their monthly PITI payment will be approximately $925. Their total monthly debt excluding the monthly PITI payment is $495. They have a credit score of 680. Do they qualify?

3. The Santana family is trying to qualify for a VA loan. Their gross monthly income is $4200. If they get the loan, their monthly PITI payment will be approximately $987. They have a credit score of 650. They have the following monthly debt payments:

$$\text{Credit card 1} = \$35$$
$$\text{Credit card 2} = \$85$$
$$\text{Car loan 1} = \$305$$
$$\text{Car loan 2} = \$289$$

Do they qualify?

4. The Rivers family is trying to qualify for a VA loan. Their gross monthly income is $2845. If they get the loan, their monthly PITI payment will be approximately $756. They have a credit score of 660. They have the following monthly debt payments:

$$\text{Credit card 1} = \$30$$
$$\text{Credit card 2} = \$45$$
$$\text{Car loan 1} = \$279$$
$$\text{Student loan} = \$93$$

Do they qualify?

5. The Bishop family is trying to qualify for an FHA loan. Their gross monthly income is $2530. If they get the loan, their monthly PITI payment will be approximately $785. They have the following monthly debt payments:

$$\text{Credit card 1} = \$40$$
$$\text{Car loan} = \$249$$
$$\text{Student loan} = \$65$$

Their credit score is 700. Do they qualify?

6. The Nicks family is trying to qualify for an FHA loan. Their gross monthly income is $2760. If they get the loan, their monthly PITI payment will be approximately $805. They have the following monthly debt payments:

$$\text{Credit card 1} = \$25$$
$$\text{Credit card 2} = \$35$$
$$\text{Car loan 1} = \$279$$

Their credit score is 650. Do they qualify?

7. Suppose a family has a gross monthly income of $4800. What is the maximum monthly PITI payment that would meet the front-end ratio qualification for a conventional loan?

8. Suppose a family has a gross monthly income of $2980. What is the maximum monthly PITI payment that would meet the front-end ratio qualification for a conventional loan?

9. Suppose a person has a gross monthly income of $1840. What is the maximum monthly PITI payment that would meet the front-end ratio qualification for a FHA loan?

10. Suppose a person has a gross monthly income of $1580. What is the maximum monthly PITI payment that would meet the front-end ratio qualification for a FHA loan?

11. Suppose a family has a gross monthly income of $3480. What is the maximum debt that would meet the back-end ratio qualification for a conventional loan?

12. Suppose a family has a gross monthly income of $2240. What is the maximum debt that would meet the back-end ratio qualification for a conventional loan?

13. Suppose a family has a gross monthly income of $1675. What is the maximum debt that would meet the back-end ratio qualification for a VA loan?

14. Suppose a family has a gross monthly income of $2110. What is the maximum debt that would meet the back-end ratio qualification for a VA loan?

15. Suppose a family has a gross monthly income of $1375. What is the maximum debt that would meet the back-end ratio qualification for a FHA loan?

16. Suppose a family has a gross monthly income of $2680. What is the maximum debt that would meet the back-end ratio qualification for a FHA loan?

17. Terese and Brian are traveling south in separate cars on the same interstate. Terese is traveling at 60 mph and Brian at 70 mph. Terese passes Exit 62 at 2:30 P.M. Brian passes the same exit at 3:30 P.M. At what time will Brian catch up to Terese?

18. Sherryl and Darrin are traveling west in separate cars on the same interstate. Darrin is traveling at 65 mph and Sherryl at 70 mph. Darrin passes a rest area at 10:05 A.M. Sherryl passes the same rest area at 10:25 P.M. At what time will Sherryl catch up to Darrin?

19. Liz and Tyler are riding bicycles in the same direction on the same trail. Liz passes a marker at 12:18 P.M. Tyler passes the same marker at 12:33 P.M. If Liz is averaging 8 mph and Tyler is averaging 10 mph, at what time will Tyler catch up to Liz?

20. Nikki and Berry are running in the same direction on the same trail. Berry passes a marker at 8:10 A.M. Nikki passes the same marker at 8:20 A.M. If Berry is averaging 4 mph and Nikki is averaging 6 mph, at what time will Nikki catch up to Berry?

21. A patient with a mass of 55 kg is to receive an antibiotic. The order is to administer 9.5 mg per kg. How much of the antibiotic should be given?

22. A patient with a mass of 81 kg is to receive 1.4 mg per kg of metroprolol tartrate (an antihypertensive drug). How much of the drug should the patient receive?

23. A patient weighing 132 lb. is to receive 0.04 mg/kg of clonazepam (an anticonvulsant drug). If each tablet contains 0.5 mg, how many tablets should the patient receive?

24. A patient weighing 165 lb. is to receive 7.2 mg/kg of azithromycin. If each capsule contains 300 mg, how many capsules should the patient receive?

25. A patient is to receive 250 ml of 5% D/W solution IV over 6 hr. The label on the box of the IV indicates that 15 drops dissipate 1 ml of the solution. How many drops should the patient receive each minute?

26. A patient is to receive 500 ml of 5% D/W solution IV over 2 hr. The label on the box of the IV indicates that 15 drops dissipate 1 ml of the solution. How many drops should the patient receive each minute?

27. A 500 ml of 5% D/W solution IV contains 20,000 units of heparin. If the patient receives 10 ml per hour, how many units is the patient receiving each hour?

28. A 500 ml of 5% D/W solution IV contains 20 units of pitocin. If the patient receives 0.002 units per minute, how many milliliters is she receiving in an hour?

REVIEW EXERCISES

1. Simplify the ratio 0.14 to 0.5.

2. What is the probability of randomly selecting a king or jack from a shuffled standard deck of cards?

3. A 16 oz. bottle of apple juice costs $2.29. What is the unit price in $/oz?

4. Convert 75 ft. to yd.

5. Convert 8.5 g to mg.

Summary

Defined Terms

Review the following terms and for those you do not know, study its definition on the page number next to it.

Section 7.1
Ratio *p. 476*
Theoretical probability *p. 478*
Unit ratio *p. 480*
Rate *p. 481*

Section 7.2
Proportion *p. 487*
Similar figures *p. 491*
Equivalent ratios *p. 487*

Section 7.3
Unit fraction *p. 499*
Capacity *p. 502*

Section 7.4
Base unit *p. 507*

Section 7.6
Front-end ratio *p. 523*
Back-end ratio *p. 523*

Procedures, Rules, and Key Examples

Procedures/Rules	Key Example(s)
Section 7.1 **To translate a phrase involving the word "ratio":** 1. Find the word *to* 2. Write the amount expressed to the left of the word *to* in the numerator. 3. Write the amount expressed to the right of the word *to* in the denominator.	A roof has a vertical rise of 8 in. for every 13 in. of horizontal length. What is the ratio of vertical rise to horizontal length? Answer: $\dfrac{8}{13}$
To simplify a ratio of two fractions or mixed numbers, divide the fractions/mixed numbers.	A recipe calls for $\frac{1}{2}$ cup of flour to $2\frac{1}{4}$ cups of milk. Write the ratio of flour to milk in simplest form. $$\frac{\frac{1}{2}}{2\frac{1}{4}} = \frac{1}{2} \div 2\frac{1}{4}$$ $$= \frac{1}{2} \div \frac{9}{4}$$ $$= \frac{1}{\overset{}{2}} \cdot \frac{\overset{2}{\cancel{4}}}{9}$$ $$= \frac{2}{9}$$

Procedures/Rules (cont.)	Key Example(s)

To clear decimals in a ratio, multiply the numerator and denominator by an appropriate power of 10 as determined by the decimal number with the most decimal places.

A small plastic tube has a diameter of 0.08 cm and a length of 12.5 cm. What is the ratio of length to diameter?

$$\frac{12.5}{0.08} = \frac{12.5(100)}{0.08(100)} = \frac{1250}{8} = \frac{625}{4}$$

To write a probability:
1. Write the number of favorable outcomes in the numerator.
2. Write the total number of possible outcomes in the denominator.
3. Simplify.

Write the probability of drawing a red 10 from a standard deck of cards in simplest form.

There are 2 red 10's in a standard deck of 52 cards.

$$P = \frac{2}{52} = \frac{1}{26}$$

To calculate a unit ratio, divide the denominator into the numerator.

The stock for a chemical company sold for $16\frac{1}{4}$ at the close of the market one day. If the company's annual profit on each share of stock was $1.85 the previous year, what is the price to earnings ratio?

$$\frac{16\frac{1}{4}}{1.85} = \frac{16.25}{1.85} = 8.78$$

Section 7.2

If ratios are equivalent then the cross products are equal.
In math language: Given two ratios: $\frac{a}{b}$ and $\frac{c}{d}$ where $b \neq 0$ and $d \neq 0$, if $\frac{a}{b} = \frac{c}{d}$ then $ad = bc$.

Determine whether the ratio is proportional.

$$9.6 \cdot 8 = 76.8 \qquad 6.4 \cdot 12 = 76.8$$

$$\frac{8}{6.4} \underset{\diagup}{\overset{\diagdown}{\times}} \frac{12}{9.6}$$

The cross products are equal, therefore the ratios are proportional.

To solve a proportion using cross products:
1. Calculate the cross products.
2. Set the cross products equal to one another.
3. Use the multiplication/division principle to isolate the variable.

Solve. $\dfrac{2.8}{-9} = \dfrac{m}{15}$

$$15(2.8) = 42 \qquad\qquad -9 \cdot m = -9m$$

$$\frac{2.8}{-9} \underset{}{\overset{}{\times}} \frac{m}{15}$$

$$42 = -9m$$
$$\frac{42}{-9} = \frac{9m}{-9}$$
$$-4.\overline{6} = 1m$$
$$-4.\overline{6} = m$$

To solve proportion problems:
1. Set up the given ratio any way you wish.
2. Set up the ratio with the unknown so that it logically corresponds to the way you set up the given ratio.
3. Solve.

A garden plan shows 30 plants to cover 96 ft.^{2}. How many plants would be needed to cover a garden area that is 140 ft.^{2}?

$$\frac{30 \text{ plants}}{96 \text{ ft.}^2} = \frac{n \text{ plants}}{140 \text{ ft.}^2}$$
$$4200 = 96n$$
$$\frac{4200}{96} = \frac{96n}{96}$$
$$43.75 = 1n$$
$$43.75 = n$$

Answer: Because one can only purchase whole plants, the larger garden would require about 44 plants.

Section 7.3

To convert units using dimensional analysis, multiply the given measurement by unit fractions so that the undesired units divide out leaving the desired units.

Tips: If an undesired unit is in a numerator, then that unit should appear in the denominator of a unit fraction.

If an undesired unit is in the denominator, then that unit should appear in the numerator of a unit fraction.

Convert:
4.5 ft. to inches

$$4.5 \text{ ft.} = \frac{4.5 \text{ ft.}}{1} \cdot \frac{12 \text{ in.}}{1 \text{ ft.}} = 54 \text{ in.}$$

20 lb. to kilograms

$$20 \text{ lb.} = \frac{20 \text{ lb.}}{1} \cdot \frac{1 \text{ kg}}{2.2 \text{ lb.}} = \frac{20}{2.2} \text{ kg} = 9.\overline{09} \text{ kg}$$

Facts for conversions involving American units:

Length:
1 ft. = 12 in.
1 yd. = 3 ft.
1 mi. = 5280 ft.

Area:
1 ft.^{2} = 144 in.^{2}
1 yd.^{2} = 9 ft.^{2}

Capacity:
1 cup = 8 oz.
1 pt. = 2 cups
1 qt. = 2 pt.
1 gal. = 4 qt.

Weight:
1 lb. = 16 oz.
1 T = 2000 lb.

Section 7.4

To convert metric units of length, capacity, or mass, use the following chart:

Kilo (k)	hecto (h)	deka (da)	Base unit	deci (d)	centi (c)	milli (m)

Memory tip:

King Hector's Data Base Decidedly Centralized Millions

1. Locate the given unit on the chart.
2. Count the number of units and note the direction (right or left) to get to the desired unit.
3. Move the decimal point in the given measurement the same number of places and in the same direction.

Notes:
In capacity, the cubic centimeter (cc) represents the same amount as the milliliter (ml).
The metric ton is 1000 kg, or 3 steps on the chart to the left of kilograms.

Convert:
2.8 m to centimeters

(Centimeters is 2 places to the right of meters on the chart so the decimal point moves right 2 places.)

2.8 m = 280 cm

97,500 dg to kilograms

(Kilograms is 4 places to the left of decigrams on the chart so the decimal point moves 4 places to the left.)

97,500 dg = 9.75 kg

Section 7.5

Facts for converting between American and metric systems:

Length:
1 in. ≈ 2.54 cm	1 m ≈ 3.3 ft.
1 yd. ≈ 0.914 m	1 m ≈ 1.094 yd.
1 mi. ≈ 1.609 km	1 km ≈ 0.621 mi.

Capacity:
1 qt. ≈ 0.946 l   1 l ≈ 1.057 qt.

Weight/Mass:
1 lb. ≈ 0.454 kg	1 kg ≈ 2.2 lb.
1 T ≈ 0.907 t	1 t ≈ 1.1 T

Formulas

Convert °C to °F:
$$F = \frac{9}{5}C + 32$$

Convert °F to °C:
$$C = \frac{5}{9}(F - 32)$$

$$\text{Front-end ratio} = \frac{\text{Monthly PITI payment}}{\text{Gross monthly income}}$$

$$\text{Back-end ratio} = \frac{\text{Total monthly debt payments}}{\text{Gross monthly income}}$$

For Exercises 1–6, answer true or false.

1. Every ratio is a rational number.

2. The ratio 0.9 to 0.5 can be expressed as $\dfrac{9}{5}$.

3. The cross products of a proportion are equal.

4. A unit price is the ratio of quantity to price.

5. The probability of rolling a 7 on a die with six sides is 0.

6. 1 cc is the same amount as 1 cl.

7. To calculate probability, write a ratio with the number of _____ outcomes in the numerator and the total number of _____ outcomes in the denominator.

8. The denominator of a unit ratio is always equal to _____.

9. To calculate a unit ratio, divide the _____ into the _____.

10. To convert a given measurement using dimensional analysis, multiply the given measurement by a unit fraction that has the given unit in the _____.

For Exercises 11 and 12, write each ratio in simplest form.

11. A company has 32 male employees and 22 female employees. Write the ratio of female to male employees.

12. The shortest side on a right triangle measures 2.5 in. The hypotenuse measures 10.8 in. Write the ratio of the shortest side to the hypotenuse.

For Exercises 13 and 14, calculate the probability in simplest form.

13. What is the probability of randomly selecting a Queen or King from a shuffled standard deck of cards?

14. What is the probability of rolling a 1 or a 2 on a die that has six sides?

For Exercises 15 and 16, calculate the unit ratio.

15. A college has 4500 students and 275 faculty. Write a unit ratio of students to faculty.

16. The price for each share of stock for a communications company closed at $12\frac{5}{8}$. The previous year's annual earning for the stock was $1.78. Calculate the price-to-earnings ratio.

For Exercises 17 and 18, calculate the rate.

17. Sekema drove 210.6 mi. in 3 hr. What was her average rate in miles per hour?

18. A phone call lasting 28 min. costs $2.52. What is the rate in dollars per minute?

For Exercises 19 and 20, calculate the unit price.

19. A 40 oz. bag of sugar costs $1.79. What is the unit price in dollars per ounce?

20. A 64 oz. bottle of juice costs $2.49. What is the unit price in dollars per ounce?

For Exercises 21 and 22, determine which is the better buy.

21. Which is the better buy?

$0.79 for a 32 oz. cup of cola

or

$0.69 for a 20 oz. cup of cola

22. Which is the better buy?

Two 12 oz. cans of name brand frozen juice for $0.99

or

four 8 oz. cans of store brand frozen juice for $1.45

For Exercises 23–26, solve.

23. Solve. $\dfrac{1.8}{-5} = \dfrac{k}{12.5}$

24. Solve. $\dfrac{2}{m} = \dfrac{3\frac{1}{4}}{15}$

25. The instructions on the back of a can of wood sealant indicate that 3.78 L will cover about 25 m² of mildly porous wood. If the can contains 4.52 L of sealant, how many m² of mildly porous wood will it cover?

26. On a map $\frac{1}{4}$ in. represents 20 mi. Suppose two cities are $3\frac{1}{2}$ in. apart on the map. How far apart are the two cities in miles?

For Exercises 27–32, use dimensional analysis to convert.

27. Convert 12 yd. to inches

28. Convert 40 oz. to pounds

29. Convert 6 qt. to ounces

30. Convert 2.25 hr. to minutes

31. A 14 ft. by 15.5 ft. room is to be carpeted. Calculate the area of the room in square yards.

32. A meteor streaks through Earth's upper atmosphere at a speed of 116,160 ft./sec. Calculate the speed in miles per hour.

For Exercises 33–36, convert.

33. Convert 5 m to decimeters

34. Convert 0.26 kg to grams

35. Convert 950,000 g to metric tons

36. Convert 0.075 l to cubic centimeters

37. A patient is to receive 450 ml of NS (normal saline) IV in 6 hr. If the IV dissipates 1 ml every 10 drops, what is the drip rate in drops per minute?

38. A patient is to receive an antibiotic. The patient has a mass of 32 kg. The order for the antibiotic is 2.9 mg per kg. How much should the patient receive?

39. Convert 20 ft. to meters

40. Convert 12 mi. to kilometers

41. Convert 145 lb. to kilograms

42. Convert 3 l to gallons

43. Convert 90°F to degrees Celsius.

44. Convert −4°C to degrees Fahrenheit.

45. A family applying for a loan has a gross monthly income of $3450. If their monthly PITI payment will be $978, what is the ratio of monthly PITI payment to gross monthly income (front-end ratio)?

46. A family has the following monthly debts:

$$\text{Monthly PITI} = \$565$$
$$\text{Credit card 1} = \$45$$
$$\text{Car loan} = \$225$$
$$\text{Student loan} = \$125$$

If their gross monthly income is $2420, what is ratio of monthly debt to gross monthly income (back-end ratio)?

47. A family with a gross monthly income of $1970 is applying for an FHA loan. To qualify, their front-end ratio must not exceed 0.29. What is the maximum monthly PITI payment that they can have and qualify for the loan?

48. A family with a gross monthly income of $4200 is applying for a VA loan. To qualify, their back-end ratio must not exceed 0.41. What is the maximum monthly debt that they can have and qualify for the loan?

49. Dale is traveling 60 mph and passes Exit 50 on a highway. $\frac{1}{2}$ hr. later Tanya passes the same exit at 70 mph. If they maintain the same rates of speed, how long will it take Tanya to catch up to Dale?

50. Vijay is walking at 3 mph in a walk-run charity event and passes a drink station at 10:42 A.M. Micah jogs past the same drink station at 11:02 A.M. at a speed of 5 mph. What time will Micah catch up with Vijay?

1. A computer screen is 9.5 in. by 13 in. Write the ratio of the shorter side to the longer side in simplest form.

2. Dianne has 22 entries in a random drawing contest that has 4578 total entries. What is the probability that Dianne will win? Write the probability in simplest form.

3. What is the probability of rolling a 9 or a 10 on a die that has 10 sides? Write the probability in simplest form.

4. A college has 3850 students and 125 faculty. Write a unit ratio of students to faculty.

5. Laurence drove 158.6 mi. in 2.25 hr. What was his average rate in miles per hour?

6. A long distance phone call lasting 36 min. costs $5.04. What is the rate in dollars per minute?

7. A 15 oz. can of mixed vegetables costs $0.59. What is the unit price in dollars per ounce?

8. Which is the better buy?

    $0.89 for a 20 oz. can of pineapple   or $0.69 for a 15 oz. can of pineapple

9. Solve. $\dfrac{-6.5}{12} = \dfrac{n}{10.8}$

10. A company charges $49 to install 20 ft.^{2} of hardwood floor. How much would they charge to install the same floor in a 210 ft.^{2} room?

11. On a map $\frac{1}{4}$ in. represents 50 mi. Suppose two cities are $4\frac{1}{2}$ in. apart on the map. How far apart are the two cities in miles?

For Exercises 12–16, use dimensional analysis to convert.

12. Convert 14 ft. to inches

13. Convert 20 lb. to ounces

14. Convert 36 pt. to gallons

15. Convert 150 min. to hours

1. _____

2. _____

3. _____

4. _____

5. _____

6. _____

7. _____

8. _____

9. _____

10. _____

11. _____

12. _____

13. _____

14. _____

15. _____

16. _____

16. A 20 ft. by 12.5 ft. room is to be carpeted. Calculate the area of the room in square yards.

17. Convert.

    a. 0.058 m to centimeters

17. _____

    b. 420 m to kilometers

18. Convert.

    a. 24 g to kilograms

18. _____

    b. 0.091 t to grams

19. Convert.

    a. 280 cm^2 to square meters

19. _____

    b. 0.8 m^2 to square decimeters

20. Convert.

20. _____

    a. 6.5 dl to liters

    b. 0.8 l to cubic centimeters

21. A patient is to receive 750 ml of NS IV in 8 hr. If the IV dissipates 1 ml every 10 drops, what is

21. _____ the drip rate in drops per minute?

22. Convert 75 ft. to meters

22. _____

23. Convert 30°F to degrees Centigrade.

24. A family has the following monthly debts:

23. _____

$$
\begin{aligned}
\text{Monthly PITI} &= \$714 \\
\text{Credit card 1} &= \$128 \\
\text{Credit card 3} &= \$78 \\
\text{Car payment} &= \$295 \\
\text{Car payment} &= \$324
\end{aligned}
$$

24. _____

If their gross monthly income is $3420, what is the ratio of monthly debt to gross monthly income?

25. Catrina is traveling 65 mph and passes mile marker 80 on a highway. 1 hr later Darryl passes the

25. _____ same mile marker at 70 mph. If they maintain the same rates, how long will it take Darryl to catch up to Catrina?

Cumulative Review Exercises

For Exercises 1–6, answer true or false.

1. $-2.4 \times 10^6 > -2.4 \times 10^5$

2. $\left(5\frac{3}{4}\right)^0 = 0$

3. 127 is a prime number.

4. $\sqrt{5}$ is a rational number.

5. $\frac{4}{9}x = 12$ is a linear equation.

6. $5^2 \cdot 5^6 = 5^{12}$

For Exercises 7–9, fill in the blank.

7. To calculate the quotient of two fractions, we write an equivalent multiplication statement using the _____ of the divisor.

8. To clear fractions from an equation, we can use the multiplication principle of equality and multiply both sides by the _____ of the denominators.

9. The _____ products of equal ratios are equal.

10. Explain in your own words how to solve a proportion.

11. Write the word name for 49,802.76.

12. Write 914,000,000 in scientific notation.

13. Graph $-3\frac{2}{3}$ on a number line.

14. Round 41.325 to the nearest hundredth.

15. What is the coefficient of $-x^3$?

16. What is the degree of $4x^2 - 12x^5 + 10x - 8$?

For Exercises 17–24, simplify.

17. $[10 - 3(7)] - [16 \div 2(7 + 3)]$

18. $5\frac{2}{3} \div \left(-2\frac{1}{6}\right)$

19. $10\frac{1}{2} - 4\frac{3}{5}$

20. $|12.5 - 18.65| \div 0.2$

21. $\left(8\frac{2}{3}\right)(-3.6)$

22. $\left(\frac{3}{4}\right)^2$

23. $\sqrt{\frac{45}{5}}$

24. $(-0.3)^4$

25. Evaluate $\frac{1}{2}mv^2$ when $m = 380$ and $v = 0.4$

26. Subtract. $(12.4y^3 - 8.2y^2 + 9) - (4.1y^3 + y - 1.2)$

27. Multiply. $\left(\frac{2}{3}a - 1\right)\left(\frac{3}{4}a - 7\right)$

28. Find the prime factorization of 2400.

29. Find the GCF of $40xy^2$ and $60x^5$

30. Divide. $18x^6 \div 3x^2$

31. Factor. $32m^4 + 24m^2 - 16m$

32. Simplify. $\dfrac{9x^2}{20y} \cdot \dfrac{5y}{12x^6}$

33. Find the LCM of $8k^2$ and $10k$.

34. Simplify. $\dfrac{2}{x} + \dfrac{5}{9}$

For Exercises 35–38, solve and check.

35. $2.5n - 16 = 30$

36. $\dfrac{3}{4}x - 5 = \dfrac{1}{5}x + 3$

37. $\dfrac{10.2}{x} = \dfrac{-12.5}{20}$

38. Convert 12.5 mi. to feet

39. Convert 0.48 kg to grams

40. Convert 60°C to degrees Fahrenheit.

For Exercises 41–50, solve.

41. Find the surface area of the shape.

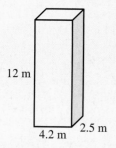

12 m

4.2 m 2.5 m

42. Calculate the area of the shape.

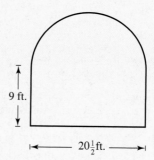

9 ft.

$20\frac{1}{2}$ ft.

43. In one month a company had a profit of $45,698, with $96,408 in revenue. Find the cost.

44. A triangle has an area of 38.75 m². If the base is 12.5 m, what is the height?

45. Write an expression in simplest form for the area.

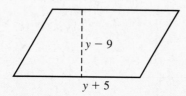

$y - 9$

$y + 5$

46. The width of a rectangle is 3 less than the length. If the perimeter is 50 m, find the length and width.

47. Don drives for 45 min., stops for a break, then drives for 1 hr. to reach his destination. If the total trip was 108.5 mi., what was his average rate in miles per hour?

48. Which is the better buy?
16 oz. of beans for $0.99   or   28 oz. of beans for $1.79

49. A family consumes 6 l of cola every 7 days. At this rate, how many liters of cola will the family consume in 1 yr?

50. Selina is traveling 55 mph and passes mile marker 42 on a highway. 1 hr later, Marissa passes the same mile marker at 65 mph. If they maintain the same rates, how long will it take Marissa to catch up to Selina? (Use a four-column table.)

Percents

Time Management

One of the most difficult tasks to do is to balance all the activities that we take on in life. Many students make the mistake of overloading themselves with too many activities and too many classes. If you have not read the *To the Student* section at the beginning of the text, take a moment to read through it. In particular, do the exercise entitled *How Do I Do It All.*

First, make sure that you have less than 60 hours per week committed to your job (if you have one) and school combined. Remember that, as a general rule, you are expected to spend about twice the number of hours outside of class as you spend in class preparing, studying, and completing assignments. This time, of course, can vary depending on the course and your natural aptitude in that subject.

In managing your time, make every minute count. Many people find it difficult to study in their home environment because there are too many distractions. One of the most effective ways to make your study time more efficient is to study in an environment that does not allow for distraction. Many people find the library or a vacant room on campus to be good environments, free of distraction. It may help to make a rule that you cannot leave until you have completed your homework and studies for the next class meeting.

The added bonus to completing all your school work on campus is that if you need help, you can easily meet with your instructor or go to tutorial services. You'll be amazed at how much more effective your study time is when you put yourself in a distraction-free environment and make every minute count.

"*Nothing valuable can be lost by taking time.*"
—ABRAHAM LINCOLN

"*Since time is the one immaterial object which we cannot influence— neither speed up nor slow down, add to nor diminish—it is an imponderably valuable gift.*"
—MAYA ANGELOU

"*Time stays long enough for those who use it.*"
—LEONARDO DA VINCI

8.1 Introduction to Percent

OBJECTIVES

1 Write a percent as a fraction.
2 Write a percent as a decimal number.
3 Write a fraction as a percent.
4 Write a decimal number as a percent.

OBJECTIVE **1** *Write a percent as a fraction.*

Think about the word *percent*. It is a compound word made from the prefix *per,* indicating a ratio, and the suffix *cent,* coming from the Latin word *centum,* which means 100. Therefore, **percent** means a ratio out of 100.

DEFINITION	**Percent:** Ratio out of 100.

The symbol for percent is %.

$$20\%$$
$$= 20 \text{ out of } 100$$

Note that the definition can be used to write percents as fractions or decimal numbers. Let's first consider writing percents as fractions.

$$20\% = 20 \text{ out of } 100$$
$$= \frac{20}{100}$$
$$= \frac{1}{5}$$

Conclusion: To write a percent as a fraction, write the numeral over 100, then simplify.

Procedure *To write a percent as a fraction:*
1. Write the numeral over 100.
2. Simplify.

EXAMPLE 1 Write each percent as a fraction in simplest form.

a. 12%

Solution: Write 12 over 100, then simplify.

$$12\% = \frac{12}{100} = \frac{3}{25}$$

> **Reminder**
> When we reduce a fraction, we divide the numerator and denominator by their GCF. In this case, the GCF for 12 and 100 is 4.

b. 20.5%

Solution: Write 20.5 over 100, then simplify.

$$20.5\% = \frac{20.5}{100}$$

$$= \frac{20.5(10)}{100(10)} \qquad \text{Clear the decimal by upscaling by 10.}$$

$$= \frac{205}{1000} \qquad \text{Reduce to lowest terms.}$$

$$= \frac{41}{200}$$

> **Reminder**
> To clear a decimal in a ratio, we upscale using an appropriate power of 10 as determined by the decimal number with the largest number of decimal places. In this case 20.5 has the most decimal places. To clear 1 place we multiply by 10.

c. $30\frac{1}{4}\%$

Solution: Write $30\frac{1}{4}$ over 100, then simplify.

$$30\frac{1}{4}\% = \frac{30\frac{1}{4}}{100} \qquad \text{Write the complex fraction in division form.}$$

$$= 30\frac{1}{4} \div 100 \qquad \text{Write an equivalent multiplication.}$$

$$= \frac{121}{4} \cdot \frac{1}{100} \qquad \text{Multiply.}$$

$$= \frac{121}{400}$$

> **Reminder**
> We can write a complex fraction as a division problem.

Do Margin 1. ▶

OBJECTIVE 2 *Write a percent as a decimal number.*

In Section 7.1 we learned to write unit ratios by dividing the denominator into the numerator. We can use this process to write a percent as a decimal. Because the denominator will always be 100, we can simply divide the numeral by 100.

$$20\% = 20 \text{ divided by } 100$$
$$= 20 \div 100$$
$$= 0.2$$

> **Reminder**
> Dividing by 100 causes a decimal point to move two places to the left from its original position in the dividend.

Conclusion: To write a percent as a decimal, divide by 100. This will cause the decimal point to move two places to the left from its original position in the dividend.

Note, this assumes that the percent is a decimal number. If the percent is a mixed number or fraction, we can write the fraction or mixed number as a decimal number and then divide by 100.

> **Procedure** *To write a percent as a decimal number:*
> 1. Write the fraction or mixed number percent in decimal form.
> 2. Divide by 100.

EXAMPLE 2 Write each percent as a decimal number.

a. 92%

Solution: Because 92 is not a fraction or mixed number we simply divide by 100, which causes the decimal point to move two places to the left.

$$92\% = 92 \div 100 = 0.92$$

b. 125%

Solution: Divide 125 by 100, which causes the decimal point to move two places to the left.

$$125\% = 125 \div 100 = 1.25$$

c. 9.2%

Solution: Divide 9.2 by 100, which causes the decimal point to move two places to the left.

$$9.2\% = 9.2 \div 100 = 0.092$$

d. $21\frac{2}{3}\%$

Solution: Because $21\frac{2}{3}$ is a mixed number we first write it as a decimal number.

$$21\frac{2}{3}\% = 21.\overline{6}\%$$

> **Reminder**
> To write a fraction as a decimal number, we divide the denominator into the numerator.

Now that we've expressed the mixed number as a decimal number, we can divide by 100, which will cause the decimal point to move two places to the left.

$$21\frac{2}{3}\% = 21.\overline{6}\%$$
$$= 21.\overline{6} \div 100$$
$$= 0.21\overline{6}$$

◀ **Do Margin 2.**

OBJECTIVE 3 *Write a fraction as a percent.*

We have seen that $20\% = \dfrac{20}{100} = \dfrac{1}{5}$ and $20\% = 20 \div 100 = 0.2$

Note that what we did to write a percent as a fraction or as a decimal was to divide by 100. To reverse this process, we should multiply by 100 and attach a percent sign.

Conclusion: To write a fraction or decimal as a percent, we multiply by 100%.

> **Procedure** *To write a fraction as a percent:*
> 1. Multiply by 100%.
> 2. Simplify.

EXAMPLE 3   Write each fraction as a percent.

a. $\dfrac{1}{2}$

Solution: $\dfrac{1}{2} = \dfrac{1}{\overset{}{\underset{1}{\cancel{2}}}} \cdot \dfrac{\overset{50}{\cancel{100}}}{1}\%$ **Multiply by 100%.**

$= 50\%$

b. $\dfrac{5}{8}$

Solution: $\dfrac{5}{8} = \dfrac{5}{\underset{2}{\cancel{8}}} \cdot \dfrac{\overset{25}{\cancel{100}}}{1}\%$ **Multiply by 100%.**

$= \dfrac{125}{2}\%$ **Simplify.**

$= 62\dfrac{1}{2}\% \text{ or } 62.5\%$

c. $\dfrac{2}{3}$

Solution: $\dfrac{2}{3} = \dfrac{2}{3} \cdot \dfrac{100}{1}\%$ **Multiply by 100%.**

$= \dfrac{200}{3}\%$ **Simplify.**

$= 66\dfrac{2}{3}\% \text{ or } 66.\overline{6}\%$

Do Margin 3. ▶

OBJECTIVE 4 *Write a decimal number as a percent.*

To write a decimal number as a percent, we follow the same procedure as for fractions: multiply by 100. Recall that multiplying a decimal number by 100 causes the decimal point to move two places to the right.

> **Procedure** *To write a decimal number as a percent, multiply by 100%.*
> **Note:** Multiplying a decimal number by 100% will cause the decimal point to move two places to the right.

♦ **Margin 3**

Write each fraction as a percent.

a. $\dfrac{3}{5}$

b. $\dfrac{5}{16}$

c. $\dfrac{4}{9}$

Answers: **a.** 60%
b. $31\dfrac{1}{4}\%$ or 31.25% **c.** $44\dfrac{4}{9}\%$ or
$44.\overline{4}\%$

EXAMPLE 4 Write each decimal number as a percent.

a. 0.7

Solution: Multiply by 100%.

$$0.7 = (0.7)(100)\%. \qquad \textbf{Multiply by 100\%.}$$
$$= 70\%$$

b. 0.018

Solution: Multiply by 100%.

$$0.018 = (0.018)(100)\% \qquad \textbf{Multiply by 100\%.}$$
$$= 1.8\%$$

c. 2

Solution: Multiply by 100%.

$$2 = (2)(100)\% \qquad \textbf{Multiply by 100\%.}$$
$$= 200\%$$

◀ **Do Margin 4.**

◆ **Margin 4**

Write each decimal number as a percent.

a. 0.49

b. 0.883

c. 1.5

Answers: **a.** 49% **b.** 88.3% **c.** 150%

8.1 Exercises

For Exercises 1–12, write each percent as a fraction in simplest form.

1. 20%   **2.** 30%   **3.** 15%   **4.** 85%

5. 14.8%   **6.** 18.6%   **7.** 3.75%   **8.** 6.25%

9. $45\frac{1}{2}\%$ **10.** $65\frac{1}{4}\%$ **11.** $33\frac{1}{3}\%$ **12.** $66\frac{2}{3}\%$

For Exercises 13–24, write each percent as a decimal number.

13. 75%   **14.** 58%   **15.** 125%   **16.** 210%

17. 12.9%   **18.** 11.6%   **19.** 1.65%   **20.** 100.5%

21. $53\frac{2}{5}\%$ **22.** $70\frac{3}{4}\%$ **23.** $16\frac{1}{6}\%$ **24.** $6\frac{2}{3}\%$

For Exercises 25–36, write each fraction as a percent.

25. $\frac{1}{2}$ **26.** $\frac{3}{4}$ **27.** $\frac{3}{5}$ **28.** $\frac{1}{5}$

29. $\frac{3}{8}$ **30.** $\frac{5}{8}$ **31.** $\frac{1}{6}$ **32.** $\frac{5}{6}$

33. $\frac{2}{3}$ **34.** $\frac{1}{3}$ **35.** $\frac{4}{9}$ **36.** $\frac{5}{11}$

For Exercises 37–48, write each decimal number as a percent.

37. 0.96   **38.** 0.42   **39.** 0.8   **40.** 0.7

41. 0.09   **42.** 0.01   **43.** 1.2   **44.** 3.58

45. 0.028   **46.** 0.065   **47.** 4.051   **48.** 0.007

45. 0.028   **46.** 0.065   **47.** 4.051   **48.** 0.007

49. $0.\overline{6}$ **50.** $0.\overline{3}$ **51.** $1.\overline{63}$ **52.** $0.\overline{47}$

REVIEW EXERCISES

1. $0.6(24) =$ **2.** $0.45(3600) =$ **3.** $0.05(76.8) =$

4. Solve. $0.15x = 200$ **5.** Solve. $0.75y = 28.5$

8.2 Solving Percent Sentences by Direct Translation

OBJECTIVE

1 Solve percent sentences by direct translation to an equation.

OBJECTIVE 1 *Solve percent sentences by direct translation to an equation.*

There are three primary elements to percent problems:

1. The percent

2. A whole amount

3. A part of the whole

All problems involving these three primary elements can be simplified to a sentence.

A **percent** of a **whole amount** is the **part of the whole.**

Consider the basic sentence:	50% of 40 is 20.
	50% is the percent
	40 is the whole amount
	20 is the part of the whole

We will be given two of the three elements in a sentence with the third element unknown. There are three primary elements, three cases to explore. Following are examples of the three cases.

20% of 80 is what number?
> Case 1: The percent and the whole amount are given and the part of the whole is the unknown amount.

32% of what number is 12?
> Case 2: The percent and the part of the whole are given and the whole amount is the unknown amount.

What percent of 240 is 64?
> Case 3: The percent is the unknown while the whole amount and the part of the whole are given.

Keep in mind that the English language allows for variations on each of the three cases. Our task will be to translate to an equation and then solve. Learning to translate the sentence to an equation is a step toward solving more complex percent problems. The idea is to relate the more complex problems to the structure of this basic sentence. We will consider two methods for translating this sentence:

  1. Direct translation to an equation.

  2. Translation to a proportion.

We will explore direct translation to equation in this section and translation to a proportion in Section 8.3.

To translate the sentence directly to an equation we use the key words in the sentence. Recall that the key word *is* translates to an equal sign. Also recall that the key word *of* indicates to multiply when it is preceded by a fraction. A percent can be written as a fraction out of 100, therefore when the word *of* is preceded by a percent, it indicates multiplication. When *of* is preceded by a whole number as in *20 of 40,* it is an abbreviation for *out of,* which indicates division.

Procedure *To translate a percent sentence directly to an equation:*

1. Select a variable to represent the unknown amount indicated by the word *what*.
2. If the word *of* is preceded by the percent, translate it to multiplication. If the word *of* is preceded by a whole number, translate it to division.
3. Translate the word *is* to an equal sign.

Once the sentence is translated, we must then solve the equation. To solve the equation, we must first write the percent as a decimal or fraction.

Let's explore Case 1. The unknown amount will be the part of the whole. For consistency we will always use the variable *c*, for *change*, to represent the part of the whole.

EXAMPLE 1 Translate to an equation. Then solve.

a. 20% of 80 is what number?

Solution: The words *what number* indicate an unknown. The unknown is the part of the whole, so we will use *c* to represent the unknown. Because *of* is preceded by the percent, it indicates multiplication. The word *is* translates to an equal sign.

20% of 80 is what number?	Translate to an equation.
$20\% \cdot 80 = c$	Write the percent as a decimal.
$(0.2)(80) = c$	Multiply.
$16 = c$	

b. What number is 105% of 60.4?

Solution: The words *what number* indicate an unknown. The unknown is the part of the whole, so we will use *c* to represent the unknown. Because *of* is preceded by the percent, it indicates multiplication. The word *is* translates to an equal sign.

Discussion Why is the result larger than 60.4?

What number is 105% of 60.4?	Translate to an equation.
$c = 105\% \cdot 60.4$	Write the percent as a decimal.
$c = (1.05)(60.4)$	Multiply.
$c = 63.42$	

c. What number is 5.4% of $34\frac{1}{3}$?

Solution: The words *what number* indicate an unknown. The unknown is the part of the whole, so we will use *c* to represent the unknown. Because *of* is preceded by the percent, it indicates multiplication. The word *is* translates to an equal sign.

What number is 5.4% of $34\frac{1}{3}$?	Translate to an equation.
$c = 5.4\% \cdot 34\frac{1}{3}$	Write the percent as a decimal number. Write the mixed number as an improper fraction.
$c = (0.054)\left(\dfrac{103}{3}\right)$	Write the decimal number over 1.
$c = \left(\dfrac{0.054}{1}\right)\left(\dfrac{103}{3}\right)$	Multiply.
$c = \dfrac{5.562}{3}$	Divide.
$c = 1.854$	

◀ **Do Margin 1.**

◆ **Margin 1**

Translate to an equation. Then solve.

a. 30% of 148 is what number?

b. What number is 6% of $20\frac{2}{5}$?

c. What number is $4\frac{1}{2}$% of 28?

Reminder
$34\frac{1}{3} = 34.\overline{3}$. To calculate with numbers that have decimal digits that repeat in an unending pattern we must round. Rounding causes the result to be inaccurate. We avoid this inaccuracy by writing the numbers in fraction form and by following the calculation procedures for fractions.

Answers: **a.** 44.4 **b.** 1.224 **c.** 1.26

In Case 2, the unknown is the whole amount. The whole amount always follows the word *of*. For consistency, we will always use the variable *w*, for *whole*, when the unknown is the whole amount.

EXAMPLE 2 Translate to an equation. Then solve.

a. 32% of what number is 12?

Solution: The words *what number* indicate an unknown. The unknown is a whole amount, so we will use *w* to represent the unknown. Because *of* is preceded by the percent, it indicates multiplication. The word *is* translates to an equal sign.

32% of what number is 12?	Translate to an equation.
$32\% \cdot w = 12$	Write the percent as a decimal number.
$0.32\,w = 12$	
$\dfrac{0.32w}{0.32} = \dfrac{12}{0.32}$	Divide both sides by 0.32 to isolate *w*.
$1w = 37.5$	
$w = 37.5$	

b. $16\dfrac{1}{2}$ is 2% of what number?

Solution: The words *what number* indicate an unknown. The unknown is a whole amount, so we will use *w* to represent the unknown. Because *of* is preceded by the percent, it indicates multiplication. The word *is* translates to an equal sign.

$16\dfrac{1}{2}$ is 2% of what?	Translate to an equation.
$16\dfrac{1}{2} = 2\% \cdot w$	Write the percent as a decimal number.
$16.5 = 0.02w$	
$\dfrac{16.5}{0.02} = \dfrac{0.02w}{0.02}$	Divide both sides by 0.02 to isolate *w*.
$825 = 1w$	
$825 = w$	

Do Margin 2. ▶

In Case 3, the percent is the unknown amount. For consistency, when the percent is the unknown, we will always use the variable *p*.

EXAMPLE 3 Translate to an equation. Then solve.

a. What percent of 240 is 60?

Solution: *What percent* indicates that the unknown is the percent, so we will use *p* to represent the unknown. Because *of* is preceded by the percent, it indicates multiplication. The word *is* translates to an equal sign.

What percent of 240 is 60?	Translate to an equation.
$p \cdot 240 = 60$	
$240p = 60$	
$\dfrac{240p}{240} = \dfrac{60}{240}$	Divide both sides by 240 to isolate *p*.
$1p = 0.25$	
$p = 0.25$	

Margin 3

*Translate to an equation.
Then solve.*

a. What percent of 400
   is 50?

b. 15.37 is what percent
   of 14.5?

c. What percent is 19 out
   of 40?

d. 32 of 72 is what percent?

Note that the answer is a decimal number. To write
the result as a percent, we multiply by 100%.

$$p = (0.25)(100)\% = 25\%$$

b. 82.4 is what percent of 164.8?

Solution: *What percent* indicates that the unknown is the percent, so we will use p to represent the unknown. Because *of* is preceded by the percent, it indicates multiplication. The word *is* translates to an equal sign.

$$82.4 \text{ is what percent of } 164.8? \quad \textbf{Translate to an equation.}$$
$$82.4 = \qquad p \cdot 164.8$$
$$82.4 = 164.8p$$
$$\frac{82.4}{164.8} = \frac{164.8p}{164.8} \quad \textbf{Divide both sides by 164.8 to isolate } p.$$
$$0.5 = 1p$$
$$0.5 = p \qquad \textbf{Write 0.5 as a percent.}$$
$$(0.5)(100)\% = p$$
$$50\% = p$$

c. What percent is 15 out of 40?

Solution: *What percent* indicates that the unknown is the percent, so we will use p to represent the unknown. The key phrase *out of* indicates a ratio or division. The word *is* translates to an equal sign. There are two ways of translating and solving.

What percent is 15 out of 40?		or	What percent is 15 out of 40?	
Divide.	$p = 15 \div 40$		$p = \dfrac{15}{40}$	**Reduce.**
	$p = 0.375$		$p = \dfrac{3}{8}$	
Multiply by 100% to write as a percent.	$p = (0.375)(100)\%$		$p = \dfrac{3}{\underset{2}{\cancel{8}}} \cdot \dfrac{\overset{25}{\cancel{100}}}{1}\%$	**Multiply by 100% to write as a percent**
	$p = 37.5\%$		$p = \dfrac{75}{2}\%$	
			$p = 37\dfrac{1}{2}\%$	

d. What percent is 12 of 18?

Solution: *What percent* indicates that the unknown is the percent, so we will use p to represent the unknown. The word *of* is preceded by a whole number, which means it is an abbreviation of *out of*. It is the same as if we had been asked: What percent is 12 out of 18? Therefore the *of* in this case indicates ratio or division.

What percent is 12 of 18?		or	What percent is 12 of 18?	
Divide.	$p = 12 \div 18$		$p = \dfrac{12}{18}$	**Reduce.**
	$p = 0.\overline{6}$		$p = \dfrac{2}{3}$	
Multiply by 100% to write as a percent.	$p = (0.\overline{6})(100)\%$		$p = \dfrac{2}{3} \cdot \dfrac{100}{1}\%$	**Multiply by 100% to write as a percent**
	$p = 66.\overline{6}\%$		$p = \dfrac{200}{3}\%$	
			$p = 66\dfrac{2}{3}\%$	

Answers: **a.** 12.5% **b.** 106%
c. 47.5% **d.** 44.$\overline{4}$%

◀ **Do Margin 3.**

8.2 Exercises

For Exercises 1–40, translate to an equation. Then solve.

1. 40% of 350 is what number?

2. 60% of 400 is what number?

3. What number is 15% of 78?

4. What number is 5% of 44?

5. 150% of 60 is what number?

6. What number is 105% of 84?

7. What number is 16% of $30\frac{1}{2}$?

8. 95% of $40\frac{3}{5}$ is what number?

9. 110% of 46.5 is what number?

10. What number is 120% of 36.82?

11. What number is $5\frac{1}{2}$% of 280?

12. $10\frac{1}{4}$% of 68 is what number?

13. $66\frac{2}{3}$% of 56.8 is what number?

14. $16\frac{1}{6}$% of 365 is what number?

15. What number is 8.5% of 54.82?

16. What number is 3.5% of 24,500?

17. 12.8% of 36,000 is what number?

18. 19.8% of 84.5 is what number?

19. 35% of what number is 77?

20. 70% of what number is 45.5?

21. 4800 is 40% of what?

22. 1,950 is 65% of what?

23. 605 is 2.5% of what number?

24. 168.3 is 19.8% of what number?

25. 105% of what number is 48.3?

26. 120% of what number is 101.52?

27. 47.25 is $10\frac{1}{2}$% of what number?

28. 2000 is $6\frac{1}{4}$% of what number?

29. What percent of 68 is 17?

30. What percent of 120 is 48?

31. 2.142 is what percent of 35.7?

32. 25.12 is what percent of 125.6?

33. What percent of $24\frac{1}{2}$ is $7\frac{7}{20}$?

34. $5\frac{5}{8}$ is what percent of $9\frac{3}{8}$?

35. What percent is 21 out of 60?

36. What percent is 114 out of 200?

37. 25 out of 35 is what percent?

38. 38 out of 42 is what percent?

39. 43 of 65 is what percent?

40. What percent is 29 of 30?

REVIEW EXERCISES

1. $12.8(100) =$

2. $5\frac{3}{4} \cdot 100 =$

3. $8\frac{1}{4} \cdot 10\frac{1}{2}$

4. Solve. $\dfrac{x}{14} = \dfrac{3}{4}$

5. Solve. $\dfrac{n}{100} = \dfrac{5}{8}$

8.3 Solving Percent Sentences by Proportion

OBJECTIVE

1 Solve percent sentences by proportion.

OBJECTIVE 1 *Solve percent sentences by proportion.*

In Section 8.2 we developed the direct translation method for solving percent sentences. Let's recall the structure of the sentence.

A percent of a whole amount is the part of the whole.

We will now consider the second method for translating percent sentences. In this method we will translate the sentence to a proportion.

Recall that a proportion is two ratios set equal. To translate the percent sentence to a proportion, we must write two ratios using the percent, the whole amount, and the part of the whole. One of the ratios in the proportion is the percent itself, because by definition, it is a ratio of a quantity to 100. The other ratio is the part of the whole to the whole amount.

To get a feel for the two ratios, let's reconsider Example 1 from Section 8.2. We discovered that 20% of 80 is 16. In terms of equivalent ratios, we can say that 16 is the part of 80 that is in the same ratio as 20 to 100. The ratios 20 to 100 and 16 to 80 are equivalent because both ratios reduce to $\frac{1}{5}$. Notice the structure of the ratios:

$$\text{Percent expressed as a ratio in fraction form.} \qquad \frac{20}{100} = \frac{16}{80} \begin{array}{l} \leftarrow \text{Part} \\ \text{to} \\ \leftarrow \text{Whole amount} \end{array}$$

Conclusion: The percent ratio is equivalent to the ratio of the part to the whole amount.

Procedure *To translate a percent sentence to a proportion, write the proportion in the following form:*

$$\text{percent} = \frac{\text{part}}{\text{whole amount}}$$

where the percent is expressed as a fraction with a denominator of 100.

> **Tip** The whole amount always follows the word *of.*

Once the sentence is translated, we must then solve the proportion. Recall that we cross multiply to solve proportions.

Consider sentences where the unknown is the part of the whole using this method of proportions. To remain consistent with Section 8.2, we will continue to use the variable c to represent the part of the whole.

EXAMPLE 1 Translate to a proportion. Then solve.

a. 60% of 40 is what number?

Solution: The first ratio is the percent expressed as a fraction with a denominator of 100, so we will write 60 over 100. The other ratio is the part to the whole amount. The number 40 follows the word *of*, therefore 40 is the whole amount and will be the denominator of the second ratio. The unknown amount is the part, so we will write *c* in the numerator of the second ratio.

$$\frac{60}{100} = \frac{c}{40}$$

$$40 \cdot 60 = 2400 \qquad\qquad 100 \cdot c = 100c$$

$$\frac{60}{100} \diagdown\!\!\!\!\diagup \frac{c}{40} \qquad\qquad \text{Cross multiply.}$$

$$2400 = 100c \qquad\qquad \text{Equate the cross products.}$$

$$\frac{2400}{100} = \frac{100c}{100} \qquad\qquad \text{Divide both sides by 100 to isolate } c.$$

$$24 = 1c$$

$$24 = c$$

b. What number is 120% of 48.5?

Solution: The first ratio is the percent expressed as a fraction with a denominator of 100, so we will write 120 over 100. The other ratio is the part to the whole amount. The number 48.5 follows the word *of*, therefore 48.5 is the whole amount and will be the denominator of the second ratio. The unknown amount is the part, so we will write *c* in the numerator of the second ratio.

$$\frac{120}{100} = \frac{c}{48.5}$$

$$(48.5)(120) = 5820 \qquad\qquad 100 \cdot c = 100c$$

$$\frac{120}{100} \diagdown\!\!\!\!\diagup \frac{c}{48.5} \qquad\qquad \text{Cross multiply.}$$

$$5820 = 100c \qquad\qquad \text{Equate the cross products.}$$

$$\frac{5820}{100} = \frac{100c}{100} \qquad\qquad \text{Divide both sides by 100 to isolate } c.$$

$$58.2 = 1c$$

$$58.2 = c \qquad\qquad \blacklozenge\ \textbf{Discussion}\ \text{Why is the result larger than 48.5?}$$

c. What number is 15.2% of $20\frac{1}{3}$?

Solution: The first ratio is the percent, so we will write 15.2 over 100. The other ratio is the part to the whole amount. Because $20\frac{1}{3}$ follows the word *of*, it is the initial amount and will be the denominator of the second ratio. The unknown amount is the part, so we will write *c* in the numerator of the second ratio.

$$\frac{15.2}{100} = \frac{c}{20\frac{1}{3}}$$ Write the mixed number as a fraction.

$$\frac{15.2}{1} \cdot \frac{61}{3} = \frac{927.2}{3} = 309.0\overline{6}$$ $$100 \cdot c = 100c$$

$$\frac{15.2}{100} \diagup\diagup \frac{c}{61}$$ Cross multiply.
$$\phantom{\frac{15.2}{100}} \diagdown 3$$

$$309.0\overline{6} = 100c$$ Equate the cross products.

$$\frac{309.0\overline{6}}{100} = \frac{100c}{100}$$ Divide both sides by 100 to isolate c.

$$3.090\overline{6} = 1c$$

$$3.090\overline{6} = c$$

Discussion How would you compare the proportion method with the direct translation method when the part is the unknown?

Do Margin 1. ▷

In Case 2, the whole amount is unknown. Recall that the whole amount always follows the word *of*. In the proportion, the whole amount is placed in the denominator because it is in logical correspondence with the 100 that is the denominator from the percent. We will continue to use the variable w to represent the unknown whole amount.

EXAMPLE 2 Translate to a proportion. Then solve.

a. 85% of what number is 352.75?

Solution: The first ratio is the percent, so we will write 85 over 100. The other ratio is the part to the whole amount. The words *what number* follow the word *of*, therefore the whole amount is the unknown. Because we have established that w will represent the whole amount, we write w in the denominator of the second ratio. The part is 352.75, so we will write 352.75 in the numerator of the second ratio.

$$\frac{85}{100} = \frac{352.75}{w}$$

$$w \cdot 85 = 85w \qquad 100 \cdot 352.75 = 35{,}275$$

$$\frac{85}{100} \diagup\diagup \frac{352.75}{w}$$ Cross multiply.

$$85w = 35{,}275$$ Equate the cross products.

$$\frac{85w}{85} = \frac{35{,}275}{85}$$ Divide both sides by 85 to isolate w.

$$1w = 415$$

$$w = 415$$

b. $18\frac{3}{8}$ is $8\frac{3}{4}$% of what number?

Solution: The first ratio is the percent, so we will write $8\frac{3}{4}$ over 100. The other ratio is the part to the whole amount. The words *what number* follow the word *of*, therefore the whole amount is the unknown. Because we have established that w will represent the whole amount, we write w in the denominator of the second ratio. The part is $18\frac{3}{8}$, so we will write $18\frac{3}{8}$ in the numerator of the second ratio.

◆ **Margin 1**

Translate to a proportion. Then solve.

a. 75% of 620 is what number?

b. What number is 8% of $46\frac{1}{2}$?

c. What number is 15.2% of 90?

$$\frac{8\frac{3}{4}}{100} = \frac{18\frac{3}{8}}{w}$$

Write the mixed numbers as fractions.

$$w \cdot \frac{35}{4} = \frac{35}{4}w \qquad \overset{25}{\underset{1}{\cancel{100}}} \cdot \frac{147}{\underset{2}{\cancel{8}}} = \frac{3675}{2}$$

$$\frac{\frac{35}{4}}{100} \diagup\diagdown \frac{\frac{147}{8}}{w}$$

Cross multiply.

$$\frac{35}{4}w = \frac{3675}{2}$$

Equate the cross products.

$$\frac{1}{\underset{1}{\cancel{35}}} \cdot \frac{\overset{1}{\cancel{35}}}{\underset{1}{\cancel{4}}}w = \frac{\overset{105}{\cancel{3675}}}{\underset{1}{\cancel{2}}} \cdot \frac{\overset{2}{\cancel{4}}}{\underset{1}{\cancel{35}}}$$

Multiply both sides by $\frac{4}{35}$ to isolate w.

$$1w = 210$$

$$w = 210$$

◀ **Do Margin 2.**

Case 3 has the percent as the unknown amount. We will continue to use the variable p when the unknown amount is the percent. Notice that the first ratio in the proportion is the percent. Even though the percent may be unknown, we know the denominator will be 100.

Conclusion: When the percent is the unknown, we write the first ratio as p to 100.

EXAMPLE 3 Translate to a proportion. Then solve.

a. What percent of 240 is 60?

Note: We have kept Example 3a the same as that in Section 8.2 to illustrate a point about the outcome.

Solution: *What percent* indicates that the unknown is the percent, so we will use p to represent the unknown. Because the first ratio in the proportion expresses the percent, we write p in the numerator and 100 in the denominator. Because 240 follows the word *of*, it is the whole amount and therefore is written in the denominator of the second ratio. This leaves 60 to be written in the numerator of the second ratio.

$$\frac{p}{100} = \frac{60}{240}$$

$$240 \cdot p = 240p \diagup\diagdown 100 \cdot 60 = 6000$$

$$\frac{p}{100} \diagup\diagdown \frac{60}{240}$$

Cross multiply.

$$240p = 6000$$

Equate the cross products.

$$\frac{240p}{240} = \frac{6000}{240}$$

Divide both sides by 240 to isolate p.

$$1p = 25$$

$$p = 25$$

Note that the answer is the correct percent. We simply need to write the percent sign.

$$p = 25\%$$

◆ **Margin 2**

Translate to a proportion. Then solve.

a. 95% of what number is 64.6?

b. $10\frac{1}{5}$ is 20% of what number?

Answers: **a.** 68 **b.** 51

With the direct translation method the result was 0.25, a decimal number, and we had to multiply by 100 to write this result as a percent. With the proportion method, the result is the correct percent. All we must do is write the percent sign. The reason the result in the proportion method is a percent is because when we cross multiply, we multiply by 100. Multiplying by 100 is already part of the procedure.

b. 17.64 is what percent of 84?

Solution: *What percent* indicates that the unknown is the percent, so we will use p to represent the unknown. Because the first ratio in the proportion expresses the percent, we write p in the numerator and 100 in the denominator. Because 84 follows the word *of*, it is the whole amount and therefore is written in the denominator of the second ratio. This leaves 17.64 to be written in the numerator of the second ratio.

$$\frac{p}{100} = \frac{17.64}{84}$$

$$84 \cdot p = 84p \qquad\qquad 100 \cdot 17.64 = 1764$$

$$\frac{p}{100} \diagdown \frac{17.64}{84} \qquad \text{Cross multiply.}$$

$$84p = 1764 \qquad \text{Equate the cross products.}$$

$$\frac{84p}{84} = \frac{1764}{84} \qquad \text{Divide both sides by 84 to isolate } p.$$

$$1p = 21\%$$

$$p = 21\%$$

c. What percent is 15 out of 40?

Note: We have kept Example 3c the same as in Section 8.2 to illustrate a point about the translation.

Solution: *What percent* indicates that the unknown is the percent, so we will use p to represent the unknown. This form of the percent sentence lends itself to the proportion method very well. 15 out of 40 translates to the ratio $\frac{15}{40}$.

$$\frac{p}{100} = \frac{15}{40}$$

$$40 \cdot p = 40p \qquad\qquad 100 \cdot 15 = 1500$$

$$\frac{p}{100} \diagdown \frac{15}{40} \qquad \text{Cross multiply.}$$

$$40p = 1500 \qquad \text{Equate the cross products.}$$

$$\frac{40p}{40} = \frac{1500}{40} \qquad \text{Divide both sides by 40 to isolate } p.$$

$$1p = 37.5\%$$

$$p = 37.5\%$$

a. What percent of 360 is 45?

b. 58.75 is what percent of 94?

c. What percent is 27 out of 162?

d. 58 of 72 is what percent?

d. What percent is 34 of 102?

Solution: *What percent* indicates that the unknown is the percent, so we will use p to represent the unknown. The word *of* is preceded by a whole number, which means it is an abbreviation of *out of*. It is the same as if we had been asked: What percent is 34 *out of* 102? 34 of 102 translates to the ratio $\frac{34}{102}$.

$$\frac{p}{100} \diagup\!\!\!\!\!\diagdown \frac{34}{102}$$

$$102 \cdot p = 102p \qquad\qquad 100 \cdot 34 = 3400$$

$$\frac{p}{100} = \frac{34}{102} \qquad\qquad \text{Cross multiply.}$$

$$102p = 3400 \qquad\qquad \text{Equate the cross products.}$$

$$\frac{102p}{102} = \frac{3400}{102} \qquad\qquad \text{Divide both sides by 102 to isolate } p.$$

$$1p = 33.\overline{3}\%$$

$$p = 33.\overline{3}\%$$

◀ **Do Margin 3.**

8.3 Exercises

FOR EXTRA HELP

 Videotape 9

 InterAct Math
Tutorial Software

 www.carsonmath.com

 AWL Math Tutor Center

InterAct MathXL www.mathxl.com

 Student's Solutions
Manual

For Exercises 1–40, translate to a proportion. Then solve.

1. 30% of 560 is what number?

2. 80% of 420 is what number?

3. What number is 12% of 85?

4. What number is 7% of 62?

5. 125% of 90 is what number?

6. What number is 106% of 77?

7. What number is 15% of $9\frac{1}{2}$?

8. 98% of $60\frac{1}{4}$ is what number?

9. 102% of 88.5 is what number?

10. What number is 150% of 45.25?

11. What number is $12\frac{1}{2}$% of 440?

12. $5\frac{1}{4}$% of 164 is what number?

13. $33\frac{1}{3}$% of 87.6 is what number?

14. $10\frac{1}{2}$% of 24.8 is what number?

15. What number is 6.5% of 22,800?

16. What number is 9.5% of 288?

17. 16.9% of 2450 is what number?

18. 14.8% of 160 is what number?

19. 65% of what number is 52?

20. 90% of what number is 133.2?

21. 4480 is 80% of what number?

22. 2400 is 75% of what number?

23. 1143 is 4.5% of what number?

24. 645 is 12.9% of what number?

25. 120% of what number is 65.52?

26. 110% of what number is 251.35?

27. 13.53 is $5\frac{1}{2}$% of what number?

28. 461.25 is $10\frac{1}{4}$% of what number?

29. What percent of 80 is 24?

30. What percent of 180 is 81?

31. 2.226 is what percent of 74.2?

32. 7.328 is what percent of 91.8?

33. What percent of $30\frac{1}{3}$ is $18\frac{1}{5}$?

34. $5\frac{1}{20}$ is what percent of $12\frac{5}{8}$?

35. What percent is 76 out of 80?

36. What percent is 186 out of 300?

37. 32 out of 45 is what percent?

38. 47 out of 60 is what percent?

39. 17 of 19 is what percent?

40. What percent is 169 of 180?

REVIEW EXERCISES

1. 0.48(560) =

2. 0.05(24.8) =

3. $\dfrac{9}{16} \cdot \dfrac{20}{27} =$

4. $\frac{5}{8}$ of the respondents in a survey disagreed with a given statement. If 400 people were surveyed, how many disagreed with the given statement?

5. $\frac{3}{4}$ of the respondents in a survey agreed with a given statement. Of these, $\frac{4}{5}$ were female. What fraction of all respondents were females that agreed with the particular statement?

8.4 Solving Percent Problems (Portions)

OBJECTIVES

1 Solve for the part in percent problems.

2 Solve for the whole in percent problems.

3 Calculate the percent.

OBJECTIVE 1 *Solve for the part in percent problems.*

Now that we have developed methods for solving the basic percent sentence, we can explore realistic problems involving percents. It is important to recognize that every percent problem can be stated in the form of a basic sentence. Let's recall the basic sentence structure:

The **percent** of a **whole** amount is the **part of the whole.**

We want to identify the pieces in the problem. You may find it helpful to then write the pieces in the basic sentence format. Then you can translate to an equation or a proportion, whichever is more appropriate or comfortable. Then solve.

> **Procedure** *To solve problems involving percents:*
> **1.** Identify the percent, whole amount, and part.
> **2.** Write these pieces in a basic sentence (if needed).
> **3.** Translate to an equation or a proportion.
> **4.** Solve.

> **Tips**
> **1.** When the percent and the whole amount are the given pieces, translate directly to an equation.
> **2.** When the percent is the missing amount, translate to a proportion.

EXAMPLE 1 30% of a solution is acetone. If the total capacity of the solution is 400 ml, how many milliliters of the solution is acetone?

Understand: We must calculate the part of the solution that is acetone. We are given the percent and the capacity of the whole solution.

Plan: Write a basic percent sentence. Translate to an equation or a proportion. Then solve.

Execute: Percent of the whole amount is the part

$$30\% \text{ of } \quad 400 \quad \text{ is what number?}$$
$$30\% \cdot 400 = c \qquad \text{Write the percent as a decimal.}$$
$$0.3\,(400) = c \qquad \text{Multiply.}$$
$$120 = c$$

Note: We chose to translate directly to an equation because we were given the percent and whole amount. Many people prefer this method for this situation because it is less involved than the proportion method.

Answer: The solution contains 120 ml of acetone.

> **Reminder**
> For consistency, we will always use the variable c to represent the part.

$$\frac{30}{100} = \frac{c}{400}$$
$$12{,}000 = 100c$$
$$\frac{12{,}000}{100} = \frac{100c}{100}$$
$$120 = 1c$$
$$120 = c$$

◀ **Do Margin 1.**

Salespeople often earn a **commission** based on sales. The commission can be a percent of the total sales amount or a percent of the profit.

DEFINITION	**Commission:** A portion of sales earnings that a salesperson receives.

EXAMPLE 2 Carry earns 15% of total sales in commission. If she sells $2485 in merchandise over a two-week period, what is her commission?

Understand: We must calculate Carry's commission. We are given the commission rate and her total sales.

Plan: Write a basic percent sentence. Translate to an equation or a proportion. Then solve.

Execute: Her commission is 15% of total sales.

What number is 15% of 2485?	Translate to an equation.
$c = 15\% \cdot 2485$	Write the percent as a decimal.
$c = 0.15(2485)$	Multiply.
$c = 372.75$	

Answer: Carry's commission will be $372.75.

Discussion If Carry receives 15% of total sales in commission, what percent does the company receive? How much of the 2485 does the company receive?

Check: We could use the proportion method as a check.

$$\frac{15}{100} = \frac{c}{2485}$$
$$37{,}275 = 100c$$
$$\frac{37{,}275}{100} = \frac{100c}{100}$$
$$372.75 = 1c$$
$$372.75 = c$$

◀ **Do Margin 2.**

OBJECTIVE 2 *Solve for the whole in percent problems.*

Sometimes the whole amount is the missing piece in a percent problem. In these cases, the percent and part will be given.

EXAMPLE 3 Tedra sells real estate and receives a commission of $5445.60. If her commission rate is 6%, what was the total sale?

Understand: We must calculate Tedra's total sale. We are given the commission rate and her commission.

◆ **Margin 1**

Sabrina scored 80% on a test with 60 total questions. How many questions did she answer correctly?

◆ **Margin 2**

a. Al is a salesperson at a computer store and earns 10% of total sales in commission. If in one month he sold $15,294 worth of merchandise, how much will he receive as commission?

b. Brin sells scanning equipment. She earns 20% of the net profit in commission. If the net profit was $16,148, what was her commission?

Answer to Margin 1: 48

Answers to Margin 2: **a.** $1529.40 **b.** $3229.60

Plan: Write a basic percent sentence. Translate to an equation or a proportion. Then solve.

Execute: 5445.60 is 6% of the total sale.

$$5445.60 \text{ is } 6\% \text{ of what?} \qquad \textbf{Translate.}$$

$$5445.60 = 6\% \cdot w \qquad \textbf{Write the percent as a decimal.}$$

$$5445.60 = 0.06w$$

$$\frac{5445.60}{0.06} = \frac{0.06w}{0.06} \qquad \textbf{Isolate } w \textbf{ by dividing both sides by 0.06.}$$

$$90,760 = 1w$$

$$90,760 = w$$

Answer: Tedra's sale was $90,760.

Check: We could use the proportion method as a check.

$$\frac{6}{100} = \frac{5445.60}{w}$$

$$6w = 544,560$$

$$\frac{6w}{6} = \frac{544,560}{6}$$

$$1w = 90,760$$

$$w = 90,760$$

Do Margin 3. ▶

◆ **Margin 3**

Todd earns 20% commission on the sale of medical equipment. If Todd receives $3857.79 as a commission, what was his total sales?

EXAMPLE 4 A reporter states that 456 people in a survey indicated that they felt their schools were safe. The reporter went on to say that this was 48% of the respondents. How many respondents were involved in the survey?

Understand: We must calculate the total respondents in a survey. We are given the number of people that represent 48% of the total respondents.

Plan: Write a basic percent sentence. Translate to an equation or a proportion. Then solve.

Execute: 456 is 48% of the total respondents.

$$456 \text{ is } 48\% \text{ of what number?}$$

$$456 = 48\% \cdot w \qquad \textbf{Write the percent as a decimal.}$$

$$456 = 0.48w$$

$$\frac{456}{0.48} = \frac{0.48w}{0.48} \qquad \textbf{Isolate } w \textbf{ by dividing both sides by 0.48.}$$

$$950 = 1w$$

$$950 = w$$

Answer: There were 950 people who responded in the survey.

Check: We could use the proportion method as a check.

$$\frac{48}{100} = \frac{456}{w}$$

$$48w = 45,600$$

$$\frac{48w}{48} = \frac{45,600}{48}$$

$$1w = 950$$

$$w = 950$$

▶ **Discussion** If 48% of the respondents felt schools are safe, what percent did not feel schools are safe? How many of the respondents did not feel schools are safe?

◆ **Margin 4**

A power amplifier is operating at 80% of its full power capability. If its current output is 400 W (watts), what is its full power capability?

Do Margin 4. ▶

OBJECTIVE 3 *Calculate the percent.*

In Case 3, the percent is missing. In this case we will need the whole and the part. Recall that if we use the direct translation method when the percent is missing, the answer must be converted to a percent. If we use the proportion method to find the percent, the answer will be in percent form.

Answer to Margin 3: $19,288.95

Answer to Margin 4: 500 W

Margin 5

Out of the 286 students taking foreign language at a certain school 234 passed. What percent passed?

EXAMPLE 5 Latasha answered 52 questions correctly on a test with a total of 60 questions. What percent of the questions did she answer correctly?

Understand: We must calculate the percent of the questions that Latasha answered correctly. We are given the total number of questions and the number of questions that she answered correctly.

Plan: Write a basic percent sentence. Translate to a proportion. Then solve.

Execute: Latasha answered 52 questions correctly out of 60 total questions.

What percent is 52 out of 60?   **Write a proportion.**

$$\frac{p}{100} = \frac{52}{60}$$ **Solve for *p*.**

$$60p = 5200$$

$$\frac{60p}{60} = \frac{5200}{60}$$

$$1p = 86.\overline{6}$$

$$p = 86.\overline{6}\%$$

Answer: Latasha answered $86.\overline{6}\%$ of the questions correctly. Her score on the test would most likely be rounded to 87%.

Check: We could use the direct translation method as a check.

What percent is 52 out of 60?

$$p = 52 \div 60$$

$$p = 0.8\overline{6}$$ **Write the decimal**

$$p = 0.8\overline{6}\,(100)\%$$ **as a percent.**

$$p = 86.\overline{6}\%$$

◄ **Do Margin 5.**

Reminder
The result of the direct translation method will not be a percent. To express the result as a percent we multiply by 100.

EXAMPLE 6 A production facility finds that out of the 485 of its products manufactured in a week's time, only 3 are defective. What percent of the products are found to be *without* defect?

Understand: We must calculate the percent of the products that are without defect. We are given the total number of products and the number of defective products.

Plan: We must first calculate the number of products without defect by subtracting the number of defective products from the total number of products. We can then write a basic percent sentence, translate to a proportion, and solve.

Execute: Number of products without defect = 485 − 3 = 482

What percent is 482 out of 485?   **Write a proportion.**

$$\frac{p}{100} = \frac{482}{485}$$ **Solve for *p*.**

$$485p = 48,200$$

$$\frac{485p}{485} = \frac{48,200}{485}$$

$$1p \approx 99.4\%$$

$$p \approx 99.4\%$$

Answer: 99.4% of the products manufactured in a week are without defect.

Check: We could use the direct translation method as a check.

What percent is 482 out of 485?

$$p = 482 \div 485$$

$$p \approx 0.994$$

$$p \approx 0.994\,(100)\%$$ **Write the decimal as a percent.**

$$p \approx 99.4\%$$

◄ **Do Margin 6.**

Margin 6

After 4 months of construction, 26 floors of the 55 floors in a skyscraper are complete. What percent remains to be completed?

Answer to Margin 5: $81.\overline{81}\%$

Answer to Margin 6: $52.\overline{72}\%$

8.4 Exercises

FOR EXTRA HELP

 Videotape 9

 InterAct Math Tutorial Software

 www.carsonmath.com

 AWL Math Tutor Center

 InterAct MathXL www.mathxl.com

 Student's Solutions Manual

For Exercises 1–32, write an equation or a proportion. Then solve.

1. 60% of a mixture is HCl (hydrochloric acid). If the total capacity of the solution is 800 ml, how many milliliters of the solution is HCl?

2. The label on a bottle of rubbing alcohol indicates that it is 70% isopropyl alcohol. If the bottle contains 473 ml, how many milliliters of isopropyl alcohol does it contain?

Trivia Bite

> Here's a case where 1 + 1 does *not* equal 2. Measure 1 cup of water and mix with 1 cup of rubbing alcohol. This will not equal exactly 2 cups of liquid. The reason is that the water and alcohol molecules are different sizes and shapes, allowing the molecules to *squeeze* together so that the total volume is less than 2 cups. The purer the alcohol, the more evident this effect is.

3. Sabrina scored 90% on a test with 30 total questions.
   a. How many questions did she answer correctly?
   b. What percent did she answer incorrectly?
   c. How many questions did she answer incorrectly?

4. Willis scored 85% on a test with 40 questions.
   a. How many questions did he answer correctly?
   b. What percent did he answer incorrectly?
   c. How many questions did he answer incorrectly?

5. According to a study by Columbia University's National Center on Addiction and Substance Abuse, drug and alcohol abuse played a part in the crimes committed by 80% of all people incarcerated. If there are about 1.7 million people in America's prisons, for how many of these did drug and alcohol abuse play a part in their crime? (*Houston Chronicle,* 1/10/98)

6. According to the Criminal Justice Institutes' 1996 Corrections Yearbook, 94% of America's inmates are men. If there are about 1.7 million inmates in America, then how many are men?

7. Wright earns 10% of total sales in commission. If he sells $3106 in merchandise over a two-week period, what is his commission?

8. Kera is a salesperson at a car dealership and earns 25% of the profit in commission. If in one month the dealership made a profit of $9,680 from her sales, how much will she receive as commission?

9. Liz earns 8% of total sales in commission and $6 per hour for working. If she works a total of 60 hours in two weeks and her total sales are $4249, what will her gross pay be for the two-week period?

10. Vastine earns 10% of total sales in commission and $5 per hour for working. If he works a total of 80 hours over two weeks and has total sales of $5827, what is his gross pay?

11. Cush sells cars and earns 25% of the profit in commission. In one month the dealership grossed $78,950. If the total cost to the company was $62,100, what was Cush's commission?

12. A writer earns 15% of the publishing company's net profit on the sale of her book in royalties (commission). If the company sells 13,000 copies of the book and makes a profit of $6 on each book, what is the writer's royalty?

13. Rosita sells real estate and receives a commission of $750. If her commission rate is 6% of the seller's net, what is the seller's net?

14. Andre earns 20% of the net profit in commission on the sale of cash registers. If he receives $2675.92 as a commission, what is the net profit on his sales?

15. A basketball team in one game made 42% of the total number of attempted shots. If they made a total of 39 baskets, how many shot attempts did they have in the game?

16. A football quarterback in one game completes 34% of his pass attempts. If he completed 14 passes, how many pass attempts did he have in the game?

17. At a certain company, 43 employees were hired in the last five years. If this represents 38% of the company's current employees, how many current employees does the company have?

18. $5,710,900 of the budget at a school comes from the state. If this represents 65% of the school's total budget, what is the school's total budget?

19. Hunter answered 58 questions correctly out of a total of 65 questions. What percent of the total questions did Hunter answer correctly?

20. Corrina answered 41 questions correctly out of a total of 45 questions. What percent of the total questions did she answer correctly?

21. A baseball player has 48 hits in a season where he has been at bat 110 times. What percent of his times at bat did he get a hit?

Trivia Bite

In baseball, a player's batting average is the ratio of the number of hits to the number of times at bat. It is expressed as a decimal number to the nearest thousandth. The batting average can be expressed as a percent by multiplying by 100. For example, a player with a batting average of 0.345 gets a hit 34.5% of the times he goes to bat.

22. In one season a football kicker scores 33 extra points in 35 attempts. What percent of his attempts did he score?

23. Of the world's 5.6 billion people, 3 billion live on less than $2 per day. What percent of the world's population live on less than $2 per day? (*Parade,* 8/17/97)

24. In 1995, Americans spent \$57.3 billion on illegal drugs. \$38 billion of that total went toward cocaine. What percent of the total money spent on illegal drugs was spent on cocaine? (*Parade,* 11/9/97)

25. Suppose a person spends 8 hours sleeping, 7 hours working, 1 hour commuting, and 1 hour eating in a 24-hour period. What percent of the day is left?

26. If Mozart composed his first music at the age of 4 and continued composing and performing until he died at the age of 35, what percent of his life was spent involved in music?

Historical Note

Wolfgang Amadeus Mozart (1756–1791) is considered to be one of the greatest classical composers. By the age of 6 he was an accomplished performer on clavier, organ, and violin and had already composed several pieces of high quality. It is said that he completed his compositions in his head and then merely copied the completed work to paper with little or no corrections. He composed nearly 600 pieces in his lifetime and performed throughout Europe for royalty as well as common folk. It is believed that he died of typhoid fever, but some believe that he may have been poisoned. Because of his poverty he was buried in a mass unmarked grave.

27. Below are the results of a survey. Respondents could respond in three ways: agree, disagree, or no opinion.

248 agreed

562 disagreed

89 said no opinion

What percent of all respondents either agreed or disagreed?

28. Below are the results of a survey. Respondents could respond in three ways: agree, disagree, or no opinion.

612 agreed

145 disagreed

94 said no opinion

What percent of all respondents either agreed or disagreed?

29. In a 1860 ft.^{2} house the living room is 16 ft. by 20 ft. What percent of the total area of the house is the living room area?

30. The Sun is a sphere with a radius of about 695,990 km. The core of the Sun is a sphere with a radius of about 170,000 km. What percent of the Sun's total volume is made up by the core?

Trivia Bite

The Sun is so large that if Earth were placed at its center, the moon would be about halfway to the surface of the Sun.

31. Below is a list of the Geiger family's net monthly income and expenses.

Income	Expenses
Mr. Geiger = $3025.85 Mrs. Gieger = $1050.72	Mortgage = $948.75 Childcare = $800 Car payment = $294.88 Car payment = $275.96 Credit card payments = $150 Utilities = $285 Groceries = $475 Entertainment = $260

Discussion What financial advice would you give the Geiger family?

   a. What percent of their income goes to each expense?
   b. What is the total percentage of the family income paid toward expenses?
   c. What percent of their income is left after expenses?

32. Below is a list of the Sharp family's net monthly income and expenses.

Income	Expenses
Mr. Sharp = $2625.85	Mortgage = $752.45 Car payment = $385.95 Credit card payments = $85 Utilities = $210 Groceries = $450 Entertainment = $360

Discussion What financial advice would you give the Sharp family?

   a. What percent of their income goes to each expense?
   b. What is the total percentage of the family income paid toward expenses?
   c. What percent of their income is left after expenses?

Puzzle Problem

There are approximately 2×10^{11} stars in our galaxy. Suppose 10% of those stars have planets orbiting them. Suppose 1% of those planetary systems have a planet with a climate that could support life. Suppose that 1% of those planets that could support life actually have living organisms. Suppose that 1% of those planets with life have intelligent life. Suppose that 1% of the intelligent life have developed civilizations with communication devices. Suppose that 1% of those civilizations with communications devices are within a 500 light year radius from Earth. Based on these suppositions how many stars might we expect to have planets with intelligent life that have communication devices within a 500 light year radius?

Discussion In the Puzzle Problem, what flaws may exist in the suppositions? If the suppositions are reasonable, then what are the implications about contact or communication with other intelligent life forms?

REVIEW EXERCISES

1. Combine like terms. $5.1x + 9.8 + x - 12.4$

2. Solve. $14.7 = x + 0.05x$

3. Solve $24.48 = y - 0.2y$

4. Solve. $\dfrac{y}{2.4} = \dfrac{5.6}{8.4}$

5. Amelia receives a raise so that her new salary is $28,350. If her former salary was $25,800, what was the amount of the raise?

8.5 Solving Problems Involving Percent of Increase or Decrease

OBJECTIVES

1 Solve problems involving a percent of increase.

2 Solve problems involving a percent of decrease.

3 Calculate the percent of increase or decrease.

OBJECTIVE 1 *Solve problems involving a percent of increase.*

In Section 8.4 we considered percent problems that were phrased in terms of parts and wholes. In this section we will explore problems where an initial amount is increased or decreased by a given percent. To help with translating the information in the problems, we will rephrase the basic percent sentence. The whole amount will be the initial amount. The part will be the amount of increase or decrease.

> The **percent** of an **initial amount** is the **amount of increase or decrease.**

Situations like sales tax and a salary raise involve increasing an initial amount by a given percent. Discounts and deductions from one's paycheck are examples of situations where a decrease by a given percent occurs.

Note that it does not matter whether we are dealing with an increase situation or a decrease situation, the percent sentence is the same. However, after calculating the amount of the increase or decrease, we must either add or subtract accordingly.

Let's first explore examples of percent of increase. Sales tax is calculated as a percent of the price of a purchase. It is paid in addition to the price of the purchase. In other words sales tax is added to the initial price of the purchase.

EXAMPLE 1 Charlie is buying a software package for $35. The sales tax rate is 5%. Calculate the sales tax and total amount of the purchase.

Understand: We must calculate the sales tax and total amount of the purchase. We are given the tax rate and initial amount.

Plan: Write a basic percent sentence, translate to an equation, and solve for the sales tax. Then to calculate the final amount, add the sales tax to the initial amount.

Execute: The sales tax is 5% of the initial price.

$$\text{What is 5\% of 35?}$$

$c = 5\% \cdot 35$	**Write the percent as a decimal.**
$c = 0.05(35)$	**Multiply.**
$c = 1.75$	

Tip

To calculate the amount of increase or decrease when the percent is a multiple of 5%, we can calculate the 10% amount, then adjust by an appropriate factor. Because 10% is 0.1, when we multiply a decimal number by 10%, the decimal point moves 1 place to the left.

> 10% of 35 is 3.5 (Move the decimal point left 1 place in 35)

Because 5% is half of 10%, the 5% amount is half of the 10% amount. Because 10% of 35 is 3.5, we can say that 5% of 35 will be half of 3.5, which is 1.75.

We could also calculate other percents that are multiples of 5%. For example 20% is twice the 10% amount. Because 10% of 35 is 3.5, we can say that 20% is twice 3.5, which is 7.

Answer: The sales tax is $1.75. Because sales tax increases the initial price, we add the tax to the initial price to get the total amount:

$$\text{Total} = 35 + 1.75 = \$36.75$$

Check: We can reverse the process as a check. We verify the tax by subtracting the initial amount from the total amount.

$$36.75 - 35 = 1.75$$

Now we can set up a proportion to verify that the tax rate is in fact 5%.

$$\frac{p}{100} = \frac{1.75}{35}$$
$$35p = 175$$
$$\frac{35p}{35} = \frac{175}{35}$$
$$1p = 5\%$$
$$p = 5\%$$

◆ **Margin 1**

a. A mountain bike is priced at $365.95. Calculate the sales tax if the tax rate is 7%.

Connection

The purpose for checking this problem this way is to give a sense of how we will handle problems of increase or decrease when the percent is the missing amount. If the initial amount and total amount are given, our first step is to calculate the amount of increase or decrease by subtracting the initial amount from the total amount. We can then set up a proportion to calculate the percent.

There is a bit of a shortcut that we can take in calculating the total amount with percent of increase problems. Consider the fact that Charlie paid the initial price in full, which is 100% of the initial price. The sales tax was an additional 5%. If he paid 100% of the initial price plus an additional 5% of the initial price in tax, then he paid a total of 105% of the initial price.

Conclusion: To calculate the final amount with percent of increase, we can add the percent of increase to 100%, then multiply the result by the initial amount.

Consider the multiplication process to get a sense of why this works.

b. A suit is priced at $128.95. Calculate the sales tax and total amount if the tax rate is 6%.

$$
\begin{array}{r}
35 \\
\times\ 1.05 \\
\hline
175 \\
000 \\
+\ 35 \\
\hline
36.75
\end{array}
$$

Calculate the tax by multiplying the 5 digit in 1.05.

The tax and initial amount are added within the multiplication procedure to end up with the total amount.

Multiplying by the digit 1 in the ones place of 1.05 places the initial amount in the sum.

◀ **Do Margin 1.**

Another example of percent of increase is a salary raise. Usually raises are given based on a percent of the salary. Note that we can use the same shortcut that we discussed for sales tax. When we get a raise, we keep 100% of our initial salary, and the percent of the raise increases that initial salary to the new salary level. We can add the percent of increase to 100%, then multiply by the initial salary to arrive at the new salary.

Answers: **a.** $25.62 **b.** $7.74; $136.69

EXAMPLE 2 Carlita's current annual salary is $23,500. If she gets a 4% raise, what is her new salary?

Understand: We must calculate Carlita's new salary. We are given her initial salary and the percent of the raise. Because the 4% is added to 100% of her current salary, her new salary will be 104% of her current salary.

Plan: Write a basic percent sentence, translate to an equation, and solve for the new salary. Because we are calculating a final amount, we will use the variable a.

Execute: The new salary is 104% of the current salary.

What is 104% of 23,500?

$a = 104\% \cdot 23{,}500$ **Write the percent as a decimal.**

$a = 1.04(23{,}500)$ **Multiply.**

$a = 24{,}440$

Answer: Carlita's new salary will be $24,440. Note that if we need to know the amount of the raise, we can subtract the initial salary from the new salary.

raise amount $= 24{,}440 - 23{,}500 = \$940$

Check: We can reverse the process and verify that the raise of $940 is in fact 4% of her initial salary. Now we can set up a proportion to verify that the raise is indeed 4%.

$$\frac{p}{100} = \frac{940}{23{,}500} \quad \begin{matrix} \leftarrow \textbf{Raise amount} \\ \leftarrow \textbf{Initial salary} \end{matrix}$$

$$23{,}500p = 94{,}000$$

$$\frac{23{,}500p}{23{,}500} = \frac{94{,}000}{23{,}500}$$

$$1p = 4\%$$

$$p = 4\%$$

> **Connection**
> Note how we are calculating the percent of the increase in the check. The percent of increase is equal to the ratio of the raise amount to the initial salary.

Do Margin 2. ▶

What if the initial amount is the missing amount? If we are given the percent with the amount of increase, we can use the basic percent sentence and solve for the initial amount. However, if we are given the percent with the amount of increase, we have a little more challenging problem.

Consider the case in which we are given the percent with the amount of increase.

EXAMPLE 3 Carmine indicates that he received a 6% raise that amounts to $1950. What was his former annual salary?

Understand: We must calculate Carmine's former salary. We are given the percent of increase and the amount of the raise.

Plan: Write a basic percent sentence, translate to an equation, and solve for the new salary. We will use the variable f to represent the former salary.

Execute: 6% of the former salary is the raise amount.

6% of what salary is 1950?   **Translate.**

$6\% \cdot \quad f \quad = 1950$ **Write the percent as a decimal.**

$0.06f = 1950$

$\dfrac{0.06f}{0.06} = \dfrac{1950}{0.06}$ **Divide both sides by 0.06 to isolate f.**

$1f = 32{,}500$

$f = 32{,}500$

◆ **Margin 2**

Benjamin has a current salary of $22,600. If he receives a 3% raise, what will be his new salary?

Answer: $23,278

Answer: Carmine's former salary was $32,500.

Check: We can verify that a 6% raise on an initial salary of $32,500 equals a raise of $1950.

$$6\% \text{ of } 32,500 \text{ is what raise?}$$
$$6\% \cdot 32,500 = r$$
$$0.06(32,500) = r$$
$$1950 = r$$

◆ It checks. The raise was $1950.

◀ **Do Margin 3.**

What if the percent and the final amount, after the increase, were the given pieces of information?

EXAMPLE 4 Suppose that Carmine (from Example 3) said that he had a 6% raise and his new salary is $34,450. What was his former salary?

Understand: We must calculate Carmine's former salary. We are given the new salary after the 6% raise.

The new salary is the former salary combined with the amount of the raise.

Plan: Write an equation, then solve.

Execute: Recall that the new salary is the former salary combined with the amount of the raise.

$$\text{new salary} = \text{former salary} + \text{amount of raise}$$

Recall that the amount of the raise is calculated by multiplying the former salary by 6%. We can incorporate this into the above equation.

$$\text{new salary} = \text{former salary} + 6\% \cdot \text{former salary}$$

If we let f represent the former salary, we can translate the above equation to:

$$34,450 = f + 0.06f$$

Now solve for f.

$34,450 = f + 0.06f$	**Combine like terms.**
$34,450 = 1.06f$	
$\dfrac{34,450}{1.06} = \dfrac{1.06f}{1.06}$	**Divide both sides by 1.06 to isolate f.**
$32,500 = 1f$	
$32,500 = f$	

Reminder

$f + 0.06f$ is the same as $1f + 0.06f$. To combine like terms, we add the coefficients:
$1 + 0.06 = 1.06$

Connection

Notice that after we combine like terms, we get $1.06f$. Recall that 1.06 is equal to 106%. This agrees with our shortcut for increase problems. 34,450 is 100% of the former salary with a 6% raise. So 34,450 is 106% of the former salary.

◆ *Answer:* Carmine's former salary was $32,500. Notice that this checks with Example 3.

◆ Margin 3

Julia indicates that she received a 3% raise in the amount of $1356. What was her former salary?

Answer: $45,200

Do Margin 4. ▶

OBJECTIVE 2 *Solve problems involving a percent of decrease.*

Now let's explore problems involving percent of decrease. The percent sentence remains the same. However, we will *subtract* the amount of decrease from the initial amount to get the final amount.

Discount prices and pay deductions are two common situations in which percent of decrease is used. A discount situation occurs when a store has a sale and marks down the prices of merchandise by a certain percent. Our task will be to calculate the discount and then subtract the discount amount from the initial price to get the final price.

EXAMPLE 5 A 20% discount is to be applied to a dress with an initial price of $49.95. What are the discount amount and final price?

Understand: We must calculate the discount amount and final price. We are given the discount rate and the initial amount.

Plan: Write a basic percent sentence, translate to an equation, and solve for the discount amount. Then to calculate the final price, subtract the discount amount from the initial price.

Execute: The discount amount is 20% of the initial price.

What amount is 20% of 49.95?   **Translate.**
$$c = 20\% \cdot 49.95 \quad \text{Write the percent as a decimal.}$$
$$c = 0.2(49.95) \quad \text{Multiply.}$$
$$c = 9.99$$

Answer: The discount amount is $9.99. Because this discount is an amount of decrease, we subtract it from the initial price to get the final price.

$$\text{final price} = 49.95 - 9.99 = \$39.96$$

Check: We can reverse the process to verify that the discount rate is correct. We verify the discount amount by subtracting the final price from the initial price.

$$49.95 - 39.96 = 9.99$$

Now we can set up a proportion to verify that the discount rate is 20%.

$$\frac{p}{100} = \frac{9.99}{49.99} \quad \begin{array}{l} \leftarrow \text{Discount amount} \\ \leftarrow \text{Initial price} \end{array}$$
$$49.95p = 999$$
$$\frac{49.95p}{49.95} = \frac{999}{49.95}$$
$$1p = 20\%$$
$$p = 20\%$$

| Connection
Notice that the pattern we've established in the checks continues. The percent of decrease is equal to the ratio of the amount of decrease (discount amount) to the initial amount (initial price).

Note that there is a shortcut with percent of decrease problems that is similar to what we did with percent of increase problems. Consider the fact that with discounts, we are deducting the discount amount from the initial amount. Instead of paying 100% of the initial price, we are deducting 20%, therefore we only pay 80% of the initial price.

◆ **Margin 4**

Josh recalls paying a total of $16.75 for a music CD. He knows the sales tax rate in his state is 5%. What was the initial price of the CD?

Answer: $15.95

Conclusion: To calculate the final amount with percent of decrease problems, we can simply subtract the percent of decrease from 100%, then multiply the resulting percent by the initial amount.

Repeating Example 2 with this method:

The final price is 80% of the initial price.

$$\text{final price} = 80\% \cdot 49.95$$
$$\text{final price} = 0.8(49.95)$$
$$\text{final price} = 39.96$$

Note that, if we need to know the amount of the discount, we can subtract the final price from the initial price.

◀ **Do Margin 5.**

Pay deductions also use percent of decrease. FICA, Medicare, retirement, etc., are deductions that are based on a percent of gross wages. A person's *net pay* is the amount that is left after all deductions, that is, take home pay.

EXAMPLE 6 Approximately 28% of Deanne's gross monthly income is deducted for Social Security, federal withholdings, state withholdings, insurance, etc. If Deanne's gross monthly salary is $2111.67, what is her net pay?

Understand: If 28% of her gross monthly salary is deducted, then her net pay will be 72% of her gross salary (100% − 28% = 72%).

Plan: Write a basic percent sentence. Translate to an equation or proportion. Then solve. We will use the variable n to represent net pay.

Execute: Her net pay is 72% of her gross monthly salary

$n = 72\% \cdot 2111.67$	Write the percent as a decimal.
Discussion What were the total deductions? $\quad n = 0.72\,(2111.67)$	Multiply.
$n = 1520.4$	

Answer: Her net pay will be $1520.40.

Check: We can reverse the process to verify that the deduction percent is correct. The deduction percent is equal to the ratio of the deduction amount to the gross pay. We first need the deduction amount by subtracting net pay from gross pay.

$$2111.67 - 1520.40 = 591.27$$

Now we can set up a proportion to verify that the deduction is 28%.

$$\frac{p}{100} = \frac{591.27}{2111.67} \quad \begin{array}{l} \leftarrow \text{Deduction amount} \\ \leftarrow \text{Gross pay} \end{array}$$
$$2111.67p = 59{,}127$$
$$\frac{2111.67p}{2111.67} = \frac{59{,}127}{2111.67}$$
$$1p \approx 28\%$$
$$p \approx 28\%$$

◀ **Do Margin 6.**

What if the initial amount is missing in a percent of decrease problem? If we are given the percent and the amount of decrease, the problem is no different than Example 3, when we were given the percent and the amount of increase.

◆ **Margin 5**

A coat with an initial price of $88.95 is marked 40% off. What are the discount amount and final price?

◆ **Margin 6**

a. 1.45% of gross monthly salary goes to Medicare. If a person's gross pay is $1187.50, what is that individual's Medicare contribution?

b. Sandra's total deductions come to about 31% of her gross pay. If her gross pay is $1337.50, what is her net pay?

Answer to Margin 5: $35.58; $53.37

Answers to Margin 6: **a.** $17.22 **b.** $922.88

EXAMPLE 7   If 27% of Petrov's gross monthly salary results in total deductions of $738, what is his gross monthly salary?

Understand: We must calculate Petrov's gross monthly salary. We are given the percent of decrease and the total deductions.

Plan: Write a basic percent sentence, translate to an equation, and solve for the salary. We will use the variable g to represent his gross monthly salary.

Execute: 27% of his gross monthly salary is total deductions.

◆ **Margin 7**

Clayton purchased a suitcase on sale for 20% off. If the discount amount was $37.09, what was the initial price?

27% of what salary is 738?		Translate.	

$$27\% \cdot \quad g \quad = 738 \qquad \text{Write the percent as a decimal.}$$

$$0.27g = 738$$

$$\frac{0.27g}{0.27} = \frac{738}{0.27} \qquad \text{Divide both sides by 0.27 to isolate } g.$$

$$1g = 2733.33$$

$$g = 2733.33$$

Answer: Petrov's gross monthly salary is $2733.33.

Check: We can verify that 27% of $2733.33 equals the total deductions of $738.

$$27\% \text{ of } 2733.33 \text{ is what amount of decrease?}$$

$$27\% \cdot 2733.33 = d$$

$$0.27(2733.33) = d$$

$$738 = d \qquad \text{It checks.}$$

Do Margin 7. ▷

What if we had been given Petrov's net salary and the percent? This situation would be similar to Example 4. However, we must remember that net salary is the amount left after deductions.

$$\text{net salary} = \text{gross salary} - \text{total deductions}$$

EXAMPLE 8   After deductions, Petrov's net pay is $1995.33. If the total of the deductions was 27% of his gross monthly salary, what was his gross monthly salary?

Understand: We must calculate Petrov's gross monthly salary. We are given the percent of decrease and the net pay.

Plan: Write an equation, then solve.

Execute: Recall that net pay is gross pay minus the total deductions.

$$\text{net pay} = \text{gross monthly salary} - \text{total deductions}$$

Also recall that the total of the deductions is 27% of the gross salary. We can incorporate this into the above equation.

$$\text{net pay} = \text{gross monthly salary} - 27\% \cdot \text{gross pay}$$

If we let g represent the gross monthly salary, we can translate the above equation to:

$$1995.33 = g - 0.27g$$

Answer: $185.45

Now solve for g.

$$1995.33 = g - 0.27g \qquad \text{Combine like terms.}$$
$$1995.33 = 0.73g$$
$$\frac{1995.33}{0.73} = \frac{0.73g}{0.73} \qquad \text{Divide both sides by 0.73 to isolate } g.$$
$$2733.33 = 1g$$
$$2733.33 = g$$

Reminder

$g - 0.27g$ is the same as $1g - 0.27g$. To combine like terms, we subtract the coefficients:
$1 - 0.27 = 0.73$

Connection

Notice that after we combine like terms, we get $0.73g$. Recall that 0.73 is equal to 73%. This agrees with our shortcut for decrease problems. 1995.33 is the net pay after a 27% deduction from 100% of the gross monthly salary. Therefore 1995.33 is 73% of the gross monthly salary.

◆ *Answer:* Petrov's gross monthly salary is $2733.33. Note that this checks with Example 7.

◀ **Do Margin 8.**

OBJECTIVE 3 *Calculate the percent of increase or decrease.*

What if the percent is the unknown? We shall see that the given information will affect how we approach the problem. Recall that the basic percent sentence is:

The **percent** of an **initial amount** is the **amount of increase or decrease.**

If we are given the initial amount and the amount of increase or decrease, we can write an equation or proportion and solve for the percent.

EXAMPLE 9 Tiera is visiting a different state and purchases a shirt that is priced $24.95. She notes that the tax on her receipt is $2.25. What is the sales tax rate?

Understand: We must calculate the sales tax rate, which is a percent of increase. We are given the initial amount (initial price) and the increase amount (sales tax).

Plan: Write a basic percent sentence. Translate to an equation or proportion. Then solve.

Execute: The percent of an initial amount is the amount of increase or decrease.

What percent of 24.95 is 2.25?

$$\frac{p}{100} = \frac{2.25}{24.95} \quad \begin{array}{l} \leftarrow \text{Amount of increase} \\ \leftarrow \text{Initial amount} \end{array}$$
$$24.95p = 225$$
$$\frac{24.95p}{24.95} = \frac{225}{24.95}$$
$$1p \approx 9\%$$
$$p \approx 9\%$$

Note: It is most likely that the actual tax rate is exactly 9%, and when the tax amount was calculated, it was rounded.

Answer: The tax rate is 9%.

◆ **Margin 8**

Clarece purchased a robe discounted by 25%. If the price after the discount was $59.96 what was the initial price?

Answer: $79.95

Check: We can verify that a tax rate of 9% applied to an initial price of $24.95 will equal a sales tax of $2.25.

$$\text{Sales tax is 9\% of 24.95.}$$
$$\text{Sales tax} = 0.09 \, (24.95) = 2.25$$

Do Margin 9. ▶

What if we were given the initial amount and final amount after the increase or decrease has been applied?

Because the percent sentence includes the initial amount and the amount of increase or decrease, we must first calculate the increase or decrease. Look back over the checks for Examples 1, 2, 5, and 6. You should notice that in each case, to find the percent, we first had to calculate the amount of the increase or decrease. We did this by calculating the difference between the final amount and initial amount. Also notice that the percent of increase or decrease is equal to the ratio of the amount of increase or decrease to the initial amount.

$$\frac{p}{100} = \frac{\text{amount of increase or decrease}}{\text{initial amount}}$$

EXAMPLE 10 Jerod was making $12.50 per hour. He is given a raise so that now he is making $14.00 per hour. What was the percent of the increase?

Understand: We must calculate the percent of the increase. We are given Jerod's initial hourly wage and his new hourly wage.

Plan: Because the percent of the increase is equal to the ratio of the amount of increase to the initial amount, we must first calculate the amount of increase. The amount can be calculated by subtracting the initial hourly wage from the new hourly wage. We then can write the proportion and solve.

Execute: amount of increase = new hourly wage − initial hourly wage
$$= 14.00 - 12.50$$
$$= \$1.50$$

Now we can put together the proportion and solve for the percent. In terms of the basic percent sentence, we are asking:

What percent of 12.50 is 1.50?

$$\frac{p}{100} = \frac{1.50}{12.50} \quad \begin{array}{l} \leftarrow \textbf{Amount of increase} \\ \leftarrow \textbf{Initial amount} \end{array}$$
$$12.50p = 150$$
$$\frac{12.50p}{12.50} = \frac{150}{12.50}$$
$$1p = 12\%$$
$$p = 12\%$$

Answer: Jerod received a 12% raise.

Check: We can verify that a 12% raise with a $12.50 initial hourly wage equals a new hourly wage of $14.50.

$$\text{The raise amount is 12\% of 12.50.}$$
$$R = 0.12 \, (12.50)$$
$$R = 1.50$$

◆ Margin 9

a. Marvin received a raise of $1728. If his former salary was $28,800 what was the percent of increase?

b. A salesperson says she can take $1280 off the price of a car. If the initial price is $25,600, what is the percent of the decrease?

Answers: a. 6% b. 5%

a. Marianne purchases a small rug on sale. The initial price was $185.79. The price after the discount was $157.92. What was the percent of decrease (discount rate)?

the new hourly wage = initial hourly wage + the raise amount

$$N = 12.50 + 1.50$$
$$N = 14.50$$

Connection

We could have used the shortcut for increase problems. Adding the percent of the increase to 100, we can say that the new hourly wage is 112% of the initial hourly wage.

$$\text{new hourly wage} = 112\% \cdot 12.50$$
$$N = 1.12(12.50)$$
$$N = 14.50$$

◀ **Do Margin 10.**

b. Quan Li manages a manufacturing plant and has just received the end-of-the-month report on production. The report indicates that the plant produced 2336 units. At the end of the previous month, the plant had produced 2540 units. What is the percent of decrease in production?

Answers: **a.** 15% **b.** 8%

8.5 Exercises

For Exercises 1–44, solve.

1. A desk is priced at $185.95. The sales tax rate is 5%. Calculate the sales tax and total amount of the purchase.

2. A guitar is priced at $585.95. Calculate the sales tax if the tax rate is 6%.

3. A car is priced at $26,450. If the sales tax rate is 6%, calculate the sales tax and total amount of the purchase.

Trivia Bite

Many states have a law that sets the maximum sales tax that can be charged on a purchase. It is sometimes beneficial to shop around in different states for large purchases.

4. A computer is priced at $1998.95. Calculate the sales tax and total amount if the tax rate is 7%.

5. Michelle's current annual salary is $29,600. If she gets a 3.5% raise, what is her new salary?

6. Lou's current annual salary is $23,800. If he receives a 6.5% raise, what is his new salary?

7. The bill in a restaurant comes to a total of $32.85. The patrons decide to tip the wait person 15%. What tip amount should be left?

8. The bill in a restaurant comes to a total of $48.75. The patrons decide to tip the wait person 20%. What tip amount should be left?

Tip

When tipping, most people round to whole dollar amounts. A convenient scale for 15% is to recognize that every $10 of the bill corresponds to a tip of $1.50. The rest of the scale would follow:

Total Bill	Tip
$5	$0.75
$10	$1.50
$20	$3.00

In our example of a total bill of $48.75, we would round to $50. $50 is two $20's and one $10. Each $20 is a $3 tip and the $10 is a $1.50 tip for a total tip of $7.50.

$$\text{Total Bill} = \$20 + \$20 + \$10 = \$50$$
$$\downarrow \qquad \downarrow \qquad \downarrow$$
$$\text{Tip} \quad = \$3 \ + \ \$3 \ + \$1.50 = \$7.50$$

9. Jack indicates that he received a 4% raise that amounts to $1008. What was his former annual salary?

10. Lucia indicates that she received a 5.5% raise in the amount of $1793. What was her former salary?

11. Huang indicates that he received a 4% raise. His new salary is $30,420. What was his former salary?

12. Denzel indicates that he received a 9% raise. His new salary is $39,828.60. What was his former salary?

13. Rachel paid a total of $45.10 for a pair of shoes. She knows the sales tax rate in her state is 5%. What was the initial price of the shoes?

14. Paul paid $635.95 for a new stereo system. If the sales tax rate in his state is 6%, what was the initial price?

15. A 30% discount is to be applied to a dress with an initial price of $65.99. What are the discount amount and final price?

16. A sweater with an initial price of $42.95 is marked 25% off. What are the discount amount and final price?

17. A sofa with an initial price of $699.99 is marked 35% off. What are the discount amount and final price?

18. A necklace with an initial price of $349.90 is marked 45% off. What are the discount amount and final price?

19. Approximately 27% of Sheila's gross monthly income is deducted for Social Security, federal withholdings, state withholdings, insurance, etc. If her gross monthly salary is $2708.33, what is her net pay?

20. Approximately 29% of Natron's gross monthly income is deducted for Social Security, federal withholdings, state withholdings, insurance, etc. If his gross monthly salary is $2487.50, what is his net pay?

21. 1.45% of gross monthly salary goes to Medicare. If a person's gross pay is $1546.67, what is the Medicare contribution?

22. 6.2% of gross pay goes to FICA. If a person's gross pay is $540.83, what is the FICA deduction?

23. If 26% of Gordon's gross monthly salary results in total deductions of $674.96, what is his gross monthly salary?

24. Marcelle's retirement deduction from her monthly pay is $187.50. If the retirement deduction is 6% of her gross monthly salary, what is her gross monthly salary?

25. Patricia purchased a table on sale for 30% off. If the discount amount was $44.09, what was the initial price?

26. A lamp is on sale for 15% off. If the discount amount is $13.49, what is the initial price?

27. After deductions, Luther's net pay is $1850.40. If the total of the deductions was 28% of his gross monthly salary, what was his gross monthly salary?

28. After deductions, Wendy's net pay is $1016.60. If the total of the deductions was 32% of her gross two-week salary, what was her gross two-week salary?

29. Hannah purchased a curio stand discounted 35%. If the price after the discount was $246.68 what was the initial price?

30. Vladimir purchased a set of golf clubs discounted 15%. If the price after the discount was $577.96 what was the initial price?

31. Melliah purchases a tennis racquet that is priced $54.95. She notes that the tax on her receipt is $3.85. What is the sales tax rate?

32. The property tax on a vehicle is $286.45. If the vehicle is assessed at a value of $8500, what is the tax rate?

33. Tina received a raise of $849.10. If her former salary was $24,260, what was the percent of the increase?

34. Jana notes that her July electricity bill is $8.96 higher than last year's amount. If last year's bill was $148.75, what is the percent of increase?

35. A salesperson says they can take $1512 off the price of a car. If the initial price is $18,900, what is the percent of decrease?

36. In 1997 a company's cost of production for one unit of a product was $328.50. After taking some cost-cutting measures, the company was able to reduce the cost of production by $14.50. What is the percent of decrease in cost?

37. Lena was making $8.50 per hour. She is given a raise so that now she is making $10.00 per hour. What was the percent of increase?

38. Gloria notes that her water bill was $18.50 in March. Her April bill is for $32.40. What is the percent of increase?

39. A social worker worked a total of 1259 cases in 1996. The next year he worked 1540 cases. What was the percent of increase in case load?

40. 1990 was the first year of operation for a business. At the end of that first year, the business recorded a profit of $15,280. In 1998 that same business recorded a profit of $880,450. What was the percent of increase in profit?

41. Van purchases a VCR on sale. The initial price was $275.90. The price after the discount was $234.52. What was the percent of decrease (discount rate)?

42. Risa's gross two-week salary is $1756.25. If her net pay is $1334.75, what is the percent of decrease?

43. By insulating their water heater, adding a storm door, and installing new double-pane windows, the Jones family decreased their electric and gas bills from a total of $1536.98 in 1997 to $1020.57 in 1998. What was the percent of decrease in their electric and gas charges?

44. Gloria is an office manager. Through her suggestions for improving efficiency and cutting excessive spending, the cost of operation dropped from $65,886.35 in 1996 to $60,245.50 in 1997. What was the percent of decrease in operation cost?

REVIEW EXERCISES

1. Solve. $0.12x = 450$

2. Solve. $20 = 9(y + 2) - 3y$

3. Simplify. $(1 + 0.08)^2$

4. Evaluate. $(5)^{-3}$

5. Will has some $5 bills and $10 bills in his wallet. If he has 16 bills worth a total of $105, then how many of each bill is in his wallet? Use a four-column table.

Categories	Value	Number	Amount

8.6 Solving Problems Involving Interest

OBJECTIVES

1 Solve problems involving simple interest.

2 Solve problems involving compound interest.

3 Solve problems involving amortization.

OBJECTIVE 1 *Solve problems involving simple interest.*

Percents can be applied to borrowing or investing money. The money that is borrowed or invested is called the **principal.** In terms of the basic percent sentence, think of principal as the initial amount.

DEFINITION	**Principal:** An initial amount of money.

For example, if we borrow or invest $100 then we say the principal is $100.

When borrowing money, the lender will assess a charge for the use of the money. This charge is a percent of the principal and is called **interest.** Think of interest as the amount of increase in the basic percent sentence. The borrower is expected to pay back the interest as well as the principal.

Similarly, when money is invested it earns interest. The receiver of the investment pays the investor interest for the use of the money.

DEFINITION	**Interest:** An amount of money that is a percent of the principal.

Interest is an amount of money calculated as a percent of the principal. This percent is called the **interest rate.**

DEFINITION	**Interest rate:** A percent used to calculate interest.

An interest rate is usually listed as an **annual percentage rate (APR).** An annual percentage rate describes the interest that would be earned on a given principal if it is allowed to earn interest for one full year.

DEFINITION	**Annual percentage rate (APR):** An interest rate that is used to calculate the interest earned on a given principal if it were allowed to earn interest for one full year.

For example, if we invested $100 principal in a savings account with a 5% APR, then we would be paid 5% of the $100 in interest at the end of a year's time. 5% of $100 is $5. So the bank would deposit $5 in our savings account at the end of the year so that we would have a total balance of $105.

All interest rates in the text will be APR. Notice this introduces the fact that time plays a part in interest. Because all interest rates will be APR, if we withdrew the money after only half a year, we would receive half of the interest. In our above example, instead of $5 we would only earn $2.50 if we withdrew the money after half a year.

Conclusion: The amount of interest depends on principal, the interest rate, and the time that the principal is allowed to earn interest.

There are two ways that interest is calculated:

1. Simple interest

2. Compound interest

Simple interest is interest that is earned only based on the original principal. Simple interest is the product of the principal, the APR, and the time in years.

DEFINITION **Simple Interest:** Interest calculated using only the original principal.

$$\text{simple interest} = \text{principal} \cdot \text{APR} \cdot \text{time}$$

$$I = Prt$$

EXAMPLE 1 Graham invests $800 in a savings account at an APR of 3%. How much interest will be earned after 1 yr? What will the final balance be after 1 yr?

Understand: We must calculate the interest and final balance after 1 yr. We are given the principal, rate, and time.

Plan: Use the simple interest formula $I = Prt$ to calculate interest. Then add the interest to the principal to calculate the final balance.

Execute: $I = Prt$

$$I = (800)(3\%)(1)$$
$$I = (800)(0.03)(1)$$
$$I = \$24$$

Answer: Graham will earn $24 in interest after 1 yr. To get the final balance, we add the interest to the principal.

$$\text{final balance} = 800 + 24 = \$824$$

◆ *Check:* Verify that 24 is in fact 3% of 800. We will leave this to the reader.

◀ **Do Margin 1.**

What if the time is other than 1 yr? Because the interest rate is an APR, the time must be expressed in terms of years. We can use dimensional analysis from Chapter 7 to convert the units of time.

Suppose the time is for 9 mo. We must convert 9 mo. to years. Because 1 yr. is 12 mo., we write:

$$9 \text{ mo.} = \frac{9 \text{ mo.}}{1} \cdot \frac{1 \text{ yr.}}{12 \text{ mo.}} = \frac{9}{12} \text{ yr.} = \frac{3}{4} \text{ yr.}$$

Suppose the time is 60 days. To simplify the calculation, financial institutions will use 365 days to a year instead of $365\frac{1}{4}$.

$$60 \text{ days} = \frac{60 \text{ days}}{1} \cdot \frac{1 \text{ yr.}}{365 \text{ days}} = \frac{60}{365} \text{ yr.}$$

Note: Because we will use the fractions in the calculation of interest, it isn't necessary to simplify them. The simplification will occur in the calculation.

Let's now consider some problems where the time is other than 1 yr.

EXAMPLE 2 Calvin borrows $2000 at 15% APR. If he negotiates to repay the loan after 90 days, how much will he have to pay back?

Understand: We must calculate the final balance after 90 days. Because the loan is for 90 days, and the time must be in terms of years, we must convert 90 days to years.

$$90 \text{ days} = \frac{90 \text{ days}}{1} \cdot \frac{1 \text{ yr.}}{365 \text{ days}} = \frac{90}{365} \text{ yr.}$$

Plan: Use the simple interest formula $I = Prt$ to calculate interest. Then add the interest to the principal to calculate the final balance.

Execute: $I = Prt$

$$I = (2000)(0.15)\left(\frac{90}{365}\right)$$

$$I = \frac{2000}{1} \cdot \frac{0.15}{1} \cdot \frac{90}{365}$$

$$I = \frac{27{,}000}{365}$$

$$I \approx 73.97$$

Answer: In 90 days Calvin will have to pay back the $2000 principal plus $73.97 in interest for a total of $2073.97.

◆ *Check:* Verify that $73.97 is in fact 15% of 2000 for 90 days. We will leave this to the reader.

Do Margin 2. ▶

What if the principal is the missing amount? Like the problems of percent of increase, our approach will depend on the given information. If we are given the percent and the interest, we can use $I = Prt$ and solve for P.

EXAMPLE 3 Sandy notes that $6.80 was deposited in her savings account after 3 mo. The APR on her account is 5.5%. What was the principal?

Understand: We must calculate the principal. Because the time is 3 mo. and the time must be in terms of years, we must convert 3 mo. to years.

$$3 \text{ mo.} = \frac{3 \text{ mo.}}{1} \cdot \frac{1 \text{ yr.}}{12 \text{ mo.}} = \frac{3}{12} \text{ yr.}$$

◆ **Margin 2**

Calculate the interest and the final balance.

a. $4500 at 12% for 1 quarter. ($\frac{1}{4}$ of a year)

b. $900 at 5.5% for 30 days.

Answers: **a.** $135; $4635
b. $4.07; $904.07

Plan: Use the simple interest formula $I = Prt$ and solve for P.

Execute: $I = Prt$

$$6.80 = P(0.055)\left(\frac{3}{12}\right)$$

$$6.80 = P\left(\frac{0.055}{1} \cdot \frac{3}{12}\right)$$

$$6.80 = P\left(\frac{0.165}{12}\right)$$

$$6.80 = 0.01375P$$

$$\frac{6.80}{0.01375} = \frac{0.01375P}{0.01375}$$

$$494.55 \approx P$$

Answer: The principal was $494.55.

Check: Verify that a principal of $494.55 at 5.5% for 3 months earns $6.80 in interest. We will ◆ leave this to the reader.

◀ **Do Margin 3.**

What if the final balance were given instead of the interest? We must remember that the final balance is a combination of principal and interest. We can say:

$$\text{final balance} = \text{principal} + \text{interest}$$

Because interest is calculated by the relationship Prt, we can write the above equation as:

$$B = P + Prt$$

Note that we rewrite the above formula by factoring out P.

$$B = P(1 + rt)$$

EXAMPLE 4 Pedro notes that his new balance at the end of a year is $1441.60. If the APR for the account is 6%, what was the principal?

Understand: We must calculate the principal. We are given the APR and the final balance.

Plan: Use the final balance formula $B = P(1 + rt)$

Execute:
$$B = P(1 + rt)$$
$$B = P[1 + 0.06(1)]$$
$$1441.60 = 1.06P$$
$$\frac{1441.60}{1.06} = \frac{1.06P}{1.06}$$
$$1360 = 1P$$
$$1360 = P$$

Divide both sides by 1.06 to isolate P.

Connection
This is the same type of problem as Example 4 in Section 8.5.

Connection
Consider the meaning of the equation $1441.60 = 1.06P$. Recall that 1.06 is 106%. This equation is saying that 1441.60 is 106% of the principal. This should make sense because 1441.60 is 100% (all) of the principal combined with the interest, where the interest is 6% of the principal.

Answer: Pedro's principal was $1360.

Check: Verify that a principal of $1360 at 6% APR will come to a final balance of 1441.60 at the end of 1 year.

◆

Do Margin 4. ▶

◆ **Margin 4**

a. The total amount that Nala must repay on a 1 yr. loan at 8% is $12,960. What was the principal?

OBJECTIVE 2 *Solve problems involving compound interest.*

Now that we've explored simple interest, we can explore the second way that interest is calculated: by compounding. Recall that simple interest was calculated based solely on the original principal. When interest is calculated using the principal as well as prior earned interest, we are dealing with **compound interest.**

DEFINITION **Compound interest:** Interest that is calculated based on principal and prior earned interest.

---

b. Ron notes that his new balance at the end of 6 mo. is $1221. If the APR for the account is 4%, what was the principal?

Let's work toward developing a formula for compound interest. As we work through Example 5 the *long* way, look for a pattern.

EXAMPLE 5 Chan invests $500 in a savings account with an APR of 8%. If the interest is compounded quarterly, how much will he have after 1 yr?

Understand: We must calculate Chan's final balance after 1 yr if the interest is compounded quarterly. Notice that this means every 3 mo. the bank will make a deposit into Chan's account. Therefore, he will start each quarter with a little more money than the prior quarter and therefore earn slightly more interest.

Plan: Calculate the final balance after each quarter in the year. This means we will perform four calculations. Recall that final balance can be calculated using the formula $B = P(1 + rt)$.

First quarter: $B = 500\left[1 + 0.08\left(\frac{1}{4}\right)\right]$

$B = 500[1 + 0.02]$

$B = 500[1.02]$

$B = 510$

Connection
1.02 is 102%. In one quarter he will earn a fourth of the APR. If the APR is 8%, then in one quarter he will earn 2%. The final balance is 100% of the principal plus the 2% in interest for a total of 102% of the principal.

Note that each quarter has the same rate of 2%. Therefore to get each final balance, we simply multiply the previous final balance by 1.02.

Second quarter: $B =$ first quarter final balance $\cdot$ 1.02

$B = 510(1.02)$

$B = 520.2$

Third quarter: $B =$ second quarter final balance $\cdot$ 1.02

$B = 520.2(1.02)$

$B = 530.604$

Note: We did not round here because we will perform another calculation with this number.

Fourth quarter: $B =$ third quarter final balance $\cdot$ 1.02
(1 year)

$B = 530.604(1.02)$

$B \approx 541.22$

We rounded here because this is the final calculation.

Answer: Chan will have $541.22 after 1 yr.

Discussion How much would Chan have earned if the situation were 8% simple interest for 1 yr?

Answers: **a.** $12,000 **b.** $1197.06

These exponent calculations are best done with a calculator. On a scientific calculator, the calculation to the right would proceed as follows:

500 [×] 1.02 [yˣ] 4 [=]

As you probably noted, compound interest calculations can be quite tedious if we follow the above approach. Consider how many calculations we would have performed if Chan had let the principal earn interest for 5 yr., or 20 yr., or more. A formula for compound interest can simplify the calculations considerably.

Notice that we multiplied 500 by four factors of 1.02. This means we can use an exponent of 4 with 1.02 as the base.

$$500 \cdot 1.02 \cdot 1.02 \cdot 1.02 \cdot 1.02$$
$$= 500 \, (1.02)^4$$

Answer: 541.21608, which is rounded to 541.22

Keep in mind that 1.02 is the $(1 + rt)$ part of the formula for figuring final balance: $B = P(1 + rt)$. Therefore, to make this formula work for compound interest, we simply introduce an exponent into the formula. But how do we determine the exponent?

In our example the exponent was the number of times that the interest was compounded.

Conclusion: The final balance in a compound interest situation can be calculated using the formula:

$$B = P(1 + rt)^n$$

Where P represents the principal

r represents the APR

t represents the time length, in years, of each compounding

n represents the number of compoundings

Some common compound time lengths are:

Annually = One compounding each year

Semiannually = Every $\frac{1}{2}$ yr., so 2 compoundings per year

Quarterly = Every $\frac{1}{4}$ yr, so 4 compoundings each year

Daily = Every $\frac{1}{365}$ yr., so 365 compoundings a year

Sometimes interest is even compounded continuously. We'll consider that situation in the next objective.

EXAMPLE 6   Bjorn invests $15,000 at 12% compounded semiannually. What will be his final balance after 4 yr?

Understand: We must calculate the final balance. We are given the principal, the APR. The fact that it is compounded semiannually means the principal is compounded every $\frac{1}{2}$ yr. or twice each year. Semiannually over 4 yr. translates to 8 compoundings.

Plan: Use the compound interest formula $B = P(1 + rt)^n$.

$$P = 15,000$$
$$r = 12\% = 0.12$$
$$t = \frac{1}{2} \text{ (com pounded every half year)}$$
$$n = 8 \text{ (compounded twice each year for 4 yr.)}$$

Execute: $B = P(1 + rt)^n$

$$B = 15{,}000\left[1 + 0.12\left(\frac{1}{2}\right)\right]^8$$

$$B = 15{,}000[1 + 0.06]^8$$

$$B = 15{,}000[1.06]^8$$

$$B \approx 23{,}907.72$$

Discussion How much would Bjorn have in 4 yr. if he had invested in a simple interest plan at 12% APR?

◆ *Answer:* Bjorn will have $23,907.72 at the end of 4 yr.

Do Margins 5 and 6. ▶

EXAMPLE 7 Paula has a balance of $685.72 on a credit card that has a 14.9% APR. The interest with this particular card is compounded daily. If she does not charge any money to this account until after she receives her next bill (30 days), what will be her balance on the next bill?

Understand: We must calculate the final balance. We are given the principal, the APR. The fact that it is compounded daily means the principal is compounded every $\frac{1}{365}$ yr. or 365 times in a year.

Plan: Use the compound interest formula $B = P(1 + rt)^n$.

$$P = \$685.72$$
$$r = 14.9\% = 0.149$$
$$t = \frac{1}{365} \qquad \text{(compounded every day)}$$
$$n = 30 \qquad \text{(compounded each day for 30 days)}$$

Execute: $B = P(1 + rt)^n$

$$B = 685.72\left[1 + 0.149\left(\frac{1}{365}\right)\right]^{30} \qquad \textbf{Divide 0.149 by 365.}$$

$$B = 685.72\,[1 + 0.000408219]^{30} \qquad \textbf{Add.}$$

$$B = 685.72\,[1.000408219]^{30} \qquad \textbf{Evaluate exponential expression, then multiply.}$$

$$B \approx 694.17$$

Discussion What is the interest or finance charge that Paula owes?

◆ *Answer:* Paula's balance will be $694.17 on the next bill.

WARNING Read the fine print of a credit card statement to find out how the interest is calculated for that card. Sometimes there are several methods listed for the same type of card. The method used will be indicated somewhere on the statement. In addition to daily compounding as we saw in Example 10, here are a few other common methods:

Method 1: Apply the monthly percentage rate to the average daily balance. Applying this method to Example 10:

Monthly percentage rate = 14.9% ÷ 12 = 1.242% = 0.01242
Finance charge (interest) = 685.72(0.01242) = $8.52
Final balance = 685.72 + 8.52 = $694.24

Method 2: Apply the daily percentage rate to the average daily balance. Then multiply the result times the number of days in the billing cycle (simple interest). Applying this method to Example 10:

Daily percentage rate = 14.9% ÷ 365 = 0.04082% = 0.0004082
Finance charge = 685.72 (0.0004082)30 = $8.40
Final balance = 685.72 + 8.40 = $694.12   **There were 30 days in the billing cycle.**

♦ **Margin 5**

Deanne invests $12,000 at 14%.

a. Suppose the interest is compounded annually. What would be her final balance after 2 yr.?

b. Suppose the interest is compounded semi-annually. What will be her final balance after 2 yr.?

♦ **Margin 6**

Kyle invests $8000 at 8%.

a. Suppose the interest is compounded semi-annually. What would be his final balance after 3 yr.?

b. Suppose the interest is compounded quarterly. What will be his final balance after 3 yr.?

Working with Scientific Calculators

We can enter the entire expression from step 1 using a scientific calculator. The key sequence is as follows:

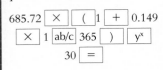

Answers to Margin 5:
a. $15,595.20 **b.** $15,729.55

Answers to Margin 6:
a. $10,122.55 **b.** $10,145.93

◀ Do Margin 7.

Marla has a balance of
$426.85 on a credit card
that has a 12.9% APR. The
interest with this particular
card is compounded daily.
If she does not charge any
money to this account un-
til after she receives her
next bill (30 days), what
will be her balance on
the next bill?

OBJECTIVE 3 *Solve problems involving amortization.*

In Example 7, notice that the interest charged during the month was $8.45 (694.17 −
685.72 = 8.45). If Paula just paid $8.45, her balance goes back to the initial balance,
$685.72, and the next month's balance will be back to $694.17. Thus, if all she does is pay
the finance charge (interest), then she will never pay off the debt.

In order to pay off a debt, we must pay more than the interest. Any amount that we pay in
addition to the interest will decrease the principal. This will then decrease the amount of interest for the next payment.

For example, suppose we borrow $1000 at 12% APR and make a $200 payment each month.
We will need to calculate the balance at the end of each month. Because the interest rate is
12% APR, the monthly rate will be 1%.

$$\text{First month balance} = 1000(1.01) = \$1010$$
$$\text{Applying the \$200 payment} = 1010 - 200 = \$810$$

> Note that $10 of the $200 payment is going toward inter-
> est while the other $190 is going toward the principal. In
> our next payment, we will owe less interest because the
> principal is now less.

$$\text{Second month balance} = 810(1.01) = \$818.10$$
$$\text{Applying the \$200 payment} = 818.10 - 200 = 618.10$$

> Note that only $8.10 is interest, so the rest of the $200
> payment, which amounts to 191.90, goes toward principal.
> In the next payment, we will pay even less interest and
> more of the $200 can go toward principal.

This process would continue until we paid off the loan. The payments we make to pay off a
loan are called *installments*. When we pay off a loan in installments, we **amortize** the loan.

DEFINITION   **Amortize:** To pay off a loan or debt in installments.

In amortizing a loan we are interested in the payment that we must make. The factors that affect this payment are the principal, percentage rate, the number of payments per year, and the
agreed upon time to pay off the loan.

Because the calculations can be rather involved, it is common to use amortization tables to
figure the payment. Most bookstores carry small amortization books that list tables by percentage rate. On page 597 is a sample from the table for loans with a 9% APR.

Notice under the 30-year column and the row for 90,000 that the payment is $724.16. Not
all principal amounts appear to be shown. However, we can use combinations of the listed
principal values to calculate a payment for a principal that is not listed.

EXAMPLE 8   Use Table 8.1 to figure the monthly payment for $72,300 at 9% APR
for 15 yr.

Understand: We must figure the monthly payment using the amortization table.

Because 72,300 is not listed we must use combinations of principals that are listed. We can use
any of the principals whose sum is 72,300.

Answer: $431.40

Table 8.1　9% APR Monthly Payment Schedule

Amount	3 Years	4 Years	5 Years	15 Years	30 Years
100	3.18	2.49	2.08	1.01	0.80
200	6.36	4.98	4.15	2.03	1.61
500	15.90	12.44	10.38	5.07	4.02
1000	31.80	24.89	20.76	10.14	8.05
2000	63.60	49.77	41.52	20.29	16.09
5000	159.00	124.43	103.79	50.71	40.23
6000	190.80	149.31	124.55	60.86	48.28
7000	222.60	174.20	145.31	71.00	56.32
8000	254.40	199.08	166.07	81.14	64.37
9000	286.20	223.97	186.83	91.28	72.42
10000	318.00	248.85	207.58	101.43	80.46
15000	477.00	373.28	311.38	152.14	120.69
20000	635.99	497.70	415.17	202.85	160.92
25000	794.99	622.13	518.96	253.57	201.16
30000	953.99	746.55	622.75	304.28	241.39
35000	1112.99	870.98	726.54	354.99	281.62
36000	1144.79	895.86	747.30	365.14	289.66
37000	1176.59	920.75	768.06	375.28	297.71
38000	1208.39	945.63	788.82	385.42	305.76
39000	1240.19	970.52	809.58	395.56	313.80
40000	1271.99	995.40	830.33	405.71	321.85
41000	1303.79	1020.29	851.09	415.85	329.90
42000	1335.59	1045.17	871.85	425.99	337.94
43000	1367.69	1070.06	892.61	436.13	345.99
44000	1399.19	1094.94	913.37	446.28	354.03
45000	1430.99	1119.83	934.13	456.42	362.08
46000	1462.79	1144.71	954.88	466.56	370.13
47000	1494.59	1169.60	975.64	476.71	378.17
48000	1526.39	1194.48	996.40	486.85	386.22
49000	1558.19	1219.37	1017.16	496.99	394.27
50000	1589.99	1244.25	1037.92	507.13	402.31
51000	1621.79	1269.14	1058.68	517.28	410.36
52000	1653.59	1294.02	1079.43	527.42	418.40
53000	1685.39	1318.91	1100.19	537.56	426.45
54000	1717.19	1343.79	1120.95	547.70	434.50
55000	1748.99	1368.68	1141.71	557.85	442.54
56000	1780.79	1393.56	1162.47	567.99	450.59
57000	1812.58	1418.45	1183.23	578.13	458.63
58000	1844.38	1443.33	1203.98	588.27	466.68
59000	1876.18	1468.22	1224.74	598.42	474.73
60000	1907.98	1493.10	1245.50	608.56	482.77
65000	2066.98	1617.53	1349.29	669.27	523.00
70000	2225.98	1741.95	1453.08	709.99	563.24
75000	2384.98	1866.38	1556.88	760.70	603.47
80000	2543.98	1990.80	1660.67	811.41	643.70
85000	2702.98	2115.23	1764.46	862.13	683.93
90000	2861.98	2239.65	1868.25	912.84	724.16
95000	3020.97	2364.08	1972.04	963.55	764.39
100000	3179.97	2488.50	2075.84	1014.27	804.62
105000	3338.97	2612.93	2179.63	1064.98	844.85
110000	3497.97	2737.35	2283.42	1115.69	885.08
120000	3815.97	2986.21	2491.00	1217.12	965.55
130000	4133.97	3235.06	2698.59	1318.55	1046.01
140000	4451.96	3483.91	2906.17	1419.97	1126.47
150000	4769.96	3732.76	3113.75	1521.40	1206.93
175000	5564.95	4354.88	3632.71	1774.97	1408.09
200000	6359.95	4977.01	4151.67	2028.53	1609.25
225000	7154.94	5599.13	4670.63	2282.10	1810.40
250000	7949.93	6221.26	5189.59	2535.67	2011.56

◆ **Margin 8**

Use Table 8.1 to figure the monthly payment.

a. $15,000 for 5 yr.

b. $125,000 for 30 yr.

c. $97,500 for 15 yr.

Plan: Locate principal amounts that are listed and sum to be 72,300. List each of their individual monthly payments. Add all the monthly payments to get the total monthly payment for 72,300.

Execute: We can make 72,300 by adding the following listed principals:

$$70{,}000 + 2000 + 200 + 100 = 72{,}300$$

Now, under the 15 yr. column, read the corresponding monthly payment for each of these listed principals. Then add the monthly payments.

$$
\begin{array}{ccccc}
70{,}000 & + & 2000 & + & 200 & + & 100 & = & 72{,}300 \\
\downarrow & & \downarrow & & \downarrow & & \downarrow & & \downarrow \\
709.99 & + & 20.29 & + & 2.03 & + & 1.01 & = & 733.32
\end{array}
$$

◆ *Answer:* The monthly payment will be 733.32.

◀ **Do Margin 8.**

8.6 Exercises

FOR EXTRA HELP

 Videotape 9

 InterAct Math
Tutorial Software

 www.carsonmath.com

 AWL Math Tutor Center

InterAct
MathXL www.mathxl.com

 Student's Solutions
Manual

For Exercises 1–10, calculate the simple interest and final balance.

1. $4000 at 3% APR for 1 yr.

2. $500 at 4% APR for 1 yr.

3. $350 at 5.5% APR for 2 yr.

4. $1200 at 6.5% APR for 3 yr.

5. $12,250 at 12.9% APR for 5 yr.

6. $18,000 at 14.5% APR for 4 yr.

7. $2400 at 8% APR for six mo. $\left(\frac{1}{2}\text{ yr.}\right)$.

8. $840 at 12% APR for three mo. $\left(\frac{1}{4}\text{ yr.}\right)$.

9. $2000 at 6.9% APR for 60 days.

10. $600 at 4.9% APR for 30 days.

For Exercises 11–14, solve for the principal.

11. Nuno notes that his savings account earned $331.12 in 1 yr. If the APR for his account is 6.9%, what was his principal?

12. Sharice notes that her savings account earned $21.28 in 1 yr. If the APR for her account is 5.5%, what was her principal?

13. Kelly notes that her savings account earned $15.57 in three mo. If the APR for her account is 3.2%, what was her principal?

14. Roger notes that his savings account earned $24.56 in nine mo. If the APR for his account is 4.8%, what was his principal?

For Exercises 15–20, calculate the final balance.

15. $5000 at 8% compounded annually for 2 yr.

16. $6400 at 12% compounded annually for 2 yr.

17. $840 at 6% compounded annually for 2 yr.

18. $1450 at 10% compounded annually for 2 yr.

19. $14,000 at 8% compounded semiannually for 1 yr.

20. $20,000 at 10% compounded semiannually for 1 yr.

For Exercises 21–34, use a calculator to solve.

21. $400 at 9% compounded annually for 3 yr.

22. $650 at 5% compounded annually for 3 yr.

23. $1600 at 6% compounded semiannually for 3 yr.

24. $960 at 14% compounded semiannually for 3 yr.

25. $290 at 9% compounded semiannually for 4 yr.

26. $500 at 11% compounded semiannually for 5 yr.

27. $1300 at 12% compounded quarterly for 2 yr.

28. $900 at 8% compounded quarterly for 3 yr.

29. $450 at 6% compounded quarterly for 4 yr.

30. $2400 at 10% compounded quarterly for 4 yr.

31. Kevin has a balance of $860.20 on a credit card that has a 16.9% APR. The interest is compounded daily. If he does not charge any money to this account until after he receives his next bill (30 days), what will be his balance on the next bill?

32. Phyllis has a balance of $285.32 on a credit card that has a 7.9% APR. The interest is compounded daily. If she does not charge any money to this account until after she receives her next bill (30 days), what will be her balance on the next bill?

33. Lindsey has a balance of $694.75 on a credit card that has a 19.8% APR. The interest is compounded daily. If she does not charge any money to this account until after she receives her next bill (30 days), what will be her balance on the next bill?

34. Scott has a balance of $1248.56 on a credit card that has a 21.9% APR. The interest is compounded daily. If she does not charge any money to this account until after she receives her next bill (30 days), what will be her balance on the next bill?

For Exercises 35–44, use Table 8.1 to figure the payment.

35. $15,000 for 5 yr.

36. $9000 for 3 yr.

37. $17,000 for 4 yr.

38. $21,000 for 5 yr.

39. $92,500 for 30 yr.

40. 78,000 for 15 yr.

41. $84,100 for 15 yr.

42. $103,500 for 30 yr.

43. $115,200 for 30 yr.

44. $147,200 for 15 yr.

REVIEW EXERCISES

1. Graph $-6\frac{3}{4}$ on a number line.

2. Write $\frac{5}{6}$ as a decimal number.

3. Simplify. $\dfrac{65 + 90 + 72 + 84}{4}$

4. Evaluate $2x + 3y$ when $x = -3$ and $y = 4$.

5. Solve. $\dfrac{3}{8}x = -\dfrac{21}{30}$

Summary

Defined Terms

Review the following terms and for those you do not know, study its definition on the page number next to it.

Section 8.1
Percent *(p. 546)*

Section 8.4
Commission *(p. 568)*

Section 8.6
Principal *(p. 589)*
Interest *(p. 589)*
Interest rate *(p. 589)*

Annual Percentage Rate
  (APR) *(p. 589)*
Simple interest *(p. 590)*
Compound interest *(p. 593)*

Amortize *(p. 596)*

Procedures, Rules, and Key Examples

Procedures/Rules	Key Example(s)
Section 8.1	Write each percent as a fraction.
To write a percent as a fraction: 1. Write the percent over 100. 2. Simplify.	$45\% = \dfrac{45}{100} = \dfrac{9}{20}$ $16.2\% = \dfrac{16.2}{100} = \dfrac{16.2(10)}{100(10)} = \dfrac{162}{1000} = \dfrac{81}{500}$ $9\dfrac{1}{2}\% = \dfrac{9\frac{1}{2}}{100} = \dfrac{19}{2} \div 100 = \dfrac{19}{2} \cdot \dfrac{1}{100} = \dfrac{19}{200}$
To write a percent as a decimal number: 1. Write the fraction or mixed number percent in decimal form. 2. Divide by 100.	Write each percent as a decimal. $62\% = 62 \div 100 = 0.62$ $4.5\% = 4.5 \div 100 = 0.045$ $33\frac{1}{3}\% = 33.\overline{3}\% = 33.\overline{3} \div 100 = 0.\overline{3}$
To write a fraction as a percent: 1. Multiply by 100%. 2. Simplify. 3. Write a percent sign to the right of the result.	Write each fraction as a percent. $\dfrac{3}{5} = \dfrac{3}{\overset{}{5}} \cdot \dfrac{\overset{20}{\cancel{100}}}{1}\% = 60\%$ $\dfrac{2}{3} = \dfrac{2}{3} \cdot \dfrac{100}{1}\% = \dfrac{200}{3}\% = 66\dfrac{2}{3}\%$
To write a decimal number as a percent multiplying by 100%. **Note:** Multiplying a decimal number by 100% causes the decimal point to move two places to the right.	Write each decimal number as a percent. $0.49 = 0.49 \cdot 100\% = 49\%$ $0.067 = 0.067 \cdot 100\% = 6.7\%$ $1.03 = 1.03 \cdot 100\% = 103\%$

Section 8.2	Translate to an equation.

To translate a percent sentence directly to an equation:

1. Select a variable to represent the unknown amount indicated by the word *what*.
2. If the word *of* is preceded by the percent, translate it to multiplication. If the word *of* is preceded by a whole number, translate it to division.
3. Translate the word *is* to an equal sign.

Note: The basic percent sentence is:
A percent of a whole amount is the part of the whole.

Note: The first step in solving any of the translated equations is to write the percent as a decimal or fraction.

Note: When using this method to calculate a missing percent, the outcome will be a decimal number or fraction. You must write the decimal or fraction as a percent.

40% of 70 is what number?
Translation: $40\% \cdot 70 = x$

15% of what number is 65?
Translation: $15\% \cdot x = 65$

What percent of 200 is 50?
Translation: $x \cdot 200 = 50$

What percent is 20 out of 30?
Translation: $x = 20 \div 30$

What percent is 15 of 20?
Translation: $x = 15 \div 20$

Section 8.3	Translate to a proportion.

To translate a percent sentence to a proportion, write the proportion in the following form:

$$\text{Percent} = \frac{\text{part}}{\text{whole amount}}$$

where the percent is written as a fraction with a denominator of 100.

Tip The whole amount always follows the word *of*.

Note: When using this method to calculate a missing percent, the result will be a percent. You must simply write the percent sign beside the result.

40% of 70 is what number?
Translation: $\dfrac{40}{100} = \dfrac{x}{70}$

15% of what number is 65?
Translation: $\dfrac{15}{100} = \dfrac{65}{x}$

What percent of 200 is 50?
Translation: $\dfrac{x}{100} = \dfrac{50}{200}$

What percent is 20 out of 30?
Translation: $\dfrac{x}{100} = \dfrac{20}{30}$

What percent is 15 of 20?
Translation: $\dfrac{x}{100} = \dfrac{15}{20}$

Sections 8.4 and 8.5

To solve problems involving percents:

1. Identify the percent, whole amount, and part.
2. Write these pieces in a basic sentence (if needed).
3. Translate to an equation or a proportion.
4. Solve.

Formulas

		Where	P represents the principal
Simple interest:	$I = Prt$		r represents the APR
Simple interest final balance:	$B = P(1 + rt)$		t represents the time length, in years, of each compound
Compound interest final balance:	$B = P(1 + rt)^n$		n represents the number of compoundings

For Exercises 1–6, answer true of false.

1. A percent cannot represent an amount more than 1.

2. 20% of 60 is greater than 30.

3. 60% of 300 is greater than 150.

4. 28 is the same as 28%.

5. 0.6 is the same as 6%.

6. The more often a given principal is compounded at the same rate, the larger the final balance will be.

7. In general, to write a percent as a decimal, we ⎯⎯⎯⎯⎯⎯⎯ by 100.

8. In general, to write a decimal or fraction as a percent, we ⎯⎯⎯⎯⎯⎯⎯ by 100 and write a percent sign by the result.

9. Explain in your own words how to translate the basic percent sentence directly to an equation.

10. Explain in your own words how to translate the basic percent sentence to a proportion.

11. Write as a fraction in simplest form.

   a. 40%   **b.** 26%   **c.** 6.5%   **d.** $24\frac{1}{2}\%$

12. Write as a decimal number.

   a. 16%   **b.** 150%   **c.** 3.2%   **d.** $40\frac{1}{3}\%$

For Exercises 13–16, write each number as a percent.

13. 0.54   **14.** 1.3   **15.** $\frac{3}{8}$ **16.** $\frac{4}{9}$

For Exercises 17–20, translate directly to an equation. Then solve.

17. What number is 15% of 90?   **18.** 12.8% of what number is 5.12?

19. 12.5 is what percent of 20?   **20.** What percent is 40 of 150?

For Exercises 21–24, translate to a proportion. Then solve.

21. What number is 40.5% of 800?

22. $10\frac{1}{2}$ is 15% of what number?

23. 8.1 is what percent of 45?

24. What percent is 16 of 30?

For Exercises 25–40, write an equation or a proportion. Then solve.

25. 35% of a mixture is soluble fat. If the total volume of the mixture is 600 ml, how many milliliters is soluble fat?

26. In a survey of 2000 adults, 540 believed the Sun still revolves around Earth. What percent of those surveyed believed the Sun revolves around Earth?

27. A stereo receiver is priced at $285.75. The sales tax rate is 6%. Calculate the sales tax and total amount of the purchase.

28. An outfit is on sale for 25% off. If the initial price is $56.95, what is the price after the discount?

29. After traveling 180 mi., Boris has completed only 35% of a trip. What is the total distance of the trip?

30. Kat indicates that she received a 3.5% raise that amounted to $1102.50. What was her former salary?

31. The Johnsons' power bill increased $27.68 from last month. If their bill was $86.50 last month, what was the percent of increase?

32. Dave's hourly wage is raised from $12 per hour to $15.50 per hour. What is the percent of increase?

33. $480 is invested at 6% APR for 1 yr. Calculate the interest.

34. $8000 is deposited into a savings account at 5.2% APR. What will be the balance after three mo.?

35. After 1 yr. at 4.5% APR, the balance in an account is $627. If no deposits or withdrawals were made during the year, what was the principal?

36. $5000 is invested at 8% compounded annually. Calculate the balance after 3 yr.

37. $1800 is invested at 12.4% compounded quarterly. Calculate the balance after a year.

38. A family borrows $90,500 at 9% APR to be amortized over 15 yr. Use Table 8.1 to figure their monthly payment.

1. Write 24% as a fraction in simplest form.

2. Write 4.2% as a decimal number.

3. Write $12\frac{1}{2}$% as a decimal number.

4. Write $40\frac{3}{4}$% as a fraction in simplest form.

For Exercises 5–8, write each number as a percent.

5. 0.26   6. 1.2   7. $\frac{2}{5}$ 8. $\frac{5}{9}$

For Exercises 9 and 10, translate directly to an equation. Then solve.

9. What number is 15% of 76?   10. 6.5% of what number is 8.32?

For Exercises 11 and 12, translate to a proportion. Then solve.

11. 14 is what percent of 60?   12. What percent is 12 of 32?

For Exercises 13–25, write an equation or a proportion. Then solve.

13. 80% of a class got an A on a particular test. If there were 30 students in the class, how many got an A?

14. Carolyn earns 25% of total sales in commission. If she sold a total of $1218 in merchandise in one week, what is her commission?

15. The Morgan family has a net monthly income of $2786.92. If they have a $345.75 car payment, what percent of their net monthly income goes toward paying for the car?

16. A microwave oven is priced at $295.75. The sales tax rate is 5%. Calculate the sales tax and total amount of the purchase.

1. _____

2. _____

3. _____

4. _____

5. _____

6. _____

7. _____

8. _____

9. _____

10. _____

11. _____

12. _____

13. _____

14. _____

15. _____

16. _____

17. _____

17. A pair of boots is on sale for 30% off. If the initial price of the boots is $84.95, what is the price after the discount?

18. _____

18. Barbara indicates that she received a 2.5% raise that amounted to $586.25. What was her former salary?

19. _____

19. Andre's hourly wage is raised from $8.75 per hour to $10.50 per hour. What is the percent of increase?

20. _____

20. $5000 is invested at 6% APR for 1 yr. Calculate the interest.

21. _____

21. $800 is deposited into a savings account at 4% APR. What will be the balance after six mo?

22. _____

22. After 1 yr. at 6% APR, the balance in an account is $2544. If no deposits or withdrawals were made during the year, what was the principal?

23. _____

23. $2000 is invested at 9% compounded annually. Calculate the balance after 3 yr.

24. _____

24. $1200 is invested at 4% compounded quarterly. Calculate the balance after a year.

25. _____

25. The Ramsey family borrows $112,000 at 9% APR to be amortized over 30 years. Use Table 8.1 to figure their monthly payment.

Cumulative Review Exercises

For Exercises 1–6, answer true or false.

1. There are two solutions for $(\quad)^2 = 81$, $+9$ and -9.

2. 91 is composite.

3. π is an irrational number.

4. $4x^2y^3$ is a monomial.

5. 12 is the LCM of 36 and 24.

6. $9\% > 0.1$

7. Explain in your own words how to multiply two polynomials.

8. Explain in your own words how to add or subtract fractions.

9. Explain in your own words how to clear decimal numbers from an equation.

10. Explain in your own words how to use dimensional analysis to convert units.

11. Write 27,500,000,000 in scientific notation.

12. Graph -5.6 on a number line.

13. Estimate $7805 \cdot 246$ by rounding so that there is only one nonzero digit in each number.

14. What is the degree of $-5y$?

For Exercises 15–21, simplify.

15. $[12 + 4(6 - 8)] - (-2)^3 + \sqrt{100-36}$

16. $8\frac{3}{4} - (-5\frac{2}{3})$

17. $10\frac{5}{8} \div \left(-\frac{3}{16}\right) - \frac{1}{2}$

18. $4.86 \div 0.8 + 58.9$

19. $\left(4\frac{3}{8}\right)(-5.6)$

20. $\frac{3}{4} - (0.6)^2$

21. $\sqrt{(9)(0.25)}$

22. Approximate $\sqrt{72}$ to the nearest hundredth.

23. Write $12\frac{1}{2}\%$ as a decimal number.

24. Write 0.71 as a percent.

25. Subtract. $(x^4 - \frac{1}{3}x^2 + 9.6) - (4x^3 + \frac{3}{5}x^2 - 14.6)$

26. Simplify. $(7.8x^6)(5.9x^8)$

27. Multiply. $(6.2x - 1)(4x + 5)$

28. Find the prime factorization of 540.

29. Find the LCM of $24y^2$ and $30x$.

30. Divide. $18x^6 \div 3x^2$

31. Factor. $32m^4 + 24m^2 - 16m$

32. Simplify. $\frac{10m}{9n^2} \div \frac{5}{12n}$

33. Simplify. $\frac{3}{5} - \frac{y}{4}$

For Exercises 34–38, solve and check.

34. $\frac{5}{8}y - 3 = \frac{3}{4}$

35. $3.5x - 12.1 = 6.8x + 3.08$

36. $\dfrac{4\frac{1}{3}}{9} = \dfrac{1\frac{2}{3}}{n}$

37. Convert 9.5 lb. to ounces.

38. Convert 80 ml to liters.

39. Convert 40°F to degrees centigrade.

For Exercises 40–50, solve.

40. A parallelogram has an area of 201.3 ft². If the base is 16.5 ft, what is the height? Convert the height to meters.

41. Write an expression in simplest form for the area. Calculate the area if $h = 9$ in.

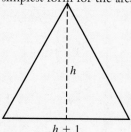

42. The sum of two consecutive integers is 123. What are the integers?

43. Find the area of the shaded region.

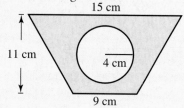

44. Find the volume of a soup can that is 4 in. tall and has a 1.2 in. radius.

45. Calculate the volume of a ball with a 16 in. diameter.

46. On a popular talk show, a studio-audience participant is selected at random by seat number. Suppose there are 480 seats. If 12 people from the same family are in the studio audience, what is the probability that someone from that family is selected? Write the probability in simplest form.

47. The following triangles are similar. Find the missing side lengths.

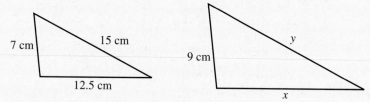

48. Liam is jogging at 6 mph. Starnell is walking at 4 mph toward Liam on the same trail. If they are 0.25 mi. apart, how long until they meet? Express the time in minutes.

49. Of the 450 people in attendance at a retirement seminar, 384 are over 50. What percent of the people in attendance are over 50?

50. $3200 is invested at 6% APR compounded semiannually. What will be the balance after 2 yr?

Statistics and Graphs

Preparing for a Final Exam

Preparing for a cumulative final exam can seem overwhelming. However, if you break up your preparation over at least several days, you'll find the task much less daunting.

First, go back through your notes, from beginning to end, to see the progression of topics. You'll find that most topics keep reappearing throughout the course. Also, review your study sheets and go through the tests you have taken. Often instructors take final exam questions directly from the tests.

In the second stage of preparation, do each cumulative review in the text. Try to review these over a few days, working through two cumulative reviews each day. Treat the last cumulative review as a practice final exam.

Try to complete all of the above preparation two days before the final exam. During the day before the final, go through the whole process one more time. Make a quick pass through your notes and study sheets, then work the final cumulative review again. Finally, be sure to get plenty of rest the night before and eat healthy.

"The will to win is important but the will to prepare is vital."
—JOE PATERNO

"You hit home runs not by chance but by preparation."
—ROGER MARIS

9.1 Mean, Median, and Mode

OBJECTIVES

1 Find the mean.
2 Find weighted means.
3 Find the median.
4 Find the mode.

OBJECTIVE **1** *Find the mean.*

Numbers can be used to describe characteristics of data. A number used in this way is called a **statistic.**

DEFINITION	**Statistic:** A number used to describe some characteristic of a set of data.

Oftentimes we use statistics that describe the *middle* or *central tendency* in a set of data. One such statistic is the **arithmetic mean,** or **average.**

DEFINITION	**Mean:** The sum of all given numbers divided by the number of numbers.

Procedure *To find the arithmetic mean, or average, of a given set of numbers:*
1. Calculate the sum of all the given numbers.
2. Divide the sum by the number of numbers.

The common variable used to indicate a mean is $\bar{x}$.

EXAMPLE 1 Below is a list of clerical specialist salaries within a company. What is the mean salary of a clerical specialist within the company?

$24,000	$21,900
$19,500	$19,500
$20,400	$25,200

Understand: We must calculate the average, or mean, salary. We are given a set of salaries.

Plan: Divide the sum of the salaries by the number of salaries in the list.

Execute: $\bar{x} = \dfrac{24,000 + 19,500 + 20,400 + 21,900 + 19,500 + 25,200}{6}$

$\bar{x} = \dfrac{130,500}{6}$

$\bar{x} = 21,750$

Answer: The average or mean salary of the clerical specialists at this company is $21,750.

◆ *Check:* We can verify the calculations by inverse operations. We will leave this to the reader.

◀ **Do Margin 1.**

◆ **Margin 1**

a. Latonya has the following test scores. What is the mean of her test scores?

82, 74, 95, 86, 90

b. The ages of the employees in a particular department of a company are listed below. What is the mean age of the employees in the department?

23, 25, 32, 40, 36, 35

Answers: **a.** 85.4 **b.** 31.8$\overline{3}$

EXAMPLE 2 Juan is taking a course in which the grade is determined by the average of four tests. He has the following test scores in the course: 68, 85, 90. To get a B, he must have a test average of 80 on the four tests. What is the minimum score he needs on the last test to get a B?

Understand: We must calculate the minimum test score on the last of four tests that will produce an 80 average for the four tests. We are given the first three test scores.

Plan: Write an equation and solve.

Execute: To calculate an average, we divide the sum of the scores by the number of scores.

$$\frac{\text{sum of the scores}}{\text{number of the scores}} = \text{average}$$

Let x represent the unknown fourth test score.

$$\frac{68 + 85 + 90 + x}{4} = 80 \qquad \text{Add in the numerator.}$$

$$\frac{243 + x}{4} = 80$$

$$4 \cdot \frac{243 + x}{4} = 80 \cdot 4 \qquad \begin{array}{l}\text{Clear the 4 by multiplying}\\ \text{both sides by 4.}\end{array}$$

$$243 + x = 320$$

$$\begin{array}{r} 243 + x = 320 \\ -243 \qquad -243 \\ \hline 0 + x = 77 \end{array}$$

$$x = 77$$

Answer: If Juan scores 77 on the last test, the average of the four tests will be exactly 80 and he will receive a B.

Check: Verify that a 77 on the last test will produce an average of 80.

$$\frac{68 + 85 + 90 + 77}{4} = \frac{320}{4} = 80$$

Discussion What if Juan knows the instructor would round a 79.5 up to 80? What would be the lowest score he could make on the last test and still receive a B? If the test is a 50-question test with each question worth 2 points and no partial credit given, how many questions can Juan miss and receive a B?

Do Margin 2. ▶

OBJECTIVE 2 *Find weighted means.*

Sometimes when we calculate a mean, some scores in the data set have more weight than others. An example of a weighted mean is **grade point average (GPA).**

DEFINITION **Grade point average (GPA):** An average of the total grade points earned divided by the total number of credit hours.

Each letter grade has a **grade point** value.

DEFINITION **Grade point:** A numerical value assigned to a letter grade.

◆ **Margin 2**

Trina's grade in a course is determined by the average of 5 tests. Her scores on the first four tests are: 80, 92, 88, 94. To get an A, she must have a final average of 90 or higher. What is the minimum score she needs on the fifth test to receive an A?

Answer: 96

The total number of grade points is weighted by the number of credit hours of the course in which the grade is received. Following is a list of grade point equivalents for each letter grade.

$$
\left.\begin{array}{ll}
A & = 4 \\
B+ & = 3.5 \\
B & = 3 \\
C+ & = 2.5 \\
C & = 2 \\
D+ & = 1.5 \\
D & = 1 \\
F & = 0
\end{array}\right\}
$$

Number of grade points per credit hour of the course.

Note: Some schools do not use + grades like B+, C+, and D+.

Suppose you receive an A in a 3 credit hour course. We can calculate the number of grade points by multiplying the number of credit hours by the grade points for the letter grade. Because an A is worth 4 grade points for each credit of the course, we multiply $4 \cdot 3$ to equal 12 grade points for that course.

To calculate the GPA, we must calculate the grade points for each course, then divide by the total number of credit hours.

$$
GPA = \frac{\text{total grade points}}{\text{total credits}}
$$

Connection

We can use dimensional analysis:

$$
\frac{4 \text{ grade points}}{\text{credit hour}} \cdot \frac{3 \text{ credit hours}}{1} = 12 \text{ grade points}
$$

EXAMPLE 3 Below is a grade report for a student. Calculate the student's GPA.

	Credits	Grade
MATH 100	5.0	B+
ENG 101	3.0	C+
PSY 101	3.0	C
BIO 101	4.0	A

Understand: We must calculate the student's GPA. We are given the credit hours for each course and the letter grade received.

Plan: GPA is the ratio of total grade points to total credit hours. To calculate the total grade points, we must multiply the credits for each course by the value of the letter grade received, then add all those results. The grade points for the grades involved are:

$$
B+ = 3.5 \text{ grade points per credit}
$$
$$
C+ = 2.5 \text{ grade points per credit}
$$
$$
C = 2.0 \text{ grade points per credit}
$$
$$
A = 4.0 \text{ grade points per credit}
$$

Execute:

	Credits	Grade	Grade Points
MATH 100	5	B+	$(5)(3.5) = 17.5$
ENG 101	3	C+	$(3)(2.5) = 7.5$
PSY 101	3	C	$(3)(2.0) = 6.0$
BIO 101	4	A	$(4)(4.0) = 16.0$
Totals	15		47.0

$$GPA = \frac{\text{total grade points}}{\text{total credits}}$$

$$GPA = \frac{47}{15} = 3.1\overline{3}$$

Answer: Rounded to the nearest thousandth, this student has a 3.133 GPA.

♦ *Check:* Verify each calculation using an inverse operation.

Do Margin 3. ▶

OBJECTIVE **3** *Find the median.*

Another statistic that describes the *middle* or *central tendency* of a set of numbers is the **median.**

DEFINITION	**Median:** The middle score in an ordered set of scores.

Consider the set of numbers: 4, 5, 9, 3, 2, 10, 9

The definition of *median* indicates that the set must be ordered. This means that we must put the numbers in order from smallest to largest. We must also write each repetition of a score.

2, 3, 4, 5, 9, 9, 10
↑
median

Notice that 5 is the middle number of the set. The median is easy to find in a set that contains an odd number of scores. What if the set contains an even number of scores?

Suppose the scores were: 2, 3, 4, 5, 9, 10

The middle of this set of ordered scores is in-between 4 and 5. In fact, the number exactly halfway between 4 and 5 is the mean of 4 and 5.

2, 3, 4, 5, 9, 10
↑
median

$$\text{median} = \text{mean of 4 and 5} = \frac{4+5}{2} = 4.5$$

Procedure *To find the median of a set of scores:*
1. Arrange the scores in order from smallest to largest.
2. Locate the middle score of the ordered set of scores.

Note: If there are an even number of scores in the set, the median will be the mean of the two middle scores.

EXAMPLE 4 Find the median salary for clerical specialists at the company described in Example 1. The salaries were:

$24,000	$21,900
$19,500	$19,500
$20,400	$25,200

♦ **Margin 3**
Below is a grade report for a student. Calculate the student's GPA. Round to the nearest thousandth.

	Credits	Grade
MATH 100	5.0	C
ENG 101	3.0	A
PSY 101	3.0	B+
BIO 101	4.0	C+

Discussion Note that the courses are the same as in Example 3. The letter grades are also the same. However, the C grades are now in BIO and MATH, which have more credit hours than ENG and PSY. How does this affect the GPA?

Answer: 2.833

Understand: We must find the median salary. We are given the salaries.

Plan: Arrange the salaries in order from least to greatest, then locate the middle salary in the ordered list. Because there are an even number of salaries, the median salary will be the mean of the middle two salaries.

Execute: 19,500   19,500   20,400   21,900   24,000   25,200

> The median is halfway between 20,400 and 21,900 so we must find the mean of 20,400 and 21,900.

$$\text{median} = \frac{20,400 + 21,900}{2} = \frac{42,300}{2} = 21,150$$

Answer: The median income for the clerical specialists at the company is $21,150.

Check: Verify the calculations by inverse operations.

Note that half of the people have salaries greater than the median salary and half have salaries less than the median salary. Median does not indicate how much greater or less the other salaries might be. Also, it is possible that no one earns exactly the mean or median salary.

◄ **Do Margin 4.**

OBJECTIVE 4 *Find the mode.*

Sometimes we are interested in a score that occurs most often. This statistic is called **mode**.

DEFINITION   **Mode:** The score that occurs most often in a set of scores.

Consider the set of scores: 2, 5, 9, 12, 9, 15, 9, 12

Notice that 9 is the score that occurs most often, so the mode is 9. If no score is repeated, then there is no mode. If there is a tie between two or three scores, then we list each as a mode.

Procedure *To find the mode of a set of scores, count the number of repetitions of each score. The score with the most repetitions is the mode.*

Note: If no score is repeated, then there is no mode. If there is a tie, then list each score as a mode.

EXAMPLE 5   Twelve players in a golf tournament post the following scores. What is the mode?

69, 66, 74, 72, 69, 70, 72, 71, 75, 65, 72, 71

Understand: We must find the mode in the set of scores. The mode is the score that occurs most frequently.

♦ **Margin 4**

a. A class of 9 students has the following final scores. What is the median score?

80, 82, 94, 78, 95, 76, 92, 95, 64

b. Below is a list of the Jones family's long distance charges for the past year. What is the median long distance charge for the year?

Jan. = $42.70
Feb. = $32.90
March = $47. 50
April = $64.85
May = $71.54
June =$41.23
July = $55.21
Aug. = $64.72
Sep. =$53.91
Oct. = $45.90
Nov. = $85.54
Dec. = $98.91

Answers: **a.** 82 **b.** $54.56

Plan: Count the number of repetitions of each score. The score with the most repetitions is the mode.

Execute: 69 occurs twice

66 appears once

74 occurs once

72 occurs three times

72 is the score with the most repetitions, so it is the mode.

70 occurs once

71 occurs twice

75 occurs once

65 occurs once

◆ *Answer:* The mode is 72.

EXAMPLE 6 Below is a list of the final averages for a class of 15 students. What is the mode?

85, 86, 72, 65, 80, 91, 62, 76, 80, 85, 76, 78, 80, 96, 85

Understand: We must find the mode in the set of scores. The mode is the score that occurs most frequently.

Plan: Count the number of repetitions of each score. The score with the most repetitions is the mode.

Execute: 85 occurs three times

86 occurs once

72 occurs once

85 and 80 both occur three times. This means they are both modes.

65 occurs once

80 occurs three times

91 occurs once

62 occurs once

76 occurs twice

78 occurs once

96 occurs once

◆ *Answer:* 85 and 80 are modes.

EXAMPLE 7 Below is a list of the birth weights for a group of 6 siblings. What is the mode?

7.4 lb.   7.5 lb.   8.2 lb.   7.6 lb.   7.8 lb.   8.1 lb.

Understand: We must find the mode in the set of weights. The mode is the weight that occurs most frequently.

Plan: Count the number of repetitions of each weight. The weight with the most repetitions is the mode.

Execute: Because no weight is repeated, there is no mode.

◆ *Answer:* There is no mode for this set of weights.

◀ **Do Margin 5.**

◆ **Margin 5**

a. Below is a list of ages for people at a work-shop. Find the mode(s).

25 31 30 38 28
25 31 36 25 35

b. Below is a list of canned vegetable prices. Find the mode(s).

$0.79 $0.89 $0.49
$0.79 $1.09 $0.89
$0.99

Answers: **a.** 25 **b.** $0.79, $0.89

9.1 Exercises

FOR EXTRA HELP

 Videotape 11

 InterAct Math Tutorial Software

 www.carsonmath.com

 AWL Math Tutor Center

InterAct MathXL www.mathxl.com

 Student's Solutions Manual

For Exercises 1–10, find the mean, median, and mode(s).

1. Below is a list of nurse salaries on the surgical floor of a hospital. Find the mean, median, and mode of the salaries.

$25,500	$28,700
$28,700	$27,500
$26,450	$24,200

2. Below is a list of the hourly wages for employees at a packaging plant. Find the mean, median, and mode of the wages.

$11.00	$12.00	$11.00
$15.00	$18.25	$12.50
$13.50	$16.25	$18.00

3. Find the mean, median, and mode for the test scores of students in a history class.

80	92	64	78	88
80	82	74	72	60
55	96	100	71	82
75	82	90	86	58

4. A basketball team has the following final scores. Find the mean, median, and mode of the scores.

76	82	80	78	85	75
78	80	72	70	84	88

5. A marine biologist studying leatherback sea turtles measures and records their length. Below is a list of lengths for the last twenty turtles. Find the mean, median, and mode of the lengths.

2.2 m	1.8 m	1.9 m	2.3 m	2.1 m
1.6 m	1.5 m	1.8 m	1.2 m	2.0 m
2.1 m	1.4 m	1.2 m	2.1 m	1.7 m
2.2 m	2.0 m	1.6 m	1.8 m	1.9 m

6. Below is a list of heights of players on a basketball team. Find the mean, median, and mode of the heights.

6.5 ft.	6.75 ft.	6.25 ft.	6.5 ft.	6.8 ft.	6.75 ft.
6.25 ft.	7 ft.	6.75 ft.	6.25 ft.	6 ft.	6.2 ft.

7. Below is a list of rainfall amounts for one month. Each amount is the total of rainfall in one 24-hour period. Calculate the mean, median, and mode.

0.4 in.	0.8 in.	1.2 in.	3.4 in.
0.6 in.	1.5 in.	0.6 in.	1.0 in.
1.4 in.	2.3 in.	1.3 in.	0.5 in.

8. Below is a list of daily high and low temperatures for the last two weeks. Find the mean, median, and mode of the high temperatures. Then find the mean, median, and mode of the low temperatures.

	Sun.	Mon.	Tues.	Wed.	Thurs.	Fri.	Sat.
High:	91	92	95	93	90	89	88
Low:	72	74	78	75	72	70	71
High:	90	94	96	95	92	90	91
Low:	72	74	78	76	73	70	72

9. Below is a list of each month's electric and gas charges for a family over a two-year period. Find the mean, median, and mode of the electric and gas charges each year.

	Year 1	Year 2
January	$158.92	$165.98
February	$147.88	$162.85
March	$125.90	$130.45
April	$108.40	$112.55
May	$87.65	$90.45
June	$114.58	$125.91
July	$145.84	$137.70
August	$142.78	$140.19
September	$90.25	$96.15
October	$104.12	$115.75
November	$136.62	$145.21
December	$158.18	$160.25

Discussion What can you conclude in comparing the mean, median, and mode for the two years?

10. Below is a list of water consumption in ft.^{3} by a family over a two-year period. Find the mean, median, and mode for each year.

	Year 1	Year 2
January	500	550
February	525	540
March	600	600
April	650	650
May	840	900
June	1100	1430
July	1000	1200
August	1400	1250
September	840	750
October	620	700
November	600	550
December	550	500

Discussion What can you conclude in comparing the mean, median, and mode for the two years?

11. Daniel's grade in a course is determined by the average of five tests. His scores on the first four tests are: 86, 96, 90, 88. To get an A he must have a final average of 90 or higher. What is the minimum score he needs on the fifth test to receive an A?

12. Heather's grade in a course is determined by the average of six tests. Her scores on the first five tests are: 83, 78, 82, 70, 76. To get a B she must have a final average of 80 or higher. What is the minimum score she needs on the sixth test to receive a B?

13. If Lonnie has a test average of 94 or higher, she will be exempt from the final exam. She has taken three out of the four tests for the course. Her scores on the three tests are: 88, 92, 90. What is the lowest she can score on the last test to be exempt from the final exam?

14. If Fernando has a test average of 90 or higher, he will be exempt from the final exam. He has taken four out of the five tests for the course. His scores on those four tests are: 85, 88, 94, 86. What is the lowest he can score on the fifth test to be exempt from the final exam?

15. Below is a grade report for a student. Calculate the student's GPA.

	Credits	Grade
MATH 101	3.0	B
ENG 101	3.0	C
HIS 102	3.0	B+
CHM 101	4.0	A

16. Below is a grade report for a student. Calculate the student's GPA.

	Credits	Grade
MATH 100	5.0	B
ENG 100	3.0	C+
PSY 101	3.0	F
BIO 101	4.0	C

Discussion What if this student had received an F in MATH 100 and a B in PSY 101?

17. Below is a grade report for a student. Calculate the student's GPA.

	Credits	Grade
MATH 102	3.0	B
ENG 100	3.0	B+
PSY 101	3.0	D+
BIO 101	4.0	C
COL 101	2.0	A

Discussion What if the student had made an A in BIO and a C in COL 101?

18. Below is a grade report for a student. Calculate the student's GPA.

	Credits	Grade
MATH 101	3.0	A
ENG 101	3.0	C
SPA 101	3.0	A
PHY 101	4.0	D+
COL 101	2.0	B

Use the grading scale below for Exercises 19 and 20.

$$A = 90\text{--}100$$
$$B = 80\text{--}89$$
$$C = 70\text{--}79$$
$$D = 60\text{--}69$$
$$F \text{ below } 60$$

19. Tamara has the following scores:

Tests: $88, 91, 90, 94$

Homework: $70, 90, 80, 70, 80, 90, 100, 90$

Final exam: 84

The instructor has decided that the test average is 40% of the final score, the homework average is 25% of the final score, and the final exam is the remaining 35% of the final score. What is Tamara's final grade?

20. Justin has the following scores:

Tests: $80, 74, 78, 65, 72$

Homework: $62, 74, 85, 91, 78, 54, 72, 81$

Project: 90

Final exam: 70

The instructor has decided that the test average is 35% of the final score, the homework average is 20% of the final score, the project is 20% of the final score, and the final exam is 25% of the final score. What is his final grade?

REVIEW EXERCISES

1. What percent of 400 is 300?

2. Write 12.5% as a decimal number.

3. Add. $4\frac{3}{5} + 6\frac{3}{4}$

4. Simplify. $-5x + 8y - 9x + 12 - 14y$

5. Solve. $\frac{4}{5}b = -12$

9.2 Tables, Charts, and Graphs

OBJECTIVES

1 Solve problems using data from tables.

2 Solve problems using data from charts.

3 Solve problems using data from bar graphs.

4 Solve problems using data from line graphs.

OBJECTIVE 1 *Solve problems using data from tables.*

We use tables, charts, and graphs to represent information in different ways. Tables are generally used to organize lists of data into common groupings. Below is a table that lists the number of crashes, injuries, and fatalities for the years 1988–1997. They are estimates based on samples.

Table 9.1   Motor Vehicle Traffic Data, 1988–1998

Year	Crashes	Injuries	Fatalities
1988	6,887,000	3,416,000	47,087
1989	6,653,000	3,284,000	45,582
1990	6,471,000	3,231,000	44,599
1991	6,117,000	3,097,000	41,508
1992	6,000,000	3,070,000	39,250
1993	6,106,000	3,149,000	40,150
1994	6,496,000	3,266,000	40,716
1995	6,699,000	3,465,000	41,817
1996	6,842,000	3,511,000	42,065
1997	6,764,000	3,399,000	42,013

Source: National Automotive Sampling System (NASS), General Estimates System (GES), and the Fatality Analysis Reporting System (FARS)

EXAMPLE 1   Use Table 9.1 to answer the following.

a. How many crashes occurred in 1994?

Answer: 6,496,000

b. How many injuries occurred in 1994?

Answer: 3,266,000

c. How many fatalities in 1994?

Answer: 40,716

d. In what year was the number of fatalities greatest?

Answer: 1988

e. What is the unit ratio of fatalities to injuries for 1997? Interpret the result.

Solution: To calculate a unit ratio, we divide the numerator (the number of fatalities) by the denominator (the number of injuries).

$$\frac{42,013}{3,399,000} \approx 0.012$$

This means there were 0.012 fatalities for every 1 injury in 1997. Or, because 0.012 is equivalent to the fraction $\frac{12}{1000}$, it means there were 12 fatalities out of every 1000 injuries.

f. What percent of the crashes in 1994 resulted in injuries?

Solution: We can write a basic percent sentence using the data from the table. There were 3,266,000 injuries out of a total of 6,496,000 crashes.

What percent of 6,496,000 is 3,266,000?

Because we are to find a percent we will use the proportion method.

$$\frac{x}{100} = \frac{3,266,000}{6,496,000}$$

$$6,496,000x = 326,600,000$$

$$\frac{6,496,000x}{6,496,000} = \frac{326,600,000}{6,496,000}$$

$$1x \approx 50.3$$

$$x \approx 50.3\%$$

g. Use the mean of the total number of crashes and the mean of the fatalities for the 10 years listed to estimate the probability that a crash results in a fatality.

Solution: To calculate the mean, we divide the sum of the numbers by 10, the number of years.

Crashes	Fatalities
6,887,000	47,087
6,653,000	45,582
6,471,000	44,599
6,117,000	41,508
6,000,000	39,250
6,106,000	40,150
6,496,000	40,716
6,699,000	41,817
6,842,000	42,065
+ 6,764,000	+ 42,013
65,035,000	424,787

$$\text{mean of the total crashes} = \frac{65,035,000}{10} = 6,503,500$$

$$\text{mean of the fatalities} = \frac{424,787}{10} = 42,478.7 \approx 42,479$$

Now we can calculate the probability using the mean.

Probability is the ratio of the number of *favorable* outcomes to the total number. In this case, *favorable* outcomes are the number of fatalities and the total number is the total number of crashes.

$$P = \frac{\text{fatalities}}{\text{crashes}}$$

$$P = \frac{42,479}{6,503,500}$$

$$P \approx 0.0065$$

Note that this means approximately 65 out of every 10,000 crashes result in fatality. Or, we could say about 6.5 out of every 1000 crashes result in a fatality.

◀ **Do Margin 1.**

◆ Margin 1

Use Table 9.1 to answer the questions.

a. How many injuries occurred in 1990?

b. Which of the listed years had the least number of injuries?

c. Calculate the unit ratio of injuries to crashes for 1992.

d. What is the median number of crashes for the 10 years listed?

e. Calculate the percent of the crashes that resulted in injury during 1992.

Answers: **a.** 3,231,000 **b.** 1992 **c.** 0.5116̄ **d.** 6,574,500 **e.** 51.16̄%

OBJECTIVE 2 *Solve problems using data from charts.*

Often data are organized in charts. One common chart is the pie chart, which is often used to show portions. The pie chart below shows how William's monthly net income was spent.

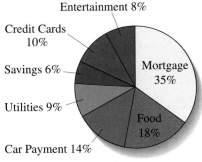

Figure 9.1

EXAMPLE 2 Use the pie chart in Figure 9.1 to answer the following.

a. If William's net income was $2458.75, how much went toward food?

Solution: William spent 18% of his net income on food. Because $2458.75 is his total income, we can say:

18% of $2458.75 is the amount spent on food

Using *f* to represent the amount spent on food, we can translate the above sentence to an equation, then solve.

$$0.18(2458.75) = f$$
$$\$442.58 = f$$

b. If William's net income was $2458.75, how much went toward mortgage and utilities combined?

Solution: 35% of his net income went toward the mortgage payment and 9% went toward utilities for a total of $9 + 35 = 44\%$ going toward mortgage and utilities combined.

Because 2458.75 is William's total income, we can say:

44% of 2458.75 is the amount spent on mortgage and utilities combined.

Using *x* to represent the amount spent on mortgage and utilities combined, we can translate the above sentence to an equation, then solve.

$$0.44(2458.75) = x$$
$$\$1081.85 = x$$

Do Margin 2. ▶

EXAMPLE 3 A company allots a budget of $50,000 each quarter for equipment/supplies, maintenance, lease, and utilities. Below is a list of actual expenditures for one quarter. Construct a pie chart showing the percent of the budget that went toward each item. Consider any remaining funds surplus.

Equipment and Supplies	$24,500
Maintenance	$ 3400
Lease	$11,845
Utilities	$ 6250

◆ **Margin 2**

Use the pie chart in Figure 9.1 to answer the following.

a. If William's net income was $2458.75, how much went toward entertainment?

b. What percent of William's income went toward credit cards and car payment combined?

c. What fraction of William's income went toward savings? Express the fraction in lowest terms.

Answers: **a.** $196.70 **b.** 24% **c.** $\frac{3}{50}$

Understand: We must construct a pie chart showing the percentage of total budget that went toward each item.

Plan: Calculate the percentage for each item. We can then develop the pie chart.

Execute:

Equipment and supplies:	Maintenance:	Lease:	Utilities:

$$\frac{E}{100} = \frac{24,500}{50,000} \qquad \frac{m}{100} = \frac{3400}{50,000} \qquad \frac{L}{100} = \frac{11,845}{50,000} \qquad \frac{u}{100} = \frac{6250}{50,000}$$

$$50,000E = 2,450,000 \quad 50,000m = 340,000 \quad 50,000L = 1,184,500 \quad 50,000u = 625,000$$

$$\frac{50,000E}{50,000} = \frac{2,450,000}{50,000} \quad \frac{50,000m}{50,000} = \frac{340,000}{50,000} \quad \frac{50,000L}{50,000} = \frac{1,184,500}{50,000} \quad \frac{50,000u}{50,000} = \frac{625,000}{50,000}$$

$$E = 49\% \qquad m = 6.8\% \qquad L \approx 23.7\% \qquad u = 12.5\%$$

What percent of the budget remains as surplus? We calculate the percent remaining by subtracting the total percent of equipment/supplies, maintenance, lease, and utilities from 100%.

$$Surplus = 100\% - (49\% + 6.8\% + 23.7\% + 12.5\%)$$
$$= 100\% - 92\%$$
$$= 8\%$$

Answer: We can now construct the pie chart.

In constructing the chart, we think about what fraction each percent represents. 49% is nearly 50%, which would be half of the chart. 23.7% is nearly 25%, which would be one quarter of the chart. 12.5% is equivalent to one eighth. 6.8% and 8% are both less than one tenth.

Check: Verify that each percentage of the $50,000 total budget equals the correct amount that was spent on that item. We will leave this to the reader.

◀ **Do Margin 3.**

OBJECTIVE 3 *Solve problems using data from bar graphs.*

Another way to organize data is in bar graphs that may have horizontal or vertical bars. Bar graphs are useful in comparing amounts or percent of data for different categories. Figure 9.2 shows an example of a vertical bar graph. Notice that two reference lines form the frame of the graph. A horizontal or vertical line used for reference in a graph is called an **axis.**

DEFINITION **Axis:** A line used for reference in a graph.

In Figure 9.2, the horizontal axis indicates the educational levels of unemployed persons. The vertical axis displays the percent of people who are unemployed.

EXAMPLE 4 Use Figure 9.2 to answer the following.

a. What was the unemployment rate for people who did not complete high school?

Answer: Shown from the left-hand bar of the graph, 8.7% of people 25 years or older who had not completed high school were unemployed.

◆ Margin 3

Below is a list of results from a survey. Construct a pie chart that shows each response as a percentage of the total responses.

Agree = 724

Disagree = 258

No opinion = 98

Answer:

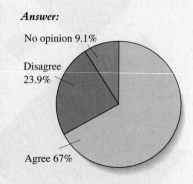

Figure 9.2 Unemployment rates of persons 25 years old and over, by highest degree attained: 1996

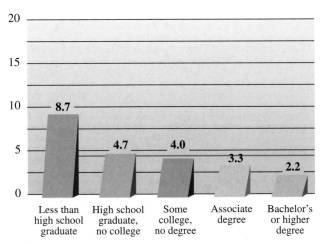

Source: U.S. Department of Labor, Bureau of Labor Statistics, Office of Employment and Unemployment Statistics, Current Population Survey, 1996.

b. Which group of people had the lowest unemployment rate?

Answer: Those with a bachelor's or higher degree.

c. Suppose about 90,000,000 people in the labor force are high school graduates with no college education and 25 years of age or older. How many of these people were unemployed in 1996?

Solution: From Figure 9.2, 4.7% of high school graduates with no college education are unemployed. If we let N represent the number of high school graduates with no college education who were unemployed, then we can write:

$$N \text{ is } 4.7\% \text{ of } 90,000,000$$
$$N = 0.047(90,000,000)$$
$$N = 4,230,000 \text{ people}$$

d. Suppose about 12,000,000 in the labor force did not complete high school and are 25 years of age or older. How many people who did not complete high school were unemployed in 1996?

Solution: 8.7% of people who did not complete high school were unemployed. If we let H represent the number of people 25 years of age or older who did not complete high school and were unemployed in 1996, then we can write:

$$H = 8.7\% \text{ of } 12,000,000$$
$$H = 0.087(12,000,000)$$
$$H = 1,044,000 \text{ people}$$

Discussion Compare the answers to questions c and d. 8.7% is a larger percentage than 4.7%, so why were there more high school graduates with no college education unemployed than people who had not completed high school?

Do Margin 4. ▶

EXAMPLE 5 Below is a list of unemployment rates for the first quarter of 1999. Construct a vertical bar graph with each category along the horizontal axis and the percentage along the vertical axis.

All workers: 4.3%

Adult men: 3.4%

Adult women: 3.8%

Teenagers: 14.6%

◆ Margin 4

Use the bar graph in Figure 9.2 to answer the following.

a. What percent of people 25 years or older with some college education but no degree were unemployed in 1996?

b. Which group had the greatest unemployment rate?

c. Suppose 35,000,000 people in the labor force were 25 years of age or older and had an associate degree as their highest level of education. How many of those people were unemployed in 1996?

Answers: **a.** 4% **b.** Less than high school graduate **c.** 1,155,000

Below is a list of scores from a test on general science that was given to grade 12 students in each country. Construct a vertical bar graph with each country listed along the horizontal axis and the scores listed along the vertical axis.

Sweden: 559

France: 487

Canada: 552

United States: 480

Australia: 527

Hungary: 471

Germany: 497

South Africa: 349

Source: Third International Mathematics and Science Study (TIMSS), p. 37.

Trivia Bite

There were actually the 21 countries involved in the study described in Margin 5. Students in 4th, 8th, and 12th grades were given four tests: general knowledge in mathematics, advanced knowledge in mathematics, general knowledge in science, and advanced knowledge in science. The United States consistently ranked near the bottom on all four tests.

Solution: Draw two axes perpendicular to one another with categories along the horizontal axis and percentages along the vertical axis. Approximate the height of each bar using the corresponding percentage along the vertical axis.

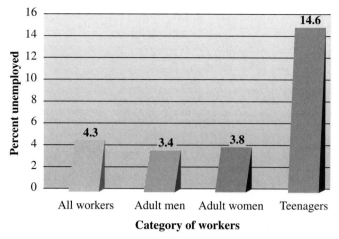

Source: Bureau of Labor Statistics, Employment situation summary 1999

◀ **Do Margin 5.**

OBJECTIVE 4 *Solve problems using data from line graphs.*

When we wish to show how data has changed over time, we generally use a line graph. The line graph in Figure 9.3 shows the change in unemployed persons from 1989 to 1999. Notice that the number of people is in millions.

Figure 9.3   Unemployed Persons, 1989–1999 (seasonally adjusted)

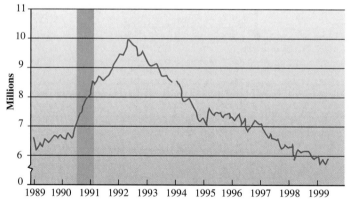

Note: Shaded area represents recession. Break in series in January 1994 is due to the redesign of the survey.

Source: Bureau of Labor Standards, Current Population Survey

Answer to Margin 5:

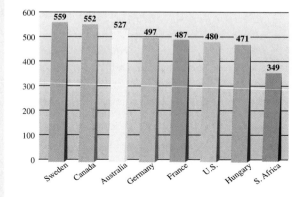

EXAMPLE 6 Use Figure 9.3 to answer the following.

a. How many people were unemployed at the beginning of 1996?

Answer: About 7,400,000 people

Explanation: If we follow a vertical line straight up from 1996 to the graph, we would reach point A on the graph. If we follow a horizontal line from point A over to the vertical axis we would reach point B. There we can read or approximate the number of people who are unemployed.

Unemployed persons, 1989–1999 (seasonally adjusted)

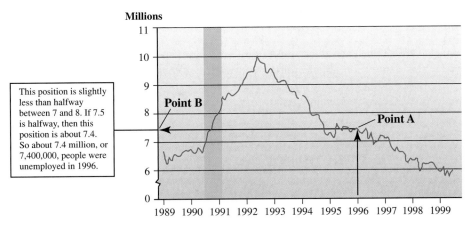

This position is slightly less than halfway between 7 and 8. If 7.5 is halfway, then this position is about 7.4. So about 7.4 million, or 7,400,000, people were unemployed in 1996.

Note: Shaded area represents recession. Break in series in January 1994 is due to the redesign of the survey.

Source: Bureau of Labor Statistics, Current Population Survey

b. In what year were the highest number of people unemployed?

Answer: During 1992

Explanation: Each mark represents the beginning of a year. Because the highest point, or peak, for the graph occurred between the marks for 1992 and 1993, the highest unemployment was during 1992.

c. If the labor force was approximately 138,000,000 at the beginning of 1996, what was the unemployment rate at the beginning of 1996?

Solution: The unemployment rate is a percent of the labor force that is unemployed. To calculate this percentage, we need the total labor force and the number of people unemployed.

The number of people who were unemployed at the beginning of 1996 is the answer from question a, 7,400,000. We can write a basic percent sentence, then solve. Because we are to find percent, we will use the proportion method.

What percent of 138,000,000 is 7,400,000?

$$\frac{p}{100} = \frac{7,400,000}{138,000,000}$$
$$138,000,000p = 740,000,000$$
$$\frac{138,000,000p}{138,000,000} = \frac{740,000,000}{138,000,000}$$
$$p \approx 5.4\%$$

Do Margin 6. ▷

◆ **Margin 6**

Use Figure 9.3 to answer the following.

a. The lowest number of unemployed people occurred during which year?

b. How many people were unemployed during June 1997? (Hint: June is the 6th month, which is halfway to 1998.)

c. If the number of people in the labor force at the beginning of 1999 was approximately 139,000,000, what was the unemployment rate at the beginning of 1999?

Answers: **a.** 1999 **b.** 6,700,000 **c.** ≈4.3%

EXAMPLE 7 Below is a list of revenue for each month during 1998 for a small business. Construct a line graph with each month listed on the horizontal axis and the revenue along the vertical axis.

January: $20,356	April: $17,450	July: $24,678	October: $19,400
February: $19,842	May: $18,540	August: $23,345	November: $24,728
March: $17,785	June: $22,258	September: $21,628	December: $28,258

Solution: Draw axes perpendicular to one another with the months along the horizontal axis and the revenue along the vertical axis. Because the revenues range from $17,450 up to $28,258, we will begin the numbers along the vertical axis at $17,000 and make each mark $1000, that is, we'll break the vertical axis and skip $1000 to $16,000 to begin at $17,000.

Imagine a vertical line extending upward for each month and a horizontal line extending out from each revenue value. Where these two imaginary lines intersect, we draw a point. We have drawn these lines for January. Notice we had to approximate where $20,356 is on the revenue line between 20,000 and 21,000. After drawing each point, we connect the points to form the line graph shown in Figure 9.4.

Figure 9.4   1998 Revenue Summary

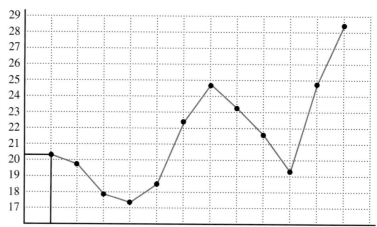

Do Margin 7.

Margin 7

Below is a list of the Wu family's electric charges for 1998. Construct a line graph with each month along the horizontal axis and the charges along the vertical axis.

January: $62.50

February: $68.25

March: $44.82

April: $56.78

May: $82.95

June: $110.80

July: $125.75

August: $138.58

September: $112.45

October: $72.35

November: $54.23

December: $56.85

Answer:

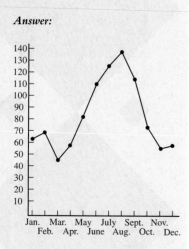

9.2 Exercises

FOR EXTRA HELP

 Videotape 11

 InterAct Math
Tutorial Software

 www.carsonmath.com

 AWL Math Tutor Center

 InterAct MathXL www.mathxl.com

 Student's Solutions Manual

For Exercises 1–10, use Table 9.2.

Table 9.2

American Housing Survey	Occupied households	Median household income	Percent owner-occupied
United States	97,693,000	31,416	65.0
Chicago	2,789,300	38,893	63.8
Columbus, OH	556,200	35,468	63.0
Denver	725,700	39,360	65.3
Detroit	1,681,100	37,638	71.1
Kansas City	659,100	35,220	67.3
Los Angeles-Long Beach	2,947,500	30,145	46.7
Miami-Fort Lauderdale	1,245,200	28,722	62.1
New Orleans	484,900	28,101	61.7
New York-Nassau-Suffolk-Orange	4,207,900	33,366	44.6
Northern NJ	2,202,500	40,368	62.1
Philadelphia	1,820,600	34,235	69.5
Pittsburgh	969,600	30,461	72.1
Portland, OR	655,100	37,178	64.7
San Antonio	484,800	29,728	61.5

Source: 1995 American Housing Survey

1. How many occupied households are in the United States?

2. How many occupied households are in New Orleans?

3. What is the median household income in Chicago?

4. What is the median household income in Denver?

5. Which of the listed cities had the highest median household income?

6. Which of the listed cities had the lowest median household income?

7. Which of the cities listed had the highest percentage of its households owner-occupied?

8. Which of the cities listed had the lowest percentage of its households owner-occupied?

9. How many households are owner-occupied in Los Angeles-Long Beach?

10. How many households are owner-occupied in Detroit?

> **Discussion** Compare your answers to 9 and 10. What are some possible reasons why the number of owner-occupied homes in LA-Long Beach is greater than the number in Detroit?

For Exercises 11–16, use the pie chart in Figure 9.5.

Figure 9.5   Sources of current-fund revenue for public institutions of higher education: 1994–1995

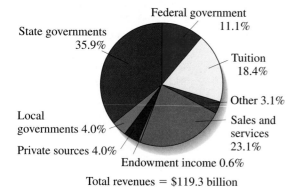

Federal government 11.1%
State governments 35.9%
Tuition 18.4%
Other 3.1%
Local governments 4.0%
Sales and services 23.1%
Private sources 4.0%
Endowment income 0.6%
Total revenues = $119.3 billion

Source: U.S. Department of Education, National Center for Education Statistics, Integrated Postsecondary Education Data System (IPEDS), "Finance FY95" survey

11. Which source contributed the smallest percentage of total revenue?

12. Which source contributed the largest percentage of total revenue?

13. What percent of the total revenue was from local governments?

14. What percent of the total revenue was from the federal government?

15. How much of the $119.3 billion total revenues was from tuition?

16. How much of the $119.3 billion total revenues was from sales and services?

17. Below is a list of deductions from a person's monthly pay. Construct a pie chart showing each deduction and the net pay as a percent of the gross salary.

Income	Deductions	
Gross monthly salary: $3220.50	FICA:	$199.67
	Medicare:	$ 46.70
	Fed W/H:	$225.44
	State W/H:	$161.03
	Health Plan:	$149.76
	Retirement:	$193.23
	Life insurance:	$ 48.31

18. Respondents in a poll could strongly agree, agree, be neutral, disagree, or strongly disagree. Below is the number of people that responded each way. Construct a pie chart showing each response as a percent of total respondents.

Strongly agree: 68

Agree: 198

Neutral: 25

Disagree: 94

Strongly disagree: 35

For Exercises 19–26, use the bar graph in Figure 9.6.

Figure 9.6 **Salaries of recent bachelor's degree recipients 1 year after graduation, by field: 1987, 1991, and 1994 (In constant 1994 dollars)**

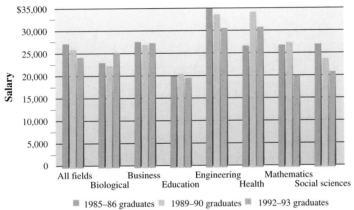

Source: U.S. Department of Education, National Center for Education Statistics, "Recent College Graduates" surveys, 1987 and 1991 and "Baccalaureate and Beyond Longitudinal Study. First Follow-up" survey

19. For the 1985–86 graduates, which field had the highest salary?

20. For the 1989–90 graduates, which field had the highest salary?

21. For graduates from 1985 through 1990, which field had the lowest salary?

22. For 1992–93 graduates, which field had the lowest salary?

23. What was the income of 1992–93 graduates in business?

24. What was the income of 1992–93 graduates in mathematics?

25. Estimate the average income of all three groups of graduates in social sciences.

26. Estimate the average income of all three groups of graduates in education.

27. In Exercise 9 from Section 9.1 we listed a family's electric and gas charges for two years. Construct a vertical bar graph with the months along the horizontal axis and the dollar amounts along the vertical axis. Show year 1 and year 2 as separate bars, side by side for each month.

	Year 1	Year 2
January	$158.92	$165.98
February	$147.88	$162.85
March	$125.90	$130.45
April	$108.40	$112.55
May	$ 87.65	$ 90.45
June	$114.58	$125.91
July	$145.84	$137.70
August	$142.78	$140.19
September	$ 90.25	$ 96.15
October	$104.12	$115.75
November	$136.62	$145.21
December	$158.18	$160.25

28. In Exercise 10 from Section 9.1 we listed a family's water consumption for two years. Construct a vertical bar graph with the months along the horizontal axis and the ft.^{3} consumed along the vertical axis. Show year 1 and year 2 as separate bars, side by side for each month.

	Year 1	Year 2
January	500	550
February	525	540
March	600	600
April	650	650
May	840	900
June	1100	1430
July	1000	1200
August	1400	1250
September	840	750
October	620	700
November	600	550
December	550	500

For Exercises 29–34, use the bar graph in Figure 9.7.

29. Which field had the largest number of degree recipients in all three groups of graduates?

30. Which field had the smallest number of degree recipients in 1994–95?

31. How many degrees were conferred in visual and performing arts in 1994–95?

32. How many degrees were conferred in education in 1989–90?

Figure 9.7 Trends in bachelor's degrees conferred in selected fields of study: 1984–85, 1989–90, and 1994–95

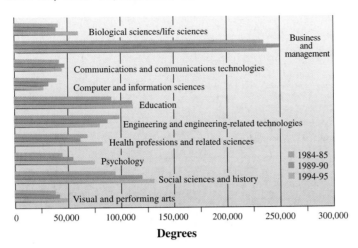

Source: U.S. Department of Education, National Center for Education Statistics, Higher Education General Information Survey (HEGIS), "Degrees and Other Formal Awards Conferred" survey, and Integrated Postsecondary Education Data System (IPEDS), "Completions" surveys

33. How many more degrees were awarded in health professions and related fields in 1994–95 than in 1989–90?

34. How many more degrees were awarded in social sciences and history fields in 1989–90 than in 1984–95?

For Exercises 35–40, use the line graph in Figure 9.8.

Figure 9.8 Items most frequently cited by the public as a major problem facing the local public schools: 1980 to 1996

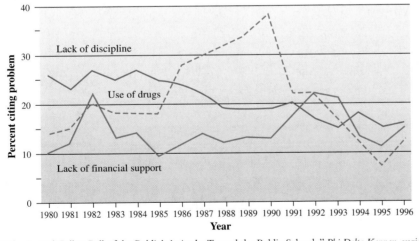

Source: "The Annual Gallup Poll of the Public's Attitudes Toward the Public Schools," *Phi Delta Kappan,* various years

35. What percent of those polled felt that use of drugs was a major problem in 1990?

36. What percent of those polled said that lack of financial support was a major problem in 1989?

37. In what year did use of drugs overtake lack of discipline as the more popular response?

38. Over what years was lack of discipline the most popular response?

39. List the percent for each response in 1995.

40. Based on the directions of the lines after 1996, what would you forecast the responses to be in 1997?

41. Below is a list of a student's GPA each semester until she graduated. Construct a line graph with each semester along the horizontal axis and GPA along the vertical axis.

Fall 1994:	3.25		Fall 1996:	2.50
Spring 1995:	3.50		Spring 1997:	2.25
Fall 1995:	3.60		Fall 1997:	3.00
Spring 1996:	2.75		Spring 1998:	3.50

42. Below is a list of revenue and cost totals at the end of each year over 6 years for a small business. Construct a line graph with each year along the horizontal axis and the dollar amounts along the vertical axis. Draw one line for assets and a different line for debts.

	Revenue	Cost
1991	$ 45,450	$ 55,000
1992	$ 65,000	$ 62,000
1993	$ 88,000	$ 64,000
1994	$112,000	$ 75,000
1995	$248,000	$ 86,000
1996	$282,000	$124,000

REVIEW EXERCISES

1. Graph -5 on a number line.

2. Graph 6 on a number line.

3. $|25 - 32| =$

4. $(6 - 10)^2 =$

5. Find the missing side length in the right triangle shown.

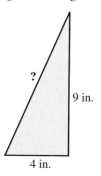

9 in.

4 in.

9.3 The Rectangular Coordinate System

OBJECTIVES

1 Determine the coordinates of a given point.

2 Plot points in the coordinate plane.

3 Determine the quadrant for a given coordinate.

4 Determine the distance between two points in the coordinate plane.

OBJECTIVE **1** *Determine the coordinates of a given point.*

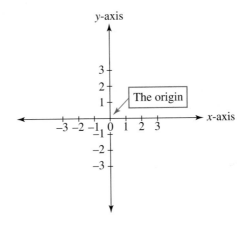

In 1619, Rene Descartes realized that positions in a plane could be described using two number lines that intersect at a right angle. These two number lines form the rectangular coordinate system or Cartesian coordinate system, named in honor of Rene Descartes. In Section 9.2 we learned that number lines used as reference in a graph are called *axes*. In the rectangular coordinate system the axes are labeled with variables. The common variables are x for the horizontal axis and y for the vertical axis. Notice that the point of intersection is 0 for both number lines. This position is called the origin. The positive numbers are to the right and up from the origin, while negative numbers are to the left and down from the origin.

To avoid confusion, we will always write the two numbers in a specific order. The number from the horizontal axis (x-axis) is stated first, then the number from the vertical axis (y-axis). Because the order in which we say or write the numbers matters, we say they form an *ordered pair*. Each number in an ordered pair is called a **coordinate.**

> **Connection**
> In Section 9.2, we defined axes in connection with bar and line graphs. When you constructed these graphs, you were developing the foundation for the rectangular coordinate system.

DEFINITION **Coordinate:** A number that refers to a number line in the rectangular coordinate system.

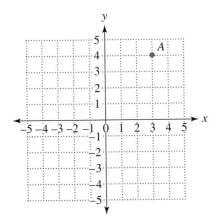

Notation for writing a coordinate pair:

$$\begin{pmatrix} \text{Horizontal} & \text{Vertical} \\ \text{coordinate,} & \text{coordinate} \end{pmatrix}$$

Consider the point labeled A in the coordinate plane to the left. The point is drawn at the intersection of the 3rd line to the right of the origin and the 4th line up from the origin. We write the coordinates that describes point A as (3, 4).

> **Tip**
> Think of points in the coordinate plane as intersections of city streets. In New York City, avenues run north/south while streets run east/west. To describe an intersection, a person would say the avenue number first, then the street number. Point A on the graph to the left is the intersection of 3rd Avenue and 4th Street. Or as New Yorkers would say, "3rd and 4th."

Procedure *To determine the coordinates for a given point in the rectangular system:*

1. Follow a vertical line from the given point to the *x*-axis (horizontal axis). The number at this position on the *x*-axis is stated as the first coordinate.

2. Follow a horizontal line from the given point to the *y*-axis (vertical axis). The number at this position on the *y*-axis is stated as the second coordinate.

Historical Note

Rene Descartes was born on March 31, 1596, at La Haye, near Tours, France. Young Rene was a sickly child and upon entering school, the headmaster allowed Rene to stay in bed every morning and rest until he felt up to coming to class. This habit of lying and thinking until late in the morning was to be Descartes' routine until near the end of his life. He became quite skeptical of the teachings of the day and began formulating his own methods for reasoning. From this early philosophizing he concluded, "Cogito ergo sum," which means "I think, therefore I am." To escape the wild antics of his friends, he entered military service. During his time in military service, after a celebration on the night of November 10, 1619, Descartes had a series of dreams. In the dreams, Descartes became aware of a new way of viewing geometry using algebra. At the age of 32, after finally tiring of military service and after the prompting of a friend, he decided to settle down and write. To avoid distractions, he went into hiding in Holland. Over the next 20 years, he stayed in small villages and out-of-the-way parts of cities and never stayed in any place very long. Only one close friend was allowed to know his whereabouts. It was through this friend that Descartes corresponded with other intellectuals of the time. In 1634, he was working on final revisions of a large treatise, *Le Monde,* when he received word that Galileo had been forced by the Inquisition to recant his belief in the Copernican system. Descartes too believed the Copernican system to be correct and had written so in his work. He therefore required that his work not be published until after his death. In 1637, after some urging by his friends, he reluctantly allowed one work, known as the *Method,* to be printed. It was in this book that the rectangular coordinate system and analytical geometry was given to the world.

EXAMPLE 1 Write the coordinates for each point shown.

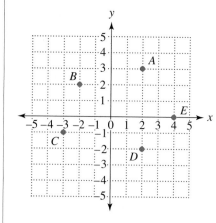

Answers: A: (2, 3)

Note that we can get to this point from the origin by moving right 2 and then up 3.

B: (−2, 2)

From the origin point B is left 2 then up 2.

C: (−3, −1)   Left 3 then down 1.

D: (2, −2)   Right 2 then down 2.

E: (4, 0)

Right 4. Because the point is on the *x*-axis the *y* coordinate is 0.

◀ **Do Margin 1.**

◆ **Margin 1**

Write the coordinates for each point shown.

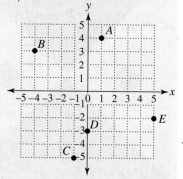

OBJECTIVE 2 *Plot points in the coordinate plane.*

Now that we have a sense of how coordinates are used to describe positions, let's graph or plot points given the coordinates. Recall what we learned about coordinates.

$$(3, 4)$$

Right 3   Up 4

Procedure *To graph or plot a point given a coordinate pair:*

1. Beginning at the origin, use the first coordinate to travel right or left along the *x*-axis.
2. From that position on the *x*-axis, use the second coordinate to travel up or down.
3. Draw a dot to represent the point described by the coordinates.

Answers: **A:** (1, 4) **B:** (−4, 3)
C: (−1, −5) **D:** (0, −3) **E:** (5, −2)

EXAMPLE 2 Plot the point described by the coordinates.

a. $(4, -2)$

Solution: Beginning at the origin, we move to the right 4, then down 2.

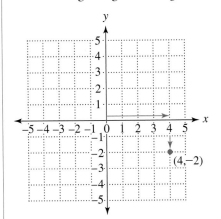

$(4, -2)$
right 4   down 2

b. $(-3, -4)$

Solution: Beginning at the origin, we move to the left 3, then down 4.

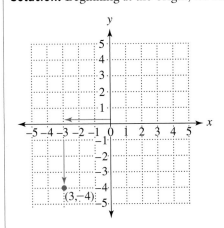

$(-3, -4)$
left 3   down 4

c. $(0, 5)$

Solution: Because the first coordinate is 0, we do not move right or left any. Because the second coordinate is 5, we move straight up 5, ending on the y-axis.

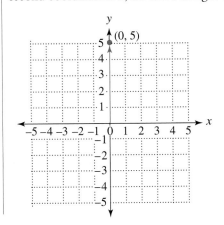

$(0, 5)$
right/left 0   up 5

Do Margin 2. ▶

◆ **Margin 2**

Plot the point described by the coordinates.

a. $(-4, 2)$

b. $(3, 1)$

c. $(0, -4)$

d. $(-2, 0)$

e. $(-1, -5)$

f. $(3, -2)$

Answer:

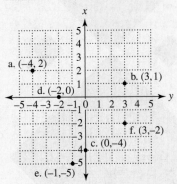

a. $(79, 109)$

b. $(-207, -12)$

c. $(-45, 0)$

d. $(-55, 72)$

OBJECTIVE 3 *Determine the quadrant for a given coordinate.*

Notice that the two axes divide the coordinate plane into four regions. We call these regions **quadrants**.

DEFINITION **Quadrant:** One of four regions created by the intersection of the axes in the coordinate plane.

The quadrants are numbered. It is customary to use roman numerals to number the quadrants. The quadrant numbers are shown below:

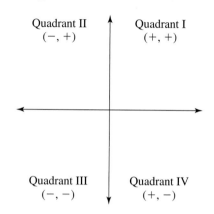

Quadrant II $(-, +)$ Quadrant I $(+, +)$

Quadrant III $(-, -)$ Quadrant IV $(+, -)$

Note that we can use the signs of the co-ordinates to determine what quadrant the point is in. Quadrant I would always have both coordinates positive. In quadrant II, the first coordinate (horizontal) is negative while the second coordinate (vertical) is positive. Quadrant III has both coordinates negative. Quadrant IV has the first coordinate positive and second coordinate negative.

Procedure *To determine the quadrant for a given coordinate, consider the signs of the numbers in the coordinate.*

$(+, +)$ = Quadrant I
$(-, +)$ = Quadrant II
$(-, -)$ = Quadrant III
$(+, -)$ = Quadrant IV

EXAMPLE 3 State the quadrant in which each point is located.

a. $(-45, 19)$

Answer: Quadrant II.

Explanation: Because the first coordinate is negative and the second coordinate positive, if we were to plot these coordinates, we would move left, then up, which places the point in Quadrant II (upper left).

b. $(102, -68)$

Answer: Quadrant IV

Explanation: Because the first coordinate is positive and the second coordinate negative, if we were to plot the coordinates, we would move right, then down, which would place the point in Quadrant IV (lower right).

c. $(0, 91)$

Answer: This point is not in a quadrant. It is on the y-axis.

Explanation: Because one of the coordinates is 0, this point will be on an axis. Therefore it ◆ is not in a quadrant. Because the x coordinate is 0, this point is on the y-axis.

◀ **Do Margin 3.**

OBJECTIVE 4 *Determine the distance between two points in the coordinate plane.*

Suppose we wanted to calculate the distance along a straight line connecting the points $(0, 0)$ and $(3, 4)$. Let's plot the points and draw the line. We'll label the unknown distance d.

Notice we can draw a right triangle with the length d as the hypotenuse. We can use the x- and y-axes to determine the lengths of the legs, and then use the Pythagorean theorem to calculate the distance d.

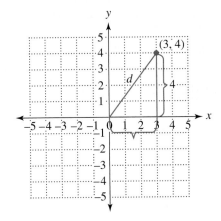

The vertical length is 4 while the horizontal length is 3. Recall the Pythagorean theorem is:

$$c^2 = a^2 + b^2$$

where c is the length of the hypotenuse.

To calculate the distance d, we replace a with 3, b with 4, and c with the variable d.

$$d^2 = 3^2 + 4^2$$
$$d^2 = 9 + 16$$
$$d^2 = 25$$
$$d = \sqrt{25}$$
$$d = 5$$

The distance between the points is 5.

Let's develop a formula that we can use to calculate the distance between any two points. We follow the same plan as above. However, we will not use numbers. Instead we will use two points labeled (x_1, y_1) and (x_2, y_2). The subscripts are used to indicate which point is which.

For convenience, we will place the two points in the first quadrant. Also, to avoid clutter, we will take out the grid and simply show the axes.

Remember, we need to find the lengths of the legs, then use the Pythagorean theorem.

Think: If y_2 were 6 and y_1 were 2, the distance between would be $6 - 2 = 4$. Therefore, to calculate the length of the vertical leg, we calculate $y_2 - y_1$.

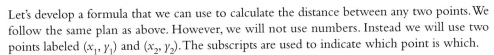

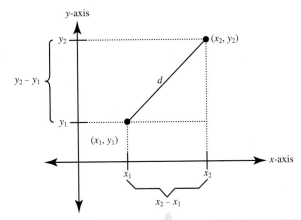

Think: If x_2 were 4 and x_1 were 1, the distance between would be $4 - 1 = 3$. Therefore, to calculate the length of the horizontal leg, we calculate $x_2 - x_1$.

Now we can use the Pythagorean theorem. We replace a with $x_2 - x_1$, b with $y_2 - y_1$, and c with d.

$$d^2 = (x_2 - x_1)^2 + (y_2 - y_1)^2$$

To isolate d, we must undo the square of d. To undo the square, we calculate the square root of both sides.

Conclusion: $d = \sqrt{(x_2 - x_1)^2 + (y_2 - y_1)^2}$

> **Procedure** *To calculate the distance between two points with coordinates* (x_1, y_1) *and* (x_2, y_2) *use the formula:*
>
> $$d = \sqrt{(x_2 - x_1)^2 + (y_2 - y_1)^2}$$

◆ **Margin 4**

Calculate the distance between the given points. Round to the nearest hundredth.

a. $(9, 1)$ and $(5, -3)$

b. $(0, -6)$ and $(8, 4)$

EXAMPLE 4 Find the distance between $(2, 6)$ and $(-4, -3)$. Round to the nearest hundredth.

Solution: Use the distance formula.

$$
\begin{array}{cc}
(2, 6) & (-4, -3) \\
\uparrow\ \uparrow & \uparrow\ \ \uparrow \\
(x_1, y_1) & (x_2, y_2)
\end{array}
$$

$d = \sqrt{(x_2 - x_1)^2 + (y_2 - y_1)^2}$ Replace the variables with the corresponding numbers.

$d = \sqrt{(-4 - 2)^2 + (-3 - 6)^2}$ Subtract within parentheses.

$d = \sqrt{(-6)^2 + (-9)^2}$ Evaluate the exponential forms.

$d = \sqrt{36 + 81}$ Add.

$d = \sqrt{117}$ Approximate the square root to the nearest hundredth.

$d \approx 10.82$

Does the order of the points matter? Does it matter which numbers are considered to be (x_1, y_1) and which are considered (x_2, y_2)? To see, we will work Example 5 with $(-4, -3)$ as (x_1, y_1) and $(2, 6)$ as (x_2, y_2).

$$
\begin{array}{cc}
(-4, -3) & (2, 6) \\
\uparrow\ \ \uparrow & \uparrow\ \uparrow \\
(x_1, y_1) & (x_2, y_2)
\end{array}
$$

$d = \sqrt{(x_2 - x_1)^2 + (y_2 - y_1)^2}$

$d = \sqrt{(2 - (-4))^2 + (6 - (-3))^2}$

$d = \sqrt{(2 + 4)^2 + (6 + 3)^2}$

$d = \sqrt{6^2 + 9^2}$

$d = \sqrt{36 + 81}$

$d = \sqrt{117}$

$d \approx 10.82$

> Recall that the results of the calculations inside the parentheses in Example 4 were -6 and -9 and here we have 6 and 9. However, when we square -6 or 6 the result is the same, 36. Likewise with -9 or 9, the result is 81. Therefore, it does not matter which coordinates are labeled as (x_1, y_1) and which are (x_2, y_2).

◆ **Conclusion:** It does not matter which coordinates you consider to be (x_1, y_1) and (x_2, y_2).

EXAMPLE 5 Find the distance between $(-4, 8)$ and $(9, -1)$. Round to the nearest hundredth.

Solution: Use the distance formula.

$d = \sqrt{(x_2 - x_1)^2 + (y_2 - y_1)^2}$ Replace the variables with the corresponding numbers.

$d = \sqrt{(-4 - 9)^2 + (8 - (-1))^2}$ Subtract within parentheses.

$d = \sqrt{(-13)^2 + (9)^2}$ Evaluate the exponential forms.

$d = \sqrt{169 + 81}$ Add.

$d = \sqrt{250}$ Approximate the square root to the nearest hundredth.

$d \approx 15.81$

Answers: **a.** 5.66 **b.** 12.81 **c.** 11.70

◆ ◁ **Do Margin 4.**

9.3 Exercises

For Exercises 1–4, write the coordinates for each point.

1.

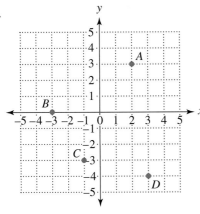

2.

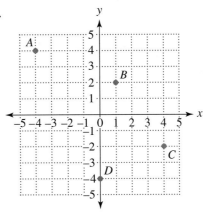

3.

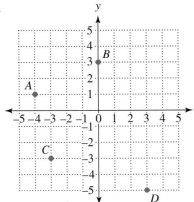

4.

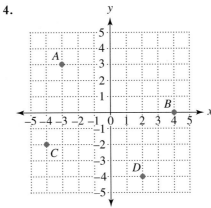

For Exercises 5–8, plot and label the points indicated by the coordinate pairs.

5. $(4, 5)$ $(3, -2)$ $(-4, -1)$ $(0, 2)$

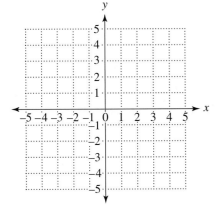

6. $(-1, 0)$ $(2, 1)$ $(-3, -3)$ $(5, -2)$

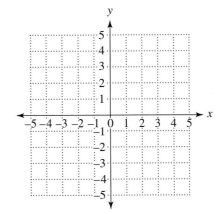

7. $(-4, 0)$ $(1, -4)$ $(-3, -5)$ $(4, 4)$ **8.** $(-2, -6)$ $(0, -3)$ $(5, -3)$ $(-4, 2)$

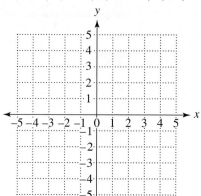   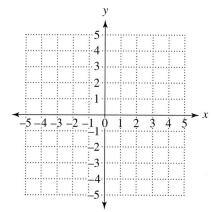

For Exercises 9–16, state the quadrant in which the point is located.

9. $(-99, 201)$ **10.** $(-61, -78)$ **11.** $(105, 50)$ **12.** $(-84, 56)$

13. $(42, -60)$ **14.** $(78, 65)$ **15.** $(-92, -106)$ **16.** $(28, -59)$

For Exercises 17–26, calculate the distance between the points. If the distance is an irrational number, use a calculator to approximate the distance to the nearest hundredth.

17. $(0, 6)$ and $(8, 0)$ **18.** $(1, 7)$ and $(9, -8)$

19. $(3, -1)$ and $(7, 2)$ **20.** $(1, -2)$ and $(10, -14)$

21. $(2, -6)$ and $(-3, 8)$ **22.** $(-4, 9)$ and $(-1, -1)$

23. $(0, -8)$ and $(-7, 11)$ **24.** $(-12, -8)$ and $(0, -3)$

25. $(-9, -3)$ and $(-5, -2)$ **26.** $(-11, -4)$ and $(-17, -7)$

27. $(1.5, -9)$ and $(-4.5, 2.2)$ **28.** $(-3.6, 12.4)$ and $(1.2, 4.8)$

29. $\left(3\frac{1}{4}, 8\right)$ and $\left(5, 2\frac{1}{2}\right)$ **30.** $\left(-14, 6\frac{2}{3}\right)$ and $\left(7, -2\frac{5}{6}\right)$

REVIEW EXERCISES

1. Graph -3 on a number line.   **2.** Evaluate $2x - 3y$, when $x = 4$ and $y = 2$.

3. Evaluate $-4x - 5y$, when $x = -3$ and $y = 0$.

4. Solve. $2x - 5 = 10$ **5.** Solve. $-\dfrac{5}{6}x = \dfrac{3}{4}$

9.4 Graphing Linear Equations

OBJECTIVES

1 Determine whether a given pair of coordinates is a solution to a given equation with two unknowns.

2 Find three solutions for an equation in two unknowns.

3 Graph linear equations.

4 Find the coordinates of the x- and y-intercepts.

OBJECTIVE 1 *Determine whether a given pair of coordinates is a solution to a given equation with two unknowns.*

In the past, we've considered equations with one variable. Now let's consider equations that have two variables.

$x + y = 5$ and $y = 3x - 4$ are examples of equations with two variables. Recall that a solution for an equation is a number that can replace the variable and make the equation a true statement. Because these equations have two variables a solution will be a pair of numbers, one number for each variable, that can replace the corresponding variables and make the equation true.

Note that the solutions for equations in two variables are ordered pairs. This means we can write a solution as a coordinate pair. The coordinate pair $(1, 4)$ is a solution for $x + y = 5$. Coordinates are written in alphabetical order, so $(1, 4)$ means that $x = 1$ and $y = 4$. We can verify this solution by replacing x with 1 and y with 4:

$$x + y = 5$$
$$1 + 4 = 5$$

This is true, so the coordinate pair checks.

> **Procedure** *To determine whether a given coordinate pair is a solution for an equation with two variables:*
>
> **1.** Replace the variables in the equation with the corresponding coordinates.
> **2.** Verify that the equation is true.

EXAMPLE 1 Determine whether the coordinate pair is a solution for the equation.

a. $(2, -5); 3x + y = 1$

Solution: Replace x with 2 and y with -5, and see if the equation is true.

$$3x + y = 1$$
$$3(2) + (-5) \text{ ? } 1$$
$$6 + (-5) \text{ ? } 1$$
$$1 = 1$$

This checks, which means that $(2, -5)$ is a solution for $3x + y = 1$.

b. $(-4, -7); y = 2x - 1.$

Solution: Replace x with -4 and y with -7, and see if the equation is true.

$$y = 2x - 1$$
$$-7 \text{ ? } 2(-4) - 1$$
$$-7 \text{ ? } -8 - 1$$
$$-7 \neq -9$$

Because the equation is not true, $(-4, -7)$ is not a solution for $y = 2x - 1$.

◄ **Do Margin 1.**

Fractions and decimal numbers can also be solutions.

EXAMPLE 2 Determine whether $\left(-\frac{1}{3}, 3.5\right)$ is a solution for $y = \frac{3}{2}x + 4$.

Solution: Replace x with $-\frac{1}{3}$ and y with 3.5, and see if the equation is true.

$$y = \frac{3}{2}x + 4$$

$$3.5 \ ? \ \frac{\overset{1}{\cancel{3}}}{2}\left(-\frac{1}{\underset{1}{\cancel{3}}}\right) + 4 \qquad \text{Divide out common factors.}$$

$$3.5 \ ? \ -\frac{1}{2} + 4 \qquad \begin{array}{l}\text{Write the fraction as an equivalent} \\ \text{decimal number.}\end{array}$$

$$3.5 \ ? \ -0.5 + 4$$

$$3.5 = 3.5 \qquad \begin{array}{l}\text{This equation is true, therefore } \left(-\frac{1}{3}, 3.5\right) \text{ is} \\ \text{a solution.}\end{array}$$

◄ **Do Margin 2.**

OBJECTIVE 2 *Find three solutions for an equation in two unknowns.*

How can we find solutions to equations in two variables? Consider the equation $x + y = 5$. We've already seen that $(1, 4)$ is a solution. Are there others?

Note that in the case of $x + y = 5$, a solution would be any pair of numbers whose sum is 5. There are lots of solutions.

$$(2, 3) \quad (3, 2) \quad (0, 5) \quad (5, 0) \quad (-1, 6) \quad (6, -1) \quad \text{and so on} \ldots$$

We could even have fraction or decimal combinations.

$$(3.5, 1.5) \quad \left(3\frac{1}{3}, 1\frac{2}{3}\right)$$

In fact, there are an infinite number of solutions. For every x value, there is a corresponding y value that will add to the x value to equal 5, and vice versa. This gives a clue as to how to find solutions. We could simply choose an x- or y-value and solve for the corresponding value of the other variable.

Procedure *To find a solution to an equation in two variables:*

1. Choose a value for one of the variables (any value).
2. Replace the corresponding variable with your chosen value.
3. Solve the equation for the value of the other variable.

Answers to Margin 1: **a.** Yes **b.** No **c.** No **d.** Yes

Answers to Margin 2: **a.** Yes **b.** No

EXAMPLE 3   Find three solutions for the equation $2x + y = 5$.

Solution: To find a solution, we replace one of the variables with a chosen value, and then solve for the value of the other variable.

For the first solution, we will choose x to be 0.	For the second solution, we will choose x to be 2.	For the third solution, we will choose x to be 3.

$$2x + y = 5$$
$$2(0) + y = 5$$
$$0 + y = 5$$
$$y = 5$$
Solution: $(0, 5)$

Note: Choosing x to be 0 makes the equation very easy to solve.

$$2x + y = 5$$
$$2(2) + y = 5$$
$$4 + y = 5$$
$$4 + y = 5$$
$$\underline{-4 \qquad -4}$$
$$0 + y = 1$$
$$y = 1$$
Solution: $(2, 1)$

$$2x + y = 5$$
$$2(3) + y = 5$$
$$6 + y = 5$$
$$6 + y = 5$$
$$\underline{-6 \qquad -6}$$
$$0 + y = -1$$
$$y = -1$$
Solution: $(3, -1)$

> **Tip**
> Zero is almost always a good choice for either x or y (or both) when solving equations with two variables because zero will eliminate one of the terms.

Keep in mind that because there are an infinite number of correct solutions, you may get different solutions than someone else solving the same equation.

Do Margin 3. ▶

EXAMPLE 4   Find three solutions for the equation $y = \frac{1}{3}x - 5$.

Solution: Notice that in this equation, y is isolated. If we select values for x, we will not have to isolate y as we did in Example 3. We will simply calculate the y value. Also notice that the coefficient for x is a fraction. Because we choose the values for x, let's choose values like 3 and 6 that will divide out nicely with the denominator of 3.

For the first solution, we will choose x to be 0.	For the second solution, we will choose x to be 3.	For the third solution, we will choose x to be 6.

$$y = \frac{1}{3}x - 5$$
$$y = \frac{1}{3}(0) - 5$$

$$y = -5$$
Solution: $(0, -5)$

$$y = \frac{1}{3}x - 5$$
$$y = \frac{1}{3}(3) - 5$$
$$y = \frac{1}{\cancel{3}}\left(\frac{\cancel{3}^{\,1}}{1}\right) - 5$$
$$y = 1 - 5$$
$$y = -4$$
Solution: $(3, -4)$

$$y = \frac{1}{3}x - 5$$
$$y = \frac{1}{3}(6) - 5$$
$$y = \frac{1}{\cancel{3}}\left(\frac{\cancel{6}^{\,2}}{1}\right) - 5$$
$$y = 2 - 5$$
$$y = -3$$
Solution: $(6, -3)$

Do Margin 4. ▶

◆ **Margin 3**

Find three solutions for each equation. (Answers may vary.)

a. $x + y = 6$

b. $3x + y = 9$

c. $2x - 3y = 12$

◆ **Margin 4**

Find three solutions for each equation. (Answers may vary.)

a. $y = x - 3$

b. $y = -4x$

c. $y = \frac{2}{3}x$

d. $y = \frac{1}{2}x + 5$

Answers to Margin 3: **a.** $(1, 5)$ $(2, 4)$ $(3, 3)$ **b.** $(1, 6)$ $(2, 3)$ $(3, 0)$ **c.** $(0, -4)$ $(6, 0)$ $\left(1, -3\frac{1}{3}\right)$

Answers to Margin 4: **a.** $(0, -3)$ $(1, -2)$ $(2, -1)$ **b.** $(-1, 4)$ $(0, 0)$ $(1, -4)$ **c.** $(0, 0)$ $(3, 2)$ $(6, 4)$ **d.** $(0, 5)$ $(2, 6)$ $(4, 7)$

Graph each equation. (Note that you found three solutions for each of these equations in Margin 3.)

a. $x + y = 6$

b. $3x + y = 9$

c. $2x - 3y = 12$

Answers:

a.
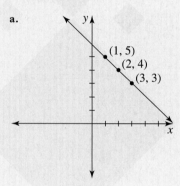

b.
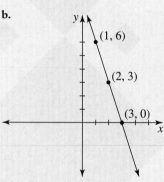

c.
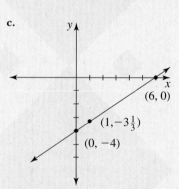

OBJECTIVE 3 *Graph linear equations.*

We've learned that equations in two variables have an infinite number of solutions. Because of this, we cannot possibly list all solutions. However, we can represent all the solutions using a graph.

Watch what happens when we plot solutions for $x + y = 5$. Recall some of the solutions: $(0, 5), (5, 0), (1, 4), (4, 1), (2, 3),$ and $(3, 2)$.

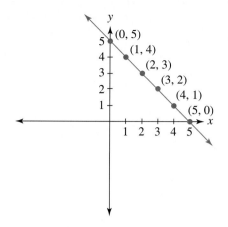

Notice that when we plot the coordinates as points in the rectangular coordinate system, they line up in a straight line. By connecting the points with a line we are graphically representing every possible solution. Placing arrows on either end of the line indicates that the solutions continue beyond our *window* in both directions. Every point on that line has coordinates with a sum equal to 5.

When we graph linear equations, we will always get a straight line. (This is why they are called *linear equations*.) For a line to be fixed in place, we need a minimum of two points, which means we must find at least two solutions in order to graph a line. However, it is wise to find three solutions, and use the third solution as a check: If we plot the three solutions as points and they cannot be connected by a straight line, we know something is wrong.

> **Procedure** *To graph a linear equation:*
> 1. Find at least two solutions to the equation.
> 2. Plot the solutions as points in the rectangular coordinate system.
> 3. Connect the points to form a straight line.

EXAMPLE 5 Graph $2x + y = 5$

Solution: We found three solutions to this equation in Example 3. Recall those solutions: $(0, 5), (2, 1),$ and $(3, -1)$.

Now we plot each solution as a point in the rectangular coordinate system, then connect the points to form a straight line.

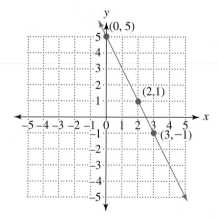

Note: Many people find it helpful to list the solutions in a table.

x	y
0	5
2	1
3	-1

◀ Do Margin 5.

EXAMPLE 6 Graph. $y = \frac{1}{3}x - 5$

Solution: We found three solutions to this equation in Example 4. Recall those solutions: $(0, -5), (3, -4)$, and $(6, -3)$.

Now we plot each solution as a point in the rectangular coordinate system, then connect the points to form a straight line.

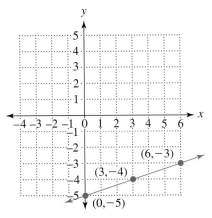

Do Margin 6. ▶

We've explored graphing linear equations in two variables, but there are also linear equations in one variable like $y = 3$ or $x = 2$. What would the graphs of these equations look like?

EXAMPLE 7 Graph.

a. $y = 3$.

Solution: Note that there is no x variable. The equation $y = 3$ indicates that y is equal to a constant, 3. In other words y is always 3 no matter what we choose for x. If we choose x to be 0, then y equals 3. If we choose x to be 2, then y is 3. If we choose x to be 4, then x is still 3.

We now have three solutions: $(0, 3), (2, 3)$, and $(4, 3)$.

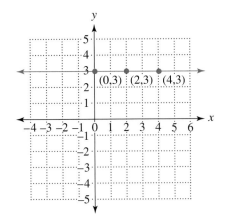

Notice that the graph of $y = 3$ is a horizontal line parallel to the x-axis that passes through the y-axis at $(0, 3)$.

b. $x = 2$

Solution: Note that there is no y variable. The equation $x = 2$ indicates that x is equal to a constant, 2. In other words x is always 2 no matter what we choose for y. If we choose y to be 0, then x equals 2. If we choose y to be 1, then x is 2. If we choose y to be 4, then x is still 2.

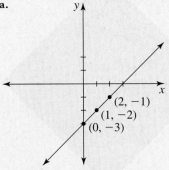
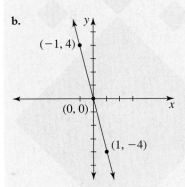
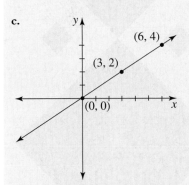
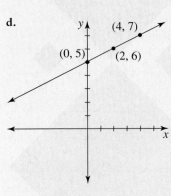

Graph.

a. $y = -4$

We now have three solutions: $(2, 0)$, $(2, 1)$, and $(2, 4)$.

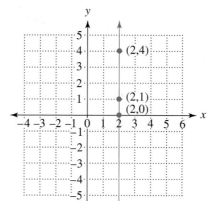

Notice that $x = 2$ is a vertical line parallel to the y-axis that passes through the x-axis at $(2, 0)$.

b. $x = -3$

◀ **Do Margin 7.**

OBJECTIVE **4** *Find the coordinates of the* **x-** *and* **y-intercepts.**

EXAMPLE 8 Graph. $x - 2y = 4$

Solution: We will find three solutions, plot each solution as a point in the rectangular coordinate system, then connect the points to form a straight line.

Recall that choosing a variable to be 0 eliminates the term with that variable, which simplifies the equation. Let's choose x to be 0 for the first solution. Then choose y to be 0 for the second solution. For the third solution, we'll choose y to be 1.

First solution: We choose x to be 0.	Second solution: We choose y to be 0.	Third solution: We choose y to be 1.

First solution:
$$x - 2y = 4$$
$$0 - 2y = 4$$
$$-2y = 4$$
$$\frac{-2y}{-2} = \frac{4}{-2}$$
$$1y = -2$$
$$y = -2$$
Solution: $(0, -2)$

Second solution:
$$x - 2y = 4$$
$$x - 2(0) = 4$$
$$x - 0 = 4$$
$$x = 4$$
Solution: $(4, 0)$

Third solution:
$$x - 2y = 4$$
$$x - 2(1) = 4$$
$$x - 2 = 4$$
$$\frac{x - 2 = 4}{+2 \quad +2}$$
$$x + 0 = 6$$
$$x = 6$$
Solution: $(6, 1)$

Answers:

a.

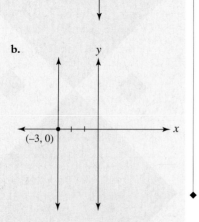

b.

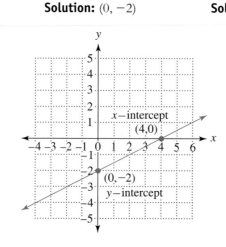

Notice when we choose x to be 0, the point is on the y-axis. We call the point where a line intersects the y-axis the **y-intercept.** When we choose y to be 0, the point is on the x-axis. We call the point where a line intersects the x-axis the **x-intercept.**

y-intercept: A point where a line intersects the y-axis.

The x-coordinate will always be 0 in the coordinates for a y-intercept.

Procedure *To find a y-intercept:*
1. Replace x with 0 in the given equation.
2. Solve for y.

x-intercept: A point where a line intersects the x-axis.

The y-coordinate will always be 0 in the coordinates for a x-intercept.

Procedure *To find an x-intercept:*
1. Replace y with 0 in the given equation.
2. Solve for x.

EXAMPLE 9 Find the coordinates for the x- and y-intercepts.

a. $5x - 2y = 15$

Solution: To find the x-intercept, we replace y with 0 and solve the equation for x. To find the y-intercept, we replace x with 0 and solve the equation for y.

For the x-intercept:

$$5x - 2y = 15$$
$$5x - 2(0) = 15$$
$$5x - 0 = 15$$
$$5x = 15$$
$$\frac{5x}{5} = \frac{15}{5}$$
$$1x = 3$$
$$x = 3$$

For the y-intercept:

$$5x - 2y = 15$$
$$5(0) - 2y = 15$$
$$0 - 2y = 15$$
$$-2y = 15$$
$$\frac{-2y}{-2} = \frac{15}{-2}$$
$$1y = -\frac{15}{2}$$
$$y = -7\frac{1}{2}$$

x-intercept coordinates: **(3, 0)** y-intercept coordinates: $\left(0, -7\frac{1}{2}\right)$

b. $y = \frac{3}{4}x$

Solution: Notice when we solve for the y-intercept by replacing x with 0, the solution for y is also 0.

Notice the intercept is the origin, $(0, 0)$. In fact the origin is the only point in the rectangular coordinate system where a line can pass through both the x-axis and y-axis simultaneously.

y-intercept: $y = \frac{3}{4}x$

$$y = \frac{3}{4}(0)$$
$$y = 0$$

x- and y-intercept coordinates: **(0, 0)**

c. $x = -5$

Solution: The graph of $x = -5$ is a vertical line parallel to the y-axis that passes through the x-axis at the point $(-5, 0)$. Notice that this point is the x-intercept. Because the line is parallel to the y-axis, it will never intersect the y-axis. Therefore there is no y-intercept.

d. $y = 7$

Solution: The graph of $y = 7$ is a horizontal line parallel to the x-axis that passes through the y-axis at the point $(0, 7)$. Notice that this point is the y-intercept. Because the line is parallel to the x-axis it will never intersect the x-axis. Therefore there is no x-intercept.

◀ **Do Margin 8.**

9.4 Exercises

FOR EXTRA HELP

 Videotape 11

 InterAct Math
Tutorial Software

 www.carsonmath.com

 AWL Math Tutor Center

InterAct MathXL www.mathxl.com

 Student's Solutions
Manual

For Exercises 1–12, determine whether the given pair of coordinates is a solution for the given equation.

1. $(2, 3); x + 2y = 8$

2. $(3, 1); 2x - y = 5$

3. $(-5, 2); y - 4x = 3$

4. $(4, -6); y = 3x - 1$

5. $(9, 0); y = -2x + 18$

6. $(0, -1); y = 2x + 1$

7. $(6, -2); y = -\dfrac{2}{3}x$

8. $(0, 0); y = \dfrac{2}{5}x$

9. $\left(-1\dfrac{2}{5}, 0\right); y - 3x = 5$

10. $\left(\dfrac{2}{3}, -4\dfrac{5}{6}\right); y = \dfrac{1}{4}x - 5$

11. $(2.2, -11.2); y + 6x = -2$

12. $(-1.5, -1.3); y = 0.2x - 1$

For Exercises 13–40, find three solutions for the given equation. Then graph.

13. $x - y = 8$

14. $x + y = -5$

15. $2x + y = 6$

16. $3x - y = 9$

17. $y = x$

18. $y = -x$

19. $y = 2x$

20. $y = -2x$

21. $y = -5x$

22. $y = 3x$

23. $y = x - 3$

24. $y = -x + 5$

25. $y = -2x + 4$

26. $y = 2x - 5$

27. $y = 3x + 2$

28. $y = -5x - 1$

29. $y = \dfrac{1}{2}x$

30. $y = -\dfrac{3}{4}x$

31. $y = -\dfrac{2}{3}x + 4$

32. $y = \dfrac{4}{5}x - 1$

33. $x - \dfrac{1}{4}y = 2$

34. $\dfrac{1}{3}x + y = -1$

35. $y = 0.4x - 2.5$

36. $1.2x + 0.5y = 6$

37. $y = -5$

38. $x = -6$

39. $x = 7$

40. $y = 4$

For Exercises 41–56, find the coordinates for the x- and y-intercepts.

41. $x + 2y = 12$

42. $x - 3y = 6$

43. $5x - 4y = 20$

44. $4x + y = 16$

45. $\frac{3}{4}x - y = -6$

46. $-x + \frac{3}{5}y = 9$

47. $6.5x + 2y = 1.3$

48. $4x - 5.6y = 2.8$

49. $y = 6x$

50. $y = -2x$

51. $y = -4x + 1$

52. $y = 2x - 3$

53. $y = \frac{1}{5}x - 5$

54. $y = \frac{2}{3}x + 6$

55. $x = -9$

56. $y = 3$

57. $y = 4$

58. $x = 2$

59. $y = -3$

60. $x = -6$

Puzzle Problem

Without lifting your pencil from the paper, connect the dots using only four straight lines.

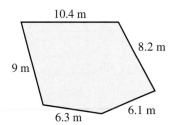

REVIEW EXERCISES

1. Evaluate $-32t + 70$, when $t = 4.2$

2. Calculate the perimeter of the shape.

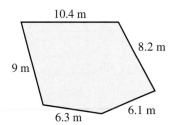

3. Calculate the area of the parallelogram with a base of $6\frac{1}{2}$ ft. and a height of 7 ft.

4. Calculate the area of the triangle with a base of 4.2 in. and a height of 2.7 in.

5. Calculate the area of the trapezoid.

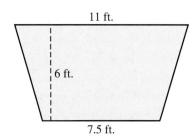

9.5 Applications with Graphing

OBJECTIVES

1 Solve problems involving linear equations in two variables.

2 Calculate the perimeter of a figure given the coordinates of its vertices.

3 Calculate the area of a figure given the coordinates of its vertices.

OBJECTIVE **1** *Solve problems involving linear equations in two variables.*

Linear equations in two variables often describe real situations. Graphing linear equations gives us insights into the situation being described.

EXAMPLE 1 The linear equation $v = -32.2t + 140$ describes the velocity of a ball thrown straight up. The variable t represents the time of the ball's flight in seconds and v represents the velocity of the ball in feet per second (ft./sec.).

a. Calculate the initial velocity of the ball.

Understand: Initial velocity means the velocity at the time the ball is released, which is 0 seconds.

Plan: Let $t = 0$ and solve for v.

Execute: $v = -32.2t + 140$

$$v = -32.2(0) + 140$$

$$v = 0 + 140$$

$$v = 140 \text{ ft./sec.}$$

Answer: The ball is traveling at 140 ft./sec. initially.

Check: Reverse the process. We will leave this to the reader.

▸**Discussion** What is this velocity in miles per hour?

> **Connection**
>
> Because one of the variables is equal to 0, this must be an intercept. Because t is the horizontal axis and v is the vertical axis, t and v values can be written as coordinates:
>
> $$(0, 140)$$
>
> This will be the v intercept on the vertical axis.

b. Calculate the velocity of the ball 2 seconds after being released.

Understand: We are given a time of 2 seconds.

Plan: Replace t with 2 and solve for v.

Execute: $v = -32.2(2) + 140$

$$v = -64.4 + 140$$

$$v = 75.6 \text{ ft./sec.}$$

> **Connection**
>
> Because $t = 2$ results in $v = 75.6$, we could write coordinates this way:
>
> $$(2, 75.6)$$

▸**Discussion** Why is the ball slowing down?

Answer: 2 seconds after being released, the ball is traveling only 75.6 ft./sec. Notice it has slowed down considerably.

Check: Reverse the process. Again, we will leave this to the reader.

c. How many seconds after the ball is released does it come to a stop before descending?

Understand: We must calculate the time at which the ball stops in midair. After stopping, the ball will then reverse direction and begin falling back toward the ground. When an object is at a stop, it is obviously not moving and therefore has no velocity. We must solve for t when the velocity, v, is 0.

Margin 1

The equation **p = 0.15r − 30,000** *describes the profit for a company, where* **r** *represents revenue.*

a. Calculate the profit if the revenue is $320,000

b. Calculate the revenue required to break even (the point at which profit is $0).

c. Graph the equation.

Plan: Replace v with 0 and solve for t.

Execute:

$$v = -32.2t + 140$$
$$0 = -32.2t + 140$$
$$0 = -32.2t + 140$$
$$\frac{-140}{-140} = \frac{-140}{-32.2t + 0}$$
$$\frac{-140}{-32.2} = \frac{-32.2t}{-32.2}$$
$$4.4 \approx 1t$$
$$4.4 \text{ sec.} \approx t$$

> **Connection**
>
> In this case, $v = 0$ and we solve for t. Because t will be shown on the horizontal axis, we have just found the t-intercept. The coordinates of the t-intercept would be written this way:
>
> $$(4.4, 0)$$

Answer: The ball will stop in midair approximately 4.4 sec. after being released.

Check: We could replace t with 4.4 in the original equation and verify that v would equal 0.

d. Graph the equation with t shown on the horizontal axis and v on the vertical axis.

Understand: We must graph the equation. To graph we need at least two solutions. In answers to questions a, b, and c, we actually found three solutions.

$$(0, 140) \qquad (2, 75.6) \qquad (4.4, 0)$$

Plan: Plot each solution in the rectangular coordinate system, then connect the points to form a straight line.

Execute:

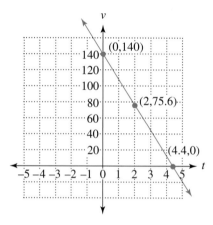

> **Discussion** Follow the line beyond the t-intercept. The line passes beyond the t-axis into quadrant IV. What does this say about the velocity?

◀ **Do Margin 1.**

OBJECTIVE 2 *Calculate the perimeter of a figure given the coordinates of its vertices.*

We can use coordinates to describe points that define the *corners* of a figure. A corner in a figure is called a *vertex*.

DEFINITION | **Vertex:** A point where two lines join to form an angle.

Points A, B, C, and D are the vertices of the figure below.

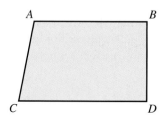

Answers:

a. $18,000

b. $200,000

c.

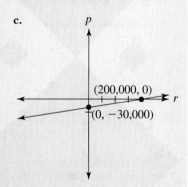

EXAMPLE 2 Calculate the perimeter of the figure with vertices $(3, 2), (-4, 2), (-5, -3)$, and $(3, -3)$.

Understand: We are to calculate the perimeter of a figure given the coordinates of its vertices.

Plan: First, we will plot the points to get a sense of the shape. Because perimeter is the total distance around the figure, we will need to determine the length of each side.

Execute:

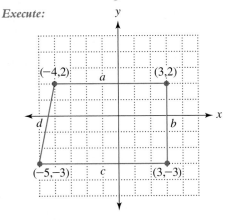

Notice the figure is a trapezoid. To aid discussion, we will label the sides $a, b, c,$ and d. Because side b is parallel to the y-axis, we can calculate its length by subtracting the y-coordinates. To avoid potential negative results for the length, we will use absolute value:

$$|2 - (-3)| = |2 + 3| = |5| = 5$$

Because sides a and c are parallel to the x-axis, we can calculate their lengths by subtracting the x-coordinates of the endpoints. Again, we find the absolute value of these differences to avoid negative lengths.

Side a: $|3 - (-4)| = |3 + 4| = |7| = 7$
Side c: $|3 - (-5)| = |3 + 5| = |8| = 8$

Note: The absolute value symbols allow us to subtract the coordinates in any order. Consider side a again. If we reverse the order of the subtraction, we still get the same length.

$$|-4 - 3| = |-7| = 7$$

Because side d is not parallel to an axis, we must use the distance formula to calculate its length. The endpoints are $(-4, 2)$ and $(-5, -3)$.

$$d = \sqrt{(-5 - (-4))^2 + (-3 - 2)^2}$$
$$d = \sqrt{(-1)^2 + (-5)^2}$$
$$d = \sqrt{1 + 25}$$
$$d = \sqrt{26}$$
$$d \approx 5.1$$

Now we can calculate the perimeter:

$$P = 5 + 7 + 8 + 5.1 = 25.1 \text{ units}$$

♦ *Answer:* The perimeter is 25.1 units.

Do Margin 2. ▶

OBJECTIVE 3 *Calculate the area of a figure given the coordinates of its vertices.*

Given the coordinates of the vertices of a figure, we can also find its area.

EXAMPLE 3 Calculate the area of the figure with vertices at $(4, 3), (-2, 3), (-4, -3)$, and $(2, -3)$.

Understand: This is a four-sided figure. We are to calculate area.

Plan: Plot the points to get a sense of the shape. Then calculate the lengths of the sides that are needed to calculate the area. Last, calculate the area.

◆ **Margin 2**

a. Calculate the perimeter of a figure with vertices $(3, 4), (3, -1), (-4, 4)$, and $(-4, -1)$.

b. Calculate the perimeter of a figure with vertices $(2, 3), (4, -5), (-4, 3)$, and $(-4, -5)$.

Answers: **a.** 24 units **b.** 30.25 units

a. Calculate the area of the figure with vertices $(5, 2), (3, -1), (-1, 2),$ and $(-3, -1)$.

Execute:

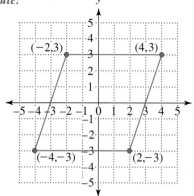

b. Calculate the area of the figure with vertices $(-1, 4), (-3, 4),$ $(-5, -5),$ and $(2, -5)$.

Notice the figure is a parallelogram. The formula for the area of a parallelogram is $A = bh$, so we need the base and height. Because the base is parallel to the x-axis, we can subtract the x-coordinates.

Base: $|2 - (-4)| = |2 + 4| = |6| = 6$

Recall that height is a vertical measurement from the base to the top of a shape. The line along which height is measured is parallel to the y-axis. Therefore, we can subtract the y-coordinates.

Height: $|3 - (-3)| = |3 + 3| = |6| = 6$

Now we can calculate the area:

$$A = 6 \cdot 6$$
$$A = 36 \text{ square units}$$

◆ *Answer:* The area is 36 square units.

◀ **Do Margin 3.**

Answers: **a.** 18 square units
b. 40.5 square units

9.5 Exercises

FOR EXTRA HELP

 Videotape 11

 InterAct Math
Tutorial Software

 www.carsonmath.com

 AWL Math Tutor Center

InterAct
MathXL www.mathxl.com

 Student's Solutions
Manual

For Exercises 1–10, use the given linear equation to answer the questions.

1. The linear equation $v = -32.2t + 600$ describes the velocity in ft./sec. of a rocket t seconds after being launched.
   a. Calculate the initial velocity of the rocket.
   b. Calculate the velocity after 3 seconds.
   c. How many seconds after launch will the rocket stop before returning to earth?
   d. Graph the equation with v as the vertical axis and t as the horizontal axis.

   Trivia Bite

   At a velocity of 7.9 km/sec., an object would just overcome the effects of gravity so that it would enter the lowest possible circular orbit around Earth. An object traveling between 7.9 km/sec. and 11.18 km/sec. would travel in an elliptical orbit around Earth. An object traveling 11.18 km/sec., or faster, would escape Earth's gravity altogether. What are these velocities in ft./sec.? mi./hr?

2. The linear equation $v = -32.2t + 86$ describes the velocity in ft./sec. of a ball t seconds after being thrown straight up.
   a. Calculate the initial velocity of the ball.
   b. Calculate the velocity after 1.5 seconds.
   c. How many seconds after launch will the ball stop before returning to Earth?
   d. Graph the equation with v on the vertical axis and t on the horizontal axis.

3. The equation $p = 0.24r - 45,000$ describes the profit for a company, where r represents revenue.
   a. Calculate the profit if the revenue is $250,000.
   b. Calculate the revenue required to break even (the point at which profit is $0).
   c. Graph the equation with p on the vertical axis and r on the horizontal axis.

4. The equation $p = 0.18r - 36,000$ describes the profit for a company, where r represents revenue.
   a. Calculate the profit if the revenue is $450,000.
   b. Calculate the revenue required to break even (the point at which profit is $0).
   c. Graph the equation with p on the vertical axis and r on the horizontal axis.

5. The equation $b = 13.5t + 300$ describes the final balance of an account t years after the initial investment is made.
   a. Calculate the initial balance (principal). (Hint: $t = 0$.)
   b. Calculate the balance after 5 years.
   c. Calculate the balance after 20 years.
   d. Graph the equation with b on the vertical axis and t on the horizontal axis.

6. The equation $b = 17.5t + 500$ describes the final balance of an account t years after the initial investment is made.
   a. Calculate the initial balance (principal). (Hint: $t = 0$.)
   b. Calculate the balance after 5 years.
   c. Calculate the balance after 20 years.
   d. Graph the equation with b on the vertical axis and t on the horizontal axis.

7. The equation $F = \frac{9}{5}C + 32$ is used to convert a temperature in °C to temperature in °F.
   a. Graph the equation with F on the vertical axis and C on the horizontal axis.
   b. What is the F-intercept?
   c. What is the C-intercept?
   d. What is F when C is 40?

8. The equation $C = \frac{5}{9}(F - 32)$ is used to convert a temperature in °F to temperature in °C.

  a. Graph the equation with C on the vertical axis and F on the horizontal axis.

  b. What is the F-intercept?

  c. What is the C-intercept?

  d. What is C when F is 68?

9. An HVAC service technician charges $9 per hour for labor plus a $58 flat fee for the visit.

  a. Write a linear equation that describes the total cost.

  b. What would be the total cost for $2\frac{1}{2}$ hr. of labor?

  c. How many hours of labor can the service technician work for a customer who only has $100 available?

  d. Graph the equation.

10. The weekly cost to produce a toy is $0.75 per unit plus a flat $4500 for lease, equipment, supplies, and other expenses. Let C represent the total cost and u represent the number of units produced.

  a. Write a linear equation that describes the total cost.

  b. What would be the total cost to produce 600 units?

  c. What would be the total cost to produce 800 units?

  d. Graph the equation.

For Exercises 11–30, calculate the perimeter and area of the figure with the given vertices. When a distance is an irrational number use a calculator to approximate the value to the nearest hundredth.

11. $(0, 0)$ $(0, 5)$ $(4, 5)$ $(4, 0)$

12. $(-2, 0)$ $(-2, 7)$ $(0, 7)$ $(0, 0)$

13. $(4, 1)$ $(-3, 1)$ $(-3, -5)$ $(4, -5)$

14. $(2, 5)$ $(-5, 5)$ $(-5, -3)$ $(2, -3)$

15. $(-1, 2)$ $(-6, 2)$ $(-6, -4)$ $(-1, -4)$

16. $(0, -3)$ $(6, -3)$ $(0, -7)$ $(6, -7)$

17. $(0, 0)$ $(2, 6)$ $(7, 6)$ $(5, 0)$

18. $(-8, 0)$ $(-5, 4)$ $(3, 4)$ $(0, 0)$

19. $(2, 5)$ $(3, -1)$ $(7, 5)$ $(8, -1)$

20. $(-5, 3)$ $(-2, 0)$ $(6, 3)$ $(9, 0)$

21. $(-3, 2)$ $(4, 2)$ $(6, -7)$ $(-1, -7)$

22. $(-6, 5)$ $(-4, -3)$ $(3, 5)$ $(5, -3)$

23. $(0, 0)$ $(0, 7)$ $(5, 7)$ $(8, 0)$

24. $(0, 0)$ $(1, 6)$ $(5, 6)$ $(5, 0)$

25. $(-3, 4)$ $(-5, 1)$ $(5, 4)$ $(5, 1)$

26. $(-2, 0)$ $(-2, -4)$ $(4, 0)$ $(7, -4)$

27. $(-3, -2)$ $(-2, 4)$ $(6, 4)$ $(7, -2)$

28. $(-1, 5)$ $(0, 0)$ $(4, 0)$ $(5, 5)$

29. $(0, 3)$ $(-3, -1)$ $(4, 3)$ $(5, -1)$

30. $(-2, -1)$ $(-4, 6)$ $(3, -1)$ $(4, 6)$

REVIEW EXERCISES

1. Calculate the mean, median, and mode of the scores.

$$88, 72, 91, 84, 88, 87, 90, 56, 71, 70, 95, 82, 70, 65$$

2. 20% of the people in a class received an A. If there were 45 people in the class, how many received an A?

3. Calculate the final balance in an account if $400 is invested at 4.8% compounded quarterly over two years.

4. A company allocates $\frac{3}{4}$ of its costs for employee wages. $\frac{2}{3}$ of the employee wages are for nonsalary wages (hourly wages). What portion of the company's total cost goes toward nonsalary wages?

5. The sum of two consecutive even integers is 106. What are the integers?

Puzzle Problem

Find the perimeter and area of the figure with vertices:

$(-2, 1)$ $(-2, -5)$ $(6, -5)$

$(6, -2)$ $(0, -2)$ $(0, 1)$

Summary

Defined Terms

Review the following terms and for those you do not know, study its definition on the page number next to it.

Section 9.1
Statistic *(p. 610)*
Mean, or average *(p. 610)*
Grade point average (GPA)
  (p. 611)

Grade point *(p. 611)*
Median *(p. 613)*
Mode *(p. 614)*

Section 9.2
Axis *(p. 624)*

Section 9.3
Coordinate *(p. 635)*
Quadrant *(p. 638)*

Section 9.4
y-intercept *(p. 649)*
x-intercept *(p. 649)*

Procedures, Rules, and Key Examples

Procedures/Rules	Key Example(s)
Section 9.1 **To find the arithmetic mean, or average of a given set of numbers:** 1. Calculate the sum of all the given numbers. 2. Divide the sum by the number of numbers.	Find the mean, or average, of the set of golf scores. 72 74 76 74 70 68 70 72 $$\bar{x} = \frac{72 + 74 + 76 + 74 + 70 + 68 + 70 + 72}{8}$$ $$\bar{x} = \frac{576}{8}$$ $$\bar{x} = 72$$
To find the median of a set of scores: 1. Arrange the scores in order from smallest to largest. 2. Locate the middle score of the ordered set of scores. **Note:** If there are an even number of scores in the set, the median will be the mean of the two middle scores.	Find the median of the test scores. 75 90 80 92 95 72 60 84 88 Arrange in order: 60 72 75 80 84 88 90 92 95 ↑ Median = 84 Suppose there were one more score: 60 72 75 80 84 88 90 92 95 98 ↑ $$\text{Median} = \frac{84 + 88}{2} = \frac{172}{2} = 86$$
To find the mode of a set of scores, count the number of repetitions of each score. The score with the most repetitions is the mode. **Note:** If no score is repeated, then there is no mode. If there is a tie, then list each score as a mode.	Find the mode of the GPAs listed. 3.5 3.875 2.75 3.0 3.5 2.25 3.5 Mode = 3.5 Find the mode of the prices listed. $24.95 $15.95 $29.99 $21.75 No mode Find the mode of the newborn babies' lengths. 21 in. 19.5 in. 18.5 in. 21 in. 19.5 in. 17 in. Modes: 21 in. and 19.5 in.

Procedures/Rules (cont.)	Key Example(s)

To determine the coordinates for a given point in the rectangular system:

1. Follow a vertical line from the given point to the *x*-axis (horizontal axis). The number at this position on the *x*-axis is stated as the first coordinate.
2. Follow a horizontal line from the given point to the *y*-axis (vertical axis). The number at this position on the *y*-axis is stated as the second coordinate.

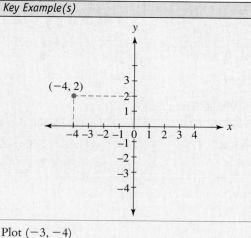

To graph or plot a point given a coordinate pair:

1. Beginning at the origin, use the first coordinate to travel right or left along the *x*-axis.
2. From that position on the *x*-axis, use the second coordinate to travel up or down.
3. Draw a dot to represent the point described by the coordinates.

Plot $(-3, -4)$

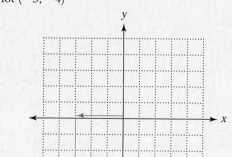

To determine the quadrant for a given coordinate, consider the signs of the numbers in the coordinate.

$(+, +)$ = Quadrant I

$(-, +)$ = Quadrant II

$(-, -)$ = Quadrant III

$(+, -)$ = Quadrant IV

Determine the quadrant in which each point is located.

$(42, 95)$ = Quadrant I

$(-65, 90)$ = Quadrant II

$(-91, -56)$ = Quadrant III

$(75, -102)$ = Quadrant IV

To calculate the distance between two points with coordinates (x_1, y_1) and (x_2, y_2), use the formula:

$$d = \sqrt{(x_2 - x_1)^2 + (y_2 - y_1)^2}$$

Find the distance between $(4, -9)$ and $(-12, -2)$.

$d = \sqrt{(x_2 - x_1)^2 + (y_2 - y_1)^2}$

$d = \sqrt{(-12 - 4)^2 + (-2 - (-9))^2}$

$d = \sqrt{(-16)^2 + (7)^2}$

$d = \sqrt{256 + 49}$

$d = \sqrt{305}$

$d \approx 17.46$

Procedures/Rules (cont.)	Key Example(s)

Section 9.4

To determine whether a given coordinate pair is a solution for an equation in two variables:

1. Replace the variables in the equation with the corresponding coordinates.
2. Verify that the equation is true.

Determine whether $(-6, 2)$ is a solution for $x + 5y = 4$.

$$x + 5y = 4$$
$$-6 + 5(2) \, ? \, 4$$
$$-6 + 10 \, ? \, 4$$
$$4 = 4 \quad \text{Yes, } (-6, 2) \text{ is a solution.}$$

Determine whether $(2.2, 4.8)$ is a solution for $y = 4x - 3$.

$$y = 4x - 3$$
$$4.8 \, ? \, 4(2.2) - 3$$
$$4.8 \, ? \, 8.8 - 3$$
$$4.8 \neq 5.8 \quad \text{No, } (2.2, 4.8) \text{ is not a solution.}$$

To find a solution to an equation in two variables:

1. Choose a value for one of the variables (any value).
2. Replace the corresponding variable with your chosen value.
3. Solve the equation for the value of the other variable.

Find a solution for $2x - y = -3$.

Choose $x = 1$

$$2(1) - y = -3$$
$$2 - y = -3$$
$$2 - y = -3$$
$$\underline{-2 \qquad -2}$$
$$0 - y = -5$$
$$\frac{-y}{-1} = \frac{-5}{-1}$$
$$1y = 5$$
$$y = 5 \qquad \text{Solution: } (1, 5)$$

To graph a linear equation:

1. Find at least two solutions to the equation.
2. Plot the solutions as points in the rectangular coordinate system.
3. Connect the points to form a straight line.

Graph $2x - y = -3$.

We found one solution above $(1, 5)$.
Two more solutions that can be found the same way as above are: $(0, 3)$ and $(-2, -1)$.

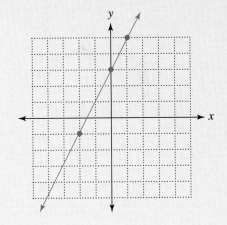

To find a *y*-intercept:
1. Replace *x* with 0 in the given equation.
2. Solve for *y*.

Find the coordinates of the *y*-intercept for the equation $3x + 5y = 8$.

Let $x = 0$.

$$3(0) + 5y = 8$$
$$0 + 5y = 8$$
$$5y = 8$$
$$\frac{5y}{5} = \frac{8}{5}$$
$$1y = \frac{8}{5}$$
$$y = 1\frac{3}{5}$$

Coordinates of the *y*-intercept: $\left(0, 1\frac{3}{5}\right)$

To find an *x*-intercept:
1. Replace *y* with 0 in the given equation.
2. Solve for *x*.

Find the coordinates of the *x*-intercept for the equation $3x + 5y = 8$.

Let $y = 0$.

$$3x + 5(0) = 8$$
$$3x + 5y = 8$$
$$3x = 8$$
$$\frac{3x}{3} = \frac{8}{3}$$
$$1x = \frac{8}{3}$$
$$y = 2\frac{2}{3}$$

Coordinates of the *x*-intercept: $\left(2\frac{2}{3}, 0\right)$.

For Exercises 1–6, answer true or false.

1. If the median income for a company is $24,000, most people in the company make around $24,000.

2. If the mode for a set of test scores is 85, 85 was the score received more often than any other score in the set of scores.

3. There can only be one mode for a set of scores.

4. $(-45, 60)$ is in quadrant IV.

5. The equation $x = 2$ is a vertical line with an x-intercept at $(2, 0)$.

6. The equation $y = -4$ has no x-intercept.

7. Explain in your own words how to find the mean of a set of numbers.

8. Explain in your own words how to find the median of a set of numbers.

9. Explain in your own words how to graph a linear equation.

10. Explain in your own words how to find the x- and y-intercepts of a linear equation.

11. Find the mean, median, and mode of the test scores below.

65	94	72	94	88	86
84	87	92	58	76	95
92	94	77	100	62	60
81	80	84	98	61	71

12. A new housing development is to have 11 different base floor plans to select from. Find the mean, median, and mode of the base prices for the houses in the development.

Savannah	$115,990	Jorden	$127,100
Charlestown	$123,190	Devon	$131,200
Richmond	$123,500	Lexington	$126,800
Dawson	$124,500	Kensington	$134,100
Maguire	$130,000	Karrington	$131,690
Cambridge	$138,250		

13. Donielle has the following test scores 86, 91, 92, 88. Her instructor will allow her to be exempt from the final exam if she has a test average of 90 or better. What must she score on the fifth test to have a test average of 90?

14. Calculate the grade point average.

	Credits	Grade
MATH 100	5.0	C
ENG 100	3.0	B+
HIS 101	3.0	A
CHM 101	4.0	A

For Exercises 15–20, use the table in Figure 9.9.

Figure 9.9 Percentage of the population in large industrialized countries who had completed secondary and higher education, by age, sex, and country: 1994

| | 25–64 years old | | 25–34 years old | | | | | |
| | Total | | Total | | Male | | Female | |
Country	Secondary ed.[1]	Higher ed.	Secondary ed.[1]	Higher ed.	Secondary ed.[1]	Higher ed.	Secondary ed.[1]	Higher ed.
Canada	74.1	16.9	82.3	18.4	80.6	18.0	84.0	18.9
France[2]	67.1	9.2	84.4	11.6	86.4	11.9	82.3	11.3
Germany	84.0	12.6	89.6	11.9	91.6	12.7	87.5	11.0
Italy	33.2	7.5	47.3	7.9	45.6	7.7	49.0	8.1
Japan[3]	69.7	13.3	90.6	22.9	89.3	34.2	91.8	11.5
United Kingdom	74.5	11.7	86.1	13.7	87.3	15.7	84.9	11.7
United States	85.1	24.4	86.4	23.4	85.2	23.4	87.5	23.5

1/ Includes individuals who have at least completed secondary education.
2/ France's definitions of ISCED levels were changed so that they are more similar to EUROSTAT definitions and are easier to compare to the definitions of other countries. As a result, data for 1994 are not directly comparable with data for other years.
3/ Data are for 1989.

Source: International comparisons of educational attainment by age; The condition of education, 1997. National Center for Education Statistics, U.S. Department of Education.

15. What percent of French people 25–64 years of age completed some form of higher education?

16. What percentage of females ages 25–34 completed secondary education in the United Kingdom?

17. Which country had the highest percentage of its total population 25–64 years of age complete some form of higher education?

18. Which country had the lowest rate of completion of secondary education for people from 25 to 64 years of age?

19. Which country had the highest percentage of its 25–34 age group complete secondary education?

20. Which country had the greatest difference in completion rates for higher education between men and women?

For Exercises 21–26, use the pie chart in Figure 9.10.

Figure 9.10 Highest level of education attained by persons 25 years and older: March 1996

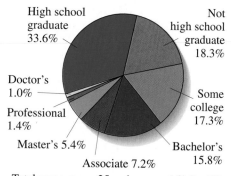

High school graduate 33.6%
Not high school graduate 18.3%
Doctor's 1.0%
Professional 1.4%
Master's 5.4%
Associate 7.2%
Bachelor's 15.8%
Some college 17.3%
Total persons age 25 and over = 168.3 million

Source: U.S. Department of Commerce, Bureau of the Census, Current Population Survey, unpublished data

21. What percent of people 25 years and older completed some college?

22. What percent of people 25 years and older did not complete high school?

23. How many people were 25 years or older in 1996?

24. How many people 25 years and older had completed a bachelor's degree as their highest level of education?

25. How many people 25 years and older had completed a master's degree as their highest level of education?

26. What is the combined number of doctor's degrees and professionals who are 25 years or older?

27. Below is an itemized list of a family's house payment. Use the data to construct a pie chart showing the percent of the total payment that goes toward each item.

Principal: $107.04
Interest: $672.58
Taxes: $129.74
Insurance: $50.08

For Exercises 28–32, use the bar graph in Figure 9.11.

Figure 9.11   Median annual income of persons 25 years old and over, by highest degree attained and gender: 1995

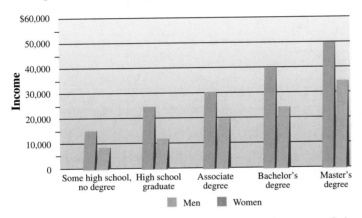

Source: U.S. Department of Commerce, Bureau of the Census, *Current Population Reports,* Series p-60, "Money Income in the United States: 1995"

28. What is the median income for men with some high school, but no degree?

29. What is the median income for women with a master's degree?

30. Which group of people had the lowest median income?

31. Which group of people had the highest median income?

32. Which group of people had a median income of about $30,000?

33. Refer to the data in Figure 9.9 for people in each country who are 25–64 years of age and who completed secondary school. Use these data to construct a bar graph with each country listed along the horizontal axis and the percentage that completed secondary education along the vertical axis.

Figure 9.12   Enrollment in institutions of higher education, by age: fall 1970 to fall 2007

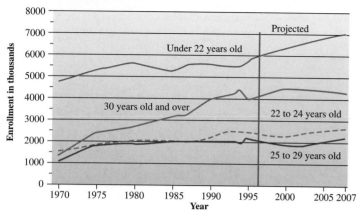

Source: U.S. Department of Education, National Center for Education Statistics, Higher Education General Information Survey (HEGIS), "Fall Enrollment in Institutions of Higher Education" survey; Integrated Postsecondary Education Data System (IPEDS), "Fall Enrollment" surveys; *Projections of Education Statistics to 2007;* and U.S. Department of Commerce, Bureau of the Census, *Current Population Reports,* Series P-20, "Social and Economic Characteristics of Students," various years

34. How many people under 22 enrolled in higher education in 1985?

35. How many people 25 to 29 years of age enrolled in higher education in 1995?

36. What is the projected number of people over 30 who will be enrolled in higher education in the year 2005?

37. Which age group had the greatest growth in enrollment from 1970 to 1990?

38. What is the difference between the number of students under 22 who were enrolled in higher education in 1970 and the number of students projected to be enrolled in 2007?

39. Below are the average daytime temperatures for each month in a certain city. Use the data to construct a line graph with each month along the horizontal axis and the temperatures along the vertical axis.

January:	35°F	May:	78°F	September:	80°F
February:	48°F	June:	83°F	October:	72°F
March:	65°F	July:	85°F	November:	54°F
April:	70°F	August:	88°F	December:	46°F

40. State the coordinates for each point shown on the graph.

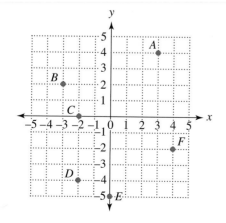

41. Plot the points in the rectangular coordinate system.

$(5, 2)$ $(2, 0)$ $(-4, -3)$ $(3, -5)$ $(0, -4)$

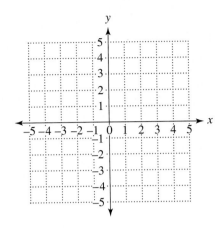

42. State the quadrant in which each point is located.

    a. $(-105, 68)$ **b.** $(-47, -158)$ **c.** $(95, 72)$ **d.** $(58, 0)$

43. Find the distance between $(2, 9)$ and $(6, 1)$. Round to the nearest hundredth.

44. Find the distance between $(-4, 2)$ and $(6, -3)$. Round to the nearest hundredth.

45. Determine whether $(1, 7)$ is a solution for $3x - y = -4$.

46. Determine whether $(2.5, -0.5)$ is a solution for $y = -x + 2$.

47. Determine whether $\left(-4\dfrac{3}{4}, -1\dfrac{9}{10}\right)$, is a solution for $y = \dfrac{2}{5}x$.

For Exercises 48–55, find three solutions, then graph.

48. $2x + y = 6$

49. $3x - 6y = 12$

50. $y = x + 3$

51. $y = -3x$

52. $y = \dfrac{2}{3}x$

53. $y = -x - 4$

54. $y = 7$

55. $x = -3$

For Exercises 56–59, find the coordinates for the x- and y-intercepts.

56. $5x + y = 10.$

57. $y = 4x - 1.$

58. $y = \dfrac{1}{5}x.$

59. $x = 6.$

60. The linear equation, $p = 0.4r - 12{,}000$ describes the profit for a company, where r is the revenue.

   a. If the company makes $430,680 in revenue, what is the profit?

   b. If the company has a profit of $80,250, what was the revenue?

   c. Graph the equation for profit.

61. The coordinates of the vertices of a figure are $(0, -1)$, $(0, 4)$, $(3, 4)$, and $(7, -1)$. Find the perimeter of the figure.

62. The coordinates of the vertices of a figure are $(-2, 5)$, $(-1, -3)$, $(4, -3)$, and $(3, 5)$. Find the area of the figure.

Practice Test

1. Calculate the mean, median, and mode of the set of test scores.

88	85	78	62	94	52
80	70	90	76	84	98
84	74	96	80	96	82

2. Below is a list of prices of houses for sale in a particular area. Calculate the mean, median, and mode of the prices.

$76,500	$88,700	$96,400
$98,200	$110,000	$105,000
$94,500	$90,250	$102,000

3. Steve has the following test scores: 84, 88, 95, and 85. His instructor will allow him to be exempt from the final exam if he has a test average of 90 or better. What must he score on the fifth test to have a test average of 90?

Use the pie chart to answer questions 4–6. The chart shows the percentage of a family's net monthly income.

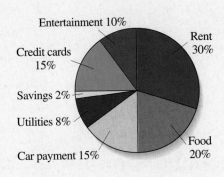

Entertainment 10%
Rent 30%
Credit cards 15%
Savings 2%
Utilities 8%
Food 20%
Car payment 15%

4. What percent of the family income goes toward credit cards?

5. If this family has a net monthly income of $2206.25, how much goes toward food?

6. How much money goes toward the credit cards and car payment combined?

Use the bar graph in Figure 9.13 to answer questions 7–9.

Figure 9.13 **Full-time–equivalent students per staff member in public and private institutions of higher education: 1976 and 1993**

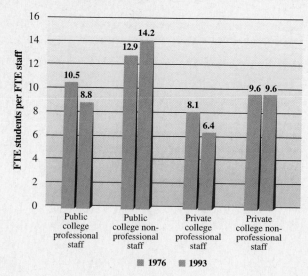

Source: U.S. Department of Education, National Center for Education Statistics, Higher Education General Information Survey (HEGIS), "Staff" survey, and Integrated Postsecondary Education Data System (IPEDS), "Staff" surveys

7. _____

7. Which type of college had the highest number of students per faculty member?

8. _____

8. How many FTE students were there per faculty member at private colleges with professional staff in 1993?

9. _____

9. How many FTE students were there per faculty member at public colleges with professional staff in 1976?

Use the line graph in Figure 9.14 to answer questions 10–12.

Figure 9.14 **Total private average weekly hours, 1988–1999 (seasonally adjusted)**

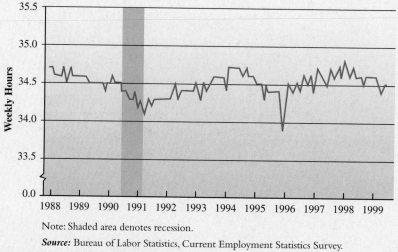

Note: Shaded area denotes recession.

Source: Bureau of Labor Statistics, Current Employment Statistics Survey.

10. What year had the lowest number of private average weekly hours?

11. What was the average number of weekly work hours at the beginning of 1999?

12. What was the average number of weekly work hours at the beginning of 1996?

13. Below are the results of a poll. Construct a pie chart showing each response as a percentage of total responses.

Strongly agree: 226

Agree: 121

Neutral: 80

Disagree: 65

Strongly disagree: 48

14. Determine the coordinates for each point in the rectangular system shown.

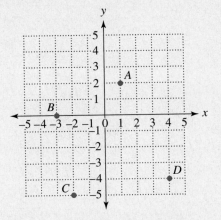

15. In which quadrant is $(-16, 48)$ located?

16. Plot the points $(-4, 2)$, $(2, 1)$, $(0, -3)$, and $(-3, -5)$.

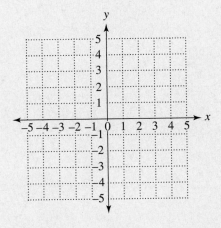

17. _____

17. Calculate the distance between the points $(4, -2)$ and $(-3, -7)$. If the distance is an irrational number, give an exact radical and a decimal approximation rounded to the nearest tenth.

18. _____

18. Determine whether $(5, -3)$ is a solution for $-x + 2y = -11$.

19. _____

19. Find the x- and y-intercepts for $3x - 5y = 6$. Find the coordinates for the x- and y-intercepts.

20. Graph $y = -3x$.

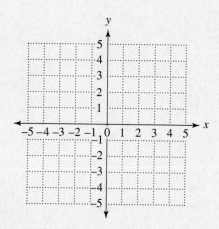

21. Graph $y = \dfrac{1}{5}x - 3$.

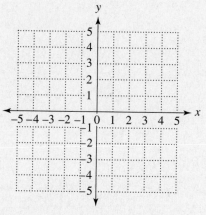

22. Graph $x + y = 7$.

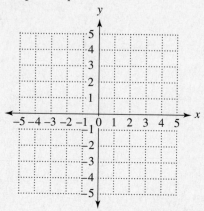

23. Graph $3x - y = 2$.

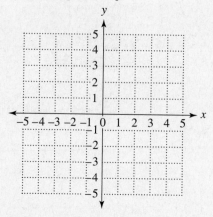

24. _____

24. The equation $v = -9.8t + 40$ describes the velocity in m/sec. of a model rocket t seconds after being launched.
   a. What is the initial velocity of the rocket?
   b. What is the velocity of the rocket 3 seconds after being launched?
   c. After how many seconds will the rocket stop in midair before returning to Earth?
   d. What is the velocity of the rocket 10 seconds after being launched?

25. _____

25. The coordinates of the vertices of a figure are $(4, 3)$, $(-2, 3)$, $(-2, -5)$, and $(4, -5)$.
   a. Find the perimeter of the figure.
   b. Find the area of the figure.

Cumulative
Review Exercises

For Exercises 1–12, answer true or false.

1. 0 is a rational number.

2. 41 is a prime number.

3. $5 - 8 = 8 - 5$

4. $6^0 = 0$

5. $5x^2 + 2$ is a monomial.

6. The degree of 9 is 1.

7. 12 is the GCF of 36 and 24.

8. $4x = 9(x + 1.5)$ is a linear equation.

9. $-9.2 \div 0 = 0$

10. $\dfrac{2.5}{9} = \dfrac{6}{21.6}$

11. $\dfrac{3}{8} > 37\%$

12. $(-5, 3)$ is in the III quadrant.

13. State the order of operations agreement.

14. Explain in your own words how to add two numbers that have the same sign.

15. Explain in your own words how to add two numbers that have different signs.

16. Explain in your own words how to write a subtraction statement as an equivalent addition statement.

17. When multiplying or dividing two numbers that have the same sign, the result is _____.

18. When multiplying or dividing two numbers that have different signs, the result is _____.

19. Explain in your own words how to solve a linear equation in one variable.

20. Explain in your own words how to graph a linear equation.

21. Write 24,607 in expanded form.

22. Write the word name for 4607.09.

23. Round 79,804.652 to the nearest:
   a. ten thousand
   b. thousand
   c. whole number
   d. tenth
   e. hundredth

24. Estimate each calculation by rounding so that there is only one nonzero digit.
   a. $21,459 + 6741$

   b. $729,105 - 4519$

   c. 86×17

   d. $2219 \div 45$

25. Write 27,500,000,000 in scientific notation.

26. Write 0.00068 in scientific notation.

27. Graph on a number line.

    a. $-(-3)$ **b.** $5\frac{1}{4}$ **c.** -7.2

28. What is the degree of each expression?

    a. $-4y$ **b.** $15x^2$ **c.** $4x^3 - 9x + 10x^4 - 7$

For Exercises 29–21, simplify.

29. $(-3)^4$ **30.** -3^4

31. $(6)^{-2}$ **32.** $(-2)^{-3}$

33. $9 - 4|3 - 8| - \sqrt{49}$ **34.** $19 - (6 - 8)^3 + \sqrt{100 - 36}$

35. $\dfrac{14 + (18 - 2 \cdot 4)}{4^2 - 10}$ **36.** $9\frac{3}{4} - \left(-2\frac{2}{3}\right)$

37. $-6\frac{1}{8} \div \dfrac{3}{16} - \dfrac{3}{4}$ **38.** $-11.7 \div 1.8 + 2.6(7.5)$

39. $\left(\dfrac{3}{4}\right)^2 (-8.4)$ **40.** $\dfrac{3}{5} - (0.2)^3$

41. $\sqrt{(27)(0.03)}$

42. Approximate $\sqrt{126}$ to the nearest hundredth.

43. Write as a decimal number.

    a. $\dfrac{3}{8}$ **b.** $\dfrac{2}{3}$ **c.** 8%

    d. $24\frac{2}{5}$ %   **e.** 4.5%

44. Write as a fraction.

    a. $5\frac{1}{6}$ **b.** 0.145   **c.** 6%

    d. $5\frac{1}{2}$%   **e.** 16.4%

45. Write as a percent.

    a. $\dfrac{1}{4}$ **b.** $\dfrac{4}{9}$ **c.** 0.65

    d. 0.035   **e.** 2.3

46. Evaluate each expression using the given values.
    a. $t^3 - 5t$ when $t = -4$

    b. $-xy + 2\sqrt{x + y}$, when $x = -4$ and $y = 20$

47. Combine like terms and write the resulting polynomial in descending order of degree.

$$9x^3 - 15.4 + 8x - 12x^3 + 4x^2 + x^4 - 4.9 - x$$

48. Add. $(5n^2 + 8n - 9.3) + (4n^2 - 12.2n - 3.51)$

49. Subtract. $\left(x^4 - \dfrac{1}{3}x^2 + 9.6\right) - \left(4x^3 + \dfrac{3}{5}x^2 - 14.6\right)$

50. Multiply.

    a. $\left(\dfrac{5}{8}a^2b\right)\left(-\dfrac{4}{9}abc\right)$ **b.** $(m - 9)(m + 9)$ **c.** $(4.5t - 3)(7.1t + 6)$

51. Find the prime factorization of 630.   **52.** Find the GCF of $30n^4$ and $45mn^2$.

53. Find the LCM of $36x^3$ and $24xy$.   **54.** Divide. $30y^5 \div 6y^2$

55. Factor. $28b^5 + 24b^3 - 32b^2$

56. Simplify.

    a. $-\dfrac{9x^3}{10} \cdot \dfrac{25y}{12x}$ **b.** $\dfrac{10m}{9n^2} \div \dfrac{5}{12n}$ **c.** $\dfrac{3}{5} - \dfrac{y}{4}$

For Exercises 57–64, solve and check.

57. $m - 25 = -31$ **58.** $-6x = 54$

59. $-4x + 19 = -1$ **60.** $\dfrac{3}{5}k - 7 = \dfrac{1}{4}$

61. $6.5h - 16.4 = 9h + 3.2$ **62.** $2.4(n - 5) + 8 = 1.6n - 14$

63. $\dfrac{u}{14} = \dfrac{-3}{8}$ **64.** $\dfrac{4}{5} = \dfrac{1\frac{3}{5}}{n}$

65. Convert.

  a. 12 ft. to inches

  b. 78 ft. to yards

  c. 2.4 mi. to feet

  d. 4.8 m to centimeters

  e. 12.5 km to meters

  f. 4500 mm to meters

66. Convert.

  a. 30 ft.^{2} to square yards

  b. 4 yd.^{2} to square feet

  c. 504 in.^{2} to square feet

  d. 0.6 m^2 to square centimeters

  e. 0.085 km^2 to square meters

  f. 42,000 cm^2 to square meters

67. Convert.

  a. 9.5 lb. to ounces

  b. 4500 lb. to tons

  c. 0.23 kg to grams

  d. 9600 hg to metric tons

68. Convert.

  a. 4 gal. to pints

  b. 68 cups to quarts

  c. 400 ml to liters

  d. 0.9 cl to cubic centimeters

69. Convert .

  a. 77°F to degrees Centigrade

  b. −20°C to degrees Fahrenheit

For Exercises 70–100, solve.

70. A parallelogram has an area of 411.4 m^2. If the base is 24.2 m, what is the height? What is the height in feet?

71. An alarm system has a three-digit code. How many possible codes are there?

72. A family has the following assets and debts. Calculate their net worth.

Assets	Debts
Savings = $1282	Credit card balance = $2942
Checking = $3548	Mortgage = $83,605
Furniture = $21,781	Automobile 1 = $4269
Jewelry = $5745	

73. A person weighs 148 lb. What is their mass in slugs? (Hint: The acceleration due to gravity is 32.2 ft./sec.^{2}.) Convert to kilograms.

74. Write an expression in simplest form for the area. Calculate the area if $h = 4.2$ in.

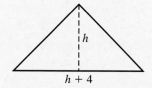

75. A company's revenue is expressed by the polynomial $3.5x + 600$, and the cost is expressed by $0.2x + 4000$, where x represents the number of units sold.
  a. Write an expression in simplest form for the profit.
  b. If the company sells 1500 units in one week, what is the net?

76. The sum of two consecutive integers is 123. What are the integers?

77. The sum of the angles in any triangle is $180°$. Suppose we have a triangle with the second angle measuring $10°$ more than the first and the third angle measuring $7°$ less than the first. What are the three angle measurements?

78. Jeremy has some \$5 bills and \$10 bills in his wallet. If he has a total of 17 bills worth a total of \$140, how many of each bill is in his wallet? (Use a four-column table.)

79. On Sandra's phone bill, 24 out of the 32 long distance calls were in state. $\frac{3}{4}$ of the in-state calls were to her parents. What fraction of all her long distance calls were to her parents? How many calls were to her parents?

80. A poll is taken to assess the president's approval rating. Respondents can answer four ways: excellent, good, fair, or poor. $\frac{1}{6}$ said "excellent," $\frac{3}{5}$ said "good," and $\frac{1}{8}$ said "fair."
  a. What fraction of the respondents said "excellent" or "good?"
  b. What fraction said "poor"?

81. Alan and Jessica pass each other going in opposite directions. If Alan walks at 3 mph and Jessica at 2 mph, how long will it take them to be $\frac{1}{4}$ mi. apart? (Use a four-column table.)

82. Find the area of the shaded region.

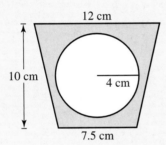

83. Calculate the surface area of a box that is 4.5 ft. by 6 ft. by 3 ft.

84. Calculate the volume of a can that is 3.5 in. tall and has a 3 in. radius.

85. Calculate the volume of a ball with a 12 in. diameter.

86. What is the volume of a cone that is formed from a blow torch flame with a 0.8 cm radius and a 12 cm height?

87. A steel support beam on a tower needs to be replaced. It is known that the connecting joint is 20 ft. above the ground, and the base of the support beam is 8 ft. from a point directly below the connecting joint. How long is the beam?

88. What is the probability of winning a drawing if you entered your name 41 times and the total entries is 650?

89. Which is the better buy?
  15.5 oz. of cereal at \$2.89   or   20 oz. of the same cereal at \$3.45

90. The triangles below are similar. Find the missing side lengths.

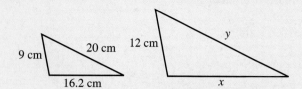

91. Because Coly is an employee of a store, he gets a 20% discount when he purchases merchandise. What would be his cost for an item that is normally $75.95?

92. A restaurant increases its price on a dish from $8.95 to $9.75. What was the percent increase?

93. Katrina answered 45 questions out of 70 correctly. What percent of the total questions did she answer correctly?

94. How much interest is earned if $680 is invested at 4.2% for half a year?

95. $3200 is invested at 6% APR compounded semiannually. What will be the balance after 2 years?

96. Calculate the mean, median, and mode for the following set of ages of participants in a survey.

$$28, 21, 45, 65, 54, 38, 39, 35, 36, 32, 40$$

97. Use the data below to construct a pie chart showing the percentages of a total house payment that was spent on each item.

Principal = $158.25

Interest = $496.78

Taxes = $125.50

Insurance = $38.50

98. Calculate the perimeter and area of a figure with vertices $(2, 4), (-5, 4), (-2, -3)$, and $(2, -3)$.

99. Graph.
   a. $x - y = 6$
   b. $y = -3x + 2$
   c. $y = \dfrac{-1}{2} x$
   d. $y = 2$

100. The linear equation $v = -32.2t + 580$ describes the velocity in ft./sec. of a rocket t seconds after being launched.
   a. Calculate the initial velocity of the rocket.
   b. Calculate the velocity after 5 seconds.
   c. How many seconds after launch will the rocket stop before returning to Earth?

Answers

Exercise Set 1.1
1. 0 **3.** 5 **5.** Thousands **7.** Hundred thousands **9.** $2 \times 10,000 + 4 \times 1000 + 3 \times 100 + 1 \times 10 + 9 \times 1$
11. $5 \times 1,000,000 + 2 \times 100,000 + 1 \times 10,000 + 3 \times 1000 + 3 \times 100 + 4 \times 1$ **13.** $9 \times 10,000,000 + 3 \times 1,000,000$ $+ 1 \times 10,000 + 4 \times 1000 + 8 \times 1$ **15.** 8792 **17.** 6,039,020 **19.** 40,980,109 **21.** Seven thousand, seven hundred sixty-eight **23.** Two hundred seventy-two million, one hundred ninety thousand **25.** One hundred eighty-six thousand, one hundred seventy-one. **27.** $599 < 899$ **29.** $4,299,308 > 4,298,308$ **31.** $609,001 = 609,001$ **33.** 5,652,992,000
35. 5,653,000,000 **37.** 6,000,000,000 **39.** 5,652,992,500 **41.** 30,000 **43.** 900,000 **45.** 9000 **47.** 93,000,000
(answers may vary)

Exercise Set 1.2
1. 9000; 8849 **3.** 101,000; 100,268 **5.** 17,000; 16,014 **7.** 14,500; 15,246 **9.** 142,500; 181,383 **11.** 5400; 5352
13. 34,200; 34,161 **15.** 45,000; 45,334 **17.** 20,000; 26,519 **19.** 160,000; 166,488 **21.** 4 **23.** 21 **25.** 18
27. 155 **29.** 225 **31.** 1810 **33.** 590 ft. **35.** 2100 K **37.** $6611 **39.** 33 A

Review Exercises:
1. 6 **2.** $3 \times 100,000,000 + 7 \times 1,000,000 + 4 \times 100,000 + 9 \times 10,000 + 1 \times 1000 + 2 \times 10 + 4 \times 1$
3. One million, four hundred seventy-two thousand, three hundred fifty-nine **4.** $12,305 < 12,350$ **5.** 23,410,000

Exercise Set 1.3
1. 352 **3.** 0 **5.** 400; 432 **7.** 4500; 4914 **9.** 80,000; 93,726 **11.** 44,940 **13.** 995,330 **15.** 8,550,756
17. 241,356,618 **19.** 330,050 **21.** 16 **23.** 1 **25.** 243 **27.** 10,000,000 **29.** 42,719,296 **31.** 130,691,232
33. 9^4 **35.** 7^5 **37.** 14^5 **39.** $2 \times 10^4 + 4 \times 10^3 + 9 \times 10^2 + 2 \times 1$ **41.** $9 \times 10^6 + 1 \times 10^5 + 2 \times 10^4 + 8 \times$ $10^3 + 2 \times 10$ **43.** $4 \times 10^8 + 7 \times 10^6 + 2 \times 10^5 + 1 \times 10^4 + 9 \times 10^2 + 2 \times 10 + 5 \times 1$ **45.** 7954 intersections; 63,632 lights **47.** 36,792,000; 2,796,192,000 **49.** 740 mg **51.** $1750 **53.** 17,280,000,000 oz **55.** 112
57. 32; No **59.** 128 **61.** No, too high; 624 sq. ft. **63.** 300 sq. ft.

Review Exercises:
1. Sixteen million, five hundred seven thousand, three hundred nine **2.** $2 \times 10,000 + 3 \times 1000 + 5 \times 100 + 6 \times 1$
3. 56,988 **4.** 831,724 **5.** 13

Exercise Set 1.4
1. 1 **3.** 0 **5.** Indeterminate **7.** Undefined **9.** No **11.** Yes **13.** Yes **15.** Yes **17.** No **19.** Yes
21. Yes **23.** Yes **25.** No **27.** 426 **29.** 207 r3 **31.** 217 **33.** 246 r3 **35.** 307 r5 **37.** 230 **39.** 1900
41. Undefined **43.** 1601 r12 **45.** 6 **47.** 8 **49.** 0 **51.** 7 **53.** 42 **55.** 20 **57.** 2854 **59.** 14
61. 125 **63.** 45 (can only buy whole stamps) **65.** 9600 boxes; 1200 bundles; 50 pallets; 2 trucks (28 in the first truck and 22 in the second) **67.** 18 ft. **69.** 22 sections **71.** 10 **73.** 13 **75.** 0 **77.** 1 **79.** 7 **81.** 14 **83.** 37
85. 214 **87.** 6 in. by 6 in. **89.** 144 ft.^{2}; 12 ft. by 12 ft.

Review Exercises:
1. 177 **2.** 60 ft. **3.** 14,800; 15,414 **4.** 625 **5.** $4 \times 10^4 + 9 \times 10^3 + 6 \times 10^2 + 2 \times 1$

Exercise Set 1.5
1. 22 **3.** 45 **5.** 141 **7.** 1 **9.** 22 **11.** 7 **13.** 4 **15.** 25 **17.** 81 **19.** 37 **21.** 47 **23.** 30
25. 51 **27.** 0 **29.** 51 **31.** 10 **33.** 13 **35.** 1354 **37.** The mistake is subtracting 6 from 48 before multiplying 6(5). Correct: 18 **39.** The mistake is squaring 3 and 5 when 3 and 5 should have been added first because of the parentheses. Correct: 50

Review Exercises:
1. 70,000 **2.** 104,187,127 **3.** 1250 r3 **4.** 56 ft.; $112 **5.** 192 tiles

Exercise Set 1.6
1. 94 cm **3.** 128 in. **5.** 64 km **7.** 288 m^2 **9.** 144 in.^{2} **11.** 112 km^2 **13.** 280 ft.^{3} **15.** 216 in.^{3}
17. 210 km^3 **19.** 9 m **21.** 62 in. **23.** 20 in. **25.** 16 ft. **27.** 18 in. **29.** 4 ft.

Review Exercises:
1. $12,833 **2.** 224 ft.^{2} **3.** $512 **4.** 120 **5.** 300

Exercise Set 1.7
1. $288 **3.** $28 **5.** $5525 **7.** $1612 **9.** 104 cm^2 **11.** 70 sheets; $1050 **13.** 648 ft.^{2}; 2 gallons
15. 14,400 ft.^{2}; $28,800; The estimate does not take into account wall thickness nor spaces that may not need carpet like bathrooms, closets, etc. **17.** 8,400,000 ft.^{3} **19.** 118,110,720 ft.^{3}

Review Exercises:
1. $2 \times 10^6 + 4 \times 10^5 + 8 \times 10^3 + 7 \times 10 + 3 \times 1$ **2.** 1929 **3.** 35,000,000; 32,183,756 **4.** 203 **5.** 53

Chapter 1 Review Exercises:
1. False **2.** True **3.** False **4.** True **5.** True **6.** False **7.** Replace x with 8, and verify that the sum is 17.
8. Cannot be checked. Any number multiplied by 0 should be 0 not 14. **9.** The mistake is adding 17 and 3 before multiplying 3(7) **10.** The square of a number is the number multiplied by itself. The square root of a number is the base that can be squared to equal the given number. **11.** $5 \times 10^6 + 6 \times 10^5 + 8 \times 10^4 + 9 \times 10^2 + 1 \times 1$ **12.** $4 \times 10^4 + 2 \times 10^3 + 5 \times 10^2 + 1 \times 10 + 9 \times 1$ **13.** 98,274 **14.** 8,020,096 **15.** 700,928,006 **16.** Forty-seven million, six hundred nine thousand, two hundred four **17.** Nine thousand, four hundred twenty-one **18.** One hundred twenty-three million, four hundred five thousand, six hundred **19.** $14 < 19$ **20.** $2930 > 2899$ **21.** 5,690,000 **22.** 3,000,000
23. 57,000; 52,721 **24.** 39,530; 39,673 **25.** 440,000; 485,716 **26.** 7700; 7707 **27.** 25 **28.** 189 **29.** $15,343
30. 22A **31.** 1,500,000; 1,409,437 **32.** 600,000; 403,500 **33.** 128 **34.** 125 **35.** 10^5 **36.** 7^8 **37.** 64
38. 288 ft.^{2} **39.** 4000; 4127 **40.** 2666 r20; 3209 r17 **41.** 24 **42.** 306 **43.** 2 ml **44.** $2438 **45.** 14
46. 15 **47.** 4 **48.** 51 **49.** 14 **50.** 10 **51.** 168 m^2 **52.** 378 ft.^{3} **53.** 16 ft. **54.** 15 ft. by 15 ft. **55.** 18 ft.
56. 89,500 ft.^{2}

Chapter 1 Practice Test:
1. 5 **2.** $4 \times 10^7 + 8 \times 10^6 + 2 \times 10^5 + 1 \times 10^4 + 9 \times 10^2 + 7 \times 1$ **3.** $19,304 > 9,204$ **4.** Sixty-seven million, one hundred ninety-four thousand, two hundred ten **5.** 2,800,000 **6.** 68,190 **7.** 425,709 **8.** 8 **9.** 246,687
10. 64 **11.** 1607 r23 **12.** 18 **13.** 14 **14.** 34 **15.** 22 **16.** 18 **17.** 28 **18.** 1 r16 **19.** 162 ft.^{2}
20. $20,612 **21.** 260 **22.** 83 classes **23.** 290 ft.; $1740 **24.** 336 ft.^{3} **25.** 123 ft.^{2}

Exercise Set 2.1
1. 450 **3.** $-13,200$ **5.** 40 **7.** -32 **9.** -78 **11.**

13. **15.** **17.** **19.** $-20 < 17$
21. $-30 < -16$ **23.** $0 > -16$ **25.** $-19 < -17$ **27.** 26 **29.** 18 **31.** 0 **33.** 14 **35.** 18 **37.** 0
39. 47 **41.** 377 **43.** 18 **45.** -61 **47.** 0 **49.** -8 **51.** 43 **53.** -6 **55.** 14 **57.** 0 **59.** -63
61. -4 **63.** 7 **65.** -87 **67.** 4 **69.** 8 **71.** 14 **73.** -5 **75.** -4

Review Exercises:
1. 94,265 **2.** 864,488 **3.** 278,313 **4.** 2031 r12 **5.** $482

Exercise Set 2.2
1. 23 **3.** -23 **5.** -15 **7.** 46 **9.** -76 **11.** 10 **13.** -14 **15.** 14 **17.** -18 **19.** 56 **21.** 138
23. -72 **25.** 61 **27.** 27 **29.** -21 **31.** -21 **33.** 0 **35.** 34 **37.** 32 **39.** 0 **41.** -59 **43.** $-$55,551
45. $-$33 **47.** $-78°$F **49.** -78 ft. **51.** 257 lbs

Review Exercises:
1. $4 \times 10^7 + 2 \times 10^6 + 5 \times 10^5 + 6 \times 10^4 + 1 \times 10^3 + 9 \times 1$ **2.** Two million, four hundred seven thousand, six
3. $7265 **4.** 55,445 **5.** 13

Exercise Set 2.3
1. $18 + (-25) = -7$ **3.** $-15 + (-18) = -33$ **5.** $20 + 8 = 28$ **7.** $-14 + 18 = 4$ **9.** $-15 + 8 = -7$
11. $0 + 5 = 5$ **13.** -40 **15.** 15 **17.** -13 **19.** 45 **21.** -12 **23.** -18 **25.** -11 **27.** -6
29. 15 **31.** 13 **33.** 14 **35.** -5 **37.** -12 **39.** $-$119; $-$139 **41.** $2,222,570 (profit) **43.** $-$21,958 (loss)
45. 46°F **47.** 55°C **49.** $67 **51.** 1340 ft.

Review Exercises:
1. 30,305 **2.** 243 **3.** 4 sq. mi. **4.** 8 **5.** 15

Exercise Set 2.4
1. -63 **3.** -65 **5.** 0 **7.** 68 **9.** -15 **11.** -380 **13.** -35 **15.** 54 **17.** 120 **19.** 49 **21.** -64
23. 81 **25.** 64 **27.** -64 **29.** $-1,000,000$ **31.** -12 **33.** -3 **35.** 8 **37.** 0 **39.** -31 **41.** -5
43. Undefined **45.** 31 **47.** -7 **49.** -2 **51.** 9 **53.** Undefined **55.** -17 **57.** 0 **59.** -5 **61.** 9
63. ±5 **65.** ±9 **67.** ±14 **69.** No integer solution **71.** 9 **73.** 8 **75.** 1 **77.** Not an integer **79.** -11
81. 0 **83.** $-$642 **85.** $-$119 **87.** -1570 N **89.** 800 kg **91.** -63 V **93.** -44 A

Review Exercises:
1. 60 ft. **2.** 138 ft.^{2} **3.** 504 ft.^{3} **4.** 9 **5.** 40

Exercise Set 2.5
1. -15 **3.** -6 **5.** 84 **7.** -29 **9.** 4 **11.** 16 **13.** -32 **15.** -9 **17.** 38 **19.** -19 **21.** -60
23. -24 **25.** 0 **27.** -5 **29.** -32 **31.** -40 **33.** -22 **35.** 23 **37.** Undefined **39.** -17 **41.** 22

43. -64 **45.** Not an integer **47.** 2 **49.** -1 **51.** 3 **53.** -2 **55.** Undefined **57.** 1 **59.** $(-2)^4$ means $(-2)(-2)(-2)(-2)$ while -2^4 means $-(2 \cdot 2 \cdot 2 \cdot 2)$. **61.** Because the exponent is odd. **63.** The mistake is subtracting 5 from 28 before multiplying $5(-6)$; correct: 58 **65.** The mistake is raising 4 to the 2nd power before calculating $9 - 4$ within the parentheses; correct: -21 **67.** The mistake is dropping the brackets before all calculations within have been completed; 41 **69.** The mistake is calculating the square root of 169 and 25 instead of subtracting first; 12

Review Exercises:
1. $-\$23$ **2.** $-\$213$ (loss) **3.** 73 ft. **4.** $-\$705$ **5.** -250 N

Exercise Set 2.6
1. $-\$15,030$ (loss) **3.** \$4576 (profit) **5.** Taxable $= \$17,683$; Tax $= \$2251$; She owes \$228. **7.** Taxable $= \$30,313$; Tax $= \$3749$; Refund of \$1054 **9.** At least 2912 lbs **11.** At least 2560 lbs **13.** -78 lbs **15.** -156 lbs **17.** -55 lbs **19.** -1110 lbs **21.** 22 ft./s **23.** -1440 ft. **25.** 65 mi./hr

Review Exercises:
1. 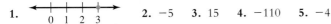 **2.** -5 **3.** 15 **4.** -110 **5.** -4

Chapter 2 Review Exercises
1. True **2.** False **3.** False **4.** True **5.** False **6.** True **7.** Subtract and keep the sign of the number with the larger absolute value. **8.** Change the operation sign from minus to plus and change the subtrahend to its additive inverse.

9. Negative **10.** Negative **11.** $-13,000$ **12.** 212 **13.** a.  ;
b. **14.** a. $-15 < 0$; b. $-26 > -35$; c. $12 > -41$ **15.** 41 **16.** 16 **17.** -27
18. 17 **19.** -26 **20.** -12 **21.** 9 **22.** -26 **23.** -12 **24.** -7 **25.** -37 **26.** 6 **27.** -5 **28.** -54
29. 60 **30.** 36 **31.** -30 **32.** 81 **33.** -1000 **34.** -6 **35.** 7 **36.** Undefined **37.** Not an integer
38. -7 **39.** -12 **40.** -32 **41.** -15 **42.** 43 **43.** 16 **44.** -16 **45.** 4 **46.** -2 **47.** -3 **48.** -41
49. 5 **50.** -9 **51.** -8 **52.** ± 12 **53.** $-\$72,083$ **54.** 0 lbs; The block is not moving. **55.** \$719,600 (profit)
56. 31°F **57.** \$70 **58.** -576 lbs **59.** 36 V **60.** \$178 **61.** Boulder weighs 2400 lbs; The bulldozer should be able to move it. **62.** 195 mi. **63.** 60 mph **64.** Taxable $= \$21,298$; Tax $= \$3191$; She owes \$675.

Chapter 2 Practice Test
1.  **2.** 26 **3.** -18 **4.** -12 **5.** -45 **6.** -14 **7.** -35 **8.** -16
9. 44 **10.** $-\$452$ **11.** \$4952 (profit) **12.** -108 **13.** -84 **14.** 4 **15.** -9 **16.** -64 **17.** ± 9
18. 138 kg **19.** 124 mi. **20.** -2 **21.** -1 **22.** 18 **23.** -27 **24.** Undefined **25.** -30

Chapters 1–2 Cumulative Review
1. True **2.** False **3.** False **4.** True **5.** False **6.** True **7.** 1) Parentheses, 2) Exponents, 3) Multiply/divide from left to right, 4) Add/subtract from left to right. **8.** The mistake is calculating the square roots of 16 and 9 when they should have first been added. **9.** Associative property of multiplication **10.** Addends can exchange places. **11.** $5 \times 10^6 + 6 \times 10^5 + 8 \times 10^4 + 9 \times 10^2 + 1 \times 1$ **12.** 50,836,009 **13.** Four hundred nine million, two hundred fifty-four thousand, six **14.** 5×10^7 **15.** $135 > -450$ **16.** $-930 > -932$ **17.** $-23,000,000$ **18.** 57,000
19. 16 **20.** 5 **21.** -8 **22.** -4 **23.** 93 **24.** -43 **25.** -50 **26.** -84 **27.** 96 **28.** -144 **29.** 243
30. 64 **31.** -64 **32.** 201 r2 **33.** -15 **34.** 11 **35.** -19 **36.** -48 **37.** 25 **38.** -37 **39.** 102
40. -7 **41.** 74 m; 300 m^2 **42.** 36 in.^{3} **43.** \$15,343 **44.** $-\$15,210$ (loss) **45.** 21,325 ft.^{2} **46.** 128
47. 5 ml **48.** 67 mph **49.** 1875 slugs **50.** 10 ft.

Exercise Set 3.1
1. Equation **3.** Expression **5.** Expression **7.** Equation **9.** Equation **11.** -7 **13.** 14 **15.** -18
17. 11 **19.** -39 **21.** 40 **23.** 6 **25.** -6 **27.** -14 **29.** 17 **31.** 5 **33.** No integer solution
35. 7 **37.** Undefined **39.** 10 **41.** 0 **43.** 7; -4

Review Exercises:
1. 2 **2.** 300,000 **3.** $2 \times 10^2 + 7 \times 10 + 9 \times 1$ **4.** 1, because 7 raised to the first power is 7. **5.** 64

Exercise Set 3.2
1. Yes **3.** No **5.** No **7.** Yes **9.** Yes **11.** Yes **13.** c. 3; d. 8 **15.** c. -9; d. 1 **17.** c. 8; d. 0 **19.** c. 1; d. 3
21. c. -1; d. 6 **23.** c. -1; d. 0 **25.** Yes **27.** No **29.** Yes **31.** Yes **33.** No **35.** Yes **37.** $5x^2$; $8x$; -7
39. $6t$; -1 **41.** $-6x^3$; x^2; $-9x$; 4 **43.** Binomial **45.** Monomial **47.** None of these **49.** Trinomial
51. Monomial **53.** Binomial **55.** 3 **57.** 6 **59.** 12 **61.** $14t^6 - 8t^4 + 9t^3 + 5t^2 - 1$ **63.** $y^5 - 18y^3 - 10y^2 + 12y + 9$ **65.** $9a^5 + 7a^3 + 2a^2 - a - 6$

Review Exercises:
1. -16 **2.** -7 **3.** -26 **4.** Yes, because of the commutative property of addition. **5.** 28 m

Exercise Set 3.3

1. $12x$ **3.** $4y$ **5.** $5m$ **7.** $-5n$ **9.** $2x$ **11.** $-16a$ **13.** $13a^2$ **15.** $-7y^2$ **17.** $2j^3$ **19.** 0 **21.** $6y^3$
23. $-8m^5$ **25.** $-xy$ **27.** $18a^2b$ **29.** Already simplest form (not like terms). **31.** m^3n **33.** $7a^2b$ **35.** $-t^3u$
37. $11x^2 + 2x + 4$ **39.** $7m^3 + m^2 - 4$ **41.** $8x^4 - 3x^2 + 8$ **43.** $-7t^5 + 6t^4 + 9t^3 - 4t - 2$ **45.** $-13y^4 - 8y^3 - 9y^2 + 18y$ **47.** $6tu^2 + 15t^3 + 7t - 12$ **49.** $2jk^3 + 4j^4 - 8jk - 5j^2 + 9$ **51.** $-3m^2n^2 - 2m^2n + 14mn - 21$
53. $3x^3y^2 + 2x^3y + 2xy^2 + x^2y - 6xy + 10$ **55.** $-13xy^2z + 19xy^2 - 10$

Review Exercises:

1. 18 ft. **2.** 6 **3.** -112 **4.** 341 **5.** Distributive Property

Exercise Set 3.4

1. $5x + 6$ **3.** $9y - 5$ **5.** $13x - 2$ **7.** $2x^2 + 6x + 10$ **9.** $15n^2 - 12n + 3$ **11.** $3a^3 + 5a^2 + a - 2$
13. $2x^3 + 3x^2 - 2$ **15.** $9t^3 - 5t^2 + t + 4$ **17.** $5m^4 + 3m^3 + 6m^2 + m - 3$ **19.** $-x^5 - 9x^4 - 6x^3 - 9x^2 + 16$
21. $12x^6 - 9x^5 + 2x^4 + 9x^3 + 3x^2 - 15x + 16$ **23.** $3x^5 - x^2y^2 - 2x^2y + 11xy - 4y^2$ **25.** $t^2u^3 - 2t^2u^2 + 6tu^2 - 14t^2u - 6tu - 1$ **27.** $4x + 7$ **29.** $8a + 4$ **31.** Mistake: Combined $4y^2 - 2y^2$ to get $-2y^2$. Correct: $10y^3 + 2y^2 - 6y + 8$ **33.** Mistake: Unlike terms were combined. Correct: $8x^6 + x^5 - 9x^4 + 7x^3 + x - 4$ **35.** $2x + 7$ **37.** $3t$
39. $8n - 6$ **41.** $3x - 7$ **43.** $6x^2 - x - 11$ **45.** $12a^2 - 11a - 6$ **47.** $-4x - 10$ **49.** $4m^2 - 18m + 4$
51. $8x^5 - 5x^4 - 12x^3 - 8x^2 - x + 19$ **53.** $-2tu^2 + 8tu - 21u^2 + 5$ **55.** $-2m^6 - 4m^3n^4 - 6m^2n^3 - 18n^2 + 16$
57. $12x^3y^4 + 7xy^3 - 6xy + 13y^2 - 17$ **59.** Mistake: Did not change all signs in the subtrahend. Correct: $2x^2 - 2x + 8$
61. Mistake: Combined $-9t^2$ and $-9t^2$ to get 0. Correct: $-19t^4 - 5t^3 - 18t^2 + 4t + 5$

Review Exercises:

1. -682 **2.** 81 **3.** -32 **4.** 375 m^2 **5.** 128

Exercise Set 3.5

1. x^7 **3.** t^7 **5.** $21a^3$ **7.** $-24u^6$ **9.** $8y^7$ **11.** $35xy^5$ **13.** $-28t^5u^3$ **15.** $56a^3b^5c^2$ **17.** $-63h^3t^5k^4$
19. $90x^7y^4$ **21.** $-36m^4n^2p^5$ **23.** $60x^3y^6z^3$ **25.** Mistake: Multiplied exponents and left out the y. Correct: $45x^7y$
27. $4x^6$ **29.** $49m^{10}$ **31.** $-32y^{30}$ **33.** $25x^2y^{12}$ **35.** $-125t^9u^3v^{18}$ **37.** $48x^{13}y^2$ **39.** Mistake: Did not raise 3 to the 4th and did not multiply x's exponent by 4. Correct: $81x^{16}y^4$ **41.** $15x - 35$ **43.** $-28a + 36$ **45.** $18u^2 + 24u$
47. $16x^3 + 24x^2 - 32x$ **49.** $-5x^4 + 6x^3 - 9x^2$ **51.** $4m^3n + 20m^2n^2 - 36mn^3$ **53.** $30t^5u^3 + 42t^3u^5 - 54t^3u^4 + 24t^2u^3$ **55.** $-4m^3n^2p^4 - 3mn^5p^2 + m^2n^2p^3 + 8m^4n^2p$ **57.** $x^2 + 6x + 8$ **59.** $m^2 + 2m - 15$ **61.** $y^2 - 7y - 8$
63. $12x^2 - 7x - 10$ **65.** $12x^2 - 23x + 5$ **67.** $12t^2 - 17tu - 5u^2$ **69.** $2a^2 - 19ab + 35b^2$ **71.** $x + 7$
73. $2x - 5$ **75.** $-2a - 8b$ **77.** $m^2 + n$ **79.** $x^2 - 9$ **81.** $4m^2 - n^2$ **83.** $36x^2 - 1$ **85.** Mistake: Did not square the initial x and the 5 should be negative. Correct: $6x^2 - 13x - 5$ **87.** $18a^4 + 6a^2b^2 - 12a^2b - 4b^3$ **89.** $2x^3 + 7x^2 - 14x + 5$ **91.** $8y^3 - 10y^2 - 15y + 18$ **93.** $4t^4 - 5t^3 - t^2 + 4t^2u - 5tu - u$ **95.** $24xyz + 8y^2z$ **97.** $3x^3 + 3x^2 - 6x$

Review Exercises:

1. 423 **2.** -4 **3.** 11 ft. **4.** 5 ft. **5.** $-4x$

Exercise Set 3.6

1. Composite **3.** Prime **5.** Neither **7.** Prime **9.** Neither **11.** Composite **13.** Composite **15.** Prime
17. No; 2 is an even prime number. **19.** $2^4 \cdot 5$ **21.** $2^2 \cdot 3 \cdot 13$ **23.** $2^2 \cdot 67$ **25.** $2^3 \cdot 5^2$ **27.** 7^3 **29.** $3 \cdot 5^2 \cdot 13$
31. $2 \cdot 3^3 \cdot 7$ **33.** $2^3 \cdot 7 \cdot 17$ **35.** 1, 2, 3, 4, 5, 6, 10, 12, 15, 20, 30, 60 **37.** 1, 3, 9, 27, 81 **39.** 1, 2, 3, 4, 5, 6, 8, 10, 12, 15, 20, 24, 30, 40, 60, 120 **41.** 12 **43.** 1 **45.** 24 **47.** 28 **49.** 26 **51.** 168 **53.** 1 **55.** 20 **57.** 32
59. 10 in. square **61.** 8 ft.; 5 in the 40 ft. trench, 4 in the 32 ft. trench, and 3 in the 24 ft. trench.

Review Exercises:

1. $5 \times 10^6 + 7 \times 10^5 + 8 \times 10^4 + 4 \times 10^3 + 2 \times 10^2 + 9 \times 1$ **2.** 2019 r5 **3.** 128 **4.** No, the ID system could only account for 260 passes. **5.** 6 ft.

Exercise Set 3.7

1. x^7 **3.** m^6 **5.** 1 **7.** $-5t^4$ **9.** $20n^8$ **11.** $3x^4y$ **13.** $-14a^5c$ **15.** $2a^2$ **17.** $-7x^4$ **19.** $4t^3u$ **21.** a^5c
23. $2x + 3$ **25.** $5a - 1$ **27.** $2t^3 - 3t^2 + 4$ **29.** $3y^3 - 5xz$ **31.** $9m$ **33.** $4t^2u$ **35.** 10 **37.** $2x^2$ **39.** $12n^2$
41. $4x^2y^5$ **43.** $6m^2n^3$ **45.** $4(2x - 1)$ **47.** $7(2y + 1)$ **49.** $2n(n + 3)$ **51.** $x^2(7x - 3)$ **53.** $4r^3(5r^2 - 6)$
55. $3xy^2(2y + 1)$ **57.** $15c^5d^3(2cf - 1)$ **59.** $3a^3(3a^4 - 4a^2 + 6)$ **61.** $7m^5n(2m^3 + 4m + 1)$ **63.** $10x^3y^3(xy^3z - 2x^2z^2 - 4y^2)$

Review Exercises:

1. 20 ft. **2.** 144m^2 **3.** 108 in.^{3} **4.** $2x^3 - 9x^2 - 17$ **5.** $x^2 - 2x - 24$

Exercise Set 3.8

1. $4x + 14$ **3.** $3n - 6$ **5.** $8b - 10$ **7.** a. 90; b. 122 **9.** a. 30; b. 18 **11.** a. 70; b. 150 **13.** $2n^2 + 8n$
15. a. 120; b. 384 **17.** $d^3 + 5d^2 - 6d$ **19.** a. 54; b. 220 **21.** 32 ft.^{2} **23.** 1536 panels **25.** 116 ft.; 36 ft.
27. 4256 ft. **29.** a. $35r + 75s - 245$; b. \$11,905 **31.** a. $5s + 9m + 15l$; b. $2s + 4m + 7l$; c. $3s + 5m + 8l$; d. \$87 profit

Review Exercises:

1. 15 **2.** -1 **3.** 0 **4.** $-x^3 - 21x^2 + 14$ **5.** 1

Chapter 3 Review Exercises

1. False **2.** True **3.** True **4.** False **5.** False **6.** False **7.** Coefficients;Variables **8.** Add; Base **9.** Subtract; Divisor's; Dividend's **10.** Multiply **11.** 40 **12.** -13 **13.** 8 **14.** -2 **15.** Yes, because it is a product of a constant with variables that are each raised to a whole number exponent. **16.** No, because it contains subtraction. **17.** c.18; d. 1 **18.** c. 1; d. 3 **19.** c. -9; d. 0 **20.** c. -3; d. 6 **21.** They have the same variables raised to the same exponents. **22.** The same variables are raised to different exponents. **23.** Binomial **24.** None of these **25.** Monomial **26.** Trinomial **27.** 6 **28.** $5a^6 + 7a^4 - a^2 - 9a + 13$ **29.** $7m^3 + 9m - 7$ **30.** $-5x^7 + 3x^2 + 13$ **31.** $4y^4 + 2y^3 - 10y - 4$ **32.** $7a^5 - 2a^2b^3 + 8ab - 22b$ **33.** $13h^3 - 5h^2 - 5h - 3$ **34.** $-m^4 + 4m^3n - 8n^2$ **35.** m^3n^4 **36.** $-10x^5y^3$ **37.** $6t^8u^5v^2$ **38.** $125x^{12}$ **39.** $-8a^{12}b^3c^{15}$ **40.** $15n^3 - 3n^2 + 21n$ **41.** $-2x^4y^3 - 5x^2y^5 + 8xy^4$ **42.** $a^2 - 2a - 35$ **43.** $10y^2 - 21y + 8$ **44.** $x^3 - x^2y + xy - y^2$ **45.** $9t^2 - 16$ **46.** $u^3 - 3u^2 - 7u + 6$ **47.** $-7x - 2$ **48.** r^6 **49.** $-4x^3$ **50.** 1 **51.** $-9m^5$ **52.** $4xy^4 - 3y^2 - 2$ **53.** $14a^2b$ **54.** Composite **55.** Prime **56.** $2^3 \cdot 3^2 \cdot 5$ **57.** $2^3 \cdot 3 \cdot 5^2 \cdot 7$ **58.** 28 **59.** 1 **60.** $12x^5$ **61.** $6a^2b^3$ **62.** $3n^2(3 - 5n)$ **63.** $6x(3x^2 + 4xy - 6z)$ **64.** $16x - 12; 132$ **65.** $m^2 - 2m - 15; 33$ **66.** $48n^3 - 32n^2; 2560$ **67.** 86 ft. **68.** a. $43n + 43b - 105$; b. \$31,414

Chapter 3 Practice Test

1. -43 **2.** 18 **3.** Binomial **4.** -1 **5.** 0 **6.** 4 **7.** $-a^4 + 5a^2$ **8.** $3y^4 + 2y^3 + y^2 - 10y - 4$ **9.** $6y^5 + 5y^3 - y^2 - 4$ **10.** $20tu^7$ **11.** $-32a^{15}$ **12.** $-2t^4u^3 + 6t^2u^4 + 14tu^3$ **13.** $b^2 - 36$ **14.** $6x + 5$ **15.** m^2 **16.** 1 **17.** $5x^3y^2$ **18.** Composite **19.** $2^2 \cdot 5 \cdot 17$ **20.** 36 **21.** $12h^5$ **22.** $8x^2(x^5 - 2)$ **23.** $5m^3(4m^3 + 3mn - 6)$ **24.** a. $2n^2 - 7n - 4$; b. 45 **25.** a. $84a + 91b$; b. \$21,217

Cumulative Review Chapters 1–3

1. False **2.** True **3.** False **4.** True **5.** True **6.** False **7.** 1) Parentheses, 2) Exponents/roots from left to right, 3) Multiply/divide from left to right, 4) Add/subtract from left to right **8.** Negative **9.** Negative **10.** Multiply every term in the second polynomial by every term in the first polynomial (FOIL). **11.** Two million, four hundred eighty thousand, forty-five **12.** 700,000,000 **13.**

14. $-110,000$ **15.** 140,000 **16.** -1 **17.** 1 **18.** 5 **19.** 6 **20.** -9 **21.** -144 **22.** -38 **23.** 108 **24.** -125 **25.** -81 **26.** -103 **27.** 20 **28.** -15 **29.** 108 **30.** 8 **31.** $-x^3 + 16x^2 - x - 12$ **32.** $4y^3 - 9y^2 + 21y - 6$ **33.** $63x^4y$ **34.** $2b^2 - 13b - 24$ **35.** $2^3 \cdot 3^2 \cdot 5$ **36.** $10x^2$ **37.** $6n^3$ **38.** $6m^2(2m^2 - 3m + 4)$ **39.** 6 **40.** 8 **41.** 255 cm^2 **42.** $16w$ **43.** $5y^3 + 5y^2$ **44.** \$282,600 **45.** 560 **46.** 10 cc **47.** $-\$279$ **48.** 548 ft.^{2} **49.** 16 ft. **50.** a. $113b + 158a - 345$; b. \$19,711

Exercise Set 4.1

1. Equation **3.** Expression **5.** Expression **7.** Yes **9.** No **11.** Yes **13.** No **15.** Yes **17.** No **19.** No

Review Exercises:

1. $5x - 12$ **2.** $-2y - 8$ **3.** $5m + 15$ **4.** $-3m + 4$ **5.** 7

Exercise Set 4.2

1. Yes **3.** No **5.** No **7.** Yes **9.** No **11.** Yes **13.** 6 **15.** 8 **17.** -4 **19.** -1 **21.** 7 **23.** 4 **25.** 0 **27.** 20 **29.** -4 **31.** -12 **33.** \$1473 **35.** 70 cc **37.** \$2465; Yes, midway through the month he's made \$8035. **39.** Yes **41.** 26 cm

Review Exercises:

1. -40 **2.** -8 **3.** -9 **4.** -2 **5.** 3

Exercise Set 4.3

1. -7 **3.** -9 **5.** 6 **7.** 7 **9.** -7 **11.** 12 **13.** 11 **15.** -3 **17.** 4 **19.** 2 **21.** -8 **23.** 2 **25.** 1 **27.** Mistake: Did not rewrite minus sign. Correct: -1 **29.** Mistake: Did not change minus sign in $x - 8$. Correct: -3 **31.** 18 in. **33.** 7 hr **35.** 72 in. **37.** 4 hrs **39.** 5 sec **41.** 666 chips

Review Exercises:

1. Six million, seven hundred eighty-four thousand, two hundred nine **2.** 1 **3.** 43 **4.** $2^4 \cdot 3 \cdot 5$ **5.** $6x^3(4x^2 - 5x + 3)$

Exercise Set 4.4

1. $n + 5 = -7; -12$ **3.** $n - 6 = 15; 21$ **5.** $x + 17 = -8; -25$ **7.** $-3y = 21; -7$ **9.** $9b = -36; -4$ **11.** $5x + 4 = 14; 2$ **13.** $-6m - 16 = 14; -5$ **15.** $39 - 5x = 8x; 3$ **17.** $17 + 4t = 6t - 9; 13$ **19.** $2(b - 8) = 5 + 9b; -3$ **21.** $6x + 5(x - 7) = 19 - (x + 6); 4$ **23.** $-8(y - 3) - 14 = -2y - (y - 5); 1$ **25.** Mistake: Subtraction order is incorrect. Correct: $n - 7 = 15$ **27.** Mistake: Multiplied 2 times x instead of the sum. Correct: $2(x + 13) = -9$ **29.** Mistake: Subtraction order is incorrect. Correct: $16 - 6n = 2(n - 4)$

Review Exercises:

1. $6xy^5$ **2.** $9b^2(2b^3 - 3b + 6)$ **3.** $8w$ **4.** 48 ft. **5.** 9; \$94

Exercise Set 4.5

1. $10w = 300; w = 30$ ft.; $l = 120$ ft. **3.** $2w + 2(20 + 2w) = 220; w = 30$ ft.; $l = 80$ ft. **5.** $2b - 5 + 2b - 5 + b = 70; b = 16$ m, both sides = 27 m **7.** $8l - 20 = 3l$; Rectangle: $l = 4$ ft., $w = 2$ ft.; Triangle sides = 4 ft. **9.** $x + 4x = 90$; Angles: 18°, 72° **11.** $15 + 5a = 180$; Angles: 33°, 147° **13.** $x + 3x + 10 + x = 180$; Angles: 34°, 112°, 34° **15.** $x + x + 10 + x - 7 = 180$; Angles: 59°, 69°, and 52° **17.** $x + x + 1 = 93$; Integers: 46, 47 **19.** $x + x + 2 = 196$; Integers: 97, 99

21. $x + x + 2 + x + 4 = 225$; Integers: 73, 75, 77 **23.** $8(x + 5) + 12x = 260$; 16 small, 11 large **25.** $10x + 5(19 - x)$ $= 125$; 13 five dollar bills, 6 ten dollar bills **27.** $18x + 36x = 108$; \$2, \$6 **29.** $13(16 - x) + 17x = 244$; \$9, \$7

Review Exercises:
1. No **2.** No **3.** -7 **4.** -5 **5.** 3

Chapter 4 Review Exercises
1. False **2.** True **3.** False **4.** True **5.** True **6.** False **7.** Add; Subtract; Same **8.** Multiply; Divide; Same
9. Simplify; Distribute; Combine; Addition/subtraction; Multiplication/division **10.** Replace the variable in the original equation with the solution and verify that it makes the equation true. **11.** Yes **12.** No **13.** Yes **14.** No **15.** Yes
16. No **17.** 8 **18.** -2 **19.** -9 **20.** 3 **21.** -7 **22.** -9 **23.** 3 **24.** -2 **25.** 1 **26.** -6 **27.** 3
28. -5 **29.** $-15 + x = 28$; 43 **30.** $-547 + x = -350$; \$197 **31.** $(50)(25)h = 5000$; 4 ft. **32.** $2(3)w + 2(3)(4) +$ $2w(4) = 52$; 2 ft. **33.** $15 - 7n = 22$; -1 **34.** $5x - 4 = 3x$; 2 **35.** $2(n + 12) = -6n - 8$; -4 **36.** $12 - 3(x - 7) =$ $6x - 3$; 4 **37.** $2(3w - 2) + 2w = 188$; Width $= 24$ m; Length $= 70$ m **38.** $b + b + 38 + b + 38 = 256$; $b = 60$ in.; Other sides $= 98$ in. **39.** $x + 2x - 15 = 180$; Angles: 65°, 115° **40.** $x + x + 16 = 90$; Angles: 37°, 53° **41.** $x +$ $x + 1 = 193$; Integers: 96, 97 **42.** $x + x + 2 = 166$; Integers: 82, 84 **43.** $10(x + 9) + 20x = 330$; 8 twenty dollar bills, 17 ten dollar bills **44.** $4(27 - x) + 6x = 146$; 19 large, 8 small

Chapter 4 Practice Test
1. Equation, because it has an equal sign. **2.** No, because there is a degree 2 term. **3.** No **4.** Yes **5.** 8 **6.** -6
7. 4 **8.** 1 **9.** 3 **10.** -21 **11.** -2 **12.** -2 **13.** $x + 375 = 458$; \$83 **14.** $(44)(26)h = 20,592$; 18 in.
15. $4n - 9 = 23$; 8 **16.** $3(x - 5) = 4x - 9$; -6 **17.** $B + 2b + 9 + 2b + 9 = 258$; Base $= 48$ in.; Other sides $= 105$ in.
18. $x + x - 30 = 180$; Angles: 105°, 75° **19.** $x + x + 2 = 76$; Integers: 37, 39 **20.** $450(12 - x) + 675x = 6300$; 4 at \$675, 8 at \$450

Cumulative Review Chapters 1–4
1. True **2.** False **3.** False **4.** False **5.** True **6.** False **7.** Positive **8.** Add **9.** Subtract **10.** Replace the variables with the given numbers, then calculate. **11.** $3 \times 10^4 + 6 \times 10^3 + 9 \times 10 + 7 \times 1$

12. 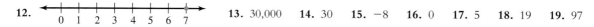 **13.** 30,000 **14.** 30 **15.** -8 **16.** 0 **17.** 5 **18.** 19 **19.** 97

20. -26 **21.** -61 **22.** -16 **23.** -22 **24.** -391 **25.** -7 **26.** $8t^3 + 14t^2 - 25t - 7$ **27.** $6x^3 - 7x^2 - x - 7$
28. $-54a^6b$ **29.** $9x^2 - 25$ **30.** $2^2 \cdot 3 \cdot 5 \cdot 7$ **31.** $36m^2$ **32.** $-7k$ **33.** $5n^2(4n^3 + 3mn - 2)$ **34.** 39 **35.** -6
36. 22 **37.** 4 **38.** -7 **39.** 222 hr **40.** 216 in.^{2} **41.** $-\$67,181$ **42.** -4 A **43.** \$50,710 **44.** 7 ft.
45. $x^2 - 4x - 12$ **46.** 5 in. **47.** 62 mph **48.** Width $= 9$ ft.; Length $= 13$ ft. **49.** 45, 47 **50.** 7 large, 9 small

Exercise Set 5.1

1. $\frac{1}{3}$ **3.** $\frac{5}{8}$ **5.** $\frac{2}{3}$ **7.** a. $\frac{5}{16}$; b. $\frac{15}{16}$; c. The sense of smell seems to contribute more information about foods than taste.;

d. If she is trying the same foods in the same order she may be guessing better the second time. **9.** $\frac{17}{800}$; $\frac{783}{800}$

11. $\frac{249}{258}$; $\frac{9}{258}$ **13.** 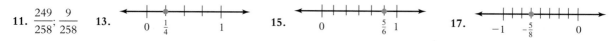 **15.** **17.**

19. **21.** 23 **23.** 0 **25.** 1 **27.** Undefined **29.** 15 **31.** 7 **33.** 30

35. -30 **37.** $\frac{4}{9} > \frac{2}{5}$ **39.** $\frac{12}{18} > \frac{9}{16}$ **41.** $-\frac{4}{15} > -\frac{6}{17}$ **43.** $-\frac{9}{12} = -\frac{15}{20}$ **45.** $4\frac{2}{7}$ **47.** $21\frac{1}{4}$ **49.** $-12\frac{4}{5}$
51. $-12\frac{7}{8}$ **53.** $\frac{31}{6}$ **55.** $\frac{11}{1}$ **57.** $-\frac{79}{8}$ **59.** $\frac{-29}{20}$

Review Exercises:
1. 4 **2.** $2^3 \cdot 3 \cdot 5 \cdot 7$ **3.** 12 **4.** $8x^2$ **5.** $8x^2(5x^3y - 7)$

Exercise Set 5.2

1. $\frac{5}{6}$ **3.** $\frac{1}{2}$ **5.** $\frac{3}{4}$ **7.** -1 **9.** $-\frac{7}{9}$ **11.** $-\frac{3}{10}$ **13.** $\frac{5}{12}$ **15.** $\frac{12}{31}$; 12 out of every 31 women can be expected to

use this product; 248 may not be a large enough sample. **17.** $\frac{212}{1001}$ **19.** $\frac{16}{77}$ **21.** $3\frac{3}{4}$ **23.** $3\frac{1}{3}$ **25.** $-4\frac{1}{7}$

27. $-5\frac{1}{6}$ **29.** $\frac{5x}{16}$ **31.** $\frac{x^2}{y}$ **33.** $\frac{2n}{5m^3}$ **35.** $\frac{1}{4t^3u}$ **37.** $\frac{-3ac}{5b}$ **39.** $\frac{-2b^4}{5a^6c^5}$

Review Exercises:
1. $-15,015$ **2.** 72 **3.** $35x^5y$ **4.** $-32x^{15}$ **5.** $99m^2$

Exercise Set 5.3

1. $\dfrac{8}{35}$ **3.** $\dfrac{1}{54}$ **5.** $\dfrac{15}{28}$ **7.** $\dfrac{49}{1000}$ **9.** $\dfrac{4}{5}$ **11.** $\dfrac{1}{3}$ **13.** $\dfrac{13}{20}$ **15.** $\dfrac{3}{10}$ **17.** $-\dfrac{2}{5}$ **19.** $\dfrac{21}{50}$ **21.** $\dfrac{1}{2}$

23. Estimate: 6; Actual: $5\dfrac{3}{5}$ **25.** Estimated: 28; Actual: $23\dfrac{1}{3}$ **27.** Estimated: 80; Actual: 74 **29.** Estimated: -3; Actual: $-3\dfrac{1}{2}$ **31.** Estimated: 42; Actual: $40\dfrac{4}{5}$ **33.** Estimated: -9; Actual: $-13\dfrac{1}{2}$ **35.** $\dfrac{2x^2}{15}$ **37.** $\dfrac{x^2}{6}$ **39.** $\dfrac{2xy^4}{45}$

41. $-\dfrac{5k^3}{12}$ **43.** $\dfrac{10x^2y}{7}$ **45.** $-\dfrac{3m}{10n^3p^2}$ **47.** $\dfrac{25}{36}$ **49.** $-\dfrac{27}{64}$ **51.** $\dfrac{1}{64}$ **53.** $\dfrac{x^3}{8}$ **55.** $\dfrac{8x^6}{27}$ **57.** $\dfrac{m^{12}n^4}{81p^8}$ **59.** 1560

61. $\dfrac{5}{8}$ **63.** $\dfrac{2}{3}$; 20 calls **65.** $\dfrac{2}{3}$ **67.** $147\dfrac{49}{50}$N **69.** 51 ft.² **71.** $4\dfrac{1}{2}$m; $28\dfrac{2}{7}$m **73.** $1\dfrac{53}{220}$mi.; $3\dfrac{9}{10}$mi.

Review Exercises:
1. -204 **2.** 16 **3.** $3x(3x-4)$ **4.** -5 **5.** 2 ft.

Exercise Set 5.4

1. $\dfrac{3}{2}$ **3.** 6 **5.** $-\dfrac{1}{15}$ **7.** $-\dfrac{4}{x}$ **9.** 3 **11.** $\dfrac{9}{25}$ **13.** $-1\dfrac{3}{5}$ **15.** $\dfrac{1}{24}$ **17.** $1\dfrac{1}{2}$ **19.** -18 **21.** Estimate: 10; Actual: $11\dfrac{5}{8}$ **23.** Estimate: 3; Actual: $2\dfrac{8}{15}$ **25.** Estimate: $-\dfrac{3}{10}$; Actual: $-\dfrac{1}{4}$ **27.** Estimate: $1\dfrac{6}{7}$; Actual: $1\dfrac{23}{27}$ **29.** $\dfrac{14x^2}{5}$

31. $\dfrac{15m^3}{16n^2}$ **33.** $-\dfrac{9x^4}{8y^5}$ **35.** $\dfrac{8}{9}$ **37.** $\dfrac{11}{6}$ **39.** 6 **41.** 9 **43.** 16 **45.** $3\dfrac{1}{3}$ **47.** $-1\dfrac{4}{5}$ **49.** $\dfrac{2}{7}$ **51.** 30 doses; There are 2 extra doses in the bottle. A little extra can account for spillage. **53.** $\dfrac{7}{8}$ cup **55.** $91\dfrac{60}{61}$ ft. **57.** $989\dfrac{109}{110}$ ft.

59. $45\dfrac{3}{4}$ mph

Review Exercises:
1. $2\cdot3^3\cdot7$ **2.** $60x^3y$ **3.** $\dfrac{5}{6}=\dfrac{10}{12}$ **4.** $9t^3+9t^2-13t-16$ **5.** $16x(2y+1)$

Exercise Set 5.5
1. 30 **3.** 36 **5.** 60 **7.** 36 **9.** 72 **11.** 252 **13.** 364 **15.** 1800 **17.** 3360 **19.** 1680 **21.** $24xy$

23. $16mn$ **25.** $30y^3z$ **27.** $36x^2y^3z$ **29.** $\dfrac{9}{30}$ and $\dfrac{25}{30}$ **31.** $\dfrac{21}{36}$ and $\dfrac{11}{36}$ **33.** $\dfrac{3}{60}$ and $\dfrac{34}{60}$ **35.** $\dfrac{27}{36}$, $\dfrac{6}{36}$, and $\dfrac{28}{36}$

37. $\dfrac{14y}{24xy}$ and $\dfrac{9x}{24xy}$ **39.** $\dfrac{9}{16mn}$ and $\dfrac{6n^2}{16mn}$ **41.** $\dfrac{21}{30y^3z}$ and $\dfrac{25y^2z^2}{30y^3z}$ **43.** $\dfrac{2y^2z^2}{36x^2y^3z}$ and $\dfrac{-15x}{36x^2y^3z}$

Review Exercises:
1. -12 **2.** 11 **3.** $4x^3+39x^2-16x-3$ **4.** -8 **5.** 53

Exercise Set 5.6
1. $\dfrac{5}{7}$ **3.** $-\dfrac{2}{3}$ **5.** $\dfrac{6}{17}$ **7.** $-\dfrac{1}{5}$ **9.** $\dfrac{12}{x}$ **11.** $\dfrac{2x^2}{3}$ **13.** $\dfrac{x+1}{6}$ **15.** $\dfrac{4x^2-3x+1}{7y}$ **17.** $\dfrac{5n^2-3}{5m}$ **19.** $1\dfrac{2}{15}$

21. $\dfrac{5}{18}$ **23.** $\dfrac{37}{60}$ **25.** $1\dfrac{25}{36}$ **27.** $\dfrac{23x}{24}$ **29.** $\dfrac{3}{16m}$ **31.** $\dfrac{13h+16}{20h}$ **33.** $\dfrac{6-7n}{9n^2}$ **35.** $10\dfrac{4}{9}$ **37.** $7\dfrac{1}{2}$ **39.** $7\dfrac{1}{2}$

41. $10\dfrac{5}{24}$ **43.** $9\dfrac{3}{4}$ **45.** $9\dfrac{1}{6}$ **47.** $3\dfrac{7}{24}$ **49.** $\dfrac{1}{2}$ **51.** $4\dfrac{5}{12}$ **53.** $-10\dfrac{1}{2}$ **55.** $-3\dfrac{5}{8}$ **57.** $-6\dfrac{5}{8}$ **59.** $\dfrac{1}{10}$

61. $1\dfrac{1}{20}$ **63.** $-\dfrac{11}{12}$ **65.** $-\dfrac{7}{10}$ **67.** $\dfrac{11}{16}$ in. **69.** $\dfrac{29}{40}$; $\dfrac{13}{120}$ **71.** $68\dfrac{1}{2}$ in. **73.** 4 ft.

Review Exercises:
1. 51 **2.** $10x^2+3x-5$ **3.** $-y^3-y^2-4y-5$ **4.** $-54a^3b^4c$ **5.** $2x^2+x-15$

Exercise Set 5.7

1. $1\dfrac{1}{2}$ **3.** 4 **5.** $1\dfrac{15}{16}$ **7.** $-2\dfrac{3}{4}$ **9.** $-2\dfrac{1}{4}$ **11.** $-\dfrac{1}{8}$ **13.** $-24\dfrac{5}{6}$ **15.** 10 **17.** $\dfrac{45}{64}$ **19.** $1\dfrac{9}{40}$

21. 4400 **23.** $52\dfrac{1}{4}$ m² **25.** $30\dfrac{1}{4}$ ft.² **27.** 616 in.² **29.** $4\dfrac{51}{56}$ in.² **31.** $240\dfrac{9}{10}$ m² **33.** $x^3-\dfrac{27}{4}x+5$

35. $6y^3+\dfrac{11}{5}y^2+y-\dfrac{5}{6}$ **37.** $\dfrac{1}{10}t^3+\dfrac{5}{12}t^2-\dfrac{2}{3}$ **39.** $-\dfrac{1}{10}m^4n^3$ **41.** $\dfrac{1}{2}t^2-\dfrac{5}{12}t-\dfrac{1}{16}$ **43.** $2x^2-\dfrac{109}{6}x+\dfrac{3}{2}$

45. $\dfrac{1}{9}$ **47.** $-\dfrac{1}{32}$ **49.** $\dfrac{10,000}{81}$ **51.** $-\dfrac{64}{27}$

1. 5   **2.** 11   **3.** -3 **4.** 4 hrs   **5.** 6 tens dollar bills, 9 five dollar bills

Exercise Set 5.8

1. $\dfrac{1}{4}$ **3.** $\dfrac{4}{9}$ **5.** $2\dfrac{1}{10}$ **7.** $10\dfrac{2}{5}$ **9.** $12\dfrac{8}{9}$ **11.** $6\dfrac{1}{5}$ **13.** $-1\dfrac{17}{24}$ **15.** $25\dfrac{1}{3}$ **17.** $4\dfrac{3}{5}$ cm   **19.** \$22,530

21. 1200 gal   **23.** $\dfrac{3}{5}$ **25.** $135°, 45°$ **27.** $\dfrac{100}{11,127}$ **29.** $\dfrac{1}{9}$ hr.

Review Exercises:

1. $\dfrac{6}{17} < \dfrac{7}{19}$ **2.** $\dfrac{3}{8x}$ **3.** $-\dfrac{14}{45}$ **4.** $\dfrac{10}{9b^3}$ **5.** $6\dfrac{13}{24}$

Chapter 5 Review Exercises

1. True   **2.** True   **3.** False   **4.** True   **5.** False   **6.** False   **7.** Multiplying; dividing   **8.** Write the numerator and denominator in prime factored form, then divide out all common primes.   **9.** Change the divisor to its reciprocal, then multiply.   **10.** Find a common denominator. Upscale the fractions to equivalent fractions with the common denominator. Add numerators and keep the common denominator. Reduce.   **11.** a. $\dfrac{1}{2}$; b. $\dfrac{1}{4}$ **12.** a.   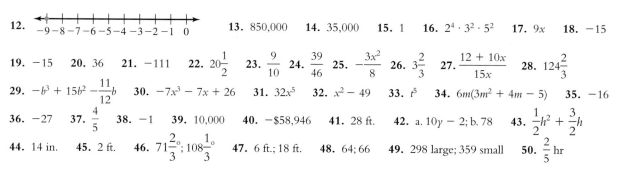;

b.   **13.** a. $4\dfrac{4}{9}$; b. $-7\dfrac{1}{4}$ **14.** a. $\dfrac{20}{3}$; b. $-\dfrac{11}{2}$ **15.** a. 18; b. Undefined; c. 0; d. 1

16. a. $\dfrac{3}{7}$; b. $\dfrac{4}{5}$ **17.** a. $-\dfrac{4m^3}{13n}$; b. $\dfrac{3x}{10y^3}$ **18.** a. $\dfrac{5}{9} > \dfrac{7}{13}$; b. $\dfrac{3}{16} < \dfrac{5}{24}$; c. $\dfrac{10}{16} = \dfrac{25}{40}$ **19.** a. $\dfrac{4}{15}$; b. -12 **20.** a. $\dfrac{35n}{12p^2}$;

b. $-\dfrac{h^2}{5}$ **21.** a. $1\dfrac{5}{9}$; b. 7   **22.** a. $\dfrac{4}{3b^2}$; b. $-\dfrac{4x^2z^2}{5}$ **23.** a. $\dfrac{5}{3}$; b. 5   **24.** a. 168; b. $60x^2y$ **25.** a. $1\dfrac{1}{30}$; b. $11\dfrac{7}{24}$

c. $-\dfrac{1}{5}$ d. $-2\dfrac{5}{6}$ **26.** a. $\dfrac{n}{2}$; b. $-\dfrac{2}{3x}$; c. $\dfrac{5}{12h}$; d. $\dfrac{15 - 14a}{24a}$ **27.** a. $3\dfrac{5}{8}$; b. $-6\dfrac{7}{12}$ **28.** a. $\dfrac{1}{64}$; b. $\dfrac{9}{25}x^2y^6$; c. $\dfrac{1}{16}$; d. $-\dfrac{27}{8}$

29. a. $\dfrac{1}{10}x^2 - 2x - 2$; b. $\dfrac{5}{8}n^2 - \dfrac{2}{3}n - 2$; c. $\dfrac{23}{2}y^3 - 2y^2 - \dfrac{5}{6}y + \dfrac{1}{15}$; d. $-\dfrac{1}{4}a^2b^4c^2$; e. $\dfrac{3}{2}x^2 + \dfrac{381}{32}x - \dfrac{3}{4}$ **30.** a. $4\dfrac{2}{15}$; b. $-\dfrac{8}{9}$;

c. $11\dfrac{1}{2}$; d. $-9\dfrac{1}{3}$ **31.** 9 ft.^{2} **32.** $110\dfrac{1}{4}$ cm^2 **33.** $1\dfrac{3}{4}$ in.   **34.** $28\dfrac{2}{5}$ cm   **35.** $7\dfrac{6}{7}$ ft.   **36.** $12\dfrac{4}{7}$ ft.^{2} **37.** 220 cm^2

38. $\dfrac{1}{6}$ **39.** 18 servings   **40.** $\dfrac{7}{20}$ **41.** $\dfrac{11}{5}$ or $2\dfrac{1}{5}$ **42.** $2\dfrac{2}{9}$ m   **43.** $10\dfrac{1}{2}$ ft.; $24\dfrac{1}{2}$ ft.   **44.** $\dfrac{7}{75}$ hr   **45.** $11\dfrac{13}{17}$ sec.

Chapter 5 Practice Test

1. $\dfrac{3}{8}$ **2.**   **3.** a. $\dfrac{3}{4} = \dfrac{9}{12}$; b. $\dfrac{14}{15} > \dfrac{5}{6}$ **4.** $6\dfrac{1}{6}$ **5.** $-\dfrac{37}{8}$ **6.** a. -16; b. Undefined;

c. 0; d. 1   **7.** a. $\dfrac{3}{5}$; b. $-\dfrac{3x^2y}{10}$ **8.** a. $-11\dfrac{5}{8}$; b. $\dfrac{4}{3b^2}$ **9.** a. $2\dfrac{1}{2}$; b. $-\dfrac{8m^3n}{15}$ **10.** $36t^3u$ **11.** a. $\dfrac{11}{20}$; b. $-1\dfrac{5}{8}$ **12.** a. $-\dfrac{x}{3}$;

b. $\dfrac{11}{12a}$ **13.** a. $\dfrac{4}{25}$; b. $\dfrac{1}{64}$ **14.** $2\dfrac{9}{16}$ **15.** a. $-\dfrac{3}{2}n - 3$; b. $2m^2 + m - \dfrac{3}{8}$ **16.** 6   **17.** a. $22\dfrac{1}{2}$ m$^2m^2$ **18.** 62 in.^{2}

19. $14\dfrac{13}{14}$ in.   **20.** $19\dfrac{9}{14}$ **21.** $\dfrac{5}{8}$ **22.** 8 pieces   **23.** $\dfrac{5}{24}$ **24.** $24\dfrac{1}{2}$ **25.** $\dfrac{1}{18}$ hr

Cumulative Review Chapters 1–5

1. False   **2.** False   **3.** False   **4.** False   **5.** True   **6.** True   **7.** Multiply   **8.** 1) Find the prime factorization. 2) Use the largest exponent of those primes that are common to all factorizations.   **9.** $-9 - 3$ is not -6.   **10.** 1) Find the LCD. 2) Upscale. 3) Addition/subtract numerators and keep the LCD. 4) Simplify.   **11.** Four million, five hundred eighty-two thousand, six hundred one

12.   **13.** 850,000   **14.** 35,000   **15.** 1   **16.** $2^4 \cdot 3^2 \cdot 5^2$ **17.** $9x$ **18.** -15

19. -15 **20.** 36   **21.** -111 **22.** $20\dfrac{1}{2}$ **23.** $\dfrac{9}{10}$ **24.** $\dfrac{39}{46}$ **25.** $-\dfrac{3x^2}{8}$ **26.** $3\dfrac{2}{3}$ **27.** $\dfrac{12 + 10x}{15x}$ **28.** $124\dfrac{2}{3}$

29. $-b^3 + 15b^2 - \dfrac{11}{12}b$ **30.** $-7x^3 - 7x + 26$ **31.** $32x^5$ **32.** $x^2 - 49$ **33.** t^5 **34.** $6m(3m^2 + 4m - 5)$ **35.** -16

36. -27 **37.** $\dfrac{4}{5}$ **38.** -1 **39.** 10,000   **40.** $-\$58,946$ **41.** 28 ft.   **42.** a. $10y - 2$; b. 78   **43.** $\dfrac{1}{2}h^2 + \dfrac{3}{2}h$

44. 14 in.   **45.** 2 ft.   **46.** $71\dfrac{2}{3}°; 108\dfrac{1}{3}°$ **47.** 6 ft.; 18 ft.   **48.** 64; 66   **49.** 298 large; 359 small   **50.** $\dfrac{2}{5}$ hr

Exercise Set 6.1

1. $\frac{1}{5}$ **3.** $\frac{1}{4}$ **5.** $\frac{3}{8}$ **7.** $\frac{6}{25}$ **9.** $1\frac{1}{2}$ **11.** $18\frac{3}{4}$ **13.** $9\frac{5}{8}$ **15.** $7\frac{9}{25}$ **17.** $-\frac{1}{125}$ **19.** $-13\frac{3}{250}$ **21.** Ninety-seven thousandths **23.** Two thousand fifteen millionths **25.** Four thousand one hundred fifty-nine and six tenths **27.** Five hundred twenty-one and six hundred eight thousandths **29.** Negative one hundred seven and ninety-nine hundredths

31. Negative fifty thousand ninety-two hundred-thousandths **33.**

35.

37.

39.

41.

43.

45. $0.81 > 0.8$ **47.** $2.891 > 2.8909$ **49.** $0.001983 < 0.001985$

51. $-1.01981 > -1.10981$ **53.** $-145.7183 < -14.57183$ **55.** 610.3 **57.** 610.2832 **59.** 610 **61.** 1 **63.** 1.0 **65.** 0.951 **67.** 408.1 **69.** 408.0626 **71.** 410

Review Exercises:

1. $\frac{41}{100}$ **2.** -13 **3.** $-7a^3 - 3a^2 + 15$ **4.** $4x^2 + 21x + 1$ **5.** -17 **6.** 57 ft.

Exercise Set 6.2

1. 159.71 **3.** 59.7511 **5.** 333.317 **7.** 841.45 **9.** 784.28 **11.** 0.01789 **13.** 10.11 **15.** -4.317 **17.** -0.9985 **19.** -190.02 **21.** -73.25 **23.** 0.9952 **25.** $-12.81y^2 + 5y - 0.93$ **27.** $4.5a^3 + 9a^2 - 1.5ab - 10.6$ **29.** $10.2x^2 - 0.9x + 8.4$ **31.** $10.01a^3 - 4.91a^2 + 7.5a + 5.99$ **33.** $0.4n^2 + 10.8n - 3.6$ **35.** $-1.42t^4 - 2.4t^3 + 0.95t^2 - 0.2t + 1.45$ **37.** 10.83 **39.** -5.81 **41.** -3.22 **43.** -46.83 **45.** $421.94 due **47.** $9.38 **49.** $1919.89 **51.** 22.3 m

Review Exercises:

1. -300 **2.** $\frac{21}{100}$ **3.** 306 **4.** $2\frac{32}{39}$ **5.** 11 **6.** $-18x^7$ **7.** $y^2 - 2y - 15$ **8.** -9

Exercise Set 6.3

1. 0.54 **3.** 10.585 **5.** 702.5922 **7.** 6.1945 **9.** 196 **11.** -15.08 **13.** -0.152 **15.** 0.00435 **17.** 0.81 **19.** 9.261 **21.** 0.00000081 **23.** -0.064 **25.** 300,000,000; Three hundred million meters per second **27.** 2,140,000; Two million one hundred forty thousand light years **29.** 273,000,000; Two hundred seventy-three million **31.** 2.3×10^6 **33.** 3.54×10^9 **35.** 5.3×10^9 **37.** $5.76x^8$ **39.** $-0.016a^3b^2c$ **41.** $0.25t^6u^4$ **43.** $0.9y^2 + 1.3y + 19$ **45.** $-0.136a^3 + 6.8ab - 12.92b$ **47.** $6.4x^2 - 18.32x - 3.66$ **49.** $33.64a^2 - 81b^2$ **51.** $14.7k^2 - 4.27k + 0.28$ **53.** $73.52 **55.** $3.84 **57.** $1827.20 **59.** $3.75 **61.** -0.0294 N **63.** -701.96 lbs **65.** 23.04 cm^2 **67.** 12,551.1 ft.^{2} **69.** 0.0081 m^3

Review Exercises:

1. 205 **2.** 13 **3.** $r = 7$ in.; C = 44 in. **4.** -9 **5.** $6x^4$

Exercise Set 6.4

1. 23.85 **3.** 0.024 **5.** 600 **7.** 4.8 **9.** -40 **11.** 125 **13.** -0.0206 **15.** 814.5 **17.** $178.\overline{18}$ **19.** 1.06 **21.** 0.6 **23.** 0.45 **25.** -0.4375 **27.** $0.4\overline{3}$ **29.** 13.25 **31.** -17.625 **33.** $104.\overline{6}$ **35.** $-216.\overline{571428}$ **37.** 0.04 **39.** 0.5 **41.** 4.90 **43.** 14.14 **45.** 1.3 **47.** 0.09 **49.** 0.00000000065 **51.** 0.00000055 **53.** 0.000000000053 **55.** 2.8×10^{-3} **57.** 1×10^{-15} **59.** 1.77×10^{-5} **61.** $25.5x^2$ **63.** $-6.4m^2n$ **65.** $0.15ac$ **67.** 6.02 **69.** -1.6 **71.** 0.6 **73.** -3.5 **75.** $613.32 **77.** 40.5 rads **79.** 400 Ω **81.** 1.71 hr. **83.** ≈ 6829.5 mi. **85.** ≈ 6.69 mi.

Review Exercises:

1. -5 **2.** 18 **3.** -180 **4.** 56 in.^{2} **5.** 680 cm^2

Exercise Set 6.5

1. 2.64 **3.** 42.99 **5.** -22.24 **7.** -254.8 **9.** -117.69 **11.** $-9.7\overline{2}$ **13.** -0.51 **15.** 0.29 **17.** -4.775 **19.** $-0.3\overline{8}$ **21.** 40 **23.** 9.75 **25.** 2.25×10^{11} **27.** 23.6 **29.** 57.6 cm^2 **31.** 1.296 m^2 **33.** ≈ 86.5 in.^{2} **35.** ≈ 0.9 cm^3 **37.** 168,750 ft.^{3} **39.** 92.10$\overline{6}$ in.^{3} **41.** $\approx 2.45 \times 10^{11}$ mi^3 **43.** ≈ 2.71 cm^2 **45.** ≈ 134.375 cm^2 **47.** 148,021.69$\overline{3}$ ft.^{3} **49.** $\approx 2,582,957.6$ m^3 **51.** 0.000027 in.^{3}; 37,037.037 grains

Review Exercises:

1. 7 **2.** $20\frac{1}{4}$ **3.** 300 ft.^{2} **4.** 720 cm^2 **5.** 60 **6.** 8 five dollar bills, 12 ten dollar bills

Exercise Set 6.6

1. 0.2 **3.** 0.6 **5.** −0.12 **7.** −6.2 **9.** $3.\overline{3}$ **11.** 0.5 **13.** −24 **15.** $6.57 **17.** $104.27 **19.** $327
21. $586.55 **23.** −3.8$\overline{3}$ **25.** 2.5$\overline{2}$ **27.** 100 km **29.** 5 mi. **31.** ≈14.1 ft. **33.** ≈59.96 ft. **35.** ≈91.9 ft.
37. 5 half dollars, 17 quarters **39.** 23 of the 16 oz., 42 of the 12 oz.

Review Exercises:

1. Twenty-four billion, nine hundred fifteen million, two hundred four **2.** 46,300,000 **3.** $13\frac{11}{24}$ **4.** $4\frac{9}{20}$ **5.** $4x^6$
6. Mistake: Calculated that dividing $-3x$ by 3 is $1x$ instead of $-1x$. Correct: 2

Chapter 6 Review Exercises

1. True **2.** False **3.** False **4.** True **5.** False **6.** True **7.** Decimal point was placed in the product incorrectly.
8. Line up the decimal points, add the digits, then bring down the decimal point directly below its position in the addends.
9. Denominator, numerator **10.** Write all the digits without the decimal point in the numerator then write the last place
value as the denominator. **11.** a. Twenty-four and thirty-nine hundredths; b. Five hundred eighty-one and four hundred

fifty-nine thousandths; c. Two thousand nine hundred seventeen ten-thousandths **12.** a. $\frac{4}{5}$; b. $\frac{3}{125}$; c. $-2\frac{13}{20}$

13. a. $2.001 > 2.0009$; b. $-0.016 < -0.008$ **14.** a. 32; b. 31.8; c. 31.81; d. 31.806 **15.** a. 88.015; b. 7.916; c. −6.2;
d. −6.12; e. 23.118; f. 20.49 **16.** a. $11.1y^4 − 8xy − 12.2y^2 + 3.8$; b. $5.4a^2 + 5.2a + 10.33$; c. $−2.7x^3 − 98.1x^2 + 2.1x − 4.2$
17. a. 17.544; b. −14.505; c. 1.44; d. −0.008; e. $0.025h^2k^3$; f. $−3.78x^8y^2c$; g. $0.12n^3 − 0.28n^2 + 0.2n − 3.6$; h. $−3.08a + 5.39b$;
i. $9a^2 − 14.45a + 3.05$; j. $1.69x^2 − 17.64$ **18.** a. 5.25; b. 0.4; c. 8.4; d. 600 e. $7.95x^2$ f. $−0.35a^3c$ **19.** a. 0.8; b. 1.1
20. a. 12.17; b. 1.26 **21.** a. 0.6; b. 0.1$\overline{6}$; c. 6.25; d. $-4.\overline{18}$ **22.** a. 3,580,000,000; b. −420,000; c. 0.000000651;
d. −0.00000219 **23.** a. 9.2×10^9; b. $−1.03 \times 10^5$; c. 1.9×10^{-5}; d. $−4.06 \times 10^{-7}$ **24.** a. −12.97; b. −3.05; c. 0.27;
d. −2.16 **25.** r = 7.5 cm; d = 15 cm; C ≈ 47.1 cm, A ≈ 176.625 cm^2 **26.** a. ≈46.12 in.^{2}; b. 0.15 m^2 **27.** a. ≈105.975
in.^{2}; b. ≈1445.43 cm^2 **28.** a. ≈42.39 in.^{3}; b. 20.096 cm^3; c. 904.32 m^3; d. 22,696.8 m^3; e. 39.978 in.^{3} **29.** 7.536 in.^{3}
30. a. 1.54; b. −0.64; c. 3; d. −0.02; e. 0.8; f. −1.05; g. 8.1; h. 5 **31.** $713.18 **32.** $47.58; Last payment = $47.60
33. 28.6 **34.** $9.82 **35.** ≈11.62 ft. **36.** 90 minutes **37.** 10 nickels, 19 dimes **38.** 9 large; 15 small

Chapter 6 Practice Test

1. Fifty-six and seven hundred eighty-nine thousandths **2.** $\frac{17}{25}$ **3.** $−0.0059 < −0.0058$ **4.** a. 2; b. 2.1; c. 2.09; d. 2.092

5. 41.991 **6.** 7.916 **7.** −38.745 **8.** 0.3 **9.** 6.02 **10.** 0.9 **11.** 3.8$\overline{3}$ **12.** a. 0.00000297; b. $−3.56 \times 10^7$

13. 5.1344 **14.** 0.256 **15.** $−2.7x^3 − 98.1x^2 + 2.1x − 4.2$ **16.** a. $1.5x^4y$; b. $9a^2 − 14.45a + 3.05$ **17.** 4.1

18. 0.6 **19.** a. 75.36 ft.; b. 452.16 ft.^{2} **20.** a. $33.49\overline{3}$ cm^3; b. 75.36 in.^{3} **21.** 785.5 ft.^{2} **22.** $7.03 **23.** $158.75

24. 19.36 ft. **25.** 8 layered, 4 sheet

Cumulative Review Chapters 1–6

1. True **2.** True **3.** False **4.** False **5.** True **6.** False **7.** Undefined **8.** Coefficients; Variables **9.** Missing

a term. Should have $5x$ term from combining $2x + 3x$. **10.** Change the divisor to its reciprocal, then multiply.

11. Twenty-nine and six thousand eighty-one ten-thousandths **12.** **13.** 2.02 **14.** 40

15. 6 **16.** $2^3 \cdot 3 \cdot 5 \cdot 7$ **17.** 168 **18.** 7 **19.** −52 **20.** $-\frac{4}{5}$ **21.** $\frac{3x}{4y^3}$ **22.** $-\frac{31}{54}$ **23.** $\frac{1}{25}$ **24.** $\frac{8n^2}{5p}$

25. $12\frac{1}{8}$ **26.** $\frac{2x + 9}{12}$ **27.** 3.98 **28.** 2.258 **29.** $1.6x^3 + \frac{3}{4}x^2 + 3.55$ **30.** $4.9m^3 − 6.7m^2 + m + 18.6$

31. $−12x^2y$ **32.** $4y^2 + 5y − 6$ **33.** 768,000,000; seven hundred sixty-eight million **34.** $5mn(6n^4 + 3n − 5)$

35. 24.7 **36.** −6.8 **37.** $\frac{7}{8}$ **38.** 7 **39.** 32°F **40.** $1,400,900 **41.** $1939.33 **42.** $8w^3 − 4w^2$; 1584 cm^3

43. 2.2 ft. **44.** ≈87.55 in.^{2} **45.** 2 **46.** $\frac{1}{2}$ **47.** $\frac{7}{30}$ **48.** Base = 15 ft.; Sides = 27.5 ft. **49.** ≈19.08 ft.

50. 9 large; 11 small

Exercise Set 7.1

1. a. $\frac{31}{66}$; b. $\frac{35}{66}$; c. $\frac{31}{35}$; d. $\frac{35}{31}$ **3.** a. $\frac{3}{2}$; b. $\frac{2}{3}$; c. $\frac{3}{5}$; d. $\frac{2}{5}$ **5.** $\frac{7}{8}$ **7.** $\frac{14}{9}$ **9.** a. $\frac{7}{30}$; b. $\frac{17}{30}$; c. 1; d. $\frac{1}{2}$ **11.** $\frac{1}{13}$ **13.** $\frac{2}{13}$

15. $\frac{1}{3}$ **17.** 0 **19.** $\frac{7}{13}$ **21.** a. $\frac{1}{5}$; b. $\frac{28}{125}$; c. $\frac{72}{125}$; d. $\frac{97}{125}$

23. ≈0.16; This means that out of every dollar of gross income, $0.16 is paid towards debt. **25.** ≈0.26; This means that out
of each dollar of gross income, $0.26 is paid towards the mortgage. **27.** ≈16.9; This means that there are 16.9 (or about 17)
students for each faculty member. **29.** ≈12.1; This means that the stock is selling at $12.1 for every dollar of annual earnings.
31. 68.2 mi./h **33.** $0.12/min. **35.** $7.50/lb **37.** 16 oz can **39.** Bag of 16 diapers **41.** Two 15.5 oz boxes

Review Exercises:

1. 120　**2.** $4\frac{1}{2}$　**3.** 12.375　**4.** 8.6　**5.** $7\frac{1}{2}$

Exercise Set 7.2

1. Yes　**3.** No　**5.** No　**7.** Yes　**9.** 5　**11.** -7.5　**13.** 25　**15.** -25　**17.** $14\frac{1}{4}$　**19.** $14\frac{2}{3}$　**21.** $\approx$34.3 gal.

23. 36 servings　**25.** 1168.8 oz　**27.** $2\frac{1}{4}$ c.　**29.** $3\frac{1}{4}$ in.　**31.** $1633\frac{1}{3}$ days　**33.** 8.96 cm　**35.** a = $3\frac{11}{18}$ in.;

b = $3\frac{1}{4}$ in.; c = $4\frac{29}{48}$ in.　**37.** $61\frac{3}{5}$ ft.　**39.** 24.8 m

Review Exercises:

1. $\frac{5}{27}$　**2.** 3.7　**3.** 1344　**4.** 5.875　**5.** $20\frac{1}{4}$

Exercise Set 7.3

1. 156 ft.　**3.** 7.5 ft.　**5.** 34,320 ft.　**7.** 66 in.　**9.** 36,608 yd.　**11.** 2.75 yd.　**13.** 270 ft.^{2}　**15.** 13,608 in.^{2}
17. 18.75 yd.^{2}　**19.** 18 c.　**21.** 20 oz.　**23.** 15 pt.　**25.** 4.25 qt.　**27.** 51.2 oz.　**29.** 5.1 gal.　**31.** 196 oz.
33. 11.25 lb.　**35.** 0.6 T　**37.** 8400 lb.　**39.** 2.5 h.　**41.** 1.5 min.　**43.** 630 sec.　**45.** 135 min.　**47.** 720 sec.
49. $\approx$0.082 y　**51.** 1440 min.　**53.** 37,869,120 min.　**55.** $12.\overline{36}$ mi./min.　**57.** No, the officer's estimate converts to a
speed of $54.\overline{54}$ mi./hr.　**59.** $25,014.\overline{27}$ mi./hr.

Review Exercises:
1. 840　**2.** 950　**3.** 4.8　**4.** 4.56　**5.** 0.009

Exercise Set 7.4
1. 450 cm　**3.** 70 m　**5.** 3.8 km　**7.** 9.5 m　**9.** 34.8 m　**11.** 1.12 hm　**13.** 120 ml　**15.** 5 l　**17.** 0.0095 dal
19. 80 cc　**21.** 12 ml　**23.** 400 cc　**25.** 12,000 mg　**27.** 4.5 kg　**29.** 50 cg　**31.** 3.6 t　**33.** 106 kg　**35.** 6.5 t
37. 18,000 cm^2　**39.** 0.22 m^2　**41.** 0.58 km^2　**43.** 0.0443 hm^2

Review Exercises:

1. 1.008　**2.** $21\frac{2}{3}$　**3.** $6.5x^2 - 18.2x - 5.6$　**4.** $481y^2 - 280y + 700$　**5.** 2.8

Exercise Set 7.5
1. 91.4 m　**3.** 42.2 km　**5.** 9.525 mm　**7.** 437.6 yd　**9.** 24.84 mi.　**11.** 17.72 in.　**13.** 0.473 l　**15.** 591.25 ml
17. 0.79 gal　**19.** 0.68 oz　**21.** 61.29 kg　**23.** 340.5 g　**25.** 0.0176 oz　**27.** 121 lb.　**29.** 37°C　**31.** $38.\overline{8}$°C
33. $-20.\overline{5}$°C　**35.** 130°F　**37.** 42.8°F　**39.** -459.67°F

Review Exercises:

1. 0.28　**2.** $0.118\overline{3}$　**3.** $-\frac{4}{3}x - \frac{7}{20}y$　**4.** $\frac{2}{3}$　**5.** 10.88

Exercise Set 7.6
1. f.e.r. = 0.22; b.e.r = 0.33; Yes　**3.** b.e.r. = 0.41; Yes　**5.** f.e.r. = 0.31; b.e.r. = 0.45; No (f.e.r. and b.e.r. too high)
7. \$1344　**9.** \$533.60　**11.** \$1252.80　**13.** \$686.75　**15.** \$563.75　**17.** 9:30 PM　**19.** 1:33 PM　**21.** 522.5 mg
23. 5 tablets　**25.** $\approx$10 drops/min　**27.** 400 units/hr.

Review Exercises:

1. $\frac{7}{25}$　**2.** $\frac{2}{13}$　**3.** $\approx$\$0.14/oz　**4.** 25 yd　**5.** 8500 mg

Chapter 7 Review Exercises
1. False　**2.** True　**3.** True　**4.** False　**5.** True　**6.** False　**7.** Favorable; Possible　**8.** 1　**9.** Denominator;
Numerator　**10.** Denominator　**11.** $\frac{11}{16}$　**12.** $\frac{25}{108}$　**13.** $\frac{2}{13}$　**14.** $\frac{1}{3}$　**15.** $\approx$16.4　**16.** $\approx$7.1　**17.** 70.2 mi./hr.

18. \$0.09/min.　**9.** \$0.045/oz.　**20.** \$0.039/oz.　**21.** 32 oz.　**22.** Two 12 oz.　**23.** -4.5　**24.** $9\frac{3}{13}$　**25.** 29.9 m^2

26. 280 mi.　**27.** 432 in.　**28.** 2.5 lb.　**29.** 192 oz.　**30.** 135 min.　**31.** $24.\overline{1}$ yd.^{2}　**32.** 79,200 mi./hr.

33. 50 dm　**34.** 260 g　**35.** 0.95 t　**36.** 75 cc　**37.** $\approx$13 drops/min.　**38.** 92.8 mg　**39.** 6.06 m　**40.** 19.3 km

41. 65.83 kg　**42.** 0.79 gal　**43.** $32.\overline{2}$°C　**44.** 24.8°F　**45.** 0.28　**46.** 0.40　**47.** \$571.30　**48.** \$1722

49. 3 hr.　**50.** 11:32 A.M.

Chapter 7 Practice Test

1. $\dfrac{19}{26}$ 2. $\dfrac{11}{2289}$ 3. $\dfrac{1}{5}$ 4. 30.8 5. $70.4\overline{8}$ mi./hr. 6. \$0.14 7. $\approx$\$0.04 8. 20 oz can 9. -5.85
10. \$514.50 11. 900 mi. 12. 168 in. 13. 320 oz. 14. 4.5 gal. 15. 2.5 hr. 16. $27.\overline{7}$ yd.^{2} 17. a. 5.8 cm;
b. 0.42 km 18. a. 0.024 kg b. 91,000 g 19. a. 0.028 m^2; b. 80 dm^2 20. a. 0.65 l; b. 800 cc 21. $\approx$16 drops/min.
22. 22.85 m 23. $-1.\overline{1}$°C 24. $\approx$0.45 25. 13 hr.

Cumulative Review Chapters 1–7

1. False 2. False 3. True 4. False 5. True 6. False 7. Reciprocal 8. LCM 9. Cross 10. Cross
multiply, then isolate the variable. 11. Forty-nine thousand, eight hundred two and seventy-six hundredths 12. 9.14×10^8

13.
number line marked -4, $-3\frac{2}{3}$, -3, -2

14. 41.33 15. -1 16. 5 17. -91 18. $-2\dfrac{8}{13}$ 19. $5\dfrac{9}{10}$

20. 30.75 21. -31.2 22. $\dfrac{9}{16}$ 23. 3 24. 0.0081 25. 30.4 26. $8.3y^3 - 8.2y^2 - y + 10.2$

27. $\dfrac{1}{2}a^2 - \dfrac{65}{12}a + 7$ 28. $2^5 \cdot 3 \cdot 5^2$ 29. $20x$ 30. $6x^4$ 31. $8m(4m^3 + 3m - 2)$ 32. $\dfrac{3}{16x^4}$ 33. $40k^2$

34. $\dfrac{18 + 5x}{9x}$ 35. 18.4 36. $14\dfrac{6}{11}$ 37. -16.32 38. 66,000 ft. 39. 480 g 40. 140°F 41. 181.8 m^2

42. $\approx$349.35 ft.^{2} 43. \$50,710 44. 6.2 m 45. $y^2 - 4y - 45$ 46. 14 m; 11 m 47. 62 mi./hr. 48. 16 oz.

49. $\approx$313 l 50. 5.5 hr.

Exercise Set 8.1

1. $\dfrac{1}{5}$ 3. $\dfrac{3}{20}$ 5. $\dfrac{37}{250}$ 7. $\dfrac{3}{80}$ 9. $\dfrac{91}{200}$ 11. $\dfrac{1}{3}$ 13. 0.75 15. 1.25 17. 0.129 19. 0.0165 21. 0.534

23. $0.161\overline{6}$ 25. 50% 27. 60% 29. $37\dfrac{1}{2}$% or 37.5% 31. $16\dfrac{2}{3}$% or $16.\overline{6}$% 33. $66\dfrac{2}{3}$% or $66.\overline{6}$% 35. $44\dfrac{4}{9}$% or

$44.\overline{4}$% 37. 96% 39. 80% 41. 9% 43. 120% 45. 2.8% 47. 405.1% 49. $66.\overline{6}$% 51. $163.\overline{63}$%

Review Exercises:

1. 14.4 2. 1620 3. 3.84 4. $1333.\overline{3}$ 5. 38

Exercise Set 8.2

1. 140 3. 11.7 5. 90 7. 4.88 9. 51.15 11. 15.4 13. $37.8\overline{6}$ 15. $\approx$4.7 17. 4608 19. 220
21. 12,000 23. 24,200 25. 46 27. 450 29. 25% 31. 6% 33. 30% 35. 35% 37. 71.4%
39. 66.15%

Review Exercises:

1. 1280 2. 575 3. $86\dfrac{5}{8}$ 4. 10.5 5. 62.5

Exercise Set 8.3

1. 168 3. 10.2 5. 112.5 7. 1.425 9. 90.27 11. 55 13. 29.2 15. 1482 17. 414.05 19. 80
21. 5600 23. 25,400 25. 54.6 27. 246 29. 30% 31. 3% 33. 60% 35. 95% 37. $71.\overline{1}$%
39. 89.47%

Review Exercises:

1. 268.8 2. 1.24 3. $\dfrac{5}{12}$ 4. 250 5. $\dfrac{3}{5}$

Exercise Set 8.4

1. 480 ml 3. a. 27; b. 10%; c. 3 5. 1.36 million 7. \$310.60 9. \$699.92 11. \$4212.50 13. \$12,500
15. 93 17. 113 19. 89% 21. $\approx$43.6% 23. $\approx$53.6% 25. $29.1\overline{6}$% 27. $\approx$90.1% 29. $\approx$17.2%
31. a. Mortgage $\approx$ 23.3%; Childcare $\approx$ 19.6%; Car Payment 1 $\approx$ 7.2%; Car Payment 2 $\approx$ 6.8%; Credit Cards $\approx$ 3.7%;
Utilities $\approx$ 7%; Groceries $\approx$ 11.7%; Entertainment $\approx$ 6.4%; b. $\approx$85.6%; c. $\approx$14.4%

Review Exercises:

1. $6.1x - 2.6$ 2. 14 3. 30.6 4. 1.6 5. \$2550

Exercise Set 8.5

1. \$9.30; \$195.25 3. \$1587; \$28,037 5. \$30,636 7. \$4.93 9. \$25,200 11. \$29,250 13. \$42.95
15. \$19.80; \$46.19 17. \$245; \$454.99 19. \$1977.08 21. \$22.43 23. \$2596 25. \$146.97 27. \$2570
29. \$379.51 31. $\approx$7% 33. 3.5% 35. 8% 37. $\approx$17.6% 39. $\approx$22.3% 41. $\approx$15% 43. $\approx$33.6%

Review Exercises:
1. 3750 **2.** $\frac{1}{3}$ **3.** 1.1664 **4.** $\frac{1}{125}$ **5.** 5 ten dollar bills; 11 five dollar bills

Exercise Set 8.6
1. $120; $4120 **3.** $38.50; $388.50 **5.** $7901.25; $20,151.25 **7.** $96; $2496 **9.** $22.68; $2022.68 **11.** $4798.84
13. $1946.25 **15.** $5832 **17.** $942.82 **19.** $15,142.40 **21.** $518.01 **23.** $1910.48 **25.** $412.41
27. $1646.80 **29.** $571.04 **31.** $872.23 **33.** $706.15 **35.** $311.38 **37.** $423.05 **39.** $744.27 **41.** $853
43. $926.92

Review Exercises:
1. **2.** $0.8\overline{3}$ **3.** 77.75 **4.** 6 **5.** $-1\frac{13}{15}$

Chapter 8 Review Exercises
1. False **2.** False **3.** True **4.** False **5.** False **6.** True **7.** Divide **8.** Multiply **9.** Translate "of" to multiplication—unless preceded by a whole number. (In this case, "of" translates to division.) Translate "is" to an equal sign. **10.** Write the percent as a fraction with a denominator of 100. Set this fraction equal to the part out of the whole. **11.** a. $\frac{2}{5}$; b. $\frac{13}{50}$; c. $\frac{13}{200}$; d. $\frac{49}{200}$

12. a. 0.16; b. 1.5; c. 0.032; d. $0.40\overline{3}$ **13.** a. 54% **14.** 130% **15.** $37\frac{1}{2}$% or 37.5% **16.** $44\frac{4}{9}$ or $44.\overline{4}$% **17.** 13.5
18. 40 **19.** 62.5% **20.** $26.\overline{6}$% **21.** 324 **22.** 70 **23.** 18% **24.** $53.\overline{3}$% **25.** 210 ml **26.** 27%
27. $17.15; $302.90 **28.** $42.71 **29.** $\approx$514.29 **30.** $31,500 **31.** 32% **32.** $29.1\overline{6}$% **33.** $28.80
34. $8104 **35.** $600 **36.** $6298.56 **37.** $2033.79 **38.** $917.91

Chapter 8 Practice Test
1. $\frac{6}{25}$ **2.** 0.042 **3.** 0.125 **4.** $\frac{163}{400}$ **5.** 26% **6.** 120% **7.** 40% **8.** $55.\overline{5}$% **9.** 11.4 **10.** 128
11. $23.\overline{3}$% **12.** 37.5% **13.** 24 **14.** $304.50 **15.** $\approx$12.4% **16.** $14.79; $310.54 **17.** $59.47 **18.** $23,450
19. 20% **20.** $300 **21.** $816 **22.** $2400 **23.** $2590.06 **24.** $1248.72 **25.** $901.17

Cumulative Review Chapters 1–8
1. True **2.** True **3.** True **4.** True **5.** True **6.** False **7.** Multiply coefficients. Add exponents of the like bases. Write unlike bases unchanged. **8.** Find a common denominator. Upscale. Add numerators and keep the common denominator. Simplify. **9.** Multiply both sides by an appropriate power of 10 as determined by the decimal number with the most decimal places.

10. Multiply the given measurement by unit fractions so that the undesired units divide out. **11.** 2.75×10^{10}
12. **13.** 1,600,000 **14.** 1 **15.** 20 **16.** $14\frac{5}{12}$ **17.** $-57\frac{1}{6}$ **18.** 64.975
19. -24.5 **20.** 0.39 **21.** 1.5 **22.** 8.49 **23.** 0.125 **24.** 71% **25.** $x^4 - 4x^3 - \frac{14}{15}x + 24.2$ **26.** $46.02x^{14}$
27. $24.8x^2 + 27x - 5$ **28.** $2^2 \cdot 3^3 \cdot 5$ **29.** $120xy^2$ **30.** $6x^4$ **31.** $8m(4m^3 + 3m - 2)$ **32.** $\frac{8m}{3n}$ **33.** $\frac{12 - 5y}{20}$
34. 6 **35.** -4.6 **36.** $3\frac{6}{13}$ **37.** 152 oz. **38.** 0.08 l **39.** $4.\overline{4}$°C **40.** 12.2 ft.; 3.7 m **41.** $\frac{1}{2}h^2 + \frac{1}{2}h$ 45 in.
42. 61; 62 **43.** $\approx$81.76 cm^2 **44.** $\approx$18.1 in.^{3} **45.** 2143.57$\overline{3}$ in.^{3} **46.** $\frac{1}{40}$ **47.** $x \approx 16.1$ cm; $y \approx 19.3$ cm
48. 0.025 hr. **49.** $85.\overline{3}$% **50.** $3601.63

Exercise Set 9.1
1. $\bar{x} = $26,841.\overline{6}$; Median = $26,975; Mode = $28,700 **3.** $\bar{x} = 78.25$; Median = 80; Mode = 82 **5.** $\bar{x} = 1.82$ m; Median = 1.85 m; Mode = 1.8 m and 2.1 m **7.** $\bar{x} = 1.25$ in.; Median = 1.1 in.; Mode = 0.6 in. **9.** Year 1: $\bar{x} = 126.76; Median = $131.26; Mode = no mode Year 2: $\bar{x} = 131.95; Median = $134.08; Mode = no mode **11.** 90 **13.** 106
15. 3.192 **17.** 2.667 **19.** B (86.6)

Review Exercises:
1. 75% **2.** 0.125 **3.** $11\frac{7}{20}$ **4.** $-14x - 6y + 12$ **5.** -15

Exercise Set 9.2
1. 97,693,000 **3.** 38,893 **5.** Northern NJ **7.** Pittsburgh **9.** 1,376,482.5 **11.** Endowment income
13. 4% **15.** 22 billion

17.

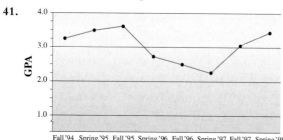

FICA: 6.2%
Medicare: 1.45%
Fed W/H: 7%
State W/H: 5%
Health Plan: 4.7%
Retirement: 6%
Life Insurance: 1.5%
Net Pay 68.2%

19. Engineering   **21.** Education   **23.** $27,000   **25.** $23,000

27.

Electric and Gas Bill (bar chart, Year 1 and Year 2, Jan.–Dec.)

29. Business and management   **31.** 48,000   **33.** 20,000   **35.** 37%   **37.** 1985   **39.** Drugs = 7%; Lack of financial support = 11%; Lack of discipline = 15%

41.

GPA vs. Semester (line graph, Fall '94 – Spring '98)

Review Exercises:

1. (number line from −5 to 0, point at −5)   **2.** (number line from 0 to 6, point at 6)   **3.** 7   **4.** 16   **5.** $\sqrt{97} \approx 9.849$

Exercise Set 9.3

1. A. (2, 3)
   B. (−3, 0)
   C. (−1, −3)
   D. (3, −4)

3. A. (−4, 1)
   B. (0, 3)
   C. (−3, −3)
   D. (3, −5)

5.

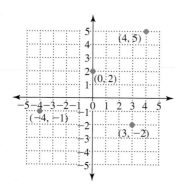

7.

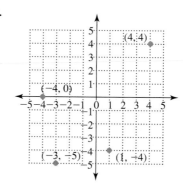

9. II   **11.** I   **13.** IV   **15.** III   **17.** 10   **19.** 5   **21.** $\sqrt{221} \approx 14.87$ **23.** $\sqrt{410} \approx 20.25$ **25.** $\sqrt{17} \approx 4.12$
27. $\sqrt{161.44} \approx 12.71$ **29.** $\sqrt{33.3125} \approx 5.77$

Review Exercises:

1. 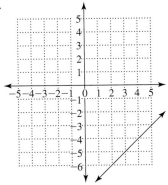 (number line with point at -3, and 0 marked)

2. 2   **3.** 12   **4.** 7.5   **5.** $-\dfrac{9}{10}$

Exercise Set 9.4

1. Yes   **3.** No   **5.** Yes   **7.** No   **9.** No   **11.** No

13.

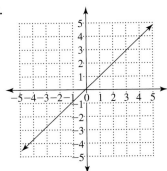

15.

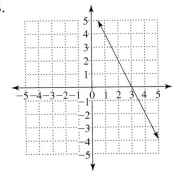

17.

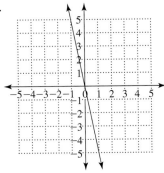

19.

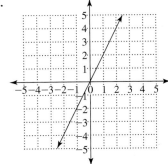

21.

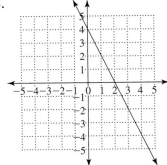

23.

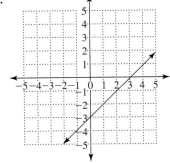

25.

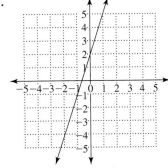

27.

29.

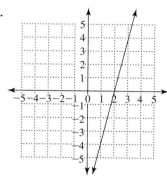

31.

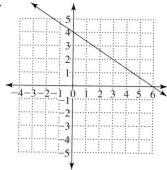

33.

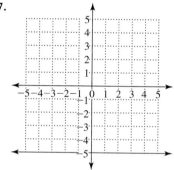

35.

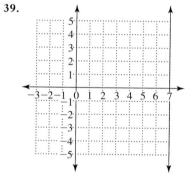

37.

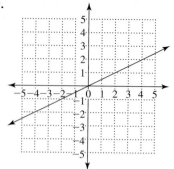

39.

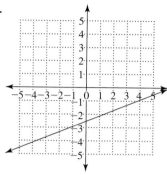

41. x: $(12, 0)$; y: $(0, 6)$ **43.** x: $(4, 0)$; y: $(0, -5)$ **45.** x: $(-8, 0)$; y: $(0, 6)$ **47.** x: $(0.2, 0)$; y: $(0, 0.65)$ **49.** x and y: $(0, 0)$

51. x: $(\frac{1}{4}, 0)$; y: $(0, 1)$ **53.** x: $(25, 0)$; y: $(0, -5)$ **55.** x: $(-9, 0)$; no y-intercept **57.** no x-intercept; y: $(0, 4)$

59. no x-intercept; y: $(0, -3)$

Review Exercises:

1. -64.4 **2.** 40 m **3.** $45\frac{1}{2}$ ft.^{2} **4.** 5.67 in.^{2} **5.** 55.5 ft.^{2}

Exercise Set 9.5

1. a. 600 ft./sec.; b. 503.4 ft./sec.; c. $\approx$18.6 s; d.

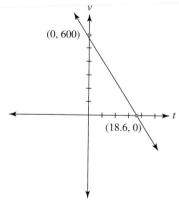

3. a. $15,000; b. $187,500; c. (18.6, 0)t

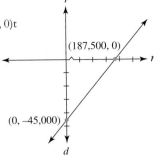

5. a. $300; b. $367.50; c. $570; d.

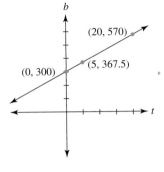

7. a.

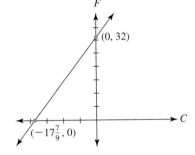

; b. (0, 32); c. $\left(-17\frac{7}{9}, 0\right)$; d. 104

9. a. $T = 9h + 58$; b. $80.50; c. $4\frac{2}{3}$hr.; d.

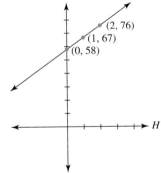

11. P = 18; A = 20 square units **13.** P = 26; A = 42 square units **15.** P = 22; A = 30 square units **17.** P = 22.64; A = 30 square units **19.** P ≈ 22.16; A = 30 square units **21.** P ≈ 32.44; A = 63 square units **23.** P ≈ 27.62; A = 45.5 square units **25.** P ≈ 24.61; A = 27 square units **27.** P ≈ 30.16; A = 54 square units **29.** P ≈ 21.12; A = 24 square units

Review Exercises:

1. $\overline{x} = 79.2$; Median = 83; Mode = 70 and 88 **2.** 9 **3.** $440.05 **4.** $\frac{1}{2}$ **5.** 52; 54

Chapter 9 Review Exercises

1. False   **2.** True   **3.** False   **4.** False   **5.** True   **6.** True   **7.** Find the sum of all the given scores, then divide the sum by the number of scores.   **8.** List the given scores in order from least to greatest, then locate the middle score in the list. When given an even number of scores, calculate the average of the middle two scores.   **9.** Find at least two solutions. Plot those solutions as points in the rectangular coordinate system. Connect the points to form a straight line.   **10.** To find the x-intercept, replace y with 0, then solve for x. To find the y-intercept, replace x with 0, then solve for y.   **11.** $\bar{x} \approx 81.3$; Median = 84; Mode = 94   **12.** $\bar{x} = \$127{,}847.\overline{27}$; Median = \$127,100; No mode   **13.** 93   **14.** 3.233   **15.** 9.2%   **16.** 84.9%   **17.** United States   **18.** Italy   **19.** Japan   **20.** Japan   **21.** 17.3%   **22.** 18.3%   **23.** 168.3 million   **24.** 26,591,400   **25.** 9,088,200   **26.** 4,039,200   **27.**

28. \$16,000   **29.** \$33,000   **30.** Women with some high school but no degree.   **31.** Men with Master's degree.   **32.** Men with an Associate's degree.

33.

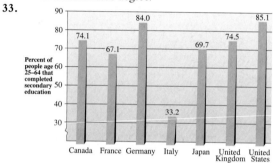

34. 5,000,000   **35.** 2,000,000   **36.** 4,000,000   **37.** 30 years old and over   **38.** 2,300,000

39.

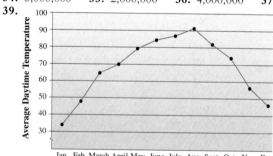

40. A. $(3, 4)$
B. $(-3, 2)$
C. $(-2, 0)$
D. $(-2, -4)$
E. $(0, -5)$
F. $(4, -2)$

41.

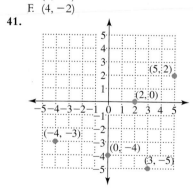

42. a. II; b. III; c. I; d. On the x-axis (not in a quadrant)   **43.** $\sqrt{80} \approx 8.94$ **44.** $\sqrt{125} \approx 11.18$ **45.** Yes   **46.** Yes
47. Yes

48.

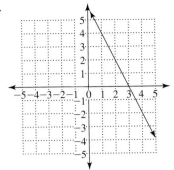

49.

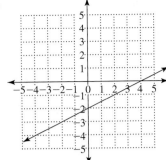

50.

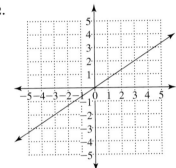

51.

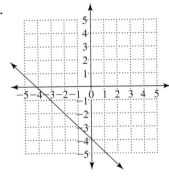

52.

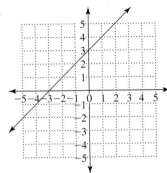

53.

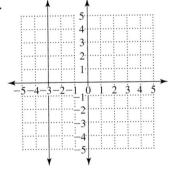

54.

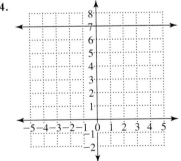

55.

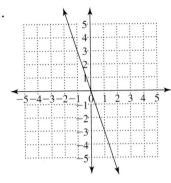

56. x: $(2, 0)$; y: $(0, 10)$ **57.** x: $(\frac{1}{4}, 0)$; y: $(0, -1)$ **58.** x and y: $(0, 0)$ **59.** x: $(6, 0)$, no y-intercept **60.** a. $\$160,272$;

b. $\$230,625$; c.

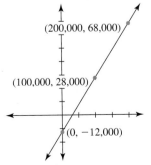

(200,000, 68,000)

(100,000, 28,000)

(0, −12,000)

61. 21.4 units **62.** 40 square units

Chapter 9 Practice Test

1. $\bar{x} = 81.61$; Median = 83; Mode = 80, 84, and 96   **2.** $\bar{x} = \$95,727.\overline{7}$, Median = \$96,400; No mode   **3.** 98   **4.** 15%
5. \$441.25   **6.** \$661.88   **7.** Public college non-professional staff   **8.** 6.4   **9.** 10.5   **10.** 1996   **11.** 34.6
12. 33.9
13.

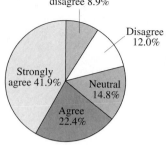

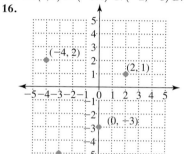

14. A. (1, 2) B. (−3.0) C. (−2, −5) D. (4, −4)   **15.** II
16.

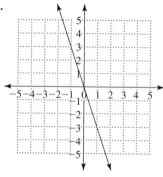

17. $\sqrt{74} \approx 8.6$ **18.** Yes   **19.** X: (2, 0); Y: $\left(0, -\dfrac{6}{5}\right)$

20.

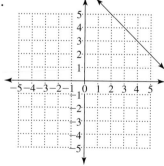

21.

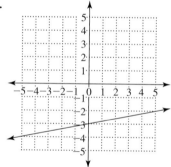

22.

23

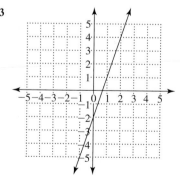

24. a. 40 m/sec.; b. 10.6 m/s; c. ≈ 4.1 sec.; d. −58 m/sec.   **25.** a. 28 units; b. 48 square units

Cumulative Review Chapters 1–9

1. True   **2.** True   **3.** False   **4.** F   **5.** False   **6.** False   **7.** True   **8.** True   **9.** False   **10.** True   **11.** True
12. False   **13.** 1. Parentheses   2. Exponents   3. Multiply/divide from left to right.   4. Add/subtract from left to right.
14. Add the absolute values and keep the same sign.   **15.** Subtract the absolute values, and keep the sign of the number with the larger absolute value.   **16.** 1. Change the operation from a minus sign to a plus sign.   2. Change the subtrahend to its additive inverse.   **17.** Positive   **18.** Negative   **19.** 1. Simplify a. Use the distributive law to clear parenthese. b. Use the multiplication principle to clear fractions or decimals. c. Combine like terms.   2. Use the addition/subtraction principle so that all variable terms are on one side of the equation and all constant terms are on the other side.   3. Use the multiplication/division principle to clear any remaining coefficient.   **20.** 1. Find at least two solutions.   2. Plot the solutions as points in the rectangular coordinate system.   3. Connect the points to form a straight line.   **21.** $2 \times 10^4 + 4 \times 10^3 + 6 \times 10^2 + 7$
22. Four thousand, six hundred seven and nine hundredths   **23.** a. 80,000; b. 80,000; c. 79,805; d. 79,804.7; e. 79,804.65

24. a. 27,000; b. 695,000; c. 1800; d. 40   **25.** 2.75×10^{10} **26.** 6.8×10^{-4} **27.** a.

b.    c. (number line from -8 to -7.2, -7)   **28.** a. 1; b. 2; c. 4   **29.** 81   **30.** -81 **31.** $\dfrac{1}{36}$

32. $-\dfrac{1}{8}$ **33.** -18 **34.** 35   **35.** 4   **36.** $12\dfrac{5}{12}$ **37.** $-33\dfrac{5}{12}$ **38.** 13   **39.** -4.725 **40.** 0.592   **41.** 0.9

42. 11.22   **43.** a. 0.375; b. $0.\overline{6}$; c. 0.08; d. 0.244; e. 0.045   **44.** a. $\dfrac{31}{6}$; b. $\dfrac{29}{200}$; c. $\dfrac{3}{50}$; d. $\dfrac{11}{200}$; e. $\dfrac{41}{250}$ **45.** a. 25%; b. $44.\overline{4}\%$;

c. 65%; d. 3.5%; e. 230%   **46.** a. -44; b. 88   **47.** $x^4 - 3x^3 + 4x^2 + 7x - 20.3$ **48.** $7n^2 - 4.2n - 12.81$

49. $x^4 - 4x^3 - \dfrac{14}{15}x^2 + 24.2$ **50.** a. $-\dfrac{5}{18}a^3b^2c$; b. $m^2 - 81$; c. $31.95t^2 + 5.7t - 18$ **51.** $2 \cdot 3^2 \cdot 5 \cdot 7$ **52.** $5n^2$

53. $72x^3y$ **54.** $5y^3$ **55.** $4b^2(7b^3 + 6b - 8)$ **56.** a. $-\dfrac{15x^2y}{8}$; b. $\dfrac{8m}{3n}$; c. $\dfrac{12 - 5y}{20}$ **57.** -6 **58.** -9 **59.** 5   **60.** $12\dfrac{1}{12}$

61. -7.84 **62.** -12.5 **63.** -5.25 **64.** 2   **65.** a. 144 in.; b. 26 yd; c. 12,672 ft.; d. 480 cm; e. 12,500 m; f. 4.5 m

66. a. $3\dfrac{1}{3}$ yd.²; b. 36 ft.²; c. 3.5 ft.²; d. 6000 cm²; e. 85,000 m²; f. 4.2 m²   **67.** a. 152 oz; b. 2.25 T; c. 230 g; d. 0.96 t

68. a. 32 pt.; b. 17 qt.; c. 0.4 l; d. 9 cc   **69.** a. 25°C; b. -4°F   **70.** 17 m; 56.1 ft.   **71.** 1000   **72.** $-\$58,460$ **73.** ≈ 4.6

slugs; 67.3 kg   **74.** $\dfrac{1}{2}h^2 + 2h$; 17.22 in.   **75.** a. $3.3x - 3400$; b. \$1550   **76.** 61, 62   **77.** 59°, 69°, and 52°   **78.** 11 tens,

6 fives   **79.** $\dfrac{9}{16}$; 18   **80.** $\dfrac{23}{30}, \dfrac{13}{120}$ **81.** $\dfrac{1}{20}$hr.   **82.** 47.26 cm²   **83.** 117 ft.²   **84.** 98.91 in.³   **85.** 904.32 in.³

86. ≈ 8.04 cm³   **87.** $\sqrt{464} \approx 21.54$ **88.** $\dfrac{41}{650}$ **89.** 20 oz. box   **90.** $x = 21.6$; $y = 26.\overline{6}$ **91.** \$60.76   **92.** 8.9%

93. 64.3%   **94.** \$14.28   **95.** \$3601.63   **96.** $\overline{x} = 39.\overline{36}$; Median $= 38$; No mode

97.

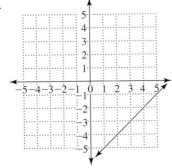

98. P $= 25.6$ units; A $= 38.5$ square units

99. a.   b.

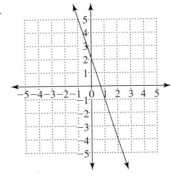

c.

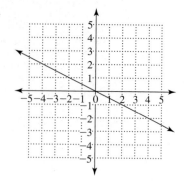

d.

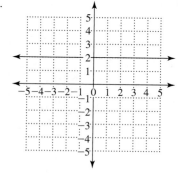

100. a. 580 ft./sec.; b. 419 ft./sec.; c. 18 sec.

Glossary

Absolute value: The absolute value of a given number is its distance from zero.

Addends: Numbers that are added.

Addition/subtraction principle of equality: We can add or subtract the same amount on both sides of an equation without affecting its solution(s).

Addition: The arithmetic operation that combines amounts.

Additive inverses: Two numbers whose sum is zero.

Amortize: To pay off a loan or debt in installments.

Annual percentage rate (APR): An interest rate that is used to calculate the interest earned on a given principal if it were allowed to earn interest for 1 full year.

Area: The total number of square units that completely fill a shape.

Associative Property of Addition: $(a + b) + c = a + (b + c)$ where a, b, and c are any numbers.

Associative Property of Multiplication: $(a \cdot b) \cdot c = a \cdot (b \cdot c)$ where a, b, and c are any numbers.

Average rate: A measure of the rate at which an object travels a total distance in a total amount of time.

Axis: A line used for reference in a graph.

Back-end ratio: The ratio of the total monthly debt payments to gross monthly income.

Balance technique: If we add or remove an amount on one side of an equation, we must add or remove the same amount on the other side to keep the equation balanced.

Base unit: A basic unit; other units are named relative to it.

Base: The number that is repeatedly multiplied.

Binomial: A polynomial that has exactly two terms.

Capacity: A measure of the amount of liquid a container holds.

Circle: A collection of points that are all equally distant from a central point, called the center.

Circumference: The distance around a circle.

Coefficient: The numerical factor in a monomial.

Combinatorics: The branch of mathematics that deals with counting total combinations or arrangements.

Commission: A portion of sales earnings that a salesperson receives.

Commutative Property of Addition: $a + b = b + a$ where a and b are any numbers.

Commutative Property of Multiplication: $a \cdot b = b \cdot a$ where a and b are any numbers.

Complementary angles: Angles whose sum is 90°.

Complex fraction: An expression that is a fraction with fractions in the numerator and/or denominator.

Composite number: A natural number that is divisible by a number other than 1 and itself.

Compound interest: Interest that is calculated based on principal and prior earned interest.

Congruent angles: Angles that have the same measurement.

Conjugates: Binomials that differ only in the sign separating the terms.

Constant: Any symbol that does not vary in value.

Coordinate: A number that refers to a number line in the rectangular coordinate system.

Cost: Money spent on production, operation, labor, and debts.

Decimal notation: A base-ten notation for expressing fractions.

Deduction: An amount that is subtracted from income.

Degree: (monomial) The sum of the exponents on all variables in a monomial; (polynomial) the largest degree of all the terms that make up the polynomial.

Denominator: The number written in the bottom position in a fraction.

Diameter: The distance across a circle along a straight line through the center.

Difference: The answer in a subtraction problem.

Distributive Property: $a(b + c) = ab + ac$ and $a(b - c) = ab - ac$ where a, b, and c are any numbers.

Dividend: The number to be divided in a division problem. The dividend is written to the left of a division sign. In fraction form, the dividend is written in the numerator (top).

Division: Repeated subtraction of the same number.

Divisor: The number that divides in a division problem. When the division sign is used, the divisor follows the division sign. In fraction form the divisor is written in the denominator (bottom).

Equation: A mathematical statement that contains an equal sign.

Equilateral triangle: A triangle with all three sides of equal length.

Equivalent fractions: Fractions that name the same number.

Equivalent ratios: Ratios that name the same number.

Exponent: A symbol written to the upper right of a base number that indicates how many times to use the base as a factor.

Expression: A mathematical relationship with constants and/or variables but no equal sign.

Factored form: A number or expression written as a product of factors.

Factorization: A multiplication equation.

Factors: Numbers that are multiplied.

Formula: An equation that describes a procedure.

Fraction: A number that describes a part of a whole.

Front-end ratio: The ratio of the house payment to gross monthly income.

GCF: (see greatest common factor)

Grade point average (GPA): An average of the total grade points earned divided by the total number of credit hours.

Grade point: A numerical value assigned to a letter grade.

Greatest common factor (GCF): The largest number that divides all given numbers evenly.

Hypotenuse: The side directly across from the 90° angle in a right triangle.

Improper fraction: A fraction in which the absolute value of the numerator is greater than or equal to the absolute value of the denominator.

Indeterminate: When a solution cannot be determined, as in $0 \div 0$.

Inequality: A mathematical statement that contains an inequality symbol.

Integers: A set of numbers that contains all whole numbers and the negative counting numbers. ..., -3, $-2, -1, 0, 1, 2, 3, ...$

Interest: An amount of money that is a percent of the principal.

Irrational number: A number that cannot be expressed in the form $\frac{a}{b}$ where a and b are integers and $b \neq 0$.

Isosceles triangle: A triangle with two sides of equal length.

LCD: See least common denominator.

LCM: See least common multiple.

Least common denominator (LCD): The LCM of the denominators.

Least common multiple (LCM): The smallest natural number that is divisible by all the given numbers.

Legs: The sides that form the 90° angle in a right triangle.

Like terms: Monomials that have the same variables raised to the same exponents.

Linear equation: An equation that is made of polynomials or monomials that are at most degree 1.

Loss: A negative net.

Lowest terms: A fraction is in lowest terms when the greatest common factor for the numerator and denominator is 1.

Mass: A measure of the amount of matter.

Mean: The sum of all given numbers divided by the number of numbers.

Median: The middle score in an ordered set of scores.

Minuend: The first number in a subtraction problem.

Mixed number: An integer combined with a fraction.

Mode: The score that occurs most often in a set of scores.

Monomial: An algebraic expression that is a product of a constant and variables, each of which is raised to a power that is a whole number.

Multiplicative inverses: Numbers whose product is 1. (see reciprocals)

Multiple: A number that is divisible by a given number.

Multiplication/division principle: We can multiply or divide both sides of an equation by the same amount without affecting its solution(s).

Multiplication: Repeated addition of the same number.

Natural numbers: The natural numbers are $1, 2, 3, ...$

Net: Money remaining after subtracting costs from revenue.

Non-linear equation: An equation that is not linear.

Numbers: Amounts or quantities.

Numerator: The number written in the top position in a fraction.

Parallel lines: Lines that, if extended forever, never intersect.

Parallelogram: A four-sided figure with two pairs of parallel sides.

Percent: Ratio out of 100.

Perfect square: A number that has a whole number square root.

Perimeter: The total distance around a shape.

Pi (π): An irrational number that is the ratio of the circumference of a circle to its diameter, usually approximated as $\frac{22}{7}$ or 3.14.

Polynomial: An expression that can be expressed as a sum of monomials.

Prime factorization: A multiplication equation that contains only prime factors.

Prime number: A number that satisfies all three of the following conditions: (1) It must be a natural number. (2) It cannot be the number 1. (3) It must be divisible only by itself and 1.

Principal: An initial amount of money.

Product: The answer in a multiplication problem.

Profit: A positive net.

Proportion: An equation with two ratios set equally.

Pythagorean theorem: The sum of the areas of the squares on the legs is the same as the area of the square on the hypotenuse; $a^2 + b^2 = c^2$, where a and b are the lengths of the legs of a right triangle, and c is the length of the hypotenuse.

Quadrant: One of four regions created by the intersection of the axes in the coordinate plane.

Quotient: The answer in a division problem.

Radius: The distance from the center to any point on the circle.

Rate: A unit ratio comparing two different measurements.

Ratio: A quotient comparison of two quantities.

Rational expression: A fraction that is a ratio of monomials or polynomials.

Rational number: A number that can be expressed in the form $\frac{a}{b}$ where a and b are integers and $b \neq 0$.

Real numbers: The set of all rational and irrational numbers.

Reciprocals: Two numbers whose product is 1. (see multiplicative inverses)

Rectangular array: A rectangle formed by a pattern of neatly arranged rows and columns.

Reduce: To create an equivalent fraction by dividing both the numerator and denominator by the same number.

Related sentence: A mathematical equation that relates the same pieces of a given equation using the inverse operation.

Remainder: The amount left over after dividing two whole numbers.

Revenue: Income.

Right triangle: A triangle that has one right angle.

Scientific notation: A number whose absolute value is greater than or equal to 1 but less than 10, and multiplied by 10 raised to an integer exponent.

Set: A group of elements.

Similar figures: Figures that have the same number of sides and all angles congruent.

Simple interest: Interest calculated using only the original principal.

Simplest form: An expression equivalent to a given expression with the fewest symbols and the smallest numbers possible.

Simplest form: An expression written with the fewest symbols possible.

Simplify: To write an equivalent expression with fewer symbols or smaller numbers.

Solution: A number that can replace the variable(s) in an equation and make the equation true.

Solve: To find a solution or solutions to an equation.

Square root: The base number that can be squared to make a given number.

Square unit: A 1×1 square.

Square: In geometry, a four-sided figure with all sides equal length; in algebra, the process of multiplying a number by itself.

Statistic: A number used to describe some characteristic of a set of data.

Subset: A set within a set.

Subtraction: An operation of arithmetic that can be interpreted as 1. take away, 2. difference, or 3. missing addend.

Subtrahend: The number following a minus sign in a subtraction problem.

Sum: The answer in an addition problem.

Supplementary angles: Angles whose sum is 180°.

Surface area: The total number of square units that completely cover the outer shell of an object.

Theoretical probability: The ratio of the number of favorable outcomes to the total number of possible outcomes.

Trapezoid: A four-sided figure with one pair of parallel sides.

Trinomial: A polynomial that has exactly three terms.

Undefined: When no numeric answer exists, as in $n \div 0$ when $n \neq 0$.

Unit fraction: A fraction with a ratio equivalent to 1.

Unit price: The price for each unit of an item.

Unit ratio: A ratio in which the denominator is 1.

Up-scale: To create an equivalent fraction by multiplying both numerator and denominator by the same number.

Variable: A symbol that can vary in value.

Vertex: A point where two lines join to form an angle.

Whole numbers: The whole numbers are 0, 1, 2, 3, ...

x-intercept: A point where a line intersects the x-axis.

y-intercept: A point where a line intersects the y-axis.

Index

A

Absolute value
 adding integers, 97, 164
 defined, 90–91
 order of operations, 92, 123–126
 perimeter graphs, 655
Absolute zero, 522
Acceleration, 100, 117–118, 139
Addend
 defined, 11
 missing, 15–16
Addition
 decimals, 399–402
 fractions, 343–347
 mixed numbers, 350
 polynomials, 169–171
 rational expressions, 345–348
Addition, integers
 different signs, 96–97
 like signs, 95–96
 order of operations, 123–126
 rules summary, 97
 solving applications, 98–99
Addition, whole numbers
 associative property, 12
 commutative property, 11–12, 165, 261, 348
 defined, 11
 estimated, 13
 key words, 13
 order of operations, 49–52
 procedures, 18
 simplifying repeated, 163
Addition/subtraction
 principle of equality, 243–248, 252–254, 351, 365–368, 403, 449
Additive inverse, 91–92, 93, 103, 104, 112, 124, 172, 350, 360, 403
Add, as key word, 13
Agreement, 49
Algebraic expression, 150
Altogether, as key word, 13
American measurement
 area, 501–502
 capacity, 502–503, 518
 length, 499–501, 517–518
 mass, 519
 speed, 504
 temperature, 519–520
 time, 503–504
 weight, 503, 519

B

Amortization, 596–598
And, as operation symbol, 368, 392
Angles
 complementary, 271–272, 370–371
 congruent, 270
 supplementary, 270, 370–371
Annual percentage rate (APR), 589–598
Apollo 11, 139
Archimedes, 444
Area problems
 circle, 358–359, 431, 440
 composite form, 443–444
 dimensional analysis, 501–502
 graph applications, 655–656
 metric conversion, 512–514
 polynomial, 215–217
 square units, 29–30
 trapezoid, 357, 439–440
 triangle, 318–319, 439
Associative property
 addition, 12
 multiplication, 22
Auditory learning, 87
Average. *See* Mean
Average rate, 133–134
Axis
 bar graphs, 624–626
 horizontal, 635–637
 vertical, 635–637

B

Back-end ratio, 523–526
Balance technique, 243–248
Bar graphs, 624–626
Base number, 27
Base-ten system
 defined, 2, 169
 metric conversion, 507
 notation, 390
Base unit
 defined, 507
 metric prefixes, 514
 order, 507–508, 510
Binary digit (bit), 27
Binomial, 158–159, 187–188
Bit (binary digit), 27
Borrowing, 15
Braces, 49–51, 123–126
Brackets, 49–51, 123–126
Brahmagupta, 88
Bringing down, 39
By, as key word, 24

C

Calculator. *See* Scientific calculator
Capacity
 metric conversion, 510–511
 unit conversion, 518
 units, 502–503
Carrying, 7
Cartesian coordinate system. *See* Rectangular coordinate system
Cavendish, Henry, 119
Celsius scale, 519–520
Centigrade scale, 519–520
Charts, 623–624
Circle
 area, 358–359, 431, 440
 circumference, 321–322, 431
 defined, 320
 diameter and radius, 320–321, 431
Circumference, circle, 321–322, 431
Coefficient
 combining like terms, 163–164
 decimal notation, 429–430
 isolating, 251–256
 monomial, 156–157
 solving equations, 247–248
Combinatorics, 27
Commission, 568–569
Commutative property
 addition, 11–12, 165, 261, 348, 359
 multiplication, 22, 111
Complementary angle, 271–272, 370–371
Complex fraction, 329
Composite form
 area and volume problems, 443–444
 problem-solving outline, 65–67
Composite number, 193–196
Compound interest, 593–595
Cone, 442
Congruent angle, 270
Conjugate, 187–188
Constant
 defined, 14, 55
 in monomial, 155–156
Coordinate, 635. *See also* Rectangular coordinate system
Coordinate pair, linear equation, 643–644

D

Cost, 105–106, 220
Cross multiplication, 297
Cubic centimeter, 59
Cubic unit, 57–58
Current, 119
Cylinder, 440–441

D

Debt problems, 523–526
Debt-to-income ratios, 523–525
Decimal equations, simplifying, 438–439, 449–452
Decimal notation
 addition, 399–403
 combining like terms, 402
 comparing, 393–394
 defined, 390–392
 division, 421–423, 429–430
 exponential forms, 411
 graphing, 392–393
 multiplication, 409–410, 413–414, 430
 ratios, 478
 rounding, 395–396
 scientific and standard forms, 411–413, 427–429
 signed, 401–402, 410
 solving applications, 403–404, 415–416, 430–432
 square root evaluation, 425–427
 subtraction, 401–403
 word names, 392
 writing fractions, 424
 writing mixed numbers, 424–425
 writing percent, 547–550
Decrease, percent of, 575, 579–584
Decreased by, as key words, 17
Deductions, income tax, 130–133
Degree
 monomial, 156–157
 polynomial, 159
Denominator, 52, 126, 290
Descartes, Rene, 635, 636
Descending order, polynomial, 159–160, 165–166
Diameter, circle, 320–321, 431
Difference, 15, 17
Digit, 2